1974

INTRODUCTION TO
NUMERICAL ANALYSIS

INTERNATIONAL SERIES IN PURE AND APPLIED MATHEMATICS

AHLFORS: Complex Analysis
BUCK: Advanced Calculus
BUSACKER AND SAATY: Finite Graphs and Networks
CHENEY: Introduction to Approximation Theory
CHESTER: Techniques in Partial Differential Equations
CODDINGTON AND LEVINSON: Theory of Ordinary Differential Equations
COHN: Conformal Mapping on Riemann Surfaces
CONTE AND DE BOOR: Elementary Numerical Analysis: An Algorithmic Approach
DENNEMEYER: Introduction to Partial Differential Equations and Boundary Value Problems
DETTMAN: Mathematical Methods in Physics and Engineering
EPSTEIN: Partial Differential Equations
GOLOMB AND SHANKS: Elements of Ordinary Differential Equations
GREENSPAN: Introduction to Partial Differential Equations
HAMMING: Numerical Methods for Scientists and Engineers
HILDEBRAND: Introduction to Numerical Analysis
HOUSEHOLDER: The Numerical Treatment of a Single Nonlinear Equation
KALMAN, FALB, AND ARBIB: Topics in Mathematical Systems Theory
LASS: Vector and Tensor Analysis
LEPAGE: Complex Variables and the Laplace Transform for Engineers
MCCARTY: Topology: An Introduction with Applications to Topological Groups
MONK: Introduction to Set Theory
MOORE: Elements of Linear Algebra and Matrix Theory
MOSTOW AND SAMPSON: Linear Algebra
MOURSUND AND DURIS: Elementary Theory and Application of Numerical Analysis
PEARL: Matrix Theory and Finite Mathematics
PIPES AND HARVILL: Applied Mathematics for Engineers and Physicists
RALSTON: A First Course in Numerical Analysis
RITGER AND ROSE: Differential Equations with Applications
RITT: Fourier Series
ROSSIER: Logic for Mathematicians
RUDIN: Principles of Mathematical Analysis
SIMMONS: Differential Equations with Applications and Historical Notes
SIMMONS: Introduction to Topology and Modern Analysis
SNEDDON: Elements of Partial Differential Equations
STRUBLE: Nonlinear Differential Equations
WEINSTOCK: Calculus of Variations

McGRAW-HILL
BOOK COMPANY
New York
St. Louis
San Francisco
Düsseldorf
Johannesburg
Kuala Lumpur
London
Mexico
Montreal
New Delhi
Panama
Rio de Janeiro
Singapore
Sydney
Toronto

F. B. HILDEBRAND
Professor of Mathematics
Massachusetts Institute of Technology

Introduction to Numerical Analysis

SECOND EDITION

This book was set in Times New Roman.
The editors were Jack L. Farnsworth and Shelly Levine Langman;
the production supervisor was Leroy A. Young.
The drawings were done by Oxford Illustrators Limited.
The printer and binder was Kingsport Press, Inc.

Library of Congress Cataloging in Publication Data

Hildebrand, Francis Begnaud.
Introduction to numerical analysis.

(International series in pure and applied mathematics)
Bibliography: P.
1. Numerical analysis. I. Title.
QA297.H54 1974 515 73-11315
ISBN 0-07-028761-9

INTRODUCTION TO
NUMERICAL ANALYSIS

1 2 3 4 5 6 7 8 9 0 KP KP 7 9 8 7 6 5 4 3

CONTENTS

PREFACE

The preface to the first edition contained the following passages:

This volume is intended to provide an introductory treatment of the fundamental processes of numerical analysis which is compatible with the expansion of the field brought about by the development of the modern high-speed calculating devices, but which also takes into account the fact that very substantial amounts of computation will continue to be effected by desk calculators (and by hand or slide rule), and that familiarity with computation on a desk calculator is a desirable preliminary to large-scale computation in any case.

* * *

The present text is based on the premise that the introductory course should provide a fairly substantial grounding in the basic operations of computation, approximation, interpolation, numerical differentiation and integration, and the numerical solution of equations, as well as in applications to such processes as the smoothing of data, the numerical summation of series, and the numerical solution of ordinary differential equations. It is believed that this course not only should exhibit techniques available for each purpose, but also should attempt to derive the relevant formulas in such a way that the underlying hypotheses are in evidence and that methods of generalization and modification are reasonably apparent, and that the problems of error analysis, convergence, and stability should be treated as adequately as time and preparation permit. Furthermore, the course desirably should be accompanied by a problem laboratory, in which enough actual computation is effected (presumably by use of desk calculators) to establish the practical significance of the theoretical developments.

Such an introduction should afford preparation for an "advanced course," dealing with certain of the somewhat more sophisticated aspects of the solution of equations and with modern methods of matrix inversion and determination of characteristic values of matrices, together with the numerical solution of partial differential equations and of integral equations, . . .

These remarks apply nearly as well to the present edition and to the philosophy underlying it. In particular, it is believed to be unrealistic to suppose that henceforth essentially all computation will be delegated to large-scale programed digital computers; that soon sufficiently efficient algorithms will exist to permit one to entrust most problems directly to a computer without preliminary or subsequent individual analyses and without a fair understanding of the rationale of the algorithms to be employed; and that, accordingly, the "numerical analysis" which is to be relevant in the classroom properly divides into the consideration of flow charts and of the technique and logic of computer programing, on the one hand, and the study of relatively profound mathematical treatments of special areas, on the other. Excellent texts for these two purposes are available and are under preparation; but, in addition, it is thought that there exists a continuing need for suitable texts which follow a more classical middle course.

This edition primarily was intended to introduce a selection of the more recent significant developments in the field and, correspondingly, to increase the focus of the text upon concepts and procedures associated with computers. At the same time, many changes have been made in phraseology, notation, and arrangement of the earlier material and a number of the treatments have been modified and amplified (or abbreviated).

The new material introduced includes sections on machine errors and on recursive calculation in Chapter 1; increased emphasis on the midpoint rule and the consideration of Romberg integration and the classical Filon integration in Chapter 3; a modified treatment of prediction-correction methods and the addition of Hamming's method in Chapter 6; the use of recursive methods in Chapters 7 and 8; brief considerations of uniform (minimax) approximation by polynomials and rational functions and four sections on spline approximation in Chapter 9; and treatments of pivoting and equilibration, ill conditioning, Crout's method for tridiagonal systems, the iterative methods of Muller, Traub, Ostrowski, and others, modifications of the method of false position and of the Lin and Bairstow procedures, and expanded treatments of the concept of the "order" of an iterative process, as well as additional methods for solving sets of nonlinear equations, in Chapter 10. Some results bearing on error analysis and on the convergence of approximation sequences, but depending upon some familiarity with analytic functions of a complex variable, were inserted as text material or as annotated problems in Sections 3.9 and 4.11. Reference lists were expanded and updated, and more than one hundred and fifty new problems were added.

As in the first edition, the chapter treating the numerical solution of equations is essentially independent of the other chapters and is placed at the end of the text (Chapter 10) so that relevant portions of its content can be inserted when they are needed in other developments at the discretion of the instructor. Thus, for example, some information relative to the practical solution of sets of linear algebraic equations should precede the consideration of least-squares methods in Chapter 7. Alternatively, it may be desirable to introduce part or all of Chapter 10 immediately following Chapter 1.

The Bibliography (Appendix B) includes only books and papers to which explicit reference was made either in the text material or in the Supplementary References sections at the ends of chapters. Obviously, it is not exhaustive in any category, and many outstanding numerical analysis texts and research contributions are omitted. In order to facilitate the use of the text for reference purposes, a Directory of Methods is included as Appendix C.

The author has profited from criticisms and encouragements by a rather long list of colleagues and students and is particularly indebted to Professor Philip Rabinowitz for many helpful suggestions relative to the present edition.

F. B. HILDEBRAND

INTRODUCTION TO
NUMERICAL ANALYSIS

1

INTRODUCTION

1.1 Numerical Analysis

The ultimate aim of the field of numerical analysis is to provide convenient methods for obtaining useful solutions to mathematical problems and for extracting useful information from available solutions which are not expressed in tractable forms. Such problems may each be formulated, for example, in terms of an algebraic or transcendental equation, an ordinary or partial differential equation, or an integral equation, or in terms of a set of such equations.

This formulation may correspond exactly to the situation which it is intended to describe; more often, it will not. Analytical solutions, when available, may be precise in themselves, but may be of unacceptable form because of the fact that they are not amenable to direct interpretation in numerical terms, in which case the numerical analyst may attempt to devise a method for effecting that interpretation in a satisfactory way, or he may prefer to base his analysis instead upon the original formulation.

More frequently, there is no known method of obtaining the solution in a precise form, convenient or otherwise. In such a case, it is necessary either to attempt to approximate the problem satisfactorily by one which *is* amenable to precise analysis, to obtain an approximate solution to the original problem by methods of numerical analysis, or to combine the two approaches.

On the other hand, the problem itself may not be clearly defined, and the analyst may be provided only with its partial solution, perhaps in the form of a table of approximate data, together with a certain amount of information with regard to its reliability, or perhaps in terms of an integral defining a function which cannot be expressed in terms of a finite number of tabulated functions. His purpose then is to obtain additional useful information concerning the function so described.

Generally the numerical analyst does not strive for exactness. Instead, he attempts to devise a method which will yield an approximation differing from exactness by less than a specified tolerance, or by an amount which has less than a specified probability of exceeding that tolerance. When the information supplied to him is inexact, he attempts both to obtain a dependable measure of the uncertainty which results from that inexactness and also to obtain an approximation which possesses a specified reliability compatible with that uncertainty.

He tries to devise a procedure which would be capable of affording an arbitrarily high degree of accuracy, in a wide class of situations, if the reliability of given information and of available calculating devices were correspondingly high. Even when successful in this attempt, he still seeks alternative procedures which may possess certain advantages in convenience or efficiency in certain situations, but which may be of less general applicability, or which may have the property that the degree of accuracy obtainable, even under ideal circumstances, cannot exceed a certain limit which depends upon the function to be analyzed. In this last case, which is of frequent occurrence, he attempts to ascertain that limit and to classify the situations in which it is not sufficiently high.

Needless to say, there are relatively few situations in which all these objectives have been, or can be, perfectly attained, as will be illustrated in the sequel. However, research with these aims in view continues to provide new procedures, as well as additional information with regard to the basic advantages and disadvantages of the older ones. Additional impetus has been afforded by the development of automatic desk calculators and, more recently, of large-scale computers. For example, certain methods had long been known to possess important theoretical advantages, but had not been *convenient*, from the point of view of the labor and time involved, for use in hand calculation or in calculation based on the use of the slide rule or of tables of logarithms, and hence had been considered as little more than mathematical curiosities. However, technological developments have promoted several of them into a much more active status and have also created additional need for detailed reexamination and modification of other existing methods and for a search for new ones.

One of the most rapidly expanding phases of numerical analysis is that which deals with the approximate solution of partial differential equations. But a basic understanding of the more involved problems which arise in that phase of the analysis depends strongly upon familiarity with similar problems which arise, in a somewhat simpler way, in connection with the solution of algebraic and transcendental equations, the processes of interpolation and approximation, numerical differentiation and integration, and the approximate solution of *ordinary* differential equations, in which only one independent variable is involved. These are the topics which are to be treated, for the most part, in what follows.

1.2 Approximation

In many of the problems which arise in numerical analysis, we are given certain information about a certain function, say $f(x)$, and we are required to obtain additional or improved information, in a form which is appropriate for interpretation in terms of numbers. Usually $f(x)$ is known or required to be *continuous* over the range of interest.

A technique which is frequently used in such cases can be described, in general terms, as follows. A convenient set of $n + 1$ *coordinate functions,* say $\phi_0(x), \phi_1(x), \ldots, \phi_n(x)$, is first selected. Then a procedure is invented which has the property that it would yield the desired additional information *simply* and *exactly* (barring inaccuracies in calculation) if $f(x)$ were a member of the set S_n of all functions which are expressible exactly as linear combinations of the coordinate functions. Next, use is made of an appropriate *selective process* which tends to choose from among all functions in S_n that one, say $y_n(x)$, whose properties are as nearly as possible identified with certain of the known properties of $f(x)$. In particular, it is desirable that the process be one which would select $f(x)$ if $f(x)$ were in S_n. The required property of $f(x)$ is then approximated by the corresponding property of $y_n(x)$. Finally, a method is devised for using additional known properties of $f(x)$, which were not employed in the selective process, for estimating the error in this approximation.

Clearly, it is desirable, first of all, to choose coordinate functions which are convenient for calculational purposes. The $n + 1$ functions $1, x, x^2, \ldots, x^n$, which generate the algebraic *polynomials* of degree n or less, are particularly appropriate, since polynomials are readily evaluated and since their integrals, derivatives, and products are *also* polynomials.

Of much greater importance, however, is the natural requirement that it be possible, by taking n sufficiently large, to be certain that the set S_n of generated functions will contain at least one member which approximates the function $f(x)$ within any preassigned tolerance, on the interval of interest. It is a most fortunate fact that the convenient set S_n, which consists of all polynomials of degree n or less, possesses this property if only $f(x)$ is continuous on that interval and the interval is of finite extent.

This fact was established in 1885 by a famous theorem of Weierstrass, which states, in fact, that any function $f(x)$ which is continuous on a closed interval $[a, b]$ can be *uniformly* approximated within any prescribed tolerance, on that interval, by some polynomial. By this statement we mean that, given any positive tolerance ε, there is a polynomial $p(x)$ such that $|f(x) - p(x)| \leqq \varepsilon$ for all x such that $a \leqq x \leqq b$.

Principally for the two reasons just given, *polynomial approximation* is of wide general use when the function to be approximated is continuous and the interval of approximation is finite, as well as in certain other cases, and accordingly is to form the basis of much of the work which follows. Other types of approximations are considered in Chap. 9.

Following the choice of the set S_n, an appropriate selective process must be chosen in accordance with the nature of the available information concerning the function $f(x)$. When the value of $f(x)$ is known for at least $n + 1$ values of x, say $x_0, x_1, \ldots, x_n$, the simplest and most often used process consists of selecting, from the members of S_n, a function $y_n(x)$ which takes on the same value as does $f(x)$ for each of those $n + 1$ values of x. Here again the choice of polynomials is convenient. For, whereas in the general case there may be no such function in S_n, or there may be several, it is a well-known fact that there exists one and only one polynomial of degree n or less which takes on prescribed values at each of $n + 1$ points. In particular, if $f(x)$ is indeed in S_n, this process will then select it. Other useful processes are described in Chaps. 7 and 9.

The final problem, that of devising an appropriate method of estimating the *error*, is a troublesome one and cannot be discussed at this point. Clearly, the precision of the estimate must depend upon the amount of available information relative to $f(x)$, and its usefulness will depend upon the form in which that information is supplied. In particular, if *all* available information is needed by the selective process, *no error estimate is possible.*

It is of some importance to notice that, if S_n is indeed taken as the set of all polynomials of degree n or less, then the Weierstrass theorem guarantees only the *existence* of a member of S_n which affords a satisfactory approximation to a continuous function $f(x)$, on any finite interval, when n is sufficiently large. This does *not* imply that the particular member chosen by a particular selective process will tend to afford such an approximation as n increases indefinitely.

Only when a dependable method of estimating the error is available can this question be resolved with certainty. Furthermore, even though it were possible to devise a selective process which had this property, it would not necessarily follow (for example) that the *derivative* of the selected polynomial $y_n(x)$ would tend to approximate the derivative of $f(x)$, even though the latter were known to exist and to be continuous. Again, recourse must be had to an error analysis.

When the selective process specifies $n + 1$ instances of exact agreement between the function $f(x)$ and its approximation and/or between certain of their derivatives, on a discrete set of points, the resultant approximation (if it exists) is called an *interpolation*. In particular, when *functional* agreement is prescribed at $n + 1$ distinct points, the interpolation process is sometimes said to be *collocative*. In some references the term interpolation is used in a somewhat more general sense. (See, for example, P. J. Davis [1963].†)

1.3 Errors

Most numerical calculations are inexact either because of inaccuracies in given data, upon which the calculations are based, or because of inaccuracies introduced in the subsequent analysis of those data. In addition to *gross errors*, occasioned by unpredictable *mistakes* (human or mechanical) and hypothetically assumed to be absent in the remainder of this discussion, it is convenient to define first a *roundoff error* as the consequence of using a number specified by n correct digits to approximate a number which requires more than n digits (generally infinitely many digits) for its exact specification.‡ Such errors are frequently present in given data, in which case they may be called *inherent* errors, due either to the fact that those data are empirical, and are known only to n digits, or to the fact that, whereas they are known exactly, they are "rounded" to n digits according to the dictates of convenience or of the capacity of a calculating device. They are introduced in subsequent analysis either because of deliberate rounding or because of the fact that a calculating device is capable of supplying only a certain number of digits in effecting operations such as addition, multiplication, division, conversion between number systems, and so forth.

It is then convenient to define a *truncation error*, by exclusion, as any error which is neither a gross error nor a roundoff error. Thus, a truncation error is one which would be present even in the hypothetical situation in which

† A name followed by a date in brackets refers to an item in Appendix B, the Bibliography.

‡ It is assumed here and elsewhere, for simplicity and consistency, that the decimal notation is used.

no "mistakes" were made, all given data were exact, and infinitely many digits were retained in all calculations. Frequently, a truncation error corresponds to the fact that, whereas an exact result would be afforded (in the limit) by an infinite sequence of steps, the process is *truncated* after a certain *finite* number of steps. However, it is rather conventional to apply the term in the more general sense defined here.

We define the error associated with an approximate value as the result of subtracting that approximation from the true value,

$$\text{True value} = \text{approximation} + \text{error} \tag{1.3.1}$$

with a remark that both this definition and that in which the algebraic sign of the error is reversed are used elsewhere in the literature.

The preceding definitions can be illustrated, for example, by calculations based on the use of power series. Thus, if a function $f(x)$ possesses $n + 1$ continuous derivatives everywhere on the interval $[a, x]$, it can be represented by a *finite* Taylor series of the form

$$f(x) = f(a) + \frac{f'(a)}{1!}(x - a) + \frac{f''(a)}{2!}(x - a)^2 + \cdots$$
$$+ \frac{f^{(n)}(a)}{n!}(x - a)^n + \frac{f^{(n+1)}(\xi)}{(n + 1)!}(x - a)^{n+1} \tag{1.3.2}$$

where ξ is *some* number between a and x. If $f(x)$ satisfies more stringent conditions, it can be represented by an *infinite* Taylor series

$$f(x) = f(a) + \frac{f'(a)}{1!}(x - a) + \frac{f''(a)}{2!}(x - a)^2 + \cdots$$
$$+ \frac{f^{(n)}(a)}{n!}(x - a)^n + \cdots \tag{1.3.3}$$

when $|x - a|$ is sufficiently small.

If $f(x)$ is *approximated* by the sum of the first $n + 1$ terms of (1.3.3), then the *error* committed is represented by the last term (*remainder*) in (1.3.2). Thus, for example, if $f(x) = e^{-x}$ and $a = 0$, we have the relation

$$e^{-x} = 1 - x + \tfrac{1}{2}x^2 - \tfrac{1}{6}x^3 + E_T(x) \tag{1.3.4}$$

where the truncation error is of the form

$$E_T = \tfrac{1}{24}e^{-\xi}x^4 \qquad (\xi \text{ between 0 and } x) \tag{1.3.5}$$

If x is positive, the same is true of ξ, and, by making use of the fact that $e^{-\xi}$ is then smaller than unity, we may deduce that the approximation

$$e^{-x} \approx 1 - x + \tfrac{1}{2}x^2 - \tfrac{1}{6}x^3 \tag{1.3.6}$$

is in error by a positive amount smaller than $\frac{1}{24}x^4$. In particular, we have

$$e^{-1/3} \approx 1 - \tfrac{1}{3} + \tfrac{1}{18} - \tfrac{1}{162} = \tfrac{116}{162} \qquad (1.3.7)$$

with an error between zero and $\frac{1}{1944}$. Since $\frac{1}{1944} \doteq 0.00051$, where the symbol $\doteq$ is used to signify "rounds to," the truncation error is smaller than 5.2×10^{-4}. If $\frac{116}{162}$ is rounded to four places, to give $e^{-1/3} \approx 0.7160$, the additional error introduced by the roundoff is less than (but here very nearly equal to) five units in the first neglected place and hence smaller than 0.5×10^{-4}. It follows finally that $e^{-1/3} \approx 0.7160$ with an error of magnitude smaller than 5.7×10^{-4}. However, whether $e^{-1/3} \doteq 0.716$ or 0.717 is not established. If *each* of the terms in (1.3.7) were rounded to four places before the terms were combined, a total roundoff error as great as 1.5×10^{-4} would be possible. Finally, if the exponent $\frac{1}{3}$ represented only an approximation to a value of x, which was not known exactly but which was known to lie, say, between 0.333 and 0.334, the approximate maximum error due to this uncertainty could be determined by noticing that the change δe^{-x} corresponding to a small change δx is approximately $(de^{-x}/dx)\,\delta x = -e^{-x}\,\delta x$. Thus, if the number $\frac{1}{3}$ is in error by an amount between -3×10^{-4} and $+7 \times 10^{-4}$, the magnitude of the maximum corresponding error in the calculated value is about $(0.716)(7 \times 10^{-4}) \approx 5 \times 10^{-4}$.

The magnitude of the roundoff errors could be reduced arbitrarily by retaining additional digits, and that of the truncation error could be reduced within any prescribed tolerance by retaining sufficiently many terms of the convergent Maclaurin expansion of e^{-x}. The effect of an inherent error could be reduced only if the uncertainty of the value of x were decreased.

It is useful to notice that, since the *sign* of the truncation error associated with (1.3.7) is known, the magnitude of the maximum possible error due to truncation can be halved by replacing the approximation $\frac{116}{162}$ by the approximation $\frac{116}{162} + \frac{1}{2}\frac{1}{1944} = \frac{2785}{3888} \doteq 0.7163$, with a corresponding truncation error accordingly known to lie between the limits $\pm 2.6 \times 10^{-4}$.

As an example of a somewhat different nature, we refer to the relation

$$\int_x^\infty \frac{e^{x-t}}{t}\,dt = \frac{1}{x} - \frac{1!}{x^2} + \frac{2!}{x^3} - \frac{3!}{x^4} + \cdots + (-1)^{n-1}\frac{(n-1)!}{x^n} + (-1)^n n! \int_x^\infty \frac{e^{x-t}}{t^{n+1}}\,dt \qquad (x > 0) \qquad (1.3.8)$$

which is readily established by successive integration by parts. If we denote the left-hand member by $F(x)$, we can thus write

$$F(x) \approx \frac{1}{x} - \frac{1!}{x^2} + \frac{2!}{x^3} - \frac{3!}{x^4} + \cdots + (-1)^{n-1}\frac{(n-1)!}{x^n} \qquad (1.3.9)$$

when $x > 0$, with a *truncation error*

$$E_T = (-1)^n n! \int_x^\infty \frac{e^{x-t}}{t^{n+1}}\,dt \qquad (1.3.10)$$

Since $x - t$ is nonpositive in the range of integration, so that $e^{x-t} \leqq 1$, we may deduce that

$$|E_T(x, n)| \leqq n! \int_x^\infty \frac{dt}{t^{n+1}}$$

or

$$|E_T(x, n)| \leqq \frac{(n-1)!}{x^n} \qquad (1.3.11)$$

Hence the truncation error is smaller in absolute value than the *last term retained* in the approximation and also is evidently of opposite sign.

Further, since $1/t^{n+1} \leqq 1/x^{n+1}$ in the integration range, we see that

$$|E_T(x, n)| \leqq \frac{n!}{x^{n+1}} \int_x^\infty e^{x-t}\,dt = \frac{n!}{x^{n+1}} \qquad (1.3.12)$$

so that the truncation error here is also smaller in absolute value than the *first term neglected*, and is of the *same* sign.

For a fixed number (n) of terms, the truncation error clearly is small when x is large and can be made *arbitrarily* small by taking x *sufficiently* large. However, for a given x, the error *cannot* be made arbitrarily small by retaining sufficiently many terms. In fact, we may notice that if the right-hand member of (1.3.9) were considered as the result of retaining the first n terms of an infinite series, then the ratio of the $(n + 1)$th term of that series to the nth would be $-n/x$. Hence, the successive terms decrease steadily in magnitude as long as $n < x$, but then *increase* unboundedly in magnitude as n increases beyond x. Thus the series does not converge for any value of x.

Nevertheless, it is useful for computation when x is fairly large. Thus, if $x = 10$, the smallest term occurs when $n = x = 10$ and is given by $-9!/10^{10} \doteq -3.6 \times 10^{-5}$. Thus, the approximation afforded by retention of 10 terms would be in error by a positive quantity smaller than 4×10^{-5}. This would be the best possible guaranteed accuracy obtainable from (1.3.9), when $x = 10$, since retention of additional terms would increase the possible magnitude of the error.

A divergent series of the type just considered, for which the magnitude of the error associated with retention of only n terms can be made arbitrarily small by taking a parameter x sufficiently near a certain fixed value x_0 (or sufficiently large in magnitude), and for which the error first decreases as n

increases but eventually increases unboundedly in magnitude with increasing n, when x is given a *fixed* value other than x_0, is often called an *asymptotic series.* An example of the former type, with $x_0 = 0$, is afforded by the relation

$$\int_0^\infty \frac{e^{-u}\,du}{1 + xu} = 1 - 1!\,x + 2!\,x^2 - 3!\,x^3 + \cdots + (-1)^n n!\,x^n + E(x, n)$$

when $x \geqq 0$, which can be obtained from (1.3.8) by replacing x by $1/x$ and making the change of variables $t = (1 + xu)/x$ in the integral, and for which it is true that $x^{-n}E(x, n) \to 0$ as x tends to zero from the positive direction, but $|E(x, n)| \to \infty$ as $n \to \infty$ for any fixed $x > 0$.

For a representation of the form

$$f(x) = a_0 + \frac{a_1}{x} + \frac{a_2}{x^2} + \cdots + \frac{a_n}{x^n} + E(x, n)$$

it is usually stipulated also (following Poincaré) that $x^n E(x, n)$ is to tend to zero as $|x| \to \infty$; for an expansion of the form

$$f(x) = a_0 + a_1(x - x_0) + a_2(x - x_0)^2 + \cdots + a_n(x - x_0)^n + E(x, n)$$

the additional requirement that $E(x, n)/(x - x_0)^n$ is to tend to zero as $x \to x_0$ is usually imposed. Equation (1.3.12) shows that (1.3.9) is thus asymptotic in the strict sense. However, the term is often applied somewhat more loosely to expansions of more general type, which are not necessarily power series.

When x is fixed, the error frequently decreases rapidly, as additional terms are taken into account, until a point of diminishing return is reached, after which the error begins to increase in magnitude. In such cases, if the error is reduced within the prescribed tolerance before that point is attained, then the approximate calculation can be successfully effected.

A great many of the expansions which are of frequent use in numerical analysis are essentially of this type. For them, the term *truncation error* generally applies only in the general sense of the definition given earlier in this section, and generally does *not* correspond to the result of truncating a convergent infinite process after a finite number of steps, but to the result of truncating a process which first tends to converge, but would ultimately diverge, at a stage before the tendency to diverge manifests itself.

The danger inherent in an unjustified *assumption* that a particular representation is of this type can be illustrated in terms of the function

$$f(x) = \frac{1 - 9.9x - 49.99x^2 - 4.999x^3}{1 - 10x - 49x^2}$$

If, for small values of x, $f(x)$ is to be approximated by a finite Taylor series in powers of x, the leading terms can be obtained in the form

$$f(x) \approx 1 + 0.1x + 0.01x^2 + 0.001x^3 + \cdots$$

by long division or otherwise. Thus, for example, when $x = 0.1$ the first four approximations to $f(0.1)$, obtained as partial sums, are 1, 1.01, 1.0101, and 1.010101. Whereas most rules of thumb would suggest that the error in the fourth approximation is positive and less than 10^{-6}, calculation shows that

$$f(0.1) \doteq 1.009998$$

and hence that the error in fact is negative, with magnitude exceeding 100 units in the sixth place.

1.4 Significant Figures

The conventional process of *rounding* or "forcing" a number to n digits (or "figures") consists of replacing that number by an n-digit approximation with minimum error.† When this requirement leads to two permissible roundings, that one for which the nth digit of the rounded number is *even* is generally selected. With this rule, the associated error is never larger in magnitude than one-half unit in the place of the nth digit of the rounded number.‡ Thus $4.05149 \doteq 4.0515$, 4.051, 4.05, 4.1, and 4. It may be noted here that whereas 4.05149 rounds to 4.0515, which in turn rounds to 4.052; nevertheless, 4.05149 rounds *directly* to 4.051. Thus rounding is not necessarily *transitive*.

The errors introduced in the rounding of a large set of numbers, which are to be combined in a certain way, usually (*but not always*) tend to be equally often positive and negative, so that their effects often tend to cancel. The slight favoring of even numbers is prompted by the fact that any subsequent operations on the rounded numbers are then somewhat less likely to necessitate additional roundoffs.

Each digit of a number, except a zero which serves only to fix the position of the decimal point, is called a *significant* digit or figure of that number. Thus, the numbers 2.159, 0.04072, and 10.00 each contain four significant figures.

† This type of abridgment is to be distinguished from the process of *chopping*, which consists of merely discarding all digits following the nth digit without modifying the nth digit and which must be used when capacity limitations of a calculating device do not permit the determination of more than n digits.

‡ It may be noticed that, when 9.95 is rounded to 10.0, the result still contains *three* correct digits; the error amounts to one-half unit in the place of the third figure of the rounded number but to five units in the place of the third figure of the original number.

Whether or not the last digit of 14620 is significant depends upon the context. If "a number known to be between 14615 and 14625" is intended, then that zero is not significant and the number would preferably be written in the form 1.462×10^4. Otherwise the form 1.4620×10^4 would be appropriate.

More generally, if any approximation $\bar{N}$ to a number N has the property that both $\bar{N}$ and N round to the same set of n significant figures, and if n is the *largest* integer for which this statement is true, then $\bar{N}$ may be said to *approximate N to n significant figures.* Thus, if $N = 34.655000\cdots$ and $\bar{N} = 34.665000\cdots$, then $n = 4$. Clearly, the error $N - \bar{N}$ cannot exceed one unit of the place of the nth digit, but, as this example illustrates, the error may take on that maximum value. In the case when $N = 38.501\cdots$ and $\bar{N} = 38.499$, while it is true that N and $\bar{N}$ both round to the same four significant digits (so that here $n = 4$), it may be noted that they do *not* round to the same *two* significant digits in spite of the fact that the error is less than three units in the place of the *fifth* digit. This point is of practical importance only in that it illustrates the fact that, no matter how accurately a calculation is to be effected, the result of rounding the calculated value to n digits cannot be guaranteed in advance to possess n *correct* digits but may differ from the rounded true value by one unit in the last digit.

It may be seen that the concept of significant figures is related more intimately to the *relative* error

$$\text{Relative error} = \frac{\text{true value} - \text{approximation}}{\text{true value}} \tag{1.4.1}$$

than to the error (or the *absolute* error) itself. In order to exhibit the relationship more specifically, it is useful to define N^* and r such that

$$N = N^* \times 10^r \qquad \text{where } 1 \leqq N^* < 10 \tag{1.4.2}$$

where r is an integer and hence is that integer for which $10^r \leqq N < 10^{r+1}$, when N is positive. Thus $N^* = N$ when $1 \leqq N < 10$, $= N/10$ when $10 \leqq N < 100$, $= 10N$ when $0.1 \leqq N < 1$, and so forth. If we write $E = N - \bar{N}$ and $R = E/N$, for the error and relative error, respectively, and suppose that $\bar{N}$ approximates N to n significant figures, so that

$$|E| \leqq 10^{r-n+1} \tag{1.4.3}$$

it then follows that

$$|R| \equiv \frac{|E|}{N} \leqq \frac{10^{r-n+1}}{N^* \times 10^r} = \frac{10^{-n+1}}{N^*} \tag{1.4.4}$$

In particular, we have $|R| \leqq 10^{-n+1}$.

Further, if $\bar{N}$ is the result of rounding N to n significant figures, (1.4.3) is then replaced by the stronger estimate

$$|E| \leqq 5 \times 10^{r-n} \tag{1.4.5}$$

and there then follows

$$|R| \leqq \frac{5}{N^*} \times 10^{-n} \tag{1.4.6}$$

In particular, $|R| \leqq 5 \times 10^{-n}$. If we also write

$$E = \omega \times 10^{r-n+1} \tag{1.4.7}$$

it follows that ω is the error expressed in units of the place of the nth digit of N and we have also

$$\omega = \frac{E}{10^{r-n+1}} = \frac{NR}{10^{r-n+1}} = N^*R \times 10^{n-1} \tag{1.4.8}$$

Suppose next that two numbers N_1 and N_2 are each rounded to n significant figures, and that the corresponding maximum error in the *product* $\bar{P} = \bar{N}_1\bar{N}_2$ of the rounded numbers is required. We notice first that, if $R(\bar{P})$ refers to $\bar{P}$ and R_1, R_2 to $\bar{N}_1$, $\bar{N}_2$, there follows

$$R(\bar{P}) = \frac{N_1N_2 - \bar{N}_1\bar{N}_2}{N_1N_2} = 1 - (1 - R_1)(1 - R_2) = R_1 + R_2 - R_1R_2$$

Thus we see that $|R(\bar{P})|$ is largest when R_1 and R_2 are *negative*, and, from (1.4.6), there follows

$$|R(\bar{P})| \leqq 5\left(\frac{1}{N_1^*} + \frac{1}{N_2^*}\right) \times 10^{-n} + \frac{25}{N_1^*N_2^*} \times 10^{-2n}$$

Hence, by using (1.4.8), we obtain

$$|\omega(\bar{P})| \leqq \frac{(N_1N_2)^*}{2}\left(\frac{1}{N_1^*} + \frac{1}{N_2^*}\right) + \frac{5}{2}\frac{(N_1N_2)^*}{N_1^*N_2^*} \times 10^{-n} \tag{1.4.9}$$

Since $(N_1N_2)^* = N_1^*N_2^* \times 10^{-\rho}$, where ρ is either 0 or 1, the right-hand member of (1.4.9) is of the form

$$\frac{10^{-\rho}}{2}(N_1^* + N_2^* + 5 \times 10^{-n})$$

and the most unfavorable cases are those for which $\rho = 0$. Under this constraint, the function $\phi(N_1^*, N_2^*) \equiv \frac{1}{2}(N_1^* + N_2^* + 5 \times 10^{-n})$ is to be considered only for $1 \leqq N_1^* < 10$, $1 \leqq N_2^* < 10$, and $1 \leqq N_1^*N_2^* < 10$, and clearly cannot take on a maximum value in the interior of this region. The maximum value of ϕ on the boundary of the region is easily seen to occur when either $N_1^* = 1$ and $N_2^* = 10$ or $N_1^* = 10$ and $N_2^* = 1$. Thus the right-hand member of (1.4.9) cannot exceed the limiting value corresponding to $(N_1 N_2)^* = 10-$ and either $N_1^* = 1$, $N_2^* = 10-$ or $N_1^* = 10-$, $N_2^* = 1$, and there follows

$$|\omega_n(\bar{N}_1 \bar{N}_2)| < \tfrac{11}{2} + \tfrac{5}{2} \times 10^{-n} < 6 \qquad (1.4.10)$$

where ω_n is the error expressed in units of the place of the nth digit of the true value.

This means that *if two numbers are rounded to n significant figures, the product of the rounded numbers differs from the true product by less than six units in the place of its nth significant digit.* In illustration, when $N_1 = 1.05+$ and $N_2 = 9.45+$ there follows $N_1 N_2 = 9.9225+$, whereas, if N_1 and N_2 are rounded to two significant figures to give $\bar{N}_1 = 1.1$ and $\bar{N}_2 = 9.5$, there follows $\bar{N}_1 \bar{N}_2 = 10.45$. Thus, in this rather extreme case, $\omega_2 = -(5.275-)$.

When $N_1 = N_2 \equiv N$, the worst (limiting) situation is that in which $(N^*)^2 = (N^2)^* = 10-$. Thus there follows

$$|\omega_n(\bar{N}^2)| < 10^{1/2} + \tfrac{5}{2} \times 10^{-n} < 4 \qquad (1.4.11)$$

so that *the square of a number rounded to n significant digits differs from the square of the unrounded number by less than four units in the place of its nth significant digit.*

More generally, if we consider $P = N_1 N_2 \cdots N_m$, we find that

$$R(\bar{P}) = 1 - [(1 - R_1)(1 - R_2) \cdots (1 - R_m)]$$

and hence

$$|\omega_n(\bar{P})| \leqq \frac{(N_1 \cdots N_m)^*}{2\alpha_n} [(1 + |R_1|)(1 + |R_2|) \cdots (1 + |R_m|) - 1]$$

$$\leqq \frac{(N_1 \cdots N_m)^*}{2\alpha_n} \left[\left(1 + \frac{\alpha_n}{N_1^*}\right)\left(1 + \frac{\alpha_n}{N_2^*}\right) \cdots \left(1 + \frac{\alpha_n}{N_m^*}\right) - 1\right] \qquad (1.4.12)$$

where

$$\alpha_n \equiv 5 \times 10^{-n} \qquad (1.4.13)$$

Here the worst situation is that in which $m - 1$ of the m numbers N_k^* are 1 and the remaining one is $10-$, such that also

$$N_1^* \cdots N_m^* = (N_1 \cdots N_m)^* = 10-$$

Thus there follows, from (1.4.12),

$$|\omega_n(\bar{N}_1 \cdots \bar{N}_m)| < \frac{5}{\alpha_n}\left[(1 + \alpha_n)^{m-1}\left(1 + \frac{\alpha_n}{10}\right) - 1\right] \qquad (1.4.14)$$

Corresponding numerical bounds on the quantity $|\omega_n(\bar{N}_1 \cdots \bar{N}_m)|$ are given in Table 1.1.

Table 1.1

n \ m	2	3	4	6	8	10
2	6	11	17	29	42	56
$\geqq 3$	6	11	16	26	36	46

In the special case when $N_1 = N_2 = \cdots = N_m \equiv N$, there follows

$$|\omega_n(\bar{N}^m)| \leqq \frac{(N^m)^*}{2\alpha_n}\left[\left(1 + \frac{\alpha_n}{N^*}\right)^m - 1\right] \tag{1.4.15}$$

and the worst situation is that in which $(N^*)^m = (N^m)^* = 10-$, when m is a positive integer, so that

$$|\omega_n(\bar{N}^m)| < \frac{5}{\alpha_n}[(1 + 10^{-1/m}\alpha_n)^m - 1] \qquad (m = 2, 3, \ldots) \tag{1.4.16}$$

Numerical bounds on $|\omega_n(\bar{N}^m)|$ are given in Table 1.2.

Table 1.2

n \ m	2	3	4	6	8	10
2	4	8	12	23	35	48
$\geqq 3$	4	7	12	21	31	41

When $P = N^m$, with $m \equiv 1/p$ the *reciprocal* of a positive integer, so that the operation involved is that of root extraction, the relation (1.4.15) again holds; but here the worst case is that in which $N = 10^{(k-m)/m}+$, and $N^* = 1+$, where k is any integer, so that $(N^m)^* = 10^{1-m}+$, in accordance with which there follows

$$|\omega_n(\bar{N}^m)| < \frac{5}{\alpha_n} \times 10^{-m}[(1 + \alpha_n)^m - 1] \qquad (m = \tfrac{1}{2}, \tfrac{1}{3}, \ldots) \tag{1.4.17}$$

Numerical bounds on $|\omega_n(\bar{N}^{1/p})|$, where p is a positive integer, are given in Table 1.3.

Table 1.3

n \ p	2	3	4	8	16	32
$\geqq 2$	0.79	0.78	0.71	0.47	0.28	0.15

Since, when $N\bar{N} \neq 0$, there follows

$$R\left(\frac{1}{\bar{N}}\right) = N\left(\frac{1}{N} - \frac{1}{\bar{N}}\right) = -\frac{N}{\bar{N}} R(\bar{N}) \approx -R(\bar{N}) \qquad (1.4.18)$$

it is seen that the bounds of Eq. (1.4.14) and Table 1.1 also apply (very nearly) to *division* if m is interpreted as the total number of factors in a ratio. Here, however, special notice should be taken of the formula

$$E\left(\frac{1}{\bar{N}}\right) = -\frac{\bar{N}}{N}\frac{E(\bar{N})}{\bar{N}^2} \approx -\frac{E(\bar{N})}{\bar{N}^2} \qquad (1.4.19)$$

relating the true errors, which is particularly significant when $|\bar{N}| \ll 1$.

Each of the given bounds applies for all N (with obvious exceptions) but may be quite conservative in any specific case. Thus, if it is known only that $|N - 9.61| \leqq 0.005$, then it can be verified that $|\sqrt{N} - 3.100| \leqq 0.0009$, whereas Table 1.3 gives a bound of 0.0078. (Here the guaranteed accuracy of the calculated result is greater than that of the basic data.) Still, none of the bounds can be appreciably lowered since each is nearly attained in some case.

In illustration, we may note that, if $N = 1.445$ and if N^6 is approximated by $(1.44)^6 \doteq 8.916$, the result differs from the true value $N^6 \doteq 9.103$ by about 19 in units of the third digit. Table 1.2 gives an upper limit of 21. The number $(106.4)^{1/3}$ should be reliable to three significant figures, according to Table 1.3, with the fourth digit in error by no more than 1. The calculated value rounds to 4.7385, whereas $(106.35)^{1/3} \doteq 4.7378$ and $(106.45)^{1/3} \doteq 4.7393$. The maximum error is thus about 0.8 units of the place of the fourth digit, as is just admitted by Table 1.3, and the *last digit* of the rounded four-place value, 4.738, is in error by not more than 1. However, whereas the value actually calculated is in error by an amount not exceeding 0.8×10^{-3}, as predicted, the rounded value may be in error by 1.3×10^{-3}.

The calculated value of the product

$$(3.658)(24.765)(1.4345)(72.43)$$

certainly will be in error by less than 16 units of the place of its fourth significant digit, by virtue of Table 1.1, under the assumptions that each factor is correctly rounded to the digits written and that sufficiently many digits are retained in the calculation itself. However, since the second and third factors each involve five significant figures, their product alone will be correct within six units of its fifth digit, so that actually the maximum error is very nearly the same as that associated with the product of *three* four-digit numbers and hence will be less than 11 units in the place of the fourth digit. Clearly (contrary to advice sometimes given) the procedure of deliberately rounding each of the factors to four digits before the multiplication would be a wasteful one, since it thus would increase the maximum possible error. Multiplication actually yields the calculated value 9.412×10^3 (to four digits), while the largest and smallest possible values of the true product are found to round to 9.415×10^3 and 9.410×10^3. Thus the maximum error here is only 3 in the fourth digit. The result of rounding each factor to four digits before multiplying rounds to 9.407×10^3, which hence *certainly* is in error by at least 3 in the fourth digit and which *may* be in error by as much as 8.

1.5 Determinacy of Functions. Error Control

When *any* differentiable function $f(N)$ is evaluated with N replaced by an approximation $\overline{N}$, the relation

$$f(N) - f(\overline{N}) = (N - \overline{N})f'(\eta) \qquad (\eta \text{ between } N \text{ and } \overline{N}) \tag{1.5.1}$$

permits us to deduce that

$$|E(f(\overline{N}))| \leqq |f'(\eta)|_{\max}|E(\overline{N})| \tag{1.5.2}$$

and

$$|R(f(\overline{N}))| \leqq \frac{|f'(\eta)|_{\max}}{|f(N)|}|E(\overline{N})| \tag{1.5.3a}$$

and also that

$$|R(f(\overline{N}))| \leqq \frac{|N|\,|f'(\eta)|_{\max}}{|f(N)|}|R(\overline{N})| \tag{1.5.3b}$$

Analogous results are readily obtained in cases when several independent variables are involved.

In illustration, if $f(N) = \log_{10} N$, there follows

$$|E(\log_{10} \overline{N})| \leqq \frac{\log_{10} e}{|\eta|}|E(\overline{N})| \qquad [\eta \text{ between } N - E(\overline{N}) \text{ and } N + E(\overline{N})]$$

and hence, if $N \geqq 1$ and $|E(\overline{N})| < \frac{1}{2}$,

$$|E(\log_{10} \overline{N})| \leqq \frac{0.44}{1 - |E(\overline{N})|}|E(\overline{N})| < |E(\overline{N})| \tag{1.5.4}$$

so that the error in the common logarithm is smaller than the error in its argument, when that argument exceeds unity. On the other hand,

$$|E(10^{\overline{N}})| = \frac{10^{\eta}}{\log_{10} e}|E(\overline{N})|$$

and hence, if $|E(\overline{N})| < \frac{1}{2}$,

$$|R(10^{\overline{N}})| \leqq 2.31 \times 10^{|E(\overline{N})|}|E(\overline{N})| < 8|E(\overline{N})| \tag{1.5.5}$$

Thus, the error in $10^{\bar{N}}$, expressed in units of the place of its nth significant figure, is less than $8|E(\bar{N})| \times 10^n$. Hence, if the error in the common logarithm is smaller in magnitude than 1 in units of the nth decimal place, then the antilogarithm is in error by less than 8 in units of its nth *significant figure* and hence is correct to at least $n - 1$ significant figures.

As a further illustration of the use of (1.5.1), we next investigate the degree of determinacy of the quantity†

$$\log \sin 1.412762$$

under the assumption that the argument is a rounded number. The use of (1.5.1) gives the bound

$$|E| \leqq (5 \times 10^{-7})|\cot \eta|_{\max} \qquad (1.4127615 \leqq \eta < 1.4127625)$$

on the inherent error E, and, since $0.17 > \cot x > 0.15$ for $1.41 < x < 1.42$, there follows $|E| < 0.85 \times 10^{-7}$, so that the desired quantity is determinate to within less than one unit in the seventh decimal place.

In the linear processes of *addition* and *subtraction*, the error in the result is merely the algebraic sum of the errors in the separate terms, and the magnitude of the maximum error is the sum of the maximum magnitudes of the component errors. Thus, whereas in multiplication and division we are concerned principally with *ratios* of errors to true quantities, and with the number of significant figures, and the absolute position of the decimal point is of importance only in fixing the magnitude of the end result, in addition and subtraction the errors themselves usually are the important quantities (see Sec. 1.6 for an important exception), significant figures are involved only incidentally, and the orientation of a digit sequence relative to the decimal point is of importance throughout the calculation.

Thus, if k numbers (positive or negative) are each rounded to n *decimal places*, so that each is in error by an amount less than $5 \times 10^{-n-1}$ in magnitude, the magnitude of the maximum error in the sum is clearly $5k \times 10^{-n-1}$, corresponding to the situation in which the signs of the errors are such that they combine without cancellation. Accordingly, the result can be in error by as much as $k/2$ units in the nth decimal place.

Formal addition assigns to the sum

$$56.434 + 251.37 - 2.6056 + 84.674 - 396.06 + 7.0228$$

the value 0.8352. However, if each number is correct only to the five significant figures given, the error in the result can have any value between the limits

† The notation $\log u$, with no base specified, is to be used consistently to denote $\log_e u$; the arguments of trigonometric functions are always to be expressed in radians unless degrees are explicitly specified.

± 0.0111, so that the result should be recorded as 0.84, with the last digit in doubt by two units,† and only one correct significant digit can be guaranteed. Rounding all of the numbers to the two decimal places which are in common, before addition, would lead to the result 0.82, and would increase the error limits to ± 0.03. Rounding each of the more accurate numbers to *one* place beyond the last place of the least-accurate one gives 0.835 with error limits ± 0.012, so that the recorded entry is again 0.84 (or 0.8_4), with the last digit in doubt by two units, and is a procedure which is generally to be recommended in such cases.

A somewhat similar situation, in which the outcome is, however, reversed, is of some importance. Tables of functions often provide a column of differences between successive entries, to facilitate linear interpolation, according to the formula

$$f(x_0 + \theta h) \approx f(x_0) + \theta[f(x_1) - f(x_0)] \qquad (0 < \theta < 1) \qquad (1.5.6)$$

where x_0 and x_1 are successive tabular arguments and $h = x_1 - x_0$. In constructing such a table, in which (say) all entries are to be rounded to a certain number of decimal places, the question arises as to whether the number tabulated for the difference $f(x_1) - f(x_0)$ should represent the rounded value of the difference or the difference between the rounded values, in those cases where these values differ. Intuition perhaps would recommend the former procedure, since it appears to make use of additional information. However, if ε represents the maximum roundoff, the maximum error in that case is clearly $(1 + \theta)\varepsilon$, whereas, since in the second case the right-hand member is properly to be considered in the form $(1 - \theta)f(x_0) + \theta f(x_1)$, and since $0 < \theta < 1$, the maximum error in that case is seen to be $(1 - \theta)\varepsilon + \theta\varepsilon = \varepsilon$. Thus the maximum error is less if the difference of the rounded values is used (for a more detailed discussion of this question, see Ostrowski [1952]). In particular, the *user* of tables which do not explicitly list differences need not regret the fact that he is *forced* to employ that procedure when using (1.5.6). The *truncation* error associated with that formula is considered in the following chapter.

The loss of significant figures in subtraction is one of the principal sources of error in numerical analysis, and it is highly desirable to arrange the sequence of calculations so that such subtractions are avoided, if possible, or so that their effects are brought into specific evidence. As a simple example, in calculating $ab - ac \equiv a(b - c)$, where b and c are very nearly equal, the products ab and ac may have many of their leading digits in common, and the number of sig-

† The notation 0.8_4 is often used to indicate that, whereas the 4 is not necessarily correct, the error associated with 0.84 is less than (or probably less than) that associated with 0.8.

nificant figures which must be retained in each product, in order that sufficiently many correct significant figures will remain in the difference, can be determined only after both products have been evaluated. This dilemma is avoided if $b - c$ is calculated first. Naturally, if a, b, and c are specified only to a certain number of significant figures, and if no roundoffs are introduced, the order of the calculations is irrelevant from this point of view, and the corresponding degree of uncertainty in the result merely must be accepted.

Frequently it is possible to exploit special properties of functions involved in the analysis. Thus, if a and b are nearly equal, it is convenient to replace $\log b - \log a$ by $\log (b/a)$, $\sin b - \sin a$ by $2 \sin \frac{1}{2}(b - a) \cos \frac{1}{2}(a + b)$, and $\sqrt{b} - \sqrt{a}$ by $(b - a)/(\sqrt{a} + \sqrt{b})$.

For example, if $4AC \ll B^2$, the *quadratic formula* is inconvenient for the determination of the smaller root of the equation $Ax^2 + Bx + C = 0$. In this case, when $B > 0$, it is desirable to replace the familiar formula $x_1 = (-B + \sqrt{B^2 - 4AC})/2A$ by the equivalent form

$$x_1 = \frac{-2C}{B + \sqrt{B^2 - 4AC}}$$

for a specific calculation, or to write

$$-B + \sqrt{B^2 - 4AC} = -B\left(1 - \sqrt{1 - \frac{4AC}{B^2}}\right)$$

in the original form and to expand the result by the binomial theorem to give

$$x_1 = -\frac{C}{B}\left(1 + \frac{AC}{B^2} + \cdots\right)$$

when the dependence of x_1 on the literal parameters is to be studied.

1.6 Machine Errors

In the preceding sections it has been implicitly assumed that the relevant sums, products, and quotients of exact numbers, or of their rounded approximations, are exactly calculable since the purpose has been to determine the extent to which the inherent *determinacy* of a function is reduced by roundoffs in its arguments.

Thus, for example, if $N_1 + N_2 + N_3$ is replaced by $\bar{N}_1 + \bar{N}_2 + \bar{N}_3$, where the barred quantities are rounded approximations, the error in the sum is the sum of the errors provided that the term *sum* has its usual meaning. However, the *machine sum* of three numbers, supplied by a digital computer, may or may not be identical with the true sum, and a similar statement applies to other operations.

In order to complement the preceding error considerations, we now suppose that the numbers supplied *to* the computer are exact and we investigate the *machine errors*.

If the computer operates in a *fixed-point* (that is, *fixed-decimal-point*) mode, then usually it deals with positive or negative numbers specified by, say, d decimal digits with the decimal point fixed to the left of the leading digit, so that each machine number has a magnitude less than 1. Although the computer usually can retain a $2d$-digit product and also can effect certain other *double-precision* operations, the necessity of choosing appropriate new units (*scaling*) in the formulation of a problem, so that all numbers lie in the interval $(-1, 1)$, is a source of considerable inconvenience. Once scaling has been accomplished, however, in such a way that all necessary sums, products, and quotients in a particular set of calculations are in $(-1, 1)$, then, assuming here that the inputs are exact d-digit numbers, the only error introduced by the computer in any single operation is the final roundoff (if it is necessary) to d digits. Thus, here the machine sum is the same as the true sum, while the d-digit machine product or quotient will differ from the true product or quotient by not more than $5 \times 10^{-d-1}$ (or by an error not exceeding some other multiple of 10^{-d} if rounding is replaced by another truncation process).

In a *floating-point* mode, a number is expressed in the form $\pm b \times 10^p$ (or often in a closely related form) where b (the *mantissa*) is expressed as an m-digit decimal such that $0.1 \leqq b < 1$ and p (the *exponent*) is a signed integer, so that the need for scaling is avoided. However, this advantage is obtained at the cost of a somewhat more complicated error analysis. To illustrate, we first consider two examples in the simple case of four-digit floating-point arithmetic, assuming that there is a double-precision eight-digit accumulator.

For the sum

$$0.2946 \times 10^4 + 0.3152 \times 10^2$$

the addition yields

$$\begin{array}{r} 0.2946\ 0000 \times 10^4 \\ +\ 0.0031\ 5200 \times 10^4 \\ \hline 0.2977\ 5200 \times 10^4 \end{array}$$

and the true sum is rounded to the machine sum 0.2978×10^4. The associated machine error is -0.48, and the *relative* error about -1.6×10^{-4}. For the product

$$(0.2946 \times 10^4)(0.3152 \times 10^2)$$

the multiplication produces $0.0928\,5792 \times 10^6$ and a shift, followed by a rounding, yields the machine product 0.9286×10^5, the machine error being -0.208×10^2 and the relative error about -2.2×10^{-4}.

In these two examples, if the accumulator had only a four-digit capacity (for the mantissa) and if "chopping" were then used so that all digits beyond the fourth would merely be dropped, the machine sum would be 0.2977×10^4 and the machine product 0.9285×10^5, the relative machine errors becoming approximately -1.7×10^{-4} and -8.5×10^{-4}.

In what follows, we will use the symbols $\oplus$, $\ominus$, $\odot$, and $\oslash$ to denote machine addition, subtraction, multiplication, and division in the floating mode. Then it can be seen that

$$\begin{aligned} N_1 \oplus N_2 &= (N_1 + N_2)(1 + \theta) \\ N_1 \odot N_2 &= N_1 N_2 (1 + \theta) \\ N_1 \oslash N_2 &= \frac{N_1}{N_2}(1 + \theta) \end{aligned} \tag{1.6.1}$$

where the value of θ may vary from one relation to another and may depend upon N_1 and N_2 but where in each case

$$|\theta| \leqq 5 \times 10^{-m} \tag{1.6.2}$$

where m is the mantissa capacity in decimal digits, assuming that the accumulator capacity permits *rounding* to m digits. Otherwise, distinct increased bounds on each of the θ's in (1.6.1) result, the increase being particularly significant for addition.

Accordingly there follows, for example,

$$\begin{aligned} (N_1 \oplus N_2) \oplus N_3 &= \{[(N_1 + N_2)(1 + \theta_1)] + N_3\}(1 + \theta_2) \\ &= (N_1 + N_2)(1 + \theta_1)(1 + \theta_2) + N_3(1 + \theta_2) \end{aligned}$$

and also

$$N_1 \oplus (N_2 \oplus N_3) = N_1(1 + \theta_3) + (N_2 + N_3)(1 + \theta_3)(1 + \theta_4)$$

so that the associative law under addition no longer holds.† We see that the maximum effect of machine error is minimized if the numbers are added in order of *increasing* magnitude, although the distinction becomes important only if *many* numbers are to be added.

† For a computer with a double-precision accumulator, the partial sum $N_1 + N_2$ would be truncated (if necessary) to $2m$ digits in the first case so that $|\theta_1|$ would not exceed a small multiple of 10^{-2m}, while the final sum would be rounded to m digits so that $|\theta_2| \leqq 5 \times 10^{-m}$. Similar statements apply to θ_3 and θ_4.

Further, there follows

$$(N_1 \odot N_2) \odot N_3 = [N_1N_2(1 + \theta_1)]N_3(1 + \theta_2)$$
$$= N_1N_2N_3(1 + \theta_1)(1 + \theta_2)$$

and

$$N_1 \odot (N_2 \odot N_3) = N_1[N_2N_3(1 + \theta_3)](1 + \theta_4)$$
$$= N_1N_2N_3(1 + \theta_3)(1 + \theta_4)$$

Although these two machine products are generally unequal, their error *bounds* (that is, bounds on their deviations from the true value $N_1N_2N_3$) are equal. No similar statement applies to the two preceding machine sums.

However, it should be reemphasized that here, for simplicity, we have supposed that the m-digit numbers supplied to the computer are exact and have then considered the error introduced by machine operations. Suppose now that two numbers N_1 and N_2 are replaced by approximations $\bar{N}_1 = N_1 - e_1$ and $\bar{N}_2 = N_2 - e_2$, where the errors e_1 and e_2 may be due to a rounding to m (or fewer) digits or to other sources, such as inaccuracies in observation or in experimental determination.

The machine sum of $\bar{N}_1$ and $\bar{N}_2$ is then given by

$$\bar{N}_1 \oplus \bar{N}_2 = [(N_1 - e_1) + (N_2 - e_2)](1 + \theta)$$

and hence

$$(N_1 + N_2) - (\bar{N}_1 \oplus \bar{N}_2) = (e_1 + e_2) - \theta(N_1 + N_2) + \theta(e_1 + e_2) \tag{1.6.3}$$

This total error is made up of the term $e_1 + e_2$, which is the *roundoff error* considered in the preceding sections, the term $-\theta(N_1 + N_2)$, which is the *machine error* currently under discussion, and the *coupling* term $\theta(e_1 + e_2)$. Whereas it could happen that the first two terms nearly cancel so that the coupling term is in fact dominant, insofar as predictable error *bounds* are concerned, if it is known that $|e_1|$ and $|e_2|$ cannot exceed ε, the best bounds on the three terms are 2ε, $(5 \times 10^{-m})|N_1 + N_2|$, and $10\varepsilon \times 10^{-m}$, and the third bound is negligible relative to at least the first one. Hence, from this point of view, the previously considered errors and the machine errors can be linearly superimposed in this case.

When the operation is multiplication, there follows

$$N_1N_2 - (\bar{N}_1 \odot \bar{N}_2) = (N_1e_2 + N_2e_1) - \theta N_1N_2 + \theta(e_1 + e_2)$$

Again the third term is negligible and the *relative* error is approximately the sum of the relative errors of the operands and the relative machine error.

In the sequel, mention seldom will be made of machine errors, as defined here. It may be noted that generally they are of importance only when the overall error tolerance is of the order of 10^{-m} so that full machine capacity is essential. Otherwise, they become of significance for the determination of *error bounds* when a specific computation requires n basic machine operations where n is so large that $n \times 10^{-m}$ is of the order of the error tolerance [or for statistical *error estimates* when the same is true of $n^{1/2} \times 10^{-m}$ (see Sec. 1.7)]. In such cases, rigorous error analysis may be unpleasant and linear superposition of machine-error effects on other errors may not be appropriate.

Finally, it should be noted that when values of a function $f(N)$, such as $N^{1/2}$ or e^N, are generated within the computer (perhaps by use of a preselected iterative process) instead of being supplied as inputs to the computer, machine errors again generally are introduced and, of course, then may be large relative to 10^{-m}.

1.7 Random Errors

If 1,000 positive numbers, each rounded to n decimal places, were added, the total error due to roundoff could amount to 500 units in the last place of the sum. Whereas this maximum error could be attained only in the case when all numbers were rounded in the same direction, by exactly one-half unit, the *possibility* of its occurrence forces us to accept its value as *the* least upper bound on the possible error.

However, the price of *certainty* in such a case is a high one, and in most situations it cannot be tolerated. Furthermore, in a great number of practical cases certainty cannot be attained. Thus each member of a set of 1,000 numbers, to be added, may itself represent the mean of a set of empirical values of a physical quantity, in which case one generally cannot *guarantee* that the error associated with it is less than, say, $5 \times 10^{-n-1}$, but can only estimate the *probability* that this is the case.

In most such cases it is assumed that the errors are symmetrically distributed about a zero mean and that, in a sufficiently large set of measurements, the probability of the occurrence of an error between x and $x + dx$ is, to a first approximation, of the form

$$\phi(x)\,dx = \frac{1}{\sqrt{2\pi}\,\sigma} e^{-x^2/2\sigma^2}\,dx \qquad (1.7.1)$$

where σ is a constant parameter, to be adjusted to the observations. The function ϕ is called the *frequency function* of the distribution. The probability that an error not exceed x algebraically is then given by the *normal distribution function*

$$\Phi(x) = \int_{-\infty}^{x} \phi(t)\,dt = \frac{1}{\sqrt{2\pi}\,\sigma} \int_{-\infty}^{x} e^{-t^2/2\sigma^2}\,dt \tag{1.7.2}$$

the numerical coefficient in (1.7.1) having been determined so that $\Phi(\infty) = 1$

$$\int_{-\infty}^{\infty} \phi(t)\,dt = \frac{1}{\sqrt{2\pi}\,\sigma} \int_{-\infty}^{\infty} e^{-t^2/2\sigma^2}\,dt = 1 \tag{1.7.3}$$

in accordance with the requirement of unit probability that any error lie *somewhere* in $(-\infty, \infty)$.

Further, the probability $P(x)$ that an error chosen at random lie between $-|x|$ and $+|x|$, that is, that its *magnitude* not exceed $|x|$, is clearly given by

$$P(x) = \Phi(|x|) - \Phi(-|x|) = \int_{-|x|}^{|x|} \phi(t)\,dt = 2\int_{0}^{|x|} \phi(t)\,dt$$

or

$$P(x) = \frac{\sqrt{2}}{\sqrt{\pi}\,\sigma} \int_{0}^{|x|} e^{-t^2/2\sigma^2}\,dt \tag{1.7.4}$$

whereas the probability that it *exceed* $|x|$ in magnitude is $Q(x) \equiv 1 - P(x)$. Equation (1.7.4) can also be written in the form

$$P(x) = \frac{2}{\sqrt{\pi}} \int_{0}^{|x|/\sqrt{2}\sigma} e^{-s^2}\,ds \equiv \operatorname{erf}\left(\frac{|x|}{\sqrt{2}\,\sigma}\right) \tag{1.7.5}$$

in terms of the error function (see Prob. 12).

Details must be omitted here with respect to the wide class of situations in which the use of this so-called *normal-distribution* law is justifiable, but the literature on this subject is extensive (for example, see Feller [1968]). In particular, even though the *frequency distribution* of the errors in a single quantity may not be capable of good approximation by a *normal* frequency distribution, of the form specified by (1.7.1), it generally is true that, when many such independent component errors are compounded, the resultant distribution *can* be so approximated.

The parameter σ is called the *standard deviation* of the distribution and its *square* σ^2 is called the *variance*. It is easily seen that the points of inflection of the curve representing $\phi(\varepsilon)$ lie at distance σ on each side of the maximum at $\varepsilon = 0$. The parameter $h = 1/(\sqrt{2\pi}\,\sigma)$ is called the *modulus of precision* and is a measure of the steepness of the frequency curve near its peak at the origin.

If ε is a random variable, the *mean value* (or *expected value*) of any function $g(\varepsilon)$, relative to the assumed distribution, is given by

$$(g(\varepsilon))_{\text{mean}} = \int_{-\infty}^{\infty} \phi(\varepsilon)g(\varepsilon)\,d\varepsilon = \frac{1}{\sqrt{2\pi}\,\sigma}\int_{-\infty}^{\infty} e^{-\varepsilon^2/2\sigma^2}g(\varepsilon)\,d\varepsilon \tag{1.7.6}$$

under the assumption that this integral exists. In particular, since $\phi(\varepsilon)$ is an even function of ε, we verify directly that the mean value of ε itself then is indeed zero, and we find also, for example, that

$$|\varepsilon|_{\text{mean}} = 2\int_0^{\infty} \varepsilon\phi(\varepsilon)\,d\varepsilon = \sqrt{\frac{2}{\pi}}\,\sigma \tag{1.7.7}$$

and

$$(\varepsilon^2)_{\text{mean}} = 2\int_0^{\infty} \varepsilon^2\phi(\varepsilon)\,d\varepsilon = \sigma^2 \tag{1.7.8}$$

Mean values of higher powers of $|\varepsilon|$ can be expressed similarly, in terms of the parameter σ.

Thus, this parameter could be determined in such a way that any one of these "moments," thus calculated for an assumed normal distribution, is made to equal the corresponding moment of the distribution actually under consideration, if that moment could be calculated or approximated. It happens that the choice of the *second* moment leads to the most convenient analysis and also is recommended by certain theoretical considerations. Thus we specify the parameter σ of the approximating normal distribution (1.7.1) in such a way that it is equal to the square root of the mean of the squared errors in the true distribution,

$$\sigma = \varepsilon_{\text{RMS}} \tag{1.7.9}$$

In general, the *root-mean-square* error ε_{RMS} for the entire distribution can be estimated only from a sample of, say, the deviations of k measurements from their mean value, and an appropriate estimate is then afforded by the formula†

$$\varepsilon_{\text{RMS}} \approx \sqrt{\frac{1}{k}(\varepsilon_1^2 + \varepsilon_2^2 + \cdots + \varepsilon_k^2)} \tag{1.7.10}$$

Having obtained such an approximation to σ, one can make use of Eq. (1.7.4) to estimate the probability that the magnitude of a random error

† A theoretically better estimate, which tends to take into account the probable deviation of the mean of the observations from the unknown true mean, is obtained by replacing $1/k$ by $1/(k - 1)$ in (1.7.10). This modification is of practical significance only when k is relatively small, in which cases the validity of the statistical analysis itself may be open to question.

exceed (or not exceed) a certain specified amount. A few useful values of $1 - P$ are listed in Table 1.4. Thus, the probability of an error of magnitude greater

Table 1.4

$\varepsilon/\varepsilon_{RMS}$	$1 - P(\varepsilon)$
0.674	0.500
0.842	0.400
1.000	0.317
1.036	0.300
1.282	0.200
1.645	0.100
2.576	0.010

than ε_{RMS} is 0.317. Only 20 percent of the errors should exceed $1.282\varepsilon_{RMS}$, 10 percent should exceed $1.645\varepsilon_{RMS}$, and 1 percent should exceed $2.576\varepsilon_{RMS}$ *if* the distribution is sufficiently nearly normal.

The number 0.67449σ is often called the *probable error* of the distribution. It should be noticed that this is merely that number which should be exceeded by the magnitude of half the errors; it is in no sense the *most* probable error, as the name tends to suggest.

If the approximation (1.7.10) were calculated for a large number of sets of samples, each containing k errors chosen at random from the same distribution, and if the mean of the estimates were selected as the best approximation to the true ε_{RMS}, the deviations of the various estimates from this best one would also be normally distributed, to a first approximation, with an RMS value of $\varepsilon_{RMS}/\sqrt{2k}$, when k is sufficiently large. This fact is often useful in estimating the reliability of the estimated value of ε_{RMS}.

Now suppose that ε is the sum of two independent errors u and v, each of which varies about a zero mean. Then the mean value of ε^2 is the sum of the mean values of u^2, $2uv$, and v^2. But, since u and v are independent, the mean of uv is the product of the means of u and v and hence is zero. Thus there follows

$$\varepsilon_{RMS} = \sqrt{u_{RMS}^2 + v_{RMS}^2} \qquad (1.7.11)$$

This argument generalizes to show that *the RMS value of the sum of n independent errors* (each having a zero mean) *is the square root of the sum of the squares of the RMS values of the component errors.*

It can be shown (see Prob. 26) that the normal-distribution law has the property that, if u and v are independent and normally distributed, with standard deviations σ_u and σ_v, then $\varepsilon = u + v$ is also normally distributed, with standard deviation $\sigma = \sqrt{\sigma_u^2 + \sigma_v^2}$. Thus if, in accordance with (1.7.9), we identify σ_u with u_{RMS} and σ_v with v_{RMS}, it will follow also that $\sigma_{u+v} = (u + v)_{RMS}$.

In illustration, if each of the numbers in the sum

$$426.44 - 43.26 + 2.72 + 9.61 - 104.26 - 218.72$$

represents the mean of a set of observations, and if the (approximate) RMS error associated with each is, say, 0.05, then the formal sum 72.53 would possess an RMS error of $\sqrt{6}\,(0.05) \doteq 0.12$. Such a result is often recorded as 72.53 ± 0.12, although some writers use the *probable* error $(0.674)(0.12) \doteq 0.07$, and write 72.53 ± 0.07, while still others use the notation $N \pm d$ to indicate that d is the *maximum* error in N (which would be undefined in the present case). *None* of these three conventions will be used here.

If we consider the error ε which arises from *rounding* a number to n decimal places, it is clear that the distribution of values of ε will *not* be well approximated by any normal distribution, since here the frequency function has the constant value $1/(2|\varepsilon|_{\max})$ when $|\varepsilon| < |\varepsilon|_{\max} = 5 \times 10^{-n-1}$ and the value zero otherwise. However, the distribution function corresponding to errors which are (exactly or approximately) linear combinations of many such errors generally will be appropriate for approximation by a normal-distribution function. Thus, in such cases, the error analysis may be based with some confidence upon the result of treating the individual errors as though they were normally distributed. (See Prob. 27.)

For this purpose, we may notice that if x takes on all values between $-\frac{1}{2}$ and $\frac{1}{2}$, and if all those values are equally likely, the RMS value of x is

$$\left(\frac{1}{1}\int_{-1/2}^{1/2} x^2\,dx\right)^{1/2} = \tfrac{1}{6}\sqrt{3} \doteq 0.2887$$

Hence, *if ε is roundoff error due to rounding to the nth decimal place, there follows*

$$\varepsilon_{\text{RMS}} = 0.2887 \times 10^{-n} \tag{1.7.12}$$

Thus, if k numbers are each rounded to n decimal places, the error in the sum of the results can be considered to be normally distributed, with an RMS value of $0.2887\sqrt{k} \times 10^{-n}$, if k is not too small.

In particular, when 1,000 such numbers are added, the RMS error in the sum is less than 10 units in the nth place. According to Table 1.4, the probability of an error of 17 units is less than 0.1, and the odds are 99 to 1 that the error will not exceed 26 units. Nevertheless, an error of 500 units in the nth place is indeed *possible.*

In accordance with such considerations, it is rather conventional to obtain a "realistic" estimate of the possible overall error due to k roundoffs, when k is fairly large, by replacing k by $\sqrt{k}$ in an expression for (or an estimate of) the *maximum* resultant error.

1.8 Recursive Computation

Frequently it is convenient to evaluate a function recursively, that is, by use of a recurrence formula. For example, if the polynomial

$$f(x) = \sum_{k=0}^{n} C_k x^k = C_n x^n + C_{n-1} x^{n-1} + \cdots + C_1 x + C_0 \tag{1.8.1}$$

were to be evaluated for a specific value of x by calculating the powers x^2, $x^3, \ldots, x^n$ and forming their linear combination, a total of $2n - 1$ multiplications generally would be needed. On the other hand, if the calculation is based on the "nested" grouping

$$f(x) = x(x\{\cdots[x(xC_n + C_{n-1}) + C_{n-2}]\cdots\} + C_1) + C_0 \tag{1.8.2}$$

it is seen that not more than n multiplications are needed.

The latter calculation can be systematized (for example) by use of the recurrence formula

$$u_k = xu_{k+1} + C_k \qquad (k = n, n-1, \ldots, 0) \tag{1.8.3}$$

with

$$u_{n+1} = 0 \tag{1.8.4}$$

so that a sequence $u_n, u_{n-1}, \ldots, u_0$ is determined recursively,

$$u_n = C_n \qquad u_{n-1} = xC_n + C_{n-1} \qquad u_{n-2} = x(xC_n + C_{n-1}) + C_{n-2}$$

and so forth, and the term u_0 provides the desired evaluation

$$\sum_{k=0}^{n} C_k x^k = u_0 \tag{1.8.5}$$

This process is often called *Horner's method* (but is due to Newton) and is equivalent to the use of so-called *synthetic division*. (See Sec. 10.14, where $C_n = 1$ and $C_k = a_{n-k}$.)

More generally, suppose that a sum of the form

$$f(x) = \sum_{k=0}^{n} C_k \phi_k(x) \tag{1.8.6}$$

is to be evaluated for a specific value of x, where ϕ_k itself satisfies a recurrence formula of the form

$$\phi_{k+1} + \alpha_k \phi_k + \beta_k \phi_{k-1} = 0 \qquad (k = 0, 1, 2, \ldots) \tag{1.8.7}$$

in which α_k and/or β_k depend upon x as well as k. (The preceding simple case, where $\phi_k = x^k$, is obtained by taking $\phi_{-1} = 0$, $\phi_0 = 1$, $\alpha_k = -x$, and

$\beta_k = 0$, so that (1.8.7) then is a *two*-term formula.) If a sequence $u_n, u_{n-1}, \ldots, u_0$ is defined by the associated recurrence formula

$$u_k + \alpha_k u_{k+1} + \beta_{k+1} u_{k+2} = C_k \qquad (k = n, n-1, \ldots, 0) \tag{1.8.8}$$

with

$$u_{n+1} = u_{n+2} = 0 \tag{1.8.9}$$

it follows that

$$\sum_{k=0}^{n} C_k \phi_k = \sum_{k=0}^{n} (u_k + \alpha_k u_{k+1} + \beta_{k+1} u_{k+2}) \phi_k$$

$$= \sum_{k=2}^{n} (\phi_k + \alpha_{k-1}\phi_{k-1} + \beta_{k-1}\phi_{k-2}) u_k + \phi_0 u_0 + (\phi_1 + \alpha_0 \phi_0) u_1$$

Hence, since the coefficient of u_k in the last sum vanishes for each relevant k, by virtue of (1.8.7), we find that

$$\sum_{k=0}^{n} C_k \phi_k = \phi_0 u_0 + (\phi_1 + \alpha_0 \phi_0) u_1 \tag{1.8.10}$$

This formula properly reduces to (1.8.5) in the preceding case.

In the special cases when $\beta_0 = 0$ or $\phi_{-1} = 0$ in (1.8.7), the sum reduces to $\phi_0 u_0$.

As an example, for the purpose of evaluating the sum

$$f(x) = \sum_{k=0}^{n} C_k \cos kx \tag{1.8.11}$$

we may note that $\phi_k = \cos kx$ satisfies the relation

$$\phi_{k+1} - 2\phi_k \cos x + \phi_{k-1} = 0 \qquad (k = 0, 1, \ldots, n) \tag{1.8.12}$$

with $\phi_{-1} = \cos x$ and $\phi_0 = 1$. Hence, if $u_n, u_{n-1}, \ldots, u_0$ are determined recursively from the relation

$$u_k - 2u_{k+1} \cos x + u_{k+2} = C_k \qquad (k = n, n-1, \ldots, 0) \tag{1.8.13}$$

with

$$u_{n+1} = u_{n+2} = 0 \tag{1.8.14}$$

there follows

$$\sum_{k=0}^{n} C_k \cos kx = u_0 - u_1 \cos x \tag{1.8.15}$$

(Other important uses of the preceding result are indicated in Chaps. 7 and 9.)

Although computation of this sort, based on a linear recurrence formula, often is a particularly efficient process, it should be noted that sometimes in *numerical* work the propagation of roundoff errors is so unfavorable that the process fails. A well-known example is that in which use is made of the relation

$$J_{k+1}(x) = \frac{2k}{x} J_k(x) - J_{k-1}(x) \qquad (1.8.16)$$

for the purpose of computing values of the Bessel function $J_n(x)$ for specific values of x from known rounded values of $J_0(x)$ and $J_1(x)$. When $x = 1$, if use is made of the five-place values

$$J_0(1) \doteq 0.76520 \qquad J_1(1) \doteq 0.44005$$

the approximate values generated are listed in the following table, together with rounded true values:

	$J_k(1)$	
k	**Approx**	**Rounded**
2	0.11490	0.11490
3	0.01955	0.01956
4	0.00240	0.00248
5	−0.00035	0.00025
6	−0.00590	0.00002

The apparent divergence might have been anticipated, assuming knowledge in advance of the fact that $J_k(1)$ rapidly tends to zero as k increases. Since, when $x = 1$, the recurrence formula becomes

$$J_{k+1}(1) = 2kJ_k(1) - J_{k-1}(1) \qquad (1.8.17)$$

it follows that if roundoff errors of magnitude ε could occur in the approximations to $J_0(1)$ and $J_1(1)$, then the corresponding absolute error in $J_2(1)$ could exceed 2ε, that error in $J_3(1)$ could exceed $4(2\varepsilon) = 8\varepsilon$, and, more generally, the magnitude of the error in $J_{k+1}(1)$ then could exceed $2^k k!\,\varepsilon$. Thus, since $J_{k+1}(1)$ itself *decreases* rapidly in magnitude as k increases, a rapid growth in the *relative* error should have been feared.

The source of the difficulty here also can be described as follows. Since the Bessel function $Y_k(x)$ of the second kind also satisfies a recurrence formula of form (1.8.16), it follows that the recurrence formula

$$\phi_{k+1} = 2k\phi_k - \phi_{k-1} \qquad (1.8.18)$$

is satisfied by

$$\phi_k = AJ_k(1) + BY_k(1) \qquad (1.8.19)$$

when $k = 0, 1, 2, \ldots$, where A and B here are determined by the specified values of ϕ_0 and ϕ_1. Unless the ratio of these two starting values is *exactly* the ratio of $J_0(1)$ and $J_1(1)$, the value of B will not vanish. Hence, since $Y_k(1)$

increases rapidly as k increases, the "parasitic" term $BY_k(1)$ eventually will dominate the desired term $AJ_k(1)$.

A third equivalent appraisal of the situation would result from writing

$$\theta_k = \frac{\phi_{k+1}}{\phi_k} \tag{1.8.20}$$

in (1.8.18) to obtain the new recurrence formula

$$\theta_k = 2k - \frac{1}{\theta_{k-1}} \tag{1.8.21}$$

This relation suggests that when k is large, there follows

$$\theta_k^2 - 2k\theta_k + 1 \approx 0 \tag{1.8.22}$$

and hence†

$$\theta_k \sim k \pm \sqrt{k^2 - 1} \qquad (k \to \infty) \tag{1.8.23}$$

Thus one solution of (1.8.18), namely, $Y_k(1)$, *grows* as k increases in such a way that $Y_{k+1}(1) \sim 2kY_k(1)$; the second solution, namely, the desired one $J_k(1)$, tends to zero as $k \to \infty$ in such a way that $J_{k+1}(1) \sim (2k)^{-1}J_k(1)$.

One method of overcoming this difficulty consists of applying the formula (1.8.17) in the opposite direction, with unknown values of $J_n(1)$ and $J_{n+1}(1)$ carried as literal parameters to be determined so that the values of $J_1(1)$ and $J_0(1)$ so generated agree appropriately with their known true values, since the parasitic solution $BY_k(1)$ then rapidly damps out as the recursion proceeds.

A somewhat simpler procedure, for numerical work, chooses N so large that $J_N(1)$ certainly is smaller than $5 \times 10^{-r-1}$, where r is the required number of significant digits in $J_n(1)$, then starts the backward recursion with $J_N(1) = 0$, $J_{N-1}(1) = A$ and determines A such that a generated value [say, $J_1(1)$] agrees suitably with a known value. Here, in fact, one can merely assign a convenient fictitious value (say, $A = 10^{-r}$) to $J_{N-1}(1)$ and scale up the generated results of interest in the proper ratio at the end of the process.

1.9 Mathematical Preliminaries

In this section, we list certain analytical results to which reference occasionally will be made in the sequel. Proofs of most are omitted. First, it is noted that in most of the following chapters it is supposed that all functions dealt with are real and continuous in the range considered and, in addition, that they possess as many continuous derivatives as the analysis may require.

† The notation $u_k \sim v_k$ $(k \to \infty)$ is used to indicate that $\lim_{k\to\infty} (v_k/u_k) = 1$.

The basic fact that *a function* $f(x)$ *which is continuous for* $a \leqq x \leqq b$ *takes on each value between* $f(a)$ *and* $f(b)$ is intuitively "obvious" but is capable of rigorous proof. Two immediate consequences of this result are the following:

Theorem 1 If $f(x)$ is continuous for $a \leqq x \leqq b$, and if $f(a)$ and $f(b)$ are of opposite sign, then $f(\xi) = 0$ for at least one number ξ such that $a < \xi < b$.

Theorem 2 If $f(x)$ is continuous for $a \leqq x \leqq b$, and if λ_1 and λ_2 are positive constants, then $\lambda_1 f(a) + \lambda_2 f(b) = (\lambda_1 + \lambda_2) f(\xi)$ for at least one ξ such that $a \leqq \xi \leqq b$.

If also $f'(x)$ exists, two additional results can be established:

Theorem 3 If $f(x)$ is continuous for $a \leqq x \leqq b$ and $f'(x)$ exists for $a < x < b$, and if $f(a) = f(b) = 0$, then $f'(\xi) = 0$ for at least one ξ such that $a < \xi < b$. (This is *Rolle's theorem.*)

Theorem 4 If $f(x)$ is continuous for $a \leqq x \leqq b$ and $f'(x)$ exists for $a < x < b$, then $f(b) - f(a) = (b - a)f'(\xi)$ for at least one ξ such that $a < \xi < b$. (This is the *mean-value theorem* for the derivative.)

In the following statements, it is assumed that the integrals involved exist and that $b > a$.

Theorem 5 If $|f(x)| \leqq M$ for $a \leqq x \leqq b$, where M is a constant, then

$$\left| \int_a^b f(x)\, dx \right| \leqq \int_a^b |f(x)|\, dx \leqq M(b - a)$$

Theorem 6 If $f(x)$ is continuous for $a \leqq x \leqq b$, then

$$\int_a^b f(x)\, dx = (b - a) f(\xi)$$

for at least one ξ such that $a < \xi < b$. (This is the *first law of the mean.*)

Theorem 7 If $m \leqq f(x) \leqq M$ and $g(x)$ is nonnegative, for $a \leqq x \leqq b$, then

$$m \int_a^b g(x)\, dx \leqq \int_a^b f(x) g(x)\, dx \leqq M \int_a^b g(x)\, dx$$

Theorem 8 If $f(x)$ is continuous for $a \leqq x \leqq b$ and $g(x)$ does not change sign in $[a, b]$, then

$$\int_a^b f(x)g(x)\,dx = f(\xi)\int_a^b g(x)\,dx$$

for at least one ξ such that $a < \xi < b$. (This is the *second law of the mean.*)

The four following theorems with relation to integrals involving a parameter are of frequent use:

Theorem 9 If a and b are finite constants and $F(x, s)$ is continuous in x and s, then

$$\lim_{x \to c} \int_a^b F(x, s)\,ds = \int_a^b F(c, s)\,ds$$

Theorem 10 If a and b are finite constants and if $\partial F/\partial x$ is continuous, then

$$\frac{d}{dx}\int_a^b F(x, s)\,ds = \int_a^b \frac{\partial F(x, s)}{\partial x}\,ds$$

Theorem 11 If a is a finite constant, u is a differentiable function of x, and $\partial F/\partial x$ is continuous, then

$$\frac{d}{dx}\int_a^u F(x, s)\,ds = \int_a^u \frac{\partial F(x, s)}{\partial x}\,ds + F(x, u)\frac{du}{dx}$$

Theorem 12 If $F_n(x)$ denotes the result of integrating $F(x)$ successively n times over $[a, x]$, then

$$F_n(x) = \frac{1}{(n-1)!}\int_a^x (x - s)^{n-1}F(s)\,ds$$

The truth of each of these assertions, except perhaps for the last two, is nearly self-evident, and the details of their proofs are rather easily supplied once the preliminary basic properties of continuous functions are established.

The validity of Theorem 11 follows from the fact that if we write

$$I(x) = \int_a^u F(x, s)\,ds$$

there follows

$$I(x + \Delta x) - I(x) = \int_a^u [F(x + \Delta x, s) - F(x, s)]\, ds + \int_u^{u+\Delta u} F(x + \Delta x, s)\, ds$$

$$= \left(\int_a^u F_x(\xi, s)\, ds\right) \Delta x + F(x + \Delta x, \eta)\, \Delta u$$

where ξ is between x and $x + \Delta x$ and η is between u and $u + \Delta u$, by virtue of Theorems 4 and 6, and hence

$$I'(x) = \lim_{\Delta x \to 0} \int_a^u F_x(\xi, s)\, ds + \lim_{\Delta x \to 0} F(x + \Delta x, \eta) \frac{\Delta u}{\Delta x}$$

$$= \int_a^u F_x(x, s)\, ds + F(x, u) \frac{du}{dx}$$

by virtue of the content of Theorem 9.

Theorem 12 can be established by successive integration by parts, or verified by the use of Theorem 11, making use of the facts that, from the definition, there follows $F_k'(x) = F_{k-1}(x)$ and $F_k(a) = 0$ for $k = 1, 2, \ldots, n$, and $F_1'(x) = F_0(x) \equiv F(x)$.

This result is useful in deriving the finite Taylor series, with an error term expressed in a form which often is more useful than that given in (1.3.2), and also in deriving the form given there. For if we write

$$F(x) = f^{(n+1)}(x) \equiv \frac{d^{n+1} f(x)}{dx^{n+1}} \tag{1.9.1}$$

and use the notation of Theorem 12, the results of integrating the equal members successively over $[a, x]$ are seen to be

$$F_1(x) = f^{(n)}(x) - f^{(n)}(a)$$

$$F_2(x) = f^{(n-1)}(x) - f^{(n-1)}(a) - (x - a) f^{(n)}(a)$$

$$F_3(x) = f^{(n-2)}(x) - f^{(n-2)}(a) - (x - a) f^{(n-1)}(a) - \frac{(x - a)^2}{2!} f^{(n)}(a)$$

and finally, after $n + 1$ integrations,

$$F_{n+1}(x) = f(x) - f(a) - (x - a) f'(a) - \frac{(x - a)^2}{2!} f''(a) - \cdots - \frac{(x - a)^n}{n!} f^{(n)}(a) \tag{1.9.2}$$

Thus, after a transposition and a reference to Theorem 12, we deduce that *if the* $(n + 1)$th *derivative of* $f(x)$ *exists and is integrable over an interval including* $x = a$, *then in that interval there follows*

$$f(x) = f(a) + f'(a)(x - a) + \frac{f''(a)}{2!}(x - a)^2 + \cdots + \frac{f^{(n)}(a)}{n!}(x - a)^n + E(x) \tag{1.9.3}$$

where

$$E(x) = \frac{1}{n!}\int_a^x (x - s)^n f^{(n+1)}(s)\, ds \tag{1.9.4}$$

Further, since $(x - s)^n$ does not change sign as s varies from a to x, we can invoke the second law of the mean (Theorem 8) to rewrite (1.9.4) in the form

$$\begin{aligned} E(x) &= \frac{f^{(n+1)}(\xi)}{n!}\int_a^x (x - s)^n\, ds \\ &= \frac{f^{(n+1)}(\xi)}{(n + 1)!}(x - a)^{n+1} \qquad (\xi \text{ between } a \text{ and } x) \end{aligned} \tag{1.9.5}$$

under the additional assumption that $f^{(n+1)}(x)$ is continuous. Whereas the form (1.9.5) has the advantage of simplicity, the form (1.9.4) often is preferable because of the fact that it is *explicit* while (1.9.5) involves a parameter which is known only to lie between a and x.

A useful generalization of the Taylor-series expansion (1.9.3) can be obtained by starting with the representation

$$F(t) = F(0) + \sum_{k=1}^{n} c_k t^k + E \tag{1.9.6}$$

where

$$c_k = \frac{1}{k!}\left[\frac{d^k F(t)}{dt^k}\right]_{t=0} \qquad E = \frac{t^{n+1}}{(n + 1)!}\left[\frac{d^{n+1} F(t)}{dt^{n+1}}\right]_{t=\tau} \tag{1.9.7}$$

with τ between 0 and t, and writing

$$t = g(x) - g(a) \qquad F(t) = f(x) \tag{1.9.8}$$

under the assumption that

$$g'(x) \neq 0 \tag{1.9.9}$$

over some interval I including $x = a$, so that $g(x) - g(a)$ increases or decreases steadily as x increases over I. The result of this substitution takes the form

$$f(x) = f(a) + \sum_{k=1}^{n} c_k[g(x) - g(a)]^k + E \qquad (1.9.10)$$

where

$$c_k = \frac{1}{k!}\left[\left\{\frac{1}{g'(x)}\frac{d}{dx}\right\}^k f(x)\right]_{x=a}$$

$$E = \frac{[g(x) - g(a)]^{n+1}}{(n+1)!}\left[\left\{\frac{1}{g'(x)}\frac{d}{dx}\right\}^{n+1} f(x)\right]_{x=\xi} \qquad (1.9.11)$$

and where ξ lies between a and x, when x is in I, under the assumption that $f^{(n+1)}(x)$ and $g^{(n+1)}(x)$ are continuous and $g'(x) \neq 0$ in I.

If we define a sequence of auxiliary functions $\alpha_0(x), \alpha_1(x), \ldots$ by the recurrence formula

$$\alpha_k(x) = \frac{\alpha'_{k-1}(x)}{g'(x)} \qquad (k = 1, 2, \ldots) \qquad (1.9.12)$$

with

$$\alpha_0(x) = f(x) \qquad (1.9.13)$$

it follows that

$$c_k = \frac{1}{k!}\alpha_k(a) \qquad E = \frac{1}{(n+1)!}\alpha_{n+1}(\xi)[g(x) - g(a)]^{n+1} \qquad (1.9.14)$$

The expansion (1.9.10) is often known as a *Bürmann series* and is useful when a certain value of a function $g(x)$ is known and the corresponding value of a second function $f(x)$ is required. The special case when $f(x)$ is identified with x itself is of most frequent occurrence.

It can be shown (see Whittaker and Watson [1927]) that the coefficient c_k can also be expressed by the formula

$$c_k = \frac{1}{k!}\left[\frac{d^{k-1}}{dx^{k-1}}\left\{f'(x)\left[\frac{x-a}{g(x) - g(a)}\right]^k\right\}\right]_{x=a} \qquad (1.9.15)$$

Although this is the form usually given, the use of the form given in (1.9.11) or (1.9.14) often leads to a somewhat less involved calculation, particularly when $(x - a)$ cannot be explicitly factored from $g(x) - g(a)$.

We conclude with a few useful basic facts relating to zeros of polynomials, recalling first the so-called *fundamental theorem* of elementary algebra, which states that *any polynomial*† *other than a constant possesses at least one zero.*

† The term *polynomial* is to be used in its common restricted sense to denote an expression of the form $a_0x^n + a_1x^{n-1} + \cdots + a_n$, where n is a nonnegative integer and the a's are constants.

The usual proofs depend upon results established in the theory of analytic functions of a complex variable. Elementary treatments *assume* the truth of this theorem and deduce easily that *any polynomial of degree n possesses exactly n zeros*, with the understanding that the zeros may be real or imaginary,† and with the convention that repeated zeros are to be counted a number of times equal to their multiplicities. In this connection, α is said to be a zero of multiplicity m if $(x - \alpha)^m$ is a factor of the polynomial but $(x - \alpha)^{m+1}$ is not.

It is now supposed that the polynomial is *monic* (that is, that the leading coefficient is 1) and that the coefficients are *real*, in which case the following theorem combines four classical results.

Theorem 13 Let $p(x) = x^n + a_1x^{n-1} + \cdots + a_n$, where the a's are real, and let the zeros of $p(x)$ be denoted by $x_1, x_2, \ldots, x_n$, with the understanding that a zero of multiplicity m is to be assigned m different subscripts. Then the following statements are valid:

DESCARTES' RULE The number of positive real zeros either equals the number of sign changes in the coefficient sequence

$$1, a_1, a_2, \ldots, a_n$$

or is smaller by an even integer, whereas the number of negative real zeros is related in the same way to the coefficient sequence associated with $p(-x)$.

BUDAN'S RULE If N_a and N_b denote the number of sign changes in the respective sequences

$$p(a), p'(a), p''(a), \ldots, p^{(n)}(a)$$

and

$$p(b), p'(b), p''(b), \ldots, p^{(n)}(b)$$

then the number of real zeros in the interval $a < x < b$ either equals $N_a - N_b$ or is smaller by an even integer.

NEWTON'S PRODUCT-SUM IDENTITIES If q_r denotes the sum of all possible products of r zeros with distinct subscripts, so that $q_1 = x_1 + x_2 + \cdots + x_n$, $q_2 = x_1x_2 + x_1x_3 + \cdots$, and so forth, then

$$q_r = (-1)^r a_r \qquad (r = 1, 2, \ldots, n)$$

† A complex number $a + ib$, with a and b real, will be said to be *imaginary* (or *nonreal*) if $b \neq 0$ and to be *pure imaginary* if also $a = 0$.

NEWTON'S POWER-SUM IDENTITIES If s_r denotes the sum of the rth powers of the n zeros

$$s_r = x_1^r + x_2^r + \cdots + x_n^r$$

then

$$s_r + s_{r-1}a_1 + \cdots + s_1 a_{r-1} + ra_r = 0 \qquad (r = 1, 2, \ldots, n)$$

and

$$s_r + s_{r-1}a_1 + \cdots + s_{r-n}a_n = 0 \qquad (r = n + 1, n + 2, \ldots)$$

The rules of Descartes and Budan (proofs omitted) are occasionally useful for the purposes of roughly locating zeros of a polynomial or of establishing their absence in a certain interval. There exists an extensive class of more powerful (and less elementary) theorems for such purposes (see Marden [1966] and Householder [1970]).

The well-known product-sum identities are easily established by exploiting the equivalence

$$(x - x_1)(x - x_2) \cdots (x - x_n) = x^n + a_1 x^{n-1} + \cdots + a_n \qquad (1.9.16)$$

In order to derive the power-sum identities, we may notice first that

$$\frac{p'(x)}{p(x)} = \frac{1}{x - x_1} + \frac{1}{x - x_2} + \cdots + \frac{1}{x - x_n} \qquad (1.9.17)$$

and that

$$\frac{1}{x - x_i} = \sum_{r=0}^{\infty} \frac{x_i^r}{x^{r+1}} \qquad (1.9.18)$$

when $|x|$ is sufficiently large. Hence substitution yields the equation

$$p(x)\left(\frac{n}{x} + \frac{s_1}{x^2} + \frac{s_2}{x^3} + \cdots + \frac{s_n}{x^{n+1}} + \cdots\right) = p'(x) \qquad (1.9.19)$$

from which the desired relations are obtained by equating coefficients of like powers of x in the expansions

$$(x^n + a_1 x^{n-1} + \cdots + a_n)\left(\frac{n}{x} + \frac{s_1}{x^2} + \cdots + \frac{s_n}{x^{n+1}} + \cdots\right)$$
$$= nx^{n-1} + (n - 1)a_1 x^{n-2} + \cdots + a_{n-1} \qquad (1.9.20)$$

1.10 Supplementary References

The Bibliography (Appendix B) lists some of the existing general texts on numerical analysis, together with a selection of collateral texts, journal references, and sources of tables. For historical references to early developments and contributions, see Whittaker and Robinson [1944], Nörlund [1954], and Kopal [1961]. Todd [1962] presents a concise critical summary of many aspects of "classical" numerical analysis. For interesting examples of challenges and pitfalls, see Stegun and Abramowitz [1956] and Forsythe [1958].

The fundamental existence proof of Weierstrass [1885] on uniform polynomial approximation was supplemented by a constructive proof by Bernstein [1912*a*], which actually *displayed* a qualifying polynomial for each specified error tolerance. For references to the more modern theory of approximation, see Sec. 9.19.

The logic of computer arithmetic is treated in references such as Flores [1960, 1963]. Numerous texts which combine treatments of theoretical aspects of numerical analysis with accounts of computer programming and related topics, in varying ratios, include Moursund and Duris [1967], Pennington [1970], and Arden and Astill [1970].

The classical paper on the effects of roundoff errors in large-scale numerical computation is Rademacher [1948]. For later developments and accounts, see chapters in Hamming [1962] and Henrici [1964] as well as the more comprehensive treatments in Wilkinson [1964] and Rall [1965]. (The last two references also contain extensive bibliographies.)

Feller [1968] and Cramér [1946] are standard references for topics in probability and statistics, and Burnside and Panton [1935] is good for classical results in the theory of equations.

PROBLEMS

Section 1.2

1 Determine A_0, A_1, and A_2 such that the function $y(x) = A_0 + A_1 x + A_2 x^2$ and the function $f(x) = 1/(1 + x)$ have each of the following sets of properties in common:

(*a*) $f(0), f(\frac{1}{2}), f(1)$

(*b*) $f(0), f'(0), f''(0)$

(*c*) $f(\frac{1}{2}), f'(\frac{1}{2}), f''(\frac{1}{2})$

(*d*) $f'(0), f(\frac{1}{2}), f'(1)$

(*e*) $\int_0^1 f(x)\,dx, \int_0^1 xf(x)\,dx, \int_0^1 x^2 f(x)\,dx$

2 Calculate three-place values of the function $f(x) = 1/(1 + x)$ and each of the parabolic approximations obtained in Prob. 1 at a spacing of 0.1 over [0, 1], and plot curves representing the errors in each approximation on a common graph.

3 Proceed as in Prob. 1 (when this is possible) with the approximation $y(x) = B_0 + B_1 \cos 2\pi x + B_2 \sin 2\pi x$.

4 Determine that member $y(x)$ of the set of all linear functions which best approximates the function $f(x) = x^2$ over $[0, 1]$ in the sense that each of the following quantities is minimized:

(*a*) $\int_0^1 [f(x) - y(x)]^2 \, dx$

(*b*) $[f(0) - y(0)]^2 + [f(\frac{1}{2}) - y(\frac{1}{2})]^2 + [f(1) - y(1)]^2$

(*c*) $\max_{0 \leqq x \leqq 1} |f(x) - y(x)|$

(*d*) $\int_0^1 x(1 - x)[f(x) - y(x)]^2 \, dx$

5 Determine c_1, c_2, and c_3 in such a way that the formula

$$\int_{-1}^{1} w(x)f(x) \, dx = c_1 f(-1) + c_2 f(0) + c_3 f(1)$$

yields an exact result when $f(x)$ is 1, x, x^2, and x^3, and hence also when $f(x)$ is any linear combination of those functions, for each of the following weighting functions:

(*a*) $w(x) = 1$; (*b*) $w(x) = \sqrt{1 - x^2}$; (*c*) $w(x) = \dfrac{1}{\sqrt{1 - x^2}}$

Section 1.3

6 Let $S = u_0 + u_1 + \cdots + u_k + R_k$ for $k = 0, 1, \ldots$. By noticing that

$$u_n + R_n = R_{n-1} \qquad u_{n+1} + R_{n+1} = R_n$$

deduce that *if R_n and R_{n-1} have opposite signs, then R_n is smaller than u_n in magnitude, and is of opposite sign, whereas if also R_n and R_{n+1} have opposite signs, then R_n is also smaller than u_{n+1} in magnitude, and is of the same sign.* (This is often known as *Steffensen's error test*.)

7 Let $S_k = v_0 - v_1 + v_2 - \cdots + (-1)^{k-1} v_{k-1}$ for $k = 1, \ldots,$ where all v's are positive. Assume also that $v_{k+1} < v_k$ for all k, and that $v_k \to 0$ as $k \to \infty$. Show that S_{2k} is positive and increasing with k, but that S_{2k} cannot exceed v_0. Hence deduce that S_{2k} tends to a limit as $k \to \infty$. Show also that S_{2k+1} tends to the same limit, and hence that the series $\sum_0^\infty (-1)^k v_k$ then converges to a limit S. Finally, show that the truncation error R_k is of the same sign as the first neglected term and is smaller than that term in magnitude. (Notice that any finite number of terms not satisfying the stated requirements may be added to the series initially, without impairing its convergence.)

8 Suppose that the alternating series

$$S = v_0 - v_1 + v_2 - \cdots \equiv \sum_{k=0}^{\infty} (-1)^k v_k$$

converges. Show that the series

$$\tfrac{1}{2}v_0 + \tfrac{1}{2}(v_0 - v_1) - \tfrac{1}{2}(v_1 - v_2) + \cdots = \tfrac{1}{2}v_0 + \tfrac{1}{2}\sum_{k=0}^{\infty}(-1)^k(v_k - v_{k+1})$$

converges to the same sum.

9 Use the transformation of Prob. 8 repeatedly to show that

$$S \equiv 1 - \tfrac{1}{2} + \tfrac{1}{3} - \tfrac{1}{4} + \cdots \equiv \sum_{k=0}^{\infty} \frac{(-1)^k}{k+1}$$

$$= \frac{1}{2} + \frac{1}{2}\sum_{k=0}^{\infty}\frac{(-1)^k}{(k+1)(k+2)} = \frac{5}{8} + \frac{1}{2}\sum_{k=0}^{\infty}\frac{(-1)^k}{(k+1)(k+2)(k+3)}$$

$$= \frac{2}{3} + \frac{3}{4}\sum_{k=0}^{\infty}\frac{(-1)^k}{(k+1)(k+2)(k+3)(k+4)} = \cdots$$

Show that the retention of five terms in the last sum given ensures that $0.69306 < S < 0.69330$ or that $S \approx 0.69318$ with a maximum error of ± 12 units in the place of the fifth digit. About how many terms of the original series would be needed to ensure this accuracy? (The true value is $S = \log 2 \doteq 0.69315$.)

10 If $f(x)$ is a positive decreasing function of x, and if $\int_K^\infty f(x)\,dx$ exists for some K, show that $\sum_1^\infty f(k)$ converges. Show also that

$$\int_K^\infty f(x)\,dx < \sum_{k=K}^{\infty} f(k) < \int_{K-1}^{\infty} f(x)\,dx$$

How many terms of the series

$$S = \sum_{k=1}^{\infty}\frac{1}{k^2+1}$$

would be required to determine the sum to four digits?

11 By making appropriate use of the known results

$$\sum_{k=1}^{\infty}\frac{1}{k^2} = \frac{\pi^2}{6} \qquad \sum_{k=1}^{\infty}\frac{1}{k^4} = \frac{\pi^4}{90} \qquad \sum_{k=1}^{\infty}\frac{1}{k^6} = \frac{\pi^6}{945}$$

evaluate the sum

$$\sum_{k=1}^{\infty}\frac{1}{k^2+1} \equiv \sum_{k=1}^{\infty}\left[\frac{1}{k^2} - \frac{1}{k^2(k^2+1)}\right] \equiv \cdots$$

correctly to four digits.

12 The *error function* is defined by the relation

$$\operatorname{erf} x = \frac{2}{\sqrt{\pi}}\int_0^x e^{-t^2}\,dt$$

It is known that $\operatorname{erf} x \to 1$ as $x \to \infty$. With the definitions

$$F_1(x) = \int_0^x e^{-t^2}\,dt \qquad F_2(x) = \int_x^\infty e^{-t^2}\,dt$$

there follows

$$\operatorname{erf} x = \frac{2}{\sqrt{\pi}} F_1(x) = 1 - \frac{2}{\sqrt{\pi}} F_2(x)$$

Show that

$$F_1(x) = \sum_{k=0}^{\infty} \frac{(-1)^k x^{2k+1}}{(2k+1)k!}$$

where the series converges for all x. About how many terms would be required for five-digit accuracy when $x = 0.2$, 1, and 2?

13 With the notation of Prob. 12, make use of repeated integration by parts to show that

$$e^{x^2} F_1(x) = x + \frac{2}{3} x^3 + \frac{2}{3} \cdot \frac{2}{5} x^5 + \cdots + \left(\frac{2}{3} \cdot \frac{2}{5} \cdots \frac{2}{2n-1} \right) x^{2n-1}$$
$$+ \left(\frac{2}{3} \cdot \frac{2}{5} \cdots \frac{2}{2n-1} \right) \cdot 2 \int_0^x e^{x^2 - t^2} t^{2n} \, dt$$

and hence that

$$F_1(x) = e^{-x^2} \left[x + \frac{2}{3} x^3 + \frac{2}{3} \cdot \frac{2}{5} x^5 + \cdots \right.$$
$$\left. + \left(\frac{2}{3} \cdot \frac{2}{5} \cdots \frac{2}{2n-1} \right) x^{2n-1} \right] + E_n(x)$$

where

$$E_n(x) = \left(\frac{2}{1} \cdot \frac{2}{3} \cdot \frac{2}{5} \cdots \frac{2}{2n-1} \right) \int_0^x e^{-t^2} t^{2n} \, dt$$

Show also that $E_n(x)$ is smaller in magnitude than the term following the last one retained in the coefficient of e^{-x^2}, and is of the same sign, that the *relative* error cannot exceed $(2x)^{2n} n!/(2n)!$, and that the infinite series obtained when $n \to \infty$ converges for all x.

14 With the notation of Prob. 12, show that

$$e^{x^2} F_2(x) = \int_x^{\infty} (t e^{x^2 - t^2}) \frac{dt}{t}$$

and, after successive integrations by parts (each followed by multiplication and division by t in the integrand), deduce that

$$e^{x^2} F_2(x) = \frac{1}{2} \left[\frac{1}{x} - \frac{1}{2} \frac{1}{x^3} + \frac{1}{2} \cdot \frac{3}{2} \frac{1}{x^5} - \cdots \right.$$
$$+ (-1)^n \left(\frac{1}{2} \cdot \frac{3}{2} \cdots \frac{2n-1}{2} \right) \frac{1}{x^{2n+1}}$$
$$\left. + (-1)^{n+1} \left(\frac{1}{2} \cdot \frac{3}{2} \cdots \frac{2n+1}{2} \right) \cdot 2 \int_x^{\infty} e^{x^2 - t^2} \frac{dt}{t^{2n+2}} \right]$$

and hence that

$$\operatorname{erf} x = 1 - \frac{e^{-x^2}}{\sqrt{\pi}}\left[\frac{1}{x} - \frac{1}{2}\frac{1}{x^3} + \frac{1}{2}\cdot\frac{3}{2}\frac{1}{x^5} - \cdots + (-1)^n\left(\frac{1}{2}\cdot\frac{3}{2}\cdots\frac{2n-1}{2}\right)\frac{1}{x^{2n+1}}\right] + E_n(x)$$

where

$$E_n(x) = (-1)^n\frac{2}{\sqrt{\pi}}\left(\frac{1}{2}\cdot\frac{3}{2}\cdots\frac{2n+1}{2}\right)\int_x^\infty e^{-t^2}\frac{dt}{t^{2n+2}}$$

Show also that the series is divergent but asymptotic (in the strict sense) and that the truncation error due to neglect of $E_n(x)$ is smaller than the last term retained and of opposite sign. Obtain the best possible approximation when $x = 2$, and give numerical bounds on the error.

15 By making use of the known expansions of e^x, $\cos x$, $\sin x$, and $(1 - x)^{1/2}$ in powers of x, obtain the coefficients of powers of x through the fourth in the corresponding expansions of the following functions:

(*a*) $e^x \cos x$; (*b*) $\dfrac{e^x}{\cos x}$; (*c*) $(\cos x)^{1/2}$; (*d*) $e^{\sin x}$

16 Under the assumption that a given series

$$y = a_0 + a_1x + a_2x^2 + \cdots \qquad (a_1 \neq 0)$$

converges for sufficiently small values of x, and that x can be expanded in a series of powers of $(y - a_0)/a_1$ which converges for y sufficiently near to a_0, in the form

$$x = u + A_2u^2 + A_3u^3 + \cdots \qquad \left(u = \frac{y - a_0}{a_1}\right)$$

show that the leading coefficients in the inverted series can be determined from the relations

$$\begin{aligned} a_1A_2 &= -a_2, \\ a_1A_3 &= -2a_2A_2 - a_3, \\ a_1A_4 &= -a_2(A_2^2 + 2A_3) - 3a_3A_2 - a_4 \\ &\cdots\cdots\cdots\cdots \end{aligned}$$

Show also that the first n terms of the inverted series can be obtained by a sequence of $n - 1$ substitutions in the right-hand member of the relation

$$x = u - \frac{a_2}{a_1}x^2 - \frac{a_3}{a_1}x^3 - \cdots$$

starting with $x^{(1)} = u$, and retaining only powers of u not exceeding the $(r + 1)$th in the rth substitution. Illustrate both methods in obtaining the first four terms in the result of inverting the series $e^x = 1 + x + \frac{1}{2}x^2 + \cdots$.

17 It is required to determine the symmetrically placed pair of nonzero roots of the equation

$$\sinh x = cx$$

where c is a real constant such that $c > 1$. Show that, with the abbreviations $s = 6(c - 1)$, $t = x^2$, the problem can be considered as that of inverting the series

$$s = t + \frac{3!}{5!}t^2 + \frac{3!}{7!}t^3 + \frac{3!}{9!}t^4 + \cdots$$

and deduce the expansion

$$x^2 = s - \tfrac{1}{20}s^2 + \tfrac{2}{525}s^3 - \tfrac{13}{37800}s^4 + \cdots$$

Sections 1.4 and 1.5

18 Show that the number $(2.46)^{1/64}$ is known within less than one unit in the place of its *fifth* significant digit if 2.46 is known only to be correctly rounded to three digits.

19 Using only five-place tables of $\sin x$ and $\cos x$, determine $\cos 0.10 - \cos 0.12$ and $\tan 0.12 - \tan 0.10$ to four significant figures.

20 Values of $\cos x$ are calculated from a five-place table of $\sin x$ by use of the formula $\cos x = (1 - \sin^2 x)^{1/2}$. What can be said about the accuracy of the calculated values?

21 If all coefficients in the definition

$$f(x) = \frac{5.03241x + 0.11095}{0.75995x + 0.014915}$$

are rounded numbers, to how many significant figures is $f(x)$ determinate when x is known only to round to 3.26?

22 If $f(x) = (\sinh x - \sin x)/(\cosh x - \cos x)$, determine $f(0.1)$ to 10 significant figures.

23 Determine bounds on the degree of indeterminacy of each of the quantities $\tan^{-1} 4.017216$, $\sin^{-1} 0.986423$, $\cos 18.4178$, and $\cos 18417.8$, under the assumption that the arguments are rounded values. To how many *significant figures* are the last two quantities determinate?

Section 1.6

24 Suppose that calculations are to be made in four-digit floating-point arithmetic, assuming a double-precision accumulator, but supposing that the computer rounds the number resulting from each operation (addition, multiplication, etc.) to four digits before effecting a subsequent operation on that number. If

$$x_1 = 0.1234 \times 10^3 \qquad x_2 = 0.3456 \times 10^2 \qquad x_3 = 0.5678 \times 10^1$$

are exact numbers, evaluate the results of each of the following machine operations and, in each case, determine the absolute and relative errors associated with the result.

(a) $(x_1 \oplus x_2) \oplus x_3$
(b) $(x_3 \oplus x_2) \oplus x_1$
(c) $(x_1 \odot x_2) \odot x_3$
(d) $(x_3 \odot x_2) \odot x_1$
(e) $x_1 \odot (x_2 \oplus x_3)$
(f) $(x_1 \odot x_2) \oplus (x_1 \oplus x_3)$
(g) $(x_1 \odot x_3) \oslash x_2$
(h) $x_1 \odot (x_3 \oslash x_2)$
(i) $\{[(x_1 \oplus x_2) \ominus x_2] \ominus x_1\} \oplus 0.1000 \times 10^1$
(j) $\{[(x_1 \oplus x_2) \ominus x_1] \ominus x_2\} \oplus 0.1000 \times 10^1$

Section 1.7

25 If $f_1(x)$ and $f_2(x)$ are the frequency functions of ε_1 and ε_2, respectively, where ε_1 and ε_2 are independent random variables, show that the distribution function of $\varepsilon_1 + \varepsilon_2$ is

$$\iint_{s+t<x} f_1(s)f_2(t)\,ds\,dt = \int_{-\infty}^{x} \left[\int_{-\infty}^{\infty} f_1(u-t)f_2(t)\,dt\right] du$$

and hence that the frequency function of $\varepsilon_1 + \varepsilon_2$ is

$$f(x) = \int_{-\infty}^{\infty} f_1(x-t)f_2(t)\,dt$$

26 Use the result of Prob. 25 to show that, if ε_1 and ε_2 are independent and are normally distributed about zero means, with standard deviations σ_1 and σ_2, then $\varepsilon_1 + \varepsilon_2$ is also normally distributed about a zero mean, with standard deviation $\sigma = (\sigma_1^2 + \sigma_2^2)^{1/2}$. [Determine constants λ_1, λ_2, and α such that

$$\frac{(x-t)^2}{\sigma_1^2} + \frac{t^2}{\sigma_2^2} \equiv \frac{x^2}{\lambda_1^2} + \frac{(t-\alpha x)^2}{\lambda_2^2}$$

and set $t - \alpha x = \sqrt{2}\,\lambda_2 v$, making use of the fact that

$$\int_{-\infty}^{\infty} e^{-v^2}\,dv = \sqrt{\pi}$$

in evaluating the integral defining the required frequency function.]

27 Suppose that $\varepsilon_1, \varepsilon_2, \ldots, \varepsilon_n$ are independent random variables with a common *uniform* frequency function

$$f(x) = \begin{cases} 1 & (-\frac{1}{2} \leqq x \leqq \frac{1}{2}) \\ 0 & (\text{otherwise}) \end{cases}$$

and denote the frequency function of $\varepsilon_1 + \varepsilon_2 + \cdots + \varepsilon_n$ by $f_n(x)$. Use the result of Prob. 25 to show that

$$f_{n+1}(x) = \int_{-\infty}^{\infty} f_1(x - t) f_n(t)\, dt = \int_{x-1/2}^{x+1/2} f_n(t)\, dt$$

In particular, deduce that $f_2(x)$ is a triangular function

$$f_2(x) = \begin{cases} 1 + x & (-1 \leqq x \leqq 0) \\ 1 - x & (0 \leqq x \leqq 1) \\ 0 & (\text{otherwise}) \end{cases}$$

and that $f_3(x)$ is defined by the relations

$$f_3(x) = \begin{cases} \frac{1}{2}(\frac{3}{2} + x)^2 & (-\frac{3}{2} \leqq x \leqq -\frac{1}{2}) \\ \frac{3}{4} - x^2 & (-\frac{1}{2} \leqq x \leqq \frac{1}{2}) \\ \frac{1}{2}(\frac{3}{2} - x)^2 & (\frac{1}{2} \leqq x \leqq \frac{3}{2}) \\ 0 & (\text{otherwise}) \end{cases}$$

Finally, plot each of the functions f_1, f_2, and f_3, and compare it graphically with the frequency function corresponding to the *normal* distribution which has the same standard deviation $\sigma_n = \sqrt{n/12}$ $(n = 1, 2, 3)$.

28 If the coefficients of the polynomial

$$f(x) = \sum_{k=0}^{n} a_k x^k$$

are independently subject to random error distributions with mean value zero and with a common RMS value δ_{RMS}, whereas x is subject to an error distribution with RMS value η_{RMS}, show that the corresponding RMS error ε_{RMS} in $f(x)$ is given approximately by

$$\varepsilon_{\text{RMS}}^2 = \frac{x^{2n+2} - 1}{x^2 - 1}\,\delta_{\text{RMS}}^2 + [f'(x)]^2 \eta_{\text{RMS}}^2$$

29 Use the result of Prob. 28 to estimate the RMS error in the calculated value of

$$f(x) = 1.47x^3 - 2.48x^2 + 2.21x - 1.65$$

when $x = 2.03$, under the assumption that the values of x and the coefficients are known only to be rounded correctly to the three digits given. Within what limits is $f(x)$ actually determinate in this case? Within what limits does its value lie with probability of about 0.9?

30 If $x_1, x_2, \ldots, x_r$ are each rounded to n decimal places, show that the corresponding RMS error in $f(x_1, x_2, \ldots, x_r)$ is approximated by

$$\varepsilon_{\text{RMS}} \approx (0.29 \times 10^{-n}) \left[\sum_{k=1}^{r} \left(\frac{\partial f}{\partial x_k} \right)^2 \right]^{1/2}$$

Show also that if

$$\left[\sum_{k=1}^{r}\left(\frac{\partial f}{\partial x_k}\right)^2\right]^{1/2} < 2K$$

and if r is not too small (say $r > 3$), then the odds are about 10 to 1 that the error in f does not exceed K units in the nth decimal place.

Section 1.8

31 Evaluate the function

$$f(x) = \sum_{k=0}^{10} \frac{\cos kx}{k+1}$$

to five decimal places (*a*) when $x = \pi/3$ and (*b*) when $\cos x = 0.2$.

32 The kth Legendre polynomial $P_k(x)$ satisfies the recurrence formula

$$P_{k+1}(x) - \frac{2k+1}{k+1}\, xP_k(x) + \frac{k}{k+1}\, P_{k-1}(x) = 0$$

with $P_0(x) = 1$ and $P_1(x) = x$. Evaluate the function

$$f(x) = \sum_{k=0}^{10} \frac{(-1)^k}{(k+1)^2}\, P_k(x)$$

to five decimal places when $x = 0.102$.

33 Determine how many decimal places should be retained in order to evaluate $P_{10}(0.102)$ safely to five decimal places by use of the recurrence formula of Prob. 32; then make the calculation. Finally, obtain $P_{10}(x)$ analytically and check your result.

34 Use a method of backward recursion to determine $J_5(1)$ to five significant figures, given that $J_0(1) \doteq 0.7651977$. (The true value rounds to 0.000249758.)

35 Indicate why the use of (1.8.21) with (1.8.20) should not be expected to be appropriate for the numerical approximation of $J_n(1)$ when n is fairly large (by forward recursion), and verify this fact numerically when $n = 5$, retaining only five decimal places. [Show that (1.4.19) can be used to explain the unfavorable error propagation.]

Section 1.9†

36 If $[a, b] = [-1, 1]$, show that the conclusions of Theorems 1 and 2 do not hold for $f(x) = 1/x$, that those of Theorems 3 and 4 do not hold for $f(x) = 1 - x^{2/3}$, and that those of Theorems 7 and 8 do not hold for $f(x) = x$ and $g(x) = x^3$. Account for each of these situations.

† The truth of the theorems stated in Sec. 1.9 may be assumed in the following problems.

37 Assuming the fact that

$$\int_0^\infty \frac{\sin t}{t}\,dt = \frac{\pi}{2}$$

show that

$$\int_0^\infty \frac{\sin xs}{s}\,ds = \begin{cases} -\dfrac{\pi}{2} & (x < 0) \\ 0 & (x = 0) \\ \dfrac{\pi}{2} & (x > 0) \end{cases}$$

Thus show that the conclusions of Theorems 9 and 10 do not hold for

$$F(x, s) = \frac{\sin xs}{s}$$

when $a = 0$ and $b = \infty$.

38 If $f(x)$ vanishes at $n + 1$ distinct points in the interval $a \leqq x \leqq b$, and if $f^{(n)}(x)$ exists for $a < x < b$, show that $f^{(n)}(x)$ vanishes at least once in (a, b).

39 If $a_r > 0$ for $r = 1, 2, \ldots, n$, show that

$$a_1 \sin t + a_2 \sin 2t + \cdots + a_n \sin nt = \sin \theta t \sum_{r=1}^{n} a_r$$

for some θ such that $1 < \theta < n$.

40 Show that

$$\int_x^\infty \frac{dt}{t^4 + 1} < \frac{1}{3x^3} \qquad (x > 0)$$

and that

$$\int_0^x t^2 e^{-t^2} \frac{dt}{t^3 + 1} < \tfrac{1}{3} \log (1 + x^3) < \tfrac{1}{3} x^3 \qquad (x > 0)$$

41 Show that

$$\int_{-1}^{1} (1 - x^2) f(x)\,dx = \tfrac{4}{3} f(\xi)$$

for some ξ in $(-1, 1)$ if $f(x)$ is continuous in that interval. Also determine ξ when $f(x) = x^2$.

42 If $F(k)$ is defined by the integral

$$F(k) = \int_0^1 \frac{x(x-1)(x-2)\cdots(x-k+1)}{k!}\,dx \qquad (k \geqq 2)$$

use the second law of the mean to show that

$$F(k) = (-1)^{k+1} \frac{(k-1-\xi)(k-2-\xi)\cdots(2-\xi)}{6k!} \qquad (0 < \xi < 1)$$

and deduce that

$$\frac{1}{6k(k-1)} < (-1)^{k+1}F(k) < \frac{1}{6k}$$

43 If $g(x)$ is continuous and $f(x)$ possesses a continuous derivative, and if

$$\phi(x) = \int_0^x f(x-t)g(t)\,dt$$

obtain an expression for $d\phi/dx$. By making an appropriate change of variables in the definition of $\phi(x)$, obtain an alternative expression for $d\phi/dx$ when the hypotheses regarding f and g are interchanged.

44 If $y'' = 2\sin y + 12x^2$ and $y(0) = y'(0) = 0$, show that

$$y(x) = x^4 + 2\int_0^x (x-s)\sin y(s)\,ds$$

and deduce that $y(x)$ lies between $x^4 - x^2$ and $x^4 + x^2$.

45 Determine the first three coefficients in the Bürmann series

$$\sin x = c_1(e^x - 1) + c_2(e^x - 1)^2 + c_3(e^x - 1)^3 + \cdots$$

and use the result to determine approximately the value of $\sin x$ when $e^x = 1.012$.

46 If $y = a_0 + a_1x + a_2x^2 + \cdots$ and if x can be expanded in a series of powers of $y - a_0$ for y near a_0, use the Bürmann expansion to show that the leading coefficients in the expansion

$$x = c_1(y - a_0) + c_2(y - a_0)^2 + c_3(y - a_0)^3 + \cdots$$

are given by

$$c_1 = \frac{1}{a_1} \qquad c_2 = -\frac{a_2}{a_1^3} \qquad c_3 = \frac{2a_2^2 - a_1a_3}{a_1^5}$$

and verify that the results agree with those of Prob. 16.

47 Derive the leading terms in the Bürmann expansion of x in powers of $\sin x$ in the form

$$x = \sin x + \tfrac{1}{6}\sin^3 x + \tfrac{3}{40}\sin^5 x + \cdots$$

and use the result to approximate $\sin^{-1}\frac{1}{2}$.

48 Show that Budan's rule reduces to Descartes' rule when $a = 0$ and $b = \infty$. Then use this fact to show that either rule is deducible from the other.

49 A polynomial $p(x)$ with real coefficients has $2m$ consecutive vanishing coefficients. Show that at least $2m$ zeros of $p(x)$ either vanish or are nonreal. Also illustrate this conclusion when $p(x) = x^5 + x^2$.

50 Show that the polynomial

$$p(x) = x^4 - 5x^3 + 5x^2 + 5x - 5$$

has one negative zero, one zero in (0, 1), either no zeros or two zeros in (2, 3), and no other real zeros.

51 Determine for what real values of c it is true that neither root of the equation

$$x^2 - 2(1 - 2c)x + 1 = 0$$

exceeds unity in magnitude.

52 Write out in detail the indicated derivation of Newton's power-sum identities.

53 Determine the sum of the rth powers of the zeros of the polynomial

$$p(x) = x^4 - 2x^3 + 3x^2 - 4x + 5$$

for $r = 1, 2, 3, 4$, and 5.

2

INTERPOLATION WITH DIVIDED DIFFERENCES

2.1 Introduction

Anyone who has had occasion to consult tables of mathematical functions is familiar with the method of *linear interpolation* and probably has encountered situations in which this method of "reading between the lines of the table" has appeared to be unreliable. If more reliable interpolates are desired, it is clearly necessary to make use of more information than that consisting of tabulated values (*ordinates*) of a function, corresponding to only *two* successive abscissas. Whereas that additional information could consist, for example, of known values of certain *derivatives* of the function at those two points, it is supposed in most of what follows (an exception is found in Sec. 8.2) that the interpolation process is to be based only on tabulated values of the function itself, with any further available information reserved for use in estimating the error involved.

There exist a number of interpolation formulas which have this property, most of which possess certain advantages in certain situations, but no one of which is preferable to all others in all respects. Whereas certain of these formulas are expressed explicitly in terms of all the ordinates on which they depend (Chap. 3), others involve only one or two of the ordinates explicitly and express their dependence upon other ordinates only in terms of differences of ordinates and successive differences of differences.

In the general case, when the abscissas are not necessarily equally spaced, the use of so-called *divided differences* affords certain conveniences. The principal purpose of this chapter is to define such differences and investigate certain of their properties, to obtain a basic interpolation formula due to Newton (Sec. 2.5), from which most of the other formulas of the type described can be deduced, and to obtain expressions for the error term (Sec. 2.6). Related methods of iterated linear interpolation (Sec. 2.7) and inverse interpolation (Sec. 2.8) are also treated.

2.2 Linear Interpolation

The assumption that a function $f(x)$ is approximately *linear*, in a certain range, is equivalent to the assumption that the ratio

$$\frac{f(x_1) - f(x_0)}{x_1 - x_0} \tag{2.2.1}$$

is approximately independent of x_0 and x_1 in that range. This ratio is called the *first divided difference* of $f(x)$, relative to x_0 and x_1, and may be designated by $f[x_0, x_1]$:†

$$f[x_0, x_1] \equiv \frac{f(x_1) - f(x_0)}{x_1 - x_0} \tag{2.2.2}$$

It is clear that $f[x_1, x_0] = f[x_0, x_1]$.

Thus the assertion of approximate linearity may be expressed in the form

$$f[x_0, x] \approx f[x_0, x_1] \tag{2.2.3}$$

which leads to the interpolation formula

$$f(x) \approx f(x_0) + (x - x_0)f[x_0, x_1] \tag{2.2.4}$$

or

$$f(x) \approx f(x_0) + \frac{x - x_0}{x_1 - x_0}[f(x_1) - f(x_0)] \tag{2.2.4'}$$

or, equivalently, to the formula

$$f(x) \approx \frac{1}{x_1 - x_0}[(x_1 - x)f(x_0) - (x_0 - x)f(x_1)] \tag{2.2.4''}$$

which can also be expressed in the convenient *determinantal* form

$$f(x) \approx \frac{1}{x_1 - x_0}\begin{vmatrix} f(x_0) & x_0 - x \\ f(x_1) & x_1 - x \end{vmatrix} \tag{2.2.5}$$

† Various other notations are used, such as $[x_0, x_1]$, $f(x_0, x_1)$, and (x_0, x_1).

It may be noticed that (2.2.4) involves one ordinate and a divided difference, (2.2.4′) involves one ordinate and an ordinary difference, and (2.2.4″) involves the two ordinates directly.

It is convenient to designate the linear function defined by the right-hand member of (2.2.4) by $p_{0,1}(x)$, the subscripts corresponding to the ordinates used in its formation. For symmetry of notation, it is desirable to write also

$$p_0(x) \equiv f[x_0] \equiv f(x_0) \qquad (2.2.6)$$

so that $f[x_0]$ is defined as the *zeroth* divided difference relative to x_0 and is merely the value of $f(x)$ at $x = x_0$, and $p_0(x)$ is the approximating polynomial of degree *zero* which agrees with $f(x)$ at $x = x_0$. With this notation, Eqs. (2.2.4) and (2.2.5) become

$$f(x) \approx p_{0,1}(x) \equiv f[x_0] + (x - x_0)f[x_0, x_1] \qquad (2.2.7)$$

and

$$p_{0,1}(x) = \frac{1}{x_1 - x_0}\begin{vmatrix} p_0(x) & x_0 - x \\ p_1(x) & x_1 - x \end{vmatrix} \qquad (2.2.8)$$

These forms are given here principally to correspond to more general forms to be obtained in following sections.

We see that the approximation $f(x) \approx p_{0,1}(x)$ is exact for *all* values of x if $f(x)$ is indeed a linear function, of the form $f(x) = A_0 + A_1x$, and further that the approximation is exact at the points $x = x_0$ and x_1 for *any* function $f(x)$.

As a numerical example, the linear interpolation of sinh x for $\bar{x} = 0.23$, from tabulated five-place values for $x_0 = 0.20$ and $x_1 = 0.30$, may be arranged as follows:

x_i	$f(x_i)$	$x_i - \bar{x}$
0.20	0.20134	−0.03
0.30	0.30452	0.07

$$f(0.23) \approx \frac{(0.07)(0.20134) - (-0.03)(0.30452)}{0.10} = 0.23229$$

Since the true five-place value is 0.23203, it is seen that linear interpolation here affords only three-place accuracy.

It is useful to notice that, since a linear interpolation merely effects a certain *weighted average* of the two ordinates involved, the result of an interpolation involving two ordinates such as 13.6340 and 13.6393 can be considered as the sum of 13.6300 and the result of effecting the same interpolation on 40 and 93, with this result added to 13.6300 in units of its last place.

Further, since the numerator and denominator of the ratio (2.2.5) are homogeneous in the abscissas, the entries x_i and $x_i - \bar{x}$ in the computational array may be multiplied by any convenient common factor. In particular, the x's in the preceding table could be replaced by 20 and 30, and the entries in the last column correspondingly by -3 and 7.

Unless $f(x)$ is truly linear, the secant slope $f[x_0, x_1]$ will depend upon the abscissas x_0 and x_1. However, if $f(x)$ were a second-degree polynomial, the secant-slope function $f[x_1, x]$ would itself be a linear function of x, for fixed x_1. That is, the ratio

$$\frac{f[x_1, x_2] - f[x_0, x_1]}{x_2 - x_0}$$

would be independent of x_0, x_1, and x_2. This ratio is called the *second divided difference*, relative to those three abscissas, and is designated here by $f[x_0, x_1, x_2]$:

$$f[x_0, x_1, x_2] \equiv \frac{f[x_1, x_2] - f[x_0, x_1]}{x_2 - x_0} \tag{2.2.9}$$

In particular, since $f[x_1, x_0, x] = f[x_0, x_1, x]$ (see Sec. 2.3), the difference between the two members of (2.2.3) can be expressed as

$$f[x_0, x] - f[x_0, x_1] = f[x_0, x] - f[x_1, x_0] = (x - x_1)f[x_0, x_1, x]$$

so that the *approximation* (2.2.4) can be replaced by the *identity*

$$f(x) = f[x_0] + (x - x_0)f[x_0, x_1] + (x - x_0)(x - x_1)f[x_0, x_1, x] \tag{2.2.10}$$

Thus the *error* committed in (2.2.7), by replacing $f(x)$ by $p_{0,1}(x)$, is given by

$$E(x) \equiv f(x) - p_{0,1}(x) = (x - x_0)(x - x_1)f[x_0, x_1, x] \tag{2.2.11}$$

Whereas knowledge of $f[x_0, x_1, x]$ is tantamount to knowledge of the exact interpolant $f(x)$, the form (2.2.11) of the error is a special case of a more general form to be obtained, which (as will be shown) is frequently useful in obtaining an *estimate* of the error in an actual calculation. For any *linear* function $f(x)$, the error term will indeed vanish identically, as may be verified directly.

Before generalizing the result just obtained, it is desirable to define divided differences of all orders, and to investigate certain of their properties.

2.3 Divided Differences

Divided differences of orders $0, 1, 2, \ldots, k$ are defined recursively by the relations

$$f[x_0] = f(x_0) \qquad f[x_0, x_1] = \frac{f[x_1] - f[x_0]}{x_1 - x_0} \qquad \cdots$$

$$f[x_0, \ldots, x_k] = \frac{f[x_1, \ldots, x_k] - f[x_0, \ldots, x_{k-1}]}{x_k - x_0} \tag{2.3.1}$$

We notice that the first $k - 1$ arguments in the first term of the numerator are the same as the last $k - 1$ arguments in the second term and that the denominator is the difference between those arguments which are not in common to the two terms. It is clear from the definition that $f[x_0, \ldots, x_k]$ is a linear combination of the $k + 1$ ordinates $f(x_0), \ldots, f(x_k)$, with the coefficients depending upon the corresponding $k + 1$ abscissas.

When $k = 1$, the divided difference obviously is a symmetric function of its arguments, that is, $f[x_1, x_0] = f[x_0, x_1]$. It is shown next that the same statement applies to divided differences of all orders. In order to establish this fact directly in the case of $k = 2$, we may write

$$f[x_0, x_1, x_2] = \frac{f[x_1, x_2] - f[x_0, x_1]}{x_2 - x_0}$$

$$= \frac{1}{x_2 - x_0}\left[\frac{f(x_2) - f(x_1)}{x_2 - x_1} - \frac{f(x_1) - f(x_0)}{x_1 - x_0}\right]$$

and the result can be put in the symmetric form

$$f[x_0, x_1, x_2] = \frac{f(x_0)}{(x_0 - x_1)(x_0 - x_2)} + \frac{f(x_1)}{(x_1 - x_0)(x_1 - x_2)} + \frac{f(x_2)}{(x_2 - x_0)(x_2 - x_1)}$$

This result suggests the truth of the more general relation

$$f[x_0, \ldots, x_k] = \frac{f(x_0)}{(x_0 - x_1)\cdots(x_0 - x_k)} + \frac{f(x_1)}{(x_1 - x_0)\cdots(x_1 - x_k)} + \cdots + \frac{f(x_k)}{(x_k - x_0)\cdots(x_k - x_{k-1})} \tag{2.3.2}$$

for any positive integer k, so that the coefficient of $f(x_i)$ is

$$\alpha_i^{(k)} = \frac{1}{(x_i - x_0)\cdots(x_i - x_k)} \qquad (i = 0, 1, \ldots, k) \tag{2.3.3}$$

where the zero factor $(x_i - x_i)$ is to be omitted in the denominator.

In order to establish this conjecture by induction, suppose that it has been proved for $k = r$. If we recall the definition

$$f[x_0, \ldots, x_{r+1}] = \frac{1}{x_{r+1} - x_0}\{f[x_1, \ldots, x_{r+1}] - f[x_0, \ldots, x_r]\} \tag{2.3.4}$$

it then follows that, for $i = 1, 2, \ldots, r$, the coefficient of $f(x_i)$ in the right-hand member is given by

$$\frac{1}{x_{r+1} - x_0}\left[\frac{1}{(x_i - x_1)\cdots(x_i - x_{r+1})} - \frac{1}{(x_i - x_0)\cdots(x_i - x_r)}\right]$$
$$= \frac{1}{(x_i - x_0)\cdots(x_i - x_{r+1})} = \alpha_i^{(r+1)} \tag{2.3.5}$$

in accordance with (2.3.3) with $k = r + 1$. When $i = 0$ or $r + 1$, only one of the terms in the right-hand member of (2.3.4) involves the ordinate $f(x_i)$, and the respective coefficient also is easily seen to be in accordance with (2.3.3) with $k = r + 1$. Thus, if (2.3.2) is valid for $k = r$, it is valid also for $k = r + 1$. Since it has been established for $k = 1$ (and $k = 2$), it is therefore valid for any positive integer k, as was to be shown.

It follows, from the symmetry of (2.3.2), that the order of the arguments is irrelevant. Hence $f[x_0, \ldots, x_k]$ can be expressed as the difference between two divided differences of order $k - 1$, having any $k - 1$ of their k arguments in common, divided by the difference between those arguments which are *not* in common. For example, there follows

$$f[x_0, x_1, x_2, x_3] = \frac{f[x_1, x_2, x_3] - f[x_0, x_1, x_2]}{x_3 - x_0}$$
$$= \frac{f[x_0, x_2, x_3] - f[x_1, x_2, x_3]}{x_0 - x_1} = \cdots$$

In those cases when two or more arguments in a divided difference become coincident, recourse must be had to appropriate limiting processes. Thus, for example, if we set $x_1 = x + \varepsilon$, there follows

$$f[x_1, x] \equiv f[x + \varepsilon, x] = \frac{f(x + \varepsilon) - f(x)}{\varepsilon}$$

and, in the limit when $\varepsilon \to 0$, we have

$$f[x, x] = f'(x) \tag{2.3.6}$$

if $f(x)$ is differentiable. A similar argument shows that

$$\frac{d}{dx} f[x_0, \ldots, x_k, x] = f[x_0, \ldots, x_k, x, x] \tag{2.3.7}$$

if $x_0, \ldots, x_k$ are constants. If $u_1, u_2, \ldots, u_n$ are differentiable functions of x, there follows also

$$\frac{d}{dx} f[x_0, \ldots, x_k, u_1, \ldots, u_n] = \sum_{\nu=1}^{n} f[x_0, \ldots, x_k, u_1, \ldots, u_n, u_\nu] \frac{du_\nu}{dx}$$

and hence, by taking $u_1 = \cdots = u_n = x$, we may deduce that

$$\frac{d}{dx} f[x_0, \ldots, x_k, \overbrace{x, \ldots, x}^{n \text{ times}}] = nf[x_0, \ldots, x_k, \overbrace{x, \ldots, x}^{n+1 \text{ times}}] \tag{2.3.8}$$

Finally, by successive differentiation of (2.3.7) combined with the use of (2.3.8) at each step, we may establish the additional useful formula

$$\frac{d^r}{dx^r} f[x_0, \ldots, x_k, x] = r! f[x_0, \ldots, x_k, \overbrace{x, \ldots, x}^{r+1 \text{ times}}] \tag{2.3.9}$$

In particular, we may deduce that *the result of allowing* $r + 1$ *arguments of a divided difference to become coincident is finite if the* rth *derivative of* $f(x)$ *is finite at the point of confluence.*

It is seen that $f[x_0, \ldots, x_k, x]$ is continuous at $x = \bar{x}$ if $\bar{x}$ is not identified with $x_0, x_1, \ldots$, or x_k, and if $f(x)$ is continuous at $\bar{x}$. If $f'(x)$ does not exist at x_0, the function $f[x_0, \ldots, x_k, x]$ generally will not tend to a finite limit as $x \to x_0$. Thus, for example, if $f(x) = \sqrt{x}$, there follows $f[0, x] = 1/\sqrt{x}$, and this function naturally becomes infinite as $x \to 0$.

However, since the product

$$(x - x_0) f[x_0, \ldots, x_k, x] \equiv P(x) \tag{2.3.10}$$

is identical with the difference

$$f[x, x_1, \ldots, x_k] - f[x_0, x_1, \ldots, x_k]$$

it follows that $P(x)$ vanishes when $x = x_0$ for any function $f(x)$ defined at $x_0, \ldots, x_k$, when these abscissas are distinct. Also $P(x)$ *tends* to zero as $x \to x_0$ (and hence is *continuous* at x_0) if $f(x)$ is continuous at x_0. Further, if x_0 and r of the other k abscissas are equal, it follows also that $P(x_0) = 0$ if $f^{(r)}(x_0)$ exists and that $P(x) \to 0$ as $x \to x_0$ (and hence is continuous at x_0) if $f^{(r)}(x)$ is continuous at x_0.

It may be expected that *the* kth *divided difference of a polynomial of degree* n *is a polynomial of degree* $n - k$ *if* $k \leqq n$, *and is identically zero if* $k > n$. The proof follows easily from the fact that the first divided difference of x^m

$$\frac{x^m - x_0{}^m}{x - x_0} = x^{m-1} + x_0 x^{m-2} + \cdots$$

is a polynomial of degree $m - 1$ in x, when m is a positive integer. In particular, it is seen that

$$f[x_0, x_1, \ldots, x_n] = 1 \qquad \text{if } f(x) = x^n \tag{2.3.11}$$

2.4 Second-degree Interpolation

If the accuracy afforded by a linear interpolation is inadequate, a generally more accurate result may be based upon the supposition that $f(x)$ may be approximated by a polynomial of *second* degree near the abscissa of the interpolate. This is equivalent to assuming that, within a certain prescribed tolerance, the first divided difference $f[x, x_0]$ is a linear function of x for fixed x_0 or, equivalently, that the second divided difference $f[x, x_0, x_1]$ is constant. The hypothesis

$$f[x, x_0, x_1] \approx f[x_2, x_0, x_1] \equiv f[x_0, x_1, x_2] \tag{2.4.1}$$

then takes the form

$$\frac{f[x, x_0] - f[x_0, x_1]}{x - x_1} \approx f[x_0, x_1, x_2]$$

or, after another reduction,

$$f(x) \approx p_{0,1,2}(x) \equiv f[x_0] + (x - x_0)f[x_0, x_1] + (x - x_0)(x - x_1)f[x_0, x_1, x_2] \tag{2.4.2}$$

Since the difference between the two members of (2.4.1) is expressible as $(x - x_2)f[x_0, x_1, x_2, x]$, the error in the approximation (2.4.2) is given by

$$E(x) = (x - x_0)(x - x_1)(x - x_2)f[x_0, x_1, x_2, x] \tag{2.4.3}$$

From this result, we may deduce that $E(x) \equiv 0$ if $f(x)$ is a polynomial of degree 2 or less, and that $E = 0$ when $x = x_0, x_1$, or x_2 for *any* function $f(x)$ which is defined for those arguments. Thus $p_{0,1,2}(x)$ is a polynomial of degree 2 which agrees with $f(x)$ when $x = x_0, x_1$, and x_2.

In order to make use of (2.4.2), one may first form a *divided-difference table* as follows:

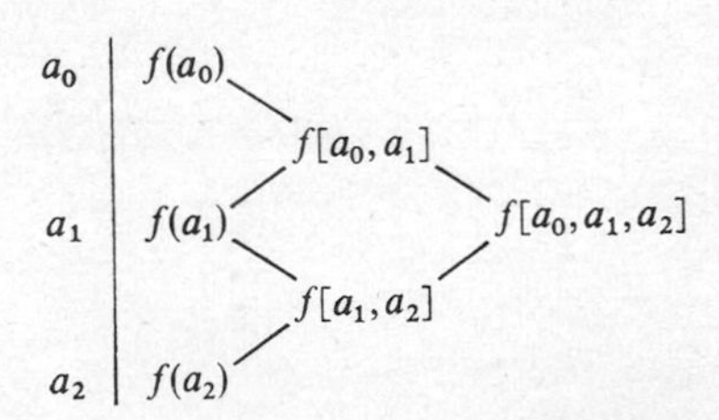

Here each entry is given by the difference between diagonally adjacent entries to its left *divided by the difference between the abscissas corresponding to the ordinates intercepted by the diagonals passing through the calculated entry.*

Thus, for $f(x) = \sinh x$, the following table may be formed, in illustration, with the abscissas 0.0, 0.2, and 0.3:

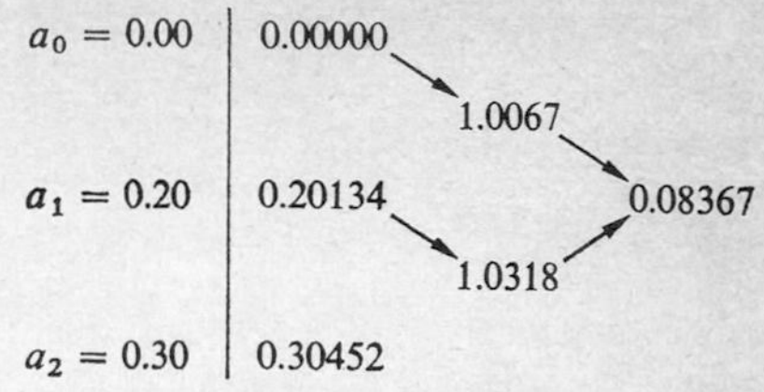

Suppose that only the given data are available and that the value of $f(0.23)$ is to be interpolated. If we take $x_i = a_i$, the calculation from (2.4.2) is of the form

$$f(0.23) \approx 0.00000 + (0.23)(1.0067) + (0.23)(0.03)(0.08367)$$
$$\doteq 0.00000 + 0.231541 + 0.000577 \doteq 0.23212$$

with an associated error of -0.00009. (One extra place was carried through the intermediate calculation, with the final result rounded to five places.)

By renumbering the x's, the calculation can be rearranged in various ways. For example, since the argument of the interpolant is nearest a_1, it may be suggested that we take $x_0 = a_1$ and, say, $x_1 = a_2$ and $x_2 = a_0$. In this case, there follows

$$f(0.23) \approx 0.20134 + (0.03)(1.0318) + (0.03)(-0.07)(0.08367)$$
$$\doteq 0.20134 + 0.030954 - 0.000176 \doteq 0.23212$$

with the same end result. The first calculation uses differences on the indicated *forward diagonal* starting from $f(a_0)$; the second uses differences on the indicated *zigzag* path starting from $f(a_1)$. By further renumbering, other paths also terminating with $f[a_0, a_1, a_2]$ could be selected, *all of which would give exactly the same value of the interpolant if no intermediate roundoff errors were present.*

The second path is the one which departs least from an imaginary horizontal line through the argument of the interpolant. Accordingly, the new information introduced at each stage of the calculation is that which may be expected to be most relevant to that interpolant, so that the *rate of approach* to the final value may be expected to be maximized at each step of the path. In addition, since the coefficients by which the successive divided differences are multiplied are smaller in magnitude along the preferred path, the effects of roundoffs introduced in the calculation of those divided differences will be somewhat reduced.†

† In this connection, it should be mentioned that if divided differences of rounded values (not rounded divided differences of true values) are used, if the results of the divisions do not require additional roundoffs, and if all following calculations are effected without roundoff, all paths which incorporate the same given data will lead to *exactly* the same end results. Thus the preferred path does *not* minimize the effects of *inherent* errors in the given data (as is sometimes argued). Those effects depend only upon the *end point* of the path and are considered in Sec. 3.2.

If the value of $f(0.10)$ were required, from the given data alone, the first path would be the preferred one from the preceding point of view and would lead to the calculation

$$\begin{aligned} f(0.10) &\approx 0.00000 + (0.10)(1.0067) + (0.10)(-0.10)(0.08367) \\ &\doteq 0.00000 + 0.10067 - 0.000837 \doteq 0.09983 \end{aligned}$$

whereas the true five-place value is 0.10017. Finally, to interpolate for $f(0.27)$, a path along the *backward diagonal* starting with $f(a_2)$ is preferable. Hence we would set $x_0 = a_2$, $x_1 = a_1$, and $x_2 = a_0$, and would obtain

$$\begin{aligned} f(0.27) &\approx 0.30452 + (-0.03)(1.0318) + (-0.03)(0.07)(0.08367) \\ &\doteq 0.30452 - 0.030954 - 0.000176 \doteq 0.27339 \end{aligned}$$

as compared with the true five-place value 0.27329.

In the preceding calculations, and in similar ones, when the number of differences to be retained has been decided in advance, and when the end point of the path is also predetermined, the reduction in loss of accuracy afforded by the *preferred path* usually is of no great consequence and the rate of approach to the final value at intermediate stages is irrelevant to the final result. Thus, the choice of paths then is relatively unimportant. However, in the more involved cases when differences of higher order are available, and when the point at which the path is to be terminated is not preassigned, it is desirable to choose that path which, when terminated after any number of steps, may be expected to afford the best result obtainable with that number of steps. The preceding examples were intended to illustrate such paths in simple cases.

2.5 Newton's Fundamental Formula

The identities (2.2.10) and (2.4.2) are special cases of a general formula, due to Newton, which may be derived as follows.

From the basic definition (2.3.1), there follows

$$\begin{aligned} f(x) &= f[x_0] + (x - x_0)f[x_0, x] \\ f[x_0, x] &= f[x_0, x_1] + (x - x_1)f[x_0, x_1, x] \\ &\cdots\cdots\cdots\cdots\cdots\cdots \\ f[x_0, \ldots, x_{n-1}, x] &= f[x_0, \ldots, x_n] + (x - x_n)f[x_0, \ldots, x_n, x] \end{aligned} \tag{2.5.1}$$

By substituting the second relation in the first, one obtains (2.2.10),

$$f(x) = f[x_0] + (x - x_0)f[x_0, x_1] + (x - x_0)(x - x_1)f[x_0, x_1, x]$$

and, by successively substituting from subsequent relations in (2.5.1), there follows finally

$$f(x) = f[x_0] + (x - x_0)f[x_0, x_1]$$
$$+ (x - x_0)(x - x_1)f[x_0, x_1, x_2]$$
$$+ \cdots + (x - x_0)\cdots(x - x_{n-1})f[x_0, \ldots, x_n] + E(x) \qquad (2.5.2)$$

where

$$E(x) = (x - x_0)\cdots(x - x_n)f[x_0, \ldots, x_n, x] \qquad (2.5.3)$$

The obvious details of the induction are omitted.

The approximate relation obtained by suppressing the error term in (2.5.2) is known as *Newton's interpolation formula with divided differences.* The resultant right-hand member, which is clearly a polynomial of degree n, may be denoted by $p_{0,\ldots,n}(x)$. An inspection of the error term then shows that $p_{0,\ldots,n}(x)$ is identical with $f(x)$ if $f(x)$ is a polynomial of degree n or less, and that it agrees with $f(x)$ at the $n + 1$ points $x = x_0, \ldots, x_n$, regardless of the form of f. Further, there exists no *other* polynomial $P(x)$ of degree n or less having this property, since, if this were the case, $P - p$ would be a polynomial of maximum degree n with $n + 1$ zeros. This situation is impossible unless $P - p$ vanishes identically.

Thus, if $f(x)$ is known at $n + 1$ distinct points $a_0, a_1, \ldots, a_n$, where $a_0 < a_1 < \cdots < a_n$, a variety of equivalent forms of the interpolation polynomial $p_{0,\ldots,n}(x)$ of degree n (or less) which agrees with $f(x)$ at these points can be obtained by identifying each of the x's in (2.5.2) with one of the a's. The various possible forms are not considered here in explicit detail. However, in Chap. 4 a more detailed consideration is given to the situation in which the abscissas $a_0, \ldots, a_n$ are *equally spaced,* so that certain simplifications are possible, and convenient use can be made of available tables of certain coefficient functions.

In illustration, we suppose that values of sinh x are given to five places for $x = 0.0$, 0.20, 0.30, and 0.50, and that sinh 0.23 is required by use of third-degree interpolation. The calculation may be arranged as follows:

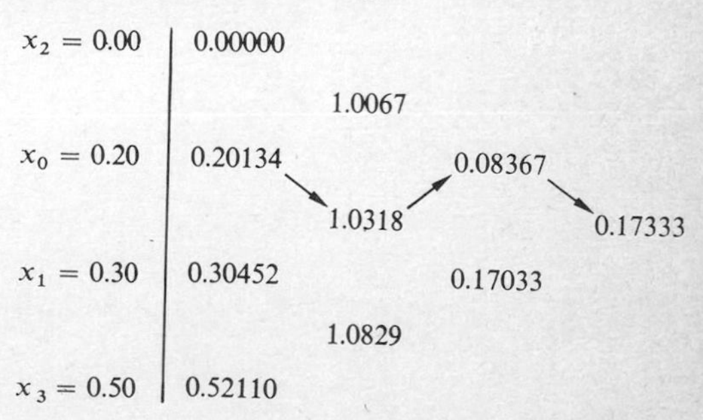

$$f(0.23) \approx 0.20134 + (0.03)(1.0318) + (0.03)(-0.07)(0.08367) + (0.03)(-0.07)(0.23)(0.17333)$$

$$\doteq 0.20134 + 0.030954 - 0.000176 - 0.000084 \doteq 0.23203$$

The initial point x_0 was taken to be as near as possible to the argument of the interpolant, and the remaining abscissas were numbered in accordance with the indicated zigzag path of differences. The same end result, which is correct to the five places given, could also have been obtained by any one of a number of other orderings of the abscissas.

Once an appropriate continuous path of differences (made up of diagonal segments, each sloping upward or downward to the right) has been selected, reference to (2.5.2) shows that *the coefficient of the* kth *difference encountered is the product of* k *factors, each of which represents the difference between the abscissa of the interpolant and the abscissa of an ordinate used in the formation of a difference previously encountered.* The instructions comprised in this statement are frequently referred to as *Sheppard's rules.*

It is convenient to speak of the data lying inside and on the boundary of the triangular region, limited by the column of ordinates (zeroth differences) in a difference table and the two diagonals passing through a specific difference in that table, as comprising the *region of determination* for that difference. It is then easily seen that *the ordinates involved in the formation of any difference are exactly those ordinates which lie in its region of determination.* Further, for a difference path of the sort considered here, the region of determination of the kth difference encountered includes the regions relevant to all differences previously encountered.

These facts permit us to write down, by inspection, the coefficient of any difference encountered in a chosen path. For example, in order to obtain the coefficient of 0.08367 in the preceding calculation, we notice that the region of determination for the *preceding* difference in the path (1.0318) includes the ordinates corresponding to the abscissas 0.20 and 0.30. Hence the desired coefficient is

$$(0.23 - 0.20)(0.23 - 0.30) = -0.0021$$

2.6 Error Formulas

It was shown in the preceding section that, if $f(x)$ is approximated by a polynomial $y(x) \equiv p_{0,\ldots,n}(x)$ of maximum degree n, which coincides with it at the $n + 1$ distinct points $x_0, \ldots, x_n$, then the error $E(x) = f(x) - y(x)$ is given by

$$E(x) = \pi(x)f[x_0, \ldots, x_n, x] \qquad (2.6.1)$$

where $\pi(x)$ is the monic polynomial of degree $n + 1$ defined as the product

$$\pi(x) = (x - x_0)(x - x_1)\cdots(x - x_n) \tag{2.6.2}$$

This form of the error term will be particularly useful in considering the accuracy of formulas for *numerical differentiation* and *integration* in subsequent chapters.

However, if $f(x)$ possesses a derivative of order $n + 1$ (and hence also continuous derivatives of order $1, 2, \ldots, n$) in the relevant interval, there exists another form of the remainder which often is more useful in other considerations.† In order to obtain it, we notice first that both $f(x) - y(x)$ and $\pi(x)$ vanish at the $n + 1$ points $x = x_0, x_1, \ldots, x_n$. We then consider a linear combination of these functions

$$F(x) \equiv f(x) - y(x) - K\pi(x) \tag{2.6.3}$$

and determine the constant K in such a way that $F(x)$ vanishes, not only at these $n + 1$ points, but also at an arbitrarily chosen point $\bar{x}$ which differs from all these points. Since $\pi(x)$ vanishes *only* at the $n + 1$ points considered previously, K certainly can be so chosen.

Let $\bar{I}$ designate the closed interval limited by the smallest and largest of the $n + 2$ values $x_0, \ldots, x_n, \bar{x}$. Then $F(x)$ vanishes at least $n + 2$ times on the interval $\bar{I}$. By Rolle's theorem (Sec. 1.9), $F'(x)$ vanishes at least $n + 1$ times inside $\bar{I}$, $F''(x)$ at least n times, . . . , and hence, finally, $F^{(n+1)}(x)$ vanishes at least once inside $\bar{I}$. Let one such point be denoted by ξ. There then follows, from (2.6.3),

$$0 = f^{(n+1)}(\xi) - y^{(n+1)}(\xi) - K\pi^{(n+1)}(\xi) \tag{2.6.4}$$

But since $y(x)$ is a polynomial of maximum degree n, its $(n + 1)$th derivative vanishes identically. Also, from the definition (2.6.2), there follows $\pi^{(n+1)}(x) \equiv (n + 1)!$. Hence (2.6.4) yields the determination

$$K = \frac{1}{(n + 1)!} f^{(n+1)}(\xi)$$

and the relation $F(\bar{x}) = 0$ becomes

$$f(\bar{x}) - y(\bar{x}) = \frac{1}{(n + 1)!} f^{(n+1)}(\xi)\pi(\bar{x})$$

for some ξ in $\bar{I}$. If $\bar{x}$ is identified with any one of the abscissas $x_0, \ldots, x_n$, both

† A third form, involving analysis in the complex plane, is derived in Chap. 4, Prob. 39.

sides of this relation vanish, so that it is valid even in that previously excluded case. The bars now may be suppressed, and there follows finally

$$E(x) = \frac{1}{(n+1)!} f^{(n+1)}(\xi)\pi(x) \qquad (2.6.5)$$

for some $\xi = \xi(x)$ in the interval I, where I is the interval limited by the largest and smallest of the numbers $x_0, x_1, \ldots, x_n, x$.

This result guarantees merely that, for any given x, there *exists* at least one corresponding number ξ in I such that the error is expressible in the given form. If $f^{(n+1)}(x)$ is *continuous* on the closed interval I (that is, the interior of I plus its end points), then $f^{(n+1)}(x)$ is *bounded* on I. In particular, there then exists a positive number M_{n+1} such that

$$|f^{(n+1)}(\xi)| \leqq M_{n+1} \qquad (2.6.6)$$

in (2.6.5) and hence

$$|E(x)| \leqq \frac{M_{n+1}}{(n+1)!} |\pi(x)| \qquad (2.6.7)$$

for all x in I. In order that this result hold, we will generally require in the sequel, not only that $f^{(n+1)}(x)$ *exist*, but also that it possess the desired continuity.

Since (2.6.1) and (2.6.5) must be equivalent, we obtain as a by-product the useful fact that

$$f[x_0, \ldots, x_n, x] = \frac{1}{(n+1)!} f^{(n+1)}(\xi) \qquad (2.6.8)$$

for some argument ξ in the interval I, whenever $f^{(n+1)}(x)$ exists in I. This fact will be needed in later developments. It can be seen that (2.6.8) continues to be valid when certain of the arguments $x_0, \ldots, x_n, x$ coalesce.

In order to illustrate the application of the error formula (2.6.5), we consider the second-degree interpolation ($n = 2$) for $f(0.23)$ effected in Sec. 2.4. Under the assumption that the analytic expression for the interpolated function is *known* to be $f(x) = \sinh x$, there follows also $f'''(x) = \cosh x$. Thus the error committed is given by

$$\begin{aligned} E(0.23) &= \frac{1}{3!}(0.23 - 0.00)(0.23 - 0.20)(0.23 - 0.30) \cosh \xi \\ &= -0.0000805 \cosh \xi \end{aligned}$$

for some ξ such that $0 < \xi < 0.30$. It happens in this case that $\cosh x$ may be computed at the tabular points from the *given data*, by use of the formula $\cosh x = (1 + \sinh^2 x)^{1/2}$, and the range of $\cosh x$ over the given interval is thus found (without the need for additional data, but with use of the fact that cosh x *increases* throughout the interval) to be between 1 and 1.04534. Thus there follows

$$-0.0000842 < E(0.23) < -0.0000805$$

so that the error in the last place retained in the calculation should be -8. Actually, the error was found to be -9 in the fifth place. The discrepancy is due, not to roundoffs in calculation (which were sufficiently controlled by retention of a sixth digit, as may be verified), but to the fact that each of the *original data* possesses a roundoff error which may be as large as 5×10^{-6}.

In other applications of interpolation, the analytic expression for $f(x)$ may not be known, and hence it may be impossible to determine the range of possible values of $f^{(n+1)}(\xi)$ in order to bound the error E. In such cases, the relation (2.6.1) may be more useful. For, if sufficient data are available to permit the evaluation of one or more sample values of the $(n + 1)$th *divided difference*, these values may be taken as estimates of the value of the divided difference which is actually relevant to (2.6.1). Thus, from the data obtained in Sec. 2.5, the divided difference

$$f[0.00, 0.20, 0.30, 0.50] \doteq 0.17333$$

may serve as an estimate of the *required* value $f[0.00, 0.20, 0.30, 0.23]$, leading to the error estimate

$$E(0.23) \approx 0.17\pi(0.23) \doteq (-0.00048)(0.17) \doteq -0.00008$$

The fact that this estimate is indeed good in this case is a consequence of the fact that the third derivative, and hence also the third divided difference, does not vary greatly in the range considered. It may be noticed that this error estimate is precisely the *correction term* which was involved in the calculation of Sec. 2.5 as a result of incorporating the contribution of the third difference. More generally, a consideration of (2.5.2) and (2.5.3) shows that if an interpolation for $f(\bar{x})$ is made, terminating with an nth difference, the error committed is given exactly by the product of the calculable number $\pi(\bar{x})$ and the $(n + 1)$th difference $f[x_0, \ldots, x_n, \bar{x}]$, which is *not* calculable unless $f(\bar{x})$ is known. On the other hand, the *first term omitted* in a calculation based on (2.5.2) is the product of $\pi(\bar{x})$ and the $(n + 1)$th difference $f[x_0, \ldots, x_n, x_{n+1}]$. If $f[x_0, \ldots, x_n, x]$ does not vary markedly over an interval including $x = \bar{x}$ and $x = x_{n+1}$, this *first term omitted* will indeed supply a good estimate of the error. This situation certainly will exist, in particular, in consequence of (2.6.8), if $f^{(n+1)}(x)$ does not vary markedly over an interval $\bar{I}$ including $x = x_0, \ldots, x_{n+1}, \bar{x}$.

It may be noticed that, as n increases without limit, the length of the interval $\bar{I}$, as well as that of the interval limited by $\bar{x}$ and x_{n+1}, generally will also increase without limit, since the later abscissas introduced generally are more remote from $\bar{x}$, so that the uncertainty of this particular error estimate may be expected to increase. In fact, in many cases the result of omitting the error term in (2.5.2), and allowing n to become infinite, leads to an infinite *interpolation series* which is itself *not convergent*. That is, the error $E(x)$ associated with retention of differences of order not greater than n very often does not tend to zero as n increases without limit.† However, if the abscissas $x_0, x_1, x_2, \ldots$ are appropriately ordered, it is usually true that the magnitude of the error E first decreases fairly rapidly with increasing n, but then increases in magnitude as n continues to increase. In most practical cases, the minimal error is extremely small, and the minimal stage occurs for a value of n so large that it is not actually encountered.

In view of this situation, the error $E(x)$ is not generally one which can be reduced in magnitude within an arbitrarily prescribed tolerance by increasing the number of differences retained. Thus, although this error is commonly known as the *truncation error*, it should be noticed again that this terminology often is somewhat misleading in that it would seem to imply an error committed by truncating a *convergent* infinite sequence of calculations after a finite number of steps.

As in Sec. 1.3, we continue to define a truncation error as any error which would be present even in the ideal case when the given data are exact and infinitely many decimal places are retained in the calculations, and we shall refer to $E(x)$ as a truncation error in this general sense. The superimposed effects of roundoff errors may be of equal or greater importance. In fact, the most efficient procedure is frequently that one in which the maximum (or RMS) errors due to truncation and to roundoff are of the *same* magnitude.

2.7 Iterated Interpolation

In Sec. 2.2, it was shown that *linear* interpolation can be conveniently effected by use of the formula

$$p_{0,1}(x) = \frac{1}{x_1 - x_0} \begin{vmatrix} p_0(x) & x_0 - x \\ p_1(x) & x_1 - x \end{vmatrix} \tag{2.7.1}$$

where $p_0(x)$ and $p_1(x)$ are two independent *interpolation polynomials of degree 0*,

$$p_0(x) = f(x_0) \qquad p_1(x) = f(x_1) \tag{2.7.2}$$

† The series obviously terminates if $f(x)$ is a polynomial, and is a convergent infinite series in certain other cases. Some information with regard to this question is given in Sec. 4.11.

In the same way, *quadratic* interpolation can be effected by linear interpolation over two independent linear interpolation polynomials, so that, for example,

$$p_{0,1,2}(x) = \frac{1}{x_2 - x_0}\begin{vmatrix} p_{0,1}(x) & x_0 - x \\ p_{1,2}(x) & x_2 - x \end{vmatrix} = \frac{1}{x_2 - x_1}\begin{vmatrix} p_{0,1}(x) & x_1 - x \\ p_{0,2}(x) & x_2 - x \end{vmatrix} \tag{2.7.3}$$

In order to verify this fact directly, we may notice, for example, that the first right-hand member of (2.7.3) is a polynomial of second degree, that it obviously takes on the values $f(x_0)$ and $f(x_2)$ when $x = x_0$ and $x = x_2$, respectively, and that when $x = x_1$ it correctly takes on the value

$$\frac{1}{x_2 - x_0}\begin{vmatrix} f(x_1) & x_0 - x_1 \\ f(x_1) & x_2 - x_1 \end{vmatrix} = f(x_1)$$

In a similar way, we may effect cubic interpolation by linear interpolation over two independent quadratic interpolation polynomials, and so forth (see Prob. 38). This procedure is particularly useful in machine calculation for the purpose of generating a *sequence* of interpolates from which the rate of effective convergence† can be estimated in cases when use cannot be made of analytic error bounds.

In *Aitken's method*, the first four stages of the calculation would be tabulated as follows for desk calculation:

x_0	p_0					$x_0 - \bar{x}$
x_1	p_1	$p_{0,1}$				$x_1 - \bar{x}$
x_2	p_2	$p_{0,2}$	$p_{0,1,2}$			$x_2 - \bar{x}$
x_3	p_3	$p_{0,3}$	$p_{0,1,3}$	$p_{0,1,2,3}$		$x_3 - \bar{x}$
x_4	p_4	$p_{0,4}$	$p_{0,1,4}$	$p_{0,1,2,4}$	$p_{0,1,2,3,4}$	$x_4 - \bar{x}$

Here, for example, the entry $p_{0,1,3}$ would be obtained by evaluating the determinant

$$\begin{vmatrix} p_{0,1} & x_1 - \bar{x} \\ p_{0,3} & x_3 - \bar{x} \end{vmatrix}$$

the elements of which are seen to be conveniently located in the above array, and dividing the result by $x_3 - x_1$. Here an additional convenience is afforded by the fact that this divisor can be obtained as the difference $(x_3 - \bar{x}) - (x_1 - \bar{x})$ between the entries in the right-hand column.

The abscissas labeled as $x_0, \ldots, x_n$ may be arranged in any algebraic order; the final value $p_{0,\ldots,n}$ is independent of that arrangement (barring

† The phrase *effective convergence* will be used in accordance with the generally asymptotic nature of the sequence.

the effects of intermediate roundoffs). However, it is often desirable to designate the abscissa nearest the argument of the interpolant $\bar{x}$ by x_0, the second nearest by x_1, and so forth. For then the entries $p_0, p_{0,1}, p_{0,1,2}$, and so forth, may be expected to represent the best possible estimates, based on the given data, which can be afforded by polynomial interpolation of degree zero, one, two, and so forth. Also, each such estimate makes use of all the information used in the preceding estimate, together with one additional datum. Thus the rate of effective convergence can be fairly confidently estimated by considering the sequence of entries in the *diagonal* of the table.

For the interpolation problem considered in Sec. 2.5, the work could be arranged as follows, through the third-degree calculation:

20	0.20134				−3
30	0.30452	0.232294			7
0	0.00000	1541	0.232118		−23
50	0.52110	3316	1936	0.232034	27

In the absence of further information, the correctness of the fourth place probably would be presumed, whereas the fifth place would be considerably in doubt. In order to decrease the uncertainty, further information would be needed. If, for example, $f(0.60)$ were also available, an additional row of entries then would be calculated, as follows:

60	0.63665	0.233988	0.231899	0.232034	0.232034	37

Thus the value 0.23203 appears to be stabilized as the five-digit interpolate corresponding to the given data.

2.8 Inverse Interpolation

It frequently happens that a variable y is given in tabular form (or analytically) as a single-valued function of x, say $y = f(x)$, and that a value of the independent variable x is required for which the dependent variable y takes on a prescribed value (frequently zero). This is the problem of *inverse interpolation*.

If $\bar{y} = f(\bar{x})$, then on any x interval including $\bar{x}$, in which $dy/dx = f'(x)$ exists and does not vanish, a unique inverse function, say $x = F(y)$, exists, such that $\bar{x} = F(\bar{y})$. Thus, if dy/dx does not vanish near the point where the inverse interpolation is to be effected (so that y increases or decreases steadily in the neighborhood of that point), it may be that $F(y)$ can be satisfactorily approximated in that neighborhood by a polynomial of moderately low degree, so that the inverse interpolation may be effected by merely tabulating x as a function of y in that neighborhood, and using the preceding methods (or any other appropriate methods) of *direct* interpolation.

In illustration, suppose that the following data are available and that the zero of $y(x)$ between $x = 1.3$ and $x = 1.4$ is required.

x	1.1	1.2	1.3	1.4	1.5
y	0.769	0.472	0.103	−0.344	−0.875

If Aitken's method is used, with the entries ordered with respect to the nearness of an ordinate to zero, the calculations may be arranged as follows:

y	x					$y - \bar{y}$
103	1.3					103
−344	1.4	1.32304				−344
472	1.2	2791	1.32509			472
769	1.1	3093	548	1.32447		769
−875	1.5	2106	432	82	1.32463	−875

Thus, a fourth-degree interpolation yields $x \approx 1.3246$, with its last place in doubt, although the uncertainty corresponding to the presence of roundoff in the given data would also remain. Actually, the given data are exact values corresponding to the algebraic relation $y = -x^3 + x + 1$, and the problem can be considered as that of determining the real zero of the algebraic equation $x^3 - x - 1 = 0$, the true value of which is 1.32472, to five places.

Evidently, if this problem were stated in its analytic form, recourse to a semianalytic method such as that of successive substitutions or the Newton-Raphson iteration (see Sec. 10.11) would also be appropriate. Even when the correspondence is given only in tabular form, it would also be possible to approximate the relation $y = f(x)$ by the relation $y = p_{0,\ldots,n}(x)$, where the equation of the approximation is expressed in explicit polynomial form, with the help of Newton's interpolation formula or of one of the other formulas to be obtained, and to solve the resultant approximating algebraic equation by such iterative methods. However, in order to estimate the accuracy obtained, it would be desirable to repeat the calculation for several values of n, each of which would lead to a distinct algebraic equation.

If dy/dx vanishes near the point $(\bar{x}, \bar{y})$ where the inverse interpolation is to be effected, then the derivative of the inverse function becomes infinite near that point, and a satisfactory approximation to the inverse function generally cannot be obtained by using a polynomial of low degree. In such a case, a simple iterative procedure is useful. For this purpose, suppose first that two abscissas x_a and x_b are available with the property that $\bar{y}$ lies between $y_a = f(x_a)$ and $y_b = f(x_b)$ (see Fig. 2.1). If y_a and y_b are sufficiently nearly equal, and if $dy/dx \neq 0$ in the interval between x_a and x_b, *linear* inverse interpolation may then be used to obtain a first approximation to $\bar{x}$, say $\bar{x}^{(1)}$. Then, by direct interpolation, using the ordinates y_a, y_b, and an appropriate number of

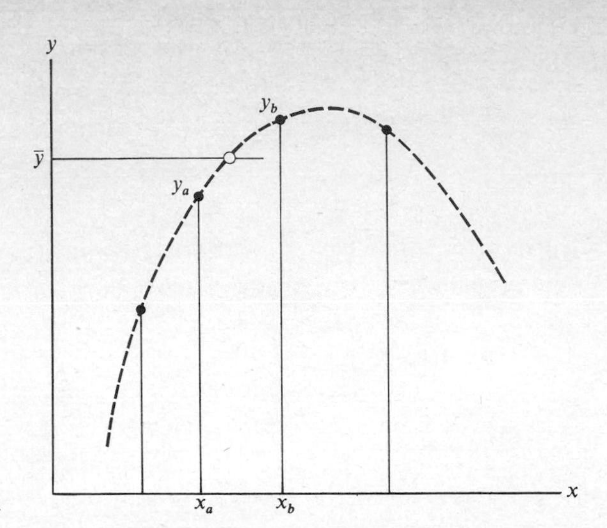

FIGURE 2.1

other known ordinates, the true value $f(\bar{x}^{(1)})$ may be approximated. Then, if that result is designated as $\bar{y}^{(1)}$, linear *inverse* interpolation based on $\bar{y}^{(1)}$ and either y_a or y_b (whichever one is separated from $\bar{y}^{(1)}$ by $\bar{y}$) is used to determine a second approximation to $\bar{x}$, say, $\bar{x}^{(2)}$, and the cycle of operations is repeated as often as necessary.

Methods of this sort, in which the only *inverse* interpolation involved is *linear*, and in which high-degree interpolation is effected only on the direct function $f(x)$, are particularly to be recommended in those cases when it is known that $f(x)$ can be satisfactorily approximated by a polynomial of reasonably low degree over an interval including $\bar{x}$, but when it is difficult to be certain that the inverse function $F(y)$, such that $\bar{x} = F(\bar{y})$, also can be fairly approximated by a polynomial in y, of comparable degree, over the corresponding interval in y. Whereas situations of this sort obviously are to be anticipated when $f'(x)$ vanishes for a real value of x near $\bar{x}$, they may also occur in the absence of such a warning.

In critical cases, in particular in the case when $dy/dx = 0$ *at* the desired point, it is usually desirable to use one of the semianalytic methods mentioned earlier, in which $f(x)$ is approximated by a polynomial $p(x)$ and the algebraic equation $p(x) = \bar{y}$ is solved by an appropriate iterative method.

2.9 Supplementary References

Standard references dealing with divided differences and with the basic Newtonian interpolation formula include Steffensen [1950], Milne-Thomson [1951], and Hartree [1958].

In Krogh [1970], algorithms are presented for the purpose of making the use of Newton's divided-difference formula for interpolation and numerical differentiation on a computer compare favorably in efficiency with the use of other formulas.

A method of iterated interpolation similar to that of Aitken [1932*b*] is due to Neville [1934]; see also Kopal [1961]. References to additional methods of inverse interpolation are included in Sec. 3.12 and Sec. 4.13.

For the use of divided differences in two-dimensional analysis see, for example, Salzer [1959] and Stancu [1964].

PROBLEMS

Section 2.2

1 Use (2.2.5) to calculate approximate values of $f(x)$ when $x = 1.1416, 1.1600$, and 1.2000 from the following rounded data:

x	1.1275	1.1503	1.1735	1.1972
$f(x)$	0.11971	0.13957	0.15931	0.17902

2 Calculate the three first divided differences relevant to successive pairs of data in Prob. 1, and use (2.2.4) to determine approximate values of $f(x)$ for

$$x = 1.1600(0.0020)1.1700\dagger$$

3 Prove that $f[x_0, x_1]$ is independent of x_0 and x_1 if and only if $f(x)$ is a linear function of x.

4 If $f(x) = u(x)v(x)$, show that

$$f[x_0, x_1] = u[x_0]v[x_0, x_1] + u[x_0, x_1]v[x_1]$$

5 If $f'(x)$ is continuous for $x_0 \leqq x \leqq x_1$, show that

$$f[x_0, x_1] = f'(\xi)$$

for some ξ between x_0 and x_1, and hence also that

$$f[x_0, x_0] \equiv \lim_{x_1 \to x_0} f[x_0, x_1] = f'(x_0)$$

Section 2.3

6 If the abscissas in Prob. 1 are numbered in increasing algebraic order, verify numerically that $f[x_0, x_1, x_2] = f[x_2, x_0, x_1]$.

† The notation $x = m(h)n$ denotes that x is to take on values between $x = m$ and $x = n$, inclusive, at increments of h units.

7 Suppose that $x_r = x_0 + rh\ (r = 1, 2, \ldots)$, so that the abscissas are at a uniform spacing h. Show that (2.3.3) then becomes

$$\alpha_i^{(k)} = \frac{(-1)^{k-i}}{i!\,(k-i)!}\,\frac{1}{h^k} = \frac{(-1)^{k-i}}{h^k k!}\binom{k}{i}$$

where $\binom{k}{i}$ is the binomial coefficient. Thus deduce that

$$f[x_0, \ldots, x_k] = \frac{1}{h^k k!}\sum_{i=0}^{k}(-1)^{k-i}\binom{k}{i}f(x_i)$$

in this case.

8 Assuming that $x_r = x_0 + rh$, verify directly (from the definition) the truth of the following special cases of the relation established in Prob. 7:

$$f[x_0, x_1] = \frac{1}{h}\,[f(x_1) - f(x_0)]$$

$$f[x_0, x_1, x_2] = \frac{1}{2!\,h^2}\,[f(x_2) - 2f(x_1) + f(x_0)]$$

$$f[x_0, x_1, x_2, x_3] = \frac{1}{3!\,h^3}\,[f(x_3) - 3f(x_2) + 3f(x_1) - f(x_0)]$$

9 If $f'(x) = df(x)/dx$, show that

$$\frac{d}{dx}f[x_0, x] \not\equiv f'[x_0, x]$$

unless $f(x)$ is linear.

10 If $f(x) = u(x)v(x)$, show that the relation established in Prob. 4 generalizes to the form

$$f[x_0, \ldots, x_n] = \sum_{k=0}^{n} u[x_0, \ldots, x_k]v[x_k, \ldots, x_n]$$

Use induction, assuming the truth of the relation for $n = N$, showing that then

$$\begin{aligned} f[x_1, \ldots, x_{N+1}] - f[x_0, \ldots, x_N] \\ = \sum_{k=0}^{N} \{(x_{N+1} - x_k)u[x_0, \ldots, x_k]v[x_k, \ldots, x_{N+1}] \\ + (x_{k+1} - x_0)u[x_0, \ldots, x_{k+1}]v[x_{k+1}, \ldots, x_{N+1}]\} \end{aligned}$$

and that this expression properly reduces to

$$\begin{aligned} (x_{N+1} - x_0)\Bigg(u[x_0]v[x_0, \ldots, x_{N+1}] \\ + \sum_{k=1}^{N} u[x_0, \ldots, x_k]v[x_k, \ldots, x_{N+1}] \\ + u[x_0, \ldots, x_{N+1}]v[x_{N+1}]\Bigg) \end{aligned}$$

11 If $f(x) = (ax + b)/(cx + d)$, obtain expressions for $f[x, y]$, $f[x, x, y]$, and $f[x, x, y, y]$ in compact forms when $x \neq y$.

12 If $f(x) = x^5$, obtain expressions for $f[a, b, c]$, $f[a, a, b]$, and $f[a, a, a]$ when $a \neq b \neq c$.

Section 2.4

13 Repeat the calculations of Prob. 2, making use of the second divided difference $f[a_0, a_1, a_2]$.

14 Compare the results of Prob. 13 with those obtained by using the second divided difference $f[a_1, a_2, a_3]$ instead.

15 Obtain the formula

$$\int_{x_0}^{x_1} f(x)\,dx = (x_1 - x_0)f(x_0) + \tfrac{1}{2}(x_1 - x_0)^2 f[x_0, x_1] - \tfrac{1}{6}(x_1 - x_0)^3 f[x_0, x_1, x_2] + E$$

where

$$E = \int_{x_0}^{x_1} (x - x_0)(x - x_1)(x - x_2) f[x_0, x_1, x_2, x]\,dx$$

16 Apply the formula of Prob. 15, neglecting the error term, to the data of Prob. 1, obtaining approximate values of the integral of $f(x)$ over each subinterval and hence obtaining also approximate values of the integral from the smallest abscissa to each of the others. Then use interpolation to approximate the integral over [1.14, 1.18].

Section 2.5

17 If $f(x) = 1/(a - x)$, show that

$$f[x_0, x_1, \ldots, x_n] = \frac{1}{(a - x_0)(a - x_1)\cdots(a - x_n)}$$

and

$$f[x_0, x_1, \ldots, x_n, x] = \frac{1}{(a - x_0)(a - x_1)\cdots(a - x_n)(a - x)}$$

and deduce that

$$\frac{1}{a - x} = \frac{1}{a - x_0} + \frac{x - x_0}{(a - x_0)(a - x_1)} + \cdots + \frac{(x - x_0)\cdots(x - x_{n-1})}{(a - x_0)\cdots(a - x_n)} + E(x)$$

where

$$E(x) = \frac{\pi(x)}{\pi(a)(a - x)}$$

with the abbreviation

$$\pi(x) = (x - x_0)(x - x_1)\cdots(x - x_n)$$

Assume throughout that $a \neq x_0, x_1, \ldots, x_n, x$.

18 The following table lists the rounded value of the probability Q that the magnitude of a normally distributed error, with mean value zero and standard deviation unity, exceed ε, for certain values of ε and Q. Calculate from it approximate values of Q for $\varepsilon = 0.7, 0.9, 1.1$, and 1.2.

ε	0.4	0.5	0.6	0.8	1.0	1.25
Q	0.68916	0.61708	0.54851	0.42371	0.31731	0.21130

19 Use the data of Prob. 18 to calculate approximate values of ε for $Q = 0.4, 0.5$, and 0.6.

20 Suppose that values of $f(x)$, $f'(x)$, and $f''(x)$ are known for $x = x_0$, values of $f(x)$ and $f'(x)$ for $x = x_1$, and the value of $f(x)$ for $x = x_2$. Show that the corresponding divided-difference table appears as follows, through third differences, where each *difference* is formed from diagonally adjacent entries to its left by the usual rule, the values of the *derivatives* being entered in advance:

x_0	$f(x_0)$			
		$f'(x_0)$		
x_0	$f(x_0)$		$\frac{1}{2}f''(x_0)$	
		$f'(x_0)$		$f[x_0, x_0, x_0, x_1]$
x_0	$f(x_0)$		$f[x_0, x_0, x_1]$	
		$f[x_0, x_1]$		$f[x_0, x_0, x_1, x_1]$
x_1	$f(x_1)$		$f[x_0, x_1, x_1]$	
		$f'(x_1)$		$f[x_0, x_1, x_1, x_2]$
x_1	$f(x_1)$		$f[x_1, x_1, x_2]$	
		$f[x_1, x_2]$		
x_2	$f(x_2)$			

Notice also that "Sheppard's rules" remain applicable to any *difference path* made up of contiguous diagonal segments, and write down the formula which introduces successively the values of $f(x_0)$, $f'(x_0)$, $f''(x_0)$, $f(x_1)$, $f'(x_1)$, and $f(x_2)$.

21 The following rounded values of $Q(\varepsilon)$ and its derivative $Q'(\varepsilon)$ are known. By appropriately modifying the procedure illustrated in Prob. 20, construct a suitable difference table and calculate approximate values of Q for $\varepsilon = 0.2(0.2)0.8$.

ε	Q	Q'
0.0	1.0000	−0.7979
0.5	0.6171	−0.7041
1.0	0.3173	−0.4839

22 Assuming that the third divided difference of $f(x)$ is constant for all x, fill in the spaces in the following divided-difference table (from right to left), and hence evaluate $f'(8)$ and $f''(8)$:

0	3			
		1		
1	4		4	
		13		1
3	30		10	
		63		1
6	219		17	
		148		—
8	515		—	
		—		—
8	515		—	
		—		
8	515			

Also use a similar procedure to obtain $f'(3)$. Determine an analytic expression for $f(x)$ and check the results.

23 If $f(x_1)$, $f(x_2)$, and $f(x_3)$ are values of $f(x)$ near a maximum or minimum point at $x = \bar{x}$, obtain the approximation

$$\bar{x} \approx \frac{x_1 + x_2}{2} - \frac{f[x_1, x_2]}{2f[x_1, x_2, x_3]}$$

and show that it can also be written in the more symmetrical form

$$\bar{x} \approx \frac{x_1 + 2x_2 + x_3}{4} - \frac{f[x_1, x_2] + f[x_2, x_3]}{4f[x_1, x_2, x_3]}$$

Show also that, when the abscissas are equally spaced, it becomes

$$\bar{x} \approx x_2 - \frac{h}{2}\left(\frac{f_3 - f_1}{f_1 - 2f_2 + f_3}\right)$$

where h is the common interval.

Section 2.6

24 Show that the truncation error associated with linear interpolation of $f(x)$, using ordinates at x_0 and x_1 with $x_0 \leqq x \leqq x_1$, is not larger in magnitude than

$$\tfrac{1}{8}M_2(x_1 - x_0)^2$$

where M_2 is the maximum value of $|f''(x)|$ on the interval $[x_0, x_1]$. Does this result hold also for extrapolation?

25 Under the assumption that the data in Prob. 1 correspond to the function $f(x) = \sin(\log x)$, show that the truncation error corresponding to linear interpolation between successive ordinates is smaller than one unit in the fourth decimal place.

26 Show that the magnitude of the truncation error, corresponding to linear interpolation of the error function

$$\text{erf}\, x = \frac{2}{\sqrt{\pi}} \int_0^x e^{-t^2}\, dt$$

between x_0 and x_1, cannot exceed

$$\frac{(x_1 - x_0)^2}{2\sqrt{2\pi e}}$$

and hence is smaller than $(x_1 - x_0)^2/8$.

27 In the special case when the abscissas are equally spaced, with separation h, show that the magnitude of the truncation error corresponding to second-degree interpolation based on ordinates at x_0, x_1, and x_2 does not exceed $(M_3h^3)/(9\sqrt{3})$, where M_3 is the maximum value of $|f'''(x)|$ on the interval $[x_0, x_2]$. Show also that, on the average, the largest errors may be expected to occur at distances of about $h/\sqrt{3} \approx 0.58h$ from the central abscissa. (Translate the origin to the point $x = x_1$.)

28 Show that the magnitude of the truncation error associated with third-degree interpolation based on ordinates at equally spaced points x_0, x_1, x_2, and x_3 does not exceed $(3M_4h^4)/128$ for interpolation between x_1 and x_2 and is, on the average, largest at the center of that interval. Show also that it does not exceed $(M_4h^4)/24$ for interpolation between x_0 and x_1 or between x_2 and x_3, with a maximum to be expected, on the average, at a distance of about $(3 - \sqrt{5})h/2 \approx 0.38h$ from x_0 or x_3, where M_4 is the maximum value of $|f^{\text{iv}}(x)|$ on $[x_0, x_4]$ in all cases. [Translate the origin to the midpoint $(x_1 + x_2)/2$.]

29 Obtain the formula

$$\begin{aligned} f(x) = f(x_0) &+ (x - x_0)f'(x_0) + (x - x_0)^2 f[x_0, x_0, x_1] \\ &+ (x - x_0)^2(x - x_1)f[x_0, x_0, x_1, x_1] + E(x) \end{aligned}$$

where

$$E(x) = \tfrac{1}{24}(x - x_0)^2(x - x_1)^2 f^{\text{iv}}(\xi) \qquad (x_0 < x, \xi < x_1)$$

and show that

$$|E(x)| \leqq \frac{h^4}{384} \max_{x_0 \leqq x \leqq x_1} |f^{\text{iv}}(x)|$$

30 If $f(x) = 1/(x + 1)$ and $y(x)$ is the polynomial approximation of degree n which agrees with $f(x)$ when $x = 0, 1, 2, \ldots, n$, show that the use of (2.6.5) leads to the error bound

$$|E(x)| < |x(x - 1) \cdots (x - n)|$$

whereas (2.6.1) permits the less conservative bound

$$|E(x)| < \frac{1}{(n + 1)!} |x(x - 1) \cdots (x - n)|$$

when $x \geqq 0$.

31 Suppose that a table presents values of $f(x)$ rounded to r decimal places at a uniform interval h in x, and that linear interpolation is employed for the calculation of

$f(\bar{x})$. Suppose also that the tabular abscissas are exact, that the abscissa $\bar{x}$ is rounded to s decimal places, and that the calculated approximate value of $f(\bar{x})$ is rounded to t decimal places. If $f''(x)$ is continuous over the tabular range, and if δ is the total error in the resultant interpolate, show that

$$|\delta| \leqq \tfrac{1}{8}M_2h^2 + 5M_1 \times 10^{-s-1} + 5 \times 10^{-r-1} + 5 \times 10^{-t-1}$$

where M_1 and M_2 are the maximum values of $|f'(x)|$ and $|f''(x)|$, respectively, in the tabular range.

32 The function $f(x) = \log_{10} \sin x$ is tabulated for $x = 0.01(0.01)2.00$ to five decimal places. If linear interpolation is employed, with the abscissa of the interpolant rounded to five decimal places, and if the calculated result is also rounded to five places, determine the portions of the table for which the results certainly will be correct within five units, and within 50 units, in the fifth place. What accuracy could be guaranteed over those ranges if the abscissa of the interpolant were rounded to four places? To three places?

33 A table of values of the function $f(x) = (x^4 - x)/12$ is to be constructed for $0 \leqq x \leqq a$ in such a way that the error in linear interpolation would not exceed ε if the effects of roundoff were negligible. Show first that, if the spacing h is to be uniform, then h should be smaller than $2\sqrt{2\varepsilon}/a$ and at least $a^2/(2\sqrt{2\varepsilon})$ entries will be required. Show also that, if the interval $[0, a]$ were divided into the subintervals $[0, \alpha]$ and $[\alpha, a]$, and if uniform spacings h_1 and h_2 were used in those respective subintervals, then the most efficient division would be such that the conditions $\alpha = a/2$, $h_2 = 2\sqrt{2\varepsilon}/a$, and $h_1 = 2h_2$ were approximately satisfied, corresponding to a reduction of about 25 percent in the number of entries.

34 From the following table of rounded values of $f(x) = (x/10)^{1/2}$, construct a divided-difference table and determine successive approximations to $f(0.5) = (0.05)^{1/2}$ corresponding to the use of one, two, three, four, and five successive ordinates, including that at $x = 0$. Compare these results with the true value. How could the existent situation have been predicted (without direct calculation) assuming knowledge of the analytical form of $f(x)$? What preliminary warning is afforded by reference to the difference table alone?

x	0	1	2	3	4
$f(x)$	0.00000	0.31623	0.44721	0.54772	0.63246

35 Form a divided-difference table based only on the ordinates of the function $f(x) = x^5 - 5x^3 + x^2 + 4x - 2$ at the points $x = -2, -1, 0, 1$, and 2. Then interpolate from this table approximate values of $f(x)$ at $x = -1.5, -0.5, 0.5$, and 1.5, and compare them with the true values. How could the possibility of the existent situation be predicted (without direct calculation) assuming knowledge of the analytical form of $f(x)$?

Section 2.7

36 Use the Aitken procedure to determine $Q(0.7)$ and $\varepsilon(0.5)$ as accurately as possible from the data of Prob. 18.

37 Use the Aitken procedure to determine $f(0.20000)$ as accurately as possible from the following rounded values of $f(x) = \sin[\sinh^{-1}(x+1)]$:

x	0.17520	0.25386	0.33565	0.42078	0.50946
$f(x)$	0.84147	0.86742	0.89121	0.91276	0.93204

38 Deduce the validity of Aitken's method by establishing the relations

$$p_{0,1,\ldots,m,n}(x) = p_{0,1,\ldots,m-1,m}(x) + [(x-x_0)\cdots(x-x_{m-1})(x-x_m)]f[x_0,\ldots,x_m,x_n]$$

$$p_{0,1,\ldots,m,n}(x) = p_{0,1,\ldots,m-1,n}(x) + [(x-x_0)\cdots(x-x_{m-1})(x-x_n)]f[x_0,\ldots,x_m,x_n]$$

and eliminating $f[x_0,\ldots,x_m,x_n]$ between them.

Section 2.8

39 If $y = f(x)$ and if $f'(x) \neq 0$ for $x_0 < x < x_1$, show that the truncation error of linear inverse interpolation based on corresponding values (x_0, y_0) and (x_1, y_1) is given by

$$-(y-y_0)(y-y_1)\frac{f''(\xi)}{2[f'(\xi)]^3}$$

where $x_0 < \xi < x_1$, if $f''(x)/[f'(x)]^3$ exists and is continuous in that interval. Show also that the magnitude of this error is limited by each of the bounds

$$\frac{(y_1-y_0)^2}{8}K \qquad \frac{h^2}{8}M_1^2K \qquad \frac{h^2}{8}\left(\frac{M_1}{m_1}\right)^2\frac{M_2}{m_1}$$

if $h = x_1 - x_0$, $|f''(x)/[f'(x)]^3| \leqq K$, $m_1 \leqq |f'(x)| \leqq M_1$, and $|f''(x)| \leqq M_2$ for $x_0 \leqq x \leqq x_1$.

40 Suppose that $f(x) = x^2$ is tabulated for $0 \leqq x \leqq 1$ with a uniform spacing of h in x. Assuming that sufficiently many significant figures are supplied and retained in the calculation to permit the neglect of the effects of roundoff errors, determine α (as a function of h and ε) so that the error of linear inverse interpolation will not exceed a specified quantity ε on the interval $[\alpha, 1]$. What spacing would be required to assure an accuracy within 0.005 for $0.1 \leqq x \leqq 1.0$?

41 Repeat the calculations of Prob. 40 when

$$f(x) = \int_0^x \sin t^2\, dt$$

[Use the inequality $\sin u > 2u/\pi$ $(0 < u < \pi/2)$ in bounding the error.]

42 Tabulate $y = x^2$ for $x = 0, 1, 2, 3$, and 4, and then interchange the roles of x and y; now considering x as a function of y (with $x \geqq 0$), use fourth-degree interpolation to obtain an approximation to x when $y = 12$. Compare the result with the true value and account for the existent situation analytically. Also plot the relevant fourth-degree interpolation polynomial $x = p(y)$, with appropriate attention to its maxima and minima, and superimpose a plot of the true relation.

43 Given the following data, use the iterative process of inverse and direct interpolation to determine, to four decimal places, the value of x between 1.50 and 1.60 for which $f(x) = 0.99800$:

x	1.40	1.50	1.60	1.70	1.80
$f(x)$	0.98545	0.99749	0.99957	0.99166	0.97385

44 Calculate an approximation to the value of x required in Prob. 43 by approximating $f(x)$ by the second-degree polynomial $p(x)$ which agrees with $f(x)$ at the points for which $x = 1.50$, 1.60, and 1.70, and solving the quadratic equation $p(x) = 0.99800$. Then use the iterative method of Prob. 43 to obtain an improved approximation which may be expected to be correct to four decimal places.

45 The following *critical table* for the function $f(x) = x(x - 1)(2x - 1)/12$ has the property that, for any x between successive tabular abscissas, the corresponding value of $f(x)$ rounds to the entry given for that interval:

x	$f(x)$
0.05667	
	0.0040
0.05844	
	0.0041
0.06025	
	0.0042
0.06208	
	0.0043
0.06394	

Construct the table, by first tabulating $f(x)$ for appropriate convenient values of x and then using inverse interpolation to obtain x when

$$f(x) = 0.00395(0.00010)0.00435$$

or otherwise.

3

LAGRANGIAN METHODS

3.1 Introduction

For many purposes, it is desirable that a formula for interpolation, numerical differentiation, or numerical integration be expressed explicitly in terms of the ordinates involved rather than in terms of their differences or divided differences. Such formulas permit a more direct consideration of the effect on the end result of a change or error in one or more of the ordinates, and their use does not require the calculation, tabulation, or storage of differences. However, it is found that these advantages are attained only at the sacrifice of others.

The basic formula, apparently due to Waring, but associated with the name of Lagrange, is derived in Sec. 3.2, and its general use in interpolation, differentiation, and integration is illustrated in Secs. 3.3 and 3.4. A number of specific formulas and techniques for numerical integration and differentiation are derived from this formula and, in the cases when the abscissas are equally spaced, these are studied in the remaining sections of the chapter.

3.2 Lagrange's Interpolation Formula

Lagrange's form of the polynomial $y(x) \equiv p_{0,\ldots,n}(x)$ of degree n, which takes on the same values as a given function $f(x)$ for the $n + 1$ distinct abscissas $x_0, x_1, \ldots, x_n$, differs from the newtonian form derived in Sec. 2.5 in that the ordinates involved are displayed explicitly in the lagrangian form, while the newtonian form explicitly involves divided differences of those ordinates. Whereas it clearly must be possible to derive Lagrange's form from (2.5.2), its importance justifies the indication of three alternative methods of approach, which are typical of methods also useful in other considerations.

As a first approach, we could write $y(x)$ in the form

$$y(x) = A_0 + A_1 x + \cdots + A_n x^n \equiv \sum_{k=0}^{n} A_k x^k \tag{3.2.1}$$

where the A's are to be determined in such a way that $y(x_i) = f(x_i)$ for $i = 0, 1, \ldots, n$. These requirements are represented by the $n + 1$ linear equations

$$\begin{aligned} A_0 + A_1 x_0 + A_2 x_0^2 + \cdots + A_n x_0^n &= f(x_0) \\ \cdots\cdots\cdots\cdots\cdots\cdots\cdots\cdots& \\ A_0 + A_1 x_n + A_2 x_n^2 + \cdots + A_n x_n^n &= f(x_n) \end{aligned} \tag{3.2.2}$$

If these equations are solved by use of determinants, the use of special properties of the determinants involved leads to rather simple expressions for the A's in terms of the ordinates, and the introduction of these results into (3.2.1) leads to the desired result (see Prob. 5). The requirement that the A's satisfy (3.2.1) *and* (3.2.2) can be expressed by the condition

$$\begin{vmatrix} y & 1 & x & x^2 & \cdots & x^n \\ f(x_0) & 1 & x_0 & x_0^2 & \cdots & x_0^n \\ \cdots & \cdots & \cdots & \cdots & \cdots & \cdots \\ f(x_n) & 1 & x_n & x_n^2 & \cdots & x_n^n \end{vmatrix} = 0 \tag{3.2.3}$$

the expanded form of which would also give the equation of the required interpolation polynomial $y = p_{0,\ldots,n}(x)$.

Alternatively, we could write $y(x)$ directly in the required form

$$y(x) = l_0(x)f(x_0) + l_1(x)f(x_1) + \cdots + l_n(x)f(x_n) \equiv \sum_{k=0}^{n} l_k(x)f(x_k) \tag{3.2.4}$$

where $l_0(x), \ldots, l_n(x)$ are polynomials of degree n or less, to be determined by the requirement that the result of replacing $y(x)$ by $f(x)$ be an *identity* when $f(x)$ is an arbitrary polynomial of degree n or less. It is clear that this situation will prevail if and only if the result of replacing $y(x)$ by $f(x)$ is an identity when

$f(x) = 1, x, x^2, \ldots,$ and x^n. These requirements are represented by the $n + 1$ equations

$$
\begin{aligned}
l_0(x) + l_1(x) + \cdots + l_n(x) &= 1 \\
x_0 l_0(x) + x_1 l_1(x) + \cdots + x_n l_n(x) &= x \\
\cdots\cdots\cdots\cdots\cdots\cdots\cdots\cdots\cdots\cdots \\
x_0^n l_0(x) + x_1^n l_1(x) + \cdots + x_n^n l_n(x) &= x^n
\end{aligned}
\tag{3.2.5}
$$

from which the coefficient functions can be determined directly as ratios of determinants which can be expanded in simple forms. The eliminant of the Eqs. (3.2.4) and (3.2.5) is merely the result of interchanging rows and columns in the matrix whose determinant appears in (3.2.3), so that the equivalence of the final forms is indeed confirmed.

Rather than pursue either of these lines, we may avoid somewhat lengthy calculation by noticing that the expression (3.2.4) will indeed take on the value $f(x_i)$ when $x = x_i$ if $l_i(x_i) = 1$ and if $l_i(x_j) = 0$ when $j \neq i$. With the convenient notation of the so-called *Kronecker delta*

$$
\delta_{ij} = \begin{cases} 0 & \text{if } i \neq j \\ 1 & \text{if } i = j \end{cases}
\tag{3.2.6}
$$

this requirement becomes merely

$$
l_i(x_j) = \delta_{ij} \qquad (i = 0, \ldots, n; j = 0, \ldots, n)
\tag{3.2.7}
$$

Since $l_i(x)$ thus is to be a polynomial of degree n which vanishes when $x = x_0, x_1, \ldots, x_{i-1}, x_{i+1}, \ldots, x_n$, there must follow

$$
l_i(x) = C_i[(x - x_0) \cdots (x - x_{i-1})(x - x_{i+1}) \cdots (x - x_n)]
\tag{3.2.8}
$$

where C_i is a constant. The final requirement $l_i(x_i) = 1$ then determines C_i in the form

$$
C_i = \frac{1}{(x_i - x_0) \cdots (x_i - x_{i-1})(x_i - x_{i+1}) \cdots (x_i - x_n)}
\tag{3.2.9}
$$

and the desired *lagrangian coefficient functions* $l_i(x)$ are obtained by introducing (3.2.9) into (3.2.8).

In order to put this result in a somewhat more compact form, we first review the notation of (2.6.2):

$$
\pi(x) = (x - x_0)(x - x_1) \cdots (x - x_n)
\tag{3.2.10}
$$

Now the *derivative* of $\pi(x)$ is clearly expressible as the sum of $n + 1$ terms, in each of which one of the factors of $\pi(x)$ is deleted. Thus, if we set $x = x_i$ in this expression, we obtain the useful result

$$\pi'(x_i) = (x_i - x_0) \cdots (x_i - x_n) = \frac{1}{C_i} \tag{3.2.11}$$

where the factor $(x_i - x_i)$ is to be omitted in the product. Thus, after introducing (3.2.8) and (3.2.9) into (3.2.4), we obtain the *lagrangian interpolation polynomial of degree n* in the form

$$y(x) = \sum_{k=0}^{n} \frac{\pi(x)}{(x - x_k)\pi'(x_k)} f(x_k) \equiv \sum_{k=0}^{n} l_k(x) f(x_k) \tag{3.2.12}$$

where

$$\begin{aligned} l_i(x) &= \frac{\pi(x)}{(x - x_i)\pi'(x_i)} \\ &= \frac{(x - x_0) \cdots (x - x_{i-1})(x - x_{i+1}) \cdots (x - x_n)}{(x_i - x_0) \cdots (x_i - x_{i-1})(x_i - x_{i+1}) \cdots (x_i - x_n)} \end{aligned} \tag{3.2.13}$$

The first expression for $l_i(x)$ is useful in theoretical considerations, the second in the actual calculation of the function.

It should be noticed that the definitions of the functions $\pi(x)$ and $l_i(x)$ involve the degree n of the interpolation polynomial. Generally, in the sequel, the value of n will be clear from the context. When a more explicit notation is necessary, we may replace $l_i(x)$ by $l_{i,n}(x)$ and $\pi(x)$ by $\pi_n(x)$.

The *direct* derivation of (3.2.12) from the newtonian form (2.5.2) is of some academic interest and may be effected by making use of (2.3.2) and comparing (2.3.3) with (3.2.11).

In view of the equivalence of (3.2.12) and (2.5.2), the error committed by replacing $f(x)$ by $y(x)$ is again given by either (2.6.1) or (2.6.5), so that we may write

$$f(x) = \sum_{k=0}^{n} l_k(x) f(x_k) + E(x) \tag{3.2.14}$$

where

$$E(x) = \pi(x) f[x_0, \ldots, x_n, x] = \pi(x) \frac{f^{n+1}(\xi)}{(n+1)!} \tag{3.2.15}$$

and where, as before, ξ is some number in the interval I spanned by x_0, $x_1, \ldots, x_n$, and x.

To illustrate the use of the lagrangian formula, we may write down the interpolation polynomial of degree 3 relevant to the data

x	-1	0	1	2
$f(x)$	1	1	1	-5

in the form

$$y = 1\,\frac{(x-0)(x-1)(x-2)}{(-1-0)(-1-1)(-1-2)} + 1\,\frac{(x+1)(x-1)(x-2)}{(0+1)(0-1)(0-2)}$$

$$+\,1\,\frac{(x+1)(x-0)(x-2)}{(1+1)(1-0)(1-2)} - 5\,\frac{(x+1)(x-0)(x-1)}{(2+1)(2-0)(2-1)}$$

$$= -\tfrac{1}{6}x(x-1)(x-2) + \tfrac{1}{2}(x+1)(x-1)(x-2)$$

$$-\,\tfrac{1}{2}(x+1)x(x-2) - \tfrac{5}{6}(x+1)x(x-1)$$

which may be reduced to

$$y = -x^3 + x + 1$$

For the purpose of actual numerical interpolation, the reduction to this final form would not be necessary.

On the other hand, whereas the newtonian method would require the formation of the divided-difference table

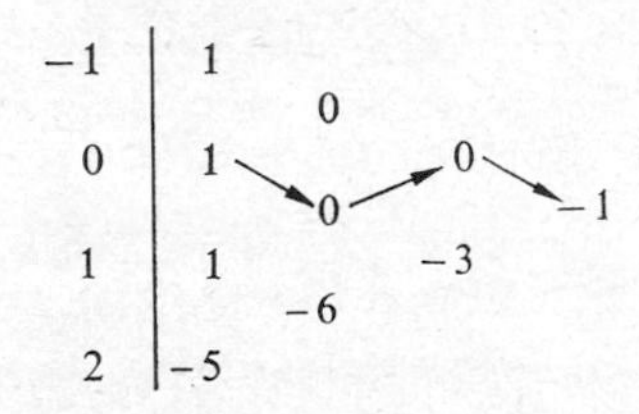

the use of the indicated difference path would involve only the following calculation:

$$y = 1 + x(0) + x(x-1)(0) + x(x-1)(x+1)(-1)$$
$$= 1 - x(x-1)(x+1) = -x^3 + x + 1$$

The lagrangian form of the interpolation formula $f(x) \approx y(x)$ possesses the advantage that its use does not involve preliminary differencing of data. However, it has the disadvantage that, unless $f(x)$ is given analytically, so that use may be made of the second form of (3.2.15), it is difficult to estimate the *truncation error* relevant to the result afforded by interpolation based on a given number of ordinates, or to estimate the number of ordinates needed to reduce the truncation error below prescribed limits. If the newtonian formula is used, a more or less dependable estimate of accuracy, based essentially on the *first* form of (3.2.15), may be obtained by sampling the first neglected higher-order difference.

Furthermore, in order to improve a certain result by taking into account one or more additional ordinates, the coefficient functions $l_i(x)$ would have to be completely redetermined in the lagrangian procedure, whereas the newtonian procedure would require merely the formation of a higher-order difference, and the addition of a multiple of that difference to the previously calculated result.

On the other hand, the lagrangian form is much better adapted to the analysis of the effects of *inherent errors* in the data. Thus, if the original data were all correctly rounded to r decimal places, so that the maximum error in each given ordinate is $5 \times 10^{-r-1}$, it is seen that the largest possible corresponding error in the interpolation for $f(x)$ would be

$$|R(x)|_{\max} = (5 \times 10^{-r-1}) \sum_{k=0}^{n} |l_k(x)| \qquad (3.2.16)$$

The corresponding calculation based on the newtonian form would be more complicated but would, of course, lead to an equivalent result. In addition to this error, the errors due to truncation and to *intermediate* roundoffs must be taken into account in either case.

3.3 Numerical Differentiation and Integration

Once an interpolation polynomial $y(x)$ has been determined so that it satisfactorily approximates a given function $f(x)$ over a certain interval I, it may be hoped that the result of differentiating $y(x)$, or of integrating it over an interval, will also satisfactorily approximate the corresponding derivative or integral of $f(x)$. However, if we visualize a curve, representing an approximating function and oscillating about the curve representing the function approximated, we may anticipate the fact that, even though the deviation between $y(x)$ and $f(x)$ be small throughout an interval, still the *slopes* of the two curves representing them may differ quite appreciably. Further, it is seen that roundoff errors (or errors of observation) of alternating sign in consecutive ordinates could affect the calculation of the derivative quite strongly if those ordinates were fairly closely spaced.

On the other hand, since *integration* is essentially a smoothing process, it would be anticipated that the error associated with integration may be small even though the interpolation polynomial itself provides only a moderately good approximation to $f(x)$.

These expectations are borne out in practice. In particular, numerical differentiation should be avoided wherever possible, particularly when the data are empirical and subject to appreciable errors of observation. When such a calculation must be made, it is desirable first to *smooth* the data to a certain extent. Certain methods of effecting such a smoothing are considered in Sec. 7.15.

From the lagrangian approximation

$$f(x) \approx \sum_{k=0}^{n} l_k(x) f(x_k) \tag{3.3.1}$$

with associated error

$$E(x) = \pi(x) \frac{f^{(n+1)}(\xi)}{(n+1)!} \tag{3.3.2}$$

we obtain the corresponding integral formula

$$\int_a^b f(x)\, dx \approx \sum_{k=0}^{n} C_k f(x_k) \tag{3.3.3}$$

where the *weighting coefficients* C_k are given by

$$C_k = \int_a^b l_k(x)\, dx \tag{3.3.4}$$

and where the associated error can be expressed in the form

$$E = \frac{1}{(n+1)!} \int_a^b \pi(x) f^{(n+1)}(\xi)\, dx \tag{3.3.5}$$

or in a related form involving a divided difference.

With regard to (3.3.5), it should be remembered that ξ depends in a specific, but generally *unknown*, way upon x, so that even though the $(n+1)$th derivative of f were known analytically, generally it would be impossible to evaluate the integral defining E exactly. However, if it is known that $|f^{(n+1)}(x)| \leqq M_{n+1}$ on I, where I is limited by the largest and smallest of $x_0, x_1, \ldots, x_n$, a, and b, and where M_{n+1} is a constant, it may be deduced that

$$|E| \leqq \frac{M_{n+1}}{(n+1)!} \int_a^b |\pi(x)|\, dx \tag{3.3.6}$$

Further, in those cases where none of the abscissas x_i lie in (a, b), the function $\pi(x)$ does not change sign in $[a, b]$ and the second law of the mean (Sec. 1.9) may be invoked to show that

$$E = \frac{f^{(n+1)}(\xi)}{(n+1)!} \int_a^b \pi(x)\, dx \tag{3.3.7}$$

where ξ is some number in I.† This last situation exists, in particular, in the

† The symbol ξ is used frequently in a generic sense to indicate an argument known only to lie in a certain interval, and often it will have different interpretations in different (possibly related) equations [as in (3.3.5) and (3.3.7)] when the resultant ambiguity is not believed to be misleading. A similar comment applies to the use of the symbol E to designate an *error term*.

frequently occurring cases when the integration is carried out over the interval *between two adjacent tabular points.*

Similarly, by differentiating (3.3.1) r times, one obtains the approximation

$$f^{(r)}(x) \approx \sum_{k=0}^{n} l_k^{(r)}(x) f(x_k) \tag{3.3.8}$$

with the associated error

$$E^{(r)}(x) = \frac{1}{(n+1)!} \frac{d^r}{dx^r} [\pi(x) f^{(n+1)}(\xi)] \tag{3.3.9}$$

However, since the dependence of ξ upon x is again unknown, the differentiation in (3.3.9) cannot be explicitly effected.

In order to obtain a somewhat more tractable form of the remainder, we replace (3.3.2) by the equivalent first form of (3.2.15), which involves the current variable x itself. The error (3.3.9) then can be expressed in the form

$$E^{(r)}(x) = \frac{d^r}{dx^r} \{\pi(x) f[x_0, \ldots, x_n, x]\} \tag{3.3.10}$$

If use is made of the *Leibnitz formula* for the rth derivative of a product,

$$\begin{aligned} \frac{d^r}{dx^r}(uv) &= u\, D^r v + r\, Du\, D^{r-1} v + \frac{r(r-1)}{2!} D^2 u\, D^{r-2} v + \cdots + D^r u\, v \\ &= \sum_{i=0}^{r} \binom{r}{i} D^i u\, D^{r-i} v \end{aligned} \tag{3.3.11}$$

where $D \equiv d/dx$ and where $\binom{r}{i}$ represents the binomial coefficient

$$\binom{r}{i} = \frac{r(r-1)\cdots(r-i+1)}{i!} = \frac{r!}{(r-i)!\, i!} \tag{3.3.12}$$

Equation (3.3.10) takes the form

$$E^{(r)}(x) = \sum_{i=0}^{r} \binom{r}{i} \pi^{(i)}(x) \frac{d^{r-i}}{dx^{r-i}} f[x_0, \ldots, x_n, x]$$

or, making use of (2.3.9),

$$E^{(r)}(x) = \sum_{i=0}^{r} \frac{r!}{i!} \pi^{(i)}(x) f[x_0, \ldots, x_n, \overbrace{x, \ldots, x}^{r-i+1 \text{ times}}] \tag{3.3.13}$$

A generalization of the relation (2.6.8) leads to the fact that

$$f[x_0, \ldots, x_n, \overbrace{x, \ldots, x}^{m \text{ times}}] = \frac{1}{(n+m)!} f^{(n+m)}(\xi_m) \tag{3.3.14}$$

where, for given n, ξ_m lies somewhere in the interval I limited by the largest and smallest of $x_0, \ldots, x_n$ and x. Hence, finally, (3.3.13) can be expressed in the form

$$E^{(r)}(x) = \sum_{i=0}^{r} \frac{r!}{(n + r - i + 1)!\, i!} \pi^{(i)}(x) f^{(n+r-i+1)}(\xi_i) \tag{3.3.15}$$

where each of the $r + 1$ numbers $\xi_0, \ldots, \xi_r$ lies in I.

The expression for the error is thus rather complicated in the general case, and when the rth derivative is calculated by differentiating an interpolation polynomial of nth degree, the estimation of the error may involve the estimation of derivatives of $f(x)$ of orders $n + 1, n + 2, \ldots, n + r$, and $n + r + 1$ in the interval I. (See, however, Prob. 10.)

It may be noticed that when $r > n$ the right-hand member of (3.3.8) vanishes identically, since $l_k(x)$ is a polynomial of degree n. Generally, at best only derivatives of order r for which r is small relative to n are given with any significant accuracy by this formula.

In the case $r = 1$, the formula (3.3.8) becomes

$$f'(x) \approx \sum_{k=0}^{n} l_k'(x) f(x_k) \tag{3.3.16}$$

and the associated error, as given by (3.3.15), is of the form

$$E'(x) = \pi'(x) \frac{f^{(n+1)}(\xi_1)}{(n + 1)!} + \pi(x) \frac{f^{(n+2)}(\xi_0)}{(n + 2)!} \tag{3.3.17}$$

where both ξ_1 and ξ_0 lie in the interval I. In particular, for numerical differentiation *at a tabular point*, there follows

$$f'(x_i) = \sum_{k=0}^{n} l_k'(x_i) f(x_k) + \pi'(x_i) \frac{f^{(n+1)}(\xi_1)}{(n + 1)!} \tag{3.3.18}$$

since $\pi(x)$ vanishes when $x = x_i$, where the factor $\pi'(x_i)$ has the simple form

$$\pi'(x_i) = (x_i - x_0) \cdots (x_i - x_n) \tag{3.3.19}$$

in accordance with (3.2.11).

It is seen that this relation is the result which would be obtained by differentiating the formula

$$f(x) = \sum_{k=0}^{n} l_k(x) f(x_k) + \pi(x) \frac{f^{(n+1)}(\xi)}{(n + 1)!}$$

with respect to x, overlooking the fact that ξ is a function of x, setting $x = x_i$ in the result, and changing ξ to a new parameter ξ_1. Except for the calculation of the *first* derivative *at a tabular point*, (3.3.15) indicates that this procedure generally would not yield the correct expression for the error term.

It can be shown, however, that the error $E^{(r)}(x)$ *can* indeed be expressed in the analogous form

$$E^{(r)}(x) = \pi^{(r)}(x)\frac{f^{(n+1)}(\eta_r)}{(n+1)!} \qquad (3.3.20)$$

for *any* positive integer r, where η_r is somewhere in I, *when x is outside or at one end of the span of the tabular values* $x_0, \ldots, x_n$ (see Prob. 8).

3.4 Uniform-spacing Interpolation

Since the coefficient function $l_i(x)$ can be expressed in the form

$$l_i(x) = \frac{(x - x_0)(x - x_1)\cdots(x - x_{i-1})(x - x_{i+1})\cdots(x - x_n)}{(x_i - x_0)(x_i - x_1)\cdots(x_i - x_{i-1})(x_i - x_{i+1})\cdots(x_i - x_n)} \qquad (3.4.1)$$

it is seen that the *form* of $l_i(x)$ is invariant under any linear change in variables

$$x = a + hs \qquad x_k = a + hs_k \qquad (3.4.2)$$

where a and h are constants:

$$l_i(x) = \frac{(s - s_0)(s - s_1)\cdots(s - s_{i-1})(s - s_{i+1})\cdots(s - s_n)}{(s_i - s_0)(s_i - s_1)\cdots(s_i - s_{i-1})(s_i - s_{i+1})\cdots(s_i - s_n)} \qquad (3.4.3)$$

It is often desirable to choose a and h in such a way that the dimensionless variable s, which measures distance from a in units of h, takes on convenient values at the tabular points used in any specific interpolation. For equally spaced abscissas, h is conveniently identified with the spacing. In the cases when the abscissas are uniformly spaced, the lagrangian coefficient functions have been tabulated rather extensively for various values of n. Formulas involving an *odd* number of ordinates are most often used, and, if that number is $n + 1 = 2m + 1$, the abscissas are then conventionally renumbered as $x_{-m}, \ldots, x_{-1}, x_0, x_1, \ldots, x_m$.

If the uniform spacing $x_{k+1} - x_k$ is denoted by h, and if s is measured from the *central* point, so that

$$x = x_0 + hs \qquad x_k = x_0 + hk \qquad (3.4.4)$$

Equation (3.4.3) then reduces to

$$l_i(x) = \frac{(s + m)(s + m - 1)\cdots(s - i + 1)(s - i - 1)\cdots(s - m + 1)(s - m)}{(i + m)(i + m - 1)\cdots(2)(1)(-1)(-2)\cdots(i - m + 1)(i - m)}$$

$$\equiv L_i(s)$$

Thus

$$L_0(s) = \frac{(1 - s^2)(4 - s^2)\cdots(m^2 - s^2)}{(m!)^2} \tag{3.4.5}$$

and

$$L_i(s) = \frac{(-1)^{i+1}s(s+i)}{(m+i)!\,(m-i)!} \times [(1-s^2)(4-s^2)\cdots(\overline{i-1}^2 - s^2)(\overline{i+1}^2 - s^2)\cdots(m^2 - s^2)] \tag{3.4.6}$$

for $i = \pm 1, \pm 2, \ldots, \pm m$.

In illustration, Table 3.1 presents exact values of the lagrangian coefficients for three-point (quadratic) interpolation to tenths, corresponding to $m = 1$ (a corresponding five-point table is included in Sec. 4.12):

Table 3.1

s	$L_{-1}(s)$	$L_0(s)$	$L_1(s)$	
0.0	0	1	0	0.0
0.1	−0.045	0.99	0.055	−0.1
0.2	−0.08	0.96	0.12	−0.2
0.3	−0.105	0.91	0.195	−0.3
0.4	−0.12	0.84	0.28	−0.4
0.5	−0.125	0.75	0.375	−0.5
0.6	−0.12	0.64	0.48	−0.6
0.7	−0.105	0.51	0.595	−0.7
0.8	−0.08	0.36	0.72	−0.8
0.9	−0.045	0.19	0.855	−0.9
1.0	0	0	1	−1.0
	$L_1(s)$	$L_0(s)$	$L_{-1}(s)$	s

From (3.4.6) it follows that $L_i(-s) = L_{-i}(s)$. This explains the fact that, for *negative* values of s, to be read from the right-hand margin, the column labels at the *foot* of the table are to be used.

Thus, for example, to interpolate the data

x	1.00	1.10	1.20	1.30
$f(x)$	0.8415	0.8912	0.9320	0.9636

for $f(1.24)$ by use of a three-point formula, the work would be centered at the nearest tabular point, $x = 1.20$. With $s = 0.04/0.10 = 0.4$, and with coefficients read from the preceding table, there would follow

$$\begin{aligned} f(1.24) &\approx (-0.12)(0.8912) + (0.84)(0.9320) + (0.28)(0.9636) \\ &= 0.945744 \doteq 0.9457 \end{aligned}$$

To interpolate for $x = 1.02$, the work would be centered at $x = 1.10$. With $s = -0.8$, there would follow

$$f(1.02) \approx (0.72)(0.8415) + (0.36)(0.8912) + (-0.08)(0.9320)$$
$$= 0.852152 \doteq 0.8522$$

The given data correspond to rounded values of $f(x) = \sin x$, and the results thus correspond to the tabulated five-place values $\sin 1.24 \doteq 0.94578$ and $\sin 1.02 \doteq 0.85211$.

Extensive tables of Lagrange coefficient functions, and of certain of their derivatives, may be found in the literature (see Sec. 3.12).

3.5 Newton-Cotes Integration Formulas

In order to obtain formulas for the approximate evaluation of an integral of the form $\int_a^b f(x)\,dx$, where a and b are finite, we may first introduce the change of variables

$$x = a + \frac{b-a}{n}s \tag{3.5.1}$$

where n is an integer, to obtain the relation

$$\int_a^b f(x)\,dx = \frac{b-a}{n}\int_0^n F(s)\,ds \tag{3.5.2}$$

where

$$F(s) = f\left(a + \frac{b-a}{n}s\right) \tag{3.5.3}$$

If now it is assumed that $f(x)$ can be approximated over $[a, b]$ by the polynomial which agrees with it at, say, $n + 1$ equally spaced points in $[a, b]$ we may obtain the approximate formula

$$\int_0^n F(s)\,ds \approx \sum_{k=0}^{n} C_k F(k) \tag{3.5.4}$$

where

$$C_k = \int_0^n \frac{s(s-1)\cdots(s-k+1)(s-k-1)\cdots(s-n)}{k(k-1)\cdots(k-k+1)(k-k-1)\cdots(k-n)}\,ds \tag{3.5.5}$$

In accordance with (2.5.3), the error term omitted on the right in (3.5.4) can be expressed in the form

$$\bar{E}_n = \int_0^n s(s-1)\cdots(s-n)F[0, 1, \ldots, n, s]\,ds \tag{3.5.6}$$

Since the coefficient of the divided difference of F is not of constant sign in $[0, n]$ unless $n = 1$, the second law of the mean cannot be applied directly. However, it is possible to prove (see Sec. 5.12 and Steffensen [1950]) that when n is *odd*, the error can be expressed in the form which would be obtained if this procedure were valid:

$$\bar{E}_n = \frac{F^{(n+1)}(\sigma)}{(n+1)!} \int_0^n s(s-1)\cdots(s-n)\,ds \qquad (n \text{ odd}) \qquad (3.5.7a)$$

whereas, when n is *even*, the error can be expressed in the form

$$\bar{E}_n = \frac{F^{(n+2)}(\sigma)}{(n+2)!} \int_0^n \left(s - \frac{n}{2}\right) s(s-1)\cdots(s-n)\,ds \qquad (n \text{ even}) \qquad (3.5.7b)$$

where $0 < \sigma < n$ in each case.

If, as before, we write $h = (b - a)/n$ and $x_i = a + hi$, the result established can be put in the more explicit form

$$\int_{x_0}^{x_n} f(x)\,dx \approx h \sum_{k=0}^{n} C_k f(x_k) \qquad (3.5.8)$$

where, for a given value of n, C_k is defined by (3.5.5). By noticing that

$$\frac{d^r}{ds^r} F(s) = h^r \frac{d^r}{dx^r} f(x)$$

and that, from (3.5.2), the error in (3.5.8) is $h\bar{E}_n$, we obtain also the expressions

$$E_n = \frac{h^{n+2} f^{(n+1)}(\xi)}{(n+1)!} \int_0^n s(s-1)\cdots(s-n)\,ds \qquad (n \text{ odd}) \qquad (3.5.9a)$$

and

$$E_n = \frac{h^{n+3} f^{(n+2)}(\xi)}{(n+2)!} \int_0^n \left(s - \frac{n}{2}\right) s(s-1)\cdots(s-n)\,ds \qquad (n \text{ even}) \qquad (3.5.9b)$$

where $x_0 < \xi < x_n$ in each case.

In illustration, we consider the case $n = 2$. Here there follows, from (3.5.5),

$$C_0 = \int_0^2 \frac{(s-1)(s-2)}{(-1)(-2)}\,ds = \frac{1}{3} \qquad C_1 = \int_0^2 \frac{s(s-2)}{(1)(-1)}\,ds = \frac{4}{3}$$

$$C_2 = \int_0^2 \frac{s(s-1)}{(2)(1)}\,ds = \frac{1}{3}$$

and (3.5.9b) gives

$$E_2 = \frac{h^5 f^{\text{iv}}(\xi)}{24} \int_0^2 s(s-1)^2(s-2)\,ds = -\frac{h^5 f^{\text{iv}}(\xi)}{90}$$

The corresponding formula (3.5.8), with the error term, then takes the form

$$\int_{x_0}^{x_2} f(x)\,dx = \frac{h}{3}(f_0 + 4f_1 + f_2) - \frac{h^5}{90} f^{\text{iv}}(\xi) \qquad (x_0 < \xi < x_2)$$

This is the celebrated formula of *Simpson's rule*.

In a similar way, the following formulas may be obtained:

$$\int_{x_0}^{x_1} f(x)\,dx = \frac{h}{2}(f_0 + f_1) - \frac{h^3}{12} f''(\xi) \tag{3.5.10}$$

$$\int_{x_0}^{x_2} f(x)\,dx = \frac{h}{3}(f_0 + 4f_1 + f_2) - \frac{h^5}{90} f^{\text{iv}}(\xi) \tag{3.5.11}$$

$$\int_{x_0}^{x_3} f(x)\,dx = \frac{3h}{8}(f_0 + 3f_1 + 3f_2 + f_3) - \frac{3h^5}{80} f^{\text{iv}}(\xi) \tag{3.5.12}$$

$$\int_{x_0}^{x_4} f(x)\,dx = \frac{2h}{45}(7f_0 + 32f_1 + 12f_2 + 32f_3 + 7f_4) - \frac{8h^7}{945} f^{\text{vi}}(\xi) \tag{3.5.13}$$

$$\int_{x_0}^{x_5} f(x)\,dx = \frac{5h}{288}(19f_0 + 75f_1 + 50f_2 + 50f_3 + 75f_4 + 19f_5) - \frac{275h^7}{12096} f^{\text{vi}}(\xi) \tag{3.5.14}$$

$$\int_{x_0}^{x_6} f(x)\,dx = \frac{h}{140}(41f_0 + 216f_1 + 27f_2 + 272f_3 + 27f_4 + 216f_5 + 41f_6) - \frac{9h^9}{1400} f^{\text{viii}}(\xi) \tag{3.5.15}$$

$$\int_{x_0}^{x_7} f(x)\,dx = \frac{7h}{17280}(751f_0 + 3577f_1 + 1323f_2 + 2989f_3 + 2989f_4 + 1323f_5 + 3577f_6 + 751f_7) - \frac{8183h^9}{518400} f^{\text{viii}}(\xi) \tag{3.5.16}$$

$$\int_{x_0}^{x_8} f(x)\,dx = \frac{4h}{14175}(989f_0 + 5888f_1 - 928f_2 + 10496f_3 - 4540f_4 + 10496f_5 - 928f_6 + 5888f_7 + 989f_8) - \frac{2368h^{11}}{467775} f^{\text{x}}(\xi) \tag{3.5.17}$$

An inspection of the error terms reveals that a formula involving an *odd* number $n + 1 = 2m + 1$ of points would yield *exact* results if $f(x)$ were a polynomial of degree $n + 1$ or less, whereas one involving an *even* number $n + 1 = 2m$ of points would be exact only if $f(x)$ were a polynomial of degree

n or less. Thus the two formulas involving $2m$ and $2m - 1$ ordinates have the same order of accuracy, so that generally no great advantage is gained by advancing from a formula involving an odd number of ordinates to one involving one more ordinate. In particular, the error in *Simpson's rule* (3.5.11) is given by $-h^5 f^{iv}(\xi_1)/90$, and that in *Newton's rule* (3.5.12) by $-3h^5 f^{iv}(\xi_2)/80$, where both ξ_1 and ξ_2 are in (a, b). In comparing these errors, when both formulas are applied to the evaluation of *the same integral*, we must notice that $h = (b - a)/2$ in the former case, whereas $h = (b - a)/3$ in the latter. Hence the coefficient of $-(b - a)^5 f^{iv}$ is $\frac{1}{2880}$ in Simpson's rule and $\frac{1}{6480}$ in Newton's rule. Thus the latter (which involves one extra ordinate) may be expected to be only slightly more accurate than the former, on the average. Clearly, the advantage may be shifted in either direction if $f^{iv}(x)$ varies strongly over $[a, b]$, so that $f^{iv}(\xi_1)$ and $f^{iv}(\xi_2)$ may differ appreciably, or if $f^{iv}(x)$ fails to exist or is discontinuous somewhere in (a, b), so that the error formulas are invalid.

Another useful set of integration formulas is obtained by dividing the interval $[a, b]$, as before, into n equal parts by inserting $n - 1$ equally spaced interior abscissas, then approximating $f(x)$ by the polynomial of degree $n - 2$ which coincides with $f(x)$ *at the* $n - 1$ *interior points*, and approximating the relevant integral by integrating the resultant polynomial over $[a, b]$. These formulas thus do not involve the ordinates at the *ends* of the interval and are said to be of *open* type, whereas those previously considered are said to be of *closed* type. The first few such formulas ($n = 2, \ldots, 6$) may be expressed as follows:

$$\int_{x_0}^{x_2} f(x)\,dx = 2hf_1 + \frac{h^3}{3} f''(\xi) \tag{3.5.18}$$

$$\int_{x_0}^{x_3} f(x)\,dx = \frac{3h}{2}(f_1 + f_2) + \frac{3h^3}{4} f''(\xi) \tag{3.5.19}$$

$$\int_{x_0}^{x_4} f(x)\,dx = \frac{4h}{3}(2f_1 - f_2 + 2f_3) + \frac{14h^5}{45} f^{iv}(\xi) \tag{3.5.20}$$

$$\int_{x_0}^{x_5} f(x)\,dx = \frac{5h}{24}(11f_1 + f_2 + f_3 + 11f_4) + \frac{95h^5}{144} f^{iv}(\xi) \tag{3.5.21}$$

$$\int_{x_0}^{x_6} f(x)\,dx = \frac{3h}{10}(11f_1 - 14f_2 + 26f_3 - 14f_4 + 11f_5) + \frac{41h^7}{140} f^{vi}(\xi) \tag{3.5.22}$$

It is sometimes more convenient to write (3.5.18) in the form

$$\int_{x_0}^{x_1} f(x)\,dx = hf_{1/2} + \frac{h^3}{24} f''(\xi) \tag{3.5.23}$$

where $f_{1/2} = f(\frac{1}{2}x_0 + \frac{1}{2}x_1)$, in which form it is often called the *midpoint formula*. (See also Prob. 28 for another family of integration formulas of which it is a member.)

The formulas of the type considered in this section are generally known as the *Newton-Cotes* (or *Cotes*) formulas. Apart from the midpoint formula (3.5.23), the formulas of open type are principally of use in the numerical integration of differential equations. In addition, however, they are needed in special cases where the values of $f(x)$ at the end points of the interval of integration are unavailable, and they tend to be somewhat preferable to comparable closed formulas when a derivative of $f(x)$ is singular at one or both of the end points (see Sec. 5.10).

Since all the integration formulas considered in this chapter must, in particular, be exact if $f(x)$ is a constant, it follows that *the sum of the weighting coefficients in any formula must equal the length of the interval of integration.* Thus, for example, that sum in (3.5.13) is $\frac{2}{45}h \cdot 90 = 4h = b - a$.

3.6 Composite Integration Formulas

In place of using a single polynomial to approximate $f(x)$ over the complete integration interval $[a, b]$, it is clearly possible to divide $[a, b]$ into subintervals and to approximate $f(x)$ by a different polynomial over each subinterval. Thus, for example, by applying the two-point formula (3.5.10) to n successive subintervals of length h, one obtains the so-called *trapezoidal rule*:

$$\int_a^b f(x)\,dx = h(\tfrac{1}{2}f_0 + f_1 + f_2 + \cdots + f_{n-2} + f_{n-1} + \tfrac{1}{2}f_n) - \frac{nh^3}{12}f''(\xi) \tag{3.6.1}$$

where $f_0 = f(a)$, $f_k = f(a + kh)$, and $f_n = f(b)$, and where now ξ is somewhere in (a, b). This formula corresponds to replacing the graph of $f(x)$ by the result of joining the ends of adjacent ordinates by line segments and is of remarkable simplicity. Whereas it is not of high accuracy, we may notice that, since here $h = (b - a)/n$, the error can be written in the form

$$E_n = -\frac{(b-a)^3}{12n^2}f''(\xi) = -\frac{b-a}{12}h^2 f''(\xi) \tag{3.6.2}$$

Hence, if only $f''(x)$ is continuous (and hence bounded) on $[a, b]$, the error will indeed tend to zero like $1/n^2$ as $n \to \infty$ (and like h^2 as $h \to 0$).

As will be seen, the accuracy afforded by a k-point Newton-Cotes formula does not *necessarily* increase as k increases, and, in fact, the accuracy may become worse and worse after a certain stage, even though $f(x)$ possess continuous derivatives of all orders for all real values of x, and even though no roundoff errors be introduced. In such cases, unless the desired accuracy is attained before this stage is attained, the use of an alternative such as a *composite* (or "compound") rule is essential as well as convenient.

Another advantage of the trapezoidal rule consists of the fact that the weighting coefficients are nearly equal to each other. For it is easily seen that, if $n + 1$ ordinates are each liable to random errors of observation (or roundoff), the RMS error in a linear combination of these ordinates, for which the sum of the constants of combination is fixed (here equal to $b - a$), is *least* when the constants of combination are equal. Newton-Cotes formulas of the open type are particularly objectionable from this point of view, since, for $n = 4$ and $n \geqq 6$, their coefficients actually fluctuate in sign. Similar sign fluctuations also occur in formulas of closed type for $n = 8$ and for $n \geqq 10$.

If the midpoint formula (3.5.23) is applied to each of the n subintervals used in deriving (3.6.1), one obtains instead the *repeated midpoint rule*:

$$\int_a^b f(x)\,dx = h(f_{1/2} + f_{3/2} + \cdots + f_{n-(1/2)}) + \frac{nh^3}{24} f''(\xi) \qquad (3.6.3)$$

where $f_{k+(1/2)} = f[a + (k + \frac{1}{2})h]$. This formula results from approximating $f(x)$ by a piecewise-constant *step function*, with a jump $f_{k+(1/2)} - f_{k-(1/2)}$ at each point $x_k = a + kh$. For a given value of n, the magnitude of its error tends (on the average) to be about half that associated with the trapezoidal formula (when f is twice differentiable) in spite of its using one less ordinate. Additional advantages follow from the *equality* of its weighting coefficients. (An offsetting computational drawback is pointed out in the next section.)

By dividing the interval $[a, b]$ into $n/2$ subintervals of length $2h$, where n is an *even* integer, and applying Simpson's rule to each subinterval [that is, by approximating the graph of $f(x)$ by a parabola in each subinterval], the so-called *parabolic rule* is obtained in the form

$$\int_a^b f(x)\,dx = \frac{h}{3}(f_0 + 4f_1 + 2f_2 + 4f_3 + \cdots + 4f_{n-3} + 2f_{n-2} + 4f_{n-1} + f_n) - \frac{nh^5}{180} f^{\text{iv}}(\xi) \qquad (3.6.4)$$

where, again, $f_0 = f(a)$, $f_k = f(a + kh)$, and $f_n = f(b)$. Here we have

$$E_n = -\frac{(b - a)^5}{180n^4} f^{\text{iv}}(\xi) \qquad (3.6.5)$$

so that, if $f^{\text{iv}}(x)$ is continuous on $[a, b]$, the error associated with the use of

$n + 1$ ordinates tends to zero like $1/n^4$ as $n \to \infty$, and like h^4 as $h \to 0$. Thus the parabolic rule usually is more accurate than the trapezoidal and repeated midpoint rules when n is sufficiently large. Since also its weighting coefficients are simple and do not fluctuate unduly in magnitude, it is perhaps the most widely used single formula for numerical integration. However, it can be used only when ordinates are available which divide $[a, b]$ into an even number of intervals of equal length h; also, in cases where a high degree of accuracy is required, a prohibitively large number of such ordinates may be needed.

It may be noted that the trapezoidal, repeated midpoint, and parabolic rules each approximate the required integral I by an associated *Riemann sum* (see Prob. 29), so that each of these approximations assuredly *converges* to I as the spacing h tends to zero if only I *exists* in the usual (Riemann) sense whether or not $f(x)$ is sufficiently differentiable to ensure the respective *rates* of convergence to which reference was made above.

It is useful to notice that when $f''(x)$ is of constant sign in $[a, b]$, the true value of I is *between* the trapezoidal and repeated midpoint approximations based on a common spacing h. Also, if these approximations are denoted by $I_T(h)$ and $I_M(h)$ in the general case, there follows

$$I_P\left(\frac{h}{2}\right) = \frac{1}{3}\left[I_T(h) + 2I_M(h)\right] \qquad (3.6.6)$$

where $I_P(h/2)$ is the *parabolic-rule approximation* based on a *halved* spacing.

The use of such formulas in two-dimensional integration over a rectangle is illustrated in Probs. 40 and 41. Other integration formulas are considered in Chaps. 5 and 8 and in Sec. 9.13.

3.7 Use of Integration Formulas

In order to illustrate the preceding formulas in a simple case, we consider first the numerical evaluation of the integral

$$\int_0^1 \frac{dx}{1+x} = \log 2 = 0.69314718\cdots \qquad (3.7.1)$$

With $f(x) = 1/(1 + x)$, there follows also

$$f^{(k)}(x) = \frac{(-1)^k k!}{(1+x)^{k+1}}$$

and hence

$$2^{-k-1}k! < (-1)^k f^{(k)}(x) < k! \qquad (0 < x < 1)$$

Thus, for example, if use were to be made of the five-point formula (3.5.13), with $h = 0.25$, the upper and lower bounds

$$0.000002 < -E_5 < 0.0004$$

would be available with regard to the truncation error. Since $f^{(k)}(x)$ is positive on [0, 1] when k is even, it follows that each error term will be negative.

The following table of upper bounds on the magnitude of the *possible* error relevant to the (closed) Newton-Cotes (NC), trapezoidal (T), and parabolic (P) rules, in the present case, is easily determined:

Ordinates	NC	T	P
3	9×10^{-3}	5×10^{-2}	9×10^{-3}
5	4×10^{-4}	2×10^{-2}	6×10^{-4}
7	3×10^{-5}	5×10^{-3}	2×10^{-4}
9	2×10^{-6}	3×10^{-3}	4×10^{-5}

More generally, it can be predicted in this case that the magnitude of the error involved in the use of $n + 1$ ordinates (where n is even) will be between $1/240n^4$ and $2/15n^4$ for the parabolic rule, between $1/48n^2$ and $1/6n^2$ for the trapezoidal rule, and between $1/96(n + 1)^2$ and $1/12(n + 1)^2$ for the repeated midpoint rule. Such preliminary upper bounds may be quite conservative, but it is difficult to obtain more precise ones.

In addition to the truncation errors, one must consider the effect of roundoff in the values of the ordinates used in the calculation. If each ordinate is rounded correctly to r decimal places, so that the maximum error in each ordinate is not greater than $5 \times 10^{-r-1}$, the maximum corresponding error in the final calculation is therefore not greater than $5 \times 10^{-r-1}$ times the sum of the *absolute values* of the relevant weighting coefficients. If those coefficients are all *positive*, this last sum must equal the length of the interval of integration (here unity). Thus, in the present case, if all weighting coefficients are positive, the error in the final result, due to inaccuracies in the original data, cannot exceed the maximum of those inaccuracies. This situation prevails in all the formulas considered in the preceding tabulation except the Newton-Cotes nine-point formula, in which a magnification factor of $\frac{41142}{28350} \approx 1.5$ would be involved. Whereas a considerable amount of cancellation in the errors of roundoffs would be expected (particularly if the weighting coefficients are nearly equal), it cannot be *guaranteed* in any particular case.

Suppose that the ordinates used in the present calculation are to be rounded to r decimal places, and that the final result is to be in error by less than one unit in the rth decimal place. If the parabolic rule is to be used, with $n + 1$ ordinates, an even integer n must then be determined such that

$$\frac{2}{15n^4} < 5 \times 10^{-r-1} \qquad \text{or} \qquad n > 0.72 \times 10^{r/4}$$

The total error, due to truncation and initial roundoff, then could not exceed 10^{-r}, under the assumption that *intermediate* roundoffs are absent or negligible. For $r = 4$, this condition gives $n > 7.2$, so that nine ordinates would do; for $r = 5$, 13 ordinates would suffice. If the trapezoidal rule were used, the need for about 58 ordinates would be indicated for guaranteed four-place accuracy. Reference to the preceding table shows that a Newton-Cotes formula using nine ordinates would lead to a result in error by less than 2×10^{-6} due to truncation. If the ordinates were rounded to five places, the effect of that roundoff here could be as large as 8×10^{-6}. Thus the final error could not exceed 10^{-5}.

Actual calculation, with the ordinates rounded to five places, shows that the error associated with Simpson's rule (three ordinates) is smaller than 2×10^{-3}, with the five-ordinate Newton-Cotes formula less than 3×10^{-5}, with the five-ordinate trapezoidal rule less than 4×10^{-3}, with the four-ordinate midpoint rule less than 2×10^{-3}, and with the five-ordinate parabolic rule less than 10^{-4}. The fact that some of the error predictions were quite conservative is a consequence of the variation of the higher derivatives of $f(x)$ over $[0, 1]$. Thus, for example, the error estimate for the five-ordinate Newton-Cotes formula assigned the maximum value 720 to $f^{vi}(x)$ on $[0, 1]$, whereas all values from $\frac{45}{8}$ to 720 are taken on. A value of about 56 would have given the proper estimate.

In those cases where $f(x)$ is given empirically or, more generally, in such a form that information with regard to bounds on higher derivatives of $f(x)$ is not readily accessible, less dependable error estimates may be based on the calculation of one or more *divided differences* of order equal to that of the derivative involved in the error estimate.

3.8 Richardson Extrapolation. Romberg Integration

Another method of estimating (or partially removing) the truncation error in an integration formula is frequently useful. Suppose that the error in a certain formula approximating

$$I = \int_a^b f(x)\, dx$$

can be expressed in the form

$$E = Ch^r f^{(r)}(\xi) \qquad (3.8.1)$$

where C is independent of h, r is a positive integer, and ξ is known only to lie in (a, b). Let two calculations be made—one with spacing h_1 and one with spacing h_2—and denote the corresponding approximations to the true integral I by I_1 and I_2, respectively.

Then, if only truncation errors are considered, there follows

$$I - I_1 = Ch_1^r f^{(r)}(\xi_1)$$
$$I - I_2 = Ch_2^r f^{(r)}(\xi_2)$$

The assumption that $f^{(r)}(\xi_1)$ and $f^{(r)}(\xi_2)$ can be (nearly) identified then leads to the *extrapolation formula*

$$I \approx \frac{h_1^r I_2 - h_2^r I_1}{h_1^r - h_2^r} = I_2 + \frac{I_2 - I_1}{(h_1/h_2)^r - 1} \qquad (3.8.2)$$

In particular, if $h_2 = h_1/2$ there follows

$$I \approx \frac{2^r I_2 - I_1}{2^r - 1} = I_2 + \frac{I_2 - I_1}{2^r - 1} \qquad (h_1 = 2h_2) \qquad (3.8.3)$$

This process, often known as *Richardson extrapolation*, generally will be effective if $f^{(r)}(x)$ does not vary rapidly and does not change sign in $[a, b]$. Usually it may be used with some confidence, in any case, if the correction to be added to I_2 is small relative to I_2 itself and if successive approximations appear to be approaching a limit from one side without oscillation.†

In the cases of the trapezoidal and repeated midpoint rules, for which $r = 2$, the divisor in (3.8.3) is 3, whereas for the parabolic rule, with $r = 4$, the divisor is 15. For example, in the case of the preceding numerical example, the parabolic rule yields the approximation $I \approx 0.694444$ with $h = \frac{1}{2}$ and the approximation $I \approx 0.693254$ with $h = \frac{1}{4}$. Use of the extrapolation formula (3.8.3), with $r = 4$, gives

$$I \approx 0.693254 - 0.000079 = 0.693175$$

Thus the error in I_2 may be estimated as about -0.00008, and the value 0.69317 may be expected to be correct within perhaps one or two units in its last place, as is indeed the case. When oscillation of the sequence of successive approximations is present, this procedure may be completely undependable, as will be illustrated in Sec. 3.9.

For the trapezoidal and parabolic rules, the process of repeatedly halving the spacing h is particularly convenient since all ordinates needed in a particular

† A procedure of this general type, in which two calculations are made, with errors of the respective forms $h_1^r \phi(h_1)$ and $h_2^r \phi(h_2)$, where $\phi(h)$ is an incompletely known function of the spacing h, and in which an extrapolation to $h = 0$ is made under the assumption that $\phi(h)$ is nearly independent of h, is also often known as *Richardson's deferred approach to the limit*. (See Richardson and Gaunt [1927].)

calculation would be used again in the next one and in all following ones. The repeated midpoint rule does not have this useful property since, in fact, no ordinate would be used more than once.

In order to treat a useful generalization of this procedure, we note next that (as will be shown in Sec. 5.8) the truncation error associated with the trapezoidal rule can be expressed in the form

$$E = C_1 h^2 + C_2 h^4 + \cdots + C_N h^{2N} - (b - a)h^{2N+2} \frac{B_{2N+2}}{(2N + 2)!} f^{(2N+2)}(\xi) \qquad (3.8.4)$$

where the C's are certain constants which depend upon f but are independent of h, and where B_{2N+2} is a *Bernoulli number* defined in Sec. 5.8. Here the integer N can be taken to be arbitrarily large, assuming only that $f(x)$ has a $(2N + 2)$th derivative in $[a, b]$. This form reduces to (3.6.2) when N is taken to be zero since $B_2 = \frac{1}{6}$.

If we now denote by $T_k^{(0)}$ the trapezoidal approximation to I with spacing $h_k = (b - a)/2^k$, then the Richardson extrapolate based on $T_k^{(0)}$ and $T_{k+1}^{(0)}$ is $[2^2 T_{k+1}^{(0)} - T_k^{(0)}]/(2^2 - 1)$, according to (3.8.2). If this number is denoted by $T_k^{(1)}$, then the derivation of (3.8.2) shows that its error, when written in a form analogous to (3.8.4), will begin with an h^4 term, the h^2 term having been eliminated.

If then we denote the Richardson extrapolate based on $T_k^{(1)}$ and $T_{k+1}^{(1)}$ by $T_k^{(2)}$, it follows that $T_k^{(2)} = [2^4 T_{k+1}^{(1)} - T_k^{(1)}]/(2^4 - 1)$ has an error of order h_k^6, when h_k is small, and this process can be iterated at pleasure if $f(x)$ is sufficiently differentiable.

In general, if we define

$$T_k^{(m)} = \frac{4^m T_{k+1}^{(m-1)} - T_k^{(m-1)}}{4^m - 1} \qquad (m = 1, 2, \ldots) \qquad (3.8.5)$$

then the approximation $T_k^{(m)}$ has an error of order h_k^{2m+2}. In fact, it is known (see Bauer, Rutishauser, and Stiefel [1963]) that

$$I - T_k^{(m)} = (-1)^{m+1}(b - a) \frac{h_k^{2m+2} B_{2m+2} f^{(2m+2)}(\xi)}{2^{m(m+1)}(2m + 2)!} \qquad (3.8.6)$$

where $h_k = (b - a)/2^k$, for some ξ in (a, b).

By recursive use of (3.8.5), we may obtain the following triangular array of approximations to the required integral I:

$$\begin{array}{lll} T_0^{(0)} & & \\ T_1^{(0)} & T_0^{(1)} & \\ T_2^{(0)} & T_1^{(1)} & T_0^{(2)} \\ \cdots & \cdots & \cdots \end{array}$$

Here the successive entries in the first column are the trapezoidal-rule approximations, with $h_k = (b - a)/2^k$ $(k = 0, 1, 2, \ldots)$, and the remaining entries can be determined recursively from them.

It can be verified that the entry $T_k^{(1)}$ is the approximation that would be obtained directly by using Simpson's rule 2^k times, and also that $T_k^{(2)}$ could be obtained directly by using the Newton-Cotes five-point formula (3.5.13) 2^k times. No similar interpretation of $T_k^{(m)}$ in terms of Newton-Cotes formulas is possible for $m \geqq 3$.

The process of estimating the required integral by use of the array of approximations $T_k^{(m)}$ is called *Romberg integration* (Romberg [1955]). It has already been noted that the sequences of elements in the first two columns of the Romberg array converge to the desired integral I. In addition, however, it is known that in fact the sequences of elements in *all* columns converge to the same limit if $f(x)$ has derivatives of all orders on $[a, b]$, the rate of convergence generally (but not always) increasing with m. The sequence of diagonal elements $T_0^{(0)}, T_0^{(1)}, \ldots$ also converges, in such cases, to the desired value and is usually the preferred one.

It is shown in Sec. 5.8 that the error associated with the repeated midpoint rule also can be expressed in a form analogous to (3.8.4), so that a completely similar array of approximations

$$\begin{array}{lll} M_0^{(0)} & & \\ M_1^{(0)} & M_0^{(1)} & \\ M_2^{(0)} & M_1^{(1)} & M_0^{(2)} \\ \cdots & \cdots & \cdots \end{array}$$

can be associated with it, where

$$M_k^{(m)} = \frac{4^m M_{k+1}^{(m-1)} - M_k^{(m-1)}}{4^m - 1} \tag{3.8.7}$$

Here $M_k^{(0)}$ is the result of applying the midpoint rule 2^k times, with the spacing $h_k = (b - a)/2^k$.

The abscissas involved in the calculation of $M_{k-1}^{(0)}$ are the *new* abscissas introduced in the determination of $T_k^{(0)}$, and, in fact, it is easily seen that

$$T_k^{(0)} = \tfrac{1}{2}[T_{k-1}^{(0)} + M_{k-1}^{(0)}] \tag{3.8.8}$$

Thus, if the midpoint-rule approximations $M_0^{(0)}, M_1^{(0)}, M_2^{(0)}, \ldots$ are calculated first, then, after $T_0^{(0)}$ is calculated, the successive trapezoidal-rule approximations $T_1^{(0)}, T_2^{(0)}, \ldots$ can be determined recursively by use of (3.8.8) in a very simple way.

More generally, since (3.8.5) and (3.8.7) are of the same form, a simple inductive argument shows that (3.8.8) generalizes to the relation

$$T_k^{(m)} = \tfrac{1}{2}[T_{k-1}^{(m)} + M_{k-1}^{(m)}] \tag{3.8.9}$$

This relation, combined with (3.8.5) for $k = 0$,

$$T_0^{(m)} = \frac{4^m T_1^{(m-1)} - T_0^{(m-1)}}{4^m - 1} \tag{3.8.10}$$

permits a recursive calculation of the entries of the T table from $T_0^{(0)}$ and the entries of the M table.

Even though the M-table entries were of no interest in themselves, this method of constructing the T table would be a particularly efficient one. In practice, the calculation frequently is terminated when the diagonal elements $T_0^{(m)}$ and $M_0^{(m)}$ agree within the prescribed error tolerance, with their mean value $T_1^{(m)}$ taken as the final approximation to I.

If a table of entries $P_k^{(m)}$ were generated from parabolic-rule approximations in correspondence with the T and M tables, it would be identical (apart from the effects of roundoff errors) with the result of suppressing the first column of the T table since $P_k^{(m)} = T_k^{(m+1)}$.

To illustrate the use of Romberg integration, the M and T tables relevant to the approximation of the previously considered integral

$$I = \int_0^1 \frac{dx}{1 + x} = \log 2 \doteq 0.69314718 \cdots$$

are presented. Six-place values of the integrand were used, and each entry was arbitrarily rounded to six places as it was determined throughout the recursive calculation.

$M_k^{(0)}$	$M_k^{(1)}$	$M_k^{(2)}$	$M_k^{(3)}$
0.666667			
0.685714	0.692063		
0.691220	0.693055	0.693121	
0.692660	0.693140	0.693146	0.693147

$T_k^{(0)}$	$T_k^{(1)}$	$T_k^{(2)}$	$T_k^{(3)}$
0.750000			
0.708333	0.694444		
0.697024	0.693254	0.693175	
0.694122	0.693155	0.693148	0.693148

3.9 Asymptotic Behavior of Newton-Cotes Formulas

This section presents some additional results which relate to the choice between the use of a single Newton-Cotes formula over an entire range of $n + 1$ points, and the use of a composite formula employing a lower-degree formula repeatedly over successive subdivisions, when n is large.

The problem consists of examining the behavior of the Newton-Cotes error term

$$E_n = \begin{cases} h^{n+2} f^{(n+1)}(\xi) \displaystyle\int_0^n \frac{s(s-1)\cdots(s-n)}{(n+1)!}\,ds & (n \text{ odd}) \quad (3.9.1) \\ h^{n+3} f^{(n+2)}(\xi) \displaystyle\int_0^n \left(s - \frac{n}{2}\right) \frac{s(s-1)\cdots(s-n)}{(n+2)!}\,ds & (n \text{ even}) \quad (3.9.2) \end{cases}$$

where $n + 1$ is the number of ordinates and ξ is somewhere in the interval of integration $[a, b]$. As may be seen by an examination of the error terms given explicitly in Sec. 3.5, the numerical factor represented by the integral in (3.9.1) or (3.9.2) decreases slowly in magnitude as n increases. Indeed, it can be shown (see Prob. 49) that *when n is sufficiently large*, the integral in (3.9.1) is approximated by $-2/[n(\log n)^2]$, and that in (3.9.2) by one-half that quantity. Thus, in either case, the numerical factor ultimately tends to zero somewhat more rapidly than $1/n$ but less rapidly than $1/n^2$.

Hence, after writing $h = (b - a)/n$, we find that†

$$E_n \sim \begin{cases} \dfrac{-2(b-a)^{n+2}}{n^{n+3}(\log n)^2} f^{(n+1)}(\xi) & (n \text{ odd}) \quad (3.9.3) \\ \dfrac{-(b-a)^{n+3}}{n^{n+4}(\log n)^2} f^{(n+2)}(\xi) & (n \text{ even}) \quad (3.9.4) \end{cases}$$

as $n \to \infty$.

If we now suppose that the function $f(z)$, where z is a *complex* variable, is *analytic* in a region of the complex plane including the interval $a \leqq x \leqq b$ on the real axis, we can use the fact that then there exists a constant K such that

$$|f^{(r)}(x)| \leqq K \frac{r!}{R^r} \qquad (a \leqq x \leqq b) \qquad (3.9.5)$$

where R is the shortest distance, in the complex plane, from a point in $[a, b]$ to a singular point of $f(z)$. (See Prob. 50.) If this relation is combined with (3.9.3) and (3.9.4), and if use is made of the *Stirling approximation to the factorial*,

$$n! \sim \sqrt{2\pi n}\, n^n e^{-n} \qquad (n \to \infty) \qquad (3.9.6)$$

we can deduce that there exists a constant C such that

$$|E_n| \leqq C \frac{[(b-a)/eR]^n}{n^{3/2}(\log n)^2} \qquad (3.9.7)$$

when n is sufficiently large, whether even or odd.

† The notation $f_n \sim g_n$ again is used to indicate that $f_n/g_n \to 1$ as $n \to \infty$.

Thus it follows that *a sufficient condition for convergence of the sequence of successive Newton-Cotes approximations is the requirement that*

$$R > \frac{b - a}{e} \tag{3.9.8}$$

That is, if $\mathscr{R}$ is the region of the complex z plane bounded by line segments of length $b - a$ parallel to the real segment $[a, b]$ at a distance $(b - a)/e$ above and below that segment and by semicircles of radius $(b - a)/e$ centered at $z = a$ and at $z = b$, then the Newton-Cotes sequence of approximations to the specified integral I will converge to I if $f(z)$ is analytic inside $\mathscr{R}$ and on its boundary.

A sharper result replaces $\mathscr{R}$ by the region $\mathscr{R}'$ bounded by a certain simple oval (with a complicated analytical specification) having its longitudinal vertices at the ends of the interval $[a, b]$ and its lateral vertices at distances of about $0.26(b - a)$ from the midpoint of that interval. (See Chap. 4, Prob. 43, and Krylov [1962].)

In the example of Eq. (3.7.1) treated in the last two sections, the only singularity of the complex function $f(z) = 1/(1 + z)$ is at $z = -1$, and hence here $R = b - a = 1$. Thus convergence is ensured. The fact that, in that case, relatively large values of n are needed to supply a specified degree of accuracy is, however, a consequence of the relative nearness of the singularity.

Nearby singularities at *nonreal* points are, of course, just as troublesome as those which occur for *real* values of z. In order to illustrate this fact, we consider the integral

$$\int_{-4}^{4} \frac{dx}{1 + x^2} = 2 \tan^{-1} 4 \doteq 2.6516 \tag{3.9.9}$$

Here, although $f(z) = 1/(1 + z^2)$ is perfectly well behaved when z is real, it possesses singularities (poles) at the imaginary points $z = \pm i$. Since both these points lie inside both $\mathscr{R}$ and $\mathscr{R}'$, convergence is in doubt and an ultimate increase in error magnitude with increasing n may be feared.

Direct calculation indicates that this undesirable situation does indeed exist, and that it evidences itself at a relatively early stage. The results of computations involving $n + 1 = 3, 5, 7, 9$, and 11 ordinates and using Newton-Cotes formulas over the entire range in each case are compared in the following table with those corresponding to the use of the parabolic rule and of the trapezoidal rule:

$n + 1$	NC	P	T
3	5.490	5.490	4.235
5	2.278	2.478	2.918
7	3.329	2.908	2.701
9	1.941	2.573	2.659
11	3.596	2.695	2.6511

It is seen that the best of the Newton-Cotes approximations corresponds to the use of only five ordinates, and that the errors associated with successive formulas of higher order oscillate with increasing amplitude about the true value.

The sequence of approximations afforded by the parabolic rule displays *damped* oscillations but is, of course, convergent. On the other hand, the trapezoidal-rule sequence is converging monotonically toward the true value at a rate which has not yet been exceeded by that of the parabolic-rule sequence, although the incorporation of additional ordinates eventually would reverse the advantage.

It is important to notice that the use of Richardson extrapolation on the P sequence may be undependable here because of the oscillation. Thus, whereas it gives a good prediction with $n_1 = 4$ and $n_2 = 8$ (the subsequence in which $n + 1 = 5, 9, 13, \ldots$ is *monotonic*), the extrapolation based on $n_1 = 6$ and $n_2 = 8$ is worse than either of the approximations upon which it is based.

The preceding example is intended, not generally to discredit the Newton-Cotes formulas which use a fairly large number of ordinates, but to serve as a warning that there exist many *nonpathological* situations in which their use is not appropriate. Such situations generally can be recognized in advance when $f(x)$ is given analytically. For example, if the Taylor-series expansions of $f(x)$ converge for all x [as for e^{-x^2}, $J_0(x)$, and so forth], convergence is ensured (when roundoff errors are suitably controlled).

In illustration, the following table compares successive Newton-Cotes, parabolic-rule, and trapezoidal-rule approximations to the integral

$$\int_0^6 \sin x \, dx \doteq 0.0398297 \qquad (3.9.10)$$

when no use is made of special properties of the integrand:

$n + 1$	NC	P	T
3	0.2850645	0.2850645	0.0042368
5	0.0250938	0.0413420	0.0320657
7	0.0405433	0.0400804	0.0364539
9	0.0398047	0.0399047	0.0379450
11	0.0398300	0.0398597	0.0386276

When only tabular values of $f(x)$ are available and no information can be obtained with respect to its analytical nature, the probability of a favorable behavior of the Newton-Cotes sequence with increasing n is difficult to estimate. In such cases, the use of a composite formula (such as the parabolic, repeated midpoint, or trapezoidal rule) probably is to be preferred. If also the data are empirical and may possess significant inherent errors, it generally is desirable to *smooth* the data before using them for numerical integration or for any other such purpose (see Sec. 7.15).

As previously noted, the presence of negative weighting coefficients in the Newton-Cotes formulas with $n = 8$ and $n \geqq 10$ makes the sum of the *magnitudes* of those coefficients exceed the minimum value $b - a$ and hence increases the possible effects of roundoff errors in those cases. Whereas the magnification factor increases unboundedly as $n \to \infty$, it has reasonably modest values rounding to 1.5 and 3.1, respectively, in the cases of the nine- and eleven-point formulas.

3.10 Weighting Functions. Filon Integration

A formula approximating an integral of the form

$$\int_a^b w(x)f(x)\,dx \tag{3.10.1}$$

by a linear combination of values of $f(x)$, rather than by a combination of values of the complete integrand, can be obtained, for example, by first approximating $f(x)$ by a polynomial and then multiplying that polynomial by $w(x)$ and carrying out the integration. As in Sec. 3.6, different polynomial approximations may be used over consecutive subintervals. Such formulas are desirable when $w(x)$ exhibits an unfavorable behavior in $[a, b]$, but the function $f(x)$ itself can be satisfactorily approximated by a polynomial of moderate degree on that interval or on each of a moderate number of subintervals. Examples are provided by the integrals

$$\int_0^1 x^{1/2}f(x)\,dx \qquad \int_0^1 f(x)\log x\,dx \qquad \int_{-1}^1 \frac{f(x)}{\sqrt{1-x^2}}\,dx$$

where, in each case, $f(x)$ is a well-behaved function.

In particular, such special formulas are desirable for the approximation of the integrals

$$\int_a^b f(x)\cos kx\,dx \qquad \int_a^b f(x)\sin kx\,dx \tag{3.10.2}$$

where k is large, because of the consequent rapid oscillation of the integrands. If, with the notation of Sec. 3.6, the interval $[a, b]$ is divided into n equal parts of length h so that

$$h = \frac{b-a}{n} \tag{3.10.3}$$

where n is *even*, and if over each double subinterval of length $2h$ the function is approximated by the second-degree polynomial agreeing with $f(x)$ at the three relevant division points, a development analogous to that which leads to Simpson's rule produces formulas associated with the name of *Filon.* (The algebraic details of the derivation are uninteresting and are omitted.)

With the previously used abbreviations

$$x_r = x_0 + rh \qquad f(x_r) = f_r \qquad (r = 0, 1, \ldots, n) \tag{3.10.4}$$

where

$$x_0 = a \qquad x_n = b$$

we first define the even and odd cosine sums

$$C_e = \tfrac{1}{2}f_0 \cos kx_0 + f_2 \cos kx_2 + f_4 \cos kx_4 + \cdots + f_{n-2} \cos kx_{n-2} + \tfrac{1}{2}f_n \cos kx_n \tag{3.10.5}$$

and

$$C_o = f_1 \cos kx_1 + f_3 \cos kx_3 + \cdots + f_{n-1} \cos kx_{n-1} \tag{3.10.6}$$

where C_e involves only ordinates with even subscripts and C_o only those with odd subscripts. It may be noted that $2hC_e$ and $2hC_o$ would afford the trapezoidal and repeated midpoint approximations, respectively, to the cosine integral relative to the spacing $2h$.

With the additional abbreviations

$$\theta = kh = \frac{k(b-a)}{n} \tag{3.10.7}$$

and

$$\alpha(\theta) = \frac{\theta^2 + \frac{1}{2}\theta \sin 2\theta - 2 \sin^2 \theta}{\theta^3} \tag{3.10.8}$$

$$\beta(\theta) = 2\,\frac{\theta(1 + \cos^2 \theta) - \sin 2\theta}{\theta^3} \tag{3.10.9}$$

$$\gamma(\theta) = 4\,\frac{\sin \theta - \theta \cos \theta}{\theta^3} \tag{3.10.10}$$

Filon's cosine formula can be written in the form

$$\int_a^b f(x) \cos kx\, dx = h[\alpha(f_n \sin kx_n - f_0 \sin kx_0) + \beta C_e + \gamma C_o] + E \tag{3.10.11}$$

where the error term E accordingly vanishes when $f(x)$ is a polynomial of degree 2 or less.†

† Unless $k = 0$, it does *not* vanish when $f(x)$ is a polynomial of degree 3 despite occasional implications to the contrary in the literature.

In fact, the error associated with a typical subinterval $[x_{2r}, x_{2r+2}]$ is expressible in the form

$$-\frac{h^5}{90}[f^{\text{iv}}(\xi_1)\cos k\xi_2 - 4kf'''(\xi_3)\sin k\xi_4]$$

where all ξ's lie in that subinterval (see Prob. 52). Since the total error E is the superposition of $n/2$ such contributions, it follows that

$$|E| \leqq \frac{b-a}{180} h^4(M_4 + 4kM_3) \qquad (3.10.12)$$

if

$$|f'''(x)| \leqq M_3 \qquad |f^{\text{iv}}(x)| \leqq M_4 \qquad (a \leqq x \leqq b) \qquad (3.10.13)$$

The bound (3.10.12) reduces when $k = 0$ to that deducible from (3.6.5) in accordance with the fact that (3.10.11) must reduce to the parabolic rule when $k = 0$. This reduction can be confirmed directly by use of the expansions

$$\alpha = \tfrac{2}{45}\theta^3 + \cdots \qquad \beta = \tfrac{2}{3} + \tfrac{2}{15}\theta^2 + \cdots \qquad \gamma = \tfrac{4}{3} - \tfrac{2}{15}\theta^2 + \cdots \qquad (3.10.14)$$

when θ is small.

In addition, *Filon's sine formula* can be written in the form

$$\int_a^b f(x)\sin kx\,dx = h[-\alpha(f_n\cos kx_n - f_0\cos kx_0) + \beta S_e + \gamma S_o] + E \qquad (3.10.15)$$

where S_e and S_o are the even and odd sine sums obtained by replacing cos by sin in (3.10.5) and (3.10.6), respectively, and where the error E differs somewhat from the corresponding term in (3.10.11) but again satisfies (3.10.12).

Since the attainable bound (3.10.12) can be written in the form

$$|E| \leqq \frac{b-a}{45} h^3\left(\theta M_3 + \frac{h}{4}M_4\right) \qquad (3.10.16)$$

it follows that $|E|$ may be expected to tend to increase linearly with θ when θ is reasonably large so that *large values of k still tend to require small values of h.* It is usually recommended that $\theta \equiv kh$ not exceed about 1.0 or 1.5.

Various modifications and generalizations are possible. In particular, it is found that when $f(x) = x^3$, the error in the cosine formula associated with the rth subinterval $[x_{2r}, x_{2r+2}]$ is expressible in the form $h^4\,\delta(\theta)\sin kx_{2r+1}$, where

$$\delta(\theta) = 4\frac{(3-\theta^2)\sin\theta - 3\theta\cos\theta}{\theta^4} \qquad (3.10.17)$$

so that $\delta(\theta)$ does not vary with r. It then follows that if one defines the sine correction sum

$$S_c = \frac{h^3}{6}[f'''(x_1)\sin kx_1 + f'''(x_3)\sin kx_3 + \cdots + f'''(x_{n-1})\sin kx_{n-1}] \qquad (3.10.18)$$

then the result of adding the term

$$h\,\delta(\theta)S_c \qquad (3.10.19)$$

to the approximation (3.10.11) and subtracting an equivalent term from its error E will be a formula in which the new error $\tilde{E}$ vanishes when $f(x)$ is any polynomial of degree 3 or less. It is found to satisfy the inequality

$$|\tilde{E}| \leqq \frac{(b-a)h^4}{180}(1+4\theta)M_4 \qquad (3.10.20)$$

with the notation of (3.10.7) and (3.10.13). (See Prob. 54.)

Similarly, with the definition

$$C_c = \frac{h^3}{6}[f'''(x_1)\cos kx_1 + f'''(x_3)\cos kx_3 + \cdots + f'''(x_{n-1})\cos kx_{n-1}] \qquad (3.10.21)$$

the addition of the term

$$h\,\delta(\theta)C_c \qquad (3.10.22)$$

to the approximation (3.10.15) leads to an error which also satisfies (3.10.20).

When θ is small, it is found that

$$\delta(\theta) = \frac{4\theta}{15} - \frac{2\theta^3}{105} + \cdots \qquad (3.10.23)$$

If only the first term is retained in (3.10.23), the correction (3.10.19) is approximated by the sum

$$\frac{2kh^5}{45}[f'''(x_1)\sin kx_1 + f'''(x_3)\sin kx_3 + \cdots + f'''(x_{n-1})\sin kx_{n-1}] \qquad (3.10.24)$$

and the correction (3.10.23) by a similar sum. These two corrections to the Filon formulas appear in the literature (see NBS [1964] and Fröberg [1969]).

3.11 Differentiation Formulas

To conclude this chapter, we list a few formulas which may be used for numerical differentiation of tabulated functions at tabular points when the need for such a calculation cannot be avoided.

By differentiating three- and five-point lagrangian interpolation formulas and evaluating the results at tabular points (see Sec. 3.4), the following symmetrical sets of derivative formulas may be obtained, with a convenient renumbering of the ordinates.

Three-point formulas:

$$f'_{-1} = \frac{1}{2h}(-3f_{-1} + 4f_0 - f_1) + \frac{h^2}{3} f'''(\xi) \qquad (3.11.1)$$

$$f'_0 = \frac{1}{2h}(-f_{-1} + f_1) - \frac{h^2}{6} f'''(\xi) \qquad (3.11.2)$$

$$f'_1 = \frac{1}{2h}(f_{-1} - 4f_0 + 3f_1) + \frac{h^2}{3} f'''(\xi) \qquad (3.11.3)$$

Five-point formulas:

$$f'_{-2} = \frac{1}{12h}(-25f_{-2} + 48f_{-1} - 36f_0 + 16f_1 - 3f_2) + \frac{h^4}{5} f^{\mathrm{v}}(\xi) \qquad (3.11.4)$$

$$f'_{-1} = \frac{1}{12h}(-3f_{-2} - 10f_{-1} + 18f_0 - 6f_1 + f_2) - \frac{h^4}{20} f^{\mathrm{v}}(\xi) \qquad (3.11.5)$$

$$f'_0 = \frac{1}{12h}(f_{-2} - 8f_{-1} + 8f_1 - f_2) + \frac{h^4}{30} f^{\mathrm{v}}(\xi) \qquad (3.11.6)$$

$$f'_1 = \frac{1}{12h}(-f_{-2} + 6f_{-1} - 18f_0 + 10f_1 + 3f_2) - \frac{h^4}{20} f^{\mathrm{v}}(\xi) \qquad (3.11.7)$$

$$f'_2 = \frac{1}{12h}(3f_{-2} - 16f_{-1} + 36f_0 - 48f_1 + 25f_2) + \frac{h^4}{5} f^{\mathrm{v}}(\xi) \qquad (3.11.8)$$

In each set of formulas, each ξ lies between the extreme values of the abscissas involved in that formula. It should be noticed that the coefficient in the truncation error is least when the derivative is calculated at the *central* point, and that the ordinate at that point is then not involved in the calculation.

The sum of the magnitudes of the coefficients (and hence the significance of the effects of *roundoff errors*) is also least for the central point, but it increases rapidly with distance from the central point for a given set of formulas as well as with increasing order of successive sets. For reference purposes, the seven- and nine-point formulas for the derivative at the central tabular point are listed as follows:

Seven-point formula:

$$f_0' = \frac{1}{60h}(-f_{-3} + 9f_{-2} - 45f_{-1} + 45f_1 - 9f_2 + f_3) - \frac{h^6}{140} f^{\text{vii}}(\xi) \qquad (3.11.9)$$

Nine-point formula:

$$f_0' = \frac{1}{840h}(3f_{-4} - 32f_{-3} + 168f_{-2} - 672f_{-1} + 672f_1 - 168f_2 + 32f_3 - 3f_4) + \frac{h^8}{630} f^{\text{ix}}(\xi) \qquad (3.11.10)$$

An inspection of the numerical differentiation formulas reveals the existence of a new problem in error control. For example, consider (3.11.2) and suppose that it is known that

$$|f'''(x)| \leqq M_3$$

on the interval $[x_0 - h, x_0 + h]$. Then if all given data were exact, the maximum possible error in the calculation of $f'(x_0)$ would be

$$|E_3|_{\max} = \frac{M_3 h^2}{6}$$

On the other hand, suppose that each of the ordinates involved could be in error by $\pm\varepsilon$. Then the magnitude of the corresponding error in the calculation of $f'(x_0)$ could be as large as

$$|R_3|_{\max} = \frac{\varepsilon}{h}$$

Whereas a reduction of the truncation error E_3 would generally require a decrease in h, a small value of h would lead to a large possible *roundoff* error R_3 and, conversely, a reduction in $|R_3|_{\max}$ would generally correspond to an increase in $|E_3|_{\max}$.

A reasonable procedure consists of determining the interval h such that the predictable upper bounds on the two errors are about equal if this is feasible. The optimum value of h and the corresponding maximum total error T_3 are then found to be

$$h_{3,\text{opt}} \approx 1.8\varepsilon^{1/3} M_3^{-1/3} \qquad |T_3|_{\max} \approx 1.1\varepsilon^{2/3} M_3^{1/3}$$

Corresponding results relevant to (3.11.6), (3.11.9), and (3.11.10) can be obtained as follows:

$$h_{5,\text{opt}} \approx 2.1\varepsilon^{1/5} M_5^{-1/5} \qquad |T_5|_{\max} \approx 1.4\varepsilon^{4/5} M_5^{1/5}$$

$$h_{7,\text{opt}} \approx 2.2\varepsilon^{1/7} M_7^{-1/7} \qquad |T_7|_{\max} \approx 1.7\varepsilon^{6/7} M_7^{1/7}$$

$$h_{9,\text{opt}} \approx 2.2\varepsilon^{1/9} M_9^{-1/9} \qquad |T_9|_{\max} \approx 1.9\varepsilon^{8/9} M_9^{1/9}$$

In illustration, suppose that empirical values were to be obtained for a function which is truly of the form $f(x) = \sin \omega x$, and that one of these formulas were to be used to approximate $f'(0) = \omega$. In this case, the relevant quantities $M_k^{1/k}$ are each equal to ω. Thus, if, say, the maximum observational error ε is 0.01, the optimum spacings for the three-, five-, seven-, and nine-point formulas are found to be about $0.39/\omega$, $0.84/\omega$, $1.14/\omega$, and $1.32/\omega$, respectively, and the corresponding maximum total errors in the calculation of $f'(0) = \omega$ are found to be about 0.051ω, 0.035ω, 0.033ω, and 0.032ω, respectively. The increase of h_{opt} with increasing n, and the fact that an increase in n affords only slight improvement in guaranteed accuracy, are both worthy of note.

The results of this example are typical of most practical situations in which the function $f(x)$ is representable by a Taylor series which converges for all values of x. When the series representations have finite radii of convergence, the quantities $M_k^{1/k}$ tend to increase with increasing k, and the incorporation of additional ordinates may lead to a *decrease* in guaranteed accuracy, at an early stage, when the inaccuracies in the given data are appreciable. (See Probs. 57 and 58.)

In practice, unless $f(x)$ is given analytically, the truncation error relevant to any lagrangian formula can be estimated only roughly by making two or more independent calculations, based on different sets of ordinates, or by determining sample values of the divided difference of order equal to the number of ordinates used. It is apparent that recourse to the latter alternative would tend to nullify the computational advantages which are inherent to the lagrangian methods. However, when *equally spaced* abscissas are used, divided differences of a given order can be calculated conveniently by use of simple formulas (see Probs. 7 and 8 of Chap. 2), without resort to the formation of a divided-difference table or to the calculation of intermediate differences of lower order. Equation (2.3.2) is available for the same purpose in the general case.

It is useful to notice that since

$$\frac{f(x_0 + h) - f(x_0 - h)}{2h} = f'(x_0) + \frac{h^2}{3!} f'''(x_0) + \frac{h^4}{5!} f^{\text{v}}(x_0) + \cdots$$

when $f(x)$ is regular at x_0, it follows that for any such function the formula (3.11.2) can be written in the form

$$f_0' = \frac{f_1 - f_{-1}}{2h} - C_1 h^2 - C_2 h^4 - \cdots \tag{3.11.11}$$

where $C_k = f^{(2k+1)}(x_0)/(2k + 1)!$. Thus the use of iterated Richardson extrapolation (Sec. 3.8) then is appropriate (*if sufficiently many significant figures can be retained to control the effects of roundoff errors*) and recursive calculation can be formulated in analogy to Romberg integration.

Specifically, if $D_0^{(0)}$ denotes the approximation to $f_0' \equiv f'(x_0)$ afforded by (3.11.2) (with the error term omitted) for a certain choice of h, and if $D_k^{(0)}$ denotes the result obtained when h has been halved k times, there follows

$$D_k^{(0)} = \frac{f(x_0 + h_k) - f(x_0 - h_k)}{2h_k} \tag{3.11.12}$$

where

$$h_k = \frac{h}{2^k} \qquad (k = 0, 1, 2, \ldots) \tag{3.11.13}$$

If then we define

$$D_k^{(m)} = \frac{4^m D_{k+1}^{(m-1)} - D_k^{(m-1)}}{4^m - 1} \qquad (m = 1, 2, \ldots) \tag{3.11.14}$$

in complete analogy to the definition (3.8.5), we can construct the array

$$\begin{array}{lll} D_0^{(0)} & & \\ D_1^{(0)} & D_0^{(1)} & \\ D_2^{(0)} & D_1^{(1)} & D_0^{(2)} \\ \cdots & \cdots & \cdots \end{array}$$

in which $D_k^{(m)}$ affords an approximation to f_0' with a truncation error of order h_k^{2m+2}. In particular, it can be verified that $D_k^{(1)}$ is identical with the result of applying (3.11.6) with h replaced by h_k.

In illustration, when $f(x) = -1/(x + 1)$ and $x_0 = 0$, so that $f_0' = 1$, an array of approximations corresponding to the choice $h = 0.5$ (for the initial spacing) is obtained as follows when each entry is rounded to eight decimal places as it is calculated:

$D_k^{(0)}$	$D_k^{(1)}$	$D_k^{(2)}$	$D_k^{(3)}$	$D_k^{(4)}$
1.33333 333				
1.06666 667	0.97777 778			
1.01587 302	0.99894 180	1.00035 273		
1.00392 157	0.99993 775	1.00000 415	0.99999 862	
1.00097 752	0.99999 617	1.00000 006	1.00000 000	1.00000 001

The effect of roundoff is visible in the last entry.

3.12 Supplementary References

Extensive tables of lagrangian interpolation coefficients relevant to equally spaced abscissas are provided in NBS [1944], Pearson [1920*a*], and in other sources listed in the comprehensive index of tables by Fletcher, Miller, Rosenhead, and Comrie [1962]. Coefficients for lagrangian polynomial differentiation are given by Salzer [1948]. Tables of coefficients for lagrangian exponential interpolation are given by Luke [1953], and tables for trigonometric interpolation (see Prob. 7) are given by Salzer [1949].

Tables for inverse lagrangian polynomial interpolation are given by Salzer [1944*a*, 1945]. Salzer [1951] also provides formulas of lagrangian type for determining the argument for which the derivative of a function takes on a prescribed value.

Comprehensive treatments of numerical integration (quadrature) are listed in Sec. 8.17.

Richardson extrapolation and Romberg integration are considered in Henrici [1964] and Ralston [1965]. Convergence of a Romberg sequence is dealt with by Bauer, Rutishauser, and Stiefel [1963], and various modifications of the Romberg process are summarized in Davis and Rabinowitz [1967].

The last reference also includes a consideration of additional iterative schemes and of other "general-purpose" algorithms which are intended to permit computer evaluation of an integral within a prescribed tolerance without the need for any analysis on the part of the user. References are provided, together with warnings of dangers inherent in blind reliance on such *automatic integrators*.

A very complete listing of formulas for numerical integration, with various weighting functions, in terms of linear combinations of ordinates at equally spaced points, is given by Miller [1960*a*].

Tables of the coefficients in the formulas of Filon [1928] may be found in the NBS [1964] handbook and in other sources listed in the index by Fletcher et al. [1962]. Generalizations of Filon's method include one obtained by Flinn [1960] by approximating $f(x)$ on each $2h$ interval by a polynomial of degree 5 such that there is agreement with both $f(x)$ and $f'(x)$ at the division points. This and other related formulas appear in Krylov and Kruglikova [1969], together with treatments of the case $b = \infty$ and with associated numerical tables.

Numerical integration over regions in two or more dimensions (see Probs. 40 and 41) is treated extensively in Stroud [1971]. See also Irwin [1923], Radon [1948], Willers [1950], Steffensen [1950], Hammer [1959], Miller [1960*b*], Stroud [1967], Davis and Rabinowitz [1967], and Haber [1970].

The fact that the usefulness of the simple midpoint formula for one-dimensional integration was long overlooked in the literature and also that it generalizes naturally to a *centroid method*, which is useful for integration in higher dimensions, is discussed by Good and Gaskins [1971]. See also Hammer [1958].

PROBLEMS

Section 3.2

1 By noticing that the zeroth lagrangian coefficient function of degree n takes on the value unity when $x = x_0$ and the value zero when $x = x_1, \ldots, x_n$, and by considering the associated divided-difference table (or otherwise), show that

$$l_0(x) = 1 + \frac{x - x_0}{x_0 - x_1} + \frac{(x - x_0)(x - x_1)}{(x_0 - x_1)(x_0 - x_2)} + \cdots + \frac{(x - x_0)\cdots(x - x_{n-1})}{(x_0 - x_1)\cdots(x_0 - x_n)}$$

and that similar expansions can be written down by symmetry for the other coefficient functions.

2 Derive the lagrangian interpolation formula directly from the newtonian divided-difference formula.

3 If $y(x)$ is the polynomial of degree n which agrees with $f(x)$ at the distinct points $x = x_0, x_1, \ldots, x_n$, and if $\pi(x) \equiv (x - x_0)(x - x_1)\ldots(x - x_n)$, obtain the lagrangian form of $y(x)$ by determining the coefficients in the partial fraction expansion of the ratio

$$\frac{y(x)}{\pi(x)} = \sum_{k=0}^{n} \frac{a_k}{x - x_k}$$

(Multiply both members by $x - x_r$ and let $x \to x_r$.)

4 Show that

$$\begin{vmatrix} 1 & a_1 & a_1^2 \\ 1 & a_2 & a_2^2 \\ 1 & a_3 & a_3^2 \end{vmatrix} = (a_2 - a_1)(a_3 - a_1)(a_3 - a_2)$$

and use this fact to express the result of expanding the left-hand member of (3.2.3) with respect to the elements of the first column, and equating the result to zero, in lagrangian form when $n = 2$.

5 Generalize the result of Prob. 4 to show that

$$\begin{vmatrix} 1 & a_1 & a_1^2 & \cdots & a_1^{n-1} \\ 1 & a_2 & a_2^2 & \cdots & a_2^{n-1} \\ 1 & a_3 & a_3^2 & \cdots & a_3^{n-1} \\ \cdots & \cdots & \cdots & \cdots & \cdots \\ 1 & a_n & a_n^2 & \cdots & a_n^{n-1} \end{vmatrix}$$

$$= (a_2 - a_1)(a_3 - a_1)(a_3 - a_2)(a_4 - a_1)(a_4 - a_2)(a_4 - a_3)\cdots(a_n - a_{n-1})$$

and to derive the lagrangian form of the interpolation polynomial from (3.2.3) in the general case. (The determinant involved here is often called *Vandermonde's determinant*.)

6 By considering the limit of the three-point lagrangian interpolation formula relative to x_0, $x_0 + \varepsilon$, and x_1, as $\varepsilon \to 0$, obtain the formula

$$f(x) = \frac{(x_1 - x)(x + x_1 - 2x_0)}{(x_1 - x_0)^2} f(x_0) + \frac{(x - x_0)(x_1 - x)}{x_1 - x_0} f'(x_0) + \frac{(x - x_0)^2}{(x_1 - x_0)^2} f(x_1) + E(x)$$

where

$$E(x) = \tfrac{1}{6}(x - x_0)^2(x - x_1)f'''(\xi)$$

7 Write down a determinantal equation analogous to (3.2.3) but corresponding to the requirement that

$$y(x) = A_0 + A_1 \cos x + A_2 \sin x$$

agree with $f(x)$ when $x = x_0, x_1$, and x_2. Then establish the identity

$$\begin{vmatrix} 1 & \cos a_1 & \sin a_1 \\ 1 & \cos a_2 & \sin a_2 \\ 1 & \cos a_3 & \sin a_3 \end{vmatrix} = 4 \sin \tfrac{1}{2}(a_2 - a_1) \sin \tfrac{1}{2}(a_3 - a_1) \sin \tfrac{1}{2}(a_3 - a_2)$$

and use this result to express $y(x)$ in the following form (due to Gauss):

$$y(x) = \frac{\sin \frac{1}{2}(x - x_1) \sin \frac{1}{2}(x - x_2)}{\sin \frac{1}{2}(x_0 - x_1) \sin \frac{1}{2}(x_0 - x_2)} f(x_0) + \frac{\sin \frac{1}{2}(x - x_0) \sin \frac{1}{2}(x - x_2)}{\sin \frac{1}{2}(x_1 - x_0) \sin \frac{1}{2}(x_1 - x_2)} f(x_1) + \frac{\sin \frac{1}{2}(x - x_0) \sin \frac{1}{2}(x - x_1)}{\sin \frac{1}{2}(x_2 - x_0) \sin \frac{1}{2}(x_2 - x_1)} (x_2)$$

Show also that the formula resulting from deleting the $\frac{1}{2}$'s in the arguments of the sines (and due to Hermite) defines the approximation

$$y = A_0 + A_1 \cos 2x + A_2 \sin 2x$$

which agrees with $f(x)$ at the same three points. Also predict the form of the generalization of the Gauss formula to the case when an approximation of the form

$$y = A_0 + A_1 \cos x + A_2 \sin x + \cdots + A_{2n-1} \cos nx + A_{2n} \sin nx$$

is to agree with $f(x)$ at the $2n + 1$ points $x_0, x_1, \ldots, x_{2n}$, and verify the correctness of this conjecture.

Section 3.3

8 Prove that (3.3.20) is valid when x is outside the span R of the values $x_0, \ldots, x_n$, by writing $F(x) \equiv f(x) - y(x) - K\pi(x)$, showing that the function

$$F^{(r)}(x) = f^{(r)}(x) - y^{(r)}(x) - K\pi^{(r)}(x)$$

vanishes at least $n - r + 1$ times inside R, showing that all zeros of $\pi^{(r)}(x)$ lie inside R and hence that K can be so chosen that $F^{(r)}(x)$ also vanishes when $x = \bar{x}$ if $\bar{x}$ is not inside R, so that

$$E^{(r)}(\bar{x}) = K\pi^{(r)}(\bar{x})$$

and proving that, with this K, there follows $0 = f^{(n+1)}(\bar{\eta}) - (n + 1)!K$ for some $\bar{\eta}$ between the smallest and the largest of $x_0, \ldots, x_n$, and $\bar{x}$.

9 Obtain the lagrangian two-point first-derivative formula in the form

$$f'(x) = \frac{1}{h}[f(x_0 + h) - f(x_0)] + E'(x)$$

where

$$E'(x) = \frac{d}{dx}\{(x - x_0)(x - x_1)f[x_0, x_1, x]\}$$

with $h = x_1 - x_0$, and hence also [by (3.3.17)]

$$E'(x) = \left(x - \frac{x_0 + x_1}{2}\right)f''(\xi_1) + (x - x_0)(x - x_1)\frac{f'''(\xi_2)}{6}$$

where ξ_1 and ξ_2 are in the interval spanned by x_0, x_1, and x. Deduce the special cases

$$f'\left(\frac{x_0 + x_1}{2}\right) = \frac{f(x_1) - f(x_0)}{h} + \frac{h^2}{24}f'''(\xi)$$

$$f'(x_0) = \frac{f(x_1) - f(x_0)}{h} - \frac{h}{2}f''(\xi)$$

$$f'(x_1) = \frac{f(x_1) - f(x_0)}{h} + \frac{h}{2}f''(\xi)$$

where each ξ is in (x_0, x_1).

10 By writing the error term in Prob. 9 in the form

$$E'(x) = \frac{d}{dx}[(x - x_1)\{f[x_1, x] - f[x_0, x_1]\}]$$

show that

$$E'(x) = (x - x_1)f[x_1, x, x] + (x - x_0)f[x_0, x_1, x]$$
$$= (x - x_1)\frac{f''(\xi_1)}{2} + (x - x_0)\frac{f''(\xi_2)}{2}$$

and hence that

$$|E'(x)| \leqq \frac{M_2}{2}(|x_1 - x| + |x - x_0|)$$

if $|f''(x)| \leqq M_2$ in the interval spanned by x_0, x_1, and x. In particular, in the case when $x_0 \leqq x \leqq x_1$, deduce that

$$|E'(x)| \leqq \frac{M_2 h}{2}$$

whereas the result of Prob. 9 [or of (3.3.17)] gives

$$|E'(x)| \leqq \frac{M_2 h}{2} + \frac{M_3 h^2}{24}$$

if also $|f'''(x)| \leqq M_3$ when $x_0 \leqq x \leqq x_1$, in that case. [Similar manipulations are possible in less simple cases for the purpose of obtaining derivative-formula error bounds, which involve only bounds on lower-order derivatives of $f(x)$, in contrast with (3.3.15). In the present case the special midpoint error bound in Prob. 9 is sacrificed, but a better "global" bound on $[x_0, x_1]$ is in fact obtained; in most cases, the added simplicity is accompanied by increased conservatism.]

11 Obtain the lagrangian three-point first-derivative formula, in the case when the abscissas are equally spaced, at spacing h, and the origin is taken at the central point, in the form

$$f'(x) = \frac{2x - h}{2h^2} f(-h) - \frac{2x}{h^2} f(0) + \frac{2x + h}{2h^2} f(h) + E'(x)$$

with

$$E'(x) = \tfrac{1}{6}(3x^2 - h^2) f'''(\xi_1) + \tfrac{1}{24} x(x^2 - h^2) f^{\text{iv}}(\xi_2)$$

where ξ_1 and ξ_2 are in $(-h, h)$ if x is in that interval. Show also that, unless hf^{iv} is large in magnitude relative to f''', the absolute error is least, on the average, at distances of about $0.6h$ from the central point.

12 By integrating the lagrangian three-point formula, when the abscissas are at equal spacing h, with the origin taken at the central point, obtain the formula

$$\int_{-h}^{x} f(t)\,dt = \frac{5h^3 - 3hx^2 + 2x^3}{12h^2} f(-h) + \frac{2h^3 + 3h^2 x - x^3}{3h^2} f(0) - \frac{h^3 - 3hx^2 - 2x^3}{12h^2} f(h) + T(x)$$

and show that the truncation error is expressible in the form

$$T(x) = \int_{-h}^{x} t(t^2 - h^2) f[-h, 0, h, t]\,dt$$

13 If the upper limit of the integration in Prob. 12 does not lie outside the interval $[-h, 0]$, show that

$$T(x) = \tfrac{1}{24}(x^2 - h^2)^2 f'''(\xi)$$

where $-h < \xi < h$. In particular, deduce the formula

$$\int_{x_0}^{x_0+h} f(x)\,dx = \frac{h}{12}[5f(x_0) + 8f(x_0 + h) - f(x_0 + 2h)] + \frac{h^4}{24} f'''(\xi)$$

where $x_0 < \xi < x_0 + 2h$, after a change in notation.

14 By integrating the expression for $T(x)$ in Prob. 12 by parts, and noticing that $x(x^2 - h^2) = \frac{1}{4}[(x^2 - h^2)^2]'$, show that

$$T(x) = \tfrac{1}{4}(x^2 - h^2)^2 f[-h, 0, h, x] - \tfrac{1}{4}\int_{-h}^{x} (t^2 - h^2)^2 f[-h, 0, h, t, t]\, dt$$

and deduce that

$$T(x) = \tfrac{1}{24}(x^2 - h^2)^2 f'''(\xi_1) - \tfrac{1}{1440}(3x^5 - 10h^2x^3 + 15h^4x + 8h^5) f^{\text{iv}}(\xi_2)$$

where ξ_1 and ξ_2 lie between the smallest and largest of $-h$, h, and x. In particular, deduce the formula of *Simpson's rule* (see Sec. 3.5):

$$\int_{x_0}^{x_0+2h} f(x)\, dx = \frac{h}{3}[f(x_0) + 4f(x_0 + h) + f(x_0 + 2h)] - \frac{h^5}{90} f^{\text{iv}}(\xi)$$

where $x_0 < \xi < x_0 + 2h$.

Section 3.4

15 Determine the lagrangian coefficient functions, in explicit polynomial form, relative to the ordinates of $f(x)$ at the four points $x = -2, -1, 1$, and 2. Use the results to obtain approximate expressions for $f(0)$, $f'(0)$, and $\int_{-2}^{2} f(x)\, dx$ in terms of those ordinates.

16 Use the results of Prob. 15 to determine the equation of the third-degree polynomial passing through the points $(-2, -5)$, $(-1, -1)$, $(1, 1)$, and $(2, 11)$.

17 Use the Lagrange interpolation formula to calculate approximate values of $f(x)$ when $x = 1.1300, 1.1500, 1.1700$, and 1.1900 from the following rounded data:

x	1.1275	1.1503	1.1735	1.1972
$f(x)$	0.11971	0.13957	0.15931	0.17902

18 Use the results of Prob. 17 and the coefficients of Table 3.1 to determine approximate values of $f(x)$ for $x = 1.1600(0.0010)1.1700$.

19 Under the assumption that the data in Prob. 17 correspond to the function $f(x) = \sin(\log x)$, obtain bounds on the truncation errors associated with the values calculated in Probs. 17 and 18.

20 Obtain bounds on the roundoff errors associated with the values calculated in Probs. 17 and 18.

21 Use the table of five-point lagrangian coefficients given in Sec. 4.12, to interpolate in that table itself for the coefficients relative to $s = 0.38, 0.05$, and 1.93, rounding the results to six places. If no roundoffs were effected, what errors would be present in the calculated coefficients?

22 Show that, if $h^3|f'''(x)|$ does not exceed 16 units in the last place to be retained in a three-point Lagrange interpolation based on equally spaced abscissas with spacing h, then the truncation error cannot exceed one unit in that place.

23 Show that, if $h^5|f^{\text{v}}(x)|$ does not exceed 32 units in the last place to be retained in a five-point Lagrange interpolation based on equally spaced abscissas with spacing h, then the truncation error cannot exceed 1 unit in that place, and also that $h^5|f^{\text{v}}(x)|$ may be as large as 84 units if the interpolation is effected only between the second and fourth of the five successive abscissas.

Section 3.5

24 Prove directly, from Eq. (3.5.6), that a Newton-Cotes formula of closed type, employing $n + 1$ ordinates, is exact when applied to any polynomial of degree $n + 1$ when $n + 1$ is *odd.* (Notice that $F[0, 1, \ldots, n, s]$ is then constant, write $s = t + (n/2)$, and show that the resultant integrand is an odd function of t.)

25 Show that the factor $s - (n/2)$ can be replaced by $s - c$ in (3.5.9*b*), where c is any constant (see Prob. 24).

26 Derive the formulas resulting from neglect of the error terms in (3.5.19) and (3.5.20).

27 Show that the truncation error associated with a Newton-Cotes formula of closed type employing $n + 1$ ordinates can be expressed in the form

$$E = h^{n+2} \int_0^n s(s-1)\cdots(s-n)f[x_0, \ldots, x_n, x_0 + hs]\, ds$$

whereas that associated with a formula of open type employing $n - 1$ ordinates is given by

$$E = h^n \int_0^n (s-1)\cdots(s-n+1)f[x_1, \ldots, x_{n-1}, x_0 + hs]\, ds$$

28 If $f_{(2k+1)/2}$ denotes the value of $f(x)$ at an abscissa midway between x_k and $x_{k+1} \equiv x_k + h$, derive the formulas

$$\int_{x_0}^{x_2} f(x)\, dx = h(f_{1/2} + f_{3/2}) + E_1$$

$$\int_{x_0}^{x_3} f(x)\, dx = \frac{3h}{8}(3f_{1/2} + 2f_{3/2} + 3f_{5/2}) + E_2$$

[These formulas, together with the midpoint formula (3.5.23), are the first members of a set due to *Maclaurin.* It can be shown that $E_1 = h^3f''(\xi)/12$ $(x_0 < \xi < x_2)$ and that $E_2 = 21h^5f^{\text{iv}}(\xi)/640$ $(x_0 < \xi < x_3)$.]

Section 3.6

29 A *Riemann sum* associated with an integral $\int_a^b f(x)\, dx$ is an approximation of the form

$$S_n = \sum_{k=0}^{n} f(t_k)(s_{k+1} - s_k)$$

where

$$a = s_0 \leqq t_0 \leqq s_1 \leqq t_1 \leqq s_2 \leqq \cdots \leqq s_n \leqq t_n \leqq s_{n+1} = b$$

Any sequence of such sums in which the subdivision of $[a, b]$ is refined in such a way that $\max (s_{k+1} - s_k) \to 0$ tends to the (Riemann) integral I if it exists.

(*a*) Show that the approximations afforded by the *repeated midpoint rule*, the *trapezoidal rule*, and the *parabolic rule* are Riemann sums. (Display the values of $s_1, s_2, \ldots, s_n$ in each case.)

(*b*) The relation

$$\int_a^b f(x)\,dx \approx \frac{h}{4}(5f_0 + f_1 + f_2 + 10f_3 + f_4 + f_5 + 10f_6 + \cdots + 10f_{n-3} + f_{n-2} + f_{n-1} + 5f_n)$$

with the notation of Sec. 3.6 is an equality when $f(x)$ is any linear function. Prove that the approximation is *not* a Riemann sum.

30 *Convergence of composite rules* Suppose that $[a, b]$ is divided into r equal parts by $a = X_0 < X_1 < \cdots < X_{r-1} < X_r = b$, and let $(b - a)/r = H$. If an m-point formula which yields exact results when integrating a *constant* is used to approximate the integral of $f(x)$ over each subinterval $[X_i, X_{i+1}]$, prove that the sum converges to the integral over $[a, b]$ as the spacing $H \to 0$. (If the result of applying the m-point formula to $[X_0, X_1]$ is of the form

$$\int_{X_0}^{X_1} f(x)\,dx \approx H \sum_{k=0}^{m-1} w_k f(X_0 + c_k) \qquad (0 \leqq c_k \leqq H)$$

show that the total approximation is given by

$$\int_a^b f(x)\,dx \approx \sum_{k=0}^{m-1} w_k \left(H \sum_{i=0}^{r-1} f(X_i + c_k)\right)$$

and that the inner sum is a Riemann sum for $f(x)$ over $[a, b]$. Then let $r \to \infty$ and complete the proof. See also Davis and Rabinowitz [1967], Sec. 2.4.

31 Show that the composite rule corresponding to the repeated use of Newton's *three-eighths* rule (3.5.12) is of the form

$$\int_{x_0}^{x_n} f(x)\,dx = \frac{3h}{8}(f_0 + 3f_1 + 3f_2 + 2f_3 + \cdots + 2f_{n-3} + 3f_{n-2} + 3f_{n-1} + f_n) - \frac{nh^5}{80} f^{\text{iv}}(\xi)$$

where n is to be an integral multiple of 3. Also, by considering the case when n is a multiple of 6, so that both this rule and the parabolic rule can be used with the same spacing h, account for the fact that the parabolic rule is nearly always preferred.

Section 3.7

32 Given the following rounded values of the function

$$f(x) = \sqrt{\frac{2}{\pi}}\, e^{-x^2/2}$$

calculate approximate values of the integral

$$P(1) = \sqrt{\frac{2}{\pi}} \int_0^1 e^{-t^2/2}\, dt \doteq 0.6826895$$

by use of the trapezoidal rule with $h = 1, \frac{1}{2}, \frac{1}{4}$, and $\frac{1}{8}$, and compare the results with the rounded true values:

x	$f(x)$	x	$f(x)$
0.000	0.7978846	0.625	0.6563219
0.125	0.7916754	0.750	0.6022749
0.250	0.7733362	0.875	0.5441100
0.375	0.7437102	1.000	0.4839414
0.500	0.7041307		

33 Repeat the calculations of Prob. 32 using instead the repeated midpoint rule with $h = 1, \frac{1}{2}$, and $\frac{1}{4}$.

34 Repeat the calculations of Prob. 32 using instead the parabolic rule with $h = \frac{1}{2}, \frac{1}{4}$, and $\frac{1}{8}$.

35 Repeat the calculations of Prob. 32 using instead the Newton-Cotes three-, five-, and nine-point formulas of closed type.

36 Calculate approximate values of the integral

$$\int_0^1 e^{\cos 2\pi x}\, dx = I_0(1) \doteq 1.2660659$$

by use of the trapezoidal rule with $h = 1, \frac{1}{2}, \frac{1}{4}$, and $\frac{1}{8}$, retaining seven decimal places, and compare the results with the rounded true value.

37 Repeat the calculations of Prob. 36 using instead the repeated midpoint rule with $h = 1, \frac{1}{2}$, and $\frac{1}{4}$.

38 Repeat the calculations of Prob. 36 using instead the parabolic rule with $h = \frac{1}{2}, \frac{1}{4}$, and $\frac{1}{8}$.

39 Repeat the calculations of Prob. 36 using instead the Newton-Cotes three-, five-, and nine-point formulas of closed type.

40 By a double application of Simpson's rule, derive the formula

$$\int_{x_0}^{x_2} \int_{y_0}^{y_2} f(x, y)\, dx\, dy = \frac{hk}{9}\,[(f_{0,0} + f_{0,2} + f_{2,0} + f_{2,2})$$
$$+ 4(f_{0,1} + f_{1,0} + f_{1,2} + f_{2,1}) + 16 f_{1,1}] + E$$

where $x_r \equiv x_0 + rh$, $y_s \equiv y_0 + sk$, and $f_{r,s} \equiv f(x_r, y_s)$, and show that

$$E = -\frac{hk}{45}\left[h^4 \frac{\partial^4 f(\xi_1, \eta_1)}{\partial x^4} + k^4 \frac{\partial^4 f(\xi_2, \eta_2)}{\partial y^4}\right]$$

where ξ_1, ξ_2 lie in (x_0, x_2) and η_1, η_2 in (y_0, y_2). [More elaborate formulas for two-way integration over a rectangle ("cubature formulas") are obtainable by double application of other one-dimensional integration formulas.]

41 By applying the formula of Prob. 40 to subrectangles and adding the results, derive the two-dimensional generalization of the parabolic rule in the form

$$\int_{x_0}^{x_m}\int_{y_0}^{y_n} f(x, y)\,dx\,dy = \frac{hk}{9}\,[(f_{0,0} + 4f_{1,0} + 2f_{2,0} + \cdots + f_{m,0})$$
$$+ 4(f_{0,1} + 4f_{1,1} + 2f_{2,1} + \cdots + f_{m,1})$$
$$+ 2(f_{0,2} + 4f_{1,2} + 2f_{2,2} + \cdots + f_{m,2}) + \cdots$$
$$+ (f_{0,n} + 4f_{1,n} + 2f_{2,n} + \cdots + f_{m,n})] + E$$

where

$$E = -\frac{hk}{90}\left[mh^4 \frac{\partial^4 f(\bar{\xi}_1, \bar{\eta}_1)}{\partial x^4} + nk^4 \frac{\partial^4 f(\bar{\xi}_2, \bar{\eta}_2)}{\partial y^4}\right]$$

for some points $(\bar{\xi}_1, \bar{\eta}_1)$ and $(\bar{\xi}_2, \bar{\eta}_2)$ inside the rectangle, when m and n are even integers.

42 Obtain an approximate evaluation of the integral

$$\int_0^1 \frac{\cos x}{\sqrt{x}}\,dx$$

(*a*) By writing it in the form

$$\int_0^1 \frac{dx}{\sqrt{x}} - \int_0^1 \frac{1 - \cos x}{\sqrt{x}}\,dx$$

evaluating the first integral analytically, and applying the parabolic rule with $h = \frac{1}{8}$ to the second one.

(*b*) By making the change of variables $x = t^2$ in the original form and applying the parabolic rule with $h = \frac{1}{8}$ directly to the result.

Also compare the approximations with a more accurate value obtained by expanding the integrand of one of the forms in a power series and integrating term by term.

Section 3.8

43 Prove that in Romberg integration $T_k^{(1)}$ is the result of applying the parabolic rule 2^k times and $T_k^{(2)}$ is the result of applying the Newton-Cotes closed five-point formula 2^k times.

44 Apply Romberg integration to the data of Prob. 32.

45 Apply Romberg integration to Prob. 36 using a total of nine ordinates.

Section 3.9

46 Show that

$$\int_0^{2m+1} s(s-1)\cdots(s-2m-1)\,ds = -\frac{2}{2m+3}\int_0^1 s(s-1)\cdots(s-2m-2)\,ds$$

when m is a nonnegative integer. (Express the left-hand integral as a sum of integrals between successive integers, translate all lower limits to zero, and show that the $2m+1$ terms in the resultant integrand can be telescoped into the sum of two terms. Then replace s by $1-s$ in the integrand of one of those terms.)

47 Show that the numerical factors in (3.9.1) and (3.9.2) can be expressed in the form

$$I_{2m} = \int_0^{2m} \frac{s(s-1)\cdots(s-2m)(s-2m-1)}{(2m+2)!}\,ds$$

when $n = 2m$, and in the form

$$I_{2m+1} = \int_0^{2m+1} \frac{s(s-1)\cdots(s-2m)(s-2m-1)}{(2m+2)!}\,ds$$

when $n = 2m+1$, and show also that

$$I_{2m+1} - I_{2m} = \int_0^1 \frac{s(s-1)\cdots(s-2m-1)}{(2m+2)!}\,ds$$

48 With the abbreviation

$$\alpha_k = \int_0^1 \frac{s(s-1)\cdots(s-k+1)}{k!}\,ds$$

show that the results of Probs. 46 and 47 lead to the relations

$$I_{2m} = -2\alpha_{2m+3} - \alpha_{2m+2}$$

$$I_{2m+1} = -2\alpha_{2m+3}$$

and deduce that the error associated with a Newton-Cotes formula of closed type, employing $n+1$ ordinates, can be expressed in the form

$$E_n = \begin{cases} -2\alpha_{n+2}h^{n+2}f^{(n+1)}(\xi) & (n \text{ odd}) \\ -(2\alpha_{n+3} + \alpha_{n+2})h^{n+3}f^{(n+2)}(\xi) & (n \text{ even}) \end{cases}$$

49 The constant α_k defined in Prob. 48 is expressible as a *generalized Bernoulli number* and is often denoted either by $B_k^{(k)}(1)/k!$ or by $B_k^{(k)}/k! + B_{k-1}^{(k-1)}/(k-1)!$. It is known (see Steffensen [1950]) that

$$\alpha_k \sim \frac{(-1)^{k+1}}{k(\log k)^2} \qquad (k \to \infty)$$

Assuming this fact, show that the numerical factor in the expression for E_n is approximated by $-2/[n(\log n)^2]$ when n is a large odd integer, and is approximated by $-1/[n(\log n)^2]$ when n is a large even integer.

50 Suppose that $f(z)$ is an analytic function of the complex variable z in a simply-connected region $\mathscr{R}$ of the complex plane which includes the segment $a \leqq x \leqq b$ of the real axis. By making use of the fact that then

$$f^{(r)}(z_0) = \frac{r!}{2\pi i} \oint_C \frac{f(z)}{(z - z_0)^{r+1}}\, dz$$

for any z_0 in $\mathscr{R}$, where C is any simple closed curve in $\mathscr{R}$ enclosing z_0, deduce that

$$|f^{(r)}(x)| \leqq \frac{r!\, ML}{2\pi d^{r+1}} \qquad (a \leqq x \leqq b)$$

where $|f(z)| \leqq M$ on C, L is the length of C, and d is the shortest distance from a point x in $[a, b]$ to a point z on C. [Notice that C can be *any* simple closed curve, enclosing the segment $a \leqq x \leqq b$ (once) but excluding all *singular points* of $f(z)$.]

Section 3.10

51 Show that

$$\int_{-1}^{1} \frac{f(x)}{\sqrt{1 - x^2}}\, dx = \sum_{k=-1}^{1} C_k f(k) + E$$

where

$$C_k = \int_{-1}^{1} \frac{L_k(x)}{\sqrt{1 - x^2}}\, dx$$

with $L_k(x)$ the Lagrange coefficient function defined by (3.4.5) and (3.4.6) when $n = 2$, and where

$$\begin{aligned} E &= \int_{-1}^{1} \frac{1}{\sqrt{1 - x^2}}\, x(x^2 - 1) f[-1, 0, 1, x]\, dx \\ &= \int_{-1}^{1} \left\{\frac{d}{dx}\, [\tfrac{1}{3}(1 - x^2)^{3/2}]\right\} f[-1, 0, 1, x]\, dx \\ &= -\frac{f^{\text{iv}}(\xi)}{72} \int_{-1}^{1} (1 - x^2)^{3/2}\, dx \end{aligned}$$

Hence deduce the formula

$$\int_{-1}^{1} \frac{f(x)}{\sqrt{1 - x^2}}\, dx = \frac{\pi}{4}\, [f(-1) + 2f(0) + f(1)] - \frac{\pi}{192} f^{\text{iv}}(\xi)$$

where $|\xi| < 1$.

52 Show that when the Filon cosine formula is applied to the rth subinterval $[x_{2r}, x_{2r+2}]$ of length $2h$, the associated error is given by

$$\int_{x_{2r}}^{x_{2r+2}} (x - x_{2r})(x - x_{2r+1})(x - x_{2r+2}) f[x_{2r}, x_{2r+1}, x_{2r+2}, x] \cos kx \, dx$$

Then show that

$$(x - x_{2r})(x - x_{2r+1})(x - x_{2r+2}) = \frac{d}{dx}\left[\tfrac{1}{4}(x - x_{2r})^2(x_{2r+2} - x)^2\right]$$

and use integration by parts to transform the error expression to the form

$$-\frac{h^5}{90}\left[f^{\text{iv}}(\xi_1)\cos k\xi_2 - 4kf'''(\xi_3)\sin k\xi_4\right]$$

where each ξ is in $[x_{2r}, x_{2r+2}]$.

53 Use the Filon cosine formula to approximate the integral

$$\int_0^{2\pi} e^{x/\pi} \cos kx \, dx = -\frac{\pi(1 - e\cos k\pi)}{1 + k^2\pi^2}$$

for $k = 1$ and $k = 4$. In each case determine the two approximations corresponding to the spacings $h = \pi/4$ and $\pi/8$, retaining five decimal places.

54 Suppose that the Filon cosine formula is applied to $f(x) = x^3$ on the subinterval $[x_{2r}, x_{2r+2}]$.

(*a*) Show that the associated error can be expressed in the form

$$\int_{x_{2r}}^{x_{2r+2}} (x - x_{2r})(x - x_{2r+1})(x - x_{2r+2}) \cos kx \, dx$$

[See (2.3.11).]

(*b*) Assuming that the error considered in part (*a*) also can be expressed in the form $h^4\,\delta(\theta)\sin kx_{2r+1}$, with the notation of (3.10.17), deduce that if the term

$$h^4 \frac{f'''(x_{2r+1})}{6}\,\delta(\theta)\sin kx_{2r+1}$$

is added to the approximation in (3.10.11), then the term

$$\int_{x_{2r}}^{x_{2r+2}} (x - x_{2r})(x - x_{2r+1})(x - x_{2r+2}) \frac{f'''(x_{2r+1})}{6} \cos kx \, dx$$

must be subtracted from the integral considered in Prob. 52.

(*c*) Use this result to deduce that when $h\,\delta(\theta)S_c$ is added to the approximation in (3.10.11), with the notation of (3.10.18), the resultant modified error $\tilde{E}$ then satisfies (3.10.20).

55 Apply the correction (3.10.19) to each of the calculated approximations in Prob. 53.

Section 3.11

56 From the following rounded values of $f(x) = (1 + x)^{-2}$, determine approximate values of $f'(x)$ for $x = 1.0$, 1.1, and 1.2 by use of appropriate three- and five-point formulas, estimate the errors, and check the validity of the estimations:

x	1.0	1.1	1.2	1.3	1.4
$f(x)$	0.2500	0.2268	0.2066	0.1890	0.1736

57 Determine the optimum values of the spacing h for the purpose of approximating $f'(0)$ when $f(x) = 1/(x + 2)$ by means of the three-, five-, and seven-point differentiation formulas considered in Sec. 3.11 with $x_0 = 0$ when the predictable bound on the magnitude of the total error $T = E + R$ is to be minimized, under the assumption that each ordinate may be in error by ± 0.01. Also calculate the corresponding bounds on $|T|$ and show that $|T_7|_{max} > |T_5|_{max}$.

58 Values of a function $f(x)$ are to be determined for $x = 0$ and for four additional positive values of x, and are to be used for the approximate determination of $f'(0)$. Assume that use is to be made of (3.11.4) and that the accuracy of the calculated values can be guaranteed within only 1 percent, and suppose that the true function is $f(x) = 1/(1 + x)$; determine the spacing for which the sum of the squares of predictable upper bounds on the truncation and roundoff errors is least, and calculate the corresponding upper bound on the total error. Also compare this situation with that in which only three ordinates are to be used.

4

FINITE-DIFFERENCE INTERPOLATION

4.1 Introduction

This chapter returns to the consideration of formulas expressed in terms of differences, rather than of the ordinates themselves, but deals only with the cases in which the abscissas are equally spaced. Here the rather cumbersome notation of *divided* differences is not needed and is replaced by other notations which are explained in Sec. 4.2.

The most important of the interpolation formulas which involve differences, together with error terms, are derived in Secs. 4.3 to 4.7, and their respective uses in connection with desk calculation or with the use of a digital computer are discussed and illustrated in Sec. 4.8. It is of some historical interest to note that the formulas bearing the names of Gauss, Stirling, and Bessel apparently were first known to Newton, while the formulas attributed to Newton (Sec. 4.3) are due to Gregory. Further, Everett's first formula is due to Laplace, and Everett's second formula apparently was first given by Steffensen.

The propagation and detection of errors in given data are considered in Sec. 4.9, whereas a useful method of taking certain higher differences into approximate account, by modifying certain earlier differences, is illustrated in Sec. 4.10. The concluding section of the chapter provides some information concerning the behavior of the error term in certain interpolation formulas, as more and more differences are retained, and indicates the practical significance of that information.

4.2 Difference Notations

When data are tabulated for uniformly spaced abscissas, with spacing h, it often is convenient to express formulations for interpolation and related processes in terms of the differences themselves, rather than the *divided* differences used in Chap. 2.

For calculation near a tabular point x_0 at the beginning of the tabulated range, it is conventional to define the *forward difference* $\Delta f(x_0)$ as

$$\Delta f(x_0) = f(x_0 + h) - f(x_0) \qquad (4.2.1)$$

If also $\Delta f(x_0 + h) = f(x_0 + 2h) - f(x_0 + h)$ is known, then the second forward difference associated with x_0 is defined as

$$\begin{aligned}\Delta^2 f(x_0) &= \Delta f(x_0 + h) - \Delta f(x_0) \\ &= f(x_0 + 2h) - 2f(x_0 + h) + f(x_0)\end{aligned} \qquad (4.2.2)$$

and succeeding forward differences are defined by iteration. More generally, we introduce the definitions

$$\Delta f(x) = f(x + h) - f(x) \qquad \Delta^{r+1} f(x) = \Delta^r f(x + h) - \Delta^r f(x) \qquad (4.2.3)$$

the spacing h being implied in Δ. If a more specific notation is needed, Δ_h may be used in place of Δ.

When forward differences are used, it is convenient to number the abscissas $x_0, x_1, \ldots$ in increasing algebraic order, so that

$$x_{k+1} = x_k + h \qquad (4.2.4)$$

Then, with the notation of Sec. 2.3, there follows

$$\begin{aligned}\Delta f(x_k) &= f(x_{k+1}) - f(x_k) = (x_{k+1} - x_k) f[x_k, x_{k+1}] \\ &= hf[x_k, x_{k+1}] \\ \Delta^2 f(x_k) &= hf[x_{k+1}, x_{k+2}] - hf[x_k, x_{k+1}] \\ &= h(x_{k+2} - x_k) f[x_k, x_{k+1}, x_{k+2}] = 2h^2 f[x_k, x_{k+1}, x_{k+2}]\end{aligned}$$

and, in general, induction shows that

$$\begin{aligned}\Delta^r f(x_k) &= (r-1)!\, h^{r-1} f[x_{k+1}, \ldots, x_{k+r}] \\ &\qquad - (r-1)!\, h^{r-1} f[x_k, \ldots, x_{k+r-1}] \\ &= (r-1)!\, h^{r-1} (x_{k+r} - x_k) f[x_k, \ldots, x_{k+r}] \\ &= r!\, h^r f[x_k, x_{k+1}, \ldots, x_{k+r}]\end{aligned} \qquad (4.2.5)$$

The beginning of the corresponding difference table is indicated in Fig. 4.1, where f_k is written for $f(x_k)$. We notice that the subscript remains constant

x_0	f_0			
		Δf_0		
x_1	f_1		$\Delta^2 f_0$	
		Δf_1		$\Delta^3 f_0$
x_2	f_2		$\Delta^2 f_1$	
		Δf_2		
x_3	f_3			
$\vdots$	$\vdots$			

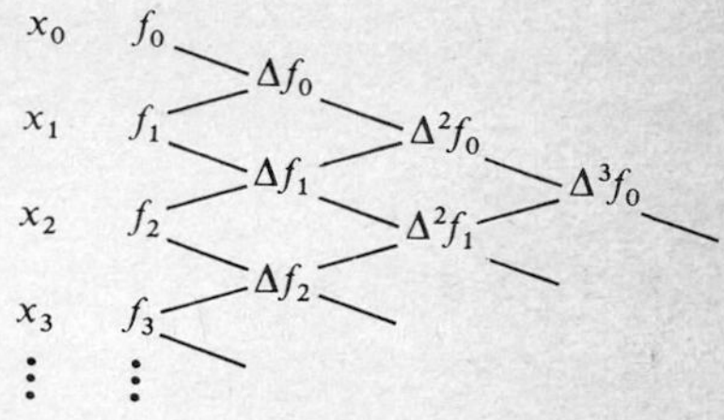

FIGURE 4.1

along each *forward* diagonal of the table, and that the region of determination of $\Delta^r f_k$ is bounded by the kth forward diagonal and the $(r + k)$th backward diagonal. Hence the difference $\Delta^r f_k$ depends upon the ordinates $f_k, f_{k+1}, \ldots, f_{k+r}$, as is also indicated by (4.2.5).

For calculation near the *end* of a tabulated range, the notation of *backward differences* is often more convenient. Here we write

$$\nabla f(x) = f(x) - f(x - h) \qquad \nabla^{r+1} f(x) = \nabla^r f(x) - \nabla^r f(x - h) \tag{4.2.6}$$

If the abscissas are again numbered in accordance with (4.2.4), there follows

$$\begin{aligned}\nabla f(x_k) &= f(x_k) - f(x_{k-1}) = (x_k - x_{k-1}) f[x_k, x_{k-1}] \\ &= h f[x_k, x_{k-1}]\end{aligned}$$

and, in general

$$\nabla^r f(x_k) = r!\, h^r f[x_k, x_{k-1}, \ldots, x_{k-r}] \tag{4.2.7}$$

in analogy with (4.2.5).

The end of the corresponding difference table is indicated in Fig. 4.2.

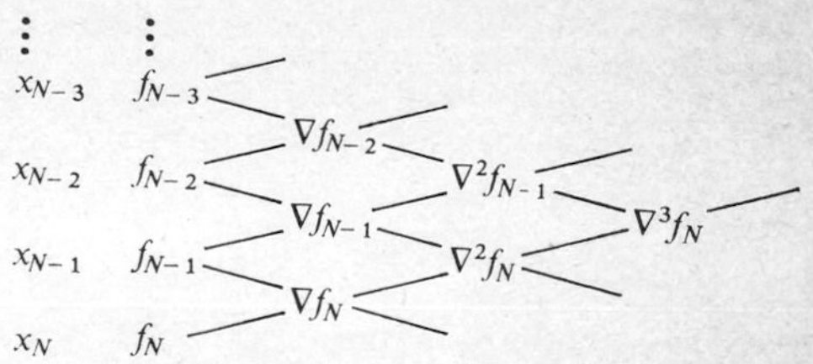

$\vdots$	$\vdots$			
x_{N-3}	f_{N-3}			
		∇f_{N-2}		
x_{N-2}	f_{N-2}		$\nabla^2 f_{N-1}$	
		∇f_{N-1}		$\nabla^3 f_N$
x_{N-1}	f_{N-1}		$\nabla^2 f_N$	
		∇f_N		
x_N	f_N			

FIGURE 4.2

Here the subscript remains constant along each *backward* diagonal. Also, it is seen that the difference $\nabla^r f_k$ depends upon the ordinates $f_{k-r}, f_{k-r+1}, \ldots, f_k$, as is also indicated by (4.2.7).

For the remaining calculation, the notation of so-called *central differences* is usually most convenient. If the calculation is to be effected near a certain interior tabular point, it is convenient to number that abscissa as x_0, and to number forward abscissas as $x_1, x_2, \ldots$ and backward abscissas as $x_{-1}, x_{-2}, \ldots$, so that (4.2.4) again holds. In the central-difference notation, one writes

$$\begin{aligned} \delta f(x) &= f(x + \tfrac{1}{2}h) - f(x - \tfrac{1}{2}h) \\ \delta^{r+1} f(x) &= \delta^r f(x + \tfrac{1}{2}h) - \delta^r f(x - \tfrac{1}{2}h) \end{aligned} \tag{4.2.8}$$

It is seen that $\delta f_k \equiv \delta f(x_k)$ generally *does not involve tabulated ordinates.* However, the *second* central difference

$$\begin{aligned} \delta^2 f_k &= \delta f(x_k + \tfrac{1}{2}h) - \delta f(x_k - \tfrac{1}{2}h) \\ &= [f(x_k + h) - f(x_k)] - [f(x_k) - f(x_k - h)] \\ &= f_{k+1} - 2f_k + f_{k-1} \end{aligned}$$

does involve tabular entries, and the same is seen to be true of all central differences $\delta^{2m} f_k$ of *even* order. Furthermore, we may notice that

$$\delta f_{k+(1/2)} = f_{k+1} - f_k$$

and, more generally, that $\delta^{2m+1} f_{k+(1/2)}$ involves only tabulated arguments.

With the notation of Sec. 2.3, we may write, for example,

$$\delta f_{1/2} = f_1 - f_0 = hf[x_0, x_1] \qquad \delta f_{-1/2} = f_0 - f_{-1} = hf[x_0, x_{-1}]$$
$$\delta^2 f_1 = \delta f_{3/2} - \delta f_{1/2} = hf[x_1, x_2] - hf[x_0, x_1] = 2!\, h^2 f[x_0, x_1, x_2]$$

and, in general,

$$\delta^{2m+1} f_{k+(1/2)} = h^{2m+1}(2m + 1)!\, f[x_{k-m}, \ldots, x_k, \ldots, x_{k+m}, x_{k+m+1}] \tag{4.2.9}$$
$$\delta^{2m+1} f_{k-(1/2)} = h^{2m+1}(2m + 1)!\, f[x_{k-m-1}, x_{k-m}, \ldots, x_k, \ldots, x_{k+m}] \tag{4.2.10}$$

and

$$\delta^{2m} f_k = h^{2m}(2m)!\, f[x_{k-m}, \ldots, x_k, \ldots, x_{k+m}] \tag{4.2.11}$$

The portion of the corresponding difference table in the neighborhood of an interior tabular point x_0, near which calculations are to be made, is indicated

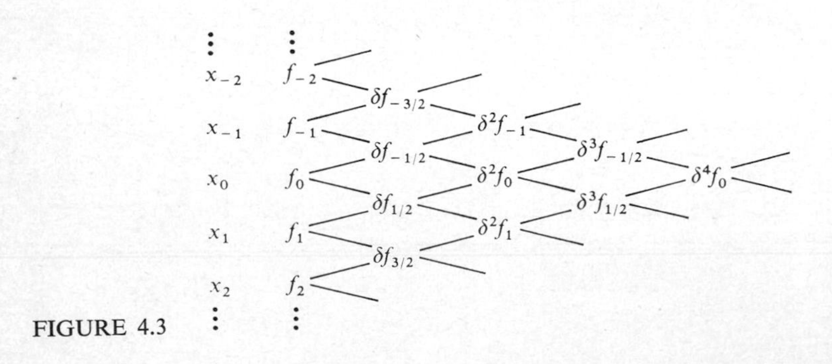

FIGURE 4.3

in Fig. 4.3. Here the subscript remains constant along *horizontal* lines of the table, which pass through differences of only even or only odd orders.

Thus, once a set of adjacent entries in a difference table has been numbered, three different sets of notations are available for the differences themselves, as may be seen from the composite Fig. 4.4. Any one of these sets of notations would suffice. However, each possesses certain advantages in certain applications, as will be seen.

$$\begin{array}{cclc} \vdots & \vdots & & \\ x_0 & f_0 & & \\ & & \Delta f_0 = \delta f_{1/2} = \nabla f_1 & \\ x_1 & f_1 & & \Delta^2 f_0 = \delta^2 f_1 = \nabla^2 f_2 \\ & & \Delta f_1 = \delta f_{3/2} = \nabla f_2 & \\ x_2 & f_2 & & \\ \vdots & \vdots & & \end{array}$$

FIGURE 4.4

4.3 Newton Forward- and Backward-difference Formulas

In order to obtain an interpolation formula such that the retention of $n + 1$ terms leads to the polynomial of degree n taking on the values of $f(x)$ at x_0, $x_1 = x_0 + h, \ldots, x_n = x_0 + nh$, we may refer to Newton's divided-difference formula (2.5.2), making use of the relation

$$f[x_0, \ldots, x_r] = \frac{1}{r!\, h^r} \Delta^r f_0 \tag{4.3.1}$$

which follows from (4.2.5), to obtain the result

$$\begin{aligned} f(x) = f_0 + (x - x_0) \frac{\Delta f_0}{1!\, h} + (x - x_0)(x - x_1) \frac{\Delta^2 f_0}{2!\, h^2} + \cdots \\ + (x - x_0)(x - x_1) \cdots (x - x_{n-1}) \frac{\Delta^n f_0}{n!\, h^n} + E(x) \end{aligned} \tag{4.3.2}$$

where

$$E(x) = (x - x_0) \cdots (x - x_n) \frac{f^{(n+1)}(\xi)}{(n + 1)!} \tag{4.3.3}$$

and where ξ is in the interval spanned by $x_0, \ldots, x_n$, and x, in accordance with (2.5.3) and (2.6.8).

The formula takes on a simpler form if we introduce a dimensionless variable s defined as distance from x_0 in units of h:

$$s = \frac{x - x_0}{h} \qquad x = x_0 + hs \tag{4.3.4}$$

Since then there follows also $x - x_k = h(s - k)$, the preceding formula takes the form

$$f_s = f_0 + s\,\Delta f_0 + \frac{s(s-1)}{2!}\Delta^2 f_0 + \cdots + \frac{s(s-1)\cdots(s-n+1)}{n!}\Delta^n f_0 + E_s \tag{4.3.5}$$

where

$$E_s = \frac{h^{n+1}}{(n+1)!} s(s-1)\cdots(s-n) f^{(n+1)}(\xi) \tag{4.3.6}$$

and where we have written

$$f_s \equiv f(x_0 + hs) = f(x) \qquad E_s \equiv E(x_0 + hs) = E(x)$$

This formula (or, more properly, the result of neglecting the error term E_s) is known as *Newton's forward-difference formula* for interpolation. It makes use of the difference path indicated in Fig. 4.5.

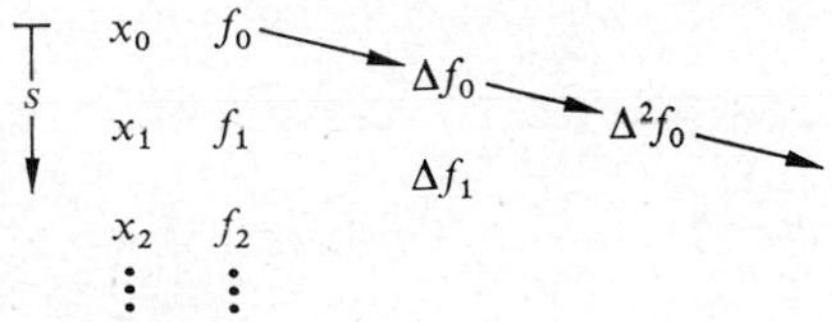

$$\begin{array}{ccccc} x_0 & f_0 & & & \\ & & \Delta f_0 & & \\ x_1 & f_1 & & \Delta^2 f_0 & \\ & & \Delta f_1 & & \\ x_2 & f_2 & & & \\ \vdots & \vdots & & & \end{array}$$

FIGURE 4.5

In a similar way, if we require a formula successively introducing the ordinates at x_N, x_{N-1}, x_{N-2}, and so forth, we may replace x_0 by x_N, x_1 by $x_{N-1}, \ldots, x_k$ by x_{N-k} in (2.5.2):

$$\begin{aligned} f(x) = f(x_N) &+ (x - x_N) f[x_N, x_{N-1}] \\ &+ (x - x_N)(x - x_{N-1}) f[x_N, x_{N-1}, x_{N-2}] + \cdots \\ &+ (x - x_N)(x - x_{N-1})\cdots(x - x_{N-n+1}) f[x_N, x_{N-1}, \ldots, x_{N-n}] \\ &+ E(x) \end{aligned}$$

and, writing here

$$s = \frac{x - x_N}{h} \qquad x = x_N + hs \tag{4.3.7}$$

we may use (4.2.7) to reduce this result to the form

$$f_{N+s} = f_N + s\,\nabla f_N + \frac{s(s+1)}{2!}\nabla^2 f_N + \cdots + \frac{s(s+1)\cdots(s+n-1)}{n!}\nabla^n f_N + E_s \tag{4.3.8}$$

where

$$E_s = \frac{h^{n+1}}{(n+1)!} s(s+1)\cdots(s+n)f^{(n+1)}(\xi) \qquad (4.3.9)$$

This formula is known as *Newton's backward-difference formula*, when E_s is neglected, and it utilizes the difference path indicated in Fig. 4.6.

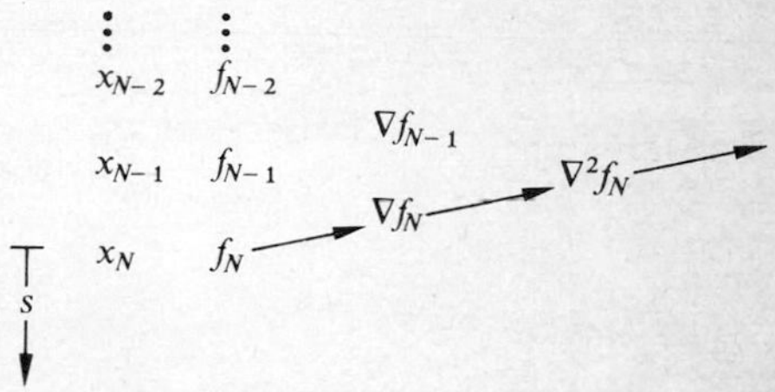

$\vdots \quad \vdots$

$x_{N-2} \quad f_{N-2}$

∇f_{N-1}

$x_{N-1} \quad f_{N-1} \qquad \nabla^2 f_N$

∇f_N

$x_N \quad f_N$

s

FIGURE 4.6

If $r + 1$ terms are retained in (4.3.5), the polynomial agreeing with $f(x)$ at $x_0, x_1, \ldots, x_r$ is obtained; the retention of $r + 1$ terms in (4.3.8) yields the polynomial agreeing with $f(x)$ at $x_N, x_{N-1}, \ldots, x_{N-r}$. If $N + 1$ terms were retained in each formula, the two formulas would involve the same ordinates and would yield the same polynomial approximation.

More generally, the former would be used near the beginning of a tabulation (at which *only* forward differences are available) and the latter would be used near the end (where only backward differences are available). In particular, the backward-difference formula is especially useful in *extending* a tabulation, and for generating other formulas useful for advancing numerical solutions of differential equations. For this reason, s was measured *forward* in the table in both formulas, so that it is positive for *extrapolation* in (4.3.8), whereas it is positive for *interpolation* in (4.3.5). Either formula can, of course, be used for either interpolation or extrapolation.

The formulas can be written in more concise form in terms of the binomial coefficients

$$\binom{r}{k} = \frac{r(r-1)\cdots(r-k+1)}{k!} \qquad (4.3.10)$$

With this notation, the *forward-difference* formula becomes merely

$$f_s \equiv f(x_0 + hs) \approx \sum_{k=0}^{n} \binom{s}{k} \Delta^k f_0 \qquad (4.3.11)$$

Further, the coefficient of $\nabla^k f_N$ in (4.3.8) is seen to be

$$\binom{s+k-1}{k} = \frac{s(s+1)\cdots(s+k-1)}{k!}$$

$$= (-1)^k \frac{(-s)(-s-1)\cdots(-s-k+1)}{k!}$$

$$= (-1)^k \binom{-s}{k} \tag{4.3.12}$$

so that the *backward-difference* formula takes the form

$$f_{N+s} \equiv f(x_N + hs) \approx \sum_{k=0}^{n} (-1)^k \binom{-s}{k} \nabla^k f_N \tag{4.3.13}$$

or, alternatively,

$$f_{N-s} \equiv f(x_N - hs) \approx \sum_{k=0}^{n} (-1)^k \binom{s}{k} \nabla^k f_N \tag{4.3.14}$$

In the form (4.3.14), s clearly is positive for *interpolation.*

Extensive tables of the coefficient functions may be found in the literature (see references in Sec. 4.13). A brief table, for interpolation or extrapolation by tenths, through fifth differences, is included in Sec. 4.12 for desk calculation.

4.4 Gaussian Formulas

For interpolation at a point $\bar{x}$, it is desirable to have available a formula in which the successively introduced ordinates correspond to abscissas which are as near as possible to $\bar{x}$. If $\bar{x}$ is near one end of the tabulation, the newtonian formulas of the preceding section serve this purpose as well as is possible. Otherwise, it is convenient to start with the abscissa x_0 nearest $\bar{x}$, then to introduce x_1 and x_{-1}, then x_2 and x_{-2}, and so forth.

If the ordinates are introduced in the order $f_0, f_1, f_{-1}, f_2, f_{-2}, \ldots$, the result of replacing $x_0, x_1, x_2, x_3, x_4, \ldots$ by $x_0, x_1, x_{-1}, x_2, x_{-2}, \ldots$ in (2.5.2), and the subsequent use of (4.2.9) and (4.2.11), with $k = 0$, leads to the form

$$f(x) = f_0 + (x - x_0)\frac{\delta f_{1/2}}{1!\,h} + (x - x_0)(x - x_1)\frac{\delta^2 f_0}{2!\,h^2}$$

$$+ (x - x_0)(x - x_1)(x - x_{-1})\frac{\delta^3 f_{1/2}}{3!\,h^3}$$

$$+ (x - x_0)(x - x_1)(x - x_{-1})(x - x_2)\frac{\delta^4 f_0}{4!\,h^4} + \cdots$$

If we write

$$s = \frac{x - x_0}{h} \qquad x = x_0 + hs \tag{4.4.1}$$

this result takes the form

$$f_s = f_0 + s\,\delta f_{1/2} + \frac{s(s-1)}{2!}\,\delta^2 f_0 + \frac{s(s^2-1^2)}{3!}\,\delta^3 f_{1/2}$$

$$+ \frac{s(s^2-1^2)(s-2)}{4!}\,\delta^4 f_0 + \cdots$$

$$+ \frac{s(s^2-1^2)\cdots(s^2-\overline{m-1}^2)(s-m)}{(2m)!}\,\delta^{2m} f_0$$

or

$$+ \frac{s(s^2-1^2)\cdots(s^2-m^2)}{(2m+1)!}\,\delta^{2m+1} f_{1/2}$$

$$+ E_s \tag{4.4.2}$$

where, if nth differences are retained, $n = 2m$ when n is even and $n = 2m + 1$ when n is odd. The error term takes the form

$$E_s = h^{2m+1}\,\frac{s(s^2-1^2)\cdots(s^2-m^2)}{(2m+1)!}\,f^{(2m+1)}(\xi) \tag{4.4.3}$$

when $n = 2m$, and the form

$$E_s = h^{2m+2}\,\frac{s(s^2-1^2)\cdots(s^2-m^2)(s-m-1)}{(2m+2)!}\,f^{(2m+2)}(\xi) \tag{4.4.4}$$

when $n = 2m + 1$. This formula employs the *forward* zigzag difference path indicated in Fig. 4.7 and is known as *Gauss' forward formula*.

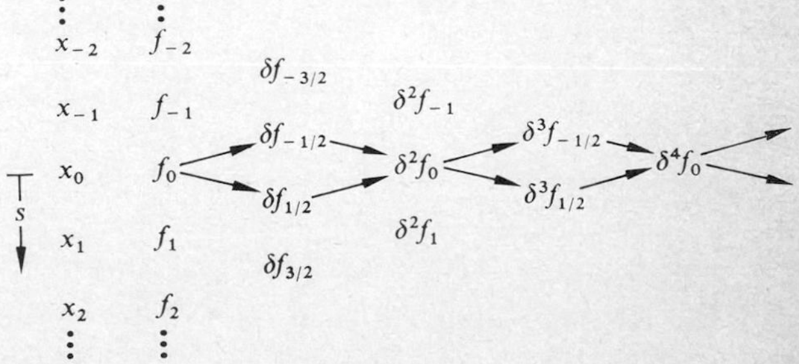

$\vdots$ $\vdots$
x_{-2} f_{-2}
$\delta f_{-3/2}$
x_{-1} f_{-1} $\delta^2 f_{-1}$
$\delta f_{-1/2}$ $\delta^3 f_{-1/2}$
x_0 f_0 $\delta^2 f_0$ $\delta^4 f_0$
$\delta f_{1/2}$ $\delta^3 f_{1/2}$
x_1 f_1 $\delta^2 f_1$
$\delta f_{3/2}$
x_2 f_2
$\vdots$ $\vdots$
s

FIGURE 4.7

In a completely similar way, by introducing the ordinates in the sequence $f_0, f_{-1}, f_1, f_{-2}, f_2, \ldots$, using (4.2.10) and (4.2.11) with $k = 0$, and again introducing the abbreviation (4.4.1), we obtain the form

$$f_s = f_0 + s\,\delta f_{-1/2} + \frac{s(s+1)}{2!}\delta^2 f_0 + \frac{s(s^2-1^2)}{3!}\delta^3 f_{-1/2} + \frac{s(s^2-1^2)(s+2)}{4!}\delta^4 f_0 + \cdots$$

$$+ \frac{s(s^2-1^2)\cdots(s^2-\overline{m-1}^2)(s+m)}{(2m)!}\delta^{2m} f_0$$

or

$$+ \frac{s(s^2-1^2)\cdots(s^2-m^2)}{(2m+1)!}\delta^{2m+1} f_{-1/2}$$

$$+ E_s \tag{4.4.5}$$

where

$$E_s = h^{2m+1}\frac{s(s^2-1^2)\cdots(s^2-m^2)}{(2m+1)!}f^{(2m+1)}(\xi)$$

or

$$E_s = h^{2m+2}\frac{s(s^2-1^2)\cdots(s^2-m^2)(s+m+1)}{(2m+2)!}f^{(2m+2)}(\xi) \tag{4.4.6}$$

according as the formula is terminated with even or odd differences. This formula utilizes the *backward* zigzag difference path in Fig. 4.7 and is known as *Gauss' backward formula.*

When terminated with an *even* difference, of order $2m$, *both* formulas yield the polynomial agreeing with $f(x)$ at $x_0, x_{\pm 1}, \ldots, x_{\pm m}$, and hence *are completely equivalent in that case.* However, when terminated with an *odd* difference, of order $2m + 1$, the forward formula gives the polynomial agreeing with $f(x)$ at $x_0, x_{\pm 1}, \ldots, x_{\pm m}$, and x_{m+1}, whereas the backward formula yields agreement at the first $2m + 1$ points and at x_{-m-1}. In this latter case, when seeking $f(\bar{x})$, the forward formula would be expected to afford somewhat better results when $\bar{x}$ is between x_0 and x_1, whereas the backward formula would generally be preferred when $\bar{x}$ is between x_0 and x_{-1}. Neither of these formulas is of frequent practical use, but from them other more useful formulas may be derived.

4.5 Stirling's Formula

When interpolations are to be effected for values of $\bar{x}$ near an interior point x_0, say, between $x_0 - \frac{1}{2}h$ and $x_0 + \frac{1}{2}h$, a formula of frequent use may be obtained by forming the *mean* of the gaussian forward and backward formulas, and so introducing a *symmetry* about the abscissa x_0:

$$f_s = f_0 + \frac{s}{2}(\delta f_{1/2} + \delta f_{-1/2}) + \frac{s}{2 \cdot 2!}[(s-1)+(s+1)]\,\delta^2 f_0$$
$$+ \frac{s(s^2-1^2)}{2 \cdot 3!}(\delta^3 f_{1/2} + \delta^3 f_{-1/2})$$
$$+ \frac{s(s^2-1^2)}{2 \cdot 4!}[(s-2)+(s+2)]\,\delta^4 f_0 + \cdots$$
$$+ \frac{s(s^2-1^2)\cdots(s^2-\overline{m-1}^2)}{2 \cdot (2m)!}[(s-m)+(s+m)]\,\delta^{2m} f_0$$

or

$$+ \frac{s(s^2-1^2)\cdots(s^2-m^2)}{2 \cdot (2m+1)!}(\delta^{2m+1} f_{1/2} + \delta^{2m+1} f_{-1/2})$$
$$+ E_s \tag{4.5.1}$$

It is then convenient to introduce symbols for the mean odd differences which appear in this formula. The notation

$$\mu f(x) \equiv \frac{1}{2}\left[f\left(x+\frac{h}{2}\right) + f\left(x-\frac{h}{2}\right)\right] \tag{4.5.2}$$

is often used, so that, for example, we may write

$$\mu\delta f_0 = \tfrac{1}{2}(\delta f_{1/2} + \delta f_{-1/2}) \qquad \mu\delta^3 f_0 = \tfrac{1}{2}(\delta^3 f_{1/2} + \delta^3 f_{-1/2}) \tag{4.5.3}$$

With this notation for the so-called *mean central differences* of odd order, (4.5.1) takes the form

$$f_s = f_0 + s\,\mu\delta f_0 + \frac{s^2}{2!}\delta^2 f_0 + \frac{s(s^2-1^2)}{3!}\mu\delta^3 f_0$$
$$+ \frac{s^2(s^2-1^2)}{4!}\delta^4 f_0 + \cdots$$
$$+ \frac{s^2(s^2-1^2)\cdots(s^2-\overline{m-1}^2)}{(2m)!}\delta^{2m} f_0$$

or

$$+ \frac{s(s^2-1^2)\cdots(s^2-m^2)}{(2m+1)!}\mu\delta^{2m+1} f_0$$
$$+ E_s \tag{4.5.4}$$

The result of omitting E_s is known as *Stirling's formula* for interpolation and corresponds to the array of Fig. 4.8.

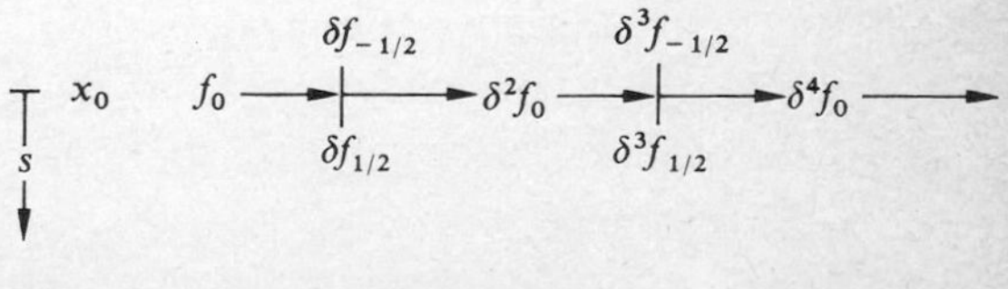

x_0 (arrow s downward) $f_0 \longrightarrow$ [$\delta f_{-1/2}$ / $\delta f_{1/2}$] $\longrightarrow \delta^2 f_0 \longrightarrow$ [$\delta^3 f_{-1/2}$ / $\delta^3 f_{1/2}$] $\longrightarrow \delta^4 f_0 \longrightarrow$

FIGURE 4.8

Since the errors associated with terminating (4.4.2) and (4.4.5) with an *even* difference are identical, there follows also

$$E_s = h^{2m+1} \frac{s(s^2 - 1^2)\cdots(s^2 - m^2)}{(2m + 1)!} f^{(2m+1)}(\xi) \qquad (4.5.5)$$

when $n = 2m$. As in the preceding cases, ξ lies between the largest and smallest of the abscissas involved in the formula (here $x_0, x_{\pm 1}, \ldots, x_{\pm m}$, and x).

However, when $n = 2m + 1$, the mean of the errors (4.4.4) and (4.4.6) takes the form

$$E_s = h^{2m+2} \frac{s(s^2 - 1^2)\cdots(s^2 - m^2)}{2(2m + 2)!} [(s - m - 1)f^{(2m+2)}(\xi_1) + (s + m + 1)f^{(2m+2)}(\xi_2)] \qquad (4.5.6)$$

where both ξ_1 and ξ_2 lie inside the interval spanned by $x_0, x_{\pm 1}, \ldots, x_{\pm(m+1)}$, and x. Thus, when Stirling's formula is terminated with an *odd* difference, the error term does not take a simple form similar to (4.5.5). It should be noticed that the interpolation polynomial of degree $2m + 1$, which is yielded by the formula in this case, agrees with $f(x)$ at the $2m + 1$ points $x_0, x_{\pm 1}, \ldots, x_{\pm m}$, but that an additional $(2m + 2)$th point of agreement (which would serve to specify the polynomial) is not known.

The Stirling formula is equivalent to either Gauss formula when terminated with even differences. But even in this case its form is more convenient because of the fact that the coefficients of the differences of even order are even functions of s, whereas the coefficients of the mean differences of odd order are odd functions of s. A brief table of the coefficients is presented in Sec. 4.12. More extensive tables can be found in the literature (see references in Sec. 4.13).

4.6 Bessel's Formula

Whereas Stirling's formula is principally intended for interpolation near a tabular entry x_0, the need frequently arises for a formula designed for interpolation over the interval, say, between x_0 and x_1. In order to obtain a formula in which the array of differences involved is symmetric about a horizontal line midway between x_0 and x_1, we again make use of the gaussian forward formula

$$f_s = f_0 + s\,\delta f_{1/2} + \frac{s(s - 1)}{2!}\delta^2 f_0 + \frac{s(s - 1)(s + 1)}{3!}\delta^3 f_{1/2} + \cdots \qquad (4.6.1)$$

which involves the differences along the forward zigzag of Fig. 4.7, and combine it with a formula which involves the differences along the backward zigzag of that figure. The latter formula may be obtained most easily by noticing that, if s were to be measured from x_1, that formula would be obtained by advancing all subscripts in the gaussian backward formula by unity. Hence, if s is to be measured from x_0 in both formulas, we must advance the subscripts in (4.4.5) by unity and, at the same time, replace s by $s - 1$, to give the result

$$f_s = f_1 + (s-1)\,\delta f_{1/2} + \frac{(s-1)s}{2!}\,\delta^2 f_1 + \frac{(s-1)s(s-2)}{3!}\,\delta^3 f_{1/2} + \cdots \tag{4.6.2}$$

The mean of (4.6.1) and (4.6.2) then takes the form

$$\begin{aligned} f_s = {} & \mu f_{1/2} + (s - \tfrac{1}{2})\,\delta f_{1/2} + \frac{s(s-1)}{2!}\,\mu\delta^2 f_{1/2} \\ & + \frac{s(s-1)(s-\frac{1}{2})}{3!}\,\delta^3 f_{1/2} + \cdots \\ & + \frac{s(s^2-1^2)\cdots(s^2-\overline{m-1}^2)(s-m)}{(2m)!}\,\mu\delta^{2m} f_{1/2} \end{aligned}$$

or

$$\begin{aligned} & + \frac{s(s^2-1^2)\cdots(s^2-\overline{m-1}^2)(s-m)(s-\frac{1}{2})}{(2m+1)!}\,\delta^{2m+1} f_{1/2} \\ & + E_s \end{aligned} \tag{4.6.3}$$

and is known as *Bessel's formula*. The associated data array is indicated in Fig. 4.9.

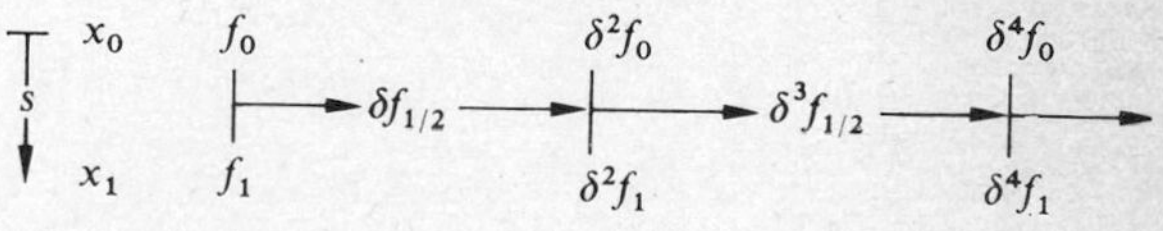

x_0 f_0 $\delta^2 f_0$ $\delta^4 f_0$

s $\delta f_{1/2}$ $\delta^3 f_{1/2}$

x_1 f_1 $\delta^2 f_1$ $\delta^4 f_1$

FIGURE 4.9

When terminated with an *odd* difference, of order $2m + 1$, both (4.6.1) and (4.6.2) yield the polynomial of degree $2m + 1$ agreeing with $f(x)$ when $x = x_0, x_{\pm 1}, \ldots, x_{\pm m}$, and x_{m+1}. Hence the same statement applies to Bessel's formula, and the error in that case is consequently identical with (4.4.4)

$$E_s = h^{2m+2}\,\frac{s(s^2-1^2)\cdots(s^2-m^2)(s-m-1)}{(2m+2)!}\,f^{(2m+2)}(\xi) \tag{4.6.4}$$

when $n = 2m + 1$. However, when Bessel's formula is terminated with even differences the error term is obtained as the mean of (4.4.3) and the first form of (4.4.6) with s replaced by $s - 1$, noticing that the parameter ξ is not generally the same in the two expressions, in the less simple form

$$E_s = h^{2m+1} \frac{s(s^2 - 1^2) \cdots (s^2 - \overline{m - 1}^2)(s - m)}{2(2m + 1)!} [(s + m) f^{(2m+1)}(\xi_1) + (s - m - 1) f^{(2m+1)}(\xi_2)] \tag{4.6.5}$$

when $n = 2m$, where ξ_1 and ξ_2 lie inside the interval spanned by $x_0, x_{\pm 1}, \ldots, x_{\pm m}, x_{m+1}$, and x.

A brief table of the coefficients appearing in (4.6.3) is given in Sec. 4.12. More extensive tables are available in the literature (see Sec. 4.13). The symmetry of this formula about the midpoint of the interval (x_0, x_1) becomes more evident if we write $s = t + \frac{1}{2}$, so that

$$t = \frac{x - \frac{1}{2}(x_0 + x_1)}{h} = s - \tfrac{1}{2} \tag{4.6.6}$$

and hence t is distance measured from that midpoint in units of h. It is readily verified that Bessel's formula then takes the equivalent form

$$f_{t+(1/2)} = \mu f_{1/2} + t\, \delta f_{1/2} + \frac{t^2 - \frac{1}{4}}{2!} \mu\delta^2 f_{1/2} + \frac{t(t^2 - \frac{1}{4})}{3!} \delta^3 f_{1/2} + \frac{(t^2 - \frac{1}{4})(t^2 - \frac{9}{4})}{4!} \mu\delta^4 f_{1/2} + \frac{t(t^2 - \frac{1}{4})(t^2 - \frac{9}{4})}{5!} \delta^5 f_{1/2} + \cdots \tag{4.6.7}$$

where the terminating term and the corresponding error term are obtainable by introducing (4.6.6) into the forms given in (4.6.3) to (4.6.5). Thus we see that the coefficients of mean even differences are even functions of t, whereas the coefficients of odd differences are odd functions of t.

An important special case results, by setting $s = \frac{1}{2}$ in (4.6.3) or $t = 0$ in (4.6.7), in the form

$$f_{1/2} = \tfrac{1}{2}(f_0 + f_1) - \tfrac{1}{16}(\delta^2 f_0 + \delta^2 f_1) + \tfrac{3}{256}(\delta^4 f_0 + \delta^4 f_1) - \tfrac{5}{2048}(\delta^6 f_0 + \delta^6 f_1) + \cdots + (-1)^m \frac{[1 \cdot 3 \cdots (2m - 1)]^2}{2^{2m+1}(2m)!} (\delta^{2m} f_0 + \delta^{2m} f_1) + E_{1/2} \tag{4.6.8}$$

where

$$E_{1/2} = (-1)^m \frac{[1 \cdot 3 \cdots (2m + 1)]^2}{2^{2m+2}(2m + 2)!} f^{(2m+2)}(\xi) \tag{4.6.9}$$

and where $x_0 - mh < \xi < x_1 + mh$. This formula is known as the formula for *interpolating to halves* and is particularly useful in subtabulation of data.

4.7 Everett's Formulas

In many tabulations, auxiliary tables of central differences of even orders (usually $\delta^2 f$ and $\delta^4 f$) are provided. In order to obtain an interpolation formula which involves only central differences of even order, we may, for example, start with the Gauss forward formula, terminated with an *odd* difference and written in the form

$$f_s = (f_0 + s\,\delta f_{1/2}) + \frac{s(s-1)}{2!}\left(\delta^2 f_0 + \frac{s+1}{3}\,\delta^3 f_{1/2}\right) + \frac{s(s^2-1^2)(s-2)}{4!}\left(\delta^4 f_0 + \frac{s+2}{5}\,\delta^5 f_{1/2}\right) + \cdots$$
$$+ \frac{s(s^2-1^2)\cdots(s^2-\overline{m-1}^2)(s-m)}{(2m)!}\left(\delta^{2m} f_0 + \frac{s+m}{2m+1}\,\delta^{2m+1} f_{1/2}\right) + E_s \qquad (4.7.1)$$

where

$$E_s = h^{2m+2}\,\frac{s(s^2-1^2)\cdots(s^2-m^2)(s-m-1)}{(2m+2)!}\,f^{(2m+2)}(\xi) \qquad (4.7.2)$$

If we now make use of the relations

$$\delta f_{1/2} = f_1 - f_0 \qquad \delta^3 f_{1/2} = \delta^2 f_1 - \delta^2 f_0 \qquad \cdots \qquad (4.7.3)$$

this formula becomes

$$f_s = (1-s)f_0 - \frac{s(s-1)(s-2)}{3!}\,\delta^2 f_0 - \frac{(s+1)s(s-1)(s-2)(s-3)}{5!}\,\delta^4 f_0 - \cdots - \frac{(s+m-1)(s+m-2)\cdots(s-m-1)}{(2m+1)!}\,\delta^{2m} f_0$$
$$+ sf_1 + \frac{(s+1)s(s-1)}{3!}\,\delta^2 f_1 + \frac{(s+2)(s+1)s(s-1)(s-2)}{5!}\,\delta^4 f_1 + \cdots + \frac{(s+m)(s+m-1)\cdots(s-m)}{(2m+1)!}\,\delta^{2m} f_1 + E_s \qquad (4.7.4)$$

where E_s is given by (4.7.2). The interpolation formula resulting from neglect of E_s is known as *Everett's first formula* (or often merely as *Everett's formula*).

In place of using the differences $\mu\delta^{2r} f_{1/2}$ and $\delta^{2r+1} f_{1/2}$ which are present in Bessel's formula, it uses the differences $\delta^{2r} f_0$ and $\delta^{2r} f_1$. However, it is seen that the result of terminating Bessel's formula with the $(2m + 1)$th difference must give a result identical with that of terminating Everett's first formula with the two $(2m)$th differences, for *both* of these formulas are equivalent to the Gauss forward formula terminated with the $(2m + 1)$th difference, as may be verified directly by comparison of the error terms.

Whereas the same number of *terms* must be evaluated in using the two formulas just mentioned, if tables are available which include differences of orders, say, two and four, then the use of the Everett formula permits a calculation taking into account all differences through the fifth without the need of differencing on the part of the computer. A brief table of the coefficients is provided in Sec. 4.12 (see references in Sec. 4.13 for more elaborate tables).

In a similar way, a formula involving only differences of *odd* order can be obtained from the Gauss forward formula terminated with an *even* difference (see Prob. 19). The result is known as *Everett's second formula* (often also as *Steffensen's formula*), but it has not found much favor in practice. The result of terminating it with the $(2m + 1)$th differences is equivalent to that of terminating *Stirling's formula* with the $(2m + 2)$th difference. A brief table of its coefficients is provided in Sec. 4.12.

If we introduce the notation of (4.3.10), Everett's first formula can be put in the form

$$f_s \approx (1 - s)f_0 + \binom{(1 - s) + 1}{3} \delta^2 f_0 + \binom{(1 - s) + 2}{5} \delta^4 f_0 + \cdots$$

$$+ sf_1 + \binom{s + 1}{3} \delta^2 f_1 + \binom{s + 2}{5} \delta^4 f_1 + \cdots \tag{4.7.5}$$

so that the coefficients of one line are obtained by replacing s by $1 - s$ in those of the other line.

4.8 Use of Interpolation Formulas

In general terms (and, for example, in relation to digital computers), the Newton forward- and backward-difference formulas (and their derivatives, integrals, and other relatives) are of importance for the purpose of generating and studying *sequences* of approximations to functions (and to their derivatives, integrals, and other properties) which correspond to collocation at successive sets of equally spaced points, each set including the preceding one, when *new* points are to be introduced in always increasing or always decreasing algebraic order. The central-difference formulas of Stirling and Bessel permit the generation and study of such sequences when successive new points of collocation are to be introduced on *both* sides of a reference point or tabular interval of interest at increasing distances therefrom.

In each of these situations, when information has been obtained relative to the number of collocation points needed for a certain purpose, it may be that the actual calculation is most efficiently effected by means of formulas of *lagrangian* type, particularly when a digital computer is to be used for that purpose.

Whereas a similar end result is provided without the use of differences by recursive processes such as Aitken interpolation and Romberg integration, the use of an appropriate difference formula may afford information with respect to a whole *set* of numerical calculations; otherwise each separate computation will require its own recursion. When only a single computation is to be made, or when available analytical information permits the use of an error term for the purpose of determining how much data will suffice, clearly the preceding considerations are of less importance.

In relation to desk calculation (for example, the use of slide rule or desk calculator), the difference formulas usually are convenient, not only for preliminary error estimation, but also for actual computation. The remainder of this section is principally oriented in this direction.

As mentioned earlier, the Newton formulas with forward or backward differences are most appropriate for calculation near the beginning or end, respectively, of a tabulation, and their use is mainly restricted to such situations. The Gauss forward and backward formulas terminated with an *even* difference are equivalent to each other and to the Stirling formula terminated with the same difference. The Gauss forward formula terminated with an *odd* difference is equivalent to the Bessel formula terminated with the same difference. The Gauss backward formula launched from x_0, and terminating with an odd difference, is equivalent to the Bessel formula launched from x_{-1}, terminated with the same difference. Thus, in place of using a Gauss formula, one may always use an equivalent Stirling or Bessel formula, for which the coefficients are extensively tabulated.

Reference to (4.5.4) shows that the coefficients of all differences of *even* order in Stirling's formula involve s^2 as a factor. Thus, for interpolation near x_0, it may be expected that the result of terminating that formula with a *mean odd difference* $\mu\delta^{2m+1}f_0$ will be nearly as accurate, on the average, as the result of retaining one additional difference. However, the relative complexity of the remainder term (4.5.6) in that situation is somewhat of a disadvantage when a precise error bound is required.

A comparison of (4.5.6) and (4.6.4) shows that, in addition to common factors, the Stirling error involves the factor

$$\tfrac{1}{2}[(m + 1 + s)f^{(2m+2)}(\xi_2) - (m + 1 - s)f^{(2m+2)}(\xi_1)]$$

whereas the Bessel error involves the factor

$$-(m + 1 - s)f^{(2m+2)}(\xi_1)$$

If it is known only that $|f^{(2m+2)}(x)| \leqq M$ for $x_{-m-1} \leqq x \leqq x_{m+1}$, the Stirling factor can be guaranteed only not to exceed $(m + 1)M$ in magnitude, whereas the Bessel factor cannot exceed $(m + 1 - s)M$ in magnitude, if extrapolation is excluded. Thus, from the point of view of *predictable* error bounds, Bessel's formula actually displays a slight advantage when the highest difference to be retained is *odd*, in spite of the fact that Stirling's formula then makes use of information afforded by an additional ordinate.† In any case, the Stirling formula is most efficient (in general) for small s, say, for $-\frac{1}{4} \leqq s \leqq \frac{1}{4}$, that is, for calculation between $x_0 - h/4$ and $x_0 + h/4$.

A similar comparison of (4.5.5) and (4.6.5) indicates that, whereas the result of truncating the Bessel formula with a *mean even difference* makes use of more information than does the Stirling formula truncated with the corresponding ordinary even difference, the use of the latter formula may actually be slightly preferable from the point of view of predictable error bounds when the highest difference to be retained is even. In any case, the Bessel formula is most efficient (in general) near $s = \frac{1}{2}$, say, for $\frac{1}{4} \leqq s \leqq \frac{3}{4}$, that is, for calculation between $x_0 + h/4$ and $x_1 - h/4$.

In a series of calculations based on a given set of data, it is inconvenient to shift from one of these two formulas to the other, and one of the two must be chosen. Given a set of data, a decision might be made first as to the highest difference which was to be retained. If that difference were of even order, Stirling's formula perhaps would be recommended; if it were of odd order, Bessel's formula might be preferred.‡ However, the difference in accuracy between the two formulas is usually small, so that the choice is usually dictated by personal preference.

Everett's first formula is particularly useful when auxiliary tables of certain even differences accompany the given data; Everett's second formula would be useful if its coefficients were tabulated and if auxiliary odd differences were available.

† If it is known (for example) that $f^{(2m+2)}(x)$ is of constant sign in the relevant interval, the advantage clearly is generally reversed.

‡ The fallibility of such generalizations is illustrated by a comparison of the results of Probs. 22 and 23.

Many tables of special functions exist, with certain even differences provided, in which the entries are determined to a large number of places, but in which a large uniform spacing h is used. Such tables store tremendous amounts of potential data in volumes of modest size for the user who employs Everett interpolation, retaining a number of differences (and a corresponding number of significant digits) consistent with his error tolerance. In addition, similar comments apply to the storage of special functions in a digital computer, a relatively small data storage sufficing for the specification of the function over a large range, in association with the use of an Everett interpolation routine. Further economizations in table formation or data storage can be effected by the use of techniques to be considered in Sec. 4.10.†

To illustrate the use of the various formulas, we consider the following difference table, based on five-place data taken from a table of $f(x) = \sin x$, where the differences are given in units of the fifth place and where the figures in parentheses are auxiliary mean central differences used in the calculations to be described:

x	$f(x)$	Δf	$\Delta^2 f$	$\Delta^3 f$	$\Delta^4 f$	$\Delta^5 f$
1.0	0.84147					
		4974				
1.1	0.89121		−891			
		4083		−40		
1.2	0.93204	(3617.5)	−931	(−36)	8	
	(0.94780)	3152	(−947)	−32	(9)	2
1.3	0.96356		−963		10	
		2189		−22		1
1.4	0.98545		−985		11	
		1204		−11		−3
1.5	0.99749		−996		8	
		208		−3		4
1.6	0.99957		−999		12	
		−791		9		
1.7	0.99166		−990			
		−1781				
1.8	0.97385					

A convenient check on the differencing effected in any difference table consists of the fact that *the sum of the entries in any column of differences should equal the difference between the last and first entries of the preceding column.* To see that this is so, suppose that the entries in a certain column, reading downward, are $u_1, u_2, u_3, \ldots, u_r$. Then the corresponding entries in the next column to the right are $(u_2 - u_1), (u_3 - u_2), \ldots, (u_{r-1} - u_{r-2})$, and

† Frequently used functions on specific intervals of more modest extent are more often stored in computers by means of coefficients specifying other types of approximations (to be considered in Chap. 9).

$(u_r - u_{r-1})$, and the sum of these quantities evidently "telescopes" into $u_r - u_1$.

Because of the irregular fluctuation of the fifth differences in the given table, we would suppose that they are not significant but that they principally reflect the propagated effects of roundoff errors present in the given data (see also Sec. 4.9). In fact, it would be suspected that the fluctuation of the fourth differences about their mean value of about 10 is also principally due to these inherent errors in the original data. Thus, not more than the first four differences are to be used here. Whether these differences are sufficient, and whether they are all *needed*, could be determined from the error term associated with the formula to be used if knowledge of the analytical form of $f(x)$ were presumed.

In order to interpolate for $f(1.02)$, we would use Newton's forward-difference formula, with $s = 0.2$:

$$f(1.02) \approx 0.84147 + 0.2(0.04974) + \frac{(0.2)(-0.8)}{2}(-0.00891)$$

$$+ \frac{(0.2)(-0.8)(-1.8)}{6}(-0.00040)$$

$$+ \frac{(0.2)(-0.8)(-1.8)(-2.8)}{24}(0.00008)$$

$$\doteq 0.84147 + 0.009948 + 0.000713 - 0.000019 - 0.000003$$

$$= 0.852109 \doteq 0.85211$$

which is correct to five places. The calculation is considerably simplified if use is made of coefficient tables (see Sec. 4.12):

$$f(1.02) \approx 0.84147 + (0.2)(0.04974) + (-0.08)(-0.00891) + (0.048)(-0.00040) + (-0.0336)(0.00008)$$

$$\doteq 0.85211$$

The interpolation for $f(1.75)$ would be accomplished by use of Newton's backward-difference formula (using coefficient tables), with $s = -0.5$:

$$f(1.75) \approx 0.97385 + (-0.5)(-0.01781) + (-0.025)(-0.00990) + (-0.0625)(0.00009) + (-0.03906)(0.00012)$$

$$\doteq 0.98398$$

the rounded value being in error by defect of one unit in the fifth place.

In order to interpolate for $f(1.22)$, we could use either Stirling's formula or Bessel's formula, with $x_0 = 1.2$ and $s = 0.2$ in either case. Since the formula is to terminate with even differences (and also since the interpolant is nearer $s = 0$ than $s = \frac{1}{2}$), Stirling's formula might be preferred. After inserting the mean odd differences indicated in parentheses in the row $x = 1.2$ of the difference table, the use of Stirling's formula gives

$$f(1.22) \approx 0.93204 + (0.2)(0.036175) + (0.02)(-0.00931) + (-0.032)(-0.00036) + (-0.0016)(0.00008)$$

$$\doteq 0.93910$$

whereas, after inserting appropriate mean even differences in the table, the use of Bessel's formula gives

$$f(1.22) \approx 0.94780 + (-0.3)(0.03152) + (-0.08)(-0.00947) + (0.008)(-0.00032) + (0.0144)(0.00009)$$

$$\doteq 0.93910$$

Both results are correct to five places. We see that both formulas would in fact give results correct to five places if only *third* differences were retained.

In a table providing $\delta^2 f$ and $\delta^4 f$, the entries used in the interpolation for $x = 1.22$ by Everett's first formula would read

x	$f(x)$	$\delta^2 f$	$\delta^4 f$
1.2	0.93204	−931	8
1.3	0.96356	−963	10

and the calculation would be of the form

$$f(1.22) \approx 0.8(0.93204) + (-0.048)(-0.00931) + (0.00806)(0.00008) + 0.2(0.96356) + (-0.032)(-0.00963) + (0.00634)(0.00010)$$

$$\doteq 0.745632 + 0.000447 + 0.000001 + 0.192712 + 0.000308 + 0.000001$$

$$= 0.939101 \doteq 0.93910$$

in agreement with the preceding results. The additional computation here is because of the fact that Everett's formula with fourth differences actually incorporates the effects of the first *five* differences. In this case, it is seen that the retention of only the two *second* differences would have been sufficient.

Since the analytical expression for $f(x)$ is known, this last situation could have been predicted by reference to the error formula (4.7.2) which, with $h = 0.1$, $s = 0.2$, and $m = 1$, gives

$$E = 10^{-4} \frac{(0.2)(-0.96)(-1.8)}{24} f^{\text{iv}}(\xi)$$

$$\doteq 1.44 \times 10^{-6} f^{\text{iv}}(\xi)$$

Since here $f^{\text{iv}}(x) = \sin x$, there follows $|f^{\text{iv}}(\xi)| \leqq 1$, so that (if no roundoff errors were present) the error resulting from terminating Everett's first formula with second differences would be less than two units in the sixth place. Similar error estimates could have been obtained, in advance, with reference to the other calculations.

Formulas for numerical differentiation and integration may be obtained by differentiating and integrating any of the interpolation formulas (see Probs. 5, 13, and 16). However, these formulas can be obtained somewhat more systematically by operational methods, and their treatment accordingly is postponed to the following chapter.

4.9 Propagation of Inherent Errors

In addition to the truncation errors, for which certain analytical expressions have been given, the effects of roundoff errors in the given data, and in the computation, must be taken into account. The latter generally can be regulated by retaining one or more extra figures in the intermediate calculation. It thus remains to investigate the way in which roundoff errors in the given data affect the interpolation process.

The error in the interpolant, corresponding to such inherent errors, clearly is merely a linear combination of the errors in the ordinates involved in the interpolation. When the interpolation polynomial is of degree n and is determined by exact fit to the given data at $n + 1$ points, the constants of combination are the Lagrange coefficients considered in Chap. 3. In particular, if the error in each given ordinate cannot exceed ε, then the error in the interpolant cannot exceed the product of ε and the sum of the absolute values of the relevant Lagrange coefficients. Three-point coefficients, corresponding to retention of second differences, are tabulated to tenths in Sec. 3.4, whereas a similar table of five-point coefficients, corresponding to retention of fourth differences, is presented in Sec. 4.12. Use of the latter table shows, for example, that the error in an interpolation at an abscissa midway between the third and fourth of the five relevant abscissas, due only to data errors not exceeding ε in magnitude, cannot exceed 1.4ε in magnitude.

The Stirling formula, when terminated with a mean odd difference, and the Bessel formula, when terminated with a mean even difference, are not based on interpolation polynomials which fit the data at $n + 1$ points, and hence must be analyzed separately. This is an additional reason for avoiding the termination of the Stirling and Bessel formulas with *mean* differences, when precise error bounds are desired.

The presence of roundoff errors in given data is also of importance in connection with the question as to the number of differences which should be *retained* in an interpolation. For this reason, it is of interest to study the *propagation* of the effects of such errors into the differences themselves.

Suppose first that a single initial entry is in error by an excess e due perhaps to rounding. Then, if all other initial entries are assumed to be exact, it is seen that the effects of this error will be propagated into the first five differences of the difference table as follows:

f	Δf	$\Delta^2 f$	$\Delta^3 f$	$\Delta^4 f$	$\Delta^5 f$
—		—		—	
	—		—		e
—		—		e	
	—		e		$-5e$
—		e		$-4e$	
	e		$-3e$		$10e$
e		$-2e$		$6e$	
	$-e$		$3e$		$-10e$
—		e		$-4e$	
	—		$-e$		$5e$
—		—		e	
	—		—		$-e$
—		—		—	

This characteristic distribution along a column, in which the successive errors alternate in sign and, indeed, vary along the column of rth differences as the binomial coefficients associated with $(1 - x)^r$, frequently serves to permit one to discover and correct a gross error in a table.

f	Δf	$\Delta^2 f$	$\Delta^3 f$
1.203		18	
	221		18
1.424		36	
	257		18
1.681		54	
	311		22
1.992		76	
	387		6
2.379		82	
	469		30
2.848		112	
	581		14
3.429		126	
	707		18
4.136		144	

Thus, for example, the third differences in the accompanying table appear to fluctuate irregularly. Their mean value is 18, and the successive deviations from the mean, reading downward, are 0, 0, 4, -12, $+12$, -4, 0. Thus an excess $e = 4$ in the last place is indicated in the entry 2.379, which occupies the row separating the maximum deviations. The corrected value is 2.375.

A fourth differencing would have given the entries 0, 4, -16, 24, -16, 4, from which the same conclusion would be drawn. When several errors are present, their discovery may be much more difficult.

Suppose now that *all* initial entries may be in error by amounts between $-e$ and e. The most unfavorable situation, with regard to effects on differences, is that in which the successive errors are as large as possible but are of alternating sign. The error-propagation table, through fourth differences, then appears as follows:

f	Δf	$\Delta^2 f$	$\Delta^3 f$	$\Delta^4 f$
e		$-4e$		$16e$
	$-2e$		$8e$	
$-e$		$4e$		$-16e$
	$2e$		$-8e$	
e		$-4e$		$16e$
	$-2e$		$8e$	
$-e$		$4e$		$-16e$
	$2e$		$-8e$	
e		$-4e$		$16e$

Thus, it follows that errors varying between $-e$ and e in the initial data will lead to errors varying between $-2^r e$ and $2^r e$ in the rth differences. Here, for example, if the initial data are correctly rounded to k decimal places, $e = 5 \times 10^{-k-1}$.

Because of this possible error growth, it usually happens in practice that calculated differences beyond a certain order are no longer significant. That is, there exists a certain "noise level" such that the effects of initial roundoffs are of the same order of magnitude as the differences which would have been obtained had the initial data been exact. If the initial data are rounded, from exactly known data, to k decimal places, then roundoff errors of magnitude $2^{r-1}/10^k$ are possible in the rth differences. Hence rth differences of magnitude appreciably smaller than $2^{r-1}/10^k$ are likely to consist largely of "noise." Thus, since $k = 5$ in the data used for the examples of the preceding section, noise of magnitude 1, 2, 4, 8, and 16 units in the fifth place *could* occur in the respective differences of order one through five, although the *probability* of noise of nearly maximum magnitude in the rth difference is clearly small and will decrease rapidly as r increases. In any case, it would be expected that, since the fifth differences in that table are small relative to the permissible noise, they are meaningless, so that the fluctuation of the fourth differences about their mean may also lack significance, in the sense that the replacement of those differences by their mean value would lead to errors in interpolation of the same order as the errors which are present in the *given* data.

4.10 Throwback Techniques

A useful procedure, due to Comrie [1956], frequently permits a table user, table maker, or computer program effectively to take into account a neglected difference by modifying certain of the differences actually retained in an interpolation formula.

In illustration, Everett's first formula, terminated with fourth differences, can be expressed in the form

$$f_s \approx (1-s)f_0 + sf_1 - \frac{s(s-1)(s-2)}{6}\left[\delta^2 f_0 + \frac{(s+1)(s-3)}{20}\delta^4 f_0\right]$$
$$+ \frac{(s+1)s(s-1)}{6}\left[\delta^2 f_1 + \frac{(s+2)(s-2)}{20}\delta^4 f_1\right] \quad (4.10.1)$$

On the interval $0 \leqq s \leqq 1$, the factors

$$\frac{(s+1)(s-3)}{20} \quad \text{and} \quad \frac{(s+2)(s-2)}{20}$$

both vary only from $-\frac{3}{20}$ to $-\frac{4}{20}$. This fact suggests that these factors be replaced by a constant value over that interval in (4.10.1). The value suggested by Comrie, -0.184, differs only slightly from the *mean* value ($-\frac{11}{60}$) of each factor.† Hence, if we define the *modified second difference*

$$\bar{\delta}^2 f_k = \delta^2 f_k - 0.184\,\delta^4 f_k \quad (4.10.2)$$

Everett's formula with fourth differences may be approximated by the formula

$$f_s \approx (1-s)f_0 + sf_1 - \frac{s(s-1)(s-2)}{6}\bar{\delta}^2 f_0 + \frac{(s+1)s(s-1)}{6}\bar{\delta}^2 f_1 \quad (4.10.3)$$

It is conventional to speak of (4.10.2) as "throwing back the fourth difference on the second."

The error associated with the introduction of this approximation is given by

$$-\left[\binom{s+1}{5} + 0.184\binom{s}{3}\right]\delta^4 f_0 + \left[\binom{s+2}{5} + 0.184\binom{s+1}{3}\right]\delta^4 f_1 \quad (4.10.4)$$

and calculation shows that, when $0 \leqq s \leqq 1$, this sum is not larger in magnitude than $0.00122M$, where M is the larger of $|\delta^4 f_0|$ and $|\delta^4 f_1|$. Hence, if $\delta^4 f_0$ and $\delta^4 f_1$ do not exceed 400 units in the last decimal place retained in the overall

† The figure $(\sqrt{2}+3)/24 \doteq 0.184$ was obtained by Comrie as that value of the factor for which the magnitude of the maximum error due to throwback is least in the case of *Bessel*'s formula. The same figure is conventionally used with the Everett formula (see also Probs. 27 and 28).

calculation, the error committed in the throwback cannot exceed $\frac{1}{2}$ unit in that place; if $\delta^4 f_0$ and $\delta^4 f_1$ are of common sign, then 1,000 units are permissible (see Prob. 28).

In such situations, only the associated modified second difference $\bar{\delta}^2 f_k$ need be tabulated (or stored, in a computer) together with each ordinate f_k, and Everett interpolation based on only those data effectively takes into account all differences through the *fifth*.

If the same throwback (4.10.2) is effected in Bessel's formula, so that $\mu\delta^2 f_{1/2}$ is replaced by $\mu\bar{\delta}^2 f_{1/2}$, and differences beyond the third are omitted, the effect of the omitted fourth difference is properly taken into account, in the same sense, if it does not exceed 1,000 units in the last decimal place retained.

Generalized techniques, relevant to higher differences or to two or more differences, have also been devised by Comrie [1956]. Similar techniques have been given (see Lidstone [1943]) for Stirling's formula.

4.11 Interpolation Series

Reference has already been made to the fact that the formal infinite series generated by the interpolation formulas considered in this chapter generally do not converge as the number n of differences retained is increased without limit while the spacing h is held fixed. In this section we consider a simple example which illustrates this fact, and we state certain known results of a more general nature.

If the Newton forward-difference formula (4.3.2) (with the error term suppressed) is considered as a formal infinite series, it can be expressed in the form

$$\begin{aligned} f(x) &\sim f(0) + x\frac{\Delta f(0)}{1!\,h} + x(x-h)\frac{\Delta^2 f(0)}{2!\,h^2} + \cdots \\ &\sim f(0) + \sum_{k=1}^{\infty} \frac{\Delta^k f(0)}{k!\,h^k}\, x(x-h)\cdots(x-\overline{k-1}h) \end{aligned} \tag{4.11.1}$$

where we have supposed that the origin has been chosen such that $x_0 = 0$. Similarly, the formal Stirling *interpolation series* is expressible in the form

$$\begin{aligned} f(x) &\sim f(0) + \frac{x}{h}\left[\mu\delta f(0) + \frac{x}{2h}\,\delta^2 f(0)\right] + \cdots \\ &\sim f(0) + \sum_{k=1}^{\infty} \frac{1}{(2k-1)!\,h^{2k-1}}\left[\mu\delta^{2k-1} f(0) + \frac{x}{2kh}\,\delta^{2k} f(0)\right] \\ &\qquad \times [x(x^2-h^2)(x^2-4h^2)\cdots(x^2-\overline{k-1}^2h^2)] \end{aligned} \tag{4.11.2}$$

and the remaining formulas correspond to similar interpolation series. The basic problem considered here is that of determining when such a series actually converges to the generating function $f(x)$.

In the special case when

$$f(x) = e^{ax} \tag{4.11.3}$$

there follows

$$\Delta f(x) = e^{a(x+h)} - e^{ax} = (e^{ah} - 1)f(x) \quad \ldots$$

$$\Delta^r f(x) = (e^{ah} - 1)^r f(x)$$

and hence also

$$\Delta^r f(0) = (e^{ah} - 1)^r \tag{4.11.4}$$

Thus the formal Newton interpolation series for e^{ax} may be obtained in the form

$$e^{ax} \sim 1 + \sum_{k=1}^{\infty} \frac{(e^{ah} - 1)^k}{k!\, h^k} [x(x - h) \cdots (x - \overline{k - 1}h)] \tag{4.11.5}$$

This series *terminates*, and represents e^{ax} correctly, when x is zero or a positive integral multiple of h. In order to investigate its convergence when the series is infinite, we notice first that the ratio of successive terms in this series is given by

$$\frac{e^{ah} - 1}{h(k + 1)} (x - kh)$$

and that, as $k \to \infty$, this ratio tends to $-(e^{ah} - 1)$, for all values of x. Thus we may deduce that the series (4.11.5) *converges* if $|e^{ah} - 1| < 1$ or $e^{ah} < 2$, whereas, if $x \neq 0, h, 2h, \ldots$, it *diverges* when $|e^{ah} - 1| > 1$ or $e^{ah} > 2$.

When $e^{ah} = 2$, the series reduces to

$$1 + \sum_{k=1}^{\infty} \frac{x(x - h) \cdots (x - \overline{k - 1}h)}{k!\, h^k}$$

$$\equiv 1 + \sum_{k=1}^{\infty} \frac{(-1)^k}{k!} \left[\left(k - \frac{x}{h} - 1\right)\left(k - \frac{x}{h} - 2\right) \cdots \left(-\frac{x}{h}\right)\right]$$

the successive terms of which alternate in sign when k is sufficiently large. Now the kth term can be written, in terms of the gamma function, in the form

$$(-1)^k \frac{\Gamma(k - x/h)}{\Gamma(-x/h)k!}$$

By making use of the fact that $\Gamma(k + u)$ is approximated by $k!\, k^{u-1}$, for large k, we find that this last ratio is approximated by

$$(-1)^k \frac{k^{-x/h-1}}{\Gamma(-x/h)}$$

when k is large, and hence that it tends to zero as $k \to \infty$ if and only if $x > -h$.

It follows that *the series* (4.11.5) *converges for all finite values of x if $ah < \log 2$ and diverges for all values of x which differ from* $0, h, 2h, \ldots$ *if $ah > \log 2$, and that, if $ah = \log 2$, the series converges when and only when $x > -h$.* It can be proved that, when the series converges, the convergence is indeed to e^{ax}.

If x is replaced by a *complex* variable z (but a is real), the preceding developments are unchanged except for the fact that, when $ah = \log 2$, the region of convergence is that half of the complex z plane for which the *real part* of z is greater than $-h$. If also a is complex, the conditions $ah \lesseqgtr \log 2$ must be replaced by $|e^{ah} - 1| \lesseqgtr 1$.

We see, therefore, that, if the Newton forward-difference formula were used for interpolating e^{ax} where $a > 0$, with a spacing h larger than $(\log 2)/a$, the successive interpolates corresponding to the incorporation of more and more data eventually would begin to oscillate with increasing amplitude about the true value. Thus, whereas the retention of an additional term of the interpolation formula generally would improve the accuracy of the interpolation up to a certain stage, there would exist a point beyond which the addition of more terms would correspond to a *loss* of accuracy, even though no roundoff errors were present. (A similar situation was encountered in Sec. 3.9.) For $ah < \log 2$ (in particular, for negative a), this situation would not arise. In the intermediate case when $ah = \log 2$, convergence would follow if and only if the formula were not used for backward *extrapolation* beyond $x = -h$. These results are particularly remarkable in view of the fact that e^{az} is such a well-behaved function that its *Taylor* series, launched from any point, converges for *all* finite values of z.

A similar analysis, in the case when Stirling's formula is used instead, leads to the fact that here the corresponding interpolation series converges for all z when $ah < 2 \log (1 + \sqrt{2})$ and diverges (when it does not terminate) for all z in all other cases, including the special case when $ah = 2 \log (1 + \sqrt{2})$. If a is complex, the corresponding condition is $|\sinh (ah/2)| < 1$.

In the general case, it is known that, if the *Stirling* series converges for *any* value of z in addition to $z = 0, \pm h, \pm 2h, \ldots$ (for which it terminates), then it converges for *all* finite values of z (real or complex). Hence, conversely, if it diverges for any finite value of z, it diverges always unless it terminates. In the language of the theory of anlaytic functions of a complex variable, the

Stirling series cannot converge to $f(z)$ for *any* z (except those for which it terminates) unless $f(z)$ is a so-called *entire function*, that is, a function which is analytic at all finite points of the complex z plane, or, equivalently, a function whose Taylor series converge everywhere.

But, even though this be the case, the series *still* may not converge (as in the preceding example, where $f(z) = e^{az}$). It is also necessary that $|f(re^{i\theta})| < Me^{ar}$ for large r, where M and a are constants and a is such that $ah \leqq \pi$. If these conditions are not satisfied, the Stirling series will *diverge* everywhere except where it terminates. On the other hand, if $f(z)$ is an entire function, and if $|f(re^{i\theta})| < Me^{ar}$ for large r, where M and a are constants and a is such that $ah < 2 \log (1 + \sqrt{2})$, then the series will *converge* everywhere. Similar statements apply to the series associated with Bessel's formula.

In the case of the *Newton* series, with forward differences, it is known that, if the series converges to $f(z)$ for any value of z, say, $\tilde{z}$, in addition to $z = 0, h, 2h, \ldots$, for which it terminates, then it converges for all values of z such that the real part of z is greater than the real part of $\tilde{z}$. Unless $f(z)$ is analytic in some right half-plane $\text{Re}\,(z) > \alpha$, and also $|f(re^{i\theta})| < Me^{ar}$ for large r, where M and a are constants and a is such that $ah \leqq \pi/2$, the series will diverge except when it terminates. If $f(z)$ is analytic in such a half-plane and if also $|f(re^{i\theta})| < Me^{ar}$ for large r, where M and a are constants and a is such that $ah < \log 2$, then the series will converge everywhere in that half-plane. (For proofs of these statements, see Nörlund [1926, 1954].)

Thus, for example, the function $f(z) = 1/(1 + z^2)$ is analytic when z is real, but it possesses poles when $z = \pm i$. Hence the Stirling series will diverge when it does not terminate. Since this function is analytic in the right half-plane $z > 0$, and since it is dominated in magnitude by *any* exponential function Me^{ar} $(a > 0)$ as $r \to \infty$, the Newton series will converge in that half-plane. Nevertheless, if both series are launched from the same point, the error in the Stirling series will at first decrease much more rapidly than that associated with the Newton series, as additional terms are incorporated into the calculation. *Eventually*, the result of adding still more terms to the Stirling series will *increase* its error, whereas the error in the Newton series will continue to decrease. However, it is of considerable practical importance that, when h is small, this point of diminishing return in the Stirling formula is likely to be preceded either by a stage at which the truncation error has decreased below the tolerance imposed or by a stage at which the "noise level" is reached, so that the effects of roundoff errors would cause the remaining higher differences to be undependable in any case.

Thus, as in many other practical situations, it is quite possible to obtain more accurate results by terminating an ultimately *divergent* process *at an appropriate stage* than by terminating a *convergent* process at a corresponding stage.

It is evident that since each partial sum of either the Newton or Stirling series represents a polynomial approximation to $f(x)$ corresponding to collocation at the points involved, the two sequences of approximations differ only in that the former results from the successive introduction of the ordinates at $x = 0, h, 2h, \ldots, kh, \ldots$, all of which lie on the half-line $0 \leqq x < \infty$, whereas the latter successively introduces the ordinates at the points $\cdots, -kh, \ldots, -h, 0, h, \ldots, kh, \ldots$, in such a way that symmetry is preserved about $x = 0$. That is, the convergence or divergence of the sequence of approximations truly depends upon the sequence of data introduced rather than upon the form in which the polynomial interpolation formula employed is written. Whereas an indication of the existence of an unfavorable situation usually is afforded by an inspection of a relevant *difference* table, such numerical evidence is not available when lagrangian methods are used.

The sequences of interpolation polynomials considered here correspond to the incorporation of successive ordinates which eventually are at unboundedly increasing distances from the point of interpolation. Whereas the sequence generated by fitting ordinates at points which divide a fixed finite interval $[a, b]$ into n equal parts and allowing n to increase without limit is more tractable, there actually exist functions which are continuous on $[a, b]$ but for which this sequence *diverges everywhere* inside that interval. For the function $f(x) = 1/(x^2 + 1)$ on the interval $[-5, 5]$, Runge [1901] established divergence when $|x| > c$, where $c \doteq 3.63$. For the simple function $f(x) = |x|$ on $[-1, 1]$, Bernstein [1912*b*] proved that the sequence diverges for all x in $[-1, 1]$ except $x = 0$ and $x = \pm 1$. However, when the function $f(z)$ is an *analytic* function of the complex variable z in a sufficiently large region including the real interval $a \leqq x \leqq b$ under consideration, convergence is guaranteed (see Probs. 38 to 43).

Indeed, it is known (see Krylov [1962]) that if $f(z)$ is analytic in the region $\mathscr{B}$ comprising the points which lie on or inside one or both of the circles with centers at $z = a$ and $z = b$ and radius $b - a$, then *any* infinite collocative interpolation sequence will converge to $f(x)$ everywhere on $[a, b]$ regardless of the distribution of the successive points of collocation, provided only that they all lie in $[a, b]$ and that their number becomes infinite.

4.12 Tables of Interpolation Coefficients

This section provides brief tables of coefficients relevant to the interpolation formulas which have been considered. For more elaborate tables, the references cited in Sec. 4.13 should be consulted.

LAGRANGE FIVE-POINT INTERPOLATION†

$f_s \approx L_{-2}(s)f_{-2} + L_{-1}(s)f_{-1} + L_0(s)f_0 + L_1(s)f_1 + L_2(s)f_2$

(for negative s, use lower column labels)

s	$L_{-2}(s)$	$L_{-1}(s)$	$L_0(s)$	$L_1(s)$	$L_2(s)$	
0.0	0.000000	0.000000	1.000000	0.000000	0.000000	0.0
0.1	0.007838	−0.059850	0.987525	0.073150	−0.008663	−0.1
0.2	0.014400	−0.105600	0.950400	0.158400	−0.017600	−0.2
0.3	0.019338	−0.136850	0.889525	0.254150	−0.026163	−0.3
0.4	0.022400	−0.153600	0.806400	0.358400	−0.033600	−0.4
0.5	0.023438	−0.156250	0.703125	0.468750	−0.039063	−0.5
0.6	0.022400	−0.145600	0.582400	0.582400	−0.041600	−0.6
0.7	0.019338	−0.122850	0.447525	0.696150	−0.040163	−0.7
0.8	0.014400	−0.089600	0.302400	0.806400	−0.033600	−0.8
0.9	0.007838	−0.047850	0.151525	0.909150	−0.020663	−0.9
1.0	0.000000	0.000000	0.000000	1.000000	0.000000	−1.0
1.1	−0.008663	0.051150	−0.146475	1.074150	0.029838	−1.1
1.2	−0.017600	0.102400	−0.281600	1.126400	0.070400	−1.2
1.3	−0.026163	0.150150	−0.398475	1.151150	0.123338	−1.3
1.4	−0.033600	0.190400	−0.489600	1.142400	0.190400	−1.4
1.5	−0.039063	0.218750	−0.546875	1.093750	0.273438	−1.5
1.6	−0.041600	0.230400	−0.561600	0.998400	0.374400	−1.6
1.7	−0.040163	0.220150	−0.524475	0.849150	0.495338	−1.7
1.8	−0.033600	0.182400	−0.425600	0.638400	0.638400	−1.8
1.9	−0.020663	0.111150	−0.254475	0.358150	0.805838	−1.9
2.0	0.000000	0.000000	0.000000	0.000000	1.000000	−2.0
	$L_2(s)$	$L_1(s)$	$L_0(s)$	$L_{-1}(s)$	$L_{-2}(s)$	s

NOTE: All coefficients become exact if each terminal 8 is replaced by 75, and each terminal 3 by 25.

† See Sec. 3.4 for three-point coefficients.

STIRLING INTERPOLATION

$f_s \approx f_0 + s\,\mu\delta f_0 + C_2(s)\,\delta^2 f_0 + C_3(s)\,\mu\delta^3 f_0 + C_4(s)\,\delta^4 f_0$

s	$C_2(s)$	$C_3(s)$	$C_4(s)$	s
0	0.00000	0.00000	0.00000	0
0.1	0.00500	−0.01650†	−0.00041	−0.1
0.2	0.02000	−0.03200†	−0.00160	−0.2
0.3	0.04500	−0.04550†	−0.00341	−0.3
0.4	0.08000	−0.05600†	−0.00560	−0.4
0.5	0.12500	−0.06250†	−0.00781	−0.5
0.6	0.18000	−0.06400†	−0.00960	−0.6
0.7	0.24500	−0.05950†	−0.01041	−0.7
0.8	0.32000	−0.04800†	−0.00960	−0.8
0.9	0.40500	−0.02850†	−0.00641	−0.9
1.0	0.50000	0.00000	0.00000	−1.0

† Change sign when reading s from right-hand column.

BESSEL INTERPOLATION

$$f_s \approx \mu f_{1/2} + (s - \tfrac{1}{2})\,\delta f_{1/2} + C_2(s)\,\mu\delta^2 f_{1/2} + C_3(s)\,\delta^3 f_{1/2} + C_4(s)\,\mu\delta^4 f_{1/2} + C_5(s)\,\delta^5 f_{1/2}$$

s	$C_2(s)$	$C_3(s)$	$C_4(s)$	$C_5(s)$	s
0	0.00000	0.00000	0.00000	0.00000	1.0
0.1	−0.04500	0.00600†	0.00784	−0.00063†	0.9
0.2	−0.08000	0.00800†	0.01440	−0.00086†	0.8
0.3	−0.10500	0.00700†	0.01934	−0.00077†	0.7
0.4	−0.12000	0.00400†	0.02240	−0.00045†	0.6
0.5	−0.12500	0.00000	0.02344	0.00000	0.5

† Change sign when reading s from right-hand column.

EVERETT INTERPOLATION

$$f_s \approx (1 - s)f_0 + C_2(s)\,\delta^2 f_0 + C_4(s)\,\delta^4 f_0 + sf_1 + C_2(1 - s)\,\delta^2 f_1 + C_4(1 - s)\,\delta^4 f_1$$

s	$C_2(s)$	$C_4(s)$
0	0.00000	0.00000
0.1	−0.02850	0.00455
0.2	−0.04800	0.00806
0.3	−0.05950	0.01044
0.4	−0.06400	0.01165
0.5	−0.06250	0.01172
0.6	−0.05600	0.01075
0.7	−0.04550	0.00890
0.8	−0.03200	0.00634
0.9	−0.01650	0.00329
1.0	0.00000	0.00000

STEFFENSEN INTERPOLATION†

$$f_s \approx f_0 + C_1(s)\,\delta f_{1/2} + C_3(s)\,\delta^3 f_{1/2} - C_1(-s)\,\delta f_{-1/2} - C_3(-s)\,\delta^3 f_{-1/2}$$

s	$C_1(s)$	$C_3(s)$
−0.5	−0.12500	0.02344
−0.4	−0.12000	0.02240
−0.3	−0.10500	0.01934
−0.2	−0.08000	0.01440
−0.1	−0.04500	0.00784
0	0.00000	0.00000
0.1	0.05500	−0.00866
0.2	0.12000	−0.01760
0.3	0.19500	−0.02616
0.4	0.28000	−0.03360
0.5	0.37500	−0.03906

† See Prob. 19.

NEWTON INTERPOLATION

$$f_s \approx f_0 + s\,\Delta f_0 + C_2(s)\,\Delta^2 f_0 + C_3(s)\,\Delta^3 f_0 + C_4(s)\,\Delta^4 f_0 + C_5(s)\,\Delta^5 f_0$$
$$f_{N-s} \approx f_N - s\,\nabla f_N + C_2(s)\,\nabla^2 f_N - C_3(s)\,\nabla^3 f_N + C_4(s)\,\nabla^4 f_N - C_5(s)\,\nabla^5 f_N$$

(s positive for interpolation)

s	$C_2(s)$	$C_3(s)$	$C_4(s)$	$C_5(s)$
−1.0	1.00000	−1.00000	1.00000	−1.00000
−0.9	0.85500	−0.82650	0.80584	−0.78972
−0.8	0.72000	−0.67200	0.63840	−0.61286
−0.7	0.59500	−0.53550	0.49534	−0.46562
−0.6	0.48000	−0.41600	0.37440	−0.34445
−0.5	0.37500	−0.31250	0.27344	−0.24609
−0.4	0.28000	−0.22400	0.19040	−0.16755
−0.3	0.19500	−0.14950	0.12334	−0.10607
−0.2	0.12000	−0.08800	0.07040	−0.05914
−0.1	0.05500	−0.03850	0.02984	−0.02447
0	0.00000	0.00000	0.00000	0.00000
0.1	−0.04500	0.02850	−0.02066	0.01612
0.2	−0.08000	0.04800	−0.03360	0.02554
0.3	−0.10500	0.05950	−0.04016	0.02972
0.4	−0.12000	0.06400	−0.04160	0.02995
0.5	−0.12500	0.06250	−0.03906	0.02734
0.6	−0.12000	0.05600	−0.03360	0.02285
0.7	−0.10500	0.04550	−0.02616	0.01727
0.8	−0.08000	0.03200	−0.01760	0.01126
0.9	−0.04500	0.01650	−0.00866	0.00537
1.0	0.00000	0.00000	0.00000	0.00000

4.13 Supplementary References

Standard references include Steffensen [1950], Milne-Thomson [1951], Nörlund [1954], and Jordan [1965].

H. T. Davis [1963] includes tables of interpolation coefficients for the formulas of Newton, Stirling, Bessel, and Everett, together with corresponding coefficients for numerical differentiation. Other tabulations are listed in the index by Fletcher et al. [1962]. Davis also lists formulas which provide approximations to the results of inverting truncated Newton and Everett formulas for the purpose of inverse interpolation [see also Prob. 30 (this chapter) and Sec. 2.9]. Salzer [1943, 1944*b*] gives tables of the relevant coefficient functions.

Miller [1950] gives a valuable discussion of the use of difference tables in the detection of errors. See Hamming [1971] for a probabilistic treatment of the effects of inherent roundoff errors on difference columns. Comrie [1956] includes additional throwback techniques.

The divergence of the interpolation sequence for $f(x) = 1/(x^2 + 1)$ in $[-5, 5]$ investigated by Runge [1901] is also considered by Steffensen [1950]. The example of $f(x) = |x|$ in $[-1, 1]$ is studied by Bernstein [1912*b*]. See also Cheney [1966]. Conditions ensuring convergence of interpolation sequences (and of their integrals) in finite intervals (see Prob. 42) are established by Krylov [1962]. General results relative to interpolation series are obtained by Nörlund [1926, 1954].

Interpolation in two-way tables [see Prob. 24 (this chapter) and Chap. 5, Probs. 4 and 5] is treated by Steffensen [1950], Willers [1950], and, in more detail, Pearson [1920*b*]. For more recent contributions, see Southard [1956], Thacher [1960], and Thacher and Milne [1960].

PROBLEMS

Section 4.2

1 Show that

$$\Delta^r f_k = \nabla^r f_{k+r} = \delta^r f_{k+r/2}$$
$$\nabla^r f_k = \Delta^r f_{k-r} = \delta^r f_{k-r/2}$$
$$\delta^r f_k = \Delta^r f_{k-r/2} = \nabla^r f_{k+r/2}$$
$$\Delta(f_{k-1}\,\Delta g_{k-1}) = \nabla(f_k\,\Delta g_k) = \Delta(f_{k-1}\,\nabla g_k) = \nabla(f_k\,\nabla g_{k+1})$$
$$\Delta\nabla f_k = \nabla\Delta f_k = \delta^2 f_k$$

2 Show that

$$\Delta(f_k g_k) = f_k\,\Delta g_k + g_{k+1}\,\Delta f_k \qquad \Delta(f_k^2) = (f_k + f_{k+1})\,\Delta f_k$$
$$\Delta\left(\frac{f_k}{g_k}\right) = \frac{g_k\,\Delta f_k - f_k\,\Delta g_k}{g_k g_{k+1}} \qquad \Delta\left(\frac{1}{f_k}\right) = -\frac{\Delta f_k}{f_k f_{k+1}}$$

3 Show that

$$\Delta^r\left(\frac{1}{x}\right) = \frac{(-1)^r r!\,h^r}{x(x+h)\cdots(x+rh)}$$
$$\Delta\cos(\omega x + \alpha) = 2\sin\frac{\omega h}{2}\cos\left(\omega x + \alpha + \frac{\omega h}{2} + \frac{\pi}{2}\right)$$
$$\Delta^r\cos(\omega x + \alpha) = \left(2\sin\frac{\omega h}{2}\right)^r\cos\left(\omega x + \alpha + \frac{r\omega h}{2} + \frac{r\pi}{2}\right)$$

Section 4.3

4 Calculate approximate values of $f(x) = \sin x$ for $x = 0.50(0.02)0.70$ and for $1.50(0.02)1.70$, by applying the appropriate newtonian formula to the following rounded data:

x	0.5	0.7	0.9	1.1	1.3	1.5	1.7
$f(x)$	0.47943	0.64422	0.78333	0.89121	0.96356	0.99749	0.99166

5 Obtain the formulas

$$hf'_s = \Delta f_0 + \tfrac{1}{2}(2s - 1)\,\Delta^2 f_0 + \tfrac{1}{6}(3s^2 - 6s + 2)\,\Delta^3 f_0 + \tfrac{1}{12}(2s^3 - 9s^2 + 11s - 3)\,\Delta^4 f_0 + \tfrac{1}{120}(5s^4 - 40s^3 + 105s^2 - 100s + 24)\,\Delta^5 f_0 + \cdots$$

and

$$\frac{1}{h}\int_{x_0}^{x_0+sh} f(x)\,dx = sf_0 + \tfrac{1}{2}s^2\,\Delta f_0 + \tfrac{1}{12}s^2(2s - 3)\,\Delta^2 f_0 + \tfrac{1}{24}s^2(s - 2)^2\,\Delta^3 f_0 + \tfrac{1}{720}s^2(6s^3 - 45s^2 + 110s - 90)\,\Delta^4 f_0 + \tfrac{1}{1440}s^2(2s^4 - 24s^3 + 105s^2 - 200s + 144)\,\Delta^5 f_0 + \cdots$$

and also obtain corresponding formulas for hf'_{N+s} and for $h^{-1}\int_{x_N-sh}^{x_N} f(x)\,dx$ in terms of backward differences.

6 Use the data of Prob. 4 and the results of Prob. 5 to obtain approximate values of $f'(0.6), f'(1.6), f''(0.6), f''(1.6)$, and of $\int_{0.5}^{0.6} f(x)\,dx$, $\int_{1.6}^{1.7} f(x)\,dx$.

Section 4.4

7 Show that Gauss' forward and backward formulas (without error terms) are truncations of the formal series

$$f_s \sim f_0 + \binom{s}{1}\delta f_{1/2} + \sum_{r=1}^{\infty}\left[\binom{s+r-1}{2r}\delta^{2r} f_0 + \binom{s+r}{2r+1}\delta^{2r+1} f_{1/2}\right]$$

and

$$f_s \sim f_0 + \binom{s}{1}\delta f_{-1/2} + \sum_{r=1}^{\infty}\left[\binom{s+r}{2r}\delta^{2r} f_0 + \binom{s+r}{2r+1}\delta^{2r+1} f_{-1/2}\right]$$

respectively.

8 Calculate approximate values of $f(1.0)$ from the data of Prob. 4, first by use of Gauss' forward formula launched from $x = 0.9$, and second by use of the backward formula launched from $x = 1.1$.

9 By specializing Sheppard's rules for the formation of interpolation formulas to the case when the relevant abscissas are at a uniform spacing h, show that *the coefficient of the kth difference encountered in a continuous difference path can be obtained by dividing by k! the product of k factors, each of which represents the distance between the abscissa of the interpolant and one of the abscissas lying in the region of determination of the preceding difference in the path, in units of the spacing,* if the result of truncating the interpolation formula with the kth difference is to yield exact results at all points involved in its formation. Also, illustrate the use of this rule by writing down the forward and backward formulas of Newton and Gauss.

10 Show that the result of truncating the Gauss forward formula with the fourth difference can be written in the form

$$f_s \approx f_0 + s\left\{\delta f_{1/2} + \frac{s-1}{2}\left[\delta^2 f_0 + \frac{s+1}{3}\left(\delta^3 f_{1/2} + \frac{s-2}{4}\,\delta^4 f_0\right)\right]\right\}$$

where the evaluation of the formula is conveniently effected from right to left, and write the backward formula, as well as the two Newton formulas, in similar forms.

Section 4.5

11 Show that Stirling's formula is a truncation of the formal series

$$f_s \sim f_0 + \binom{s}{1}\mu\delta f_0 + \sum_{r=1}^{\infty}\left[\frac{s}{2r}\binom{s+r-1}{2r-1}\delta^{2r} f_0 + \binom{s+r}{2r+1}\mu\delta^{2r+1} f_0\right]$$

12 Use Stirling's formula to calculate approximate values of $f(x)$ for the points $x = 1.00(0.02)1.20$ from the data of Prob. 4.

13 Obtain the formulas

$$hf_s' = \mu\delta f_0 + s\,\delta^2 f_0 + \tfrac{1}{6}(3s^2 - 1)\,\mu\delta^3 f_0 + \tfrac{1}{12}s(2s^2 - 1)\,\delta^4 f_0$$
$$+ \tfrac{1}{120}(5s^4 - 15s^2 + 4)\,\mu\delta^5 f_0 + \tfrac{1}{360}s(3s^4 - 10s^2 + 4)\,\delta^6 f_0 + \cdots$$

and

$$\frac{1}{h}\int_{x_0 - hs}^{x_0 + hs} f(x)\,dx = 2sf_0 + \tfrac{1}{3}s^3\,\delta^2 f_0 + \tfrac{1}{180}s^3(3s^2 - 5)\,\delta^4 f_0$$
$$+ \tfrac{1}{7560}s^3(3s^4 - 21s^2 + 28)\,\delta^6 f_0 + \cdots$$

and use them to calculate approximate values of $f'(1.1)$, $f'(1.0)$, $f''(1.1)$, $f''(1.0)$, and of $\int_{1.0}^{1.2} f(x)\,dx$, $\int_{0.9}^{1.3} f(x)\,dx$ from the data of Prob. 4.

Section 4.6

14 Show that Bessel's formula is a truncation of the formal series

$$f_0 \sim \mu f_{1/2} + (s - \tfrac{1}{2})\,\delta f_{1/2} + \sum_{r=1}^{\infty}\left[\binom{s+r-1}{2r}\mu\delta^{2r} f_{1/2}\right.$$
$$\left. + \frac{s-\frac{1}{2}}{2r+1}\binom{s+r-1}{2r}\delta^{2r+1} f_0\right]$$

15 Use Bessel's formula to calculate approximate values of $f(x)$ for the points $x = 0.90(0.02)1.10$ from the data of Prob. 4.

16 Obtain the formulas

$$hf_{t+(1/2)}' = \delta f_{1/2} + t\,\mu\delta^2 f_{1/2} + \tfrac{1}{24}(12t^2 - 1)\,\delta^3 f_{1/2} + \tfrac{1}{24}t(4t^2 - 5)\,\mu\delta^4 f_{1/2}$$
$$+ \tfrac{1}{1920}(80t^4 - 120t^2 + 9)\,\delta^5 f_{1/2} + \cdots$$

and

$$\frac{1}{h}\int_{x_{1/2}-th}^{x_{1/2}+th} f(x)\,dx = 2t\,\mu f_{1/2} + \tfrac{1}{12}t(4t^2 - 3)\,\mu\delta^2 f_{1/2}$$
$$+ \tfrac{1}{2880}t(48t^4 - 200t^2 + 135)\,\mu\delta^4 f_{1/2} + \cdots$$

where $x_{1/2} \equiv (x_0 + x_1)/2$, and use them to calculate approximate values of $f'(1.1)$, $f'(1.0)$, $f''(1.1)$, $f''(1.0)$, and of $\int_{0.9}^{1.1} f(x)\,dx$, $\int_{0.7}^{1.3} f(x)\,dx$ from the data of Prob. 4.

Section 4.7

17 Use Everett's first formula to obtain approximate values of $f(x) = \sin x$ for $x = 1.00(0.02)1.20$ from the following data:

x	$f(x)$	$\delta^2 f$	$\delta^4 f$
0.9	0.78333	−3123	125
1.1	0.89121	−3553	141
1.3	0.96356	−3842	155

18 By integrating Everett's first formula, obtain the formula

$$\frac{1}{h}\int_{x_0}^{x_1} f(x)\,dx = \mu f_{1/2} - \tfrac{1}{12}\,\mu\delta^2 f_{1/2} + \tfrac{11}{720}\,\mu\delta^4 f_{1/2} - \tfrac{191}{60480}\,\mu\delta^6 f_{1/2} + \cdots$$

and verify that it follows also from the result of Prob. 16.

19 Derive *Steffensen's formula* (Everett's second formula) in the form

$$f_s = f_0 + \frac{(s+1)s}{2!}\,\delta f_{1/2} + \frac{(s+2)(s+1)s(s-1)}{4!}\,\delta^3 f_{1/2} + \cdots$$
$$+ \frac{(s+m+1)(s+m)\cdots(s-m)}{(2m+2)!}\,\delta^{2m+1} f_{1/2}$$
$$- \frac{s(s-1)}{2!}\,\delta f_{-1/2} - \frac{(s+1)s(s-1)(s-2)}{4!}\,\delta^3 f_{-1/2} - \cdots$$
$$- \frac{(s+m)(s+m-1)\cdots(s-m-1)}{(2m+2)!}\,\delta^{2m+1} f_{-1/2} + E_s$$

where

$$E_s = h^{2m+3}\,\frac{s(s^2-1^2)\cdots(s^2-\overline{m+1}^2)}{(2m+3)!}\,f^{(2m+3)}(\xi)$$

Also show that it can be put in the form

$$f_s \approx f_0 + \binom{s+1}{2}\delta f_{1/2} + \binom{s+2}{4}\delta^3 f_{1/2} + \cdots$$
$$- \binom{-s+1}{2}\delta f_{-1/2} - \binom{-s+2}{4}\delta^3 f_{-1/2} - \cdots$$

so that one set of coefficients is obtained from the other by replacing s by $-s$.

20 Use Steffensen's formula to obtain approximate values of $f(x) = \sin x$ for $x = 1.00(0.02)1.20$ from the following data:

x	$f(x)$	$\delta^3 f$
0.9	0.78333	
		−430
1.1	0.89121	
		−289
1.3	0.96356	

Section 4.8

21 Sketch the function $\pi(x) = x(x - 1)(x - 2)(x - 3)(x - 4)$ over the interval $-1 \leqq x \leqq 5$. Noticing that the error associated with the approximation of $f(x)$ by the result of retaining fourth differences in either of Newton's or Gauss' interpolation formulas is of the form $\pi(x)f^{\text{v}}(\xi)/120$ for some ξ, if $f^{\text{v}}(x)$ is continuous, when the ordinates at $x = 0$, 1, 2, 3, and 4 are employed, and assuming that Newton's formulas would be used principally in [0, 1] and [3, 4], whereas central-difference formulas would be used principally in [1, 3], account for the fact that the former are sometimes erroneously said to be "less accurate" than the latter. If interpolations were effected in the interval [1, 3] by both Newton and Gauss formulas, based on the five ordinates at $x = 0$, 1, 2, 3, and 4, and if no roundoffs were committed, how would the results actually compare in accuracy? What evidence is afforded by the graph with respect to the general relative dependability of interpolation and extrapolation?

22 If $f(x) = (1 + x)^5$, determine the Stirling and Bessel approximations over [0, 1] corresponding to a spacing $h = 1$, with $x_0 = 0$ and $x_1 = 1$, and corresponding to the successive retention of differences through the first, second, third, and fourth. Then calculate the error in each of these eight approximations for $x = 0.0(0.2)1.0$, retaining only one decimal place, and plot the error curves in a common graph. Thus show that, in this example, the following facts are true over [0, 1]:

(*a*) The Stirling mean first-difference approximation is better than that which also incorporates the second difference over most of the interval, and the mean third-difference approximation is better than that which also incorporates the fourth difference over the entire interval.

(*b*) The Bessel mean second-difference approximation is better than that which also incorporates the third difference over half the interval.

(*c*) The Stirling mean first-difference approximation is better than the three Bessel approximations which employ the first difference, the mean second difference, and/or the third difference, near $x = \frac{1}{2}$ as well as near $x = 0$.

(*d*) The Bessel fourth-difference approximation is much better than all the others and is followed successively by the Stirling third-difference and the Stirling fourth-difference approximations.

(Compare the results of Prob. 23.)

23 Proceed as in Prob. 22 with $f(x) = \cos(\pi x/8)$, retaining five decimal places. Thus show that, in this example, the following facts are true over $[0, 1]$:

(*a*) The Stirling zeroth-difference approximation is better than the Bessel first-difference approximation over half the interval, whereas the Bessel mean zeroth-difference approximation is better than the Stirling mean first-difference approximation over the remainder of the interval.

(*b*) The Bessel mean second-difference approximation is better than that which also incorporates the third difference over half the interval.

(*c*) The Stirling second-difference approximation is better than the Bessel third-difference approximation over most of the interval.

(*d*) The Stirling fourth-difference approximation is better than all the others and is followed successively by the Bessel fourth-difference and the Stirling second-difference approximations.

(Compare the results of Prob. 22.)

24 The following data represent rounded four-place values of the elliptic-integral function $E(x, y) = \int_0^y \sqrt{1 - \sin^2 x \sin^2 t}\, dt$:

y \ x	50°	54°	58°	62°
50°	0.8134	0.8060	0.7988	0.7920
52°	0.8414	0.8332	0.8251	0.8174
54°	0.8690	0.8598	0.8508	0.8422
56°	0.8962	0.8859	0.8759	0.8663

Determine an approximation to $E(52°, 51°)$ by (*a*) interpolating horizontally to obtain $E(52°, y)$ for $y = 50°(2°)56°$ and then interpolating these values vertically, (*b*) interpolating vertically then horizontally, and (*c*) interpolating directly along a diagonal. Also interpolate as accurately as possible for $E(55.4°, 53.1°)$ by any method.

Section 4.9

25 Construct a difference table, corresponding to the results of rounding true values of $f(x) = x^3$ for $x = 1.0(0.1)3.0$ to two decimal places, and study the propagated effects of the roundoff errors. Also compare the mean absolute values of the third, fourth, and fifth differences with the ideal values, and show that a more regular difference array would result from an *improper* rounding of the values corresponding to $x = 1.4$ and $x = 2.6$ by one unit.

26 Certain of the following 20 consecutive values, corresponding to equally spaced arguments, are incorrect because of typical copying errors. Locate the errors and correct them.

17278	48818	79779	112630
23424	54440	86249	119398
29585	60723	92752	126246
35764	67041	99318	133180
41964	73398	105937	140206

Section 4.10

27 Show that the additional error $R(s)$ introduced into Bessel's formula, by replacing $\mu\delta^2 f_{1/2}$ by $\mu(\delta^2 f_{1/2} - k\,\delta^4 f_{1/2})$ and neglecting $\mu\delta^4 f_{1/2}$ otherwise, can be expressed in the form

$$R(s) = \tfrac{1}{24}[(s^2 - s)^2 + (12k - 2)(s^2 - s)]\,\mu\delta^4 f_{1/2}$$

and that the extreme values of the coefficient of $\mu\delta^4 f_{1/2}$ for $0 \leqq s \leqq 1$ occur when $s = \frac{1}{2}$ and when $s^2 - s = 1 - 6k$ and are given by $(3 - 16k)/128$ and $-(1 - 6k)^2/24$, respectively. Show also that the requirement that the extreme values be of equal magnitude and opposite sign (so that the maximum additional error is minimized) gives $k = (3 + \sqrt{2})/24 \doteq 0.184$, and that $R(s)$ then varies between the limits $\pm(3 - 2\sqrt{2})\mu\delta^4 f_{1/2}/384 \doteq \pm 0.00045\mu\delta^4 f_{1/2}$.

28 Show that the additional error $R(s)$ introduced into Everett's first formula, by replacing $\delta^2 f_0$ and $\delta^2 f_1$ by $\delta^2 f_0 - k\,\delta^4 f_0$ and $\delta^2 f_1 - k\,\delta^4 f_1$ and neglecting fourth differences otherwise, is identical with that associated with Bessel's formula when $\delta^4 f_0 = \delta^4 f_1$; hence deduce that if k is assigned the same value as for Bessel's formula (Prob. 27), then the additional error cannot exceed 0.00045 times the larger of $|\delta^4 f_0|$ and $|\delta^4 f_1|$ if the two fourth differences have the same sign. Show also that if those differences are equal in magnitude and of opposite sign, then $R(s)$ is given by

$$R(s) = \pm\frac{2s - 1}{15}\left[3\binom{s+1}{4} + 5k\binom{s}{2}\right]\delta^4 f_0$$

$$= \pm\tfrac{1}{1920}[t^5 - 10t^3 + 9t + 80k(t^3 - t)]\,\delta^4 f_0 \qquad (t = 2s - 1)$$

and show that the maximum additional error for $0 \leqq s \leqq 1$ is smaller than $0.00122\,|\delta^4 f_0|$ in magnitude. Thus deduce that $|R(s)| < 0.00122M$ in the general case, where M is the larger of $|\delta^4 f_0|$ and $|\delta^4 f_1|$.

29 Solve Prob. 17 by using the throwback technique.

30 Show that Everett's modified second-difference formula can be expressed in the form

$$s \approx \frac{1}{f_1 - f_0}\left\{(f_s - f_0) + \frac{s(s-1)}{6}[(s - 2)\,\bar{\delta}^2 f_0 - (s + 1)\,\bar{\delta}^2 f_1]\right\}$$

for *inverse interpolation* between f_0 and f_1, when f_s is given. Use it iteratively, first replacing s by zero in the coefficients in the right-hand member to calculate an initial approximation to the required value of s, and then successively introducing each new approximation into the right-hand member to obtain the next one, to determine approximately the value of x for which $f(x) = 0.9$ with the data of Prob. 17.

Section 4.11

31 Show that the error corresponding to the truncation of the series in (4.11.5) with the nth term is of the form

$$E_n(x) = \frac{a^{n+1}}{(n+1)!} x(x-h)\cdots(x-nh)e^{a\xi_n}$$

for some ξ_n, and deduce that, if $x < nh$, the errors $E_n(x)$ and $E_{n+1}(x)$ are of opposite sign, so that the error is smaller than the first term omitted and of the same sign (see Prob. 6 of Chap. 1). Under the assumption that $e^{ah} > 2$, so that (4.11.5) diverges, show also that the term corresponding to the $(k+1)$th difference is smaller than the preceding one so long as k does not exceed k_0, where k_0 is the integral part of $[(e^{ah}-1)x + h]/[h(e^{ah}-2)]$.

32 Illustrate the results of Prob. 31 by calculating successive approximations to e^x from successive partial sums of the Newton interpolation series (4.11.5) with $a = 1$ and $h = 1$, when $x = 0.5$. In particular, show that the best approximation is afforded by retention of only two differences, that a consideration of the first neglected term gives the result $1.49008 < e^{0.5} < 1.80716$, and that the *mean* of these limits differs from the true value by one unit in the fourth decimal place.

33 If $f(x) = e^{ax}$, show that

$$\delta^{2r} f(0) = \left(2 \sinh \frac{ah}{2}\right)^{2r} \qquad \mu\delta^{2r+1} f(0) = \left(2 \sinh \frac{ah}{2}\right)^{2r} \sinh ah$$

and deduce that the formal Stirling series centered at $x = 0$ is of the form

$$\begin{aligned} e^{ax} \sim 1 + \frac{x}{h}\sinh ah &+ \frac{x^2}{2!\,h^2}\left(2\sinh\frac{ah}{2}\right)^2 \\ &+ \frac{x(x^2-h^2)}{3!\,h^3}\left(2\sinh\frac{ah}{2}\right)^2 \sinh ah \\ &+ \frac{x^2(x^2-h^2)}{4!\,h^4}\left(2\sinh\frac{ah}{2}\right)^4 + \cdots \end{aligned}$$

34 Calculate six successive approximations to the value of e^x when $x = 0.5$ by use of Stirling's formula centered at $x = 0$, with $h = 1$, and investigate the trend of the successive deviations from the true value. (Notice that the infinite Stirling series itself is convergent in this case.)

35 By successively equating the even and odd parts of the two members of the expansion of Prob. 33 and taking $h = 1$, deduce the formal expansions

$$\cosh ax \sim 1 + \frac{x^2}{2!}\beta^2 + \frac{x^2(x^2-1^2)}{4!}\beta^4 + \frac{x^2(x^2-1^2)(x^2-2^2)}{6!}\beta^6 + \cdots$$

and

$$\begin{aligned} \frac{\sinh ax}{\sinh a} &= \frac{1}{\beta}\frac{\sinh ax}{\cosh (a/2)} \\ &\sim x + \frac{x(x^2-1^2)}{3!}\beta^2 + \frac{x(x^2-1^2)(x^2-2^2)}{5!}\beta^4 + \cdots \end{aligned}$$

where $\beta = 2 \sinh (a/2)$. Show also that these series converge when $|\beta| < 2$.

36 Show that the formal Bessel-series representation of $f(x) = e^{ax}$, centered at $x = h/2$, is of the form

$$e^{ax} \sim e^{ah/2}\left[\cosh\frac{ah}{2} + \frac{2x-h}{2h}\beta + \frac{x(x-h)}{2!\,h^2}\beta^2\cosh\frac{ah}{2} + \frac{x(x-h)(2x-h)}{2\cdot 3!\,h^3}\beta^3 + \cdots\right]$$

where $\beta = 2\sinh(ah/2)$. Also, by taking $h = 1$, replacing x by $x + \frac{1}{2}$, and successively equating the even and odd parts of the result, deduce the expansions

$$\frac{\cosh ax}{\cosh(a/2)} \sim 1 + \frac{x^2 - \frac{1}{4}}{2!}\beta^2 + \frac{(x^2-\frac{1}{4})(x^2-\frac{9}{4})}{4!}\beta^4 + \cdots$$

and

$$\sinh ax = \frac{\beta}{2}\frac{\sinh ax}{\sinh(a/2)} \sim x\beta + \frac{x(x^2-\frac{1}{4})}{3!}\beta^3 + \frac{x(x^2-\frac{1}{4})(x^2-\frac{9}{4})}{5!}\beta^5 + \cdots$$

where $\beta = 2\sinh(a/2)$, and show that these series converge when $|\beta| < 2$.

37 Let $f(x)$ be a function such that $f(kh) = 0$ ($k = \pm 1, \pm 2, \ldots$) and $f(0) = 1$.

(*a*) Obtain the formal Newton series

$$f(x) \sim 1 - \frac{x}{1!\,h} + \frac{x(x-h)}{2!\,h^2} - \cdots + (-1)^n\frac{x(x-h)\cdots(x-\overline{n-1}h)}{n!\,h^n} + \cdots$$

and the corresponding Stirling series

$$f(x) \sim 1 - \frac{x^2}{(1!\,h)^2} + \frac{x^2(x^2-h^2)}{(2!\,h^2)^2} - \cdots + (-1)^n\frac{x^2(x^2-h^2)\cdots(x^2-\overline{n-1}^2h^2)}{(n!\,h^n)^2} + \cdots$$

(*b*) Show that the Newton series converges for $x \geqq 0$ and diverges when $x < 0$, and that the Stirling series converges for *all* x.†

† Use may be made of a test associated with Raabe and Gauss, which states that if all terms of a real series are of the same sign after a certain point, and if the ratio of the $(n+1)$th term to the nth term can be expressed in the form

$$1 - \frac{c}{n} + \frac{\theta_n}{n^2}$$

where c is independent of n and θ_n is bounded as $n \to \infty$, then the series converges if $c > 1$ and diverges otherwise unless it terminates. (See Knopp [1956].)

(*c*) Verify that the Newton series can be obtained as the limit as $t \to 1-$ of the binomial expansion of $(1 - t)^{x/h}$ in powers of t, and deduce that the series converges to a function $f(x)$ such that $f(x) = 0$ when $x > 0$ and $f(0) = 1$. [The Stirling series, on the other hand, can be shown to converge to the function $f(x) = (\sin \pi x/h)/(\pi x/h)$.]

(Note, for example, that accordingly the effect on the value of the nth interpolation polynomial at $x = h/2$, due to an isolated error ε in the ordinate at $x = 0$, would tend toward $2\varepsilon/\pi \approx 0.64\varepsilon$ as n increases in the Stirling sequence, but would tend to *zero* in the Newton sequence, when both sequences are "launched" from $x = 0$.)

38 If $f(z)$ is analytic in a simply-connected region $\mathscr{R}$ of the complex plane which includes the segment of the real axis spanned by the points $x_0, x_1, \ldots, x_n$, and x, and if C is a simple closed curve in $\mathscr{R}$ which surrounds this segment, use the fact that

$$f(x) = \frac{1}{2\pi i} \oint_C \frac{f(z)\,dz}{z - x}$$

to deduce that

$$f[x_0, x_1, \ldots, x_n, x] = \frac{1}{2\pi i} \oint_C \frac{f(z)\,dz}{\pi_n(z)(z - x)}$$

where

$$\pi_n(z) = (z - x_0)(z - x_1)\cdots(z - x_n)$$

(Review Prob. 17 of Chap. 2.)

39 Apply the calculus of residues to the result of Prob. 38 to derive Lagrange's polynomial interpolation identity in the familiar form

$$f(x) - \sum_{k=0}^{n} \frac{\pi_n(x)}{(x - x_k)\pi_n'(x_k)} f(x_k) = E_n(x)$$

but with the error term

$$E_n(x) = \pi_n(x) f[x_0, x_1, \ldots, x_n, x]$$

expressed in the alternative form

$$E_n(x) = \frac{1}{2\pi i} \oint_C \frac{\pi_n(x) f(z)\,dz}{\pi_n(z)(z - x)}$$

Deduce also that

$$|E_n(x)| \leqq \frac{ML}{2\pi d} \max \left| \frac{\pi_n(x)}{\pi_n(z)} \right|$$

where $|f(z)| \leqq M$ on C, L is the length of C, d is the shortest distance from the real point x to a point on C, and where, for a given real value of x, $|\pi_n(x)/\pi_n(z)|$ is to be maximized for all z on C.

40 In Prob. 39, suppose that the points $x_0, x_1, \ldots, x_n$ are equally spaced and span a fixed interval $[a, b]$ so that, with $h_n = (b - a)/n$, there follows $x_k = x_0 + kh_n$, with $x_0 = a$ and $x_n = b$. Suppose also that x is in $[a, b]$. Show that

$$h_n \log |\pi_n(x)| = \frac{b-a}{n} \sum_{k=0}^{n} \log |x - x_k|$$

and hence that

$$\lim_{n\to\infty} [h_n \log |\pi_n(x)|] = \int_a^b \log |x - t| \, dt$$

41 With the abbreviation

$$v(x) = \frac{1}{b-a} \int_a^b \log \left(\frac{b-a}{|x-t|} \right) dt$$

deduce from the result of Prob. 40 that

$$\left| \frac{\pi_n(z)}{\pi_n(x)} \right| \sim e^{-n[v(x)-v(z)]} \qquad (n \to \infty)$$

and hence, with the notation of Prob. 39,

$$|E_n(x)| \leqq K_n$$

where

$$K_n \sim \frac{ML}{2\pi d} e^{-n[v(x)-v(z)]} \qquad (n \to \infty)$$

so that the interpolation error $E_n(x)$ tends to zero as $n \to \infty$ for all x in $[a, b]$ if C can be so chosen that $v(z) < v(x)$ for all such x and for all z on C.

42 With the notation of Prob. 41, verify that

$$\begin{aligned} v(z) &= \operatorname{Re} \left[\frac{1}{b-a} \int_a^b \log \left(\frac{b-a}{z-t} \right) dt \right] \\ &= \operatorname{Re} \left\{ 1 + \log (b-a) - \frac{1}{b-a} [(z-a) \log (z-a) \right. \\ &\qquad\qquad \left. + (b-z) \log (b-z)] \right\} \\ &= 1 + \log (b-a) - \frac{1}{b-a} \left[(x-a) \log \sqrt{(x-a)^2 + y^2} \right. \\ &\qquad\qquad + (b-x) \log \sqrt{(b-x)^2 + y^2} \\ &\qquad\qquad \left. - y \left(\tan^{-1} \frac{y}{b-x} + \tan^{-1} \frac{y}{x-a} \right) \right] \end{aligned}$$

Also show that the locus $v(z) = 1$ in the complex plane passes through the ends of the real interval $[a, b]$ and that at the midpoint of that interval $v(z) = 1 + \log 2$.

43 It can be shown that the locus $v(z) = 1$ in Prob. 42 is an oval curve Γ (somewhat resembling an ellipse) with longitudinal vertices at the ends of the real interval $[a, b]$, center at its midpoint, and lateral vertices at a distance of about $0.26(b - a)$ from the center. Also it is true that $v(z) > 1$ everywhere *inside* Γ

and $v(z) < 1$ everywhere *outside* Γ. Assuming these facts, deduce from the results of Probs. 38 to 42 that *if* $f(z)$ *is analytic in a region* $\mathcal{R}$ *including the curve* Γ, *then the polynomial interpolation sequence generated by fitting* $f(x)$ *at* $n + 1$ *equally spaced points in* $[a, b]$ *converges to* $f(x)$ *everywhere in* $[a, b]$ *and also the Newton-Cotes sequence of approximations converges to the integral* $\int_a^b f(x)\,dx$ *as* $n \to \infty$. (Notice also that it is sufficient to require that $f(z)$ be analytic *inside and on* Γ since then, by the definition of analyticity, $f(z)$ also would be analytic in a region $\mathcal{R}$ including Γ. To deduce convergence of the *integration* sequence, use must be made of the *uniformity* of the convergence of the interpolation sequence. For a detailed treatment, see Krylov [1962].)

5

OPERATIONS WITH FINITE DIFFERENCES

5.1 Introduction

The purpose of this chapter is twofold: first, to indicate the power and simplicity of operational methods in deriving a variety of formulas which are useful in various aspects of numerical analysis, and, second, to display certain such formulas for convenient reference and for use in following chapters.

The operational methods which are illustrated supply only the formulas themselves and do not furnish the relevant error terms, which therefore must be obtained independently. Many of the formulas also could be obtained by differentiating or integrating appropriate interpolation formulas, although it often would be somewhat more difficult to obtain the rule of formation of the general term in an expansion. However, in such cases, it is clearly possible to deduce the desired *error term* by differentiating or integrating the known error term relevant to the parent formula.

In addition to formulas for numerical differentiation and integration, generally expressed in terms of forward, backward, or central differences, there are included certain formulas which are useful in subtabulation (Sec. 5.7) and approximate summation of series (Secs. 5.8, 5.9). The concluding sections (Secs. 5.10 and 5.11) deal with the problem of determining an expression for the error term relevant to a formula, when the coefficients in the formula are known.

5.2 Difference Operators

For many purposes, it is convenient to think of the symbols, Δ, ∇, and δ, defined in the preceding chapter, as *operators*, which transform a given function $f(x)$ into related functions, according to the laws

$$\Delta f(x) = f(x+h) - f(x) \qquad \nabla f(x) = f(x) - f(x-h)$$

$$\delta f(x) = f\left(x + \frac{h}{2}\right) - f\left(x - \frac{h}{2}\right) \tag{5.2.1}$$

Also, in addition to the *averaging operator* μ, such that

$$\mu f(x) = \frac{1}{2}\left[f\left(x + \frac{h}{2}\right) + f\left(x - \frac{h}{2}\right)\right] \tag{5.2.2}$$

we define the *shifting operator* $\mathbf{E}$ such that

$$\mathbf{E}f(x) = f(x+h) \tag{5.2.3}$$

and *differential* and *integral operators* $\mathbf{D}$ and $\mathbf{J}$ with the properties

$$\mathbf{D}f(x) = f'(x) \tag{5.2.4}$$

and

$$\mathbf{J}f(x) = \int_x^{x+h} f(t)\, dt \tag{5.2.5}$$

In all these operators except $\mathbf{D}$, the *spacing* h is implied. When a more explicit notation is needed, the spacing may be indicated as a subscript, so that, for example, we could write $\delta_{2h} f(x)$ for $f(x+h) - f(x-h)$. In addition to the operator $\mathbf{J}$, which is associated with Δ, one can define integral operators which correspond similarly to δ and to ∇; this will not be done here.

Positive integral powers of the operators are defined by iteration. Also, we define the *zero*th power of any operator as the *identity operator* 1, which leaves any function unchanged. For the operator $\mathbf{E}$, the power $\mathbf{E}^\alpha$ is defined for *any* α so that

$$\mathbf{E}^\alpha f(x) = f(x + \alpha h) \tag{5.2.6}$$

assuming the existence of $f(x + \alpha h)$.

We say that two operators, say, $\mathbf{L}_1$ and $\mathbf{L}_2$, are *equal* if $\mathbf{L}_1 f(x) = \mathbf{L}_2 f(x)$ for any function $f(x)$ for which the operations are defined. It is easily verified that the seven operators defined here possess the *commutative*, *distributive*, and *associative* properties shared by real numbers, so that if $\mathbf{L}_1$, $\mathbf{L}_2$, and $\mathbf{L}_3$ are any of these operators, there follows

$$\mathbf{L}_1 + \mathbf{L}_2 = \mathbf{L}_2 + \mathbf{L}_1 \qquad \mathbf{L}_2\mathbf{L}_1 = \mathbf{L}_1\mathbf{L}_2$$

$$\mathbf{L}_1(\mathbf{L}_2 + \mathbf{L}_3) = \mathbf{L}_1\mathbf{L}_2 + \mathbf{L}_1\mathbf{L}_3$$

$$(\mathbf{L}_1\mathbf{L}_2)\mathbf{L}_3 = \mathbf{L}_1(\mathbf{L}_2\mathbf{L}_3) \qquad (\mathbf{L}_1 + \mathbf{L}_2) + \mathbf{L}_3 = \mathbf{L}_1 + (\mathbf{L}_2 + \mathbf{L}_3) \tag{5.2.7}$$

The exponential law $\mathbf{L}^m\mathbf{L}^n = \mathbf{L}^{m+n}$ is also readily established for each of these operators.

In particular, to show that $\mathbf{D}$ and $\mathbf{J}$ are commutative, we make the calculations

$$\mathbf{DJ}f(x) = \frac{d}{dx}\int_x^{x+h} f(t)\,dt = f(x+h) - f(x) = \int_x^{x+h} \frac{df(t)}{dt}\,dt = \mathbf{JD}f(x)$$

and so deduce also that

$$\mathbf{DJ} = \mathbf{JD} = \Delta \tag{5.2.8}$$

We may define $\mathbf{L}^{-1}$ as an operator such that

$$\mathbf{LL}^{-1} = 1 \tag{5.2.9}$$

so that, if $\mathbf{L}^{-1}g = f$, then $\mathbf{LL}^{-1}g = \mathbf{L}f$ or $g = \mathbf{L}f$, and refer to $\mathbf{L}^{-1}$ as an *inverse* of $\mathbf{L}$. It is important, however, to notice that the inverse operator $\mathbf{L}^{-1}$ may not be uniquely defined. For if $\omega(x)$ is any function which is *annihilated* by $\mathbf{L}$, so that $\mathbf{L}\omega(x) \equiv 0$, and if one interpretation of $\mathbf{L}^{-1}g(x)$ is $f(x)$, so that $\mathbf{L}f(x) = g(x)$, then another one is $f(x) + \omega(x)$, since also $\mathbf{L}[f(x) + \omega(x)] = g(x)$. That is, we may write $\mathbf{L}^{-1}\mathbf{L}f = f + \omega$, where ω is *any* function annihilated by $\mathbf{L}$. Conversely, if $\mathbf{L}^{-1}\mathbf{L}f = f + \omega$, then it must follow that $\mathbf{LL}^{-1}\mathbf{L}f = \mathbf{L}f + \mathbf{L}\omega$ or $\mathbf{L}f = \mathbf{L}f + \mathbf{L}\omega$, so that $\mathbf{L}\omega$ *must* vanish.

If no function is annihilated by an operator $\mathbf{L}$, there follows

$$\mathbf{L}^{-1}\mathbf{L} = \mathbf{LL}^{-1} = 1$$

and $\mathbf{L}^{-1}$ is then said to be a *proper* inverse. Thus, whereas no function is annihilated by $\mathbf{E}$, the operators Δ, ∇, and δ annihilate any function of period h, $\mathbf{J}$ annihilates the derivative of any such function, and $\mathbf{D}$ annihilates any constant. Further, μ annihilates any so-called *odd-harmonic* function of period $2h$, that is, any function $f(x)$ for which $f(x + h) = -f(x)$. Hence, care must be taken with respect to the *order* of operations involving the inverses of those operators.

In the case of the operator $\mathbf{D}$, it is seen that $\mathbf{D}^{-1}$ corresponds to the formation of an indefinite integral, or *antiderivative*, and the situation described corresponds to the fact that, whereas the derivative of that integral is the original function, the integral of the derivative involves an arbitrary additive constant. On the other hand, it should be noticed that $\mathbf{\Delta D}^{-1}f(x)$ *is* uniquely determined since Δ annihilates the arbitrary constant. In fact, if we write

$$\mathbf{D}^{-1}f(x) = F(x) + C$$

we see that

$$\mathbf{\Delta D}^{-1}f(x) = F(x + h) - F(x) = \mathbf{J}f(x)$$

so that we may write also

$$\mathbf{\Delta D}^{-1} = \mathbf{J} \qquad (5.2.10)$$

This result follows also by using (5.2.8) to deduce that

$$\mathbf{\Delta D}^{-1} = \mathbf{JDD}^{-1} = \mathbf{J}$$

In the present chapter, we will be concerned principally with applying operators to *polynomials*. In this connection, we may notice that *each of the operators* Δ, ∇, δ, *and* $\mathbf{D}$ *reduces the degree of any polynomial*, and that the same statement is true of any positive integral powers of these operators. We will refer to such operators as *reductive operators*. (They are sometimes called *delta* or *theta* operators.) It should be noticed that $\mathbf{E}$, μ, and $\mathbf{J}$ are *not* reductive.

From the definitions given, we may obtain the relations

$$\Delta = \mathbf{E} - 1 \qquad \nabla = 1 - \mathbf{E}^{-1}$$
$$\delta = \mathbf{E}^{1/2} - \mathbf{E}^{-1/2} \qquad \mu = \tfrac{1}{2}(\mathbf{E}^{1/2} + \mathbf{E}^{-1/2}) \qquad (5.2.11)$$

whereas (5.2.8) leads also to the relation

$$\mathbf{DJ} = \mathbf{JD} = \mathbf{E} - 1 \qquad (5.2.12)$$

so that these operators are simply expressed in terms of $\mathbf{E}$.

Further, if r is any nonnegative integer, there follows

$$\begin{aligned}\Delta^r &= \mathbf{E}^r\nabla^r = \mathbf{E}^{r/2}\delta^r \\ &= (\mathbf{E} - 1)^r = \mathbf{E}^r - \frac{r}{1!}\mathbf{E}^{r-1} + \frac{r(r-1)}{2!}\mathbf{E}^{r-2} - \cdots \\ &\qquad + (-1)^{r-1}\frac{r}{1!}\mathbf{E} + (-1)^r\end{aligned}$$

and hence, by applying these equal operators to $y(x_k) \equiv y_k$, we obtain the useful formulas

$$\Delta^r y_k = y_{k+r} - \binom{r}{1} y_{k+r-1} + \binom{r}{2} y_{k+r-2} - \cdots + (-1)^{r-1}\binom{r}{1} y_{k+1} + (-1)^r y_k$$

$$\nabla^r y_k = y_k - \binom{r}{1} y_{k-1} + \binom{r}{2} y_{k-2} - \cdots + (-1)^{r-1}\binom{r}{1} y_{k-r+1} + (-1)^r y_{k-r}$$

$$\delta^r y_k = y_{k+r/2} - \binom{r}{1} y_{k+r/2-1} + \cdots + (-1)^{r-1}\binom{r}{1} y_{k-r/2+1} + (-1)^r y_{k-r/2} \qquad (5.2.13)$$

These relations permit the calculation of an arbitrary difference as a linear combination of ordinates, without the formation of a difference table or the calculation of differences of lower order, the coefficients of successive ordinates being merely binomial coefficients prefixed by alternating signs.

From the relations of (5.2.11), we may properly deduce the relations

$$\mathbf{E} - \Delta = 1 \qquad \mathbf{E}(1 - \nabla) = 1$$

$$(\mathbf{E}^{1/2} - \tfrac{1}{2}\delta)^2 - \tfrac{1}{4}\delta^2 = 1 \qquad \mathbf{E}^{1/2} - \tfrac{1}{2}\delta - \mu = 0$$

after which the formal symbolism of elementary algebra suggests the forms

$$\mathbf{E} = 1 + \Delta \qquad \mathbf{E} = \frac{1}{1 - \nabla}$$

$$\mathbf{E}^{1/2} = (1 + \tfrac{1}{4}\delta^2)^{1/2} + \tfrac{1}{2}\delta \qquad \mu = (1 + \tfrac{1}{4}\delta^2)^{1/2} \qquad (5.2.14)$$

While the first form requires no explanation, the term $1/(1 - \nabla)$ can be interpreted at this stage only as representing the *inverse* of the operator $1 - \nabla$, that is, as an alternative notation for the operator $(1 - \nabla)^{-1}$ such that

$$(1 - \nabla)(1 - \nabla)^{-1} = 1 \qquad (5.2.15)$$

whereas the derivation of the third form shows that $(1 + \frac{1}{4}\delta^2)^{1/2}$ is to represent an operator such that its *iterate* is the operator $1 + \frac{1}{4}\delta^2$

$$[(1 + \tfrac{1}{4}\delta^2)^{1/2}]^2 = 1 + \tfrac{1}{4}\delta^2 \qquad (5.2.16)$$

If we now suppose that the function upon which the operations are to be effected is a *polynomial* $p(x)$, of degree n, we may obtain a more useful interpretation of these operators. For, if t is a variable, we have the identity

$$(1 - t)(1 + t + t^2 + \cdots + t^n) = 1 - t^{n+1}$$

for any nonnegative integer n. Clearly, it is proper to replace t by ∇ (or by the symbol representing any other distributive operator), to give

$$(1 - \nabla)(1 + \nabla + \nabla^2 + \cdots + \nabla^n) = 1 - \nabla^{n+1} \qquad (5.2.17)$$

Since the operator ∇^{n+1} will *annihilate* $p(x)$, the operator in (5.2.17) is equivalent to the unit operator for any such $p(x)$. Since the inverse of $1 - \nabla$ is uniquely defined by (5.2.15), it follows that we may write

$$(1 - \nabla)^{-1} = 1 + \nabla + \nabla^2 + \cdots + \nabla^n$$

when only polynomials of degree n or less are to be affected by the operator. More generally, we are justified in writing

$$\mathbf{E} = (1 - \nabla)^{-1} = 1 + \nabla + \nabla^2 + \cdots = \sum_{k=0}^{\infty} \nabla^k \qquad (5.2.18)$$

when the class of *all* polynomials is included, since the finite number of required terms, for which the exponent of ∇ does not exceed the degree of the polynomial, is present, and the remaining terms each annihilate that polynomial.

It is useful to notice that the coefficients in (5.2.18) are those which would be present if the superscript -1, denoting the *inverse*, were interpreted as a *power* and if formal use then were made of the binomial expansion of the *reciprocal* of $1 - \nabla$.

In a similar way, it can be seen that if we retain only the terms which involve powers of δ which do not exceed n in the formal expansion

$$(1 + \tfrac{1}{4}\delta^2)^{1/2} = 1 + \tfrac{1}{8}\delta^2 - \tfrac{1}{128}\delta^4 + \cdots$$

the resultant *polynomial* in δ possesses the property that its square differs from $1 + \frac{1}{4}\delta^2$ by the product of δ^{n+1} and a polynomial in δ, and hence is equivalent to $1 + \frac{1}{4}\delta^2$ for any nth-degree $p(x)$. Clearly the *negative* of the indicated expansion also has this property. However, the result of applying the expanded form of the third relation in (5.2.14) to any arbitrarily chosen function (say a constant) shows that the former alternative is the proper one, so that we are justified in writing

$$\mathbf{E}^{1/2} = (1 + \tfrac{1}{4}\delta^2)^{1/2} + \tfrac{1}{2}\delta = 1 + \tfrac{1}{2}\delta + \tfrac{1}{8}\delta^2 - \cdots \qquad (5.2.19)$$

when we deal only with polynomials.

It then follows that we may write

$$\mathbf{E}^s = (1 + \Delta)^s = (1 - \nabla)^{-s} = [(1 + \tfrac{1}{4}\delta^2)^{1/2} + \tfrac{1}{2}\delta]^{2s} \tag{5.2.20}$$

when s is any integer, where each right-hand member may be expanded in a series of powers of the relevant reductive operator when we are dealing only with polynomials. The extension to the more general case when s is any rational number offers no difficulties. It is also possible to give a rigorous direct justification of the use of these expansions when s is irrational although the required arguments are somewhat more subtle.

The first two equivalences in (5.2.20) are, in fact, seen to be symbolic representations of the relations

$$p(x_0 + sh) = \sum_{k=0}^{\infty} \binom{s}{k} \Delta^k p(x_0) = \sum_{k=0}^{\infty} (-1)^k \binom{-s}{k} \nabla^k p(x_0) \tag{5.2.21}$$

to which the previously established Newton forward- and backward-difference formulas (4.3.11) and (4.3.13) reduce when $f(x)$ is replaced by a polynomial $p(x)$, since only a finite number of terms then do not vanish and since the remainder term also vanishes. This fact can be considered as constituting an indirect proof of the validity of the first two relations of (5.2.20) for a polynomial, when s is unrestricted.

As was discussed in Sec. 4.11, the series in (5.2.21) frequently do not converge when $p(x)$ is replaced by a function $f(x)$ other than a polynomial; they must be truncated, say, after $n + 1$ terms, and the appropriate error term (4.3.6) or (4.3.9) then must be added. However, the *coefficients* in the formula are not dependent upon the nature of $f(x)$, and the present operational methods serve to determine those coefficients in a simple and systematic way.

The equivalence of the extreme members of (5.2.20) can be expressed in a variety of forms, such as

$$\begin{aligned}\mathbf{E}^s &= [(1 + \tfrac{1}{4}\delta^2)^{1/2} + \tfrac{1}{2}\delta]^{2s} = [1 + \tfrac{1}{2}\delta^2 + \delta(1 + \tfrac{1}{4}\delta^2)^{1/2}]^s \\ &= (1 + \tfrac{1}{2}\delta^2 + \mu\delta)^s = (1 + \mathbf{E}^{1/2}\delta)^s = (1 - \mathbf{E}^{-1/2}\delta)^{-s}\end{aligned} \tag{5.2.22}$$

The operational formula obtained by expanding the first or second of these expressions would correspond to an interpolation employing the central differences $\delta^{2m+1}f(x_0)$ as well as the central differences of even order. Since the former differences generally are not available in *tabular* work, this formula would be of limited use. The Stirling formula could be obtained by expanding the third expression and afterward replacing μ^{2m} by $(1 + \delta^2/4)^m$ and μ^{2m+1} by $\mu(1 + \delta^2/4)^m$. The two gaussian formulas could be obtained from the remaining two expressions. Since the results have been obtained in the preceding chapter (see also Prob. 5), these calculations are omitted here. However, the use of operational methods in deriving formulas for interpolation in *two-way* tables is illustrated in Probs. 3 to 6. The derivations of formulas for other purposes are indicated in the following sections.

In the remainder of this chapter we shall proceed, in general, as though we were concerned only with polynomials, and often we shall emphasize this fact by writing $p(x)$ in place of $f(x)$. Formulas to be obtained then will have been established as *identities* for any polynomial $p(x)$, in which case all relevant series of operations by reductive operators will terminate. The determination of the error term to be introduced when the formula is applied to a function of more general type, after a *truncation* of the series involved, is then to be considered in each case as a separate problem.

5.3 Differentiation Formulas

In order to obtain formulas for numerical differentiation, by operational methods, it is necessary to relate $\mathbf{D}$ to other reductive operators. For this purpose, we notice that the familiar formula of the Taylor-series expansion

$$p(x + h) = p(x) + \frac{h}{1!}\,p'(x) + \frac{h^2}{2!}\,p''(x) + \cdots + \frac{h^n}{n!}\,p^{(n)}(x) + \cdots \qquad (5.3.1)$$

(which certainly is valid for any polynomial) can be written in the operational form

$$\mathbf{E}p(x) = \left(1 + \frac{h\mathbf{D}}{1!} + \frac{h^2\mathbf{D}^2}{2!} + \cdots + \frac{h^n\mathbf{D}^n}{n!} + \cdots\right) p(x)$$

Since the series in parentheses is the formal expansion of the function $e^{h\mathbf{D}}$, we deduce the curious relationship

$$\mathbf{E} = e^{h\mathbf{D}} \qquad (5.3.2)$$

which is to be interpreted as an abbreviation of the statement that the operators $\mathbf{E}$ and $1 + h\mathbf{D}/1! + \cdots + (h\mathbf{D})^n/n!$ are equivalent when applied to any polynomial $p(x)$ of degree n, for any n.

Further, we may obtain the additional relations

$$\begin{aligned} h\mathbf{D} &= \log \mathbf{E} = \log (1 + \Delta) = -\log (1 - \nabla) \\ &= 2 \log \left[(1 + \tfrac{1}{4}\delta^2)^{1/2} + \tfrac{1}{2}\delta\right] \equiv 2 \sinh^{-1} \frac{\delta}{2} \end{aligned} \qquad (5.3.3)$$

Here, for example, the symbolic relation $h\mathbf{D} = \log (1 + \Delta)$ asserts that the operators $h\mathbf{D}$ and $P_n(\Delta) \equiv \Delta - \Delta^2/2 + \cdots + (-1)^{n+1}\Delta^n/n$ are equivalent for any nth-degree $p(x)$. Its validity can be verified directly by noticing that since Δ and $h\mathbf{D}/1! + \cdots + (h\mathbf{D})^n/n!$ have been shown to be equivalent for any such $p(x)$, we may replace Δ by this last operator in the polynomial $P_n(\Delta)$. The result will differ from $h\mathbf{D}$ by a polynomial of the form $a_1(h\mathbf{D})^{n+1} + \cdots + a_n(h\mathbf{D})^{2n}$, which will annihilate $p(x)$.

In terms of *forward* differences, we thus deduce the formula

$$p_0' = \frac{1}{h}\log(1+\Delta)\,p_0 = \frac{1}{h}(\Delta - \tfrac{1}{2}\Delta^2 + \tfrac{1}{3}\Delta^3 - \cdots)p_0 \tag{5.3.4}$$

By iteration, there follows also

$$\begin{aligned} p_0^{(r)} &= \frac{1}{h^r}[\log(1+\Delta)]^r p_0 \\ &= \frac{1}{h^r}(1 - \tfrac{1}{2}\Delta + \tfrac{1}{3}\Delta^2 - \cdots)^r \Delta^r p_0 \\ &= \frac{1}{h^r}\Big[\Delta^r - \frac{r}{2}\Delta^{r+1} + \frac{r(3r+5)}{24}\Delta^{r+2} \\ &\qquad - \frac{r(r+2)(r+3)}{48}\Delta^{r+3} + \cdots\Big]p_0 \end{aligned} \tag{5.3.5}$$

The coefficients in this expansion are expressible in terms of the so-called *Stirling numbers of the first kind*, which may be denoted by $S_k^{(r)}$, and which are then defined by the relation

$$\frac{[\log(1+t)]^r}{r!} = \sum_{k=r}^{\infty} \frac{S_k^{(r)}}{k!} t^k \tag{5.3.6}$$

when $|t|$ is small, so that (5.3.5) can be written in the form

$$p_0^{(r)} = \frac{1}{h^r}\left[\frac{S_r^{(r)}}{1} + \frac{S_{r+1}^{(r)}}{r+1}\Delta + \frac{S_{r+2}^{(r)}}{(r+1)(r+2)}\Delta^2 + \cdots\right]\Delta^r p_0 \tag{5.3.7}$$

In a completely similar way, the corresponding *backward*-difference formulas are obtained in the form

$$p_N' = -\frac{1}{h}\log(1-\nabla)\,p_N = \frac{1}{h}(\nabla + \tfrac{1}{2}\nabla^2 + \tfrac{1}{3}\nabla^3 + \cdots)p_N \tag{5.3.8}$$

and

$$\begin{aligned} p_N^{(r)} &= \frac{1}{h^r}(1 + \tfrac{1}{2}\nabla + \tfrac{1}{3}\nabla^2 + \cdots)^r \nabla^r p_N \\ &= \frac{1}{h^r}\Big[\nabla^r + \frac{r}{2}\nabla^{r+1} + \frac{r(3r+5)}{24}\nabla^{r+2} \\ &\qquad + \frac{r(r+2)(r+3)}{48}\nabla^{r+3} + \cdots\Big]p_N \\ &\equiv \frac{1}{h^r}\left[\frac{S_r^{(r)}}{1} - \frac{S_{r+1}^{(r)}}{r+1}\nabla + \frac{S_{r+2}^{(r)}}{(r+1)(r+2)}\nabla^2 - \cdots\right]\nabla^r p_N \end{aligned} \tag{5.3.9}$$

From the last form of (5.3.3), there follows symbolically

$$p_0' = \left[\frac{2}{h}\sinh^{-1}\frac{\delta}{2}\right]p_0 \qquad (5.3.10)$$

and several types of *central*-difference expansions are possible. Since the right-hand member is an *odd* function of δ, its expansion in powers of δ would involve odd central differences, which are not generally useful in tabular interpolation.

To obtain a result equivalent to that of differentiating the Stirling formula, and evaluating the result at x_0, we require an expansion involving *mean* odd central differences. Hence, by multiplying the right-hand member by μ and dividing by its equivalent $\sqrt{1 + \delta^2/4}$, we obtain the form†

$$p_0' = \frac{2\mu}{h}\frac{\sinh^{-1}\delta/2}{\sqrt{1+\delta^2/4}}\,p_0$$

$$= \frac{\mu}{h}\left(\delta - \frac{1^2}{3!}\delta^3 + \frac{1^2\cdot 2^2}{5!}\delta^5 - \cdots\right)p_0 \qquad (5.3.11)$$

This formula is useful for calculating the derivative at interior tabular points, whereas (5.3.4) and (5.3.8) would be required at end points of the tabulation. Intermediate values are then conveniently interpolated from these values.

Higher derivatives of *even* order $2m$ are obtained by use of the operator $\mathbf{D}^{2m}$, where $\mathbf{D}$ is expressed as in (5.3.10),

$$\mathbf{D} = \frac{2}{h}\sinh^{-1}\frac{\delta}{2}$$

$$= \frac{1}{h}\left(\delta - \frac{1^2}{2^2\cdot 3!}\delta^3 + \frac{1^2\cdot 3^2}{2^4\cdot 5!}\delta^5 - \frac{1^2\cdot 3^2\cdot 5^2}{2^6\cdot 7!}\delta^7 + \cdots\right) \qquad (5.3.12)$$

whereas higher derivatives of *odd* order $2m + 1$ are obtained by multiplying the operator in (5.3.11) by the expansion of $\mathbf{D}^{2m}$. Thus, for example, we may obtain the formula

$$p_0'' = \frac{1}{h^2}\left(\delta - \frac{1^2}{2^2\cdot 3!}\delta^3 + \frac{1^2\cdot 3^2}{2^4\cdot 5!}\delta^5 - \frac{1^2\cdot 3^2\cdot 5^2}{2^6\cdot 7!}\delta^7 + \cdots\right)^2 p_0$$

$$= \frac{1}{h^2}(\delta^2 - \tfrac{1}{12}\delta^4 + \tfrac{1}{90}\delta^6 - \tfrac{1}{560}\delta^8 + \cdots)p_0 \qquad (5.3.13)$$

† More properly, we should *operate* on the right-hand member by $1 = \mu\mu^{-1} = \mu(1 + \delta^2/4)^{-1/2}$ to obtain

$$p_0' = \left[\frac{2}{h}\mu\left(1 + \frac{\delta^2}{4}\right)^{-1/2}\sinh^{-1}\frac{\delta}{2}\right]p_0$$

and *then* effect the desired expansion in powers of δ. But since the *coefficients* in that expansion are independent of the *order* of the factors, the fact that *some* ordering is justifiable permits us to ignore the noncommutativity of μ^{-1} with other factors when we actually determine those coefficients. Similar comments are applicable to other such operational manipulations in what follows.

Other formulas obtainable in this way may be listed as follows:

$$p_0''' = \frac{\mu}{h^3}(\delta^3 - \tfrac{1}{4}\delta^5 + \tfrac{7}{120}\delta^7 - \cdots)p_0 \qquad (5.3.14)$$

$$p_0^{\text{iv}} = \frac{1}{h^4}(\delta^4 - \tfrac{1}{6}\delta^6 + \tfrac{7}{240}\delta^8 - \cdots)p_0 \qquad (5.3.15)$$

$$p_0^{\text{v}} = \frac{\mu}{h^5}(\delta^5 - \tfrac{1}{3}\delta^7 + \cdots)p_0 \qquad (5.3.16)$$

The error term to be introduced in each case, when $p(x)$ is replaced by a function $f(x)$ which is not a polynomial, so that the series must be terminated with, say, nth differences, can be determined by use of the results of Sec. 3.3 if it is noticed that the result of this truncation corresponds to the differentiation of a polynomial of degree n which agrees with $f(x)$ at $x_0, x_1, \ldots, x_n$ in the case of (5.3.5), at $x_N, x_{N-1}, \ldots, x_{N-n}$ in the case of (5.3.9), and at $x_0, x_{\pm 1}, \ldots, x_{\pm m}$ in the case of the central-difference formulas, where $m = n/2$ if n is even, and $m = (n + 1)/2$ if n is odd. In the case of the forward- and backward-difference formulas, when the differentiation is effected at an end point of the range, formula (3.3.20) is valid. However, in the central-difference formulas, the more complicated formula (3.3.15) cannot be avoided. In practice, unless the analytical definition of $f(x)$ is known and is of sufficiently simple form to permit the determination and estimation of corresponding analytical expressions for the higher derivatives involved in those error terms, one generally must estimate the error by considering the magnitude of the first term omitted, realizing that this estimate is not necessarily a dependable one. The importance of *inherent* errors in numerical differentiation has already been emphasized in Sec. 3.8.

Formulas for differentiation at a point midway between two tabular points are obtained by writing **D** in the form (5.3.12) and operating on $p_{1/2}$. In addition, in order to obtain *ordinary* odd central differences and mean *even* central differences at $s = \frac{1}{2}$, we must multiply the expansion by the unit operator $\mu/\sqrt{1 + \delta^2/4}$ in calculating derivatives of *even* order, whereas in the preceding case this device was necessary when calculating derivatives of *odd* order. Thus we have, symbolically,

$$p_{1/2}^{(2m)} = \frac{\mu}{\sqrt{1 + \delta^2/4}}\left[\frac{2}{h}\sinh^{-1}\frac{\delta}{2}\right]^{2m} p_{1/2} \qquad (5.3.17)$$

and

$$p_{1/2}^{(2m+1)} = \left[\frac{2}{h}\sinh^{-1}\frac{\delta}{2}\right]^{2m+1} p_{1/2} \qquad (5.3.18)$$

In particular, when $m = 0$ in (5.3.17), we thus rederive (4.6.8) in the form

$$p_{1/2} = \mu(1 - \tfrac{1}{8}\delta^2 + \tfrac{3}{128}\delta^4 - \tfrac{5}{1024}\delta^6 + \tfrac{35}{32768}\delta^8 - \cdots)p_{1/2} \tag{5.3.19}$$

and obtain also derivative formulas which may be listed as follows:

$$p'_{1/2} = \frac{1}{h}(\delta - \tfrac{1}{24}\delta^3 + \tfrac{3}{640}\delta^5 - \tfrac{7}{7168}\delta^7 + \cdots)p_{1/2} \tag{5.3.20}$$

$$p''_{1/2} = \frac{\mu}{h^2}(\delta^2 - \tfrac{5}{24}\delta^4 + \tfrac{259}{5760}\delta^6 - \tfrac{3229}{322560}\delta^8 + \cdots)p_{1/2} \tag{5.3.21}$$

$$p'''_{1/2} = \frac{1}{h^3}(\delta^3 - \tfrac{1}{8}\delta^5 + \tfrac{37}{1920}\delta^7 - \cdots)p_{1/2} \tag{5.3.22}$$

$$p^{\text{iv}}_{1/2} = \frac{\mu}{h^4}(\delta^4 - \tfrac{7}{24}\delta^6 + \tfrac{47}{640}\delta^8 - \cdots)p_{1/2} \tag{5.3.23}$$

$$p^{\text{v}}_{1/2} = \frac{1}{h^5}(\delta^5 - \tfrac{5}{24}\delta^7 + \cdots)p_{1/2} \tag{5.3.24}$$

In certain applications, it is desirable to express differences at a point in terms of derivatives at that point. This is the inverse of the problem just considered. Thus, in order to express forward differences in terms of derivatives, we again refer to (5.3.3) and obtain the relation

$$\Delta^r = (e^{h\mathbf{D}} - 1)^r = \left(\frac{h\mathbf{D}}{1!} + \frac{h^2\mathbf{D}^2}{2!} + \cdots\right)^r \tag{5.3.25}$$

Thus there follows

$$\Delta^r p_0 = \left[(h\mathbf{D})^r + \frac{r}{2}(h\mathbf{D})^{r+1} + \frac{r(3r+1)}{24}(h\mathbf{D})^{r+2} + \frac{r^2(r+1)}{48}(h\mathbf{D})^{r+3} + \cdots\right]p_0 \tag{5.3.26}$$

The coefficients in this formula are expressible in terms of the so-called *Stirling numbers of the second kind*, which may be denoted by $\mathscr{S}_k^{(r)}$, and are then defined by the relation

$$\frac{(e^t - 1)^r}{r!} = \sum_{k=r}^{\infty} \frac{\mathscr{S}_k^{(r)}}{k!} t^k \tag{5.3.27}$$

when $|t|$ is small, so that (5.3.26) can be written in the form

$$\Delta^r p_0 = \left[\frac{\mathscr{S}_r^{(r)}}{1} + \frac{\mathscr{S}_{r+1}^{(r)}}{r+1}(h\mathbf{D}) + \frac{\mathscr{S}_{r+2}^{(r)}}{(r+1)(r+2)}(h\mathbf{D})^2 + \cdots\right](h\mathbf{D})^r p_0 \tag{5.3.28}$$

By comparing the relation

$$(-\nabla)^r = (e^{-h\mathbf{D}} - 1)^r \qquad (5.3.29)$$

with (5.3.25), we see that a corresponding formula for *backward* differences can be obtained by replacing Δ by $-\nabla$ and $\mathbf{D}$ by $-\mathbf{D}$ in (5.3.26) or (5.3.28).

Similar formulas involving *central* differences are readily obtained from the relations

$$\delta = 2 \sinh \frac{h\mathbf{D}}{2} \qquad \mu = \cosh \frac{h\mathbf{D}}{2} \qquad \mu\delta = \sinh h\mathbf{D} \qquad (5.3.30)$$

Thus, for example, there follows

$$\mu\delta p_0 = [(h\mathbf{D}) + \tfrac{1}{6}(h\mathbf{D})^3 + \tfrac{1}{120}(h\mathbf{D})^5 + \cdots]p_0 \qquad (5.3.31)$$

and

$$\delta^2 p_0 = [(h\mathbf{D})^2 + \tfrac{1}{12}(h\mathbf{D})^4 + \tfrac{1}{360}(h\mathbf{D})^6 + \cdots]p_0 \qquad (5.3.32)$$

5.4 Newtonian Integration Formulas

For the purpose of obtaining formulas for numerical integration, we may make use of (5.2.10)

$$\mathbf{J} = \Delta\mathbf{D}^{-1} \qquad (5.4.1)$$

combined with one of the relations of (5.3.3). Thus, to obtain a formula involving *forward* differences for the approximation of the integral

$$\int_{x_0}^{x_0+rh} f(x)\,dx$$

we may notice first that, when $f(x)$ is a polynomial $p(x)$, this integral can be expressed as

$$(1 + \mathbf{E} + \mathbf{E}^2 + \cdots + \mathbf{E}^{r-1})\mathbf{J}p_0 = \frac{\mathbf{E}^r - 1}{\mathbf{E} - 1}\mathbf{J}p_0$$

where it is assumed that the operators are destined to be expanded in series of powers of a reductive operator. Hence, by expressing $\mathbf{E}$ and $\mathbf{D}$ in terms of Δ, there follows, symbolically,

$$\int_{x_0}^{x_0+rh} p(x)\,dx = h\left[\frac{(1+\Delta)^r - 1}{\Delta}\right]\left[\frac{\Delta}{\log(1+\Delta)}\right]p_0 \qquad (5.4.2)$$

The expansion of the first operator is easily found to be

$$\frac{(1+\Delta)^r - 1}{\Delta} = \sum_{i=0}^{\infty}\binom{r}{i+1}\Delta^i \qquad (5.4.3)$$

whereas the expansion of the second factor may be written in the form

$$\frac{\Delta}{\log(1+\Delta)} = \sum_{j=0}^{\infty} c_j \Delta^j \tag{5.4.4}$$

the first nine coefficients of which are

$$\begin{array}{lll} c_0 = 1 & c_1 = \frac{1}{2} & c_2 = -\frac{1}{12} \\ c_3 = \frac{1}{24} & c_4 = -\frac{19}{720} & c_5 = \frac{3}{160} \\ c_6 = -\frac{863}{60480} & c_7 = \frac{275}{24192} & c_8 = -\frac{33953}{3628800} \end{array} \tag{5.4.5}$$

Hence the operator involved in (5.4.2) can be expressed in the form

$$\sum_{i=0}^{\infty}\sum_{j=0}^{\infty} c_j \binom{r}{i+1} \Delta^{i+j} = \sum_{k=0}^{\infty}\left[\sum_{i=0}^{k} c_{k-i}\binom{r}{i+1}\right]\Delta^k \tag{5.4.6}$$

Thus, if we write

$$\alpha_k^{(r)} \equiv \sum_{i=0}^{k} c_{k-i}\binom{r}{i+1} = c_k\binom{r}{1} + c_{k-1}\binom{r}{2} + \cdots \tag{5.4.7}$$

where the series terminates when the subscript of c vanishes *or* when the arguments of the binomial coefficient become equal, the required formula becomes

$$\int_{x_0}^{x_0+rh} p(x)\,dx = h\left(\sum_{k=0}^{\infty} \alpha_k^{(r)}\Delta^k\right) p_0 \tag{5.4.8}$$

In particular, in the case $r = 1$, there follows $\alpha_k^{(1)} = c_k$ and (5.4.8) becomes

$$\int_{x_0}^{x_0+h} p(x)\,dx = h\left(1 + \tfrac{1}{2}\Delta - \tfrac{1}{12}\Delta^2 + \tfrac{1}{24}\Delta^3 - \tfrac{19}{720}\Delta^4 + \tfrac{3}{160}\Delta^5 + \cdots\right)p_0 \tag{5.4.9}$$

In the case $r = 2$, there follows $\alpha_k^{(2)} = 2c_k + c_{k-1}$, and we may obtain the formula

$$\int_{x_0}^{x_0+2h} p(x)\,dx = 2h\left(1 + \Delta + \tfrac{1}{6}\Delta^2 + 0\Delta^3 - \tfrac{1}{180}\Delta^4 + \tfrac{1}{180}\Delta^5 + \cdots\right)p_0 \tag{5.4.10}$$

Further, in the case $r = -1$, there follows

$$\alpha_k^{(-1)} = -c_k + c_{k-1} - c_{k-2} + \cdots + (-1)^{k+1}c_0$$

and (5.4.8) becomes

$$\int_{x_0-h}^{x_0} p(x)\,dx = h\left(1 - \tfrac{1}{2}\Delta + \tfrac{5}{12}\Delta^2 - \tfrac{3}{8}\Delta^3 + \tfrac{251}{720}\Delta^4 - \tfrac{95}{288}\Delta^5 + \cdots\right)p_0 \tag{5.4.11}$$

This formula amounts to the result of using the Newton formula to extrapolate $f(x) = p(x)$ backward over the interval $[x_0 - h, x_0]$ and integrating the result over that interval.

Similar formulas are easily obtained in terms of *backward* differences, for the purpose of integrating a function over r intervals terminating at the *end* of a tabulation. It may be seen that *the formula for integrating from* $x_N - rh$ *to* x_N *can be obtained from* (5.4.8) *by replacing* Δ *by* $-\nabla$ *and* p_0 *by* p_N. Thus, for example, one has

$$\int_{x_N-h}^{x_N} p(x)\,dx$$
$$= h(1 - \tfrac{1}{2}\nabla - \tfrac{1}{12}\nabla^2 - \tfrac{1}{24}\nabla^3 - \tfrac{19}{720}\nabla^4 - \tfrac{3}{160}\nabla^5 - \cdots)p_N \qquad (5.4.12)$$

and

$$\int_{x_N}^{x_N+h} p(x)\,dx$$
$$= h(1 + \tfrac{1}{2}\nabla + \tfrac{5}{12}\nabla^2 + \tfrac{3}{8}\nabla^3 + \tfrac{251}{720}\nabla^4 + \tfrac{95}{288}\nabla^5 + \cdots)p_N \qquad (5.4.13)$$

the last formula being useful for integration over an interval beyond the range of tabulation and playing an important role in the numerical solution of differential equations.

In each case, the error term to be introduced when $p(x)$ is replaced by $f(x)$ and the series is truncated with the nth difference can be expressed in the form given by (3.3.5). Thus, in the case of (5.4.8), there follows

$$E_n = \frac{1}{(n+1)!}\int_{x_0}^{x_r} (x - x_0)(x - x_1)\cdots(x - x_n)f^{(n+1)}(\xi)\,dx \qquad (5.4.14)$$

where ξ depends upon x and lies between the smaller of x_0 and x_r and the larger of x_r and x_n, and an analogous term applies to the formula with backward differences.

When $r = 1$, or when r is a negative integer, the coefficient of $f^{(n+1)}$ in (5.4.14), which has been denoted by $\pi(x)$, does not change sign in the interval of integration. Thus the second law of the mean can then be applied to give the more useful form

$$E_n = \frac{f^{(n+1)}(\xi)}{(n+1)!}\int_{x_0}^{x_r} \pi(x)\,dx \qquad (r \leqq 1) \qquad (5.4.15)$$

where now ξ is a *constant*, such that $\min(x_0, x_r) < \xi < \max(x_r, x_n)$. A reference to the form of Newton's interpolation formula, from which the preceding formulas may be obtained by integration, shows that, *when* (5.4.15) *applies, the error term is obtained by replacing* $\Delta^k p_0$ *or* $\nabla^k p_N$ *by* $h^k f^{(k)}(\xi)$ *in the first nonvanishing term omitted.* Thus, for example, we may deduce from (5.4.12) that

$$\int_{x_N-h}^{x_N} f(x)\,dx = h(1 - \tfrac{1}{2}\nabla - \tfrac{1}{12}\nabla^2)f_N - \frac{h^4}{24} f'''(\xi)$$

where $x_N - 2h < \xi < x_N$.

In those cases when $n = r$, so that the number of differences retained is equal to the number of h intervals in the range of integration, the formulas reduce to Newton-Cotes formulas when expressed in terms of the ordinates, and the error terms can be supplied by reference to the results of Sec. 3.5. Thus, for example, we may deduce from (5.4.10) that

$$\int_{x_0}^{x_0+2h} f(x)\,dx = 2h(1 + \Delta + \tfrac{1}{6}\Delta^2)f_0 - \frac{h^5}{90} f^{\text{iv}}(\xi)$$

where $x_0 < \xi < x_0 + 2h$, and the formula is equivalent to *Simpson's rule.*

If the terms involving Δ and Δ^2 in (5.4.10) are expressed explicitly in terms of the ordinates p_0, p_1, and p_2, the result takes the form

$$\int_{x_0}^{x_0+2h} p(x)\,dx = \frac{h}{3}(p_0 + 4p_1 + p_2) - \frac{h}{90}(\Delta^4 - \Delta^5 + \tfrac{37}{42}\Delta^6 - \cdots)p_0 \tag{5.4.16}$$

which may be considered as Simpson's rule with "correction terms" expressed in terms of forward differences, for use at the beginning of a tabulation. The corresponding formula with backward differences is

$$\int_{x_N-2h}^{x_N} p(x)\,dx = \frac{h}{3}(p_N + 4p_{N-1} + p_{N-2}) - \frac{h}{90}(\nabla^4 + \nabla^5 + \tfrac{37}{42}\nabla^6 + \cdots)p_N \tag{5.4.17}$$

5.5 Newtonian Formulas for Repeated Integration

It frequently happens that the *second* derivative of a function $F(x)$ is known

$$F''(x) = f(x) \tag{5.5.1}$$

and that $F(x)$, and perhaps also $F'(x)$, are required at a set of equally spaced points $x_0, x_1, \ldots, x_N$, with the values $F(x_0) \equiv F_0$ and $F'(x_0) \equiv F'_0$ prescribed in advance. In order to treat this problem operationally, without being concerned with remainder terms, we again imagine that F and f are replaced by polynomials and denote this fact by writing P and p for F and f, respectively.

If $P''(x)$ is tabulated at the points $x_0, \ldots, x_N$, it is seen that use may be made, say, of (5.4.9), written in the form

$$P'_{k+1} = P'_k + h(1 + \tfrac{1}{2}\Delta - \tfrac{1}{12}\Delta^2 + \tfrac{1}{24}\Delta^3 - \tfrac{19}{720}\Delta^4 + \tfrac{3}{160}\Delta^5 - \cdots)P''_k \tag{5.5.2}$$

where h is the spacing, to obtain a corresponding tabulation of $P'(x)$, after which the same formula may be used again in the form

$$P_{k+1} = P_k + h(1 + \tfrac{1}{2}\Delta - \tfrac{1}{12}\Delta^2 + \tfrac{1}{24}\Delta^3 - \tfrac{19}{720}\Delta^4 + \tfrac{3}{160}\Delta^5 - \cdots)P'_k \tag{5.5.3}$$

to determine the desired tabulation of $P(x)$. Clearly, the formula (5.4.12) could be used instead and would be *needed* near the end of the tabulation if the value of P''_k were not available for $k > N$.

This procedure involves the formation of difference arrays relative to both $P'(x)$ and $P''(x)$. In order to avoid the necessity of two such arrays, we may transform (5.5.1) into the form

$$P_{k+1} = P_k + hP'_k + \int_{x_k}^{x_{k+1}} \int_{x_k}^{x} p(t)\, dt\, dx \tag{5.5.4}$$

and seek an operator $\boldsymbol{\theta}$ such that

$$\int_{x_k}^{x_{k+1}} \int_{x_k}^{x} p(t)\, dt\, dx = \boldsymbol{\theta} p_k \tag{5.5.5}$$

Thus $\boldsymbol{\theta}$ must be such that

$$(\mathbf{E} - 1 - h\mathbf{D})P_k = \boldsymbol{\theta} p_k = \boldsymbol{\theta}\mathbf{D}^2 P_k$$

and hence

$$\boldsymbol{\theta} = \frac{\mathbf{E} - 1 - h\mathbf{D}}{\mathbf{D}^2} = h^2 \frac{\mathbf{E} - 1 - \log \mathbf{E}}{(\log \mathbf{E})^2} \tag{5.5.6}$$

In terms of the operator Δ, there then follows

$$\begin{aligned}\boldsymbol{\theta} = h^2 \frac{\Delta - \log(1+\Delta)}{[\log(1+\Delta)]^2} &= h^2 \left[\frac{\Delta - \log(1+\Delta)}{\Delta^2}\right]\left[\frac{\Delta}{\log(1+\Delta)}\right]^2 \\ &= h^2(\tfrac{1}{2} - \tfrac{1}{3}\Delta + \tfrac{1}{4}\Delta^2 - \cdots)(1 + \tfrac{1}{2}\Delta - \tfrac{1}{12}\Delta^2 + \cdots)^2 \\ &= h^2(\tfrac{1}{2} + \tfrac{1}{6}\Delta - \tfrac{1}{24}\Delta^2 + \tfrac{1}{45}\Delta^3 - \tfrac{7}{480}\Delta^4 + \tfrac{107}{10080}\Delta^5 + \cdots)\end{aligned} \tag{5.5.7}$$

so that (5.5.4) takes the form

$$\begin{aligned}P_{k+1} = P_k + hP'_k + h^2(\tfrac{1}{2} + \tfrac{1}{6}\Delta - \tfrac{1}{24}\Delta^2 + \tfrac{1}{45}\Delta^3 \\ - \tfrac{7}{480}\Delta^4 + \tfrac{107}{10080}\Delta^5 + \cdots)P''_k\end{aligned} \tag{5.5.8}$$

Thus, if (5.5.2) and (5.5.8) are used for the calculation of values of $P'(x)$ and $P(x)$, only the differences of $P''(x)$ are needed. In a similar way, the formula

$$P_{k+1} = P_k + hP'_{k+1} - h^2(\tfrac{1}{2} - \tfrac{1}{6}\nabla - \tfrac{1}{24}\nabla^2 - \tfrac{1}{45}\nabla^3 - \tfrac{7}{480}\nabla^4 - \tfrac{107}{10080}\nabla^5 + \cdots)P''_{k+1} \qquad (5.5.9)$$

can be derived for use near the end of a tabulation of $P''(x)$, in conjunction with (5.4.12), written in the form

$$P'_{k+1} = P'_k + h(1 - \tfrac{1}{2}\nabla - \tfrac{1}{12}\nabla^2 - \tfrac{1}{24}\nabla^3 - \tfrac{19}{720}\nabla^4 - \tfrac{3}{160}\nabla^5 - \cdots)P''_{k+1} \qquad (5.4.12')$$

In those cases when values of P' are not required, we may derive a more useful formula by noticing that

$$\Delta^2\mathbf{D}^{-2}P''(x) = \Delta^2 P(x) \qquad (5.5.10)$$

where the factor Δ^2 is inserted to annihilate the arbitrary linear function of x which would correspond to the (improper) inverse operator $\mathbf{D}^{-2}$. Hence there follows also

$$\Delta^2 P_k = h^2\left[\frac{\Delta}{\log(1+\Delta)}\right]^2 P''_k$$
$$= h^2(1 + \Delta + \tfrac{1}{12}\Delta^2 + 0\Delta^3 - \tfrac{1}{240}\Delta^4 + \tfrac{1}{240}\Delta^5 - \cdots)P''_k \qquad (5.5.11)$$

Thus, since $\Delta^2 P_k = P_{k+2} - 2P_{k+1} + P_k$, this formula permits the determination of P_{k+2} from two preceding values of P.

A corresponding expansion involving backward differences is obtained by replacing Δ by $-\nabla$ in the form

$$\nabla^2 P_k = h^2(1 - \nabla + \tfrac{1}{12}\nabla^2 + 0\nabla^3 - \tfrac{1}{240}\nabla^4 - \tfrac{1}{240}\nabla^5 - \cdots)P''_k \qquad (5.5.12)$$

This formula determines P_k from P_{k-1} and P_{k-2} and makes use of P''_k. Another formula, in which only *preceding* values of P'' are needed, is obtained by operating on both sides of (5.5.12) by $\mathbf{E}$, and replacing $\mathbf{E}$ by $(1 - \nabla)^{-1}$ in the right-hand member, to give

$$\nabla^2 P_{k+1} = h^2(1 + \nabla + \nabla^2 + \cdots)(1 - \nabla + \tfrac{1}{12}\nabla^2 + \cdots)P''_k$$
$$= h^2(1 + 0\nabla + \tfrac{1}{12}\nabla^2 + \tfrac{1}{12}\nabla^3 + \tfrac{19}{240}\nabla^4 + \tfrac{3}{40}\nabla^5 + \cdots)P''_k \qquad (5.5.13)$$

In fact, a whole series of formulas of either type can be obtained, for example, by operating on both members of (5.5.12) or (5.5.13) by $a_0 + a_1\nabla + a_2\nabla^2 + \cdots$, where the a's are arbitrary constants. Such formulas are particularly useful in the numerical solution of differential equations (see Sec. 6.10), which include (5.5.1) as a very special case.

In order to illustrate the use of these formulas in connection with (5.5.1), we consider a simple example. It is supposed that the values of F'' listed in the following table are known and that the values $F(1) = 0$ and $F'(1) = 1$ are prescribed:

x	F	F'	F''	$\Delta F''$	$\Delta^2 F''$	$\Delta^3 F''$
1.0	0	1.000	1.000			
				331		
1.1	0.1055	1.1160	1.331		66	
				397		6
1.2	0.2244		1.728		72	
				469		6
1.3	0.3606		2.197		78	
				547		6
1.4	.		2.744		84	
	.			631		
1.5	.		3.375			

In order to determine $F_1 \equiv F(1.1)$ and $F_1' \equiv F'(1.1)$, we use the approximate relations resulting from replacing P by F in (5.5.8) and (5.5.2):

$$F_1 \approx 0 + (0.1)(1) + 0.01[\tfrac{1}{2}(1.000) + \tfrac{1}{6}(0.331) - \tfrac{1}{24}(0.066) + \tfrac{1}{45}(0.006)] \doteq 0.1055$$

$$F_1' \approx 1 + 0.1[1.000 + \tfrac{1}{2}(0.331) - \tfrac{1}{12}(0.066) + \tfrac{1}{24}(0.006)] \doteq 1.1160$$

Formula (5.5.8) is then used again to determine F_2. For the evaluation of F_3, sufficiently many *backward* differences are available for the use of (5.5.9) or (5.5.12). Hence, unless values of F' are required, F_2' need not be calculated, and F_3 may be determined by (5.5.12):

$$F_3 \approx 2F_2 - F_1 + h^2(F_3'' - \nabla F_3'' + \tfrac{1}{12}\nabla^2 F_3'')$$
$$= 0.3433 + 0.01[2.197 - 0.469 + \tfrac{1}{12}(0.072)] \doteq 0.3606$$

From this stage onward, use may be made exclusively of (5.5.12).

In this example, the given data are exact values of F'' corresponding to $F''(x) = x^3$, from which there follows $F(x) = 0.05x^5 + 0.75x - 0.8$, and the results are correct to the places given. Since here the third difference of $F''(x)$ is constant, *exact* values would have been obtained if no intermediate roundoffs had been effected. A check on the calculation, which would be useful if the last difference retained were *not* constant, would be afforded by the use of (5.5.13).

5.6 Central-Difference Integration Formulas

The most useful integration formulas involving *central* differences are those in which the differences are evaluated at the center of the range of integration, and the integral is expressed in the form

$$\int_{x_0 - mh}^{x_0 + mh} f(x)\, dx$$

In terms of the operator **J** defined in (5.2.5), this integral can be expressed in the symbolic form

$$(\mathbf{E}^{-m} + \mathbf{E}^{-m+1} + \cdots + \mathbf{E}^{m-1})\mathbf{J}p_0 = \frac{\mathbf{E}^m - \mathbf{E}^{-m}}{\mathbf{E} - 1}\mathbf{J}p_0$$
$$= \frac{e^{mh\mathbf{D}} - e^{-mh\mathbf{D}}}{\Delta}\Delta\mathbf{D}^{-1}p_0$$

when $f(x)$ is a polynomial $p(x)$, and hence we may write

$$\int_{x_0-mh}^{x_0+mh} p(x)\,dx = 2\frac{\sinh mh\mathbf{D}}{\mathbf{D}}p_0 \qquad (5.6.1)$$

In order to obtain an expansion in central differences, we may first obtain the expansion

$$2\frac{\sinh mh\mathbf{D}}{\mathbf{D}} = 2mh\left[1 + \frac{m^2(h\mathbf{D})^2}{6} + \frac{m^4(h\mathbf{D})^4}{120} + \frac{m^6(h\mathbf{D})^6}{5040} + \cdots\right]$$

and then replace $h\mathbf{D}$ by its expansion given in (5.3.12), to give

$$2\frac{\sinh mh\mathbf{D}}{\mathbf{D}} = 2mh\left[1 + \frac{m^2\delta^2}{6}(1 - \tfrac{1}{24}\delta^2 + \tfrac{3}{640}\delta^4 + \cdots)^2 + \frac{m^4\delta^4}{120}(1 - \tfrac{1}{24}\delta^2 + \cdots)^4 + \frac{m^6\delta^6}{5040} + \cdots\right]$$

if, say, only coefficients of differences of order less than eight are desired. Hence we may obtain the formula

$$\int_{x_0-mh}^{x_0+mh} p(x)\,dx = 2mh\left[1 + \frac{m^2}{6}\delta^2 + \frac{m^2(3m^2-5)}{360}\delta^4 + \frac{m^2(3m^4 - 21m^2 + 28)}{15120}\delta^6 + \cdots\right]p_0 \qquad (5.6.2)$$

In the special cases $m = 1$ and $m = 2$, the relevant formulas are of the forms

$$\int_{x_0-h}^{x_0+h} p(x)\,dx = 2h(1 + \tfrac{1}{6}\delta^2 - \tfrac{1}{180}\delta^4 + \tfrac{1}{1512}\delta^6 - \tfrac{23}{226800}\delta^8 - \cdots)p_0 \qquad (5.6.3)$$

and

$$\int_{x_0-2h}^{x_0+2h} p(x)\,dx = 4h(1 + \tfrac{2}{3}\delta^2 + \tfrac{7}{90}\delta^4 - \tfrac{2}{945}\delta^6 + \tfrac{13}{56700}\delta^8 - \cdots)p_0 \qquad (5.6.4)$$

Formula (5.6.3) can also be expressed in the form

$$\int_{x_0-h}^{x_0+h} p(x)\,dx = \frac{h}{3}(f_{-1} + 4f_0 + f_1) - \frac{h}{90}(\delta^4 - \tfrac{5}{42}\delta^6 + \tfrac{23}{1260}\delta^8 - \cdots)p_0 \tag{5.6.5}$$

and so considered as Simpson's rule with "correction terms" expressed in terms of central differences.

It is known (see Steffensen [1950]) that, if $p(x)$ is replaced by $f(x)$ in (5.6.2) and the formula is truncated with the difference of order $2k$, then the error term to be introduced can be expressed in the convenient form

$$E_{2k} = h^{2k+3}\frac{f^{(2k+2)}(\xi)}{(2k+2)!}\int_{-m}^{m} s^2(s^2-1^2)\cdots(s^2-k^2)\,ds \tag{5.6.6}$$

where $x_0 - mh < \xi < x_0 + mh$ if $k \leqq m$ and $x_0 - kh < \xi < x_0 + kh$ if $k \geqq m$. Reference to the Stirling interpolation formula, from which the preceding formulas may be obtained by integration, shows that *this error term is obtained by replacing* $\delta^{2k+2}p_0$ *by* $h^{2k+2}f^{(2k+2)}(\xi)$ *in the first nonvanishing term omitted.*

An important formula, relating to repeated integration, is obtained by noticing that since

$$\delta^2\mathbf{D}^{-2}P''(x) = \delta^2 P(x)$$

and since we have the expansion

$$\frac{\delta^2}{\mathbf{D}^2} = h^2(1 - \tfrac{1}{12}\delta^2 + \tfrac{1}{90}\delta^4 - \tfrac{1}{560}\delta^6 + \cdots)^{-1}$$

$$= h^2[1 + (\tfrac{1}{12}\delta^2 - \tfrac{1}{90}\delta^4 + \tfrac{1}{560}\delta^6 + \cdots) + (\tfrac{1}{12}\delta^2 - \tfrac{1}{90}\delta^4 + \cdots)^2 + (\tfrac{1}{12}\delta^2 + \cdots)^3 + \cdots]$$

from (5.3.12) and (5.3.13), there follows

$$\delta^2 P_k = h^2(1 + \tfrac{1}{12}\delta^2 - \tfrac{1}{240}\delta^4 + \tfrac{31}{60480}\delta^6 - \cdots)P_k'' \tag{5.6.7}$$

Because of the fact that only differences of even order are involved, this formula is usually preferable to (5.5.12) for advancing a step-by-step double integration of a given tabulated function, over the portion of the range in which the requisite central differences are available. The formula (5.6.7) also will be used in the numerical solution of boundary-value problems governed by certain second-order differential equations (Sec. 6.15), whereas the analogous formulas (5.5.12) and (5.5.13) are to be used for corresponding initial-value problems (Sec. 6.10).

5.7 Subtabulation

In some situations it is desirable to determine, from a given difference table based on the spacing h, a new set of differences based on a new spacing ρh. This problem would occur, for example, if a function were initially tabulated for increments of 0.1 in x and it were required to *subtabulate* the function for increments of 0.01, in which case $\rho = 0.1$. Whereas this subtabulation clearly could be effected by the use of an appropriate interpolation formula, it is often more convenient to form certain new differences, based on the new spacing, and to build up the table by addition, as will be illustrated at the end of this section. The problem also arises when a finite-difference method is used in a step-by-step numerical solution of a differential equation, in which case a halving of the interval is desirable when a derivative of the solution being determined begins to change too rapidly.

In order to obtain formulas for such purposes, we designate the shifting operator relative to the new spacing ρh by $\mathbf{E}_1$, and notice that since $\mathbf{E}_1$ effects an h shift ρ times, there follows

$$\mathbf{E}_1 = \mathbf{E}^\rho$$

If we designate the forward-difference operator relative to ρh by Δ_1, there then follows

$$1 + \Delta_1 = (1 + \Delta)^\rho \tag{5.7.1}$$

and hence we obtain the desired transformation in the symbolic form

$$\begin{aligned} \Delta_1^r &= [(1 + \Delta)^\rho - 1]^r \\ &= \left[\rho\Delta + \frac{\rho(\rho - 1)}{2!}\Delta^2 + \frac{\rho(\rho - 1)(\rho - 2)}{3!}\Delta^3 + \cdots\right]^r \end{aligned} \tag{5.7.2}$$

The leading terms in this expansion can be obtained in the form

$$\begin{aligned} \Delta_1^r = \rho^r \Bigg\{ \Delta^r &+ \frac{r(\rho - 1)}{2}\Delta^{r+1} \\ &+ \frac{r(\rho - 1)}{24}[4(\rho - 2) + 3(r - 1)(\rho - 1)]\Delta^{r+2} \\ &+ \frac{r(\rho - 1)}{48}[2(\rho - 2)(\rho - 3) + 4(r - 1)(\rho - 1)(\rho - 2) \\ &\qquad + (r - 1)(r - 2)(\rho - 1)^2]\Delta^{r+3} + \cdots \Bigg\} \end{aligned} \tag{5.7.3}$$

In particular, in the important case $\rho = \frac{1}{2}$, where the spacing is *halved*, this formula becomes

$$\Delta_1^r = \left(\frac{1}{2}\Delta - \frac{1}{2^2 \cdot 2!}\Delta^2 + \frac{1 \cdot 3}{2^3 \cdot 3!}\Delta^3 - \frac{1 \cdot 3 \cdot 5}{2^4 \cdot 4!}\Delta^4 + \cdots\right)^r$$

$$= 2^{-r}\left[\Delta^r - \frac{r}{4}\Delta^{r+1} + \frac{r(r+3)}{32}\Delta^{r+2} - \frac{r(r+4)(r+5)}{384}\Delta^{r+3} + \cdots\right] \tag{5.7.4}$$

whereas the formula reduces in the case $\rho = \frac{1}{10}$ to

$$\Delta_1^r = 10^{-r}\left[\Delta^r - \frac{9r}{20}\Delta^{r+1} + \frac{3r(27r+49)}{800}\Delta^{r+2} - \frac{3r(81r^2 + 441r + 580)}{16000}\Delta^{r+3} + \cdots\right] \tag{5.7.5}$$

In order to illustrate an appropriate technique, we again consider the data tabulated in Sec. 4.8, where a difference table is constructed with spacing $h = 0.1$, and suppose that the data are to be subtabulated by tenths, that is, with a new spacing 0.01. Here, with $\rho = 0.1$, Eq. (5.7.5) gives the formulas

$$\begin{aligned} \Delta_1 &= 0.1\Delta - 0.045\Delta^2 + 0.0285\Delta^3 - 0.0206625\Delta^4 + \cdots \\ \Delta_1^2 &= 0.01\Delta^2 - 0.009\Delta^3 + 0.007725\Delta^4 - \cdots \\ \Delta_1^3 &= 0.001\Delta^3 - 0.00135\Delta^4 + \cdots \\ \Delta_1^4 &= 0.0001\Delta^4 + \cdots \end{aligned} \tag{5.7.6}$$

through fourth differences, where the coefficients have been expressed exactly, for convenient reference. In units of the fifth place, the new forward differences relative to $x = 1.0$ and $x = 1.1$ are found as follows:

$$x = 1.0: \quad \Delta_1 f = 536.2 \qquad \Delta_1^2 f = -8.5$$
$$\Delta_1^3 f = -0.05 \qquad \Delta_1^4 f = 0.0008 \approx 0$$
$$x = 1.1: \quad \Delta_1 f = 449.1 \qquad \Delta_1^2 f = -8.9$$
$$\Delta_1^3 f = -0.05 \qquad \Delta_1^4 f = 0.001 \approx 0$$

Thus, we may suppose that the third differences are constant (within the accuracy indicated) over the first range, and we may set up the *underlined* entries in the following table:

x	f	Δf	$\Delta^2 f$	$\Delta^3 f$
1.00	0.84147			
		536.2		
1.01	0.846832		-8.5	
		527.7		-0.05
1.02	0.852109		-8.6	
		519.1		-0.05
1.03	0.857300		-8.6	
		510.5		-0.05
1.04	0.862405		-8.6	
		501.9		-0.05
1.05	0.867424		-8.7	
		493.2		-0.05
⋮	⋮			

The remaining entries are then filled in by addition, proceeding from right to left, and the results rounded, correctly to five places, to known rounded values of $f(x) = \sin x$.

The decimal parts of units in the fifth place are retained in order to reduce the danger of propagated effects of roundoff errors. Since here the errors are propagated to the left, and since (see Sec. 4.9) then errors of magnitude e in the rth difference could lead to errors of magnitude $2^r e$ in the calculated values of f, it follows that if no errors of one-half unit are to be so introduced, the roundoff errors in the rth differences should be smaller than 2^{-r-1} units in that place. Hence, for this reason alone, at least one extra place should be retained in the intermediate subtabulation of f and in the first two differences, two extra places in each of the next three differences, and so forth.

If *backward* differences are used, we see that (5.7.1) must be replaced by

$$1 - \nabla_1 = (1 - \nabla)^\rho$$

and hence *all the formulas of this section are transformed to corresponding formulas for backward differences by replacing* Δ_1 *by* $-\nabla_1$ and Δ by $-\nabla$. Formulas using *central* differences may be derived similarly (see Prob. 16).

5.8 Summation and Integration. The Euler-Maclaurin Sum Formula

The problem of evaluating a sum

$$\sum_{\nu=m}^{k-1} f(x_0 + \nu h) = f_m + f_{m+1} + \cdots + f_{k-1} \qquad (k > m)$$

where $f_\nu \equiv f(x_0 + \nu h)$, is closely related to the problem of determining a function $F(x)$ such that

$$\Delta F(x) = f(x) \qquad (5.8.1)$$

since, if any such function $F(x)$ is known, there follows immediately

$$\sum_{\nu=m}^{k-1} f_\nu = (F_{m+1} - F_m) + (F_{m+2} - F_{m+1}) + \cdots + (F_{k-1} - F_{k-2}) + (F_k - F_{k-1})$$

$$= F_k - F_m \equiv [F_\nu]_m^k \tag{5.8.2}$$

It should be noticed that the upper limit in the last term exceeds by unity the upper limit in the original sum.

If we invert (5.8.1) in the symbolic form $F_k = \Delta^{-1} f_k$, it follows that we may write

$$\Delta^{-1} f_k = C + \sum_{\nu=M}^{k-1} f_\nu \quad \text{and} \quad \sum_{\nu=m}^{k-1} f_\nu = [\Delta^{-1} f_\nu]_m^k \tag{5.8.3}$$

where C is an arbitrary constant and M is an arbitrarily fixed integer such that $M \leqq m \leqq k$, with the usual convention that

$$\sum_{\nu=m}^{m-1} f_\nu = 0$$

Thus we may refer to $\Delta^{-1} f_k$ as an *indefinite sum* of f_k and may correspondingly consider indefinite summation to be the inverse of the process of differencing, just as indefinite integration is the inverse of differentiation. As was noted previously, to any one inverse $\Delta^{-1} f(x)$ we may add any function $\omega_h(x)$ which is of period h since any such function is annihilated by Δ. However, if only values of x which differ from some fixed value x_0 by integral multiples of h are involved, then, for that set of values of x, the additive function $\omega_h(x)$ reduces to the constant C, which itself disappears in *definite* summation between limits.

There exist extensive tables of sum functions (analogous to tables of integrals) which facilitate the evaluation of many special sums by use of (5.8.3). We consider next some other techniques for evaluating or approximating the value of a sum. A simple formula for summing any *polynomial* $p(x)$ is obtained by writing

$$\sum_{k=0}^{r-1} p_k = (1 + \mathbf{E} + \mathbf{E}^2 + \cdots + \mathbf{E}^{r-1}) p_0 = \frac{\mathbf{E}^r - 1}{\mathbf{E} - 1} p_0$$

$$= \frac{(1 + \Delta)^r - 1}{\Delta} p_0$$

$$= \left[r + \frac{r(r-1)}{2!} \Delta + \frac{r(r-1)(r-2)}{3!} \Delta^2 + \cdots \right] p_0 \tag{5.8.4}$$

Thus, for example, in order to sum the series $1^2 + 2^2 + \cdots + r^2$, we may take $p(x) = (x + 1)^2$, $x_0 = 0$, and $h = 1$. With $p_0 = 1$, $\Delta p_0 = 3$, $\Delta^2 p_0 = 2$, $\Delta^3 p_0 = \cdots = 0$, Eq. (5.8.4) gives

$$1^2 + 2^2 + \cdots + r^2 = r \cdot 1 + \frac{r(r-1)}{2!} \cdot 3 + \frac{r(r-1)(r-2)}{3!} \cdot 2$$

$$= \tfrac{1}{6} r(r+1)(2r+1)$$

The formula (5.8.4) is principally useful for the finite summation of a polynomial of degree n small relative to the number of terms r, so that the number of terms in the transformed series is small relative to the original number. In order to obtain a formula which is of more general usefulness in finite or infinite summation, as well as in numerical integration, we again first restrict attention to a polynomial $p(x)$. From the operational identity

$$\mathbf{D}\mathbf{J}\Delta^{-1} = 1 \tag{5.8.5}$$

we may deduce that

$$hp(x) = \left(\frac{h\mathbf{D}}{e^{h\mathbf{D}} - 1}\right) \mathbf{J}p(x) \tag{5.8.6}$$

The coefficients B_k in the expansion

$$\frac{t}{e^t - 1} = \sum_{\nu=0}^{\infty} \frac{B_\nu}{\nu!} t^\nu \tag{5.8.7}$$

for small $|t|$, are the so-called *Bernoulli numbers*, which occur in many fields of mathematics.† It is found that $B_3 = B_5 = B_7 = \cdots = 0$, and the following additional values may be listed:

$$\begin{array}{llll} B_0 = 1 & B_1 = -\frac{1}{2} & B_2 = \frac{1}{6} & B_4 = -\frac{1}{30} \\ B_6 = \frac{1}{42} & B_8 = -\frac{1}{30} & B_{10} = \frac{5}{66} & B_{12} = -\frac{691}{2730} \\ B_{14} = \frac{7}{6} & B_{16} = -\frac{3617}{510} & B_{18} = \frac{43867}{798} & B_{20} = -\frac{174611}{330} \end{array} \tag{5.8.8}$$

Hence, with this notation, (5.8.6) can be expressed in the form

$$hp(x) = \sum_{\nu=0}^{\infty} \frac{B_\nu}{\nu!} h^\nu \mathbf{D}^\nu \mathbf{J}\, p(x)$$

or

$$hp(x) = \int_x^{x+h} p(t)\, dt + \sum_{\nu=1}^{\infty} \frac{B_\nu}{\nu!} h^\nu \mathbf{D}^\nu \mathbf{J}\, p(x) \tag{5.8.9}$$

† The notation B_k is sometimes used for the present B_{2k}.

By using (5.8.5) to replace $\mathbf{D}^\nu \mathbf{J} p(x)$ by $\mathbf{D}^{\nu-1}[p(x+h) - p(x)]$, we may express this result in the more explicit form

$$p(x) = \frac{1}{h}\int_x^{x+h} p(t)\,dt + \sum_{\nu=1}^{\infty} \frac{B^\nu}{\nu!} h^{\nu-1}[p^{(\nu-1)}(x+h) - p^{(\nu-1)}(x)] \tag{5.8.10}$$

If we write (5.8.10) for

$$x = x_0, \quad x_1\ (\equiv x_0 + h), \quad \ldots, \quad x_{r-1}\ [\equiv x_0 + (r-1)h]$$

and sum the results, noticing the "telescoping" of the resultant terms in brackets, we deduce the identity

$$\sum_{k=0}^{r-1} p_k = \frac{1}{h}\int_{x_0}^{x_r} p(x)\,dx + \sum_{\nu=1}^{\infty} \frac{B_\nu}{\nu!} h^{\nu-1}[p_r^{(\nu-1)} - p_0^{(\nu-1)}] \tag{5.8.11}$$

where $p_k \equiv p(x_k)$ and $p_k^{(\nu-1)} \equiv p^{(\nu-1)}(x_k)$. This result is usually known as the *Euler-Maclaurin sum formula* for a polynomial, although that name is also sometimes applied instead to (5.8.10), which leads to (5.8.11), or to still another formula, which generalizes (5.8.11).

It can be written in a somewhat more convenient form by making use of the fact that all Bernoulli numbers with odd subscripts greater than unity are zero. Thus, if we extract the term corresponding to $\nu = 1$, and afterward replace ν by $2i$, we obtain the form

$$\sum_{k=0}^{r} p_k = \frac{1}{h}\int_{x_0}^{x_r} p(x)\,dx + \tfrac{1}{2}(p_0 + p_r) + \sum_{i=1}^{\infty} \frac{B_{2i}}{(2i)!} h^{2i-1}[p_r^{(2i-1)} - p_0^{(2i-1)}] \tag{5.8.12}$$

If the degree of the polynomial $p(x)$ is $2m$ or $2m + 1$, the series on the right terminates when $i = m$.

When $f(x)$ is *not* a polynomial, the result of replacing $p(x)$ by $f(x)$ in the series (5.8.12) must be terminated, say, with $i = m$, and an appropriate error term must be introduced, so that we write

$$\sum_{k=0}^{r} f_k = \frac{1}{h}\int_{x_0}^{x_r} f(x)\,dx + \tfrac{1}{2}(f_0 + f_r) + \sum_{i=1}^{m} \frac{B_{2i}}{(2i)!} h^{2i-1}[f_r^{(2i-1)} - f_0^{(2i-1)}] + E_m \tag{5.8.13}$$

It is known (see Prob. 28) that this error term is expressible in the form

$$E_m = r\,\frac{B_{2m+2}h^{2m+2}}{(2m+2)!} f^{(2m+2)}(\xi) \tag{5.8.14}$$

where $x_0 < \xi < x_r$, when r is finite. When $r \to \infty$ and also $x_r \to \infty$, this form becomes indeterminate and must be replaced by a somewhat more elaborate one.

Frequently it is possible to avoid the use of the error formula (5.8.14) or its substitute by making use of the fact (see Steffensen [1950]) that if $f^{(2m+2)}(x)$ and $f^{(2m+4)}(x)$ do not change sign for $x_0 < x < x_r$, then E_m is numerically smaller than the first neglected term and is of the same sign. More generally, if it is known only that $f^{(2m+2)}(x)$ does not change sign, then E_m is numerically smaller than *twice* the first neglected term and of the same sign (see Prob. 28 and Steffensen [1950]).

The fact that rules of this type apply rather frequently to interpolation series and to allied series makes the procedure of using the *first omitted term* as a basis for estimating the order of magnitude of the error somewhat less hazardous in connection with such series than with *convergent* series more often encountered in other fields.

A formula similar to (5.8.13), but summing instead the ordinates midway between the successive ordinates involved in (5.8.13), is sometimes called the *second Euler-Maclaurin formula* (see Steffensen [1950]) and is of the form

$$\sum_{k=0}^{r-1} f_{k+(1/2)} = \frac{1}{h}\int_{x_0}^{x_r} f(x)\,dx - \sum_{i=1}^{m} \frac{(1-2^{1-2i})B_{2i}h^{2i-1}}{(2i)!}\left[f_r^{(2i-1)} - f_0^{(2i-1)}\right] + E_m \qquad (5.8.15)$$

where

$$E_m = -r\,\frac{(1-2^{-1-2m})B_{2m+2}h^{2m+2}}{(2m+2)!}\,f^{(2m+2)}(\xi) \qquad (5.8.16)$$

Here, again, if $f^{(2m+2)}(x)$ and $f^{(2m+4)}(x)$ are of constant sign in (x_0, x_r), the error is numerically smaller than the first neglected term and is of the same sign. If only $f^{(2m+2)}(x)$ is known to be of constant sign, then E_m can be shown to be numerically smaller than *three times* the first neglected term and of the same sign.

It may be seen that the *correction terms* in both (5.8.13) and (5.8.15) all vanish if, say, $f(x)$ is *periodic*, of period $x_r - x_0$, although the error term naturally remains (see, for example, Prob. 36 of Chap. 3).† Here it is of interest to notice that m can be assigned *any* positive integral value in E_m, assuming only that $f^{(2m+2)}(x)$ is continuous on $[x_0, x_r]$. Generally there is one such value, in correspondence with a specified $f(x)$, for which the corresponding error *bound* is minimized.

† For significant applications of these formulas in such cases, see Fettis [1955, 1958] and Luke [1956].

We see that these formulas each relate a given *sum* and an *integral* in terms of an associated sum of m terms and a corresponding error term, where m can be chosen at pleasure. While the first formula is useful both for numerical integration and for numerical summation of series, the second is used chiefly for integration.

Before considering the use of these formulas for approximate summation, we note that the Euler-Maclaurin formula (5.8.12) can be written in the form

$$\int_{x_0}^{x_r} f(x)\,dx = h(\tfrac{1}{2}f_0 + f_1 + f_2 + \cdots + f_{r-1} + \tfrac{1}{2}f_r) - \frac{h^2}{12}(f_r' - f_0') + \frac{h^4}{720}(f_r''' - f_0''') - \frac{h^6}{30240}(f_r^{\text{v}} - f_0^{\text{v}}) + \cdots - \frac{B_{2m}}{(2m)!}h^{2m}[f_r^{(2m-1)} - f_0^{(2m-1)}] - hE_m \tag{5.8.17}$$

where E_m is defined by (5.8.14), and hence can be considered as the *trapezoidal rule* with correction terms expressed in terms of derivatives. Similarly, the formula (5.8.15) can be written in the form

$$\int_{x_0}^{x_r} f(x)\,dx = h(f_{1/2} + f_{3/2} + \cdots + f_{r-(1/2)}) + \frac{h^2}{24}(f_r' - f_0') - \frac{7h^4}{5760}(f_r''' - f_0''') + \frac{31h^6}{967680}(f_r^{\text{v}} - f_0^{\text{v}}) - \cdots + \frac{(1 - 2^{1-2m})B_{2m}}{(2m)!}h^{2m}[f_r^{(2m-1)} - f_0^{(2m-1)}] - hE_m \tag{5.8.18}$$

where E_m here is defined by (5.8.16), and hence can be considered as the *repeated midpoint rule* with derivative correction terms. A comparison of (5.8.14) and (5.8.16) shows that the second formula tends to be slightly more accurate than the first, on the average, when truncated with the same number of correction terms.

A useful modification of (5.8.17) is obtained when the derivatives at x_0 are expressed in terms of forward differences, by using (5.3.5), and those at x_r are expressed in terms of backward differences, by using (5.3.9):

$$\begin{aligned}
hf_0' &= \Delta f_0 - \tfrac{1}{2}\Delta^2 f_0 + \tfrac{1}{3}\Delta^3 f_0 - \tfrac{1}{4}\Delta^4 f_0 + \tfrac{1}{5}\Delta^5 f_0 - \cdots \\
hf_r' &= \nabla f_r + \tfrac{1}{2}\nabla^2 f_r + \tfrac{1}{3}\nabla^3 f_r + \tfrac{1}{4}\nabla^4 f_r + \tfrac{1}{5}\nabla^5 f_r + \cdots \\
h^3 f_0''' &= \Delta^3 f_0 - \tfrac{3}{2}\Delta^4 f_0 + \tfrac{7}{4}\Delta^5 f_0 - \cdots \\
h^3 f_r''' &= \nabla^3 f_r + \tfrac{3}{2}\nabla^4 f_r + \tfrac{7}{4}\nabla^5 f_r + \cdots \\
h^5 f_0^{\text{v}} &= \Delta^5 f_0 - \cdots \\
h^5 f_r^{\text{v}} &= \nabla^5 f_r + \cdots
\end{aligned}$$

The result of this substitution is of the form

$$\int_{x_0}^{x_r} f(x)\,dx = h(\tfrac{1}{2}f_0 + f_1 + f_2 + \cdots + f_{r-1} + \tfrac{1}{2}f_r)$$
$$- \frac{h}{12}(\nabla f_r - \Delta f_0) - \frac{h}{24}(\nabla^2 f_r + \Delta^2 f_0) - \frac{19h}{720}(\nabla^3 f_r - \Delta^3 f_0)$$
$$- \frac{3h}{160}(\nabla^4 f_r + \Delta^4 f_0) - \frac{863h}{60480}(\nabla^5 f_r - \Delta^5 f_0) - \cdots \tag{5.8.19}$$

and is known as *Gregory's formula*. If no differences beyond the rth are retained, only values of the integrand in the interval of integration are involved. It appears that no tractable expression is known for the implied error term to be inserted after an appropriate truncation of the series. This formula can also be derived directly by operational methods (see Prob. 29). A similar modification of (5.8.18) is clearly possible.

In addition, instead the derivatives at the end points can be replaced by mean *central* differences of odd order. The formula so obtained from (5.8.17) is associated with Gauss and is derived in Prob. 18. Although the leading coefficients of the correction terms decrease more rapidly in magnitude than those in Gregory's formula, Gauss' formula has the disadvantage that it always involves ordinates which lie outside the interval of integration.

5.9 Approximate Summation

The formulas (5.8.17) and (5.8.19) are expressed in a form suitable for approximate evaluation of the relevant integral. When the formulas are to be used instead, say, for approximate *summation* of an infinite series, under the assumption that the integral can be evaluated (or suitably approximated) otherwise, they may be expressed in the form

$$\sum_{k=0}^{\infty} f_k = \frac{1}{h}\int_{x_0}^{\infty} f(x)\,dx + \begin{cases} \tfrac{1}{2}(f_0 - \tfrac{1}{6}hf_0' + \tfrac{1}{360}h^3 f_0''' - \cdots) & (5.9.1) \\ \tfrac{1}{2}(f_0 - \tfrac{1}{6}\Delta f_0 + \tfrac{1}{12}\Delta^2 f_0 - \cdots) & (5.9.2) \end{cases}$$

where $f_k \equiv f(x_0 + kh)$, when applied formally to a function $f(x)$, under the assumptions that $f(x)$ and its derivatives vanish as $x \to \infty$, and that the series and integral are convergent.

If the terms f_k are of constant sign and decrease slowly in magnitude, so that the given series converges slowly, the successive terms in the transformed series generally decrease rapidly in magnitude, at least up to a certain stage. Thus these series, while generally asymptotic, are often useful for calculation in such cases. In illustration, for $f(x) = 1/x^2$ and $h = 1$, the relation (5.9.1) becomes

$$\frac{1}{a^2} + \frac{1}{(a+1)^2} + \frac{1}{(a+2)^2} + \cdots = \frac{1}{a} + \frac{1}{2a^2} + \frac{1}{6a^3} - \frac{1}{30a^5} + \cdots + \frac{B_{2m}}{a^{2m+1}} + E_m \tag{5.9.3}$$

since here $f_0^{(2m-1)} = -(2m)!/a^{2m+1}$. Whereas the series on the left converges rather slowly, the terms on the right decrease rapidly when a is fairly large. Thus, if we take $a = 100$, there follows

$$\frac{1}{100^2} + \frac{1}{101^2} + \frac{1}{102^2} + \cdots = 10^{-2} + \tfrac{1}{2} \times 10^{-4} + \tfrac{1}{6} \times 10^{-6} - \tfrac{1}{30} \times 10^{-10} + E_4$$

and the retention of only the first three terms on the right gives

$$\frac{1}{100^2} + \frac{1}{101^2} + \frac{1}{102^2} + \cdots = 0.0100501667$$

correctly to 10 places. Nearly 2×10^{10} terms of the *original* series would be needed to supply this accuracy!

It is of interest to notice that B_{2i} was shown by Euler to be expressible in the form

$$B_{2i} = \frac{2(-1)^{i-1}(2i)!}{(2\pi)^{2i}}\left(1 + \frac{1}{2^{2i}} + \frac{1}{3^{2i}} + \cdots\right) \qquad (i \geqq 1) \tag{5.9.4}$$

Thus, since $(2i)!$ ultimately grows more rapidly than a^{2i} for *any* fixed a, it follows that, whereas B_{2i} at first decreases with i, ultimately B_{2i} increases more rapidly than a^{2i} for any fixed a, as i increases without limit. Hence it is evident that the result of omitting E_m in the right-hand member of (5.9.3) will not *converge* as $m \to \infty$. The expression (5.8.14) is of no use when $r \to \infty$. However, the test described following that equation shows indeed that here E_m is smaller in magnitude than the first neglected term, and that it decreases in magnitude until m is approximately equal to πa, after which it begins to increase unboundedly in magnitude and to oscillate in sign. In the case $a = 100$, this would mean that the retention of additional terms would continue to improve the approximation until more than 300 terms were taken. However, in the case $a = 1$, for which the left-hand member of (5.9.3) has the known value $\pi^2/6 \doteq 1.64493$, in accordance with (5.9.4), the right-hand member becomes

$$1 + \tfrac{1}{2} + \tfrac{1}{6} - \tfrac{1}{30} + \tfrac{1}{42} - \tfrac{1}{30} + \tfrac{5}{66} - \tfrac{691}{2730} + \tfrac{7}{6} - \tfrac{3617}{510} + \cdots$$

Here the error E associated with the truncation of this series after n terms varies with n as is indicated in the following table.

n	E
1	0.645
2	0.145
3	−0.022
4	0.012
5	−0.012
6	0.021
7	−0.055
8	0.198
9	−0.968
10	6.124

This type of phenomenon, in which the successive members of a sequence of approximations first approach nearer and nearer to the desired result, and then begin to oscillate about it with ever-increasing amplitude, arises very frequently in numerical analysis. Whereas such a situation *can* often be brought about by prolonged propagation of *roundoff* errors (*and is too often attributed to this cause by computers!*), we have seen here, and in Secs. 3.9 and 4.11, that it can also result from successively progressing to procedures of "higher-order accuracy," when this progress leads eventually to using *too many* terms of a divergent (but asymptotic) series, even though it be assumed that *no* roundoff errors are introduced. (Additional situations of this type will be encountered in other chapters.)

The preceding transformations usually are not useful when the terms in the given series fluctuate in sign. However, in those situations when the signs of successive terms steadily alternate, there exist more appropriate transformations, of similar type, which possess the additional advantage that their use does not involve the evaluation of an integral. Their formal derivation is simply effected by noticing that the operational relation

$$p_0 - p_1 + p_2 - p_3 + \cdots \equiv (1 - \mathbf{E} + \mathbf{E}^2 - \cdots)p_0 = \frac{1}{1 + \mathbf{E}}\, p_0 \tag{5.9.5}$$

is valid for any polynomial $p(x)$, and that we have also

$$\frac{1}{1 + \mathbf{E}} = \frac{1}{2}\left(1 - \tanh \frac{h\mathbf{D}}{2}\right) = \frac{1}{2 + \Delta} \tag{5.9.6}$$

Hence, by formally replacing p by f in (5.9.5) and expanding the operator $1/(1 + \mathbf{E})$ in accordance with (5.9.6), we obtain the relations

$$\sum_{k=0}^{\infty} (-1)^k f_k = \begin{cases} \frac{1}{2}(f_0 - \frac{1}{2}hf_0' + \frac{1}{24}h^3 f_0''' - \frac{1}{240}h^5 f_0^{\text{v}} + \frac{17}{40320}h^7 f_0^{\text{vii}} - \cdots) & (5.9.7) \\ \frac{1}{2}(f_0 - \frac{1}{2}\Delta f_0 + \frac{1}{4}\Delta^2 f_0 - \cdots + (-1)^r 2^{-r}\Delta^r f_0 + \cdots) & (5.9.8) \end{cases}$$

The second relation (5.9.8), expressed in terms of forward differences, is often known as *Euler's transformation.*† It is known (see Hardy [1949]) that the transformed series in (5.9.8) will converge whenever the given series does so, and to the same sum. (A transformation having this property sometimes is said to be *regular*.) Indeed, the transformed series may converge when the parent series does not, in which case the sum of the transformed series is often called the *Euler sum* of the parent series. The other transformed series (5.9.7) generally is asymptotic, but the rate of effective convergence of the leading terms often is more rapid.

In illustration, if only the first four terms of the series

$$S = 1 - \frac{1}{2} + \frac{1}{3} - \frac{1}{4} + \cdots + (-1)^{n+1}\frac{1}{n} + \cdots (= \log 2) \tag{5.9.9}$$

are summed initially, to give

$$S = \tfrac{7}{12} + (\tfrac{1}{5} - \tfrac{1}{6} + \tfrac{1}{7} - \cdots)$$

the use of (5.9.7) and (5.9.8), with $f(x) = 1/x$, $x_0 = 5$, and $h = 1$, is found to yield the transformed series

$$S = \tfrac{7}{12} + \begin{cases} \tfrac{1}{10} + \tfrac{1}{100} - \tfrac{1}{5000} + \tfrac{1}{62500} - \tfrac{17}{6250000} + \cdots & (5.9.10) \\ \tfrac{1}{10} + \tfrac{1}{120} + \tfrac{1}{840} + \tfrac{1}{4480} + \tfrac{1}{20160} + \cdots & (5.9.11) \end{cases}$$

after an appropriate tabulation and differencing of the ordinates f_k in the second case.

Retention of five terms of the transformed series in (5.9.10) yields an approximation to $\log 2 \doteq 0.693147183$ with an error smaller than 6×10^{-7}, whereas the same truncation of the Euler series (5.9.11) is in error by about 2×10^{-5}. If additional terms were retained in these two series, the second would continue to converge indefinitely, whereas the oscillation of the first series eventually (after about nine terms) would begin to increase unboundedly. More efficient transformations would have been effected by summing more than four terms of the given series in advance.

A useful variant of the Euler transformation (5.9.8), which also yields a convergent series when the parent series converges, is expressible in the form‡

$$\sum_{k=0}^{\infty} (-1)^k f_k = \frac{1}{2} \sum_{k=0}^{n} \frac{(-1)^k}{2^k} \Delta^k f_0 + \frac{(-1)^{n+1}}{2^{n+1}} \sum_{k=0}^{\infty} (-1)^k \Delta^{n+1} f_k \tag{5.9.12}$$

† This transformation is closely related to that considered in Probs. 8 and 9 of Chap. 1.

‡ This formula can be obtained by operational methods or, rigorously, by n iterations of the transformation considered in Prob. 8 of Chap. 1.

The right-hand member can be interpreted as the result of truncating the Euler formula with nth differences and expressing the error term as an infinite series of $(n + 1)$th differences. In particular, if $\Delta^{n+1} f_k$ is of constant sign for $k \geqq 0$ and tends steadily to zero as $k \to \infty$, we may deduce that the truncation error in the Euler formula (5.9.8) is smaller in magnitude than *twice* the first omitted term and is of the same sign. This situation will exist if $f^{(n+1)}(x)$ is of constant sign when $x \geqq x_0$ and if it tends steadily to zero as $x \to \infty$.

The Euler transformation is most efficient when the alternating series $f_0 - f_1 + f_2 - \cdots$ converges very slowly, so that f_k tends to zero, say, like $1/k$ as $k \to \infty$. When f_k tends to zero, say, like r^k $(r < 1)$, so that the series simulates an alternating geometric series, a useful generalization results from writing

$$f_k = r^k g_k \tag{5.9.13}$$

where r may be identified, for example, with a representative value of f_{k+1}/f_k or with its limit as $k \to \infty$. The formal symbolic relation

$$f_0 - f_1 + f_2 - f_3 + \cdots = (1 - \mathbf{E} + \mathbf{E}^2 - \mathbf{E}^3 + \cdots) f_0$$

then becomes

$$\begin{aligned} f_0 - f_1 + f_2 - f_3 + \cdots &= (1 - r\mathbf{E} + r^2\mathbf{E}^2 - r^3\mathbf{E}^3 + \cdots) g_0 \\ &= \frac{1}{1 + r\mathbf{E}} g_0 = \frac{1}{(1 + r) + r\Delta} g_0 \end{aligned}$$

and yields the formula

$$\sum_{k=0}^{\infty} (-1)^k r^k g_k = \frac{1}{1 + r}\left[g_0 - \frac{r}{1 + r}\Delta g_0 + \left(\frac{r}{1 + r}\right)^2 \Delta^2 g_0 - \cdots\right] \tag{5.9.14}$$

which is equivalent to (5.9.8) when r is taken to be unity. For any fixed $r > 0$, the right-hand member will converge when the left-hand member does so.

Other generalizations of a similar nature are readily devised. Thus, if we write $f_k = c_k g_k$, we may derive the formal relation

$$\sum_{k=0}^{\infty} (-1)^k c_k g_k = \left[\phi(1) + \frac{\phi'(1)}{1!}\Delta + \frac{\phi''(1)}{2!}\Delta^2 + \cdots\right] g_0 \tag{5.9.15}$$

where $\phi(t)$ is the function possessing the Taylor expansion

$$\phi(t) = \sum_{k=0}^{\infty} (-1)^k c_k t^k \tag{5.9.16}$$

under the assumption that the interval of convergence includes $t = 1$. Here c_k is to be determined so that g_k tends to vary slowly with increasing k and, desirably, so that $\phi(t)$ is identifiable in closed form.

A related class of transformations, which frequently accelerate the convergence of alternating series, deals directly with the sequence of *partial sums* S_k, such that

$$S_k = f_0 - f_1 + f_2 - f_3 + \cdots + (-1)^k f_k \tag{5.9.17}$$

and replaces the sequence $S_0, S_1, \ldots, S_k, \ldots$ by a new sequence $T_0, T_1, \ldots, T_k, \ldots$, where

$$T_k = \frac{w_0 S_k + w_1 S_{k-1} + \cdots + w_k S_0}{w_0 + w_1 + \cdots + w_k} \tag{5.9.18}$$

with a suitable definition of the weighting coefficients $w_0, w_1, \ldots, w_k$, after which the transformation may be iterated. It is known (see Hardy [1949]) that the T sequence will converge to the same limit as does the S sequence if the conditions

$$w_0 > 0 \qquad w_r \geqq 0\ (1 \leqq r \leqq k) \qquad \lim_{k\to\infty} \frac{w_k}{w_0 + w_1 + \cdots + w_k} = 0 \tag{5.9.19}$$

are satisfied.

The choices

$$w_0 = w_1 = \cdots = w_k = 1 \tag{5.9.20}$$

and

$$w_0 = w_1 = 1 \qquad w_2 = w_3 = \cdots = w_k = 0 \tag{5.9.21}$$

are most often used, the latter often being particularly efficient (when the f's are positive), and they are associated with the names of Cesàro and Hutton, respectively.

As in the case of the Euler transformation, it happens that the Cesàro sequence may converge to a limit C when the parent series $\sum (-1)^k f_k$ is divergent, in which case C may be called the *Cesàro sum* of that series. A similar statement applies to the Hutton sequence.

5.10 Error Terms in Integration Formulas

This section presents methods of obtaining expressions for the error term to be inserted in a formula for numerical integration, obtained (by operational methods or otherwise) in such a way that it reduces to an identity when applied to a *polynomial* of sufficiently low degree, in those cases when the formula is applied to a function of more general type. The methods are readily modified to the consideration of formulas for interpolation or for numerical differentiation or other linear processes. For simplicity, we deal specifically with integral formulas involving only ordinates.

For present purposes, it is convenient to suppose that the formula at hand is expressed explicitly in terms of the ordinates involved rather than differences or divided differences. Also, in order to include formulas considered in Sec. 3.10 and a class of formulas to be developed in Chap. 8, as well as all others considered so far, we suppose that the formula is of the rather general form

$$\int_a^b w(x)f(x)\,dx = \sum_{k=0}^{n} W_k f(x_k) + R \qquad (5.10.1)$$

where $w(x)$ is a prescribed *weighting function*, which is *unity* in most of the formulas considered so far and which is *nonnegative* in $[a, b]$ in most other applications; where $x_0, x_1, \ldots, x_n$ are $n + 1$ abscissas, not necessarily equally spaced; and where $W_0, W_1, \ldots, W_n$ are the corresponding so-called *weighting coefficients*.

It is supposed that the required error R is zero when $f(x)$ is any polynomial of degree N or less. If also R is *not* zero when $f(x)$ is a polynomial of degree $N + 1$, then N is called the *degree of precision* of the integration formula. However, we suppose here only that the degree of precision is *at least* N, where N is a known positive integer. We also assume explicitly that $w(x) \geqq 0$ in $[a, b]$.

We may transpose Eq. (5.10.1) into the form

$$\mathbf{R}[f(x)] = \int_a^b w(x)f(x)\,dx - \sum_{k=0}^{n} W_k f(x_k) \qquad (5.10.2)$$

where the notation $\mathbf{R}[f(x)]$ is used to indicate that the *operation* involved in the right-hand member has been effected on $f(x)$. Our hypothesis, therefore, is that

$$\mathbf{R}[x^r] = 0 \qquad (r = 0, 1, 2, \ldots, N) \qquad (5.10.3)$$

In order to treat situations in which some of the abscissas lie outside the integration interval $[a, b]$, we suppose that the abscissas are ordered in increasing algebraic order and denote the smaller of x_0 and a by A and the larger of x_n and b by B, so that all relevant values of x lie in the interval $I \equiv [A, B]$. Attention is restricted to those functions which possess $N + 1$ continuous derivatives on $[A, B]$.

Then, for any values of x and $\bar{x}$ in $[A, B]$, we can write

$$f(x) = f(\bar{x}) + \frac{f'(\bar{x})}{1!}(x - \bar{x}) + \frac{f''(\bar{x})}{2!}(x - \bar{x})^2 + \cdots$$
$$+ \frac{f^{(N)}(\bar{x})}{N!}(x - \bar{x})^N + \frac{f^{(N+1)}(\xi)}{(N+1)!}(x - \bar{x})^{N+1} \qquad (5.10.4)$$

where, for any fixed $\bar{x}$ in I, ξ depends upon x, but lies in (A, B). Since the first

$N + 1$ terms in the right-hand member comprise a polynomial of degree N, which is *annihilated* by the operator in (5.10.2), the error $\mathbf{R}[f(x)]$ is the same as the error term corresponding to the remainder term

$$E_N(x) = \frac{f^{(N+1)}(\xi)}{(N+1)!}(x - \bar{x})^{N+1} \qquad (5.10.5)$$

and hence

$$(N + 1)!\,\mathbf{R}[f(x)] = \int_a^b w(x)(x - \bar{x})^{N+1} f^{(N+1)}(\xi)\,dx - \sum_{k=0}^{n} W_k(x_k - \bar{x})^{N+1} f^{(N+1)}(\xi_k) \qquad (5.10.6)$$

where $\xi, \xi_0, \xi_1, \ldots, \xi_n$ all lie in (A, B).

This form of the error term is generally not a very useful one. However, if we denote the maximum value of $|f^{(N+1)}(x)|$ on $[A, B]$ by M, and notice that $|x - \bar{x}| \leqq (B - A)/2$ in $[A, B]$ when $\bar{x} = (A + B)/2$, it permits the crude estimate

$$|R| \leqq \frac{ML^{N+1}}{2^{N+1}(N+1)!}\left[\int_a^b w(x)\,dx + \sum_{k=0}^{n} |W_k|\right] \qquad (5.10.7)$$

where

$$L = B - A \qquad \text{and} \qquad |f^{(N+1)}(x)| \leqq M \text{ on } [A, B] \qquad (5.10.8)$$

Since $R = 0$ in (5.10.1) when $f(x) = 1$, there follows

$$\int_a^b w(x)\,dx = \sum_{k=0}^{n} W_k \qquad (5.10.9)$$

Hence, in those cases *when all the weights W_i are nonnegative*, the error bound (5.10.7) can be expressed in the simpler form

$$|R| \leqq \frac{ML^{N+1}}{2^N(N+1)!}\int_a^b w(x)\,dx \qquad (5.10.10)$$

where $L = b - a$ when none of the abscissas lies outside $[a, b]$.

This error bound, while of simple form, is often extremely conservative. In order to obtain a more useful form, we may replace the remainder (5.10.5) by the integral form

$$E_N(x) = \frac{1}{N!}\int_{\bar{x}}^{x} (x - s)^N f^{(N+1)}(s)\,ds \qquad (5.10.11)$$

which possesses the advantage that no unknown parameter, corresponding to the ξ in (5.10.5), appears (see Sec. 1.9). If we identify $\bar{x}$ with A, the relation (5.10.6) is then replaced by the form

$$N!\,\mathbf{R}[f(x)] = \int_a^b w(x) \int_A^x (x - s)^N f^{(N+1)}(s)\, ds\, dx - \sum_{k=0}^{n} W_k \int_A^{x_k} (x_k - s)^N f^{(N+1)}(s)\, ds \qquad (5.10.12)$$

In order to express this result in more convenient form, it is useful to introduce the notation

$$(x - s)_+^k = \begin{cases} (x - s)^k & \text{when } x > s \\ 0 & \text{when } x \leqq s \end{cases} \qquad (5.10.13)$$

in accordance with which (5.10.12) can be written in the form

$$N!\,\mathbf{R}[f(x)] = \int_a^b w(x) \int_A^B (x - s)_+^N\, f^{(N+1)}(s)\, ds\, dx - \sum_{k=0}^{n} W_k \int_A^B (x_k - s)_+^N\, f^{(N+1)}(s)\, ds \qquad (5.10.14)$$

Since the integration limits are now constant, the order of integration is readily reversed to give

$$N!\,\mathbf{R}[f(x)] = \int_A^B f^{(N+1)}(s) \left[\int_a^b (x - s)_+^N\, w(x)\, dx - \sum_{k=1}^{n} W_k (x_k - s)_+^N\right] ds$$

or, equivalently,

$$\mathbf{R}[f(x)] = \int_A^B G(s) f^{(N+1)}(s)\, ds \qquad (5.10.15)$$

where $G(s)$ is defined by the equation

$$N!\, G(s) = \int_a^b (x - s)_+^N\, w(x)\, dx - \sum_{k=0}^{n} W_k (x_k - s)_+^N \qquad (5.10.16)$$

and may be called the *influence function* (or *kernel function*) for the integration formula (5.10.1) relevant to N.†

It is useful to notice that $G(s)$ *can be considered as the error in* (5.10.1) *when* $f(x)$ *is identified with* $(x - s)_+^N/N!$. The definition can also be expressed in the more explicit form

$$N!\, G(s) = \left.\begin{cases} \displaystyle\int_a^b (x - s)^N w(x)\, dx & (s \leqq a) \\ \displaystyle\int_s^b (x - s)^N w(x)\, dx & (a \leqq s \leqq b) \\ 0 & (s \geqq b) \end{cases}\right\} - \sum_{x_k > s} W_k (x_k - s)^N \qquad (5.10.17)$$

† This form appears to be due to Peano [1913, 1914].

where the notation in the right-hand member indicates that the sum is to be taken over those values of k for which $x_k > s$. It is easily seen that $G(s)$ vanishes for all values of s outside the interval $[A, B]$ over which the integration is effected in (5.10.15).

In illustration, we consider the simple integration formula

$$\int_{-1}^{1} f(x)\,dx = f(\alpha) + f(-\alpha) + R \qquad (0 \leqq \alpha \leqq 1) \qquad (5.10.18)$$

where α is a fixed constant. It is seen that $R = 0$ for $f(x) = 1$ and for $f(x) = x$, but that $R \neq 0$ for $f(x) = x^2$ unless $\alpha^2 = \frac{1}{3}$. Thus we have always $N = 1$, and also $N > 1$ when $\alpha^2 = \frac{1}{3}$. Here $[A, B] = [a, b] = [-1, 1]$ and $w(x) = 1$. The use of (5.10.16) or (5.10.17) gives

$$\begin{aligned} 1!\,G(s) &= \int_{-1}^{1} (x - s)_+\,dx - (-\alpha - s)_+ - (\alpha - s)_+ \\ &= \left[\frac{(x - s)_+^2}{2}\right]_{-1}^{1} - (-\alpha - s)_+ - (\alpha - s)_+ \\ &\equiv \int_{s}^{1} (x - s)\,dx - \sum_{x_k > s} (x_k - s) \end{aligned} \qquad (5.10.19)$$

when $|s| \leqq 1$, so that

$$G(s) = \frac{(1 - s)^2}{2} - \sum_{x_k > s} (x_k - s) \qquad (5.10.20)$$

where $x_0 = -\alpha$ and $x_1 = +\alpha$. Hence there follows

$$G(s) = \begin{cases} \dfrac{(1 - s)^2}{2} - (-\alpha - s) - (\alpha - s) = \dfrac{(1 + s)^2}{2} & (-1 \leqq s \leqq -\alpha) \\ \dfrac{(1 - s)^2}{2} - (\alpha - s) = \dfrac{s^2 + (1 - 2\alpha)}{2} & (-\alpha \leqq s \leqq \alpha) \\ \dfrac{(1 - s)^2}{2} & (\alpha \leqq s \leqq 1) \end{cases} \qquad (5.10.21)$$

and, with $G(s)$ so defined, the error R in (5.10.18) can be expressed in the form

$$R = \int_{-1}^{1} G(s) f''(s)\,ds \qquad (5.10.22)$$

We may notice that this function $G(s)$ is made up of the arcs of three parabolas which join continuously at the transition points, coinciding with the abscissas employed in (5.10.18). However, the slope $G'(s)$ decreases abruptly by unity as each such point is crossed in the positive direction. Also, in each subinterval we have $G''(s) = 1$.

If $\alpha \leqq \frac{1}{2}$, $G(s)$ vanishes only at the ends $s = \pm 1$, and at $s = 0$ in the special case $\alpha = \frac{1}{2}$, and is otherwise *positive* in $[-1, 1]$. Hence, in this case, the second law of the mean may be invoked to permit (5.10.22) to be written in the form

$$R = f''(\xi) \int_{-1}^{1} G(s)\, ds = \frac{1 - 3\alpha^2}{3} f''(\xi) \qquad (0 \leqq \alpha \leqq \tfrac{1}{2}) \qquad (5.10.23)$$

where $|\xi| < 1$. The formula (5.10.18) reduces to the *midpoint rule* over $[-1, 1]$ when $\alpha = 0$. When $\alpha = 1$, and the formula becomes the *trapezoidal rule* over $[-1, 1]$, there follows merely $G(s) = (s^2 - 1)/2$ for $-1 \leqq s \leqq 1$. In this case $G(s)$ is *negative* throughout the interior of the interval, so that the law of the mean again can be applied, and (5.10.23) also holds in this case:

$$R = f''(\xi) \int_{-1}^{1} \frac{s^2 - 1}{2}\, ds = -\tfrac{2}{3} f''(\xi) \qquad (\alpha = 1) \qquad (5.10.24)$$

If $\frac{1}{2} < \alpha < 1$, $G(s)$ changes sign at $s = \pm\sqrt{2\alpha - 1}$, and (5.10.22) cannot be transformed in this way. However, in any case it *can* be deduced that

$$|R| \leqq |f''|_{\max} \int_{-1}^{1} |G(s)|\, ds \qquad (5.10.25)$$

In the special case in which $\alpha = \sqrt{3}/3$ in (5.10.18), R vanishes also for $f(x) = x^2$ and for $f(x) = x^3$, but does not vanish for $f(x) = x^4$. Hence the degree of precision is then 3, and we may obtain a more useful formula by taking $N = 3$, in accordance with which

$$24G(s) = \begin{cases} (1 + s)^4 & (-1 \leqq s \leqq -\alpha) \\ s^4 + 6(1 - 2\alpha)s^2 + (1 - 4\alpha^3) & (-\alpha \leqq s \leqq \alpha) \\ (1 - s)^4 & (\alpha \leqq s \leqq 1) \end{cases} \qquad (5.10.26)$$

where $\alpha = \sqrt{3}/3$. It is easily verified that $G(s)$ is continuous and that it vanishes only at the ends of the interval, so that the second law of the mean may be invoked to give

$$R = f^{\text{iv}}(\xi) \int_{-1}^{1} G(s)\, ds = \tfrac{1}{135} f^{\text{iv}}(\xi)$$

and hence there follows

$$\int_{-1}^{1} f(x)\, dx = f\left(-\frac{\sqrt{3}}{3}\right) + f\left(\frac{\sqrt{3}}{3}\right) + \tfrac{1}{135} f^{\text{iv}}(\xi) \qquad (5.10.27)$$

where $|\xi| < 1$.†

† This remarkable formula is a member of the class of so-called *Gauss quadrature* formulas (to be considered in Sec. 8.5) as well as the class of *Chebyshev quadrature* formulas (to be treated in Sec. 8.14).

This example may serve to indicate the use of the influence function in other cases. From the definition (5.10.17), it is easily seen that $G(s)$ and its first $N - 1$ derivatives are continuous at the transition points and that they all vanish at the end points, $x = A$ and $x = B$, of the interval of integration in (5.10.15). Further, it is found from (5.10.17) that

$$(-1)^N G^{(N)}(s) = \int_s^b w(x)\,dx - \sum_{x_k > s} W_k \qquad (5.10.28)$$

and

$$G^{(N+1)}(s) = (-1)^{N+1} w(s) \qquad (5.10.29)$$

in each subinterval, with the convention that $w(x)$ is to be taken as zero when x is outside $[a, b]$ in both (5.10.28) and (5.10.29). Thus, $(-1)^N G^{(N)}(s)$ increases abruptly by W_i as s increases through the ith abscissa, but is continuous inside each subinterval.

In the usual cases when none of the relevant ordinates correspond to abscissas outside (a, b), so that $(A, B) \equiv (a, b)$, it follows that G and its first $N - 1$ derivatives vanish at the end points of that interval. If, in addition, the ordinates at the end points are not involved, so that the formula is of *open* type, it follows that $G^{(N)}(s)$ also vanishes at those points.†

It may be seen that, if $G(s)$ does not change sign in $[A, B]$, the use of the second law of the mean shows that (5.10.15) is expressible in the form

$$\mathbf{R}[f(x)] = K f^{(N+1)}(\xi) \qquad (A < \xi < B) \qquad (5.10.30)$$

where K is independent of $f(x)$. In particular, if we take $f(x) = x^{N+1}$ there follows

$$\mathbf{R}[x^{N+1}] = (N + 1)!\,K$$

Thus K is determined, and, from (5.10.30), we deduce that

$$\mathbf{R}[f(x)] = \frac{f^{(N+1)}(\xi)}{(N + 1)!}\,\mathbf{R}[x^{N+1}] \qquad (5.10.31)$$

if $G(s)$ does not change sign in $[A, B]$.

In illustration, we have seen that the function $G(s)$ associated with (5.10.18) does not change sign in the cases when $0 \leqq \alpha \leqq \frac{1}{2}$ or when $\alpha = 1$, and that then $N = 1$. Thus, in place of evaluating the integral involved in (5.10.23) *in those cases*, we can use (5.10.31) to obtain the same result more easily:

$$\mathbf{R}[f(x)] = \frac{f''(\xi)}{2!}\left[\int_{-1}^{1} x^2\,dx - \alpha^2 - (-\alpha)^2\right] = \frac{1 - 3\alpha^2}{3} f''(\xi)$$

† For this reason, the presence of a singular derivative of $f(x)$ at an end point tends to be somewhat less troublesome for an open formula than for a comparable closed one.

However, the initial labor of determining $G(s)$ and actually investigating whether or not it changes sign in $[A, B]$ may be appreciable when N is moderately large. The preceding simple example shows that the requirement that the *weights* W_i be positive is *not* sufficient to guarantee that $G(s)$ will be of constant sign.

A third form of the error term, complementing the alternatives (5.10.6) and (5.10.15), can be obtained by replacing $f(x)$ by the sum of the polynomial $y_n(x)$, which agrees with it at the $n + 1$ points $x_0, x_1, \ldots, x_n$ involved in the integration formula, and the appropriate remainder term (2.6.1), so that we write

$$f(x) = y_n(x) + \pi(x)f[x_0, x_1, \ldots, x_n, x] \qquad (5.10.32)$$

where, as before,

$$\pi(x) = (x - x_0)(x - x_1)\cdots(x - x_n) \qquad (5.10.33)$$

If we suppose that the degree of precision of (5.10.1) is at least equal to n, as is true for most of the useful formulas, the polynomial $y_n(x)$ is annihilated by the operator in (5.10.2). Since also the remainder term in (5.10.32) vanishes when $x = x_i$, for $i = 0, 1, \ldots, n$, there follows simply

$$\mathbf{R}[f(x)] = \int_a^b w(x)\pi(x)f[x_0, x_1, \ldots, x_n, x]\,dx \qquad (5.10.34)$$

In many cases of interest, there exists a function $V(x)$ such that

$$w(x)\pi(x) = \frac{d^r V(x)}{dx^r} \qquad (5.10.35)$$

where $V(x)$ and its first $r - 1$ derivatives vanish for both $x = a$ and $x = b$, for some positive integer r. Under this assumption, the result of integrating (5.10.34) by parts r times is seen to be

$$\mathbf{R}[f(x)] = (-1)^r \int_a^b V(x)\frac{d^r}{dx^r} f[x_0, \ldots, x_n, x]\,dx$$

and, after making use of (2.3.9) and (3.3.14), combined in the form

$$\frac{d^r}{dx^r} f[x_0, x_1, \ldots, x_n, x] = \frac{r!}{(n + r + 1)!} f^{(n+r+1)}(\eta) \qquad (5.10.36)$$

where η is interior to the interval spanned by the $n + 2$ arguments on the left, there follows

$$\mathbf{R}[f(x)] = \frac{(-1)^r r!}{(n + r + 1)!}\int_a^b V(x)f^{(n+r+1)}(\eta)\,dx \qquad (5.10.37)$$

If also $V(x)$ is of constant sign in $[a, b]$, this result can be further simplified to the form

$$\mathbf{R}[f(x)] = \frac{(-1)^r r!\, f^{(n+r+1)}(\xi)}{(n+r+1)!} \int_a^b V(x)\, dx \qquad (5.10.38)$$

where ξ lies between the smaller of a and x_0 and the larger of b and x_n. In addition, by integrating by parts r times, and again making use of (5.10.35) and of the assumed properties of $V(x)$, we find that

$$\int_a^b V(x)\, dx = \frac{(-1)^r}{r!} \int_a^b [x^r + u_{r-1}(x)] V^{(r)}(x)\, dx$$
$$= \frac{(-1)^r}{r!} \int_a^b [x^r + u_{r-1}(x)] w(x)\pi(x)\, dx$$

where $u_{r-1}(x)$ is an arbitrary polynomial of degree $r - 1$ or less, which can be taken to be identically zero. Hence (5.10.38) is also expressible in the equivalent form

$$\mathbf{R}[f(x)] = \frac{f^{(n+r+1)}(\xi)}{(n+r+1)!} \int_a^b x^r w(x)\pi(x)\, dx \qquad (5.10.39)$$

where x^r can be replaced by any convenient monic polynomial of degree r (in which the coefficient of x^r is unity) if so desired.

This result will be of particular usefulness in Chap. 8. In the case of the formula (5.10.18), it is found that

$$w(x)\pi(x) = x^2 - \alpha^2 = \frac{d}{dx}\left[\tfrac{1}{3}x^3 - \alpha^2 x + (\tfrac{1}{3} - \alpha^2)\right]$$

where the constant of integration is determined so that the function in brackets vanishes when $x = -1$. That function will also vanish when $x = +1$ if $\alpha^2 = \frac{1}{3}$, in which case there follows further

$$w(x)\pi(x) = \frac{d^2}{dx^2}\left[\tfrac{1}{12}(1 - x^2)^2\right]$$

so that we may take $V(x) = (1 - x^2)^2/12$ in that case. The use of (5.10.38) or (5.10.39), with $n = 1$ and $r = 2$, leads again to the result given in (5.10.27).

It may be noticed that *if* (5.10.38) *or* (5.10.39) *is valid, the degree of precision of the relevant integration formula is $n + r$.*

In order to express in a different form the conditions permitting the use of (5.10.38) or (5.10.39), we may make use of Theorem 12 of Sec. 1.9, to show that, if $V^{(r)}(x) = w(x)\pi(x)$ and if $V, V', \ldots, V^{(r-1)}$ vanish at $x = a$, there follows

$$V(x) = \frac{1}{(r-1)!} \int_a^x (x - s)^{r-1} w(s)\pi(s)\, ds \qquad (5.10.40)$$

and also the requirements that $V, V', \ldots, V^{(r-1)}$ also vanish at $x = b$ take the form

$$\int_a^b (b - s)^k w(s)\pi(s)\, ds = 0 \qquad (k = 0, 1, 2, \ldots, r - 1) \qquad (5.10.41)$$

Further, if we assume that the degree of precision of (5.10.1) is $n + r$, where $r \geqq 1$, it follows that the right-hand member of (5.10.34) will vanish when $f(x)$ is any polynomial of degree $n + r$ or less, or, equivalently, when the divided difference $f[x_0, x_1, \ldots, x_n, x]$, of order $n + 1$, is any polynomial of degree $r - 1$ or less. But this situation implies the truth of (5.10.41).

Hence we may deduce that *if the degree of precision of the integration formula* (5.10.1) *is* $n + r$, *where* $r \geqq 1$, *and if the function*

$$V(x) = \frac{1}{(r-1)!} \int_a^x (x - s)^{r-1} w(s)\pi(s)\, ds$$

does not change sign in $[a, b]$, *then the error* R *is given by* (5.10.38) *or* (5.10.39).

5.11 Other Representations of Error Terms

If the degree of precision of (5.10.1) is exactly n, where $n + 1$ ordinates are used, the function $V(x)$ defined by (5.10.35) will not vanish at both ends of the interval $[a, b]$ when $r \geqq 1$, so that (5.10.38) and (5.10.39) then are not valid. Whereas the use of the G function of the preceding section generally involves the individual consideration of each of the segments $[x_k, x_{k+1}]$, and whereas the vanishing of $\pi(x)$ at each abscissa x_k would require the same subdivision of $[a, b]$ before the second law of the mean could be used in connection with (5.10.34), it may be possible to define V functions which are appropriate to subintervals comprising several such segments, and so to obtain a more useful form of the remainder with decreased labor.

In illustration, a formula approximating the integral of $f(x)$ over $[0, 3]$ by a linear combination of the three ordinates at $x = 0, 1, 2$, with $w(x) = 1$, would possess the error term

$$R = \int_0^3 \pi(x) f[0, 1, 2, x]\, dx \qquad \pi(x) = x(x - 1)(x - 2) \qquad (5.11.1)$$

if its degree of precision were at least 2, by (5.10.34). Here we have

$$\pi(x) = x^3 - 3x^2 + 2x = \tfrac{1}{4}(x^4 - 4x^3 + 4x^2)' = \tfrac{1}{4}[x^2(x - 2)^2]'$$

so that the function $V(x) = x^2(x - 2)^2/4$ is appropriate for the subinterval

[0, 2]. In the remaining subinterval [2, 3], $\pi(x)$ does not change sign. Hence we may deduce that

$$R = -\int_0^2 V(x)f[0, 1, 2, x, x]\, dx + \int_2^3 \pi(x)f[0, 1, 2, x]\, dx$$

$$= -\frac{f^{\text{iv}}(\xi_1)}{4!}\int_0^2 V(x)\, dx + \frac{f'''(\xi_2)}{3!}\int_2^3 \pi(x)\, dx$$

$$= -\tfrac{1}{90}f^{\text{iv}}(\xi_1) + \tfrac{3}{8}f'''(\xi_2) \tag{5.11.2}$$

where both ξ_1 and ξ_2 lie in (0, 3).

In other cases, the function $Q(x)$ defined by the relations

$$Q'(x) = \frac{w(x)\pi(x)}{x - x_k} \qquad Q(A_k) = 0 \tag{5.11.3}$$

or, equivalently,

$$Q(x) = \int_{A_k}^{x} \frac{w(t)\pi(t)}{t - x_k}\, dt \tag{5.11.4}$$

where x_k is one of the abscissas, may have the property that it does not change sign in the subinterval $[A_k, x_k]$ of $[a, b]$, when A_k is suitably chosen. In view of the identity

$$(x - x_k)f[x_0, \ldots, x_k, \ldots, x_n, x]$$
$$= f[x_0, \ldots, x_{k-1}, x_{k+1}, \ldots, x_n, x] - f[x_0, \ldots, x_n] \tag{5.11.5}$$

where the second term on the right is independent of x, we can write

$$\int_{A_k}^{x_k} w(x)\pi(x)f[x_0, \ldots, x_n, x]\, dx$$

$$= \int_{A_k}^{x_k} Q'(x)\{f[x_0, \ldots, x_{k-1}, x_{k+1}, \ldots, x_n, x] - f[x_0, \ldots, x_n]\}\, dx$$

$$= [Q(x)\{f[x_0, \ldots, x_{k-1}, x_{k+1}, \ldots, x_n, x] - f[x_0, \ldots, x_n]\}]_{A_k}^{x_k}$$

$$- \int_{A_k}^{x_k} Q(x)f[x_0, \ldots, x_{k-1}, x_{k+1}, \ldots, x_n, x, x]\, dx \tag{5.11.6}$$

after an integration by parts. Now $Q(x)$ vanishes when $x = A_k$, and its coefficient in the integrated term vanishes when $x = x_k$. Since also $Q(x)$ is assumed not to change sign in $[A_k, x_k]$, the second law of the mean is applicable to the second term, and there follows

$$\int_{A_k}^{x_k} w(x)\pi(x)f[x_0, \ldots, x_n, x]\, dx = -\frac{f^{(n+1)}(\xi)}{(n+1)!}\int_{A_k}^{x_k} Q(x)\, dx \tag{5.11.7}$$

Also, if we notice that $\int Q(x)\,dx = \int Q(x)\,d(x - x_k)$, and integrate by parts, there follows

$$\int_{A_k}^{x_k} Q(x)\,dx = [(x - x_k)Q(x)]_{A_k}^{x_k} - \int_{A_k}^{x_k} (x - x_k)Q'(x)\,dx$$
$$= -\int_{A_k}^{x_k} w(x)\pi(x)\,dx$$

so that (5.11.7) becomes

$$\int_{A_k}^{x_k} w(x)\pi(x)f[x_0, \ldots, x_n, x]\,dx = \frac{f^{(n+1)}(\xi)}{(n+1)!}\int_{A_k}^{x_k} w(x)\pi(x)\,dx \qquad (5.11.8)$$

Thus, in spite of the fact that $\pi(x)$ may change sign in $[A_k, x_k]$, it follows that the result of *formally* applying the law of the mean to the left-hand member of (5.11.8), and *then* using (5.10.36), with $r = 0$, yields a correct result *when the function $Q(x)$ defined by* (5.11.3) *or* (5.11.4) *does not change sign in* $[A_k, x_k]$.

We may notice also that if, instead, $Q(x)$ does not change sign between $x = A_k$ and $x = B_k$, and if $Q(B_k) = 0$, there follows also

$$\int_{A_k}^{B_k} w(x)\pi(x)f[x_0, \ldots, x_n, x]\,dx = \frac{f^{(n+1)}(\xi)}{(n+1)!}\int_{A_k}^{B_k} w(x)\pi(x)\,dx \qquad (5.11.9)$$

by a slight modification of the same argument.

As a first example, we notice that the error term relevant to the Newton-Cotes four-point formula of closed type with $h = 1$

$$\int_0^3 f(x)\,dx = \tfrac{3}{8}[f(0) + 3f(1) + 3f(2) + f(3)] + R \qquad (5.11.10)$$

is of the form

$$R = \int_0^3 \pi(x)f[0, 1, 2, 3, x]\,dx \qquad \pi(x) = x(x-1)(x-2)(x-3) \qquad (5.11.11)$$

Here the use of the function $V(x)$ is found to be inappropriate. However, we find that

$$\frac{\pi(x)}{x-3} = \tfrac{1}{4}[x^2(x-2)^2]'$$

so that the function $Q(x) = x^2(x-2)^2/4$, corresponding to the choice $A_k = 0$ in (5.11.3), is nonnegative for $0 \leqq x \leqq 3$ (as well as for all other real values of x). Hence (5.11.8) applies, with $A_k = 0 \equiv a$ and $x_k = 3 \equiv b$, and it yields

$$R = \frac{f^{\text{iv}}(\xi)}{4!}\int_0^3 \pi(x)\,dx = -\tfrac{3}{80}f^{\text{iv}}(\xi) \qquad (5.11.12)$$

in accordance with (3.5.12).

As a second example, we consider the Newton-Cotes two-point formula of open type with $h = 1$

$$\int_0^3 f(x)\,dx = \tfrac{3}{2}[f(1) + f(2)] + R \qquad (5.11.13)$$

for which we may write

$$R = \int_0^3 \pi(x) f[1, 2, x]\,dx \qquad \pi(x) = (x-1)(x-2) \qquad (5.11.14)$$

Again the use of $V(x)$ is inappropriate. However, we have

$$\frac{\pi(x)}{x-2} = x - 1 = \tfrac{1}{2}[x(x-2)]'$$

corresponding to the choice $x_k = 2$, $A_k = 0$ in (5.11.3), so that (5.11.8) applies over $[0, 2]$. Since $\pi(x)$ does not change sign in $[2, 3]$, we may write

$$\begin{aligned} R &= \frac{f''(\xi_1)}{2!}\int_0^2 \pi(x)\,dx + \frac{f''(\xi_2)}{2!}\int_2^3 \pi(x)\,dx \\ &= \tfrac{1}{3}f''(\xi_1) + \tfrac{5}{12}f''(\xi_2) \end{aligned}$$

and, since the numerical coefficients are of the same sign, we may combine the terms in the form

$$R = \tfrac{3}{4}f''(\xi) \qquad (5.11.15)$$

in accordance with (3.5.18).†

The V and Q methods, *when applicable*, are usually considerably more convenient than the more general G method of Sec. 5.10, which generally entails the determination and analysis of n or more distinct functions [each a polynomial of degree $N + 1$ if $w(x) = 1$] when $n + 1$ ordinates are involved. However, it must be noticed that the V and Q methods are not applicable in those cases when the degree of precision of the integration formula is less than n.

Formulas which involve values of certain *derivatives* of $f(x)$ as well as the value of $f(x)$ itself, at certain points, may be considered as limits of formulas in which $r + 1$ abscissas coalesce into a single abscissa, corresponding to which the values of $f, f', \ldots,$ and $f^{(r)}$ are used. Thus, for example, if the coefficients W_0, W_1, W_2, and C_1 are determined in such a way that the formula

$$\int_{-1}^1 w(x) f(x)\,dx \approx W_0 f(-1) + W_1 f(0) + W_2 f(1) + C_1 f'(0) \qquad (5.11.16)$$

† The same methods apply, in particular, to all Newton-Cotes formulas which employ an even number of ordinates, whereas the V method succeeds when an odd number of ordinates is used. The methods are based on analyses given by Steffensen in those cases.

is exact for $f(x) = 1, x, x^2$, and x^3, and so for any polynomial of degree 3 or less, the error term will be of the form

$$R = \int_{-1}^{1} w(x)(x+1)x^2(x-1)f[-1, 0, 0, 1, x]\,dx \qquad (5.11.17)$$

Here the second law of the mean applies directly and gives the simpler result

$$R = \frac{f^{\text{iv}}(\xi)}{4!}\int_{-1}^{1} w(x)x^2(x^2-1)\,dx \qquad (5.11.18)$$

which yields

$$R = -\tfrac{1}{90}f^{\text{iv}}(\xi) \qquad (5.11.19)$$

in the special case $w(x) = 1$.

However, for the formula

$$\int_{-1}^{1} f(x)\,dx \approx W_0 f(-1) + W_1 f(0) + W_2 f(1) + C_2 f'(1) \qquad (5.11.20)$$

with the weighting coefficients determined by the same requirements, there follows

$$R = \int_{-1}^{1} (x+1)x(x-1)^2 f[-1, 0, 1, 1, x]\,dx \qquad (5.11.21)$$

and, since here $\pi(x)$ changes sign at $x = 0$, another approach is needed. Since also the function $\int_{-1}^{x} \pi(t)\,dt$ does not vanish when $x = 1$, the V method fails. On the other hand, since

$$\frac{\pi(x)}{x-1} = x(x^2-1) = \tfrac{1}{4}[(x^2-1)^2]'$$

the function $Q(x) = (x^2-1)^2/4$ is appropriate with $A_k = -1$, $x_k = 1$, and Eq. (5.11.8) gives

$$R = \frac{f^{\text{iv}}(\xi)}{4!}\int_{-1}^{1} \pi(x)\,dx = -\tfrac{1}{90}f^{\text{iv}}(\xi) \qquad (5.11.22)$$

The fact that (5.11.19) and (5.11.22) are both identical with the error term relevant to Simpson's rule (for which $W_0 = W_2 = \frac{1}{3}$, $W_1 = \frac{4}{3}$, and $C_1 = 0$ or $C_2 = 0$) suggests that both (5.11.16) and (5.11.20) will reduce to Simpson's rule in the case $w(x) = 1$, when the weights are determined in such a way that the degree of precision is at least 3, that is, that the weights C_1 and C_2 will be required to vanish. A direct derivation will confirm this suspicion.

The direct derivation of the error formula relevant to Simpson's rule itself, over $[-1, 1]$, is effected most easily by the V method since here

$$R = \int_{-1}^{1} \pi(x)f[-1, 0, 1, x]\,dx$$

where

$$\pi(x) = x(x^2 - 1) = \tfrac{1}{4}[(x^2 - 1)^2]' \equiv V'(x)$$

Thus there follows

$$R = -\tfrac{1}{4}\int_{-1}^{1} (x^2 - 1)^2 f[-1, 0, 1, x, x]\, dx = -\frac{f^{\text{iv}}(\xi)}{4 \cdot 4!}\int_{-1}^{1} (x^2 - 1)^2\, dx$$
$$= -\tfrac{1}{90} f^{\text{iv}}(\xi)$$

5.12 Supplementary References

The use of symbolic methods essentially dates from Boole [1970 (1860)]. See also Michel [1946], Bickley [1948], and Steffensen [1950]. Useful tables of the coefficients in many finite-difference integration and differentiation formulas, with nonoperational derivations, are given in Singer [1964].

For Comrie's method of *bridging differences* in subtabulation, see Hartree [1958].

The polynomials and numbers of Bernoulli, Euler, and Stirling are treated in Fort [1948], which also lists many other sources.

Techniques for accelerating the convergence of series or sequences are developed by Bickley and Miller [1936], Szasz [1950], Cherry [1950], Rosser [1951], Shanks [1955], Wynn [1956], and Hamming [1962]. Series with known sums are listed by Jolley [1961] and by Mangulis [1965].

General expressions for remainder (error) terms are given by Peano [1913, 1914], Rémès [1940], and Sard [1948*a*]. See also Birkhoff [1906], Radon [1935], von Mises [1936], Daniell [1940], Householder [1953], and Kuntzmann [1959].

PROBLEMS

Section 5.2

1 Obtain the formal relations

$$\mu = \tfrac{1}{2}(\mathbf{E}^{1/2} + \mathbf{E}^{-1/2}) = \frac{2 + \Delta}{2\sqrt{1 + \Delta}} = \frac{2 - \nabla}{2\sqrt{1 - \nabla}} = \sqrt{1 + \tfrac{1}{4}\delta^2}$$

and construct a table expressing each of the operators $\mathbf{E}$, Δ, ∇, δ, and μ similarly in terms of each of the operators $\mathbf{E}$, Δ, ∇, and δ.

2 Establish the relations

$$\Delta = \mathbf{E}\nabla \qquad \nabla = \mathbf{E}^{-1}\Delta \qquad \mathbf{E}^{-1/2}\Delta = \mathbf{E}^{1/2}\nabla = \delta \qquad \Delta\nabla = \nabla\Delta = \Delta - \nabla = \delta^2$$

$$\mu\delta = \tfrac{1}{2}(\Delta + \nabla) \qquad \mathbf{E}^{\pm 1/2} = \mu \pm \tfrac{1}{2}\delta \qquad \mu^2 = 1 + \tfrac{1}{4}\delta^2$$

3 Let $\mathbf{E}_x$, $\mathbf{E}_y$, Δ_x, Δ_y, and so forth, designate operators which affect only the variable indicated by the subscript, with uniform spacings h and k implied in the x and y directions, respectively, so that, for example, $\delta_x^2 f_{0,0} \equiv f_{1,0} - 2f_{0,0} + f_{-1,0}$, where $f_{r,s} \equiv f(x_0 + rh, y_0 + sk)$. By writing

$$f_{r,s} = \mathbf{E}_x^r \mathbf{E}_y^s f_{0,0}$$

and referring to the interpolation formulas of Newton, Stirling, Bessel, and Everett, deduce that a variety of two-dimensional interpolation formulas can be obtained by substituting one of the following indicated expansions for each operator, and truncating the result:

$$\begin{aligned}
\mathbf{E}^p &= 1 + p\Delta + \frac{p(p-1)}{2!}\Delta^2 + \cdots \\
&= 1 + p\nabla + \frac{p(p+1)}{2!}\nabla^2 + \cdots \\
&= 1 + p\mu\delta + \frac{p^2}{2!}\delta^2 + \cdots \\
&= \left[\mu + (p - 0.5)\delta + \frac{p(p-1)}{2!}\mu\delta^2 + \cdots\right]\mathbf{E}^{1/2} \\
&= \left[(1-p) - \frac{p(p-1)(p-2)}{3!}\delta^2 + \cdots\right] \\
&\qquad + \left[p + \frac{p(p^2-1)}{3!}\delta^2 + \cdots\right]\mathbf{E}
\end{aligned}$$

Which pairs of expansions would be appropriate for interpolation near corners of a table? Near the borders? At interior points?

4 By using the Newton forward-difference expansion in both directions in Prob. 3 and retaining only differences through the first in each direction, deduce the approximate formula

$$\begin{aligned}
f_{r,s} &\approx (1 + r\Delta_x)(1 + s\Delta_y)f_{0,0} \\
&= (1-r)(1-s)f_{0,0} + r(1-s)f_{1,0} + s(1-r)f_{0,1} + rsf_{1,1}
\end{aligned}$$

and show that this formula would yield exact results if $f(x, y)$ were of the form $A + Bx + Cy + Dxy$. Also obtain the formula which neglects the mixed second difference $\Delta_x\Delta_y f_{0,0}$, show that it would yield exact results if f were of the form $A + Bx + Cy$, and specialize both formulas when $r = s = \frac{1}{2}$.

5 By using the Everett expansion in both directions in Prob. 3 and neglecting differences and mixed differences of order greater than 3, deduce the approximate formula

$$f_{r,s} \approx (1-r)(1-s)\left[1 - \frac{r(2-r)}{6}\delta_x^2 - \frac{s(2-s)}{6}\delta_y^2\right] f_{0,0}$$

$$+ r(1-s)\left[1 - \frac{1-r^2}{6}\delta_x^2 - \frac{s(2-s)}{6}\delta_y^2\right] f_{1,0}$$

$$+ (1-r)s\left[1 - \frac{r(2-r)}{6}\delta_x^2 - \frac{1-s^2}{6}\delta_y^2\right] f_{0,1}$$

$$+ rs\left[1 - \frac{1-r^2}{6}\delta_x^2 - \frac{1-s^2}{6}\delta_y^2\right] f_{1,1}$$

Show also that it would yield exact results for

$$f(x, y) = A + B_1x + B_2y + C_1x^2 + C_2xy + C_3y^2 + D_1x^3$$
$$+ D_2x^2y + D_3xy^2 + D_4y^3 + E_1x^3y + E_2xy^3$$

and specialize the formula when $r = s = \frac{1}{2}$.

6 A table includes the following ordinates and differences, together with a statement that differences of order 4 or greater are negligible. Use the formula of Prob. 5 to interpolate for $f(6.55, 1.05)$ and for $f(6.524, 1.042)$.

	$y = 1.0$			$y = 1.1$		
x	$f(x, y)$	$\delta_x^2 f$	$\delta_y^2 f$	$f(x, y)$	$\delta_x^2 f$	$\delta_y^2 f$
6.5	0.9989623	−168	−31	0.9989783	−171	−28
6.6	0.9990866	−147	−28	0.9991026	−150	−26

Section 5.3

7 Express each of the operators $\mathbf{E}$, Δ, ∇, δ, μ, and $\mu\delta$ in terms of $h\mathbf{D}$.

8 Express the operator $h^{-1}\mathbf{J}$ in terms of $\mathbf{E}$, Δ, ∇, δ, and $h\mathbf{D}$.

9 Show that the interpolation formulas of Stirling, Bessel, and Everett can be obtained operationally by rewriting the relation $\mathbf{E}^s = e^{sh\mathbf{D}}$ in the forms

$$\mathbf{E}^s = \cosh sh\mathbf{D} + \frac{\sinh sh\mathbf{D}}{\cosh \frac{1}{2}h\mathbf{D}}\mu$$

$$\mathbf{E}^s = \mathbf{E}^{t+(1/2)} = \left(\frac{\cosh th\mathbf{D}}{\cosh \frac{1}{2}h\mathbf{D}}\mu + \sinh th\mathbf{D}\right)\mathbf{E}^{1/2}$$

and

$$\mathbf{E}^s = \frac{\sinh sh\mathbf{D}}{\sinh h\mathbf{D}}\mathbf{E} + \frac{\sinh (1-s)h\mathbf{D}}{\sinh h\mathbf{D}}$$

respectively, and expanding the right-hand members in powers of $\delta = 2 \sinh \frac{1}{2}h\mathbf{D}$

by using the results of Probs. 35 and 36 of Chap. 4 with a replaced by $h\mathbf{D}$ and x by s or t. Why would the corresponding expansion of the simpler relation

$$\mathbf{E}^s = e^{sh\mathbf{D}} = \cosh sh\mathbf{D} + \sinh sh\mathbf{D}$$

be of limited usefulness?

10 From the following rounded values of the function $f(x) = \sin x$, calculate approximate values of $f'(x)$ and $f''(x)$ at each tabular point and compare the results with rounded true values:

x	0.5	0.7	0.9	1.1	1.3	1.5	1.7
$f(x)$	0.47943	0.64422	0.78333	0.89121	0.96356	0.99749	0.99166

Section 5.4

11 If c_j is defined by (5.4.4), show that c_j can be determined recursively by use of the formula

$$c_j = \tfrac{1}{2}c_{j-1} - \tfrac{1}{3}c_{j-2} + \cdots + (-1)^{k+1}\frac{1}{k+1}c_{j-k} + \cdots$$

with $c_0 = 1$ and $c_{-1} = c_{-2} = \cdots = 0$. [Clear fractions in (5.4.4), replace $\log(1 + \Delta)$ by its expansion, and equate coefficients.]

12 Using the data of Prob. 10, calculate the approximate value of $\int_{0.5}^{x} f(t)\,dt$ for $x = 0.7, 0.9$, and 1.1, and the approximate value of $\int_{x}^{1.7} f(t)\,dt$ for $x = 1.1, 1.3$, and 1.5. From these results determine approximate values of the integral taken over each tabular interval.

Section 5.5

13 Using the data of Prob. 10, calculate approximate values of the quantities

$$\int_{0.5}^{0.7}\int_{0.5}^{x} f(t)\,dt\,dx \qquad \int_{1.5}^{1.7}\int_{1.5}^{x} f(t)\,dt\,dx$$

14 If $F''(x) = \log \tan x$, and if $F(1) = 0$ and $F'(1) = 1$, calculate approximate values of $F(x)$ for $x = 1.00(0.02)(1.10)$, using only tabulated five-place values of $\log \tan x$ [$\equiv (\log 10)(\log_{10} \tan x)$] for $x \geqq 1$.

15 Show that, if the operator $\boldsymbol{\theta}$ is defined by the relation

$$\int_{x_k}^{x_k+rh}\int_{x_k}^{x} p(t)\,dt\,dx = \boldsymbol{\theta} p_k$$

then

$$\boldsymbol{\theta} = \frac{\mathbf{E}^r - 1 - rh\mathbf{D}}{\mathbf{D}^2} = h^2\left[\frac{(1+\Delta)^r - 1 - r\log(1+\Delta)}{\Delta^2}\right]\left[\frac{\Delta}{\log(1+\Delta)}\right]^2$$

and determine the first three coefficients in the expansion of the operator $\boldsymbol{\theta}$ in powers of Δ, as functions of r.

16 Show that the right-hand member of the result of operating on the equal members of (5.5.13) by $1 + a_1\nabla + a_2\nabla^2$ is independent of ∇^3 if and only if $a_1 = -1$, that the result is equivalent to (5.5.12) if also $a_2 = 0$, and that a particularly convenient choice is that for which $a_2 = \frac{1}{3}$, leading to the formula

$$P_{k+1} - P_k - P_{k-2} + P_{k-3} = 3h^2(1 - \nabla + \tfrac{5}{12}\nabla^2 + 0\nabla^3 + \tfrac{17}{720}\nabla^4 + \tfrac{17}{720}\nabla^5 + \cdots)P_k''$$

(This formula is used in Sec. 6.10.)

Section 5.6

17 Using the data of Prob. 10, calculate approximate values of

$$\int_{1.1-0.2m}^{1.1+0.2m} f(x)\,dx$$

for $m = 1$, 2, and 3.

18 Derive the operational relation

$$\int_{x_0}^{x_0+h} p(x)\,dx = h\mu\,\frac{\tanh \frac{1}{2}h\mathbf{D}}{\frac{1}{2}h\mathbf{D}}\,p_{1/2}$$

and obtain the expansion in powers of δ in the form

$$\int_{x_0}^{x_0+h} p(x)\,dx = h[\mu - \tfrac{1}{12}\mu\delta^2 + \tfrac{11}{720}\mu\delta^4 - \tfrac{191}{60480}\mu\delta^6 + \cdots]p_{1/2}$$

(Compare with Prob. 18 of Chap. 4.) Show also that it can be expressed in the alternative form

$$\frac{1}{h}\int_{x_0}^{x_1} f(x)\,dx = \tfrac{1}{2}(f_0 + f_1) - \tfrac{1}{12}(\mu\delta f_1 - \mu\delta f_0) + \tfrac{11}{720}(\mu\delta^3 f_1 - \mu\delta^3 f_0) - \tfrac{191}{60480}(\mu\delta^5 f_1 - \mu\delta^5 f_0) + \cdots$$

and deduce *Gauss' sum formula* in the form

$$\frac{1}{h}\int_{x_0}^{x_r} f(x)\,dx = \tfrac{1}{2}f_0 + f_1 + f_2 + \cdots + f_{r-1} + \tfrac{1}{2}f_r - \tfrac{1}{12}(\mu\delta f_r - \mu\delta f_0) + \tfrac{11}{720}(\mu\delta^3 f_r - \mu\delta^3 f_0) - \tfrac{191}{60480}(\mu\delta^5 f_r - \mu\delta^5 f_0) + \cdots$$

19 Use the result of Prob. 18 and the data of Prob. 10 to calculate approximate values of the integral $\int_x^{x+0.2} f(t)\,dt$ for $x = 0.9$ and 1.1.

Section 5.7

20 Subtabulate the data of Prob. 10 for $x = 0.50(0.02)0.70$ and $x = 1.50(0.02)1.70$.

21 If δ' represents the central-difference operator relative to the spacing $h' = ph$, show that

$$\frac{\delta'}{\delta} = \frac{\sinh \frac{1}{2}p h\mathbf{D}}{\sinh \frac{1}{2}h\mathbf{D}}$$

and obtain the expansion of the right-hand member in powers of $\delta = 2 \sinh \frac{1}{2}h\mathbf{D}$ (see Prob. 36 of Chap. 4, with $x = p/2$, $a = h\mathbf{D}$, and $\beta = \delta$), thus deducing the relation

$$\delta' = 2p\left[\frac{\delta}{2} - \frac{1}{3!}(1^2 - p^2)\left(\frac{\delta}{2}\right)^3 + \frac{1}{5!}(1^2 - p^2)(3^2 - p^2)\left(\frac{\delta}{2}\right)^5 - \cdots\right]$$

Show also that

$$\frac{\mu'\delta'}{\mu\delta} = \frac{\sinh p h\mathbf{D}}{\sinh h\mathbf{D}}$$

and obtain the expansion of the right-hand member in powers of δ (see Prob. 35 of Chap. 4, with $x = p$, $a = h\mathbf{D}$, and $\beta = \delta$), thus deducing the relation

$$(\mu\delta)' \equiv \mu'\delta' = p\left[\mu\delta - \frac{1}{3!}(1^2 - p^2)\mu\delta^3 + \frac{1}{5!}(1^2 - p^2)(2^2 - p^2)\mu\delta^5 - \cdots\right]$$

22 In the case of subtabulation to tenths ($p = \frac{1}{10}$), deduce from the results of Prob. 21 the formulas

$$\begin{aligned}
(\mu\delta)' &= 0.1\mu\delta - 0.0165\mu\delta^3 + 0.00329175\mu\delta^5 - \cdots \\
\delta'^2 &= 0.01\delta^2 - 0.000825\delta^4 + \cdots \\
(\mu\delta^3)' &= 0.001\mu\delta^3 - 0.0002475\mu\delta^5 + \cdots \\
\delta'^4 &= 0.0001\delta^4 - \cdots \\
(\mu\delta^5)' &= 0.00001\mu\delta^5 - \cdots
\end{aligned}$$

when differences of order greater than five are neglected, and use these formulas to subtabulate the data of Prob. 10 for $x = 0.90(0.02)1.10$.

23 Suppose that mean values of $f(x)$ are known over each of the subintervals $(x_k - h/2, x_k + h/2)$ $(k = 0, 1, 2, \ldots)$, where $x_{k+1} - x_k = h$, and that approximate mean values over subintervals of length $2ph$, again centered about the points x_k, are required. With the notations

$$m_k = \frac{1}{h}\int_{x_k - h/2}^{x_k + h/2} f(x)\,dx \qquad m_k' = \frac{1}{2ph}\int_{x_k - ph}^{x_k + ph} f(x)\,dx$$

derive the operational relation

$$m_k' = \frac{1}{2p}\frac{\sinh p h\mathbf{D}}{\sinh \frac{1}{2}h\mathbf{D}}\, m_k$$

and deduce the formula

$$m_k' = \left[1 - \frac{1}{3!}(\tfrac{1}{4} - \rho^2)\delta^2 + \frac{1}{5!}(\tfrac{1}{4} - \rho^2)(\tfrac{9}{4} - \rho^2)\delta^4 - \cdots\right] m_k$$

(See Prob. 36 of Chap. 4, with $x = \rho$, $a = h\mathbf{D}$, and $\beta = \delta$.) In particular, deduce the formula

$$f_k = \left[1 - \frac{1^2}{3!}\left(\frac{\delta}{2}\right)^2 + \frac{1^2 \cdot 3^2}{5!}\left(\frac{\delta}{2}\right)^4 - \cdots\right] m_k$$

Section 5.8

24 The *Bernoulli polynomial* $B_k(x)$, of kth degree, is defined as the coefficient of $u^k/k!$ in the expansion

$$\frac{ue^{xu}}{e^u - 1} = \sum_{\nu=0}^{\infty} \frac{u^\nu}{\nu!} B_\nu(x)$$

(*a*) By differentiating the equal members of this relation, deduce the differential recurrence formula

$$B_k'(x) = kB_{k-1}(x) \qquad (k = 1, 2, \ldots)$$

and show also that $B_0(x) = 1$.

(*b*) By making use of the identity

$$\frac{(-u)e^{-xu}}{e^{-u} - 1} = \frac{ue^{(1-x)u}}{e^u - 1}$$

prove that

$$B_k(1 - x) = (-1)^k B_k(x)$$

Also, by integrating the equal members of the defining relation over [0, 1], deduce that

$$\int_0^1 B_k(x)\, dx = 0 \qquad (k > 0)$$

and use this result, together with the recurrence formula of (*a*), to show that

$$B_0(x) = 1 \qquad B_1(x) = x - \tfrac{1}{2} \qquad B_2(x) = x^2 - x + \tfrac{1}{6}$$
$$B_3(x) = x^3 - \tfrac{3}{2}x^2 + \tfrac{1}{2}x$$

and so forth.

(*c*) In accordance with (5.8.7), the kth *Bernoulli number* B_k is defined by the relation $B_k \equiv B_k(0)$. Show that

$$\frac{u}{e^u - 1} + \frac{u}{2} = \frac{u}{2}\coth\frac{u}{2}$$

is an even function of u, and hence deduce that $B_1 = -\frac{1}{2}$ and that $B_{2m+1} = 0$ when $m \geqq 1$.

(*d*) Use the identity

$$\frac{ue^{u/2}}{e^u - 1} = 2\frac{u/2}{e^{u/2} - 1} - \frac{u}{e^u - 1}$$

to deduce that

$$B_k(\tfrac{1}{2}) = (2^{1-k} - 1)B_k$$

25 Use appropriate results of Prob. 24 to show that $B_{2m+1}(x)$ vanishes when $x = 0$, $\frac{1}{2}$, and 1. Show also that, if it vanishes at any point inside [0, 1] in addition to $x = \frac{1}{2}$, then it must vanish at at least two such points. Then deduce that this situation is impossible by using Rolle's theorem to show that its existence would imply that $B'_{2m+1}(x) = (2m + 1)B_{2m}(x)$ vanishes at least four times inside [0, 1], that $B_{2m-1}(x)$ vanishes at at least two points inside [0, 1], in addition to $x = \frac{1}{2}$, and hence that $B_{2m-3}(x), \ldots, B_3(x)$ have the same property, thus establishing a contradiction since $B_3(x) = x(x - \frac{1}{2})(x - 1)$. Show further that the function $\beta_{2m+2}(x) \equiv B_{2m+2}(x) - B_{2m+2}$ vanishes at the ends of the interval [0, 1], and that its vanishing anywhere inside [0, 1] would contradict the preceding result. Hence deduce that *the function $\beta_{2m+2}(x) \equiv B_{2m+2}(x) - B_{2m+2}$ vanishes at $x = 0$ and at $x = 1$, is of constant sign in* [0, 1], *and takes on its extreme value in that interval at $x = \frac{1}{2}$.*

26 Use successive integrations by parts to show that

$$\int_0^1 [B_{2m+2}(s) - B_{2m+2}]F^{(2m+2)}(s)\,ds$$

$$= [\{B_{2m+2}(s) - B_{2m+2}\}F^{(2m+1)}(s) - B'_{2m+2}(s)F^{(2m)}(s + \cdots$$

$$+ B^{(2m)}_{2m+2}(s)F'(s) - B^{(2m+1)}_{2m+2}(s)F(s)]_0^1 + \int_0^1 B^{(2m+2)}_{2m+2}(s)F(s)\,ds$$

Then by using results of Prob. 24, deduce the formula

$$\tfrac{1}{2}[F(1) + F(0)] = \int_0^1 F(s)\,ds + \sum_{i=1}^{m} \frac{B_{2i}}{(2i)!}[F^{(2i-1)}(1) - F^{(2i-1)}(0)] + E$$

where

$$E = -\frac{1}{(2m + 2)!}\int_0^1 [B_{2m+2}(s) - B_{2m+2}]F^{(2m+2)}(s)ds$$

27 By summing the results of increasing the argument of F successively by 0, 1, 2, ..., and $r - 1$, in Prob. 26, obtain the formula

$$\sum_{k=0}^{r} F(k) = \int_0^r F(s)\,ds + \tfrac{1}{2}[F(0) + F(r)]$$

$$+ \sum_{i=1}^{m} \frac{B_{2i}}{(2i)!}[F^{(2i-1)}(r) - F^{(2i-1)}(0)] + E_m(r)$$

where

$$E_m(r) = -\frac{1}{(2m + 2)!}\int_0^1 [B_{2m+2}(s) - B_{2m+2}]\left[\sum_{k=0}^{r-1} F^{(2m+2)}(s + k)\right] ds$$

28 Show that the error term in Prob. 27 can be written in the form

$$E_m(r) = -r\frac{F^{(2m+2)}(\sigma)}{(2m+2)!}\int_0^1 [B_{2m+2}(s) - B_{2m+2}]\,ds = r\frac{B_{2m+2}}{(2m+2)!}F^{(2m+2)}(\sigma)$$

for some σ such that $0 < \sigma < r$, if $F^{(2m+2)}$ is continuous in that interval, and also that, *if* $F^{(2m+2)}(s)$ *does not change sign for* $0 < s < r$, the error term can be expressed in the form

$$E_m(r) = -\frac{B_{2m+2}(\eta) - B_{2m+2}}{(2m+2)!}\sum_{k=0}^{r-1}[F^{(2m+1)}(k+1) - F^{(2m+1)}(k)]$$

$$= -\frac{B_{2m+2}(\eta) - B_{2m+2}}{(2m+2)!}[F^{(2m+1)}(r) - F^{(2m+1)}(0)]$$

for some η such that $0 < \eta < 1$. Further, use the results of Probs. 25 and 24*d* to show that this term is numerically smaller than twice the first term neglected in the expansion of Prob. 27 and is of the same sign. [Notice that this expansion is reduced to that of (5.8.13) if $F(s)$ is identified with $f(x_0 + hs)$, with the substitution $x_0 + hs = x$.]

29 Show that

$$\int_{x_0}^{x_r} p(x)\,dx - h(p_0 + p_1 + \cdots + p_r)$$

$$= \frac{h}{1-\mathbf{E}}[(1-\mathbf{E}^r)h^{-1}\mathbf{J} - (1-\mathbf{E}^{r+1})]p_0$$

$$= h\left[\frac{h^{-1}\mathbf{J} - 1}{1-\mathbf{E}}p_0 - \frac{h^{-1}\mathbf{J} - \mathbf{E}}{1-\mathbf{E}}p_r\right]$$

and, by expressing the operator affecting p_0 in terms of Δ and that affecting p_r in terms of ∇, deduce the Gregory summation formula in the operational form

$$\frac{1}{h}\int_{x_0}^{x_r} p(x)\,dx = (p_0 + p_1 + \cdots + p_{r-1} + p_r)$$

$$-\frac{\phi(\Delta) - 1}{\Delta}p_0 - \frac{\phi(-\nabla) - 1}{(-\nabla)}p_r$$

where

$$\phi(u) = \frac{u}{\log(1+u)} = \sum_{k=0}^{\infty} c_k u^k$$

with the notation of (5.4.4).

30 Use the data given in Prob. 32 of Chap. 3 to obtain approximate values of the integral

$$\sqrt{\frac{2}{\pi}}\int_0^1 e^{-t^2/2}\,dt$$

by means of the Euler-Maclaurin and Gregory formulas.

31 By replacing the derivatives at x_0 and at x_r in (5.8.17) by combinations of mean central differences using formulas such as

$$hf' = \mu\delta f - \tfrac{1}{6}\mu\delta^3 f + \tfrac{1}{30}\mu\delta^5 f - \cdots \qquad h^3 f''' = \mu\delta^3 f - \tfrac{1}{4}\mu\delta^5 f + \cdots$$

deduce *Gauss' sum formula*

$$\int_{x_0}^{x_r} f(x)\,dx = h(\tfrac{1}{2}f_0 + f_1 + f_2 + \cdots + f_{r-1} + \tfrac{1}{2}f_r)$$
$$- \frac{h}{12}(\mu\delta f_r - \mu\delta f_0) + \frac{11h}{720}(\mu\delta^3 f_r - \mu\delta^3 f_0) - \cdots$$

[See also Prob. 18. When the series is truncated, with differences of order $2k - 1$, the error term is the result of replacing the contents of the parentheses in the *first omitted term* by $rh^{2k+3}f^{(2k+2)}(\xi)$, where ξ is between the extreme relevant values of x. (See Steffensen [1950].)]

32 Approximate the value of the integral

$$\int_{-4}^{4} \frac{dx}{1 + x^2} \doteq 2.6516353$$

by use of the Euler-Maclaurin and Gregory formulas with $r = 8$, investigating the effects of successive corrections through those of third order and retaining six decimal places. (See also Sec. 3.9.)

33 Use Gauss' formula (Prob. 31) in Prob. 32.

34 Use the second Euler-Maclaurin formula (5.8.15) in Prob. 32.

35 Express the first two pairs of end corrections in the second Euler-Maclaurin formula (5.8.15) in terms of forward and backward differences, as in Gregory's formula, and use the result in Prob. 32.

Section 5.9

36 Show that if the first $N - 1$ terms of the series

$$S \equiv 1 + \frac{1}{2^2} + \frac{1}{3^2} + \cdots + \frac{1}{n^2} + \cdots = \frac{\pi^2}{6} \doteq 1.644934$$

are summed directly, and if the Euler-Maclaurin sum formula is used to approximate the remainder, there follows

$$S = \left[1 + \frac{1}{2^2} + \frac{1}{3^2} + \cdots + \frac{1}{(N-1)^2}\right]$$
$$+ \left[\frac{1}{N} + \frac{1}{2N^2} + \frac{1}{6N^3} - \frac{1}{30N^5} + \frac{1}{42N^7} - \cdots + \frac{B_{2m}}{N^{2m+1}}\right] + E_m(N)$$

Then determine N and m in such a way that the number of terms to be retained is minimized, assuming successively that approximations which round correctly to 5, 10, and 20 decimal places are required.

37 Suppose that neither $\sum_0^r F(k)$ nor $\int_0^r F(s)\,ds$ necessarily converges as $r \to \infty$, but that their difference tends to a limit C, so that

$$C = \lim_{r\to\infty} \left[\sum_{k=0}^{r} F(k) - \int_0^r F(s)\,ds\right]$$

and that $F(s)$ and all its derivatives tend to zero as $s \to \infty$. Show that the Euler-Maclaurin expansion of Prob. 27 then can be written in the form

$$\sum_{k=0}^{r} F(k) = \int_0^r F(s)\,ds + C + \tfrac{1}{2}F(r) + \sum_{i=1}^{m} \frac{B_{2i}}{(2i)!} F^{(2i-1)}(r) + \bar{E}_m(r)$$

where

$$\begin{aligned}\bar{E}_m(r) &= E_m(r) - E_m(\infty)\\ &= \frac{1}{(2m+2)!}\int_0^1 [B_{2m+2}(s) - B_{2m+2}]\left[\sum_{k=r}^{\infty} F^{(2m+2)}(s+k)\right] ds\end{aligned}$$

and also obtain results analogous to those of Prob. 28 in this case. Show further that

$$C = \tfrac{1}{2}F(0) - \sum_{i=1}^{m} \frac{B_{2i}}{(2i)!} F^{(2i-1)}(0) + E_m(\infty)$$

where

$$E_m(\infty) = \frac{B_{2m+2}(\eta) - B_{2m+2}}{(2m+2)!} F^{(2m+1)}(0) \qquad (0 < \eta < 1)$$

if $F^{(2m+2)}(s)$ does not change sign for $0 < s < \infty$.

38 Use the result of Prob. 37 to deduce the asymptotic expansion

$$1 + \frac{1}{2} + \frac{1}{3} + \cdots + \frac{1}{n} = \log n + C + \frac{1}{2n} - \frac{1}{12n^2} - \cdots - \frac{B_{2m}}{(2m)n^{2m}} + \cdots$$

where

$$C = \lim_{n\to\infty}\left(\sum_{k=1}^{n} \frac{1}{k} - \log n\right)$$

assuming the existence of this limit. Also show that

$$C = \frac{1}{2} + \frac{1}{12} - \frac{1}{120} + \cdots + \frac{B_{2m}}{2m} + E_m$$

where

$$E_m = -\frac{B_{2m+2}(\eta) - B_{2m+2}}{2m+2} \qquad (0 < \eta < 1)$$

and that E_m is of the same sign as the first neglected term and is less than twice as large. Finally, determine the best approximation to C obtainable from this expansion and determine C to five places by equating the two members of the former expansion when $n = 10$. (The constant C involved here is known as *Euler's constant* and is known to round to 0.5772156649.)

39 Use an appropriate modification of the result of Prob. 37 to deduce the asymptotic expansion

$$\log n! \equiv \log 1 + \log 2 + \cdots + \log n$$
$$= (n + \tfrac{1}{2}) \log n + K - n + \frac{1}{12n} - \frac{1}{360n^3} + \cdots$$
$$+ \frac{B_{2m}}{2m(2m-1)n^{2m-1}} + \cdots$$

where

$$K = \lim_{n\to\infty} [\log n! - (n + \tfrac{1}{2}) \log n + n]$$

assuming the existence of this limit, and show that

$$K = 1 - \frac{1}{12} + \frac{1}{360} - \cdots - \frac{B_{2m}}{2m(2m-1)} + E_m$$

where E_m is of the same sign as the first neglected term and is less than twice as large. Also, calculate an approximate value of K from this expansion, and determine K to five places by setting $n = 10$ in the former one and using the fact that $\log 10! \doteq 15.104412$. The true value of K is known to be $\frac{1}{2} \log 2\pi \doteq 0.91894$. Assuming this fact, deduce *Stirling's asymptotic formula for the factorial*, in the form

$$n! = \sqrt{2\pi n}\, n^n e^{-n} \left(1 + \frac{1}{12n} + \frac{1}{288n^2} - \cdots\right)$$

40 Apply the Gregory formula to the approximate summation of the series

$$\sum_{k=1}^{\infty} \frac{k}{(2k+1)^3}$$

to five places, after summing an appropriate number of terms in advance.

41 Use each of the formulas (5.9.7) and (5.9.8) to sum the series

$$\sum_{k=0}^{\infty} \frac{(-1)^k}{k^2 + 4}$$

to five places, after summing an appropriate number of terms in advance.

42 Determine the *Euler sum* of each of the following divergent series:

(*a*) $1 - 1 + 1 - 1 + \cdots + (-1)^n + \cdots$

(*b*) $1 - 2 + 3 - 4 + \cdots + (-1)^{n-1} n + \cdots$

(*c*) $1 - 2 + 4 - 8 + \cdots + (-1)^n 2^n + \cdots$

Also verify that the three series can be obtained formally by setting $x = 1$ in the power-series expansions of $(1 + x)^{-1}$, $(1 + x)^{-2}$, and $(1 + 2x)^{-1}$, respectively, and that the Euler sum in each case is the value taken on by the generating function when $x = 1$.

43 Derive (5.9.12) by means of n iterations of the transformation considered in Prob. 8 of Chap. 1.

44 Show that if N terms of the original series in (5.9.12) are summed in advance and if r terms of the second sum in the right-hand member are retained, there follows

$$\sum_{k=0}^{\infty}(-1)^k f_k = \sum_{k=0}^{N-1}(-1)^k f_k + (-1)^N\left[\frac{1}{2}\sum_{k=0}^{n}\frac{(-1)^k}{2^k}\Delta^k f_N + \frac{(-1)^{n+1}}{2^{n+1}}\sum_{k=0}^{r-1}(-1)^k\,\Delta^{n+1}f_{N-n}\right] + E$$

where

$$E = \frac{(-1)^{N+r+n+1}}{2^{n+1}}\sum_{k=0}^{\infty}(-1)^k\,\Delta^{n+1}f_{N+r+k}$$

Thus, noting that E depends on $N + r$, but not on N or r separately, deduce that *the same result is obtained (1) by summing N terms of the original series in advance, then applying* (5.9.12) *to the remainder, with a chosen value of n, and retaining r terms of the second sum and (2) by the process described by replacing N by N − m and r by r + m, when* $-r \leqq m \leqq N$. Also verify this fact numerically in the case of the series

$$S = 1 - \tfrac{1}{2} + \tfrac{1}{3} - \tfrac{1}{4} + \cdots$$

taking first $N = 2$, $n = 4$, and $r = 2$, and then $N = 3$, $n = 4$, and $r = 1$, and showing that both calculations yield the same result as does the retention of five terms in the transformed series in (5.9.11) (with $N = 4$, $n = 4$, and $r = 0$). (Thus this procedure is useful for remedying a situation in which it is found that an insufficient number N of terms was summed initially to make the Euler transformation effective, without sacrificing the calculation already completed.)

45 Use the formula (5.9.14) to sum the Taylor series

$$\sum_{k=0}^{\infty}(-1)^k\frac{x^k}{k+1} = \frac{1}{x}\log(1 + x) \qquad (|x| < 1)$$

to five decimal places when $x = \tfrac{1}{2}$.

46 Derive (5.9.15) by writing

$$\sum_{k=0}^{\infty}(-1)^k c_k g_k = \sum_{k=0}^{\infty}(-1)^k c_k(1 + \Delta)^k g_0$$

$$= \sum_{k=0}^{\infty}(-1)^k c_k\left[\sum_{r=0}^{\infty}\binom{r}{k}\Delta^r\right] g_0$$

and formally interchanging the order of summation.

47 (*a*) Show that

$$\sum_{k=0}^{\infty} (-1)^k \frac{t^k}{a^k} \sin (k+1)\theta = \frac{a^2 \sin \theta}{a^2 + 2at \cos \theta + t^2} \qquad (|t| < |a|)$$

by taking the imaginary part of $e^{i\theta} \sum (-te^{i\theta}/a)^k$.

(*b*) Deduce from (5.9.15) the formal summation formula

$$\sum_{k=0}^{\infty} (-1)^k \frac{g_k}{a^k} \sin (k+1)\theta = g_0 U(1, \theta) + \frac{\Delta g_0}{1!} U_t(1, \theta) + \frac{\Delta^2 g_0}{2!} U_{tt}(1, \theta) + \cdots$$

where $U(t, \theta) = (a^2 \sin \theta)/(a^2 + 2at \cos \theta + t^2)$. (The parameter a can be taken to be 1 if the series on the left then converges.)

48 (*a*) Show that the Cesàro transformation is not very effective for the sum $S = 1 - \frac{1}{2} + \frac{1}{3} - \cdots$, but that repeated application of the Hutton transformation to the partial sums $S_0, S_1, \ldots, S_8$ yields a five-place approximation to S.

(*b*) Verify that the Cesàro and Hutton sums of the divergent series $1 - 1 + 1 - \cdots + (-1)^n + \cdots$ are equal to the Euler sum (Prob. 42*a*).

Section 5.10

49 Show that $(x - s)_+^n$ is a continuous function of x and s if $n > 0$, and that

$$\int_a^b (x - s)_+^n \, dx = \left. \frac{(x - s)_+^{n+1}}{n+1} \right|_a^b \qquad (n \neq -1)$$

$$\frac{\partial}{\partial x} (x - s)_+^n = n(x - s)_+^{n-1}$$

50 Derive (5.10.17) from (5.10.16) and, under the assumption that the degree of precision of (5.10.1) is at least N, show also that $G(s)$ vanishes when s is outside (A, B), where A and B are the smallest and largest of $x_0, x_1, \ldots, x_n, a$, and b.

51 Obtain the influence function $G(s)$ for which

$$\int_{-1}^{1} F(x)\, dx = \tfrac{1}{3}[F(-1) + 4F(0) + F(1)] + \int_{-1}^{1} G(s) F^{\text{iv}}(s)\, ds$$

in the form

$$G(s) = -\tfrac{1}{72}(1 - |s|)^3(1 + 3|s|) \qquad (|s| \leqq 1)$$

and show that

$$\int_{-1}^{1} G(s) F^{\text{iv}}(s)\, ds = -\tfrac{1}{90} F^{\text{iv}}(\xi) \qquad (|\xi| < 1)$$

Also, by writing $x = (t - x_0 - h)/h$ and $F(x) = f(t)$, deduce Simpson's rule in the form (3.5.11).

52 Apply integration by parts to the result of Prob. 51, to show that the error in Simpson's rule, as applied to $F(x)$ over the interval $[-1, 1]$, can be expressed in the alternative forms

$$R = -\frac{1}{72}\int_{-1}^{1} (1 - |s|)^3(1 + 3|s|)F^{\text{iv}}(s)\, ds$$

$$= -\frac{1}{6}\int_{-1}^{1} s(1 - |s|)^2 F'''(s)\, ds$$

$$= \frac{1}{6}\int_{-1}^{1} (1 - |s|)(1 - 3|s|)F''(s)\, ds$$

and deduce that when the rule is applied to $f(x)$ over an interval $[x_0, x_0 + 2h]$, there follows

$$|R| \leqq \frac{h^5}{90} M_4 \qquad |R| \leqq \frac{h^4}{36} M_3 \qquad |R| \leqq \frac{8h^3}{81} M_2$$

where M_k is the maximum value of $|f^{(k)}(x)|$ on $[x_0, x_0 + 2h]$ under the assumption that $f^{(k)}(x)$ exists and is integrable over that interval.

53 Determine W_0, W_1, and W_2, as functions of α, in such a way that the error term in the formula

$$\int_{-1}^{1} F(x)\, dx = W_0 F(-\alpha) + W_1 F(0) + W_2 F(\alpha) + R \qquad (0 < \alpha \leqq 1)$$

vanishes when $F(x)$ is an arbitrary polynomial of degree 3 or less, showing that the resultant formula is of the form

$$\int_{-1}^{1} F(x)\, dx = \frac{1}{3\alpha^2}\,[F(-\alpha) + 2(3\alpha^2 - 1)F(0) + F(\alpha)] + R$$

and that its degree of precision is 3 unless $\alpha = \sqrt{\frac{3}{5}}$ and is 5 in that case. Also, show that the influence function corresponding to $N = 3$ is given by

$$G(s) = \begin{cases} \frac{1}{24}(1 - |s|)^4 - \dfrac{1}{18\alpha^2}(\alpha - |s|)^3 & (|s| \leqq \alpha) \\ \frac{1}{24}(1 - |s|)^4 & (\alpha \leqq |s| \leqq 1) \end{cases}$$

(Compare with Prob. 58.)

54 Show that the function $G(s)$ obtained in Prob. 53 does not change sign in $[-1, 1]$ when $\alpha = \frac{1}{2}$, and deduce the formula

$$\int_{-1}^{1} F(x)\, dx = \frac{2}{3}[2F(-\tfrac{1}{2}) - F(0) + 2F(\tfrac{1}{2})] + \frac{7}{720}F^{\text{iv}}(\xi)$$

where $|\xi| < 1$. Also transform this result to the Newton-Cotes three-point formula (3.5.20), of open type.

55 Show that the degree of precision of the formula

$$\int_{-1}^{1} F(x)\, dx = \tfrac{1}{15}[7F(1) + 16F(0) + 7F(-1)] - \tfrac{1}{15}[F'(1) - F'(-1)] + R$$

is 5, obtain the influence function relative to $N = 5$ in the form

$$G(s) = \tfrac{1}{3600}(1 - |s|)^4(1 + 4|s| + 5s^2)$$

and deduce that

$$R = \tfrac{1}{4725}F^{\text{vi}}(\xi) \qquad (|\xi| < 1)$$

Also, generalize this result by writing $x = (t - x_0 - h)/h$ and $F(x) = f(t)$.

56 Show that the degree of precision of the formula

$$F(1) - 2F(0) + F(-1) = \tfrac{1}{12}[F''(1) + 10F''(0) + F''(-1)] + R$$

is 5, and that R can be expressed in the form

$$R = \tfrac{1}{360}\int_{-1}^{1} (1 - |s|)^3(3s^2 - 6|s| - 2)F^{\text{vi}}(s)\, ds = -\tfrac{1}{240}F^{\text{vi}}(\xi) \qquad (|\xi| < 1)$$

57 Assuming that $x_0 \leqq x \leqq x_1$, obtain $g(x, s)$ such that

$$f(x) = f(x_0) + (x - x_0)\frac{f(x_1) - f(x_0)}{h} + \int_{x_0}^{x_1} g(x, s)f''(s)\, ds \qquad (x_1 - x_0 = h)$$

in the form

$$hg(x, s) = \begin{cases} -(s - x_0)(x_1 - x) & (x_0 \leqq s \leqq x) \\ -(x - x_0)(x_1 - s) & (x \leqq s \leqq x_1) \end{cases}$$

and deduce the more familiar form of the error term

$$R = \tfrac{1}{2}(x - x_0)(x - x_1)f''(\xi) \qquad (x_0 < \xi < x_1)$$

58 Show that the error term relevant to the formula of Prob. 53 can be written in the form

$$R = \int_{-1}^{1} V'(x)F[-\alpha, 0, \alpha, x]\, dx$$

where $V(x) = \tfrac{1}{4}[(x^2 - \alpha^2)^2 - (1 - \alpha^2)^2]$, and deduce that

$$R = \frac{3 - 5\alpha^2}{180}F^{\text{iv}}(\xi) \qquad (|\xi| < 1)$$

when $0 < \alpha^2 \leqq \tfrac{1}{2}$ or $\alpha^2 = 1$. Also show that this result reduces to the results of Probs. 51 and 54 when $\alpha = 1$ and $\tfrac{1}{2}$, respectively, and, by determining α such that the weighting coefficients are equal, deduce the additional formula

$$\int_{-1}^{1} F(x)\, dx = \tfrac{2}{3}\left[F\left(-\frac{\sqrt{2}}{2}\right) + F(0) + F\left(\frac{\sqrt{2}}{2}\right)\right] + \tfrac{1}{360}F^{\text{iv}}(\xi)$$

59 Determine α such that the error term relevant to the formula of Prob. 53 can be written in the form

$$R = \int_{-1}^{1} V'''(x)F[-\alpha, 0, \alpha, x]\, dx$$

where V, V', and V'' vanish for $x = \pm 1$, show that then $V(x)$ is nonpositive in $[-1, 1]$, and deduce the formula

$$\int_{-1}^{1} F(x)\, dx = \tfrac{1}{9}[5F(-\sqrt{\tfrac{3}{5}}) + 8F(0) + 5F(\sqrt{\tfrac{3}{5}})] + \tfrac{1}{15750}F^{\text{vi}}(\xi) \qquad (|\xi| < 1)$$

Section 5.11

60 By specializing the Newton-Cotes four-point formula of open type to the interval $[-2, 3]$ with $h = 1$, in the form

$$\int_{-2}^{3} F(x)\, dx = \tfrac{5}{24}[11F(-1) + F(0) + F(1) + 11F(2)] + \int_{-2}^{3} \pi(x)F[-1, 0, 1, 2, x]\, dx$$

where $\pi(x) = x(x^2 - 1)(x - 2)$, and considering the function

$$Q(x) = \int_{-2}^{x} \frac{\pi(t)}{t - 2}\, dt$$

show that the error can be expressed in the form

$$E = \frac{F^{\text{iv}}(\xi_1)}{4!}\int_{-2}^{2} \pi(x)\, dx + \frac{F^{\text{iv}}(\xi_2)}{4!}\int_{2}^{3} \pi(x)\, dx = \tfrac{95}{144}F^{\text{iv}}(\xi)$$

where ξ_1, ξ_2, and ξ are in the interval $(-2, 3)$.

61 Determine W_1, W_2, and W_3 such that the formula

$$\int_{0}^{2} xF(x)\, dx = W_1F(0) + W_2F(1) + W_3F(2) + R$$

possesses a degree of precision of at least 2, and show that the resultant formula takes the form

$$\int_{0}^{2} xF(x)\, dx = \tfrac{2}{3}[2F(1) + F(2)] - \tfrac{2}{45}F'''(\xi) \qquad (0 < \xi < 2)$$

62 Derive the formula

$$\int_{-1}^{1} \frac{F(x)}{\sqrt{1 - x^2}}\, dx = \frac{\pi}{4}[F(-1) + 2F(0) + F(1)] - \frac{\pi}{192}F^{\text{iv}}(\xi) \qquad (|\xi| < 1)$$

63 Show that the error R in Prob. 55 can be written in the form

$$R = \int_{-1}^{1} x^2(1 - x^2)^2 F[-1, -1, 0, 0, 1, 1, x]\, dx$$

and that this form leads again to the result

$$R = \tfrac{1}{4725} F^{\text{vi}}(\xi) \qquad (|\xi| < 1)$$

64 Show that the error R relevant to the Newton-Cotes five-point formula of closed type, as applied to $F(x)$ over $[-2, 2]$, can be expressed in the form

$$R = \int_{-2}^{2} V'(x) F[-2, -1, 0, 1, 2, x]\, dx$$

where

$$V(x) = \int_{-2}^{x} t(t^2 - 1)(t^2 - 4)\, dt$$

Show also that $V(x)$ is an even function, so that $V(-x) = V(x)$, and that $V(2) = 0$. Show further that $V(x)$ increases to a positive maximum value as x increases from -2 to -1, that it then decreases steadily as x increases from -1 to 0, and that

$$\begin{aligned} V(0) &= V(-1) + \int_{-1}^{0} t(t^2 - 1)(t^2 - 4)\, dt \\ &= V(-1) - \int_{-2}^{-1} \frac{3 + t}{2 - t} [t(t^2 - 1)(t^2 - 4)]\, dt \\ &= \left(1 - \frac{3 + \eta}{2 - \eta}\right) V(-1) \qquad (-2 < \eta < -1) \end{aligned}$$

Hence deduce that $V(0)$ is positive, that $V(x)$ does not change sign in $[-2, 2]$, and therefore that

$$R = \frac{F^{\text{vi}}(\xi)}{6!} \int_{-2}^{2} x^2(x^2 - 1)(x^2 - 4)\, dx = -\tfrac{8}{945} F^{\text{vi}}(\xi) \qquad (|\xi| < 2)$$

(A similar analysis, due to Steffensen [1950], applies to all Newton-Cotes formulas, of closed type, employing an odd number of ordinates.)

65 For the Newton-Cotes six-point formula of closed type, as applied to $F(x)$ over $[-2, 3]$, show that the function $V(x)$ of Prob. 64 serves as an appropriate Q function over $[-2, 2]$, so that the error R can be expressed in the form

$$\begin{aligned} R &= \frac{F^{\text{vi}}(\xi_1)}{6!} \int_{-2}^{2} x(x^2 - 1)(x^2 - 4)(x - 3)\, dx \\ &\qquad + \frac{F^{\text{vi}}(\xi_2)}{6!} \int_{2}^{3} x(x^2 - 1)(x^2 - 4)(x - 3)\, dx \\ &= -\tfrac{8}{945} F^{\text{vi}}(\xi_1) - \tfrac{863}{60480} F^{\text{vi}}(\xi_2) = -\tfrac{275}{12096} F^{\text{vi}}(\xi) \quad (-2 < \xi < 3) \end{aligned}$$

6

NUMERICAL SOLUTION OF DIFFERENTIAL EQUATIONS

6.1 Introduction

Many techniques are available for the approximate solution of ordinary differential equations, or of sets of such equations, by numerical methods. This chapter presents a selection of frequently used procedures of various types and illustrates their application. In addition, an indication is given of the troublesome problem of error propagation in stepwise integration processes, and overall error bounds are obtained in illustrative cases.

Some comments relative to the problem of selecting an appropriate technique are included in the concluding section (Sec. 6.17). Whereas most of the treatments deal with initial-value problems, brief considerations of boundary-value problems (Sec. 6.15) and characteristic-value problems (Sec. 6.16) are also included.

6.2 Formulas of Open Type

We consider first the problem in which it is desired to obtain a numerical approximate solution of the first-order equation

$$\frac{dy}{dx} = F(x, y) \tag{6.2.1}$$

which takes on a prescribed value y_0 when $x = x_0$,

$$y(x_0) = y_0 \tag{6.2.2}$$

Starting with the known ordinate, it is proposed to calculate successively the ordinates

$$y_1 \equiv y(x_0 + h) \equiv y(x_1), \quad y_2 \equiv y(x_0 + 2h) \equiv y(x_2), \quad \ldots, \quad y_n \equiv y(x_0 + nh) \equiv y(x_n), \quad \ldots \tag{6.2.3}$$

where h is a suitably chosen spacing.

For this purpose, we may, in particular, make use of the relation

$$y_{n+1} = y_n + \int_{x_n}^{x_n+h} y'(x)\,dx \tag{6.2.4}$$

Suppose that the ordinates $y_n, y_{n-1}, \ldots, y_1$, and y_0 are known. Then the corresponding values of $y'(x)$ are calculable from the formula

$$y'_k \equiv y'(x_k) = F(x_k, y_k) \tag{6.2.5}$$

If we approximate $y'(x)$ by the polynomial of degree N which takes on the calculated values at the $N + 1$ points $x_n, x_{n-1}, \ldots,$ and x_{n-N}, by making use of the Newton backward-difference formula (4.3.8),

$$y'_{n+s} \approx y'_n + s\,\nabla y'_n + \frac{s(s+1)}{2!}\nabla^2 y'_n + \cdots + \frac{s(s+1)\cdots(s+N-1)}{N!}\nabla^N y'_n \tag{6.2.6}$$

where

$$s = \frac{x - x_n}{h} \tag{6.2.7}$$

we may use this polynomial to *extrapolate* $y'(x)$ forward over the interval $[x_n, x_n + h]$, for the purpose of approximately effecting the integration indicated in (6.2.4).

The result of this calculation is

$$y_{n+1} = y_n + h\int_0^1 y'_{n+s}\,ds \approx y_n + h\sum_{k=0}^{N} a_k \nabla^k y'_n \tag{6.2.8}$$

where

$$a_k = \int_0^1 \frac{s(s+1)\cdots(s+k-1)}{k!}\,ds \qquad (6.2.9)$$

the leading terms of (6.2.8) being of the form

$$y_{n+1} \approx y_n + h(1 + \tfrac{1}{2}\nabla + \tfrac{5}{12}\nabla^2 + \tfrac{3}{8}\nabla^3 + \tfrac{251}{720}\nabla^4 + \tfrac{95}{288}\nabla^5 + \cdots)y_n' \qquad (6.2.10)$$

in accordance with (5.4.13).

The error term corresponding to truncation with the Nth difference of y_n' is given by h times the integral of the right-hand member of (4.3.9) with $f = y'$, in the form

$$E = h^{N+2}\int_0^1 \frac{s(s+1)\cdots(s+N)}{(N+1)!}\,y^{(N+2)}(\xi)\,ds$$

or, since the coefficient of $y^{(N+2)}$ does not change sign in [0, 1], in the form

$$E = a_{N+1}h^{N+2}y^{(N+2)}(\xi) \qquad (6.2.11)$$

where $x_{n+1} > \xi > x_{n-N}$. Thus, for example, if only third differences are retained, the error is given by $\frac{251}{720}h^5y^{\mathrm{v}}(\xi)$ where $x_{n+1} > \xi > x_{n-3}$.

More generally, we may use (6.2.6) in the relation

$$y_{n+1} = y_{n-p} + h\int_{-p}^{1} y_{n+s}'\,ds \qquad (6.2.12)$$

where p is any positive integer, to express the ordinate following the nth one in terms of the ordinate calculated p steps previously and in terms of, say, $N + 1$ already calculated values of y'. The formulas most frequently used, in addition to (6.2.10) with $p = 0$, are those for which $p = 1$, 3, and 5, the leading terms of which are of the form

$$y_{n+1} \approx y_{n-1} + h(2 + 0\nabla + \tfrac{1}{3}\nabla^2 + \tfrac{1}{3}\nabla^3 + \tfrac{29}{90}\nabla^4 + \tfrac{14}{45}\nabla^5 + \cdots)y_n' \qquad (6.2.13)$$

$$y_{n+1} \approx y_{n-3} + h(4 - 4\nabla + \tfrac{8}{3}\nabla^2 + 0\nabla^3 + \tfrac{14}{45}\nabla^4 + \tfrac{14}{45}\nabla^5 + \cdots)y_n' \qquad (6.2.14)$$

and

$$y_{n+1} \approx y_{n-5} + h(6 - 12\nabla + 15\nabla^2 - 9\nabla^3 + \tfrac{33}{10}\nabla^4 + 0\nabla^5 + \cdots)y_n' \qquad (6.2.15)$$

Whereas the error associated with terminating one of these formulas with the Nth difference can be expressed in the form

$$E = h^{N+2}\int_{-p}^{1} \frac{s(s+1)\cdots(s+N)}{(N+1)!}\,y^{(N+2)}(\xi)\,ds \qquad (6.2.16)$$

where ξ now depends upon s and lies between x_{n+1} and the smaller of x_{n-p} and x_{n-N}, the fact that the coefficient of $y^{(N+2)}$ changes sign in the integration range when $p > 0$ makes it impossible to apply the law of the mean directly in order to obtain a simple form similar to (6.2.11). Somewhat more complicated forms are obtainable by subdividing the range of integration and applying the law of the mean to each subinterval, or, better, by using one of the methods of Secs. 5.10 and 5.11.

The formulas for which p is an *odd* integer are of particular interest because of the fact that, in each such formula, the coefficient of the pth difference is found to be zero. In these cases, the retention of $p - 1$ differences thus affords the same accuracy as the retention of p differences. Indeed, the cases in which $N = p$ correspond to the use of *Newton-Cotes* formulas of *open* type, employing an *odd* number of ordinates, in the integration indicated in (6.2.12). Further, the error terms in those cases can be expressed in a form similar to (6.2.11) and are given for $p = 3$ and $p = 5$ in Eqs. (3.5.20) and (3.5.22). Thus, in particular, we have the special formulas

$$y_{n+1} = y_{n-1} + 2hy'_n + \frac{h^3}{3}\, y'''(\xi) \tag{6.2.17}$$

$$y_{n+1} = y_{n-3} + 4h(y'_n - \nabla y'_n + \tfrac{2}{3}\nabla^2 y'_n) + \frac{14h^5}{45}\, y^{\text{v}}(\xi) \tag{6.2.18}$$

and

$$y_{n+1} = y_{n-5} + 6h(y'_n - 2\nabla y'_n + \tfrac{5}{2}\nabla^2 y'_n - \tfrac{3}{2}\nabla^3 y'_n + \tfrac{11}{20}\nabla^4 y'_n) + \frac{41h^7}{140}\, y^{\text{vii}}(\xi) \tag{6.2.19}$$

where, in each case, ξ lies between the largest and smallest of the arguments involved in that formula. These formulas, and corresponding ones for $p = 7, 9, \ldots$, have the property that, in each case, the retention of differences through the Nth leads to a formula with "accuracy of order $N + 2$," that is, to an error term proportional to h^{N+3}, whereas for the other formulas of the type considered here the accuracy corresponding to the retention of differences through the Nth is of order $N + 1$.†

It is clear that, since a formula employing Nth differences depends upon knowledge of $N + 1$ successive values of y_k, and since initially only y_0 is known, such a formula cannot be used until N additional ordinates have been determined by another method. Before illustrating the use of such formulas, it is desirable to consider a class of related formulas.

† It is seen that the terminology here is also such that a formula with "accuracy of order m" would yield exact results if the required solution $y(x)$ were a polynomial of degree m or less. When $y(x)$ is *not* such a function, it is not *necessarily* true that an increase in m corresponds to an improvement in the approximation afforded, as was seen in Sec. 3.9.

6.3 Formulas of Closed Type

The formulas derived in the preceding section express y_{n+1} in terms only of previously calculated ordinates and slopes. A set of similar formulas which involve also the unknown slope y'_{n+1} is obtained by replacing the right-hand member of (6.2.6) by the interpolation polynomial agreeing with $y'(x)$ at $x_{n+1}, x_n, \ldots, x_{n-N+1}$:

$$y'_{n+s} \approx y'_{n+1} + (s-1)\,\nabla y'_{n+1} + \frac{(s-1)s}{2!}\nabla^2 y'_{n+1} + \cdots + \frac{(s-1)s(s+1)\cdots(s+N-2)}{N!}\nabla^N y'_{n+1} \tag{6.3.1}$$

where s is again defined by (6.2.7). If this approximation is introduced into (6.2.12), the results in the cases $p = 0$, 1, and 3 are obtained in the forms

$$y_{n+1} \approx y_n + h(1 - \tfrac{1}{2}\nabla - \tfrac{1}{12}\nabla^2 - \tfrac{1}{24}\nabla^3 - \tfrac{19}{720}\nabla^4 - \tfrac{3}{160}\nabla^5 - \cdots)y'_{n+1} \tag{6.3.2}$$

$$y_{n+1} \approx y_{n-1} + h(2 - 2\nabla + \tfrac{1}{3}\nabla^2 + 0\nabla^3 - \tfrac{1}{90}\nabla^4 - \tfrac{1}{90}\nabla^5 - \cdots)y'_{n+1} \tag{6.3.3}$$

and

$$y_{n+1} \approx y_{n-3} + h(4 - 8\nabla + \tfrac{20}{3}\nabla^2 - \tfrac{8}{3}\nabla^3 + \tfrac{14}{45}\nabla^4 - 0\nabla^5 - \cdots)y'_{n+1} \tag{6.3.4}$$

The error associated with retaining only Nth differences in a formula relating y_{n+1} and y_{n-p} can be expressed in the form

$$E = h^{N+2}\int_{-p}^{1} \frac{(s-1)s(s+1)\cdots(s+N-1)}{(N+1)!}\, y^{(N+2)}(\xi)\, ds \tag{6.3.5}$$

where ξ lies between x_{n+1} and the smaller of x_{n-p} and x_{n-N}. When $p = 0$, the law of the mean can be used, as in the preceding section, to show that the error is expressible in the form (6.2.11), where a_{N+1} is the numerical coefficient in the first neglected term. In the cases for which p is an odd integer, it is found that retention of $p + 1$ differences is equivalent to the retention of $p + 2$ differences and that the use of these special formulas corresponds to the use of Newton-Cotes formulas of *closed* type, employing an odd number of ordinates, for which the error terms are obtainable from Sec. 3.5. Thus, when $p = 1$ and $p = 3$, we have the special formulas

$$y_{n+1} = y_{n-1} + 2h(y'_{n+1} - \nabla y'_{n+1} + \tfrac{1}{6}\nabla^2 y'_{n+1}) - \frac{h^5}{90}y^{\text{v}}(\xi) \tag{6.3.6}$$

and

$$y_{n+1} = y_{n-3} + 4h(y'_{n+1} - 2\nabla y'_{n+1} + \tfrac{5}{3}\nabla^2 y'_{n+1} - \tfrac{2}{3}\nabla^3 y'_{n+1} + \tfrac{7}{90}\nabla^4 y'_{n+1}) - \frac{8h^7}{945} y^{\text{vii}}(\xi) \qquad (6.3.7)$$

for which the retention of Nth differences yields an accuracy of order $N + 2$, whereas the other formulas of the type considered generally yield $(N + 1)$th-order accuracy.

Formulas of the sort derived in this section are said to be of *closed* type, since the expressions for the required ordinate y_{n+1}, at the point x_{n+1}, involve the unknown slope y'_{n+1} at that point, whereas those of the preceding section involve only *known* slopes at preceding points and are accordingly said to be of *open* type. A comparison of corresponding formulas employing a like number of differences shows that the error terms associated with formulas of closed type possess smaller numerical coefficients. However, since the unknown y_{n+1} is involved (explicitly and implicitly) in *both* members of formulas of closed type, it would appear that this advantage must be weighed against the fact that, unless $y' \equiv F(x, y)$ is a *linear* function of y, the equation relevant to such a formula generally must be solved for y_{n+1} by iterative methods. (This point is considered further in Sec. 6.6.)

6.4 Start of Solution

Except for the special formula

$$y_{n+1} = y_n + hy'_n + \frac{h^2}{2} y''(\xi) \qquad (6.4.1)$$

obtained by omitting all differences in (6.2.10), and for simple closed formulas obtained from (6.3.2) by retaining not more than one difference, each of the formulas obtained in the preceding sections can be applied only after the calculation of a number of ordinates $y_1, y_2, \ldots, y_r$, in addition to the prescribed ordinate y_0, where r is the number of differences retained in an open formula and is one less than that number in a closed formula.

One method of starting the solution of the problem

$$\frac{dy}{dx} = F(x, y) \qquad y(x_0) = y_0 \qquad (6.4.2)$$

consists of determining the coefficients of a finite Taylor expansion

$$y_s \equiv y(x_0 + hs) = y_0 + \frac{hy'_0}{1!} s + \frac{h^2 y''_0}{2!} s^2 + \cdots + \frac{h^r y_0^{(r)}}{r!} s^r + \frac{h^{r+1} y^{(r+1)}(\xi)}{(r+1)!} s^{r+1} \qquad (6.4.3)$$

where $y_0^{(k)} \equiv (d^k y/dx^k)_{x=x_0}$ and $x_0 < \xi < x_0 + hs$, by successively differentiating the basic differential equation or otherwise, under the assumption that a representation of this type exists when s is sufficiently small. Thus, recalling that $d/dx = \partial/\partial x + y' \, \partial/\partial y$, we obtain the relations

$$\begin{aligned} y' &= F(x, y) \\ y'' &= F_x(x, y) + y'F_y(x, y) \\ y''' &= F_{xx}(x, y) + 2y'F_{xy}(x, y) + y'^2 F_{yy}(x, y) + y''F_y(x, y) \end{aligned}$$

and so forth, and hence there follows

$$y_0' = F(x_0, y_0) \qquad y_0'' = F_x(x_0, y_0) + y_0' F_y(x_0, y_0) \tag{6.4.4}$$

and so forth.

Whereas these *general* expressions become quite involved as the order of the required derivative increases, they are not actually needed in practice. In order to illustrate this fact, we consider the specific example

$$\frac{dy}{dx} = x^2 - y \qquad y(0) = 1 \tag{6.4.5}$$

for which the *exact* solution is readily found to be

$$y = 2 - 2x + x^2 - e^{-x} \tag{6.4.6}$$

From the given equation, we obtain successively

$$y' = x^2 - y, \quad y'' = 2x - y', \quad y''' = 2 - y'', \\ y^{\text{iv}} = -y''', \quad y^{\text{v}} = -y^{\text{iv}}, \quad \ldots \tag{6.4.7}$$

and hence, with $x_0 = 0$, there follows

$$y_0 = 1, \quad y_0' = -1, \quad y_0'' = 1, \quad y_0''' = 1, \quad y_0^{\text{iv}} = -1, \quad y_0^{\text{v}} = 1, \quad \ldots$$

Thus, if we take $h = \frac{1}{10}$, Eq. (6.4.3) gives

$$y_s = 1 - \frac{s}{10} + \frac{1}{2}\left(\frac{s}{10}\right)^2 + \frac{1}{6}\left(\frac{s}{10}\right)^3 - \frac{1}{24}\left(\frac{s}{10}\right)^4 + \frac{1}{120}\left(\frac{s}{10}\right)^5 + \cdots \tag{6.4.8}$$

and, with $s = 1$, 2, and 3, we obtain

$$y_1 = 0.90516 \qquad y_2 = 0.82127 \qquad y_3 = 0.74918 \tag{6.4.9}$$

to five places. Since the successive terms of (6.4.8) alternate in sign from the fourth term onward and decrease steadily in magnitude, the error due to truncation is smaller in magnitude than the first neglected term and is of the same sign. Additional ordinates could be obtained to this accuracy by retaining sufficiently many terms in the expansion. Alternatively, a new expansion could be launched from the point x_3, in the form

$$y_{3+s} = y_3 + \frac{hy_3'}{1!} s + \frac{h^2 y_3''}{2!} s^2 + \cdots$$

with y_3 known and $y_3', y_3'', \ldots$ calculable in terms of y_3 from (6.4.7).

It is obvious that the linear example chosen illustrates a particularly simple case, because of the simplicity of the relations (6.4.7) and because of the fact that (6.4.8) is an alternating series and hence is amenable to a precise truncation-error analysis. More usually, the relations (6.4.7) are replaced by successive equations which increase fairly rapidly in complexity, so that it is usually desirable to abandon this procedure in favor of a more convenient one when sufficiently many starting values have been obtained.†

Discussion of the existence of (6.4.3) in the general case of the problem (6.4.2), as well as consideration of other types of representations which can be used when (6.4.3) cannot, must be omitted here (see Sec. 6.18). In some cases it is preferable to determine the coefficients in an assumed expansion of the form

$$y(x) = \sum_{k=0}^{\infty} A_k (x - x_0)^k$$

by inserting that expansion in the differential equation and obtaining a recurrence formula to be satisfied by the A's.

A method similar to the preceding one, which has the advantage that the order of the highest derivative required is about half that needed in (6.4.3), but has the disadvantage that each forward step involves an iterative process, is treated in Sec. 6.12.

Mention should also be made of *Picard's method*, in which the problem (6.4.2) is first transformed into the *integral equation*

$$y(x) = y_0 + \int_{x_0}^{x} F(x, y(x))\, dx$$

and successive *functions* approximating $y(x)$ near $x = x_0$ are generated by the iteration

$$y^{[k+1]}(x) = y_0 + \int_{x_0}^{x} F(x, y^{[k]}(x))\, dx \qquad (6.4.10)$$

The initial approximation $y^{[0]}(x)$ is conveniently taken to be the constant y_0

† For adaptations of the "Taylor-series method" to computers, see, for example, references given in Lapidus and Seinfeld [1971], p. 81.

or the linear function $y_0 + y_0'(x - x_0)$, where y_0' is determined from the differential equation.

Thus, in the preceding example, we would write

$$y^{[k+1]}(x) = 1 + \int_0^x [x^2 - y^{[k]}(x)]\, dx \tag{6.4.11}$$

and, with $y^{[0]}(x) = 1$, there would then follow

$$y^{[1]}(x) = 1 - x + \tfrac{1}{3}x^3 \qquad y^{[2]}(x) = 1 - x + \tfrac{1}{2}x^2 + \tfrac{1}{3}x^3 - \tfrac{1}{12}x^4 \tag{6.4.12}$$

and so forth. The accuracy afforded by a member of the sequence of approximations at a certain number of points $x_1, x_2, \ldots$ could be estimated by comparing calculated values at those points with values calculated from the preceding approximation, or by use of appropriate analytical methods.

While Picard's method is of great theoretical importance, the explicit evaluation of the integral in (6.4.10) is often impracticable in cases which are less simple than the preceding one. Thus, for the problem $y' = \cos(x + y)$, $y(0) = 1$, the first iteration with $y^{[0]}(x) = 1$ gives $y^{[1]}(x) = 1 - \sin 1 + \sin(x + 1)$, and the second iteration would involve the evaluation of the form

$$y^{[2]}(x) = 1 + \int_0^x \cos[1 - \sin 1 + x + \sin(x + 1)]\, dx$$

Also, when $F(x, y)$ is not given analytically, neither this procedure nor the Taylor-series method is directly applicable.

A frequently used class of procedures consists of evaluating the integral of $y' \equiv F(x, y)$ in (6.4.10) approximately by use of numerical methods. Thus, in particular, if y' is approximated by the Newton forward-difference polynomial-interpolation formula, the results of the integration are obtained by replacing $p(x)$ by $y'(x)$ in (5.4.8), and we have the formulas

$$\begin{aligned} y_1 &= y_0 + h[1 + \tfrac{1}{2}\Delta - \tfrac{1}{12}\Delta^2 + \tfrac{1}{24}\Delta^3 - \tfrac{19}{720}\Delta^4 + \cdots]y_0' \\ y_2 &= y_0 + h[2 + 2\Delta + \tfrac{1}{3}\Delta^2 + 0\Delta^3 - \tfrac{1}{90}\Delta^4 + \cdots]y_0' \\ y_3 &= y_0 + h[3 + \tfrac{9}{2}\Delta + \tfrac{9}{4}\Delta^2 + \tfrac{3}{8}\Delta^3 - \tfrac{3}{80}\Delta^4 + \cdots]y_0' \\ y_4 &= y_0 + h[4 + 8\Delta + \tfrac{20}{3}\Delta^2 + \tfrac{8}{3}\Delta^3 + \tfrac{14}{45}\Delta^4 + \cdots]y_0' \end{aligned} \tag{6.4.13}$$

and so forth. Here y_0 is given, and if, say, y_1, y_2, y_3, and y_4 are *estimated*, the corresponding values of $y_0', \ldots, y_4'$ can be calculated from the differential equation and introduced into the right-hand members to give the new approximations to $y_1, \ldots, y_4$, after which the process may be iterated.

In the case of (6.4.5) we may notice that, since the value $y_0' = -1$ is obtained from the differential equation, the initial approximation $y^{[0]}(x_k) =$

$1 - x_k$ is appropriate. With $h = 0.1$, the following initial array then may be formed when three additional ordinates are incorporated:

x	y	y'	$\Delta y'$	$\Delta^2 y'$	$\Delta^3 y'$
0.0	1.00000	−1.00000			
			11000		
0.1	0.90000	−0.89000		2000	
			13000		0
0.2	0.80000	−0.76000		2000	
			15000		
0.3	0.70000	−0.61000			

The use of the first three of Eqs. (6.4.13), retaining third differences, leads to the new array

0.0	1.00000	−1.00000			
			10467		
0.1	0.90533	−0.89533		799	
			11266		−198
0.2	0.82267	−0.78267		601	
			11867		
0.3	0.75400	−0.66400			

Three additional iterations yield results which are unchanged to five places by further iteration and which are correct to those places. The correctness of those values would be checked, in practice, by considering the effects of the neglected fourth differences, sample values of which become available as the calculation is advanced from this stage.

A useful variation of this procedure consists of using *central* differences and of determining ordinates on both sides of the initial point x_0. Thus, by appropriate integration of the Stirling interpolation formula, we obtain the relations

$$\begin{aligned} y_{-2} &= y_0 + h[-2 + 2\mu\delta - \tfrac{4}{3}\delta^2 + \tfrac{1}{3}\mu\delta^3 - \tfrac{7}{45}\delta^4 + \cdots]y_0' \\ y_{-1} &= y_0 + h[-1 + \tfrac{1}{2}\mu\delta - \tfrac{1}{6}\delta^2 - \tfrac{1}{24}\mu\delta^3 + \tfrac{1}{180}\delta^4 + \cdots]y_0' \\ y_1 &= y_0 + h[1 + \tfrac{1}{2}\mu\delta + \tfrac{1}{6}\delta^2 - \tfrac{1}{24}\mu\delta^3 - \tfrac{1}{180}\delta^4 + \cdots]y_0' \\ y_2 &= y_0 + h[2 + 2\mu\delta + \tfrac{4}{3}\delta^2 + \tfrac{1}{3}\mu\delta^3 + \tfrac{7}{45}\delta^4 + \cdots]y_0' \end{aligned} \tag{6.4.14}$$

Since here the calculated ordinates are taken as close to the given one as is possible, the convergence of the iterative process is generally more rapid than that associated with the use of (6.4.13) unless the solution displays unfavorable characteristics to the left of the point x_0.† Since truncation with a mean odd difference would not correspond to true collocation, it is desirable to use a symmetrical array of abscissas.

† The calculation of ordinates on both sides of the starting point is also frequently convenient when use is made of Taylor series.

In the case of the preceding example, we may start with the array

x	y	y'	$\delta y'$	$\delta^2 y'$	$\delta^3 y'$	$\delta^4 y'$
−0.2	1.20000	−1.16000				
			7000			
−0.1	1.10000	−1.09000		2000		
			9000		0	
0.0	1.00000	−1.00000	(10000)	2000	(0)	0
			11000		0	
0.1	0.90000	−0.89000		2000		
			13000			
0.2	0.80000	−0.76000				

when four additional ordinates are used. The mean odd central differences are entered in parentheses. After four iterations, using (6.4.14), we obtain the array

−0.2	1.21860	−1.17860				
			8377			
−0.1	1.10483	−1.09483		1106		
			9483		−105	
0.0	1.00000	−1.00000		1001		9
			10484		−96	
0.1	0.90516	−0.89516		905		
			11389			
0.2	0.82127	−0.78127				

which is unchanged, to the five places retained, by further iteration. Thus five-place values of y_{-2}, y_{-1}, y_0, y_1, and y_2 are now available for the advancing calculation. Here the fact that the effect of the fourth differences is negligible supplies fair evidence that sufficiently many ordinates were used.

It is important to notice that the formulas of (6.4.13) or (6.4.14) can also be expressed explicitly in terms of the slopes y'_k, if the use of differences is undesirable, once the number of slopes to be retained has been decided (the corresponding *five*-slope formulas are given in Milne [1970]).

Another class of self-starting methods, which are also useful when $F(x, y)$ is not defined analytically, but which are *noniterative*, is treated in Secs. 6.13 and 6.14.

6.5 Methods Based on Open-type Formulas

Once at least N additional ordinates, say, $y_1, y_2, \ldots, y_N$, are determined, the calculation may be continued by use of one of the formulas, derived in Sec. 6.2 or 6.3, which involves Nth differences. In the case of the example (6.4.5), with the calculated data (6.4.9), the preliminary tabulation may be arranged as in Table 6.1 where, for compactness, the backward difference $\nabla^k y'_n$ is written in the same line as the entry y'_n.

Table 6.1

x	y	y'	$\nabla y'$	$\nabla^2 y'$	$\nabla^3 y'$	$\nabla^4 y'$
0.0	1.00000	−1.00000				
0.1	0.90516	−0.89516	10484			
0.2	0.82127	−0.78127	11389	905		
0.3	0.74918	−0.65918	12209	820	−85	

In particular, the *Adams* (or *Adams-Bashforth*) *method* (Bashforth and Adams [1883]) uses formula (6.2.10), truncated to a suitable number of terms, for advancing the calculation. (The simplest such procedure, in which *no* differences are retained, is often known as *Euler's method.*) Thus, if third differences are retained, the Adams method next yields

$$y_4 \approx 0.74918 + \tfrac{1}{10}[-0.65918 + \tfrac{1}{2}(0.12209) + \tfrac{5}{12}(0.00820) - \tfrac{3}{8}(0.00085)]$$
$$\doteq 0.68968$$

after which an additional line

0.4	0.68968	−0.52968	12950	741	−79	6	(6.5.1)

is entered for the purpose of advancing to y_5. If again only third differences are retained, the next line appears as follows:

0.5	0.64347	−0.39347	13621	671	−70	9	(6.5.2)

The fourth difference is carried along as a partial check column. Since the truncation error in each step is of the form

$$\tfrac{251}{720}h^5 y^{\mathrm{v}}(\xi)$$

for some ξ, and since $h^4 y^{\mathrm{v}}(\xi)$ is given by $\nabla^4 y'(\eta)$, for some η, the two available sample values of $\nabla^4 y'$ indicate that $h^4 y^{\mathrm{v}}$ probably does not vary strongly over the relevant range, so that a fairly dependable estimate of the truncation error committed in each of the steps can be obtained by calculating the contribution $\tfrac{251}{720}h\,\nabla^4 y'_n$ of the first neglected difference. With $h = 0.1$, this contribution will amount to less than one-half unit in the fifth place if $\nabla^4 y'_n$ does not exceed 14 units in that place.

If use is made instead of formula (6.2.18), in which only *second* differences are retained, the same results are obtained. Here the error estimate again depends upon the fourth difference, the factor $\tfrac{14}{45} \doteq 0.31$ replacing the factor $\tfrac{251}{720} \doteq 0.35$ relevant to the Adams formula with third differences. Thus, as compared with the Adams method, this method here possesses the advantage that one less difference is needed in the calculation (but not in the error check) and that the coefficients in the formula are somewhat simpler.

It should be emphasized that the errors so far considered are those which would arise in a single step from x_n to x_{n+1} if $y_0, y_1, \ldots, y_n$ were exactly correct and if no roundoff errors were introduced in that step. In addition, however, one must consider the cumulative effect of the errors introduced in *preceding* steps. Whereas consideration of the *propagation* of errors is postponed to Secs. 6.7 and 6.8, it may be remarked here that the advantage in stability generally lies with the Adams method. This situation is related to the fact that the ordinates themselves are "loosely coupled" by (6.2.18), in that the ordinate y_k is linked directly only with ordinates of the form y_{k-4i}, where i is an integer, whereas in (6.2.10) all ordinates are directly linked together.

6.6 Methods Based on Closed-type Formulas. Prediction-Correction Methods

A possible method of employing one of the formulas of Sec. 6.3 to calculate y_{n+1} consists of first *estimating* y_{n+1}, calculating $y'_{n+1} = F(x_{n+1}, y_{n+1})$ corresponding to this estimate, forming the requisite corresponding differences $\nabla^k y'_{n+1}$, and then calculating an improved estimate of y_{n+1} by use of the formula. The cycle then is to be repeated, if necessary, until two successive estimates agree within the prescribed tolerance, assuming convergence of this iterative process. The *initial* estimate, say, $y_{n+1}^{(0)}$, may be obtained by use of a formula of open type.

In illustration, returning to the example considered in the preceding section, line (6.5.1) can be considered as the result of using the Adams method, with third differences, as a *predictor*. If now the data in this line are used in (6.3.2), truncated also with third differences, the first *correction* $y_4^{(1)}$ is given by

$$y_4^{(1)} = 0.74918 + \tfrac{1}{10}[-0.52968 - \tfrac{1}{2}(0.12950) - \tfrac{1}{12}(0.00741) + \tfrac{1}{24}(0.00079)]$$
$$\doteq 0.68968$$

which agrees with the initial prediction to the five places retained, so that iteration is not needed.

This specific process of prediction and successive correction is considered now in somewhat more detail. For this purpose, we here denote by Y_{n+1} the *true* ordinate at x_{n+1} and by y_{n+1} the approximation which would be afforded by the chosen truncation of (6.3.2) if the resultant equation could be solved exactly. Further, we again denote by $y_{n+1}^{(0)}$ the initial prediction yielded by the corresponding truncation of (6.2.10) and by $y_{n+1}^{(1)}, y_{n+1}^{(2)}, \ldots,$ the results of the successive *iterations*, or *corrections*, just described.

In order to simplify the analysis, we suppose that the errors in all the previously calculated ordinates $y_0, y_1, \ldots, y_n$ and slopes $y'_0, y'_1, \ldots, y'_n$ are negligible. If only third differences are retained in (6.2.10) and (6.3.2), there then follows

$$y_{n+1}^{(0)} = y_n + \frac{h}{24}(55y'_n - 59y'_{n-1} + 37y'_{n-2} - 9y'_{n-3}) \qquad (6.6.1)$$

and also

$$y_{n+1} = y_n + \frac{h}{24}(9y'_{n+1} + 19y'_n - 5y'_{n-1} + y'_{n-2}) \qquad (6.6.2)$$

where y_{n+1} and y'_{n+1} are related by the equation

$$y'_{n+1} = F(x_{n+1}, y_{n+1}) \qquad (6.6.3)$$

In addition, the true ordinate Y_{n+1} satisfies the equations

$$Y_{n+1} = y_n + \frac{h}{24}(55y'_n - 59y'_{n-1} + 37y'_{n-2} - 9y'_{n-3}) + \tfrac{251}{720}h^5 y^{\text{v}}(\xi_1) \qquad (6.6.4)$$

and

$$Y_{n+1} = y_n + \frac{h}{24}(9Y'_{n+1} + 19y'_n - 5y'_{n-1} + y'_{n-2}) - \tfrac{19}{720}h^5 y^{\text{v}}(\xi_2) \qquad (6.6.5)$$

where both ξ_1 and ξ_2 lie between x_{n-3} and x_{n+1}, and where

$$Y'_{n+1} = F(x_{n+1}, Y_{n+1}) \qquad (6.6.6)$$

Finally, we have

$$y_{n+1}^{(i+1)} = y_n + \frac{h}{24}[9F(x_{n+1}, y_{n+1}^{(i)}) + 19y'_n - 5y'_{n-1} + y'_{n-2}] \qquad (6.6.7)$$

Accordingly, there follows from (6.6.1) and (6.6.4)

$$Y_{n+1} - y_{n+1}^{(0)} = \tfrac{251}{720}h^5 y^{\text{v}}(\xi_1) \qquad (6.6.8)$$

and, from (6.6.2) and (6.6.5),

$$Y_{n+1} - y_{n+1} = \frac{3h}{8}[F(x_{n+1}, Y_{n+1}) - F(x_{n+1}, y_{n+1})] - \tfrac{19}{720}h^5 y^{\text{v}}(\xi_2) \qquad (6.6.9)$$

Assuming the existence of $F_y \equiv \partial F/\partial y$ in the relevant region of the xy plane, the mean-value theorem gives

$$F(x_{n+1}, Y_{n+1}) - F(x_{n+1}, y_{n+1}) = (Y_{n+1} - y_{n+1})F_y(x_{n+1}, \eta_{n+1}) \qquad (6.6.10)$$

for some η_{n+1} between y_{n+1} and Y_{n+1}.

If now it is assumed that h is sufficiently small to ensure that

$$\frac{3h}{8}|F_y(x_{n+1}, \eta_{n+1})| \ll 1 \qquad (6.6.11)$$

so that the first term on the right in (6.6.9) is relatively negligible, and also that $y^{\text{v}}(x)$ does not vary strongly for $x_{n-3} < x < x_{n+1}$, so that $y^{\text{v}}(\xi_1)$ and $y^{\text{v}}(\xi_2)$ can be equated to a first approximation, Eqs. (6.6.8) and (6.6.9) lead to the useful approximate relation

$$Y_{n+1} - y_{n+1} \approx \frac{-19}{251 + 219}(y_{n+1} - y_{n+1}^{(0)})$$

or

$$Y_{n+1} - y_{n+1} \approx -\tfrac{19}{270}\gamma_{n+1} \tag{6.6.12}$$

where

$$\gamma_{n+1} = y_{n+1} - y_{n+1}^{(0)} \tag{6.6.13}$$

Thus, if a column of the differences $\gamma_k \equiv y_k - y_k^{(0)}$ is carried along in the calculation, the error in the final iterate y_{n+1} which is due to truncation error in the step from x_n to x_{n+1} can be estimated as $-19\gamma_{n+1}/270 \approx -\gamma_{n+1}/14$. The reliability of this estimate depends upon the validity of the assumption that errors propagated from preceding calculations are negligible and upon the smallness of the magnitudes of hF_y and hy^{vi} in the relevant region. In this connection, it may be noted that if the first neglected difference $\nabla^4 y'_{k+1}$ does not vary strongly with k, it can be expected that the same is true of y^{v}, so that hy^{vi} probably is small relative to y^{v}.

The magnitude of hF_y near (x_{n+1}, y_{n+1}) turns out to be the governing factor relative to the convergence of the iteration leading from $y_{n+1}^{(0)}$ to y_{n+1} since, from (6.6.3) and (6.6.7), there follows

$$\begin{aligned} y_{n+1} - y_{n+1}^{(i+1)} &= \frac{3h}{8}[F(x_{n+1}, y_{n+1}) - F(x_{n+1}, y_{n+1}^{(i)})] \\ &= \left[\frac{3h}{8} F_y(x_{n+1}, \eta_{n+1}^{(i)})\right](y_{n+1} - y_{n+1}^{(i)}) \end{aligned}$$

Hence, if $|F_y| \leqq K_{n+1}$ near (x_{n+1}, y_{n+1}), we have

$$|y_{n+1} - y_{n+1}^{(i+1)}| \leqq \frac{3h}{8} K_{n+1} |y_{n+1} - y_{n+1}^{(i)}|$$

and accordingly

$$|y_{n+1} - y_{n+1}^{(i)}| \leqq \left(\frac{3h}{8} K_{n+1}\right)^i |y_{n+1} - y_{n+1}^{(0)}|$$

Thus convergence is ensured if

$$\frac{3h}{8} K_{n+1} < 1 \tag{6.6.14}$$

The *rate* of convergence is specified by the ratio of the magnitudes of the errors in successive iterates, and this ratio here is approximated by the absolute value of the *convergence factor* ρ_{n+1} such that

$$\rho_n = \frac{3h}{8} F_y(x_n, y_n) \qquad (6.6.15)$$

In the case of the example in which

$$y' = x^2 - y^2 \qquad y(0) = 1$$

we would notice *in advance* that, near the beginning of the calculation,

$$F_y = -2y \approx -2$$

Thus the convergence factor in the early steps would be about $-3h/4$. The choice $h = 0.1$ then would appear to be acceptable since each iterate then would tend to deviate from the limiting value y_{n+1} by about one-thirteenth of the deviation possessed by the preceding iterate.

Usually, in practice, the spacing h is taken to be so small that in fact *no* iterations beyond the *first* correction $y_{n+1}^{(1)}$ are needed, so that $y_{n+1}^{(1)}$ itself is an acceptable approximation to Y_{n+1}. In this connection, the following consideration is useful.

If, temporarily, we assume that $y^{\text{v}}(x) > 0$ for $x_{n-3} < x < x_{n+1}$, we conclude from (6.6.8) that

$$Y_{n+1} > y_{n+1}^{(0)} \qquad (6.6.16)$$

Also, from (6.6.9) it follows that, at least when h is sufficiently small to ensure iteration convergence, we have

$$Y_{n+1} < y_{n+1} \qquad (6.6.17)$$

and, accordingly,

$$y_{n+1} > Y_{n+1} > y_{n+1}^{(0)} \qquad (6.6.18)$$

Next, by comparing (6.6.5) with (6.6.7) when $i = 0$, we obtain the relation

$$Y_{n+1} - y_{n+1}^{(1)} = \left[\frac{3h}{8} F_y(x_{n+1}, \eta'_{n+1})\right](Y_{n+1} - y_{n+1}^{(0)}) - \tfrac{19}{720}h^5 y^{\text{v}}(\xi_1)$$

and hence, since the first term on the right is small of order h^6, it follows that, at least for sufficiently small values of h, we have

$$Y_{n+1} < y_{n+1}^{(1)} \qquad (6.6.19)$$

Finally, by comparing (6.6.2) with (6.6.7), we obtain

$$\begin{aligned} y_{n+1} - y_{n+1}^{(1)} &= \frac{3h}{8}[F(x_{n+1}, y_{n+1}) - F(x_{n+1}, y_{n+1}^{(0)})] \\ &= \left[\frac{3h}{8} F_y(x_{n+1}, \eta''_{n+1})\right](y_{n+1} - y_{n+1}^{(0)}) \end{aligned} \qquad (6.6.20)$$

Thus, if $F_y > 0$ near (x_{n+1}, y_{n+1}), it follows that (6.6.18) and (6.6.20) imply the desired relation

$$y_{n+1} > y_{n+1}^{(1)} > Y_{n+1} > y_{n+1}^{(0)} \qquad (6.6.21)$$

which states, in particular, that the *first* correction $y_{n+1}^{(1)}$ affords a better approximation to Y_{n+1} than does the limit of an infinite sequence of iterates,† the difference being small of order h^6 when h is small. In the alternative situation when $y^{\text{v}}(x) < 0$ for $x_{n-3} < x < x_{n+1}$, all inequalities are reversed and the same conclusion follows.

Accordingly, unless it is established that $F_y < 0$ near (x_{n+1}, y_{n+1}), iteration is as likely to worsen the approximation afforded by $y_{n+1}^{(1)}$ as to improve it, on the average. It is probably best to calculate γ_{n+1} as $y_{n+1}^{(1)} - y_{n+1}^{(0)}$, to continue the calculation from point to point as long as, say, $|\gamma_{n+1}|$ is smaller than seven units in the last place retained (so that the additional truncation error introduced per step does not dominate the current roundoff-error contributions) and to decrease the spacing h when this situation no longer holds.

The preceding analysis generalizes immediately to include any predictor-corrector process in which the predictor has a truncation error of the form $c_1 h^p y^{(p)}(\xi_1)$ and the corrector has one of the form $-c_2 h^p y^{(p)}(\xi_2)$, where the order p is the same in both error terms and where c_1 and c_2 are positive. Then the factor $\frac{270}{19}$ in (6.6.12) is to be replaced by $C \equiv (c_1 + c_2)/c_2$, and the factor $\frac{3}{8}$ in (6.6.11), (6.6.14), and (6.6.15) is to be replaced by the coefficient α_{-1} of hy'_{n+1} in the corrector formula. It is easily seen that $h\alpha_{-1}$ *is the algebraic sum of the coefficients of all backward differences retained* (including the zeroth) when a truncation of one of the formulas in Sec. 6.3 is used.

In particular, if (6.2.10) is used as a predictor for (6.3.2), the factor $C = \frac{270}{19} \approx 14$ corresponding to retention of *third* differences is to be replaced by 2 for no differences, 6 for first differences, 10 for second differences, $\frac{270}{19} \approx 14$ for third differences, and $\frac{502}{27} \approx 18$ when fourth differences are retained. That is, if the difference between the prediction and the corrected value does not exceed half the value listed, in units of the last place retained, then the truncation error *in each step* probably does not exceed one-half unit in that place.

Some computers treat $-\gamma_{n+1}/C$ as a *modification*, to be added to the calculated y_{n+1} in each step, while others prefer merely to *control* the magnitude of this quantity and to use it only to provide warning or assurance with respect to stepwise truncation errors. Likewise, since under the preceding assumptions the approximate error in the *prediction* $y_{n+1}^{(0)}$ is $+(C - 1)\gamma_{n+1}/C$, some computers add $(C - 1)\gamma_n/C$ to $y_{n+1}^{(0)}$, under the assumption that $\gamma_{n+1} \approx \gamma_n$, to define a "modified" prediction.

† The occurrence of such situations was pointed out in a note by D. D. Wall [1956]; see also Henrici [1962].

We will refer to the preceding predictor-corrector method as the *modified Adams method* (it is ascribed to *Moulton* [1926]). The procedure based on retaining only the *first* difference in (6.3.2) is often called the *modified Euler method.*

Milne's method differs from the methods just described in that it uses (6.3.6) for correction and (6.2.18) for prediction, retaining second differences. The truncation error in the nth step can be estimated as $-(y_{n+1} - y_{n+1}^{(0)})/29$. This method possesses the advantage that the truncation error in each step is proportional to h^5, whereas retention of only second differences in the modified Adams method leads to stepwise truncation errors of order h^4. On the other hand (as will be indicated in Sec. 6.7), it compares unfavorably with the Adams method in regard to possible *error growth.*

It is obvious that each of the formulas considered could be expressed explicitly in terms of the values of the derivative y', in place of differences of those values, once a decision was made as to the number of differences which were to be effectively retained. Thus, for example, the Milne second-difference procedure can be based on Eqs. (6.2.18) and (6.3.6) or, equivalently, on the equations

$$y_{n+1} = y_{n-3} + \frac{4h}{3}(2y'_n - y'_{n-1} + 2y'_{n-2}) + \frac{14h^5}{45}y^{\text{v}}(\xi) \tag{6.6.22}$$

and

$$y_{n+1} = y_{n-1} + \frac{h}{3}(y'_{n+1} + 4y'_n + y'_{n-1}) - \frac{h^5}{90}y^{\text{v}}(\xi) \tag{6.6.23}$$

where, of course, the values of ξ in the two equations are generally unequal. The second equation is seen to be equivalent to *Simpson's rule.*

This procedure not only avoids the calculation and storage of differences, but also possesses an additional advantage in that then only the entries y_{n+1} and y'_{n+1} are modified in successive steps of any *iteration* process. However, whereas these advantages are of particular significance when large-scale computing devices are used, so that simplicity in programming and minimization of storage requirements are of prime importance, they may compare unfavorably in other cases with the advantages which follow from the possibility of considering the regularity and the trend of the difference columns.

6.7 The Special Case $F = Ay$

Each of the formulas treated in the preceding sections is expressible in the form

$$y_{n+1} = y_{n-p} + h(\alpha_{-1}y'_{n+1} + \alpha_0 y'_n + \alpha_1 y'_{n-1} + \cdots + \alpha_r y'_{n-r}) \tag{6.7.1}$$

where $\alpha_{-1} = 0$ for the formulas of open type, and where

$$y'_k = F(x_k, y_k) \tag{6.7.2}$$

and, accordingly, is a member of the more extensive class of so-called $(p + 1)$-*step formulas*. If, in fact, it is the result of retaining r differences in one of the formulas considered in Sec. 6.2, or $r + 1$ differences in one of the formulas in Sec. 6.3, then it will reduce to an identity if $y(x)$ is a polynomial of degree $r + 2$ or less, when $\alpha_{-1} \neq 0$, and of degree $r + 1$ or less, when $\alpha_{-1} = 0$.

In the case when the differential equation is of the very special form

$$\frac{dy}{dx} = Ay \tag{6.7.3}$$

so that $F(x, y) = Ay$, where A is a constant, the relation (6.7.1) takes the form

$$(1 - \alpha_{-1}Ah)y_{n+1} = y_{n-p} + Ah(\alpha_0 y_n + \alpha_1 y_{n-1} + \cdots + \alpha_r y_{n-r}) \tag{6.7.4}$$

and is subject to a simple analysis, the results of which are helpful in understanding the propagation of errors in the more general case. It may be noticed that the exact solution of (6.7.3), subject to the condition $y(x_0) = y_0$, is

$$y(x) = y_0 e^{A(x-x_0)} \tag{6.7.5}$$

In order to fix ideas, we suppose that $r \geqq p$ and so include most of the commonly used formulas , such as formulas (6.2.10) and (6.3.2), for which $p = 0$, and the formulas (6.3.6) and (6.3.7), for which $r = p$. As will be seen, this restriction is easily removed.

The relation (6.7.4) then affords a linear relation among the $r + 2$ ordinates $y_{n+1}, y_n, y_{n-1}, \ldots,$ and y_{n-r} (one of which is identical with y_{n-p}) and can be considered as a *difference equation* of order $r + 1$, under the assumption that $\alpha_{-1}Ah \neq 1$ and $\alpha_r \neq 0$. It holds only for $n \geqq r$, the ordinate y_0 being prescribed, and the remaining r initial ordinates $y_1, y_2, \ldots, y_r$ supposedly being supplied by an independent calculation.

We may notice that $y_n = \beta^n$ will satisfy (6.7.4), with β constant, if β is determined such that

$$(1 - \alpha_{-1}Ah)\beta^{n+1} = \beta^{n-p} + Ah(\alpha_0\beta^n + \alpha_1\beta^{n-1} + \cdots + \alpha_r\beta^{n-r})$$

or, after removing the common factor β^{n-r}, such that β satisfies the *characteristic equation*

$$(1 - \alpha_{-1}Ah)\beta^{r+1} - Ah(\alpha_0\beta^r + \alpha_1\beta^{r-1} + \cdots + \alpha_r) - \beta^{r-p} = 0 \tag{6.7.6}$$

Since p and r are nonnegative integers, such that $r - p \geqq 0$, this relation is an algebraic equation of degree $r + 1$ in β and hence possesses $r + 1$ roots $\beta_0, \beta_1, \ldots, \beta_r$, which may be real or imaginary.†

† When $p > r$, as when one of the open formulas (6.2.17) to (6.2.19) is used (without a corrector), the characteristic equation is of degree $p + 1$ and corresponding minor changes must be made in the analysis which follows.

If no roots are repeated, then, from the linearity and homogeneity of the difference equation (6.7.4), it follows that

$$y_n = c_0\beta_0^n + c_1\beta_1^n + \cdots + c_r\beta_r^n \qquad (6.7.7)$$

satisfies (6.7.4) for arbitrary values of the $r + 1$ independent constants c_0, $c_1, \ldots, c_r$, which are available for satisfying the $r + 1$ initial conditions which prescribe $y_0, y_1, \ldots, y_r$. It can be shown that (6.7.7) then represents the *most general* solution of (6.7.4), when n is restricted to integral values.

If $\beta_1 = \beta_2$, the terms $c_1\beta_1^n + c_2\beta_2^n$ are to be replaced by $\beta_1^n(c_1 + c_2 n)$, as is easily verified. Furthermore, if β_1 and β_2 are conjugate complex, so that

$$\beta_1 = \rho e^{i\phi} \qquad \beta_2 = \rho e^{-i\phi}$$

where $\rho = |\beta_1| = |\beta_2|$, we may replace c_1 and c_2 by $\frac{1}{2}(c_1 - ic_2)$ and $\frac{1}{2}(c_1 + ic_2)$ and rewrite the corresponding two terms in (6.7.7) in the more convenient form

$$\rho^n(c_1 \cos n\phi + c_2 \sin n\phi)$$

It remains to investigate the roots of the characteristic equation (6.7.6). We may notice first that, when $h = 0$, the equation reduces to

$$\beta^{r-p}(\beta^{p+1} - 1) = 0$$

so that $\beta = 0$ is then a root of multiplicity $r - p$, and the remaining $p + 1$ roots are the $(p + 1)$th roots of unity. In the complex β plane (Fig. 6.1), $r - p$ roots coincide at the origin, whereas the remaining $p + 1$ roots are

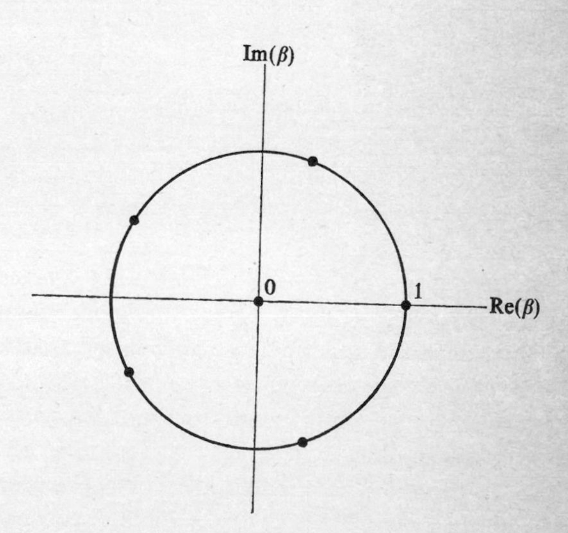

FIGURE 6.1

equally spaced about the unit circle $|\beta| = 1$, with one root at the point $\beta = 1$. When h is small, the $r + 1$ roots will generally be distinct, with $r - p$ roots *near* the origin, and $p + 1$ roots in the *neighborhood* of the unit circle.

In particular, if we denote by β_0 that root which tends to unity as h tends to zero, we may write

$$\beta_0 = 1 + m_1 h + m_2 h^2 + \cdots \tag{6.7.8}$$

where the coefficients $m_1, m_2, \ldots$ are to be determined in such a way that the result of replacing β by β_0 in (6.7.6), and expanding the result in powers of h, reduces identically to zero. A simple calculation then shows that the result of that substitution is of the form

$$h[(p + 1)m_1 - A(\alpha_{-1} + \alpha_0 + \alpha_1 + \cdots + \alpha_r)] + h^2[\cdots] + \cdots = 0$$

and hence, in particular, that we must have

$$m_1 = \frac{A}{p + 1}(\alpha_{-1} + \alpha_0 + \alpha_1 + \cdots + \alpha_r) \tag{6.7.9}$$

But, under the assumption that the integration formula which led to (6.7.1) gives exact results when applied to the integration of a *constant*, that is, for $y'(x) \equiv 1$, we can deduce from (6.7.1) the relation

$$\alpha_{-1} + \alpha_0 + \alpha_1 + \cdots + \alpha_r = p + 1 \tag{6.7.10}$$

and so find that $m_1 = A$. Accordingly, one root of (6.7.6) can be expressed in the form

$$\beta_0 = 1 + Ah + O(h^2) \tag{6.7.11}$$

where, as always, the symbol $O(h^2)$ represents a term which is small, of the order of h^2, when h is small.

The corresponding part of the solution (6.7.7) is thus of the form

$$c_0(1 + Ah + \cdots)^n = c_0(1 + Ah + \cdots)^{(x_n - x_0)/h}$$

and is approximated by $c_0 e^{A(x_n - x_0)}$ when h is small. Thus we see that this part of the general solution of the *difference* equation tends toward the general solution of the approximated *differential* equation as $h \to 0$ and, indeed, tends toward the *required* solution, for which $y(x_0) = y_0$, if $c_0 \to y_0$ as $h \to 0$.

The remaining r terms in (6.7.7) represent so-called *parasitic solutions*, which correspond to the fact that the order of the difference equation exceeds the order of the simulated differential equation by r. For small values of h, we have seen that $r - p$ of the roots β_i will be small in magnitude, relative to unity, and hence that the corresponding terms β_i^n will tend rapidly to zero as the calculation proceeds and n increases.

However, if $p > 0$, there are p roots in addition to β_0 which are of unit absolute value when $h = 0$. If, for $h > 0$, any one of these roots, say, β_k, has a magnitude *greater* than unity, then (unless the coefficient c_k happens to vanish) the corresponding term $c_k\beta_k^n$ will increase unboundedly in magnitude as n increases.

In illustration, if use is made of the simplest formula of open type (Euler's formula),

$$y_{n+1} = y_n + hy'_n = (1 + Ah)y_n \qquad (6.7.12)$$

with $p = r = 0$, the only root is $\beta_0 = 1 + Ah$, and hence the solution is

$$y_n = y_0(1 + Ah)^n = y_0(1 + Ah)^{(x_n - x_0)/h} \qquad (6.7.13)$$

which does indeed approximate (6.7.5) when h is small.

For the open formula with $p = 0$ and $r = 1$,

$$\begin{aligned} y_{n+1} &= y_n + h(y'_n + \tfrac{1}{2}\nabla y'_n) = y_n + \frac{h}{2}(3y'_n - y'_{n-1}) \\ &= (1 + \tfrac{3}{2}Ah)y_n - \tfrac{1}{2}Ahy_{n-1} \end{aligned} \qquad (6.7.14)$$

the characteristic equation (6.7.6) becomes

$$\beta^2 - (1 + \tfrac{3}{2}Ah)\beta + \tfrac{1}{2}Ah = 0$$

and yields

$$\beta_0 = \tfrac{1}{2} + \tfrac{3}{4}Ah + \tfrac{1}{2}\sqrt{1 + Ah + \tfrac{9}{4}A^2h^2} = 1 + Ah + \tfrac{1}{2}A^2h^2 + \cdots$$

and

$$\beta_1 = \tfrac{1}{2} + \tfrac{3}{4}Ah - \tfrac{1}{2}\sqrt{1 + Ah + \tfrac{9}{4}A^2h^2} = \tfrac{1}{2}Ah(1 - Ah + \cdots)$$

Thus, for small h, the solution is of the form

$$y_n = c_0(1 + Ah + \cdots)^{(x_n - x_0)/h} + c_1(\tfrac{1}{2}Ah + \cdots)^{(x_n - x_0)/h} \qquad (6.7.15)$$

If c_0 and c_1 are determined such that $y_n = c_0\beta_0^n + c_1\beta_1^n$ reduces to y_0 and y_1 for $n = 0$ and 1, respectively, there follows

$$c_0 = \frac{y_1 - \beta_1 y_0}{\beta_0 - \beta_1} \qquad c_1 = \frac{\beta_0 y_0 - y_1}{\beta_0 - \beta_1} \qquad (6.7.16)$$

The ordinate y_1 is assumed to be supplied by another method. If we assume that y_1 differs from the true value $y_0e^{Ah} = y_0(1 + Ah + \cdots)$, at worst, by an amount of order h, it is easily seen that c_0 differs from y_0 and c_1 from zero by an amount at worst of order h. Hence here the parasitic solution is small when h is small, and also it tends to zero as $n \to \infty$ for any fixed value of h which is sufficiently small to make $|\beta_1| < 1$.†

As an example in which $p > 0$, we may notice that if the Simpson's rule formula (6.6.23) is used in the form

$$y_{n+1} = y_{n-1} + \frac{Ah}{3}(y_{n+1} + 4y_n + y_{n-1}) \qquad (6.7.17)$$

Eq. (6.7.6) becomes

$$\left(1 - \frac{Ah}{3}\right)\beta^2 - \frac{4Ah}{3}\beta - \left(1 + \frac{Ah}{3}\right) = 0$$

with roots expressible in the forms

$$\beta_0 = 1 + Ah + \cdots \qquad \beta_1 = -1 + \tfrac{1}{3}Ah + \cdots$$

when h is small. Thus the solution of (6.7.17) is expressible in the form

$$\begin{aligned} y_n &= c_0(1 + Ah + \cdots)^{(x_n - x_0)/h} + (-1)^n c_1(1 - \tfrac{1}{3}Ah + \cdots)^{(x_n - x_0)/h} \\ &\approx c_0 e^{A(x_n - x_0)} + (-1)^n c_1 e^{-(A/3)(x_n - x_0)} \end{aligned} \qquad (6.7.18)$$

when h is small.

When A is *positive*, so that the *exact* solution grows exponentially with x, the root β_1 lies inside the circle $|\beta| = 1$, and the parasitic solution accordingly damps out exponentially in magnitude, as the calculation proceeds with increasing n. However, when A is *negative*, so that the exact solution tends exponentially to zero as x increases, the parasitic solution *increases* exponentially in magnitude and, in addition, alternates in sign from step to step in an advancing calculation. To a first approximation, c_1 is found to be half the difference between the value of y_1 used in the calculation and the true value $y_0 e^{Ah}$. However, the value which should be assigned to y_1 in order that c_1 vanish exactly is *not* the true value, but a value which *tends* to the true value as $h \to 0$. That is, the parasitic solution would be present even though y_0 and y_1 were *exactly* correct. It is important to notice also that each *roundoff* committed at *any* stage of the advancing calculation will initiate a *new* parasitic solution, of the same type.

In the somewhat more general case when the differential equation is of the form $y' = Ay + Bx + C$, where A, B, and C are constants, a linear function of x accordingly is added to the exponential term present in the true solution when $B = C = 0$. It is found that the same modification occurs in the solution of the approximating difference equation, so that a linear function of n is

† If $A \geqq 0$, the requirement $|\beta_1| < 1$ is satisfied for all h. However, if $A < 0$, the spacing h must be such that $h < 1/|A|$.

merely added to the right-hand member of (6.7.7). Thus the same parasitic solutions are present, and the preceding discussion again applies, except for the fact that here the true solution will not decrease in magnitude as x increases and when A is negative, but will grow linearly, while the parasitic solutions may grow exponentially.

Finally, in the general case, when we are concerned with an equation of the form $y' = F(x, y)$, we may imagine that $F(x, y)$ is replaced by the linear approximation

$$F(x, y) \approx F(x_n, y_n) + (x - x_n)F_x(x_n, y_n) + (y - y_n)F_y(x_n, y_n)$$

in the neighborhood of a point (x_n, y_n), and so imagine that the differential equation is replaced by the linear equation $y' = A_n y + B_n x + C_n$, where

$$A_n = F_y(x_n, y_n) \qquad B_n = F_x(x_n, y_n)$$
$$C_n = F(x_n, y_n) - x_n F_x(x_n, y_n) - y_n F_y(x_n, y_n)$$

It is then plausible (but not *always* true) that the nature of the error propagated in the numerical solution of the true equation will be simulated by that for the linearized equation, over a short range near x_n. The situations in which no one of the parasitic terms tends to dominate the term simulating the true solution, as the calculation proceeds from that point, are often said to be characterized by *short-range stability*.†

In order to illustrate the occurrence of instability, we present in Table 6.2 the results of calculations based on the problem

$$y' + 2y = 2 \qquad y(0) = 2$$

The entries in the column headed y_M are values determined by the Milne method, using (6.6.22) for prediction and (6.6.23) (Simpson's rule) for correction.

† Various definitions of numerous types of stability and instability occur in the literature.

Table 6.2

x	y_M	γ_M	y_A	γ_A	y_T
0.0	2.000	—	2.000	—	2.000
0.5	1.368	—	1.368	—	1.368
1.0	1.132	—	1.135	—	1.135
1.5	1.052	—	1.046	—	1.050
2.0	1.014	−42	1.016	−64	1.018
2.5	1.012	−36	1.005	−17	1.007
3.0	0.995	15	1.002	−10	1.002
3.5	1.011	−33	1.001	−2	1.001
4.0	0.986	40	1.000	−2	1.000
4.5	1.020	−57	1.000	−1	1.000

The entries γ_M represent the differences between the final results and the initial predictions, and $-\gamma_M/29$ affords the Milne estimate of the truncation error in each step. The entries in the column headed y_A were obtained by the modified Adams method, using (6.2.10) with third differences as a predictor and (6.3.2) with third differences as a corrector. The estimated truncation error in each step is afforded by $-\gamma_A/14$. The entries in the y_T column are values of the true solution $y(x) = e^{-2x} + 1$, rounded to three decimal places. As has been noted, both numerical procedures introduce truncation errors of order h^5 in each step.

A large spacing is chosen deliberately, and all calculations are rounded to three decimal places, in order to cause the effects of the error propagation to become evident at a relatively early stage of the process. The requisite starting values (above the broken lines) are correct to the places retained.

The tabulation is intended, not only to show the increasing oscillation of the first solution about the true solution, but also to serve as a reminder that the quantity $-\gamma_M/29$ affords at best only an estimate of the *truncation* error introduced *in each step* and does not, in itself, indicate the manner in which the effects of that error (and other errors) are propagated. Thus, for example, the fact that $-\gamma_M/29$ is smaller than 2 in each step must not be interpreted as indicating that the *accumulated* error at each step is less than 2 units in the last place. In fact, that error is seen to amount to -20 units in $y_M(4.5)$ in the present case, and its magnitude exceeds the *sum* of the magnitudes of the individual error estimates.

In addition to formulas which are specializations of (6.7.1), it is possible to derive an extensive class of other $(p + 1)$-step simulations to the differential equation

$$y' = F(x, y)$$

in the more general form

$$\begin{aligned} y_{n+1} &+ \beta_0 y_n + \beta_1 y_{n-1} + \cdots + \beta_p y_{n-p} \\ &= h(\alpha_{-1} y'_{n+1} + \alpha_0 y'_n + \alpha_1 y'_{n-1} + \cdots + \alpha_r y'_{n-r}) \end{aligned} \tag{6.7.19}$$

where $\beta_p \neq 0$. In particular, when $p = 2$ and $r = 1$, the six adjustable constants can be made to yield a one-parameter family of *correctors* with accuracy of order 4 (that is, with truncation-error terms of order h^5), including, for example, the third-difference Adams corrector (6.6.2) and the Milne (Simpson's rule) corrector (6.6.23).

These formulas can be obtained operationally by requiring that the relation

$$(\mathbf{E} + \beta_0 + \beta_1 \mathbf{E}^{-1} + \beta_2 \mathbf{E}^{-2})y = h(\alpha_{-1}\mathbf{E} + \alpha_0 + \alpha_1 \mathbf{E}^{-1})F$$

become equivalent to the relation

$$\mathbf{D}y = F$$

with $\mathbf{E} = e^{h\mathbf{D}}$, when an *error term* of order h^5 is introduced in the former relation. Thus the α's and β's are to be such that

$$e^t + \beta_0 + \beta_1 e^{-t} + \beta_2 e^{-2t} = t(\alpha_{-1}e^t + \alpha_0 + \alpha_1 e^{-t}) + O(t^5)$$

so that there follow five conditions relating the six available parameters.

No study of this set is undertaken here. However, we note that the member for which $\beta_1 = 0$ is of the form

$$y_{n+1} = \frac{1}{8}(9y_n - y_{n-2}) + \frac{3h}{8}(y'_{n+1} + 2y'_n - y'_{n-1}) - \frac{h^5}{40}y^{\text{v}}(\xi) \qquad (6.7.20)$$

This formula was first advocated by *Hamming* [1959], who showed that it is a commendable member of a set of formulas which (apart from other considerations) combine the *stability* properties of the third-difference Adams corrector with the advantage of requiring only three evaluations of $F(x, y)$ per calculation, as does the Milne corrector, whereas the third-difference Adams corrector requires four such evaluations. It can be used with the fourth-order Milne predictor (6.6.22), which also requires only three F evaluations, in which case the stepwise truncation-error estimate considered in Sec. 6.6 is found to be $-9(y_{n+1} - y_{n+1}^{(0)})/121$.

6.8 Propagated-Error Bounds

In actual calculation, the calculated value of y_{n+1} generally will not be given exactly by the right-hand member of the relevant formula (6.7.1), because of the necessity of effecting roundoffs. If we replace y_{n+1} by $y_{n+1} + R_n$, where R_n is inserted to account for the effects of roundoff in the nth step, Eqs. (6.7.1) and (6.7.2) can be combined in the form

$$y_{n+1} = y_{n-p} + h\sum_{k=-1}^{r} \alpha_k F(x_{n-k}, y_{n-k}) - R_n \qquad (6.8.1)$$

On the other hand, if we denote the value of the *true* solution of the given problem when $x = x_k$ by Y_k, we have also the relation

$$Y_{n+1} = Y_{n-p} + h\sum_{k=-1}^{r} \alpha_k F(x_{n-k}, Y_{n-k}) + T_n \qquad (6.8.2)$$

where we here denote the *truncation* error corresponding to the nth step by T_n.

If we subtract (6.8.1) from (6.8.2) and write

$$\varepsilon_n = Y_n - y_n \qquad E_n = T_n + R_n \qquad (6.8.3)$$

we find that the error ε_n associated with the calculated value y_n satisfies the difference equation

$$\varepsilon_{n+1} = \varepsilon_{n-p} + h \sum_{k=-1}^{r} \alpha_k [F(x_{n-k}, Y_{n-k}) - F(x_{n-k}, y_{n-k})] + E_n \qquad (6.8.4)$$

In order to obtain a bound on the magnitude of ε_n, we notice first that we may write

$$\begin{aligned} F(x_i, Y_i) - F(x_i, y_i) &= (Y_i - y_i)F_y(x_i, \eta_i) \\ &= \varepsilon_i F_y(x_i, \eta_i) \end{aligned} \qquad (6.8.5)$$

if $F_y \equiv \partial F/\partial y$ exists, where η_i is between y_i and Y_i, so that (6.8.4) can be written in the form

$$\begin{aligned} &[1 - h\alpha_{-1}F_y(x_{n+1}, \eta_{n+1})]\varepsilon_{n+1} \\ &\qquad = \varepsilon_{n-p} + h \sum_{k=0}^{r} \alpha_k \varepsilon_{n-k} F_y(x_{n-k}, \eta_{n-k}) + E_n \end{aligned} \qquad (6.8.6)$$

Suppose now that, for the range of values of x and y involved in the overall calculation, we have

$$|F_y(x, y)| \leqq K \qquad (6.8.7)$$

where K is a known constant, and consider the related difference equation

$$(1 - Kh|\alpha_{-1}|)e_{n+1} = e_{n-p} + Kh \sum_{k=0}^{r} |\alpha_k| e_{n-k} + E \qquad (6.8.8)$$

where E is such that

$$|E_n| \leqq E \qquad (n = r, r+1, \ldots) \qquad (6.8.9)$$

If $\alpha_{-1} \neq 0$, so that the formula is of closed type, suppose also that h is sufficiently small to ensure that

$$Kh|\alpha_{-1}| < 1 \qquad (6.8.10)$$

From (6.8.6) and (6.8.7), we have

$$|1 - h\alpha_{-1}F_y(x_{n+1}, \eta_{n+1})|\,|\varepsilon_{n+1}| \leqq |\varepsilon_{n-p}| + Kh \sum_{k=0}^{r} |\alpha_k|\,|\varepsilon_{n-k}| + |E_n|$$

and hence, if $|\varepsilon_n| \leqq e_n$, $|\varepsilon_{n-1}| \leqq e_{n-1}, \ldots, |\varepsilon_{n-r}| \leqq e_{n-r}$, there follows also

$$|1 - h\alpha_{-1}F_y(x_{n+1}, \eta_{n+1})|\,|\varepsilon_{n+1}| \leqq (1 - Kh|\alpha_{-1}|)e_{n+1}$$

and thus $|\varepsilon_{n+1}| \leqq e_{n+1}$. That is, if $|\varepsilon_i| \leqq e_i$ for $r + 1$ successive integral values of i, then, by induction, the same is true for all succeeding integral values of i.

Now the error ε_0 vanishes except for roundoff since $y_0 = Y_0$ is prescribed. Also, again assuming that $r \geqq p$, the errors $\varepsilon_1, \ldots, \varepsilon_r$ are errors associated with the starting values $y_1, \ldots, y_r$, supplied by an independent analysis. Let $\bar{e}$ be a positive number which is not exceeded in magnitude by any of these initial errors. Then, if e_n is a solution of (6.8.8) which is not smaller than $\bar{e}$ for $n = 0, 1, \ldots, r$, it follows that

$$|\varepsilon_n| \leqq e_n$$

for all relevant values of n. That is, any such solution of (6.8.8) will "dominate" the solution of (6.8.4).

Since the nonhomogeneous term E in (6.8.8) is a constant, a *particular* solution of (6.8.8) may be assumed in the form $e_n = -\lambda$, where λ is a constant, and the introduction of this assumption leads to the determination

$$\lambda = \frac{E}{Kh\sigma} \tag{6.8.11}$$

with the additional abbreviation

$$\sigma \equiv \sum_{k=-1}^{r} |\alpha_k| \tag{6.8.12}$$

It may be noticed that $\sigma = p + 1$ when all the α's are positive, and also that $\sigma \geqq p + 1$ in any case, by virtue of (6.7.10).

To this particular solution may be added any multiple of β^n, where β is determined such that β^n satisfies the homogeneous difference equation obtained by replacing E by zero in (6.8.8), and where β accordingly must satisfy the characteristic equation

$$\begin{aligned}(1 - Kh|\alpha_{-1}|)\beta^{r+1} \\ - Kh(|\alpha_0|\beta^r + |\alpha_1|\beta^{r-1} + \cdots + |\alpha_{r-1}|\beta + |\alpha_r|) - \beta^{r-p} = 0\end{aligned} \tag{6.8.13}$$

Since the left-hand member is negative when $\beta = 1$ and tends to $+\infty$ as $\beta \to +\infty$, there is a positive real root β_0 which is larger than unity. Indeed, for small values of h, it is found to be expressible in the form

$$\beta_0 = 1 + \frac{Kh\sigma}{p + 1} + O(h^2) \tag{6.8.14}$$

With this value of β, and the value of λ defined by (6.8.11), it follows that $e_n = c\beta_0^n - \lambda$ satisfies (6.8.8) for any constant value of c. In addition, since it increases steadily with n, if we determine c in such a way that $e_0 = \bar{e}$, so that

$$e_n = \bar{e}\beta_0^n + \lambda(\beta_0^n - 1) \tag{6.8.15}$$

then we will have $e_n \geqq \bar{e}$ for all $n \geqq 0$. Thus this particular solution of (6.8.8) will dominate the solution of (6.8.4).

Hence, in summary, we deduce that *the error* ε_n *association with the value of* y_n *determined by step-by-step calculation based on the formula*

$$y_{n+1} \approx y_{n-p} + h \sum_{k=-1}^{r} \alpha_k F(x_{n-k}, y_{n-k}) \qquad (n \geqq r) \tag{6.8.16}$$

is limited by the inequality

$$|\varepsilon_n| \leqq \bar{e}\beta_0^n + \lambda(\beta_0^n - 1) \tag{6.8.17}$$

where $\bar{e}$ *is the absolute value of the largest error associated with the* $r + 1$ *starting values* $y_0, y_1, \ldots, y_r$; λ *is defined by the equations*

$$\lambda = \frac{E}{Kh\sigma} \qquad \sigma = \sum_{k=-1}^{r} |\alpha_k| \tag{6.8.18}$$

K *is the maximum value of* $|\partial F/\partial y|$ *for the range of values of* x *and* y *involved in the calculation;* E *is the absolute value of the maximum total error introduced in each step; and* β_0 *is the positive real root of the equation*

$$\beta^{n+1} = \beta^{n-p} + Kh \sum_{k=-1}^{r} |\alpha_k|\beta^{n-k} \tag{6.8.19}$$

which exceeds unity.†

In those cases when the coefficients α_k are all *positive*, reference to (6.7.10) shows that (6.8.11) reduces to $\lambda = E/[Kh(p + 1)]$ and that the expansion (6.8.14) becomes $\beta_0 = 1 + Kh + O(h^2)$. Also, the characteristic equation (6.8.13) is then equivalent to Eq. (6.7.6) with A replaced by K, and finally $\beta_0^n \approx \exp [K(x_n - x_0)]$.

Of the three constants $\bar{e}$, E, and K needed for the application of this error estimate, the first may be estimated initially; and it represents the maximum roundoff error associated with the initial values determined before the stepwise calculation is begun, if those values are correctly determined to the number of places retained.

The constant E comprises the maximum error introduced in one step because of roundoff and truncation. The latter effect cannot be estimated in advance unless $F(x, y)$ is of a particularly convenient form, but it can be estimated in the course of the calculation by approximating the factor $h^m y^{(m)}(\xi)$ by $h\,\Delta^{m-1}y'$ in the truncation-error term, or by making use of the quantity γ_n defined in Sec. 6.6 if one of the methods described in that section is employed.

† If $\partial F/\partial y$ is known to be negative throughout the calculation, a less conservative bound often can be obtained in a correspondingly simple form. For example, see Probs. 21 and 22.

The constant K can be calculated in advance if the equation is linear, since then $\partial F/\partial y$ is independent of y, and it can be estimated in advance (assuming that an analytical expression for $\partial F/\partial y$, in terms of x and y, can be obtained) if the range of values of y can be estimated initially. Otherwise, sample values of $\partial F/\partial y$ can be tabulated as the calculation proceeds. Thus, for example, in the case of (6.4.5) we have $\partial F/\partial y \equiv -1$, and hence $K = 1$. For the equation $y' = x^2 + y^2$, K would be estimated as the largest value of $2|y|$ encountered in the calculation.†

The maximum effects of errors due to truncation and to roundoff can be treated separately. However, because of the more or less random fluctuation in sign of errors due to roundoff, any upper bound on the overall effect of a large number of roundoffs, no matter how precise, is likely to be extremely conservative in any actual calculation. On the other hand, the *statistical* analysis of such effects in stepwise integration is rather involved (see Henrici [1962]) and, in any case, can afford only the *probability* that the overall effect of roundoff errors will not exceed a certain amount.

6.9 Equations of Higher Order. Sets of Equations

In order to apply one of the preceding methods to a differential equation of higher order, it is often convenient first to replace that equation by an equivalent *set* of equations of the first order. We here illustrate the procedure only in the case of a second-order equation, after which the generalization to higher-order equations, or to sets of simultaneous equations of more general type, will be obvious.

The problem

$$y'' = G(x, y, y') \qquad y(x_0) = y_0 \qquad y'(x_0) = y'_0 \tag{6.9.1}$$

is equivalent to the problem

$$\begin{aligned} y' &= u & y(x_0) &= y_0 \\ u' &= G(x, y, u) & u(x_0) &= y'_0 \end{aligned} \tag{6.9.2}$$

which is, in turn, a specialization of the more general problem in which u is replaced by, say, $F(x, y, u)$ in the right-hand member of the first equation.

† It should be noticed that the estimate $\partial F/\partial y \approx \Delta y'/\Delta y$ (which has been suggested in the literature) is *not* generally significant. Whereas we do have the relation

$$\begin{aligned} \Delta y'_n &\equiv [F(x_{n+1}, y_{n+1}) - F(x_{n+1}, y_n)] + [F(x_{n+1}, y_n) - F(x_n, y_n)] \\ &\approx F_y(x_{n+1}, y_n)\,\Delta y_n + hF_x(x_n, y_n) \end{aligned}$$

there is no reason to suppose that the last term is small relative to $\Delta y'_n$.

It is usually convenient, but not necessary, to use the same formula in dealing with the two equations in (6.9.2). The approximate formulation then comprises two relations which are expressible in the general form

$$y_{n+1} = y_{n-p} + h(\alpha_{-1} y'_{n+1} + \alpha_0 y'_n + \alpha_1 y'_{n-1} + \cdots + \alpha_r y'_{n-r}) \tag{6.9.3}$$

and

$$u_{n+1} = u_{n-p} + h(\alpha_{-1} u'_{n+1} + \alpha_0 u'_n + \alpha_1 u'_{n-1} + \cdots + \alpha_r u'_{n-r}) \tag{6.9.4}$$

or in equivalent forms in terms of backward differences, with

$$y'_n = u_n \tag{6.9.5}$$

and

$$u'_n = G(x_n, y_n, u_n) \tag{6.9.6}$$

In any case, the relations (6.9.3) and (6.9.4) apply only for $n \geqq r$, the values y_0 and u_0 being given and the values $y_1, \ldots, y_r$ and $u_1, \ldots, u_r$ being obtained by another method (such as the use of Taylor series or of one of the methods to be given in Sec. 6.15). The values of $u'_0, u'_1, \ldots, u'_r$ are calculated in advance from (6.9.6). If the formula is of *open* type, so that $\alpha_{-1} = 0$, y_{r+1} and u_{r+1} are then calculated directly by use of (6.9.3) and (6.9.4). Next y'_{r+1} is given immediately as u_{r+1}, and u'_{r+1} is calculated as $G(x_{r+1}, y_{r+1}, u_{r+1})$, so that data are then available for advancing by another step.

If the formula is of *closed* type, so that $\alpha_{-1} \neq 0$, an initial prediction $u_{r+1}^{(0)}$ is first obtained (by use of a supplementary formula of open type, by pure estimation, or otherwise) and $y_{r+1}^{(0)}$ is obtained by replacing y'_{r+1} by $u_{r+1}^{(0)}$ in (6.9.3). Next $u'^{(0)}_{r+1}$ is obtained from (6.9.6), with y and u replaced by their zeroth approximations, and the cycle is closed by calculating $u_{r+1}^{(1)}$ from (6.9.4). If the calculated value differs from the initial prediction $u_{r+1}^{(0)}$, the cycle may be iterated until agreement is obtained, when the iteration converges.† The next step is then taken in the same way.

The iteration is thus described by the equations

$$\begin{aligned} y_{n+1}^{(i)} &= y_{n-p} + h(\alpha_{-1} u_{n+1}^{(i)} + \cdots) \\ u'^{(i)}_{n+1} &= G(x_{n+1}, y_{n+1}^{(i)}, u_{n+1}^{(i)}) \\ u_{n+1}^{(i+1)} &= u_{n-p} + h(\alpha_{-1} u'^{(i)}_{n+1} + \cdots) \end{aligned} \tag{6.9.7}$$

where y_{n-p}, u_{n-p}, and all omitted terms remain fixed throughout the iteration. There then follows also

$$y - y^{(i)} = h\alpha_{-1}(u - u^{(i)}) \tag{6.9.8}$$

† As in the first-order case, it is true here that, on the average, iteration beyond a *first correction* is as likely to worsen the approximation as to improve it when the iteration converges.

and

$$u - u^{(i+1)} = h\alpha_{-1}[G(x, y, u) - G(x, y^{(i)}, u^{(i)})] \qquad (6.9.9)$$

where the common subscript $n + 1$ is suppressed throughout. Now if, near $(x_{n+1}, y_{n+1}, u_{n+1})$, it is true that

$$|G_y(x, y, u)| \leqq K_{n+1} \qquad |G_u(x, y, u)| \leqq L_{n+1} \qquad (6.9.10)$$

then we may deduce from (6.9.9) the inequality

$$|u - u^{(i+1)}| \leqq h|\alpha_{-1}|[K_{n+1}|y - y^{(i)}| + L_{n+1}|u - u^{(i)}|] \qquad (6.9.11)$$

and hence, making use of (6.9.8),

$$|u - u^{(i+1)}| \leqq h|\alpha_{-1}|(h|\alpha_{-1}|K_{n+1} + L_{n+1})|u - u^{(i)}| \qquad (6.9.12)$$

Thus, convergence will attain if h is so chosen that

$$h|\alpha_{-1}|(h|\alpha_{-1}|K_{n+1} + L_{n+1}) < 1 \qquad (6.9.13)$$

and the convergence factor ρ_n in the nth step is such that

$$|\rho_n| \leqq h|\alpha_{-1}|(h|\alpha_{-1}|K + L) \qquad (6.9.14)$$

where K and L are upper bounds on $|G_y|$ and $|G_u|$. If G does not explicitly involve $u \equiv y'$, it is seen that the convergence factor is of second order in h.

Whether or not iteration beyond a first correction is contemplated (and especially if it is not!), the spacing h should be chosen so that $|\rho_n| \ll 1$ in the initial computation (and modified as necessary in following steps) since otherwise the stepwise truncation-error estimates are invalid and, more importantly, unfavorable propagation of truncation errors may be anticipated.

In order to illustrate the procedures, we consider the simple problem

$$y'' = y + xy' \qquad y(0) = 1 \qquad y'(0) = 0 \qquad (6.9.15)$$

For the purpose of starting the calculation, we first obtain the expressions

$$y'' = y + xy', \quad y''' = 2y' + xy'', \quad y^{\text{iv}} = 3y'' + xy''', \quad \ldots$$

and hence, with $x_0 = 0$,

$$y_0 = 1, \quad y_0' = 0, \quad y_0'' = 1, \quad y_0''' = 0, \quad y_0^{\text{iv}} = 3, \quad y_0^{\text{v}} = 0, \quad y_0^{\text{vi}} = 15, \quad \ldots$$

so that, with $h = 0.1$, there follows

$$y_s = 1 + \frac{1}{2}\left(\frac{s}{10}\right)^2 + \frac{1}{8}\left(\frac{s}{10}\right)^4 + \frac{1}{48}\left(\frac{s}{10}\right)^6 + \cdots$$

$$y_s' = \frac{s}{10} + \frac{1}{2}\left(\frac{s}{10}\right)^3 + \frac{1}{8}\left(\frac{s}{10}\right)^5 + \cdots$$

Thus we may obtain, in particular,

$$\begin{aligned} y_1 &= 1.0050, \quad y_2 = 1.0202, \quad y_3 = 1.0461, \quad \ldots \\ y_1' &= 0.1005, \quad y_2' = 0.2040, \quad y_3' = 0.3138, \quad \ldots \end{aligned} \tag{6.9.16}$$

if only four places are retained. For the purpose of simplicity, we here make use only of the calculated values of y_1 and y_1', and proceed by using a formula involving only *first* differences. The preliminary calculation then can be arranged as follows:

x	y	$y' = u$	$\nabla y'$	$\nabla^2 y'$	$y'' = y + xu$	$\nabla y''$	$\nabla^2 y''$
0.0	1.0000	0.0000	—	—	1.0000	—	—
0.1	1.0050	0.1005	1005	—	1.0150	150	—

If the Adams formula of the open type is used with first differences, there follows

$$y_2 \approx 1.0050 + 0.1[0.1005 + \tfrac{1}{2}(0.1005)] \doteq 1.0201$$

$$u_2 \approx 0.1005 + 0.1[1.0150 + \tfrac{1}{2}(0.0150)] \doteq 0.2028$$

and the third line of the calculation is

0.2	1.0201	0.2028	1023	18	1.0607	457	307

(6.9.17)

Since a second difference $\nabla^2 y''$ of about 300 units would contribute about $\frac{5}{12} \cdot \frac{1}{10} \cdot 300 \doteq 12$ units to y', while a second difference $\nabla^2 y'$ of about 18 units would contribute about 0.7 units to y, if we suppose that the neglected second differences relative to x_1 are of the same order of magnitude as those calculated here, we may consider these quantities as rough estimates of the truncation errors in y_2 and u_2, introduced in the step from x_1 to x_2. (Further information with regard to the reliability of these estimates would be afforded, in succeeding steps, by a consideration of the extent to which the second differences remain constant.) If such errors are not tolerable, and this method is to be used, it is then necessary to calculate one or more additional starting values of y and u, and to retain at least one more difference. In this connection, it should be kept in mind that the errors introduced in each step are propagated into *succeeding* calculations, as was seen in the first-order case in Secs. 6.7 and 6.8, in a manner which depends both upon the problem involved and the integration formula employed.

If, instead, the Adams formula of closed type with first differences is used as a corrector, with the open-type formula as a predictor, the value 0.2028 is obtained, as in the preceding method, as the zeroth approximation $u_2^{(0)} \equiv$

$y_2'^{(0)}$. The corresponding difference $\nabla y_2'^{(0)}$ is then entered, the prediction $y_2^{(0)}$ is determined by the formula

$$y_2^{(0)} = 1.0050 + 0.1[0.2028 - \tfrac{1}{2}(0.1023)] \doteq 1.0202$$

and $u_2'^{(0)}$ is determined as $y_2^{(0)} + x_2 y_2'^{(0)} \doteq 1.0608$, so that the third line takes the form

0.2	1.0202	0.2028	1023	—	1.0608	458	—

(6.9.18)

Next the cycle is closed by calculating

$$u_2^{(1)} = 0.1005 + 0.1[1.0608 - \tfrac{1}{2}(0.0458)] \doteq 0.2043$$

Since this corrected result differs from the initial prediction, the entry 0.2028 in the third line is altered to 0.2043 and the cycle may be repeated, at the end of which the third line has been changed to

0.2	1.0202	0.2043	1038	33	1.0611	461	311

(6.9.19)

Finally, the value $u_2^{(2)}$ is calculated and is found to agree with $u_2^{(1)}$ to four places, so that the iteration is completed.

The fact that $h = 0.1$ is an appropriate spacing here could have been established *in advance* by noticing that since $G_u = x$ and $G_y = 1$, and since $\alpha_{-1} = \frac{1}{2}$ for the Adams one-difference corrector, the initial convergence factor ρ_0 (with $x = 0$) is approximately $h/2$.

Reference to (6.3.2) shows that incorporation of the second differences would contribute $-\frac{1}{10} \cdot \frac{1}{12} \cdot 311 \approx -3$ units to y_2' and $-\frac{1}{10} \cdot \frac{1}{12} \cdot 33 \approx 0$ units to y_2. A somewhat more dependable estimate of the truncation error introduced in a single step can be obtained by calculating the prediction $y_{n+1}^{(0)}$ by use of the open formula, in place of the closed one, but using the closed formula to supply at least one corrected value of y_{n+1}. Since also $u_{n+1}^{(0)}$ is calculated by the open formula, it then follows that if we write

$$\gamma_{n+1} \equiv y_{n+1} - y_{n+1}^{(0)} \qquad \gamma_{n+1}' \equiv u_{n+1} - u_{n+1}^{(0)}$$

where y_{n+1} and u_{n+1} are corrected values provided by the closed formula, then the desired truncation errors in y_{n+1} and u_{n+1} are approximated respectively by $-\gamma_{n+1}/C$ and $-\gamma_{n+1}'/C$, where C is the numerical factor considered in Sec. 6.6, here equal to 6 (see also Prob. 27). It is convenient to tabulate γ_n and γ_n' in place of the two first neglected differences, so that the line (6.9.19) then is replaced by

				γ			γ'
0.2	1.0202	0.2043	1038	1	1.0611	461	15

The estimated truncation errors introduced into the calculated values of y_2' and y_2 are then $-\frac{15}{6} \approx -3$ units and $-\frac{1}{6} \approx 0$ units, as before.

It may be noticed that the actual errors, obtainable by reference to the rounded true values given in (6.9.16), are indeed correctly predicted, in this case, by the estimates afforded by both procedures. In a fairly lengthy sequence of steps, however, the *propagation* of errors becomes important. A treatment analogous to that of Sec. 6.8 is rather unpleasant in the case of a second-order equation and is omitted.

In this connection, however, it may be remarked that an elementary analysis quite similar to that of Sec. 6.7 permits a simple stability study of the use of (6.9.3) and (6.9.4) in the numerical solution of an equation of the special form

$$y'' = Ly' + Ky \tag{6.9.20}$$

where L and K are *constants*. Here the exact solution is of the form

$$y(x) = c_1 e^{A_1 x} + c_1 e^{A_2 x} \tag{6.9.21}$$

where A_1 and A_2 are the roots of $A^2 - LA - K = 0$ and c_1 and c_2 are determined by the initial conditions, and it is again found that the use of formulas (6.9.3) and (6.9.4) with $p > 0$ introduces *parasitic solutions* which may dominate the part of the solution which simulates the exact solution when A_1 and A_2 are negative or have negative real parts. When $p = 0$, this situation can exist only when excessively large spacings are employed.

In particular, if use is made of Milne's method, based on (6.6.23) and corresponding to $p = 1$, the generated numerical solution is found to be approximated by

$$C_1 e^{A_1 x_k} + C_2 e^{A_2 x_k} + (-1)^k [C_3 e^{-A_1 x_k/3} + C_4 e^{-A_2 x_k/3}] \tag{6.9.22}$$

when $x = x_k$, if the spacing h is small and if roundoff errors are neglected, where the C's are determined by the starting values. Thus, for example, if the true solution is of the form

$$y(x_k) = ce^{-ax_k} \cos (bx_k + \omega)$$

where $a > 0$, then the parasitic part of the numerical solution will be approximated by

$$(-1)^k c' e^{ax_k/3} \cos \left[\left(\frac{bx_k}{3} \right) + \omega' \right]$$

and will tend to dominate the desirable part of the numerical solution when x is large.

It is also important to notice that the propagated error generally will possess components simulating both the terms $e^{A_1 x}$ and $e^{A_2 x}$, even when the parasitic solutions are not troublesome. This situation is of particular disadvantage when the initial conditions require that the *exact* solution involve only the term which grows least rapidly (or decays most rapidly).

When the governing differential equation $y'' = G(x, y, y')$ is less simple than (6.9.20), a qualitative analysis of short-range error propagation near a point (x_k, y_k) generally can be obtained by identifying L and K in (6.9.20) with the values of $\partial G/\partial y'$ and $\partial G/\partial y$, respectively, at that point, if those partial derivatives do not vary excessively near that point.

The generalization of the preceding treatments to a pair of equations of the form

$$\begin{aligned} y' &= F(x, y, u) \qquad y(x_0) = y_0 \\ u' &= G(x, y, u) \qquad u(x_0) = u_0 \end{aligned} \tag{6.9.23}$$

or to a more general set of simultaneous differential equations is straightforward in principle and in numerical interpretation. However, the complexity of the relevant analysis of propagated errors and stability may be forbidding.

6.10 Special Second-Order Equations

Second-order equations of the special form

$$y'' = G(x, y) \tag{6.10.1}$$

in which y' is not explicitly involved, arise rather frequently in practice. If values of y' are not required, it is desirable to have available methods which do not entail their calculation.

Two formulas having this property were derived, as formulas (5.5.12) and (5.5.13) for repeated integration, and may be written with the present notation in the forms

$$\begin{aligned} y_{n+1} = 2y_n - y_{n-1} + h^2(1 + 0\nabla + \tfrac{1}{12}\nabla^2 + \tfrac{1}{12}\nabla^3 + \tfrac{19}{240}\nabla^4 \\ + \tfrac{3}{40}\nabla^5 + \cdots)y''_n \end{aligned} \tag{6.10.2}$$

and

$$\begin{aligned} y_{n+1} = 2y_n - y_{n-1} + h^2(1 - \nabla + \tfrac{1}{12}\nabla^2 + 0\nabla^3 - \tfrac{1}{240}\nabla^4 \\ - \tfrac{1}{240}\nabla^5 + \cdots)y''_{n+1} \end{aligned} \tag{6.10.3}$$

The former is of open type, the latter of closed type. In order to use either, a suitable number of preliminary ordinates must be calculated by another method, which takes into account the fact that y and y' are prescribed at $x = x_0$, after which the technique of the ensuing calculation, often known as *Störmer's method* (Störmer [1921]), is evident.

Formulas (6.10.2) and (6.10.3) are each representative of a whole class of similar formulas, one class of open type and the other of closed type, which are analogous to the formulas given in Secs. 6.2 and 6.3. In particular, an additional formula of open type,

$$y_{n+1} = y_n + y_{n-2} - y_{n-3} + 3h^2(1 - \nabla + \tfrac{5}{12}\nabla^2 + 0\nabla^3 + \tfrac{17}{720}\nabla^4 + \tfrac{17}{720}\nabla^5 + \cdots)y_n'' \qquad (6.10.4)$$

may be listed (see Prob. 16 of Chap. 5).

Formulas (6.10.3) and (6.10.4) comprise a pair of formulas for both of which the coefficient of the third difference vanishes. If only second differences are retained in these formulas, they become

$$y_{n+1} = y_n + y_{n-2} - y_{n-3} + \left\{\begin{matrix} 3h^2(y_n'' - \nabla y_n'' + \tfrac{5}{12}\nabla^2 y_n'') \\ \dfrac{h^2}{4}(5y_n'' + 2y_{n-1}'' + 5y_{n-2}'') \end{matrix}\right\} + \frac{17h^6}{240} y^{\mathrm{vi}}(\xi) \qquad (6.10.5)$$

and

$$y_{n+1} = 2y_n - y_{n-1} + \left\{\begin{matrix} h^2(y_{n+1}'' - \nabla y_{n+1}'' + \tfrac{1}{12}\nabla^2 y_{n+1}'') \\ \dfrac{h^2}{12}(y_{n+1}'' + 10y_n'' + y_{n-1}'') \end{matrix}\right\} - \frac{h^6}{240} y^{\mathrm{vi}}(\xi) \qquad (6.10.6)$$

where the coefficients of the remainder terms are the same as those of the omitted fourth differences in the formulas (6.10.3) and (6.10.4). The error term when one of the formulas is truncated with a difference not preceding one with a zero coefficient is of more complicated form.

Milne's method employs (6.10.5) to afford an initial prediction $y_{n+1}^{(0)}$, and (6.10.6) as the corrector. If the factor $\gamma_{n+1} = y_{n+1} - y_{n+1}^{(0)}$ is calculated (as in Sec. 6.6), the estimated truncation error in the nth step is $T_n \approx -\gamma_n/18$. Also, the convergence factor in the iteration at the nth step (see Sec. 6.6) is easily found to be approximately $\rho_n \approx \frac{1}{12}h^2 G_y(x_n, y_n)$, so that h should be sufficiently small to ensure that $\frac{1}{12}h^2|G_y| \ll 1$.

In order to illustrate the relevant analysis of error *propagation*, we consider the special case in which (6.10.6) is the basis of the method.† With the same notation as was used in earlier developments, it is easily seen that the error ε_n associated with the calculated value y_n satisfies a relation of the form

$$\varepsilon_{n+1} = 2\varepsilon_n - \varepsilon_{n-1} + \frac{h^2}{12}(G_{y_{n+1}}\varepsilon_{n+1} + 10G_{y_n}\varepsilon_n + G_{y_{n-1}}\varepsilon_{n-1}) + E_n \qquad (6.10.7)$$

† The method used for prediction is irrelevant to this analysis.

for $n \geqq 1$, where G_{y_r} is an appropriate value of G_y. This relation can also be written in the form

$$\left(1 - \frac{h^2}{12} G_{y_{n+1}}\right)(\varepsilon_{n+1} - \varepsilon_n) = \varepsilon_n - \varepsilon_{n-1} + \frac{h^2}{12}[(G_{y_{n+1}} + 10G_{y_n})\varepsilon_n + G_{y_{n-1}}\varepsilon_{n-1}] + E_n \qquad (6.10.8)$$

so that, if we have

$$|G_y(x, y)| \leqq K \qquad (6.10.9)$$

for all relevant values of x and y, and if h is sufficiently small to ensure that

$$\frac{Kh^2}{12} < 1 \qquad (6.10.10)$$

there follows

$$\left(1 - \frac{Kh^2}{12}\right)|\varepsilon_{n+1} - \varepsilon_n| \leqq |\varepsilon_n - \varepsilon_{n-1}| + \frac{Kh^2}{12}(11|\varepsilon_n| + |\varepsilon_{n-1}|) + |E_n| \qquad (6.10.11)$$

If e_n satisfies the relation

$$\left(1 - \frac{Kh^2}{12}\right)(e_{n+1} - e_n) = e_n - e_{n-1} + \frac{Kh^2}{12}(11e_n + e_{n-1}) + E \qquad (6.10.12)$$

where

$$E = |E_n|_{\max} \qquad (6.10.13)$$

and if

$$e_0 \geqq |\varepsilon_0| \qquad e_1 - e_0 \geqq |\varepsilon_1 - \varepsilon_0| \qquad (6.10.14)$$

there follows also $e_1 - e_0 \geqq |\varepsilon_1| - |\varepsilon_0|$ and hence $e_1 - |\varepsilon_1| \geqq e_0 - |\varepsilon_0|$ or $e_1 \geqq |\varepsilon_1|$. Then, by comparing (6.10.11) and (6.10.12), we find also that $e_2 - e_1 \geqq |\varepsilon_2 - \varepsilon_1|$ and hence $e_2 - e_1 \geqq |\varepsilon_2| - |\varepsilon_1|$ or $e_2 \geqq |\varepsilon_2|$. By induction, there follows $e_n - e_{n-1} \geqq |\varepsilon_n - \varepsilon_{n-1}|$ and $e_n \geqq |\varepsilon_n|$ for $n = 0, 1, 2, \ldots$, so that the error ε_n is then dominated by e_n.

The general solution of (6.10.12) is readily found to be of the form

$$e_n = A_0\beta_0^n + A_1\beta_1^n - \frac{E}{Kh^2} \qquad (6.10.15)$$

where β_0 and β_1 are the roots of the equation

$$\left(1 - \frac{Kh^2}{12}\right)\beta^2 - 2\left(1 + \frac{5Kh^2}{12}\right)\beta + \left(1 - \frac{Kh^2}{12}\right) = 0 \qquad (6.10.16)$$

and hence

$$\beta_0 = \frac{1}{\beta_1} = \frac{1 + \frac{5}{12}Kh^2 + \sqrt{Kh^2 + \frac{1}{6}K^2h^4}}{1 - \frac{1}{12}Kh^2} = 1 + \sqrt{K}h + O(h^2) \quad (6.10.17)$$

When A_0 and A_1 are determined by the conditions $e_0 = 0$ and $e_1 = |\varepsilon_1|$, under the assumption that $\varepsilon_0 = 0$, there follows finally

$$|\varepsilon_n| \leqq e_n$$

where

$$e_n = \frac{E}{Kh^2}\frac{1}{\beta_0 + 1}(\beta_0^n + \beta_0^{1-n} - 1 - \beta_0) + \frac{\beta_0|\varepsilon_1|}{\beta_0^2 - 1}(\beta_0^n - \beta_0^{-n}) \quad (6.10.18)$$

If roundoff errors are ignored, we have $E \leqq \frac{1}{240}h^6|y^{\text{vi}}|_{\max}$. Also, since $\beta_0 = 1 + \sqrt{K}h + O(h^2)$, and $n = (x_n - x_0)/h$, there follows

$$e_n \approx \frac{h^4}{240K}|y^{\text{vi}}|_{\max}\{\cosh[\sqrt{K}(x_n - x_0)] - 1\} + \frac{|\varepsilon_1|}{\sqrt{K}h}\sinh[\sqrt{K}(x_n - x_0)] \quad (6.10.19)$$

when h is small. Whereas $|y^{\text{vi}}|_{\max}$ generally is not easy to estimate directly, the factor $[h^4|y^{\text{vi}}|_{\max}]/240$ can be estimated as $h^{-2}|T_n|_{\max} \approx |\gamma_n|_{\max}/18h^2$ or as $|\nabla^4 y_k''|_{\max}/240$. (Similar but more involved error bounds can be derived in the more general case.)

In order to illustrate the calculation and to provide a basis for the considerations of the following section, we apply Milne's method, based on (6.10.5) and (6.10.6), to the problem

$$y'' = xy \qquad y(0) = 0 \qquad y'(0) = 1 \quad (6.10.20)$$

for which the exact solution is expressible in the form

$$y = 3^{1/3}\Gamma(\tfrac{4}{3})I_{1/3}(\tfrac{2}{3}x^{3/2}) \quad (6.10.21)$$

where $I_{1/3}$ is the *modified Bessel function of the first kind of order* $\frac{1}{3}$. With $h = 0.1$, the calculation can be arranged as in Table 6.3, if differences are used and if five places are retained. The first five lines are easily calculated in advance (only *three* lines are *needed*), the ordinates being determined by use of a single Taylor series, and the values of y'', determined from the equation $y'' = xy$, and of the differences are entered as shown. If (6.10.5) is used to predict y_5, the prediction is found to be 0.50522; the remainder of the sixth line is then filled in, after which (6.10.6) gives the corrected value 0.50523, and the resultant slight modification in the remainder of the line does not call for additional iteration. The value $\gamma_5 = 0.50523 - 0.50522$ is then listed as $+1$ unit in the fifth place, and the calculation proceeds in the same way in succeeding steps.

Table 6.3

x	y	y''	$\nabla y''$	$\nabla^2 y''$	γ
0.0	0.00000	0.00000			
0.1	0.10001	0.01000	1000		
0.2	0.20013	0.04003	3003	2003	
0.3	0.30068	0.09020	5017	2014	
0.4	0.40214	0.16086	7066	2049	
0.5	0.50523	0.25262	9176	2110	1
0.6	0.61086	0.36652	11390	2214	−1
0.7	0.72017	0.50412	13760	2370	−1
0.8	0.83454	0.66763	16351	2591	−1

Since the truncation error in each step may be estimated as $-\gamma/18$, we may be reasonably confident of the calculated values of y to the places retained (except for the usual uncertainty of one unit in the last place, due to roundoff). In fact, the small values of γ may be expected to correspond to the effects of roundoff.

Clearly, the alternative forms of (6.10.5) and (6.10.6) may be used instead, without the need for calculation and recording (or storing) of differences. Further, in place of using (6.10.5) to obtain an initial prediction for y_{n+1}, it is possible to *estimate* the second difference $\nabla^2 y''_{n+1}$ and then to fill in the remainder of that line from right to left through y''_{n+1}, after which use may be made of (6.10.6) to initiate the correction. Thus, in Table 6.3, a glance at the $\nabla^2 y''$ column would suggest the estimate $\nabla^2 y''_8 \approx 0.026$ after the calculation of y_7. However, this procedure would not supply data for the γ column.

It may be noticed that any *linear* equation of the second order of the form

$$Y'' + P(x)Y' + Q(x)Y = F(x) \qquad (6.10.22)$$

can be reduced to the form (6.10.1) by the change of variables

$$Y(x) = e^{-(1/2)\int P\,dx} y(x) \qquad (6.10.23)$$

in accordance with which (6.10.22) takes the form

$$y'' + f(x)y = g(x) \qquad (6.10.24)$$

where

$$f(x) = \tfrac{1}{4}\{4Q(x) - 2P'(x) - [P(x)]^2\} \qquad g(x) = e^{(1/2)\int P\,dx} F(x) \qquad (6.10.25)$$

6.11 Change of Interval

In many situations it is desirable to, say, double or halve the spacing at a certain stage of the advancing calculation. Doubling the spacing presents no difficulties, since it involves only the use of alternate values of previously calculated data, together with a direct calculation of modified differences relevant to the new spacing, if differences are used.

Thus, in illustration, the smallness of the entries in the γ column of Table 6.3 suggests that the same accuracy may be obtained with a doubled spacing $h' = 2h = 0.2$. In fact, reference to the error expression in (6.10.6) shows that the truncation error in each step can be estimated roughly by

$$T_n \approx -\frac{h^6}{240}\frac{\nabla^4 y_n''}{h^4} \approx -0.00004\nabla^4 y_n''$$

and it is found that, for the data of Table 6.3, $\nabla^4 y''$ varies from 0.00024 to 0.00065, so that the largest single *truncation* error in the range covered is probably less than about three units in the eighth decimal place. Doubling the spacing h will multiply the truncation error by a factor of the order 2^6 and hence may be expected to lead to a truncation error of less than about one unit in the sixth place in each step. The calculation following the work of Table 6.3, with doubled spacing, is given in Table 6.4.

Table 6.4

x	y	y''	$\nabla y''$	$\nabla^2 y''$	γ
0.0	0.00000	0.00000			
0.2	0.20013	0.04003	4003		
0.4	0.40214	0.16086	12083	8080	
0.6	0.61086	0.36652	20566	8483	
0.8	0.83454	0.66763	30111	9545	
1.0	1.08531	1.08531	41768	11657	1
1.2	1.38000	1.65600	57069	15301	3
1.4	1.74164	2.43830	78230	21161	7

After three lines of calculation, the γ column serves a warning that the truncation error per step may have increased at that stage to about one-half unit in the last place retained. Thus (as might have been anticipated in advance from the increasing rate of growth of $\nabla^2 y''$) the advantages of the more rapid calculation were short-lived, and the doubling of the spacing was in fact ill-advised in the present case. However, the results of Table 6.4 may serve to illustrate the somewhat more complicated transition to a *halved* spacing.†

† An obvious alternative consists of merely retaining additional differences in the relevant integration formulas (6.10.3) and (6.10.4). In the present case, however, it is assumed that retention of the advantages of the special formulas (6.10.5) and (6.10.6) is considered to be desirable.

In the present analysis, knowledge of $y(1.3)$ would permit the determination of $y''(1.3)$ and, consequently, $\nabla y''(1.4)$ and $\nabla^2 y''(1.4)$, relative to the *new* spacing $h = 0.1$. Then an iteration, based on (6.10.6), could be initiated by estimating $\nabla^2 y''(1.5)$ and proceeding as was outlined in the preceding section. The value of $y(1.3)$ could be obtained by an interpolation involving certain of the available calculated ordinates. Clearly, care should be taken to obtain this ordinate to the same degree of accuracy as the other ordinates. The use of a difference formula, for this purpose, would entail the calculation of differences of the ordinates themselves, but would be desirable, in order that the accuracy of the interpolation could be estimated.

Another procedure consists of using the formulas derived in Sec. 5.7, to transform the tabulated differences $\nabla y''(1.4)$ and $\nabla^2 y''(1.4)$ to corresponding differences relative to the halved spacing. The ordinate $y(1.3)$ can then be determined, for example, by rewriting (6.10.6) in the form

$$y_n \approx \tfrac{1}{2}(y_{n+1} + y_{n-1}) - \frac{h^2}{2}(y''_{n+1} - \nabla y''_{n+1} + \tfrac{1}{12}\nabla^2 y''_{n+1}) \qquad (6.11.1)$$

If the difference operators corresponding to the halved spacing are denoted by ∇' and ∇'^2, Eq. (5.7.4) yields the formulas

$$\nabla' = \tfrac{1}{2}\nabla + \tfrac{1}{8}\nabla^2 + \tfrac{1}{64}\nabla^3 + \tfrac{5}{128}\nabla^4 + \tfrac{7}{256}\nabla^5 + \tfrac{21}{1024}\nabla^6 + \tfrac{33}{2048}\nabla^7 + \cdots \qquad (6.11.2)$$

and

$$\nabla'^2 = \tfrac{1}{4}\nabla^2 + \tfrac{1}{8}\nabla^3 + \tfrac{5}{64}\nabla^4 + \tfrac{7}{128}\nabla^5 + \tfrac{21}{512}\nabla^6 + \tfrac{33}{1024}\nabla^7 + \cdots \qquad (6.11.3)$$

when Δ is replaced by $-\nabla$ in (5.7.4), in accordance with the results of Sec. 5.7. The use of these formulas permits the calculation of the differences relative to $x = 1.4$ in the third line of Table 6.5, after which $\nabla y''$ and y'' are obtained in line two and y'' in line one. A useful check on the accuracy of the modified differences is then afforded by a comparison of the value of $y''(1.2)$ so obtained with that previously obtained in the direct calculation of Table 6.4. The ordinate $y(1.3)$ is next calculated from (6.11.1).

Table 6.5

x	y	y''	$\nabla y''$	$\nabla^2 y''$	y
1.2	1.38000	1.65600			
1.3	1.55071	2.01592	35992		
1.4	1.74164	2.43830	42238	6246	

Then if, say, $\nabla^2 y''(1.5)$ is estimated roughly as being *equal* to $\nabla^2 y''(1.4)$, the line

1.5	1.95701	2.92314	48484	6246	

(6.11.4)

is obtained (from right to left), the first approximation to $y(1.5)$ being obtained by use of (6.10.6). When $y''(1.5)$ and its differences are recalculated (from left to right), the next approximation to $y(1.5)$ is obtained from (6.10.6) as 1.95702, and the final form of this line of the calculation reads as follows:

$$1.5 \;\; | \;\; 1.95702 \;\; | \;\; 2.93553 \;\; | \;\; 49723 \quad 7487 \;\; | \tag{6.11.5}$$

Sufficient data are now available for the use of (6.10.5) as a predictor in the next step, if this is desired, after which entries in the y column are again calculable. (Variations and modifications of this procedure are easily devised here and in other cases.)

6.12 Use of Higher Derivatives

It is possible to derive a variety of formulas, for the numerical integration of differential equations, which involve values of certain higher derivatives of the unknown function. In particular, the Euler-Maclaurin sum formula (5.8.17) can be expressed in the form

$$\begin{aligned} y_{n+1} - y_{n-p} = h(\tfrac{1}{2}y'_{n+1} + y'_n + y'_{n-1} + \cdots \\ + y'_{n-p+2} + y'_{n-p+1} + \tfrac{1}{2}y'_{n-p}) \\ - \frac{h^2}{12}(y''_{n+1} - y''_{n-p}) + \frac{h^4}{720}(y^{\text{iv}}_{n+1} - y^{\text{iv}}_{n-p}) \\ - \frac{h^6}{30240}(y^{\text{vi}}_{n+1} - y^{\text{vi}}_{n-p}) + \cdots \end{aligned} \tag{6.12.1}$$

where the error committed by truncation with the term of order h^{2k} is $(p + 1)h$ times the term of order h^{2k+2} with the contents of the relevant parentheses replaced by $y^{(2k+3)}(\xi)$, where $x_{n-p} < \xi < x_{n+1}$. Thus, for example, with $p = 0$ we have the formula

$$y_{n+1} = y_n + \frac{h}{2}(y'_{n+1} + y'_n) - \frac{h^2}{12}(y''_{n+1} - y''_n) + \frac{h^5}{720}y^{\text{v}}(\xi) \tag{6.12.2}$$

of closed type, which may be used with any convenient predictor formula (preferably also with an error of order h^5) as in the methods discussed previously.

Formula (6.12.2) can be obtained also as a special case of the so-called *Hermite interpolation formula* (to be discussed in Sec. 8.2) and can also be derived by a method of *undetermined coefficients*, in which we write

$$y_{n+1} = y_n + h(\alpha_0 y'_{n+1} + \alpha_1 y'_n) + h^2(\beta_0 y''_{n+1} + \beta_1 y''_n) + E$$

so that $E = 0$ if $y(x)$ is a constant, and determine α_0, α_1, β_0, and β_1 in such a way that $E = 0$ also when $y(x) = x, x^2, x^3$, and x^4, and hence for *any* polynomial of degree 4 or less. For this purpose, it is convenient and nonrestrictive to take $h = 1$ and $x_n = 0$, so that the relevant equations become

$$\alpha_0 + \alpha_1 = 1 \qquad 2\alpha_0 + 2(\beta_0 + \beta_1) = 1$$

$$3\alpha_0 + 6\beta_0 = 1 \qquad 4\alpha_0 + 12\beta_0 = 1$$

and yield $\alpha_0 = \alpha_1 = \frac{1}{2}$ and $\beta_0 = -\beta_1 = -\frac{1}{12}$, in accordance with (6.12.2). The error term can then be determined by the methods of Secs. 5.10 and 5.11, if the formula is first rewritten in the equivalent form

$$\int_0^1 f'(s)\,ds = \tfrac{1}{2}[f'(0) + f'(1)] + \tfrac{1}{12}[f''(0) - f''(1)] + E$$

where

$$f(s) \equiv y(x_n + sh)$$

Reference to Sec. 5.11 then gives

$$E = \int_0^1 \pi(s) f'[0, 0, 1, 1, s]\,ds \qquad \pi(s) = s^2(s-1)^2$$

and, since $\pi(s)$ does not change sign, there follows

$$E = \frac{f^{\mathrm{v}}(\eta)}{4!}\int_0^1 \pi(s)\,ds = \tfrac{1}{720} f^{\mathrm{v}}(\eta) = \tfrac{1}{720} h^5 y^{\mathrm{v}}(x_n + \eta h) = \tfrac{1}{720} h^5 y^{\mathrm{v}}(\xi)$$

where $0 < \eta < 1$, and hence $x_n < \xi < x_{n+1}$.

In the same way, a formula of open type, involving only y_{n+1}, y_{n-1}, y'_{n-1}, y''_n, and y''_{n-1}, can be obtained in the form

$$y_{n+1} = y_{n-1} + 2hy'_{n-1} + \frac{2h^2}{3}(2y''_n + y''_{n-1}) + \frac{2h^5}{45} y^{\mathrm{v}}(\xi) \tag{6.12.3}$$

and can be used as a predictor in connection with (6.12.2). Since (6.12.2) affords an accuracy which is generally better than that associated with the result of retaining fourth-order terms in the Taylor expansion

$$y_{n+1} = y_n + hy'_n + \frac{h^2}{2} y''_n + \frac{h^3}{6} y'''_n + \frac{h^4}{24} y^{\mathrm{iv}}_n + \frac{h^5}{120} y^{\mathrm{v}}(\xi)$$

it is often useful in *starting* the solution when a procedure of fourth order is appropriate and when the calculation of values of y''' and y^{iv} is to be avoided. The formula $y_1 \approx y_0 + hy'_0 + \frac{1}{2}h^2 y''_0$ can be used for a *prediction* in the first step, after which (6.12.3) is available.

When the differential equation is of second order, Eqs. (6.12.2) and (6.12.3) are to be supplemented by the two equations obtained by replacing y

by u, where $u = y'$. Formulas of higher-order accuracy may be obtained if derivatives of order three or more are also employed.

A useful class of formulas, associated with the name of Obrechkoff [1942], can be derived by an inverse method in which we first seek a formula for $\int_0^h \phi(x)\,dx$ with an error expressed in the form

$$E = \frac{1}{(2r)!}\int_0^h x^r(x-h)^r\phi^{(2r)}(x)\,dx \qquad (6.12.4)$$

where r is an arbitrarily prescribed integer. If we integrate by parts r times, there follows immediately

$$E = \frac{1}{(2r)!}\int_0^h \phi^{(r)}(x)\,\frac{d^r}{dx^r}\,[x^r(h-x)^r]\,dx$$

since the integrated terms vanish at both limits, and r additional integrations by parts yield the result

$$E = \int_0^h \phi(x)\,dx - \frac{r!}{(2r)!}\sum_{k=1}^{r}(-1)^{k-1}\frac{(2r-k)!}{(r-k)!}\frac{h^k}{k!}\,[\phi^{(k-1)}(h) + (-1)^{k-1}\phi^{(k-1)}(0)] \qquad (6.12.5)$$

which supplies the required formula after a transposition.

If we write $\phi(x) = y'(x)$, and translate the origin to x_n, the result takes the form

$$y_{n+1} = y_n + \frac{r!}{(2r)!}\sum_{k=1}^{r}(-1)^{k-1}\frac{(2r-k)!}{(r-k)!}\frac{h^k}{k!}\,[y_{n+1}^{(k)} + (-1)^{k-1}y_n^{(k)}] + E \qquad (6.12.6)$$

where, after an application of the second mean-value theorem, (6.12.4) becomes

$$E = \frac{y^{(2r+1)}(\xi)}{(2r)!}\int_0^h x^r(x-h)^r\,dx = (-1)^r\frac{h^{2r+1}}{2r+1}\left[\frac{r!}{(2r)!}\right]^2 y^{(2r+1)}(\xi) \qquad (6.12.7)$$

with $x_n < \xi < x_{n+1}$.

When $r = 2$, this result becomes identical with (6.12.2). However, when $r = 3$ we obtain the formula

$$y_{n+1} = y_n + \frac{h}{2}(y'_{n+1} + y'_n) - \frac{h^2}{10}(y''_{n+1} - y''_n) + \frac{h^3}{120}(y'''_{n+1} + y'''_n) - \frac{h^7}{100800}\,y^{\text{vii}}(\xi) \qquad (6.12.8)$$

which possesses obvious advantages (in general) over the corresponding formula

$$y_{n+1} = y_n + \frac{h}{2}(y'_{n+1} + y'_n) - \frac{h^2}{12}(y''_{n+1} - y''_n) + \frac{h^4}{720}(y^{\text{iv}}_{n+1} - y^{\text{iv}}_n) - \frac{h^7}{30240}y^{\text{vii}}(\xi) \qquad (6.12.9)$$

obtained from (6.12.1). An appropriate predictor formula can be obtained in the form

$$y_{n+1} = y_{n-1} + 2h(4y'_n - 3y'_{n-1}) - \frac{2h^2}{5}(8y''_n + 7y''_{n-1}) + \frac{2h^3}{15}(7y'''_n - 3y'''_{n-1}) + \frac{13h^7}{6300}y^{\text{vii}}(\xi) \qquad (6.12.10)$$

An infinite variety of other formulas can be derived by employing data relevant to more than two points for correction, and to more than three points for prediction. Thus, for example, the three-point formula of highest precision, using first and second derivatives, is readily found to be

$$y_{n+1} - 2y_n + y_{n-1} = \frac{3h}{8}(y'_{n+1} - y'_{n-1}) - \frac{h^2}{24}(y''_{n+1} - 8y''_n + y''_{n-1}) + \frac{h^8}{60480}y^{\text{viii}}(\xi) \qquad (6.12.11)$$

6.13 A Simple Runge-Kutta Method

The methods associated with the names of Runge [1895], Kutta [1901], Heun [1900], and others as applied to the numerical solution of the problem

$$y' = F(x, y) \qquad y(x_0) = y_0 \qquad (6.13.1)$$

effectively replace the result of truncating a Taylor-series expansion of the form

$$y_{n+1} = y_n + hy'_n + \frac{h^2}{2}y''_n + \frac{h^3}{6}y'''_n + \cdots \qquad (6.13.2)$$

by an approximation in which y_{n+1} is calculated from a formula of the type

$$y_{n+1} = y_n + h[\alpha_0 F(x_n, y_n) + \alpha_1 F(x_n + \mu_1 h, y_n + b_1 h) + \alpha_2 F(x_n + \mu_2 h, y_n + b_2 h) + \cdots + \alpha_p F(x_n + \mu_p h, y_n + b_p h)] \qquad (6.13.3)$$

Here the α's, μ's, and b's are so determined that, if the right-hand member of (6.13.3) were expanded in powers of the spacing h, the coefficients of a certain number of the leading terms would agree with the corresponding coefficients in (6.13.2).

They possess the advantages that they are self-starting but do not require the evaluation of derivatives of $F(x, y)$ and hence can be used (even at the beginning of the solution) when $F(x, y)$ is not given by an analytical expression, and also that a change in spacing is easily effected at any intermediate stage of the calculation. On the other hand, each step involves several evaluations of $F(x, y)$, which may be excessively laborious and/or time-consuming, and also the estimation of errors is less simply accomplished than in the previously described methods.

It is convenient, in order both to simplify the derivation and also to systematize the formulation, to express each of the b's in (6.13.3) as a linear combination of the preceding values of F. Thus, in place of using the notation of (6.13.3), it is desirable to write the approximation in the form

$$y_{n+1} = y_n + \alpha_0 k_0 + \alpha_1 k_1 + \cdots + \alpha_p k_p \tag{6.13.4}$$

with

$$\begin{aligned}
k_0 &= hF(x_n, y_n) \\
k_1 &= hF(x_n + \mu_1 h, y_n + \lambda_{10} k_0) \\
k_2 &= hF(x_n + \mu_2 h, y_n + \lambda_{20} k_0 + \lambda_{21} k_1) \\
&\cdots\cdots\cdots\cdots\cdots\cdots\cdots\cdots\cdots\cdots \\
k_p &= hF(x_n + \mu_p h, y_n + \lambda_{p0} k_0 + \lambda_{p1} k_1 + \cdots + \lambda_{p,p-1} k_{p-1})
\end{aligned} \tag{6.13.5}$$

where the coefficients α_i, μ_i, and λ_{ij} are to be determined.

Since the actual derivation of such formulas involves considerable algebraic manipulation, we consider in detail only the very simple case $p = 1$, which may serve to illustrate the procedure in the more general case. Thus, writing μ for μ_1 and λ for λ_{10}, we proceed to determine α_0, α_1, μ, and λ such that

$$y_{n+1} = y_n + \alpha_0 k_0 + \alpha_1 k_1 \tag{6.13.6}$$

with

$$k_0 = hF(x_n, y_n) \qquad k_1 = hF(x_n + \mu h, y_n + \lambda k_0) \tag{6.13.7}$$

possesses an expansion in powers of h whose leading terms agree, insofar as possible, with the leading terms of (6.13.2).

We first obtain the expansion

$$\begin{aligned} k_1 &= h[F + (\mu h F_x + \lambda k_0 F_y) \\ &\qquad + \tfrac{1}{2}(\mu^2 h^2 F_{xx} + 2\mu\lambda h k_0 F_{xy} + \lambda^2 k_0^2 F_{yy}) + O(h^3)] \\ &= hF + h^2(\mu F_x + \lambda F F_y) \\ &\qquad + \frac{h^3}{2}(\mu^2 F_{xx} + 2\mu\lambda F F_{xy} + \lambda^2 F^2 F_{yy}) + O(h^4) \end{aligned} \tag{6.13.8}$$

where $F \equiv F(x_n, y_n)$, $F_x \equiv F_x(x_n, y_n)$, and so forth. Hence (6.13.6) becomes

$$\begin{aligned} y_{n+1} &= y_n + h(\alpha_0 + \alpha_1)F + h^2\alpha_1(\mu F_x + \lambda F F_y) \\ &\qquad + \frac{h^3}{2}\alpha_1(\mu^2 F_{xx} + 2\mu\lambda F F_{xy} + \lambda^2 F^2 F_{yy}) + O(h^4) \end{aligned} \tag{6.13.9}$$

On the other hand, with the same abbreviated notation, we obtain from (6.13.1) the relations

$$\begin{aligned} y' &= F \\ y'' &= F_x + F F_y \\ y''' &= F_{xx} + 2F F_{xy} + F^2 F_{yy} + F_y(F_x + F F_y) \end{aligned} \tag{6.13.10}$$

so that (6.13.2) becomes

$$\begin{aligned} y_{n+1} &= y_n + hF + \frac{h^2}{2}(F_x + F F_y) \\ &\qquad + \frac{h^3}{6}[F_{xx} + 2F F_{xy} + F^2 F_{yy} + F_y(F_x + F F_y)] + O(h^4) \end{aligned} \tag{6.13.11}$$

Thus, if we identify the coefficients of hF, h^2F_x, and h^2FF_y in (6.13.9) and (6.13.11), we obtain the three conditions

$$\alpha_0 + \alpha_1 = 1 \qquad \mu\alpha_1 = \tfrac{1}{2} \qquad \lambda\alpha_1 = \tfrac{1}{2} \tag{6.13.12}$$

involving the four adjustable parameters, which are satisfied if and only if

$$\alpha_0 = 1 - c \qquad \alpha_1 = c \qquad \mu = \frac{1}{2c} \qquad \lambda = \frac{1}{2c}$$

where c is an arbitrary nonzero constant. The expansion (6.13.9) then reduces to

$$\begin{aligned} y_{n+1} &= y_n + hF + \frac{h^2}{2}(F_x + F F_y) \\ &\qquad + \frac{h^3}{8c}(F_{xx} + 2F F_{xy} + F^2 F_{yy}) + O(h^4) \end{aligned} \tag{6.13.13}$$

and reference to (6.13.10) shows that (6.13.13) or, equivalently, (6.13.6) would then be brought into agreement with (6.13.11) or (6.13.2) if a truncation-error term of the form

$$T_n = -\left(\frac{h^3}{8c} - \frac{h^3}{6}\right)[(F_{xx} + 2FF_{xy} + F^2F_{yy}) + F_y(F_x + FF_y)] + \frac{h^3}{8c}F_y(F_x + FF_y) + O(h^4)$$

or

$$T_n = -\frac{h^3}{24c}[(3 - 4c)y_n''' - 3F_y(x_n, y_n)y_n''] + O(h^4) \qquad (6.13.14)$$

were added to its right-hand member.

The remaining free parameter c clearly cannot be determined so that T_n is of order h^4, except in trivial special cases. One convenient choice is $c = \frac{1}{2}$, in which case the second abscissa involved in (6.13.6) and (6.13.7) is x_{n+1}, and the formula becomes

$$y_{n+1} = y_n + \tfrac{1}{2}(k_0 + k_1) + T_n, \qquad (6.13.15)$$

with

$$k_0 = hF(x_n, y_n) \qquad k_1 = hF(x_n + h, y_n + k_0) \qquad (6.13.16)$$

where

$$T_n = -\frac{h^3}{12}[y_n''' - 3F_y(x_n, y_n)y_n''] + O(h^4) \qquad (6.13.17)$$

Stepwise calculation based on this formula is sometimes known as *Heun's method.*

If, for all values of x and y involved in the calculation, it is known that

$$|F_y(x, y)| \leqq K \qquad (6.13.18)$$

then, as in earlier developments, it is readily shown that the propagated error ε_n in the nth step is dominated by the solution of the difference equation

$$e_{n+1} = e_n + \frac{hK}{2}e_n + \frac{hK}{2}(e_n + hKe_n) + E$$

or

$$e_{n+1} = \left(1 + hK + \frac{h^2K^2}{2}\right)e_n + E \qquad (6.13.19)$$

where

$$e_0 = 0 \qquad E = |T_n + R_n|_{\max} \qquad hK + \frac{h^2K^2}{2} < 1 \qquad (6.13.20)$$

Further, it can be shown that (6.13.17) can be replaced by

$$T_n = -\frac{h^3}{12}[y'''(\xi_1) - 3F_y(x_{n+1}, \eta)y''(\xi_2)] \quad (6.13.21)$$

where ξ_1 and ξ_2 are intermediate between x_n and x_{n+1}, and η between y_{n+1} and $y_n + hy'_n$. Thus, if the roundoff error R_n is ignored, and if

$$|y''(x)| \leqq M_2 \qquad |y'''(x)| \leqq M_3 \quad (6.13.22)$$

it follows after a simple calculation that

$$|\varepsilon_n| \leqq \frac{h^2(M_3 + 3KM_2)}{12K(1 + \frac{1}{2}hK)}\left[\left(1 + hK + \frac{h^2K^2}{2}\right)^n - 1\right] \quad (6.13.23)$$

The formula (6.13.15), using (6.13.16), is of limited accuracy. Indeed, it can be considered to be a modification of the result of retaining only the first difference in (6.3.2)

$$y_{n+1} = y_n + \frac{h}{2}(y'_n + y'_{n+1}) - \frac{h^3}{12}y'''(\xi) \quad (6.13.24)$$

in which the unknown derivative $y'_{n+1} \equiv F(x_{n+1}, y_{n+1})$ is replaced by the approximation $y'_{n+1} \approx F(x_{n+1}, y_n + hy'_n)$. This consideration is useful in deriving (6.13.21). The details of the analysis were presented here principally to illustrate the similar but more complicated analysis relevant to formulas of higher-order accuracy, certain of which are listed in the following section.

It is of some importance to notice that the error (6.13.21), associated with (6.13.15) and (6.13.16), depends upon the form of the function $F(x, y)$ as well as upon the solution y itself. This situation is characteristic of formulas of the Runge-Kutta type. For example, whereas the equations $y' = 2(x + 1)$ and $y' = 2y/(x + 1)$ both define the function $y = (x + 1)^2$ when the condition $y(0) = 1$ is imposed, the formula defined by (6.13.15) and (6.13.16) would yield this solution *exactly* when applied to the first equation, if no roundoffs were committed, but would not do so when applied to the second form. On the other hand, the formula (6.13.24) would yield exact results when applied to *either* form, or to any *other* first-order equation whose required solution is a polynomial of degree 2 or less (see also Milne [1950, 1970]).

At the same time, the mere fact that (6.13.15), with (6.13.16), does not have this last property does not imply that its interpretation as a weakened modification of (6.13.24) is proper in the more general case when the true solution is not such a polynomial. For example, it is easily seen that the use of (6.13.15) and (6.13.16) would yield exact results when applied to the problem $y' = -y/(x + 1)$, $y(0) = 1$, for which the solution is $y = 1/(x + 1)$, whereas the use of (6.13.24) would lead only to an approximation.

6.14 Runge-Kutta Methods of Higher Order

When k_0, k_1, and k_2 are employed in (6.13.4), corresponding to $p = 2$, it is found that the requirement that the expansion of the right-hand member be correct through h^3 terms imposes only six conditions on the eight arbitrary parameters involved, so that a doubly infinite set of such formulas with third-order accuracy can be obtained.

One such formula, due to Kutta, is of the form

$$y_{n+1} = y_n + \tfrac{1}{6}(k_0 + 4k_1 + k_2) + O(h^4) \qquad (6.14.1)$$

with

$$\begin{aligned} k_0 &= hF(x_n, y_n) \\ k_1 &= hF(x_n + \tfrac{1}{2}h, y_n + \tfrac{1}{2}k_0) \\ k_2 &= hF(x_n + h, y_n + 2k_1 - k_0) \end{aligned} \qquad (6.14.2)$$

A second formula, due to Heun, is of the form

$$y_{n+1} = y_n + \tfrac{1}{4}(k_0 + 3k_2) + O(h^4) \qquad (6.14.3)$$

with

$$\begin{aligned} k_0 &= hF(x_n, y_n) \\ k_1 &= hF(x_n + \tfrac{1}{3}h, y_n + \tfrac{1}{3}k_0) \\ k_2 &= hF(x_n + \tfrac{2}{3}h, y_n + \tfrac{2}{3}k_1) \end{aligned} \qquad (6.14.4)$$

These two formulas are generally of about equal accuracy, with each possessing certain obvious computational advantages. Kutta's form is seen to be analogous to the formula of Simpson's rule and would reduce to that formula if F were independent of y.

It is also possible to derive a two-parameter family of formulas of fourth-order accuracy, by retaining an additional k in (6.13.4). The simplest such formula, due to Kutta, is of the form

$$y_{n+1} = y_n + \tfrac{1}{6}(k_0 + 2k_1 + 2k_2 + k_3) + O(h^5) \qquad (6.14.5)$$

with

$$\begin{aligned} k_0 &= hF(x_n, y_n) \\ k_1 &= hF(x_n + \tfrac{1}{2}h, y_n + \tfrac{1}{2}k_0) \\ k_2 &= hF(x_n + \tfrac{1}{2}h, y_n + \tfrac{1}{2}k_1) \\ k_3 &= hF(x_n + h, y_n + k_2) \end{aligned} \qquad (6.14.6)$$

and would also reduce to Simpson's rule if F were independent of y.

Such formulas can also be generalized to the treatment of *simultaneous* equations of the form

$$\frac{dy}{dx} = F(x, y, u)$$
$$\frac{du}{dx} = G(x, y, u) \tag{6.14.7}$$

where y and u are prescribed when $x = x_0$. In particular, the preceding formula generalizes as follows:

$$y_{n+1} = y_n + \tfrac{1}{6}(k_0 + 2k_1 + 2k_2 + k_3) + O(h^5)$$
$$u_{n+1} = u_n + \tfrac{1}{6}(m_0 + 2m_1 + 2m_2 + m_3) + O(h^5) \tag{6.14.8}$$

with

$$k_0 = hF(x_n, y_n, u_n)$$
$$k_1 = hF(x_n + \tfrac{1}{2}h, y_n + \tfrac{1}{2}k_0, u_n + \tfrac{1}{2}m_0)$$
$$k_2 = hF(x_n + \tfrac{1}{2}h, y_n + \tfrac{1}{2}k_1, u_n + \tfrac{1}{2}m_1) \tag{6.14.9}$$
$$k_3 = hF(x_n + h, y_n + k_2, u_n + m_2)$$

and

$$m_0 = hG(x_n, y_n, u_n)$$
$$m_1 = hG(x_n + \tfrac{1}{2}h, y_n + \tfrac{1}{2}k_0, u_n + \tfrac{1}{2}m_0)$$
$$m_2 = hG(x_n + \tfrac{1}{2}h, y_n + \tfrac{1}{2}k_1, u_n + \tfrac{1}{2}m_1) \tag{6.14.10}$$
$$m_3 = hG(x_n + h, y_n + k_2, u_n + m_2)$$

A consideration of this form indicates the way in which other formulas are so generalized.

In particular, when $F = u$, so that (6.14.7) is equivalent to

$$\frac{d^2y}{dx^2} = G(x, y, y') \tag{6.14.11}$$

with $u \equiv y'$, (6.14.9) gives

$$k_0 = hy'_n \qquad k_1 = hy'_n + \frac{h}{2}m_0 \qquad k_2 = hy'_n + \frac{h}{2}m_1 \qquad k_3 = hy'_n + hm_2$$

and hence (6.14.8) and (6.14.10) reduce to

$$y_{n+1} = y_n + hy'_n + \frac{h}{6}(m_0 + m_1 + m_2) + O(h^5)$$
$$y'_{n+1} = y'_n + \tfrac{1}{6}(m_0 + 2m_1 + 2m_2 + m_3) + O(h^5) \tag{6.14.12}$$

with

$$
\begin{aligned}
m_0 &= hG(x_n, y_n, y'_n) \\
m_1 &= hG(x_n + \tfrac{1}{2}h, y_n + \tfrac{1}{2}hy'_n, y'_n + \tfrac{1}{2}m_0) \\
m_2 &= hG(x_n + \tfrac{1}{2}h, y_n + \tfrac{1}{2}hy'_n + \tfrac{1}{4}hm_0, y'_n + \tfrac{1}{2}m_1) \\
m_3 &= hG(x_n + h, y_n + hy'_n + \tfrac{1}{2}hm_1, y'_n + m_2)
\end{aligned}
\tag{6.14.13}
$$

The use of this formula is clearly simplified in those cases when G is independent of y'.

Many variations and generalizations of these formulas are present in the literature, some of which afford certain computational advantages in certain situations. One such modification, due to Gill [1951], is of particular usefulness when the computation is to be effected by a computer in which it is desirable to minimize the *storage* of data.

No simple expressions are known for the precise truncation errors in the preceding formulas. An *estimate* of the error can be obtained, in practice, in the following way. Let the truncation error associated with a formula of rth-order accuracy, in progressing from the ordinate at x_n to that at $x_{n+1} = x_n + h$, in a single step, be denoted by $C_n h^{r+1}$, and suppose that C_n varies slowly with n and is nearly independent of h when h is small. Then if the true ordinate at x_{n+1} is denoted by Y_{n+1}, the value obtained by two steps starting at x_{n-1} by $y_{n+1}^{(h)}$, and the value obtained by a single step with *doubled* spacing $2h$ by $y_{n+1}^{(2h)}$, there follows approximately

$$
\begin{aligned}
Y_{n+1} - y_{n+1}^{(h)} &\approx 2C_n h^{r+1} \\
Y_{n+1} - y_{n+1}^{(2h)} &\approx 2^{r+1} C_n h^{r+1}
\end{aligned}
\tag{6.14.14}
$$

when h is small. The result of eliminating C_n from these approximate relations is then the *extrapolation* formula†

$$
Y_{n+1} \approx y_{n+1}^{(h)} + \frac{y_{n+1}^{(h)} - y_{n+1}^{(2h)}}{2^r - 1} \tag{6.14.15}
$$

Thus if, at certain stages of the advancing calculation, the newly calculated ordinate y_{n+1} is recomputed from y_{n-1} with a doubled spacing, the truncation error in the originally calculated value is approximated by the result of dividing the difference between the two values by the factor $2^r - 1$, that is, by 3 in (6.13.15), by 7 in (6.14.1) or (6.14.3), and by 15 in the formulas of fourth-order accuracy.

It is apparent that an arbitrary change in spacing can be introduced at any stage of the forward progress, when a method of the Runge-Kutta type is used, without introducing any appreciable complication.

† This is another example of so-called *Richardson extrapolation* (see Sec. 3.8).

6.15 Boundary-Value Problems

Problems in which the conditions to be satisfied by the solution of a differential equation, of order two or greater, are specified at both ends of an interval in which the solution is required are known as *boundary-value* problems and are generally much less amenable to numerical analysis than are *initial-value* problems, in which all conditions are imposed at one point. In this section, we consider briefly the application of certain elementary methods to the numerical solution of such problems. More efficient methods often can be based upon the result of reformulating the problem as an *integral equation* or as a problem in the *calculus of variations*, the treatment of both of which falls outside the scope of this work.

For a *linear* problem, such as one governed by a second-order equation of the form

$$y'' + P(x)y' + Q(x)y = F(x) \qquad (a < x < b) \tag{6.15.1}$$

and by the end conditions

$$y(a) = A \qquad y(b) = B \tag{6.15.2}$$

where A and B are prescribed, the analysis can be based on the principle of superposition. Thus, if $u(x)$ is *any* solution of the equation

$$u'' + Pu' + Qu = F \tag{6.15.3}$$

which satisfies the initial condition

$$u(a) = A \tag{6.15.4}$$

and $v(x)$ is any *nontrivial* solution of the equation

$$v'' + Pv' + Qv = 0 \tag{6.15.5}$$

which satisfies the initial condition

$$v(a) = 0 \tag{6.15.6}$$

then the function

$$y(x) = u(x) + cv(x) \tag{6.15.7}$$

satisfies (6.15.1) and the condition $y(a) = A$ for *any* constant value of c. Further, if P, Q, and F are continuous on $[a, b]$, there cannot exist additional functions having this property. Thus, a solution is found if c can be determined such that

$$u(b) + cv(b) = B \tag{6.15.8}$$

Unless c can be so determined, no solution exists.

If P, Q, and F are continuous on $[a, b]$, the initial slopes $u'(a)$ and $v'(a)$ can be chosen arbitrarily, so long as $v'(a) \neq 0$; the choices $u'(a) = 0$ and $v'(a) = 1$ are frequently convenient. It may be noticed that, if $u(b) = B$

and $v(b) = 0$, then c is arbitrary and infinitely many solutions exist; if $u(b) \neq B$ and $v(b) = 0$, then *no* solution exists. Unless it happens that $v(b) = 0$, the solution exists uniquely. Any of the previously discussed methods can be used in determining u and v.

In the case of a corresponding nonlinear problem, such as that governed by an equation of the more general form

$$y'' = G(x, y, y') \qquad (a < x < b) \tag{6.15.9}$$

and the end conditions

$$y(a) = A \qquad y(b) = B \tag{6.15.10}$$

superposition generally is not valid. One possible procedure consists of defining $u(x, \alpha)$ as the solution of the initial-value problem

$$\begin{gathered} u'' = G(x, u, u') \\ u(a) = A \qquad u'(a) = \alpha \end{gathered} \tag{6.15.11}$$

and attempting to determine α such that

$$u(b, \alpha) = B \tag{6.15.12}$$

For this purpose, $u(b)$ could be determined for two or more trial values of α. Then, by linear (or higher-order) inverse interpolation, an "improved" value of α would be obtained, and the process would be iterated until (6.15.12) is satisfactorily approximated if the iteration converges.

This "shooting" process is apt to be tedious, and is complicated by the fact that (even in the linear case) small changes in α do not necessarily correspond to small changes in $u(b, \alpha)$. Further, the basic questions of existence and uniqueness of the solution are particularly troublesome in themselves, in the general nonlinear case. There exists no completely satisfactory general method (numerical or otherwise) for dealing with such problems. (Problems in which the end conditions prescribe y' or a linear combination of y and y', or which are expressed in a more complicated way, involve more or less obvious modifications.)

Another class of methods, which usually are convenient only when the problem is *linear*, consists of approximating the differential equation by a difference equation and of solving the *simultaneous* set of algebraic equations resulting from the requirement that this equation be satisfied at each of a set of equally spaced points in the relevant interval.

In illustration, any linear second-order equation can be transformed to an equation of the form

$$y'' + f(x)y = g(x) \tag{6.15.13}$$

as was pointed out at the end of Sec. 6.10. Reference to Eq. (5.6.7) yields the relation

$$y_{n+1} - 2y_n + y_{n-1} = h^2(1 + \tfrac{1}{12}\delta^2)y_n'' + T_n \quad (6.15.14)$$

where the truncation error T_n is expressible in the form

$$T_n = -\frac{h^6}{240}\, y^{\text{vi}}(\xi) \qquad (x_{n-1} < \xi < x_{n+1}) \quad (6.15.15)$$

Thus, if we use (6.15.13) to replace y_n'' by $g_n - f_n y_n$, (6.15.14) takes the form

$$\left(1 + \frac{h^2}{12} f_{n+1}\right) y_{n+1} - 2\left(1 - \frac{5h^2}{12} f_n\right) y_n + \left(1 + \frac{h^2}{12} f_{n-1}\right) y_{n-1}$$
$$= \frac{h^2}{12}(g_{n+1} + 10g_n + g_{n-1}) + T_n \quad (6.15.16)$$

Now, if the interval $[a, b]$ is divided into $N + 1$ equal parts, in such a way that $x_0 = a$, $x_1 = a + h, \ldots, x_N = a + Nh$, $x_{N+1} = b$, where $h = (b - a)/(N + 1)$, we may require the result of ignoring T_n in (6.15.16) to hold for $n = 1, 2, \ldots, N$, and so obtain a set of N simultaneous linear algebraic equations in $y_1, y_2, \ldots, y_N$, of the form

$$\left(1 + \frac{h^2}{12} f_0\right) y_0 - 2\left(1 - \frac{5h^2}{12} f_1\right) y_1 + \left(1 + \frac{h^2}{12} f_2\right) y_2 = \frac{h^2}{12} G_1$$
$$\left(1 + \frac{h^2}{12} f_1\right) y_1 - 2\left(1 - \frac{5h^2}{12} f_2\right) y_2 + \left(1 + \frac{h^2}{12} f_3\right) y_3 = \frac{h^2}{12} G_2$$
$$\cdots\cdots\cdots\cdots\cdots\cdots\cdots\cdots\cdots\cdots\cdots\cdots\cdots\cdots$$
$$\left(1 + \frac{h^2}{12} f_{N-1}\right) y_{N-1} - 2\left(1 - \frac{5h^2}{12} f_N\right) y_N$$
$$+ \left(1 + \frac{h^2}{12} f_{N+1}\right) y_{N+1} = \frac{h^2}{12} G_N \quad (6.15.17)$$

where

$$G(x) = g(x + h) + 10g(x) + g(x - h)$$

supplemented by the prescribed conditions $y_0 = A$ and $y_{N+1} = B$.

By virtue of (6.15.15), this procedure is of fifth order, in the sense that it would afford exact results if $y(x)$ were a polynomial of degree 5 or less. A simpler procedure, of third order, corresponds to the neglect of second differences in (6.15.14) and hence consists of solving the simultaneous equations

$$y_{n+1} - 2\left(1 - \frac{h^2}{2} f_n\right) y_n + y_{n-1} = h^2 g_n \qquad (n = 1, 2, \ldots, N) \quad (6.15.18)$$

supplemented by the boundary conditions.

On the other hand, if fourth differences are retained in (5.6.7), the corresponding equations are easily obtained in the form

$$-\frac{h^2}{240} f_{n+2} y_{n+2} + \left(1 + \frac{h^2}{10} f_{n+1}\right) y_{n+1} - 2\left(1 - \frac{97h^2}{240} f_n\right) y_n$$
$$+ \left(1 + \frac{h^2}{10} f_{n-1}\right) y_{n-1} - \frac{h^2}{240} f_{n-2} y_{n-2}$$
$$= \frac{h^2}{240}(-g_{n+2} + 24g_{n+1} + 194g_n + 24g_{n-1} - g_{n-2}) \qquad (6.15.19)$$

and would reduce to identities for $n = 2, 3, \ldots, N - 1$ if $y(x)$ were a polynomial of degree 7 or less. For $n = 1$ and $n = N$, Eq. (6.15.19) would involve the irrelevant quantities y_{-1} and y_{N+2}. Two additional "off-center" relations, which would also be satisfied exactly by any polynomial solution of degree 7 or less, are thus needed. They may be obtained, for example, by retaining *fifth* differences in the backward-difference formula (5.5.12), relative to $\nabla^2 y_{N+1}$, and in the corresponding forward-difference formula relative to $\Delta^2 y_0$, in the forms†

$$\left(1 + \frac{3h^2}{40} f_0\right) y_0 - 2\left(1 - \frac{209h^2}{480} f_1\right) y_1 + \left(1 + \frac{h}{60} f_2\right) y_2$$
$$+ \frac{7h^2}{120} f_3 y_3 - \frac{h^2}{40} f_4 y_4 + \frac{h^2}{240} f_5 y_5$$
$$= \frac{h^2}{240}(18g_0 + 209g_1 + 4g_2 + 14g_3 - 6g_4 + g_5) \qquad (6.15.20)$$

and

$$\frac{h^2}{240} f_{N-4} y_{N-4} - \frac{h^2}{40} f_{N-3} y_{N-3} + \frac{7h^2}{120} f_{N-2} y_{N-2} + \left(1 + \frac{h^2}{60} f_{N-1}\right) y_{N-1}$$
$$- 2\left(1 - \frac{209h^2}{480} f_N\right) y_N + \left(1 + \frac{3h^2}{40} f_{N+1}\right) y_{N+1}$$
$$= \frac{h^2}{240}(g_{N-4} - 6g_{N-3} + 14g_{N-2} + 4g_{N-1} + 209g_N + 18g_{N+1}) \qquad (6.15.21)$$

The $N - 2$ equations (6.15.19), the two equations (6.15.20) and (6.15.21), and the two prescribed end conditions serve to determine (approximately) the values of the $N + 2$ ordinates $y_0, y_1, \ldots, y_N, y_{N+1}$. Here the interval $[a, b]$ must be divided into at least five equal parts.

† The same relations can be obtained by using the approximate relations

$$\Delta^6 y''_{-1} = \Delta^6(g_{-1} - f_{-1}y_{-1}) = 0 \qquad \nabla^6 y''_{N+2} = \nabla^6(g_{N+2} - f_{N+2}y_{N+2}) = 0$$

to eliminate the ordinates y_{-1} and y_{N+2} from the equations which correspond to setting $n = 1$ and $n = N$ in (6.15.19).

If the prescribed end condition at $x = a \equiv x_0$ involves $y'(a)$ in place of (or in combination with) $y(a)$, that condition can be replaced by an appropriate approximate one, involving y_0 and y_1, by means of the result of retaining terms involving powers of h through h^n (where n is the order of the procedure used) in the expansion

$$\Delta y_0 = \left(h\mathbf{D} + \frac{1}{2!}h^2\mathbf{D}^2 + \frac{1}{3!}h^3\mathbf{D}^3 + \frac{1}{4!}h^4\mathbf{D}^4 + \cdots\right) y_0$$

combined with (6.15.13) to give

$$y_1 - y_0 = hy_0' + \frac{h^2}{2}(g_0 - f_0 y_0) + \frac{h^3}{6}(g_0' - f_0' y_0 - f_0 y_0') + \frac{h^4}{24}(g_0'' - f_0'' y_0 - 2f_0' y_0' - f_0 g_0 + f_0^2 y_0) + \cdots$$

or

$$\begin{aligned} y_1 = y_0 \Bigg[1 &- \frac{h^2}{2} f_0 - \frac{h^3}{6} f_0' - \frac{h^4}{24}(f_0'' - f_0^2) - \frac{h^5}{120}(f_0''' - 4f_0' f_0) - \cdots\Bigg] \\ &+ y_0' \left[h - \frac{h^3}{6} f_0 - \frac{h^4}{12} f_0' - \frac{h^5}{120}(3f_0'' - f_0^2) - \cdots\right] \\ &+ \Bigg[\frac{h^2}{2} g_0 + \frac{h^3}{6} g_0' + \frac{h^4}{24}(g_0'' - f_0 g_0) + \frac{h^5}{120}(g_0''' - 3f_0' g_0 - f_0 g_0') + \cdots\Bigg] \end{aligned} \tag{6.15.22}$$

A similar relation, for use at $x = b \equiv x_{N+1}$, is obtainable from the expansion

$$\nabla y_{N+1} = \left(h\mathbf{D} - \frac{1}{2!}h^2\mathbf{D}^2 + \frac{1}{3!}h^3\mathbf{D}^3 - \frac{1}{4!}h^4\mathbf{D}^4 + \cdots\right) y_{N+1}$$

6.16 Linear Characteristic-Value Problems

When $g(x) \equiv 0$ in (6.15.13), and the prescribed end conditions are of the special form $y(a) = y(b) = 0$, one solution of the problem is clearly the *trivial solution* $y(x) \equiv 0$. It frequently happens that an arbitrary constant parameter λ is linearly involved in the definition of the function $f(x)$ and that it is then desired to determine values of λ for which the problem also admits a *nontrivial solution.* Such values of λ are known as its *characteristic values* (or *eigenvalues*), and the corresponding solutions are called the *characteristic functions* of the problem.

The study of their properties and applications comprises an important field of mathematics, and a great variety of methods have been (and are being) devised for their approximate numerical treatment.

One such method can be based on the result of appropriately specializing (6.15.17). Thus, if the problem is of the form

$$\begin{aligned} y'' + [q(x) + \lambda r(x)]y &= 0 \\ y(a) = y(b) &= 0 \end{aligned} \tag{6.16.1}$$

we may replace f by $q + \lambda r$ in (6.15.17), to obtain a set of N equations of the form

$$\begin{aligned}
&-2\left[\left(1 - \frac{5h^2}{12} q_1\right) - \frac{5h^2}{12} \lambda r_1\right] y_1 \\
&\qquad + \left[\left(1 + \frac{h^2}{12} q_2\right) + \frac{h^2}{12} \lambda r_2\right] y_2 = 0 \\
&\left[\left(1 + \frac{h^2}{12} q_1\right) + \frac{h^2}{12} \lambda r_1\right] y_1 - 2\left[\left(1 - \frac{5h^2}{12} q_2\right) - \frac{5h^2}{12} \lambda r_2\right] y_2 \\
&\qquad + \left[\left(1 + \frac{h^2}{12} q_3\right) + \frac{h^2}{12} \lambda r_3\right] y_3 = 0 \\
&\cdots\cdots\cdots\cdots\cdots\cdots\cdots\cdots\cdots\cdots \\
&\left[\left(1 + \frac{h^2}{12} q_{N-1}\right) + \frac{h^2}{12} \lambda r_{N-1}\right] y_{N-1} \\
&\qquad - 2\left[\left(1 - \frac{5h^2}{12} q_N\right) - \frac{5h^2}{12} \lambda r_N\right] y_N = 0
\end{aligned} \tag{6.16.2}$$

This set of homogeneous linear equations will admit a nontrivial solution for $y_1, y_2, \ldots, y_N$ if and only if the determinant of the matrix of coefficients vanishes (see Sec. 10.2), a requirement which demands that λ be a root of an algebraic equation of degree N if no one of the values of r_k vanishes, as is generally the case in practice. For each such value of λ, this set of equations becomes redundant and (at least) one equation can be ignored, after which the remaining equations can be solved for the *ratios* of certain of the ordinates to the remaining one or ones. Except in unusual cases, only one of the ordinates (but, usually, *any* one) can be chosen arbitrarily, and the ratios of the remaining ones to that one are determinate. In this way, approximations to N of the characteristic numbers (generally the N *smallest* ones) of the true problem are obtained, together with ordinates of the corresponding characteristic functions, defined within a common arbitrary multiplicative factor.

The crudest approximation is obtained by taking $N = 1$, so that only the central ordinate y_1, at $x = (a + b)/2$, is involved, and $y_0 = y_2 = 0$. Thus only the equation

$$-2\left[\left(1 - \frac{5h^2}{12} q_1\right) - \frac{5h^2}{12} \lambda r_1\right] y_1 = 0$$

is obtained, and the requirement $y_1 \neq 0$ leads to the approximation

$$\lambda = \frac{12}{5h^2 r_1}\left(1 - \frac{5h^2 q_1}{12}\right) \equiv \lambda_1^{(1)} \qquad \left(h = \frac{b - a}{2}\right) \tag{6.16.3}$$

to the smallest characteristic number λ_1. The ordinate y_1 is then indeterminate.

When $N = 2$, the two permissible values of λ are found to be the roots of the determinantal equation

$$\begin{vmatrix} -2\left[\left(1 - \frac{5h^2}{12} q_1\right) - \frac{5h^2}{12} \lambda r_1\right] & \left[\left(1 + \frac{h^2}{12} q_2\right) + \frac{h^2}{12} \lambda r_2\right] \\ \left[\left(1 + \frac{h^2}{12} q_1\right) + \frac{h^2}{12} \lambda r_1\right] & -2\left[\left(1 - \frac{5h^2}{12} q_2\right) - \frac{5h^2}{12} \lambda r_2\right] \end{vmatrix} = 0 \tag{6.16.4}$$

where $h = (b - a)/3$, and may be denoted by $\lambda_1^{(2)}$ and $\lambda_2^{(1)}$. For each of these calculated values of λ, there follows also, from the first equation,

$$\frac{y_2}{y_1} = 2\,\frac{(1 - 5h^2 q_1/12) - (5h^2 r_1/12)\lambda}{(1 + h^2 q_2/12) + (h^2 r_2/12)\lambda} \tag{6.16.5}$$

with y_1 arbitrary. The use of the second equation would lead to an equivalent result.

In illustration, the problem

$$Y'' + 2Y' + \lambda x Y = 0 \qquad Y(0) = Y(1) = 0 \tag{6.16.6}$$

is transformed to the problem

$$y'' - y + \lambda x y = 0 \qquad y(0) = y(1) = 0 \tag{6.16.7}$$

with the change of variables

$$Y = e^{-x} y \tag{6.16.8}$$

in accordance with (6.10.23). With $q(x) = -1$ and $r(x) = x$, Eq. (6.16.3) yields

$$\lambda_1^{(1)} = \tfrac{96}{5}\left(1 + \tfrac{5}{48}\right) = \tfrac{106}{5} = 21.2$$

Equation (6.16.4) becomes

$$\begin{vmatrix} -2\left(\dfrac{339 - 5\lambda}{324}\right) & \dfrac{321 + 2\lambda}{324} \\ \dfrac{321 + \lambda}{324} & -2\left(\dfrac{339 - 10\lambda}{324}\right) \end{vmatrix} = 0$$

and expands into the relevant *characteristic equation*

$$22\lambda^2 - 2367\lambda + 39627 = 0$$

to yield a second approximation $\lambda_1^{(2)} \doteq 20.74$ to λ_1 and a first approximation $\lambda_2^{(1)} \doteq 89.38$ to λ_2. Equation (6.16.5) becomes

$$\frac{y_2}{y_1} = \frac{678 - 10\lambda}{321 + 2\lambda}$$

and yields $y_2/y_1 \doteq 1.30$ for $\lambda_1^{(2)}$ and $y_2/y_1 \doteq -0.432$ for $\lambda_2^{(1)}$. Thus, from (6.16.8) there follows $Y(\frac{2}{3})/Y(\frac{1}{3}) \approx 0.93$ in the first "mode" and -0.31 in the second one.

If three interior ordinates were used, to afford improved approximations to λ_1 and λ_2, and a first approximation to λ_3, it would seem to be necessary to expand a determinant of third order and to determine the roots of a cubic equation. However, various iterative techniques for determining the roots of the relevant characteristic equation without explicitly expanding the determinant, in such cases, exist in the literature (see Sec. 6.18).

A simpler procedure would be based on the use of (6.15.18), rather than (6.15.16), whereas a more elaborate procedure could be based on (6.15.19) to (6.15.21). Modifications, which are appropriate to situations when a linear combination of y and y' is required to vanish at each end of the interval, may be based on the use of (6.15.22) and a similar equation relevant to $x = b$, with $f = q + \lambda r$ and $g = 0$, in place of the conditions $y_0 = y_{N+1} = 0$.

In the simple special case when $q(x) = 0$ and $r(x) = 1$, so that (6.16.1) reduces to $y'' + \lambda y = 0$, where $y(a) = y(b) = 0$, the exact value of the rth characteristic value is easily found to be

$$\lambda_r = \frac{r^2\pi^2}{(b - a)^2}$$

The approximation $\bar{\lambda}_r$ afforded by use of the simpler procedure is found to be

$$\bar{\lambda}_r = \frac{4}{h^2}\sin^2\left(\frac{h}{2}\sqrt{\lambda_r}\right)$$

whereas that afforded by use of (6.16.2) can be shown to be

$$\bar{\lambda}_r = \frac{4}{h^2} \frac{\sin^2 [(h/2)\sqrt{\lambda_r}]}{1 - \frac{1}{3}\sin^2 [(h/2)\sqrt{\lambda_r}]}$$

where $h = (b - a)/(N + 1)$, from which results the nature of the approximations to the first N characteristic numbers can be determined in the two cases. In particular, the error in the former case is found to be positive and less than $h^2\lambda_r^2/12$, whereas that in the latter case is positive and less than $h^4\lambda_r^3/240$. The error associated with the use of the more elaborate seventh-order procedure would be of the order of $h^6\lambda_r^4$. These facts permit crude preliminary estimates of the requisite number of subdivisions in similar (but less simple) cases.

6.17 Selection of a Method

Whereas a rather large number of methods for dealing with initial-value problems have been outlined in this chapter, it should be remarked that a very substantial number of additional variations may also be found in the literature. The problem of deciding which one of these methods is most appropriate in a specific situation is rather difficult to discuss because of the large number of factors which may affect the decision.

First of all, the choice will depend upon the nature of the computational device to be used. Thus, for example, a method which is well adapted to the use of a desk calculator may be inconvenient for hand calculation because of the fact that it involves too many operations with multidigit numbers; or a procedure which involves a large number of iterations of a relatively simple technique may be remarkably well adapted to an automatic high-speed calculator, but its use may entail a prohibitive amount of time when the calculations are to be made on a desk calculator.

A procedure which involves a large number of evaluations of a certain function $F(x, y)$ may not be objectionable for desk calculation if $F(x, y)$ is well tabulated but may require an excessive amount of time in machine calculation. On the other hand, the situation may be completely reversed if the function is of complicated analytical form but if a subroutine is available for generating it directly in the computer.

Stability considerations may be of great importance, for a given problem, when a large number of steps is to be taken, but may be much less significant when only a relatively small number of ordinates is required, or when a different problem is dealt with. The computational advantages associated with a simple procedure with relatively large truncation error must be weighed against the fact that the use of that procedure generally requires a small spacing and a correspondingly large number of steps, and hence increases the importance of the effects of roundoff errors. At the same time, such a procedure may be preferred to a more elaborate one when its use permits a fairly confident estimate of an upper bound on the total propagated error, whereas such an estimate corresponding to the use of the more elaborate procedure is not readily available.

One may be faced with the problem of choosing the procedure which is most appropriate in the solution of a single differential equation, or that which appears to be best on the average for a wide class of equations.

The methods described in this chapter, for advancing the solution, fall into three broad classes: (1) methods which express the future ordinate as a linear combination of present and/or past ordinates and slopes (Secs. 6.5, 6.6, 6.9, and 6.10); (2) methods which also involve the calculation of certain higher derivatives (Sec. 6.12); and (3) methods in which the determination of the future ordinate does not involve memory of the past (Secs. 6.13 and 6.14).

The Euler procedure and its modification of closed type are the simplest of the procedures in the first class. The modified Adams method and the special fourth-order Milne and Hamming methods are perhaps the most frequently used procedures of higher-order accuracy in this class. In each case, the relevant formulas for prediction and correction may be expressed either in terms of differences of slopes or in terms of the slopes themselves. The Milne procedure requires fewer values of the derivative than the corresponding Adams procedure, but it compares unfavorably with the latter and with Hamming's method from the point of view of stability. Except for the simple Euler methods, the procedures in this class are not self-starting.

The methods of the second class are highly efficient when and only when the differential equation is of such a form that analytical relations between higher derivatives of the unknown function and the function itself are readily obtained. If derivatives at only two points are used, the special methods actually treated here are effectively self-starting.

The Runge-Kutta methods, of the third class, possess the advantage that, since their use at each stage of the advancing calculation does not require information relevant to past stages, they are completely self-starting and are particularly appropriate when memory requirements are to be minimized. Furthermore, these procedures are inherently stable and are such that a change in spacing is easily effected at any stage of the advance. The principal disadvantage consists of the fact that each forward step entails several evaluations of the "right-hand member" of the differential equation, a fact which may be of considerable importance when each such evaluation is difficult or time-consuming. In addition, the nonexistence of a tractable expression for the associated truncation error, in most cases, is a source of some inconvenience.

For the purpose of starting a solution, when a method of the first class is to be used for advancing the solution, one may choose among the methods of the second and third classes as well as the methods of Sec. 6.4. When the right-hand member of the differential equation is of such a form that the formation of higher derivatives is readily effected, the use of Taylor series (Sec. 6.4) or of series which involve values of higher derivatives at two points (Sec. 6.12) is often convenient. Otherwise, resort may be had to one of the iterative methods of Sec. 6.4 or to the methods of Runge-Kutta type (Secs. 6.13 and 6.14).

6.18 Supplementary References

General texts on the numerical solution of ordinary differential equations include Milne [1970], Collatz [1960], Fox [1962], Henrici [1962, 1963], Babuska, Prager, and Vitasek [1966], and Lapidus and Seinfeld [1971], each of which provides a useful bibliography. For theoretical considerations and analytical methods, see texts such as Ince [1952], Coddington and Levinson [1955], and Birkhoff and Rota [1969]. Daniel and Moore [1970] combine numerical and analytical considerations in a rather unusual way.

The basic papers on the stability of step-by-step processes are those of Rutishauser [1952] and Dahlquist [1956]. For references to later contributions, see the texts listed above. Roundoff-error propagation is studied in Henrici [1962].

Numerical techniques for dealing with boundary-value problems governed by ordinary differential equations are dealt with in Fox [1957] and Keller [1968].

PROBLEMS

Section 6.2

1 Show that the operator affecting y_n' in the open formula (6.2.12) relating y_{n+1} and y_{n-p} can be obtained by multiplying the one corresponding to $p = 0$ in (6.2.10) by

$$\frac{1 - (1 - \nabla)^{p+1}}{\nabla} = (p + 1) - \frac{(p + 1)p}{2!}\nabla + \cdots$$

and use this method to derive (6.2.13) to (6.2.15), as well as the formulas corresponding to $p = 2$ and $p = 4$.

2 Verify that the results of terminating (6.2.13) to (6.2.15) with the zeroth, second, and fourth differences, respectively, are Newton-Cotes formulas of open type.

Section 6.3

3 Show that the method used in Prob. 1 also applies to the closed formulas, and derive the formulas relating y_{n+1} and y_{n-p} from (6.3.2) for $p = 1, 2, 3, 4$, and 5 in this way.

4 Verify that the results of terminating (6.3.3) and (6.3.4) with the second and fourth differences, respectively, are Newton-Cotes formulas of closed type.

Section 6.4

5 Obtain additional values of y, corresponding to $x = \pm 0.1$ and ± 0.2, for each of the following problems, by use of power series, rounding the results to five decimal places:
(*a*) $y' + y = 0, y(0) = 1$
(*b*) $y' + 2xy = 2x^3, y(0) = 0$
(*c*) $y' + y + xy^2 = 0, y(0) = 1$
(*d*) $xy' = 1 - y + x^2y^2, y(0) = 1$
[The respective analytical solutions are e^{-x}, $e^{-x^2} - 1 + x^2$, $(2e^x - 1 - x)^{-1}$, and $(\tan x)/x$.]

6 (*a*) to (*d*) Proceed as in Prob. 5 by Picard's method.

7 (*a*) to (*d*) Proceed as in Prob. 5 by use of Eqs. (6.4.14).

8 Obtain additional starting values of y when $x = 0.1, 0.2, 0.3$, and 0.4 for the following problem, by use of Eqs. (6.4.13), assuming that only the tabulated values of $\phi(x)$ are available and rounding the results to five decimal places:

$$y' + xy = \phi(x) \qquad y(0) = 1$$

x	$\phi(x)$	x	$\phi(x)$
0.0	1.00000	0.6	1.16412
0.1	1.00499	0.7	1.21579
0.2	1.01980	0.8	1.27059
0.3	1.04399	0.9	1.32660
0.4	1.07683	1.0	1.38177
0.5	1.11730		

[Here $\phi(x) \approx \cos x + x \sin x$, and hence $y(x) \approx e^{-x^2/2} + \sin x$.]

Section 6.5

9 (*a*) to (*d*) Advance the calculation of the solutions of Prob. 5 to $x = 1$ with $h = 0.1$ by use of the Adams method, rounding all calculated ordinates to five decimal places and estimating the errors.

10 (*a*) to (*d*) Proceed as in Prob. 9, using Eq. (6.2.18).

11 Advance the calculations of Prob. 8 to $x = 1$ with $h = 0.1$, (*a*) by use of the Adams method and (*b*) by use of (6.2.18), rounding all calculated ordinates to five decimal places and estimating the errors.

Section 6.6

12 (*a*) to (*e*) Recalculate the ordinates required in Probs. 9*a* to *d* and 11 by use of the modified Adams method, again retaining five places and estimating the errors.

13 (*a*) to (*e*) Proceed as in Prob. 12 by use of Milne's method.

Section 6.7

14 Show that the approximate solution of the problem

$$y' + y = 0 \qquad y(0) = 1$$

afforded by the result of retaining first differences in the formula (6.3.2) would be of the form

$$y_n^{(1)} = \left(\frac{2-h}{2+h}\right)^n$$

if no roundoffs were effected, and that corresponding to retention of first differences in (6.3.3) would be of the form

$$y_n^{(2)} = \frac{1}{2\sqrt{1+h^2}}[(\sqrt{1+h^2} + h + y_1)(\sqrt{1+h^2} - h)^n$$
$$+ (-1)^n(\sqrt{1+h^2} - h - y_1)(\sqrt{1+h^2} + h)^n]$$

where y_1 is the independently calculated approximation to the true value e^{-h}, whereas the exact solution is given by $y_n = e^{-nh}$.

15 Show that, when h is small, the solutions obtained in Prob. 14 can be expressed in the forms

$$y_n^{(1)} = (e^{-h} - \tfrac{1}{12}h^3 + \cdots)^n$$

and

$$y_n^{(2)} = [(1 - \tfrac{1}{12}h^3 + \cdots) - \tfrac{1}{2}\varepsilon(1 - \tfrac{1}{2}h^2 + \cdots)](e^{-h} + \tfrac{1}{6}h^3 + \cdots)^n$$
$$+ (-1)^n[(\tfrac{1}{12}h^3 + \cdots) + \tfrac{1}{2}\varepsilon(1 - \tfrac{1}{2}h^2 + \cdots)](e^h - \tfrac{1}{6}h^3 + \cdots)^n$$

where ε represents the error associated with the value employed for y_1, and where omitted terms in each expansion are small, of order h^4.

16 Suppose that a spacing $h = 0.1$ is used in the approximations of Prob. 14, and that the value of y_1 used in the second calculation is assumed to be free of error. Calculate the errors and relative errors in the two approximations for values of n in the neighborhood of 10, 50, and 100, neglecting the effects of roundoff errors.

17 Show that when Milne's method is used, the parasitic part $u_n \equiv u(x_n)$ of the approximate solution of the problem

$$y' = Ay \qquad y(0) = y_0$$

where A is a constant, is approximated by

$$u_n \approx (-1)^n \left[\frac{A^5h^5}{360}\, y_0 + \tfrac{1}{2}\varepsilon\right] e^{-Ax_n/3}$$

when $|Ah|$ is small, where ε is the error inherent in y_1.

18 For Hamming's corrector (6.7.20), show that the characteristic equation corresponding to (6.7.6) is

$$(1 - \tfrac{3}{8}Ah)\beta^3 - (\tfrac{9}{8} + \tfrac{3}{4}Ah)\beta^2 + \tfrac{3}{8}Ah\beta + \tfrac{1}{8} = 0$$

Verify that when $h = 0$ the roots are 1 and $(1 \pm \sqrt{33})/16$. Then show that the Hamming simulation to the solution of (6.7.3) is approximately of the form

$$y_n = c_0(1 + Ah + \cdots)^n + c_1(0.4216 - 0.0080Ah + \cdots)^n$$
$$+ c_2(-0.2966 + 0.1798Ah + \cdots)^n$$

when h is small, where the c's are determined by starting values and are such that c_1 and c_2 are small relative to c_0. Thus deduce that the parasitic solutions cannot be troublesome when Ah is of reasonably small magnitude and hence that Hamming's method displays short-range stability whether $\partial F/\partial y$ is positive or negative unless h is abnormally large.

19 (*a*) to (*e*) Proceed as in Prob. 12 by use of Hamming's method.

Section 6.8

20 If the formula

$$y_{n+1} = y_n + h(\alpha_{-1}y'_{n+1} + \alpha_0 y'_n)$$

is used for the numerical solution of the problem

$$y' = F(x, y) \qquad y(x_0) = y_0$$

if $\alpha_{-1} \geqq 0$, $\alpha_0 \geqq 0$, if $|F_y(x, y)| \leqq K$ throughout the calculation leading to y_n, and if $Kh\alpha_{-1} < 1$, show that the error ε_n in y_n is bounded by the inequality

$$|\varepsilon_n| \leqq \frac{E}{Kh}\left[\left(\frac{1 + Kh\alpha_0}{1 - Kh\alpha_{-1}}\right)^n - 1\right] \approx \frac{E}{Kh}(e^{nKh} - 1)$$

where E is the largest error introduced in a single step. Also specialize to the cases ($\alpha_{-1} = 0, \alpha_0 = 1$), ($\alpha_{-1} = 1, \alpha_0 = 0$), and ($\alpha_{-1} = \alpha_0 = \tfrac{1}{2}$), showing that E can be taken to be $\tfrac{1}{2}h^2M_2 + R$ in the first two cases and to be $\tfrac{1}{12}h^3M_3 + R$ in the third case, where M_k is the maximum value of $|y^{(k)}(x)|$ for $x_0 \leqq x \leqq x_n$, and where R is the maximum roundoff error introduced in a single step.

21 Suppose that $F_y(x, y)$ is known to be negative throughout the calculation considered in Prob. 20, and also that

$$0 < \omega \leqq -F_y(x, y)$$

where ω is a constant. Show that then ε_n is dominated by e_n, where

$$(1 + h\omega\alpha_{-1})e_{n+1} = (1 - h\omega\alpha_0)e_n + E \qquad (n = 0, 1, 2, \ldots)$$

and $e_0 = 0$, if $h\omega\alpha_0 < 1$, and deduce the more useful bound

$$|\varepsilon_n| \leqq \frac{E}{\omega h}\left[1 - \left(\frac{1 - h\omega\alpha_0}{1 + h\omega\alpha_{-1}}\right)^n\right] \approx \frac{E}{\omega h}(1 - e^{-n\omega h})$$

in this case.

22 Suppose that the formula of the Adams method, written in the form

$$y_{n+1} = y_n + h\alpha_{-1}y'_{n+1} + h\alpha_0 y'_n + \sum_{k=1}^{r} \alpha_k y'_{n-k}$$

is applied to the problem

$$y' = F(x, y) \qquad y(x_0) = y_0$$

where it is known that

$$0 < \omega \leqq -F_y(x, y) \leqq K$$

where ω and K are constants, and assume also that $\alpha_0 \geqq 0$, $\alpha_{-1} \geqq 0$, and $h\omega\alpha_0 < 1$. Show that, if the maximum error introduced in a single step is E, then the error ε_n in y_n is dominated by e_n, where

$$(1 + h\omega\alpha_{-1})e_{n+1} = (1 - h\omega\alpha_0)e_n + hK\sum_{k=1}^{r}|\alpha_k|e_{n-k} + E$$

if $|\varepsilon_k| \leqq e_k$ for $k = 0, 1, 2, \ldots, r$. Show that one solution of this equation is of the form

$$e_n = \frac{E}{h\delta} - c\beta_0^n$$

where

$$\delta = \omega(\alpha_{-1} + \alpha_0) - K\sum_{k=1}^{r}|\alpha_k|$$

that it is possible to take β_0 such that $0 < \beta_0 < 1$ if $\delta > 0$, and that then

$$|\varepsilon_n| \leqq \bar{e}\beta_0^{n-r} + \frac{E}{h\delta}(1 - \beta_0^n)$$

where $\bar{e}$ is the absolute value of the largest of the errors $\varepsilon_0, \varepsilon_1, \ldots, \varepsilon_r$.

23 Show that the absolute values of the errors in the calculations of Prob. 14 are dominated by

$$\left(\frac{h^2}{12} + \frac{R}{h}\right)\left[1 - \left(\frac{2 - h}{2 + h}\right)^n\right] \approx \left(\frac{h^2}{12} + \frac{R}{h}\right)(1 - e^{-nh})$$

and

$$\left(\frac{h^2}{3} + \frac{R}{h}\right)[(\sqrt{1 + h^2} + h)^n - 1] + |\varepsilon|(\sqrt{1 + h^2} + h)^n$$

$$\approx \left(\frac{h^2}{3} + \frac{R}{h}\right)(e^{nh} - 1) + |\varepsilon|e^{nh}$$

respectively, where R is the magnitude of the maximum roundoff error introduced in a single step and ε is the inherent error in y_1, and compare these bounds with the direct error calculations of Prob. 16. (Use the results of Prob. 21 in the first case.)

24, 25 (*a*) to (*e*) Calculate values of F_y at appropriate stages of the calculations of Probs. 12 and 13, and obtain corresponding approximate error bounds, considering separately the effects of truncation and roundoff errors and using the result of Prob. 22 when it is appropriate. Estimate the truncation error in each step by approximating $h^m y^{(m)}$ by $h\,\nabla^{m-1} y'_m$ in the appropriate error term.

26 Use any numerical step-by-step method for the calculation of approximate values of the solution of each of the following problems for $x = 0.0(0.1)1.0$, with an error which can be reasonably confidently expected to be less than one unit in the fifth decimal place:

(*a*) $y' = x - y^2,\ y(0) = 1$

(*b*) $y' = x + \sin y,\ y(0) = \pi/2$

(*c*) $y' = e^{-xy},\ y(0) = 1$

Section 6.9

27 Suppose that closed formulas of the type (6.9.3) and (6.9.4) are used for the numerical integration of (6.9.1), that the relevant truncation errors in the nth step are E_2 and E'_2, respectively, and that another pair of formulas of open type is used for prediction of y_{n+1} and u_{n+1}, with truncation errors E_1 and E'_1, respectively. If the predicted values are denoted by $y^{(0)}_{n+1}$ and $u^{(0)}_{n+1}$, the corrected values by y_{n+1} and u_{n+1}, and the true values by Y_{n+1} and U_{n+1}, and if the notation

$$\frac{E_2 - E_1}{E_2} = C \qquad \frac{E'_2 - E'_1}{E'_2} = C'$$

is introduced, show that there follows

$$Y_{n+1} - y^{(0)}_{n+1} = -(C-1)E_2$$

$$Y_{n+1} - y_{n+1} = h\alpha_{-1}(U_{n+1} - u_{n+1}) + E_2$$

$$U_{n+1} - u^{(0)}_{n+1} = -(C'-1)E'_2$$

$$U_{n+1} - u_{n+1} = h\alpha_{-1}[G_{y_{n+1}}(Y_{n+1} - y_{n+1}) + G_{y'_{n+1}}(U_{n+1} - u_{n+1})] + E'_2$$

where $G_{y_{n+1}}$ and $G_{y'_{n+1}}$ are appropriate values of G_y and $G_{y'}$, respectively. By eliminating E_2 and E'_2, express $T_{n+1} \equiv Y_{n+1} - y_{n+1}$ and $T'_{n+1} \equiv U_{n+1} - u_{n+1}$ as linear combinations of $\gamma_{n+1} \equiv y_{n+1} - y^{(0)}_{n+1}$ and $\gamma'_{n+1} \equiv u_{n+1} - u^{(0)}_{n+1}$. Thus show that if

$$h|\alpha_{-1}|[h|\alpha_{-1}||G_{y_{n+1}}| + |G_{y'_{n+1}}|] \ll 1$$

so that also the convergence factor ρ_{n+1} is such that $|\rho_{n+1}| \ll 1$, and if it is true that $C' \approx C \gg 1$, then the approximations

$$T_{n+1} \approx -\frac{1}{C}(\gamma_{n+1} + h\alpha_{-1}\gamma'_{n+1}) \qquad T'_{n+1} \approx -\frac{1}{C}(\gamma'_{n+1} + h\alpha_{-1}G_{y_{n+1}}\gamma_{n+1})$$

generally provide better estimates than the usual simpler approximations $T_{n+1} \approx -\gamma_{n+1}/C$ and $T'_{n+1} \approx -\gamma'_{n+1}/C$. In particular, obtain the estimates

$$T_{n+1} \approx -\tfrac{1}{6}(\gamma_{n+1} + \tfrac{1}{20}\gamma'_{n+1}) \qquad T'_{n+1} \approx -\tfrac{1}{6}(\gamma'_{n+1} + \tfrac{1}{20}\gamma_{n+1})$$

for the calculations of the illustrative example of Sec. 6.9.

28 Obtain approximate values of the solution of each of the following problems for $x = 0.0(0.1)1.0$, determining appropriate starting values by power-series methods or otherwise, and proceeding by use of the modified Adams method, retaining only *first* differences, estimating the errors introduced in each step, and retaining an appropriate number of decimal places in the calculations:

(*a*) $y'' - y = 0,\ y(0) = 1,\ y'(0) = -1$

(*b*) $y'' + 2y' + 2y = 0,\ y(0) = 1,\ y'(0) = -1$

(*c*) $xy'' + y' + xy = 0,\ y(0) = 1,\ y'(0) = 0$

(*d*) $y'' + y' + y^2 = x,\ y(0) = 1,\ y'(0) = 0$

(*e*) $\begin{cases} u' = x + u - v^2,\ u(0) = 0 \\ v' = x^2 - v + u^2,\ v(0) = 1 \end{cases}$

29 (*a*) to (*e*) Repeat the calculations of Prob. 28, retaining differences through the third.

30 (*a*) to (*e*) Use Milne's method in Prob. 28.

31 (*a*) to (*e*) Use Hamming's method in Prob. 28.

Section 6.10

32 Suppose that the formula (6.10.3) is used with *no* differences to generate an approximation to e^{-x} as the solution of the problem

$$y'' = y \qquad y(0) = 1 \qquad y'(0) = -1$$

with spacing h, and that the value used for y_1 is in error by ε, so that $y_1 = e^{-h} - \varepsilon$. Show that, if all subsequent calculations were effected without roundoff, then the nth calculated ordinate would be given exactly by

$$\begin{aligned} y_n \equiv y(x_n) = {} & \frac{1-h^2}{2h}\left(\frac{1}{1-h} - e^{-h} + \varepsilon\right) e^{-n\log(1+h)} \\ & - \frac{1-h^2}{2h}\left(\frac{1}{1+h} - e^{-h} + \varepsilon\right) e^{n\log[1/(1-h)]} \end{aligned}$$

where $x_n = nh$, and that the approximation

$$\begin{aligned} y(x_n) &\approx \left(1 + \frac{h}{4} + \frac{\varepsilon}{2h}\right) e^{-x_n} - \left(\frac{h}{4} + \frac{\varepsilon}{2h}\right) e^{x_n} \\ &= e^{-x_n} - \left(\frac{h}{2} + \frac{\varepsilon}{h}\right) \sinh x_n \approx e^{-x_n} - \left(\frac{h}{4} + \frac{\varepsilon}{2h}\right) e^{x_n} \end{aligned}$$

would hold when h is small and n large, so that the *relative* error in $y(x_n)$ would then be approximated by $(h^2 + 2\varepsilon)e^{2x_n}/(4h)$. Show also that the corresponding relative error in the approximation to e^x, with the modified condition $y'(0) = +1$, would be approximated by the constant $(h^2 + 2\varepsilon)/(4h)$ when h is small and n large, again neglecting roundoffs.

33 Obtain approximate values of the solution of each of the following problems for $x = 0.0(0.1)1.0$, using the Milne procedure (6.10.5) and (6.10.6), and estimating the error introduced in each step:

(*a*) $y'' - y = 0,\ y(0) = 1,\ y'(0) = -1$

(*b*) $y'' + xy = 0,\ y(0) = 0,\ y'(0) = 1$

(*c*) $y'' + xy + \frac{1}{6}xy^3 = 0,\ y(0) = 0,\ y'(0) = 1$

(*d*) $y'' + \sin y = x,\ y(0) = \pi/2,\ y'(0) = 0$

34 (*a*) to (*d*) Obtain approximate error bounds for the calculations of Prob. 33.

35 Obtain approximate values of the solution of each of the following problems at the points noted, using the Milne procedure (6.10.5) and (6.10.6) after introducing the transformation (6.10.23), and estimating the error introduced in each step:

(*a*) $xY'' + Y' + xY = 0,\ Y(1) = 0.76520,\ Y'(1) = -0.44005\quad [x = 1.0(0.1)2.0]$

(*b*) $Y'' + 2Y' + x^2Y = 0,\ Y(0) = 0,\ Y'(0) = 1\quad [x = 0.0(0.1)1.0]$

(*c*) $Y'' + 2Y' + xY^2 = 0,\ Y(0) = 1,\ Y'(0) = -1\quad [x = 0.0(0.1)1.0]$

36 Verify that the substitution

$$Y = \rho y \qquad \rho = e^{-(1/2)\int P\,dx}$$

reduces the equation

$$Y'' + P(x)\,Y' = H(x, Y)$$

to the form

$$y'' = \tfrac{1}{4}(2P' + P^2)y + \frac{1}{\rho}\,H(x, \rho y)$$

which is a special case of (6.10.1), and verify also that this transformation includes the reduction of (6.10.22) to (6.10.25) when H is linear in Y.

37 Show that the equation $y'' + f(x)y = 0$ is satisfied by

$$y(x) = A(x)\cos\theta(x)$$

where

$$\theta(x) = \int_{x_0}^{x} v(t)\,dt + \omega$$

if A and v satisfy the equations $A'' - Av^2 + fA = 0$ and $2A'v + Av' = 0$, or hence if

$$A'' + fA = \frac{c^2}{A^3} \qquad v \equiv \theta' = \frac{c}{A^2}$$

where c and ω are arbitrary constants. Show also that the conditions

$$A(x_0) = A_0 \qquad A'(x_0) = 0 \qquad A''(x_0) = 0$$

which tend to require that $A(x)$ remain constant near $x = x_0$, are consistent with the conditions $y(x_0) = y_0$ and $y'(x_0) = y_0'$ if A_0 and ω satisfy the relations

$$A_0 \cos \omega = y_0 \qquad A_0 \sin \omega = -\frac{y_0'}{f_0^{1/2}}$$

and if

$$c = f_0^{1/2} A_0^2$$

under the assumption that $f_0 \equiv f(x_0) > 0$. {This procedure, attributed to Madelung [1931], is often useful when $f(x)$ is large and *positive*, so that $y(x)$ is strongly oscillatory, since $A(x)$ often varies much less rapidly. A similar transformation, which is often useful when $f(x)$ is large and *negative*, and $y(x)$ increases or decreases rapidly, may be obtained analogously by replacing $\cos\theta$ by $\cosh\theta$, $\sinh\theta$, or e^θ in the expression assumed for y, according as the ratio of $|y_0|$ to $|y_0'|/(-f_0)^{1/2}$ is greater than, less than, or equal to unity, respectively.}

38 Use the results of Prob. 37 to show that the solution of the problem

$$y'' + (16 - x^2)y = 0 \qquad y(0) = 1 \qquad y'(0) = 0$$

can be expressed in the form

$$y(x) = A(x) \cos \theta(x)$$

where $A(x)$ is the solution of the problem

$$A'' + (16 - x^2)A = \frac{16}{A^3} \qquad A(0) = 1 \qquad A'(0) = 0$$

and where

$$\theta(x) = 4 \int_0^x \frac{dt}{[A(t)]^2}$$

Also determine $A(x)$, and hence $\theta(x)$ and $y(x)$, for $x = 0.0(0.1)1.0$ to five places, by a numerical method.

Section 6.11

39 to *41* (*a*) to (*e*) Advance the calculations of Probs. 12, 13, and 19 to $x = 1.2$ with $h = 0.05$, given that $\phi(1.1) \doteq 1.43392$ and $\phi(1.2) \doteq 1.48080$ in Prob. 8.

42 to *44* (*a*) to (*e*) Advance the calculations of Probs. 28 to 30 to $x = 1.2$ with $h = 0.05$.

45 (*a*) to (*d*) Advance the calculation of Prob. 33 to $x = 1.2$ with $h = 0.05$.

Section 6.12

46 (*a*) to (*d*), *47*(*a*) to (*e*), *48*(*a*) to (*d*) Obtain approximate solutions of Probs. 5, 28, and 33 for $x = 0.0(0.1)1.0$ by use of (6.12.2) and (6.12.3).

Section 6.13

49 (*a*) to (*e*) Obtain approximate values of the solutions of Probs. 5*a* to *d* and 8 for $x = 0.0(0.1)1.0$ by use of (6.13.15) and (6.13.16), and estimate the errors.

Section 6.14

50 (*a*) to (*e*) Obtain approximate values of the solutions of Probs. 5*a* to *d* and 8 for $x = 0.0(0.1)1.0$ by use of (6.14.5) and (6.14.6), and estimate the errors.

51 (*a*) to (*e*), *52*(*a*) to (*d*) Obtain approximate values of the solutions of Probs. 28 and 33 for $x = 0.0(0.1)1.0$ by use of (6.14.12) and (6.14.13), and estimate the errors.

Section 6.15

53 Use an appropriate step-by-step method to determine approximate five-place values of $u(x)$ such that $u'' + u = 1$, $u(0) = 0$, $u'(0) = 0$, and of $v(x)$ such that $v'' + v = 0$, $v(0) = 0$, $v'(0) = 1$, for $x = 0.0(0.1)1.0$. Then use these results to determine approximate values of $y(x)$ for $x = 0.0(0.1)1.0$ satisfying the conditions $y''(x) + y(x) = 1$, $y(0) = 0$, $y(1) = 1$, and compare the results with exact values.

54 Use an appropriate step-by-step method to determine approximate five-place values of the solution of the problem

$$u'' + u = 1 \qquad u(0) = 0 \qquad u'(0) = \alpha$$

for $x = 0.0(0.1)1.0$, taking successively $\alpha = 0$ and $\alpha = 1$. Then use linear interpolation to estimate the value of α for which $u(1) = 1$, and investigate the correctness of this estimate by making another corresponding step-by-step calculation (see Prob. 55).

55 Prove that the procedure described in connection with Eqs. (6.15.11) and (6.15.12) would yield an exact result with *linear* interpolation on α if (6.15.11) were a *linear* equation and if no errors were committed in the determination of solutions corresponding to two trial values of α.

56 Obtain approximate values of the solution of the problem

$$y'' + y = 0 \qquad y(0) = 0 \qquad y(1) = 1$$

for $x = 0.0(0.1)1.0$ by use of (6.15.18) with $h = 0.2$.

57 Repeat the calculation of Prob. 56 using (6.15.17) with $h = 0.2$, and compare the two approximations.

58 Repeat the calculation of Prob. 56 using (6.15.19) with $h = 0.2$, together with (6.15.20) and (6.15.21).

59 Use the method of Prob. 56, together with (6.15.22), to deal with the modification of that problem in which the condition $y(0) = 0$ is replaced by the condition $y'(0) = y(0)$.

60 Repeat the calculation of Prob. 59 using the method of Prob. 57, and compare the results with those obtained in Prob. 59.

Section 6.16

61 Determine approximate values of the smallest characteristic value of λ for the problem

$$y'' + \lambda y = 0 \qquad y(0) = y(1) = 0$$

by use of (6.16.3) and (6.16.4), and compare those approximations with the true value π^2, and with the corresponding approximations based on the use of (6.15.18) with $N = 1$ and 2.

62 Repeat the calculations of Prob. 61 when the condition $y(0) = 0$ is replaced by the condition $y'(0) = 0$, making use of (6.15.22), in each case, in such a way that the order of the procedure is not reduced; compare the results which correspond to the use of the approximate condition $\Delta y_0 \equiv y_1 - y_0 = 0$.

63, 64 Repeat the calculations of Probs. 61 and 62, making use of (6.16.2) with $N = 3$.

65 to *68* Deal as in Probs. 61 to 64 with the corresponding modified formulations involving the equation $y'' + \lambda xy = 0$. [The true characteristic numbers in Probs. 65 and 67 are the zeros of the function $J_{1/3}(2\lambda^{1/2}/3)$, the smallest of which rounds to 18.956, whereas those of Probs. 66 and 68 are the zeros of the function $J_{-1/3}(2\lambda^{1/2}/3)$, the smallest of which rounds to 7.8373.]

7

LEAST-SQUARES POLYNOMIAL APPROXIMATION

7.1 Introduction

There are two classes of situations in which the process of determining an approximation (polynomial or otherwise) to a function by fitting given data exactly on a certain set of discrete points often is a particularly inefficient one.

First, when the function $f(x)$, to be approximated, is specified for *all* values of x in an interval, it is clearly desirable to take many or all of the known values into account, rather than to select an arbitrary set, consisting of the least possible number of discrete values which leads to a determinate set of conditions. This is especially true when $f(x)$ or one of its derivatives of low order possesses known finite discontinuities, or "jumps."

Second, and on the opposite extreme, when only a discrete set of approximate values of $f(x)$ is provided, and when the degree of reliability of those values is not well established, it is foolish (and, indeed, inherently dangerous) to attempt to determine a polynomial of high degree which fits the vagaries of such data exactly and hence, in all probability, is represented by a curve which oscillates violently about the curve which represents the true function. In particular, the use of the result for numerical differentiation would be hard to justify.

The so-called *method of least squares*, which is designed for the treatment of both these classes of problems, is introduced in the present chapter, and its application to the analysis of typical situations is treated. Several of the classical sets of orthogonal polynomials, which are particularly useful in these applications and which also will be needed in Chap. 8, are introduced, and certain of their properties are discussed.

7.2 The Principle of Least Squares

In place of determining a polynomial approximation $y(x)$, of degree n, to a certain function $f(x)$, by requiring that the values of $y(x)$ on a set of $n + 1$ points agree with known exact or approximate values of $f(x)$ at those points, as was done in preceding chapters, it is often preferable to require that $y(x)$ and $f(x)$ agree *as well as possible* (in some sense) over a domain D of greater extent. This domain may be taken as a continuous interval, when $f(x)$ is specified analytically, or as a discrete set, say, of $N + 1$ points, where $N > n$.

When the available data in D are either exact or of equal reliability, it is frequently agreed that the "best approximation" over D is that one for which the aggregate (sum or integral) of the squared error in D is least. This postulate is often known as *Legendre's principle of least squares*. More generally, if $w(x_i)$ is a measure of the dependability of the value assigned to $f(x)$ when $x = x_i$, the criterion is modified by requiring that the squared error at x_i be multiplied by the *weight* $w(x_i)$ before the aggregate is calculated.

Suppose first that *exact* values of $f(x)$ are known over a certain domain D, which may consist of a discrete set of points $x_0, x_1, \ldots, x_N$ or of a continuous interval $[a, b]$, and that the approximation is to be of the form

$$f(x) \approx \sum_{k=0}^{n} a_k \phi_k(x) \equiv y(x) \qquad (7.2.1)$$

where $\phi_0(x), \ldots, \phi_n(x)$ are $n + 1$ appropriately chosen linearly independent functions. In particular, in order to obtain a *polynomial* approximation of degree n, we could take $\phi_0 = 1, \phi_1 = x, \ldots, \phi_n = x^n$, although other choices of the coordinate functions, which would also afford a *basis* for the generation of all polynomials of degree n, are often more convenient, as will be seen. It is supposed that the specified *weighting function* $w(x)$ is nonnegative in D,

$$w(x) \geqq 0 \qquad (7.2.2)$$

If we define the *residual* $R(x)$ by the equation

$$R(x) = f(x) - \sum_{k=0}^{n} a_k \phi_k(x) \equiv f(x) - y(x) \qquad (7.2.3)$$

the best approximation (7.2.1), in the least-squares sense, is defined to be that for which the a's are determined so that the *aggregate* (sum or integral) of $w(x)R^2(x)$ over D is as small as possible. It is convenient to denote this aggregate here by $\langle wR^2 \rangle$. The requirement

$$\langle wR^2 \rangle \equiv \left\langle w \left[f - \sum_{k=0}^{n} a_k \phi_k \right]^2 \right\rangle = \min \qquad (7.2.4)$$

then imposes the necessary conditions

$$\frac{\partial}{\partial a_r} \left\langle w \left[f - \sum_{k=0}^{n} a_k \phi_k \right]^2 \right\rangle = 0 \qquad (r = 0, 1, \ldots, n) \qquad (7.2.5)$$

or

$$\left\langle w\phi_r \left[f - \sum_{k=0}^{n} a_k \phi_k \right] \right\rangle \equiv \langle w\phi_r(f - y) \rangle = 0 \qquad (7.2.6)$$

or

$$\sum_{k=0}^{n} a_k \langle w\phi_r \phi_k \rangle = \langle w\phi_r f \rangle \qquad (r = 0, 1, \ldots, n) \qquad (7.2.7)$$

and hence leads to $n + 1$ simultaneous linear equations in the $n + 1$ unknown parameters $a_0, a_1, \ldots, a_n$. These equations are called the *normal equations* of the process.

It is useful to notice that these conditions can be expressed also in the form

$$\langle w(x)\phi_r(x)R(x) \rangle = 0 \qquad (r = 0, 1, \ldots, n) \qquad (7.2.8)$$

Hence, since we have also

$$\langle wR^2 \rangle = \langle wR \cdot R \rangle = \left\langle wR \left[f - \sum_{k=0}^{n} a_k \phi_k \right] \right\rangle$$

$$= \langle wRf \rangle - \sum_{k=0}^{n} a_k \langle w\phi_k R \rangle \qquad (7.2.9)$$

it follows that, when the coefficients $a_0, \ldots, a_n$ satisfy (7.2.7), the corresponding aggregate squared residual reduces to

$$\Delta_n \equiv \langle wR^2 \rangle_{\min} = \langle wRf \rangle \equiv \langle wf(f - y) \rangle$$

$$\equiv \langle wf^2 \rangle - \sum_{k=0}^{n} a_k \langle w\phi_k f \rangle \qquad (7.2.10)$$

The smallness of the quantity Δ_n can be used as a criterion for the efficiency of the approximation over D.

In particular, if the domain D consists only of $n + 1$ discrete points, and if the set of functions S_n generated by the coordinate functions $\phi_0, \ldots, \phi_n$ comprises, say, all polynomials of degree not exceeding n, it is possible to reduce $R(x)$ to *zero* at each point of the domain. Thus here the least-squares procedure reduces to the determination of the polynomial $y(x)$ of degree n which agrees exactly with $f(x)$ at $n + 1$ points, and the minimum value of $\langle wR^2 \rangle$ is zero. If the domain consists of $N + 1$ points, where $N > n$, or of a continuous interval, exact fit over all of D is generally impossible and the procedure gives the function of the class considered which affords the best approximate fit under the criterion (7.2.4), in which the weighting function $w(x)$ must be specified.

It is seen from (7.2.7) that the coefficients of the unknowns in the left-hand members of the normal equations are independent of the function $f(x)$ to be approximated, so that they may be precalculated, once the coordinate functions and the weighting function have been selected. Also, since $\langle w\phi_i\phi_j \rangle \equiv \langle w\phi_j\phi_i \rangle$, the coefficient of a_i in the jth equation is equal to that of a_j in the ith equation, so that the array of the coefficients of the a's is symmetrical with respect to its principal diagonal. This fact appreciably reduces the labor in both the formation and the solution of the set of equations (see Sec. 10.4).

Clearly, these equations are greatly simplified if the coordinate functions are chosen, in advance, in such a way that

$$\langle w\phi_i\phi_j \rangle = 0 \qquad (i \neq j) \tag{7.2.11}$$

A set of ϕ's having this property over D is said to be an *orthogonal set*, relative to the weighting function $w(x)$, over D. For such a set of coordinate functions, the corresponding set of normal equations (7.2.7) becomes "uncoupled" and takes the form

$$a_r\langle w\phi_r^2 \rangle = \langle w\phi_r f \rangle \qquad (r = 0, 1, \ldots, n) \tag{7.2.12}$$

Since $w(x)$ is nonnegative, the coefficient of a_r cannot vanish, except in a very special situation where $w(x)$ vanishes at all points of D for which $\phi_r(x)$ does not. Henceforth we exclude such cases and accordingly obtain the result

$$a_r = \frac{\langle w\phi_r f \rangle}{\langle w\phi_r^2 \rangle} \qquad (r = 0, 1, \ldots, n) \tag{7.2.13}$$

Further, reference to (7.2.10) and (7.2.12) shows that the corresponding minimum value of $\langle wR^2 \rangle$ can be expressed in the more convenient alternative form

$$\Delta_n \equiv \langle wR^2 \rangle_{\min} = \langle wf^2 \rangle - \sum_{k=0}^{n} a_k^2 \langle w\phi_k^2 \rangle \tag{7.2.14}$$

in this case.

In theoretical work it is often convenient to suppose that the ϕ's have also been *normalized* in such a way that $\langle w\phi_i^2 \rangle = 1$, so that (7.2.13) and

(7.2.14) are still further simplified. However, this normalization is rarely convenient in practice.

The root-mean-square (RMS) error in an approximation over D, relative to $w(x)$, is defined to be

$$\varepsilon_{\text{RMS}} \equiv (f - y)_{\text{RMS}} = \sqrt{\frac{\langle wR^2 \rangle}{\langle w \rangle}} \tag{7.2.15}$$

Here, in particular, when $w(x) \equiv 1$ the quantity $\langle 1 \rangle$ represents the length of the interval in the continuous case and the number $(N + 1)$ of points in D in the discrete case.

In the discrete case, it frequently happens that the given data are empirical and correspond accordingly to an *observed function* $\bar{f}(x)$, and that the *true function* $f(x)$ is not known. Here we must replace $f(x)$ and $y(x)$ by $\bar{f}(x)$ and $\bar{y}(x)$ in the preceding developments, and we then are in position to calculate only $\bar{\varepsilon}_{\text{RMS}} = (\bar{f} - \bar{y})_{\text{RMS}}$ over D. The subsequent estimation of the *desired* quantity $(f - \bar{y})_{\text{RMS}}$ is considered in Sec. 7.4.

7.3 Least-Squares Approximation over Discrete Sets of Points

Before exploiting the convenience afforded by the use of orthogonal functions, we here consider the application of the general least-squares method to the case when the domain D comprises a discrete set of points. The case when D is a continuous interval is treated in a completely analogous way.

In accordance with the results of the preceding section, if an approximation to the true function f of the form

$$f(x) \approx \sum_{k=0}^{n} a_k \phi_k(x) \tag{7.3.1}$$

is to hold over a set S_N of $N + 1$ points $x_0, x_1, \ldots, x_N$, where $N \geqq n$, in the sense that the aggregate weighted squared error is to be a minimum,

$$\sum_{i=0}^{N} w(x_i)[f(x_i) - \sum_{k=0}^{n} a_k \phi_k(x_i)]^2 = \min \tag{7.3.2}$$

the set of $n + 1$ normal equations (7.2.7) becomes

$$\begin{aligned} a_0 \sum_{i=0}^{N} w(x_i)\phi_r(x_i)\phi_0(x_i) + a_1 \sum_{i=0}^{N} w(x_i)\phi_r(x_i)\phi_1(x_i) + \cdots \\ + a_n \sum_{i=0}^{N} w(x_i)\phi_r(x_i)\phi_n(x_i) \\ = \sum_{i=0}^{N} w(x_i)\phi_r(x_i)f(x_i) \qquad (r = 0, 1, \ldots, n) \end{aligned} \tag{7.3.3}$$

These equations can be obtained quite simply by first writing down the $N + 1$ equations which would require that (7.3.1) be an *equality* at the $N + 1$ points x_i,

$$\begin{aligned} a_0\phi_0(x_0) + a_1\phi_1(x_0) + \cdots + a_n\phi_n(x_0) &= f(x_0) \\ a_0\phi_0(x_1) + a_1\phi_1(x_1) + \cdots + a_n\phi_n(x_1) &= f(x_1) \\ \cdots\cdots\cdots\cdots\cdots\cdots\cdots\cdots\cdots \\ a_0\phi_0(x_N) + a_1\phi_1(x_N) + \cdots + a_n\phi_n(x_N) &= f(x_N) \end{aligned} \tag{7.3.4}$$

The rth normal equation is then obtained by multiplying each equation by the coefficient of a_r in that equation, and by the weight associated with that equation, and summing the results. Unless there is a reason for proceeding otherwise, the weights are generally taken to be equal and then may be assigned the value unity.

When $N = n$, the problem reduces to that of satisfying the $n + 1$ equations (7.3.4) in $n + 1$ unknowns, and the normal equations are equivalent to the original ones.

As a very simple example, suppose that the problem is that of fitting the equation of a straight line as well as possible (in the least-squares sense) to the following data:

x	0	1	2	3	4
$f(x)$	1.00	3.85	6.50	9.35	12.05

In place of writing out the equations (7.3.4), corresponding to the substitution of these corresponding values into the equation

$$a_0 + a_1 x = f(x) \tag{7.3.5}$$

we may merely write down the array of the coefficients of a_0 and a_1 and the right-hand members (the *augmented matrix* of the system) in the form

$$\begin{matrix} 1 & 0 & 1.00 \\ 1 & 1 & 3.85 \\ 1 & 2 & 6.50 \\ 1 & 3 & 9.35 \\ 1 & 4 & 12.05 \end{matrix} \tag{7.3.6}$$

Under the assumption that all the data are of equal significance, we take all weights equal to unity. The first normal equation then corresponds to the result of adding the elements of the respective columns of (7.3.6), to give the array [5 10 32.75], and the second corresponds to the result of multiplying the elements in each row by the element of that row which lies in the second column, and adding the results, to give the array [10 30 93.10], so that the normal equations are

$$\begin{aligned} 5a_0 + 10a_1 &= 32.75 \\ 10a_0 + 30a_1 &= 93.10 \end{aligned} \tag{7.3.7}$$

yielding the solution $a_0 = 1.03$, $a_1 = 2.76$, and hence determining the linear approximation

$$f(x) \approx y(x) = 1.03 + 2.76x \qquad (7.3.8)$$

The values obtained from this approximation at the points $x = 0, 1, 2, 3$, and 4 are 1.03, 3.79, 6.55, 9.31, and 12.07, respectively, and the sum of the squared errors is found to be 0.0090. Thus the RMS error over these five points is $\sqrt{0.0090/5} \doteq 0.042$.

The interpretation of this result must depend upon the context. If the given values are considered to be exact values of a *true* function, then the figure 0.042 represents the RMS departure between the true function $f(x)$ and the approximation $y(x)$ over the five points for which information is available. In the absence of any further information, this figure would afford the only available estimate of the RMS error over the continuous range $0 \leqq x \leqq 4$. On the other hand, if the given ordinates are empirical and hence properly correspond to an *observed function* $\bar{f}(x)$, the figure 0.042 represents only the RMS departure between $\bar{f}(x)$ and *its* least-squares approximation $\bar{y}(x)$ over the five relevant points. Unless additional information is supplied or additional assumptions are made, no conclusions can be drawn with respect to the RMS value of the true error $f - \bar{y}$.

However, if it is postulated that the *true* function is such that the residuals at each of the $N + 1$ points *can* be reduced to zero (or, more realistically, can be made *negligible*), but that the impossibility of achieving this end in the case at hand is due to the presence of independent random errors in the several observed values, then it is possible to obtain a certain amount of additional information. It is also frequently desirable to estimate the errors in the calculated *coefficients*. For both these purposes, we examine the general problem in greater detail in the following section.

7.4 Error Estimation

Suppose that the right-hand members of (7.3.4) are replaced by values of an observed function $\bar{f}(x)$, and that the *calculated* coefficients then are denoted by $\bar{a}_0, \ldots, \bar{a}_n$, so that those relations become

$$\sum_{k=0}^{N} \bar{a}_k \phi_k(x_i) \approx \bar{f}(x_i) \qquad (i = 0, 1, \ldots, N) \qquad (7.4.1)$$

whereas the proper equations are

$$\sum_{k=0}^{n} a_k \phi_k(x_i) = f(x_i) \equiv \bar{f}(x_i) + E(x_i) \qquad (7.4.2)$$

where $E(x_i)$ is the error associated with the "observed value" $\bar{f}(x_i)$. The normal equations (7.3.3), associated with (7.4.1), then can be written in the form

$$\sum_{k=0}^{n} c_{rk}\bar{a}_k = \bar{v}_r \qquad (r = 0, 1, \ldots, n) \tag{7.4.3}$$

where

$$c_{rk} = c_{kr} = \sum_{i=0}^{N} w(x_i)\phi_r(x_i)\phi_k(x_i) \tag{7.4.4}$$

and

$$\bar{v}_r = \sum_{i=0}^{N} w(x_i)\phi_r(x_i)\bar{f}(x_i) \tag{7.4.5}$$

The corresponding approximation $\Sigma\bar{a}_k\phi_k(x)$ will be denoted by $\bar{y}(x)$, and the residual $\bar{f}(x_i) - \bar{y}(x_i)$ by $\bar{R}(x_i)$.

If we denote by C_{rs} the cofactor of c_{rs} in the coefficient matrix of (7.4.3),

$$\mathbf{C} \equiv \begin{bmatrix} c_{00} & c_{01} & \cdots & c_{0n} \\ c_{10} & c_{11} & \cdots & c_{1n} \\ \cdots & \cdots & \cdots & \cdots \\ c_{n0} & c_{n1} & \cdots & c_{nn} \end{bmatrix} \tag{7.4.6}$$

and define the *reduced cofactor* $\tilde{C}_{rs} = C_{rs}/D$, where D is the determinant of $\mathbf{C}$, noticing that $\tilde{C}_{sr} = \tilde{C}_{rs}$ because of the symmetry of the array, the solution of the set (7.4.3) can be expressed in the form (see Sec. 10.2)

$$\bar{a}_r = \sum_{k=0}^{n} \tilde{C}_{rk}\bar{v}_k \qquad (r = 0, 1, \ldots, n) \tag{7.4.7}$$

In order to express the $\bar{a}$'s directly in terms of the given ordinates, we introduce (7.4.5) into (7.4.7), to obtain

$$\bar{a}_r = \sum_{k=0}^{n} \tilde{C}_{rk} \sum_{i=0}^{N} w(x_i)\phi_k(x_i)\bar{f}(x_i) = \sum_{i=0}^{N}\left[\sum_{k=0}^{n} \tilde{C}_{rk}\phi_k(x_i)\right] w(x_i)\bar{f}(x_i)$$

Thus, if we introduce the abbreviation

$$\Phi_r(x) = \sum_{k=0}^{n} \tilde{C}_{rk}\phi_k(x) \tag{7.4.8}$$

this relation takes the form

$$\bar{a}_r = \sum_{i=0}^{N} w(x_i)\Phi_r(x_i)\bar{f}(x_i) \tag{7.4.9}$$

Accordingly, if a_r denotes the corresponding coefficient calculated from the *true* ordinates, there follows also

$$a_r - \bar{a}_r = \sum_{i=0}^{N} w(x_i)\Phi_r(x_i)E(x_i) \tag{7.4.10}$$

This relation gives the difference between the rth coefficient actually obtained and that which would have been obtained if no observational errors (or roundoff errors) were present. If the hypothetical assumption is made that $f(x)$ is actually a member of the set of all functions expressible as linear combinations of $\phi_0, \ldots, \phi_n$, then $f(x)$ truly is specified by the constants $a_0, \ldots, a_n$, and we then have $f(x) - \bar{y}(x) \equiv \Sigma(a_r - \bar{a}_r)\phi_r(x)$.

Generally, only *bounds* on the observational errors $E(x_i)$, or *mean values of their squares* over a set of observations, are available in practice. In the latter case, the weights $w(x_0), \ldots, w(x_N)$ desirably are so chosen that $w(x_i)$ *is inversely proportional to the mean value of* $E^2(x_i)$, so that the mean values of $w(x_0)E^2(x_0), \ldots, w(x_N)E^2(x_N)$ over the set of observations are (approximately) equal.

Assuming that this has been done, we may first obtain from (7.4.10) the relation

$$(a_r - \bar{a}_r)^2 = [w(x_0)\Phi_r^2(x_0)][w(x_0)E^2(x_0)] + \cdots$$
$$+ [w(x_N)\Phi_r^2(x_N)][w(x_N)E^2(x_N)] + \cdots \qquad (7.4.11)$$

where the omitted terms at the end involve products of the form $E(x_i)E(x_j)$ where $i \neq j$. If both sides of this equation are averaged over the available set of observations, if the mean value of the product $E(x_i)E(x_j)$ is assumed to be zero when $i \neq j$, and if $w(x_0), \ldots, w(x_N)$ are assigned values such that

$$w(x_0)[E^2(x_0)]_m = \cdots = w(x_N)[E^2(x_N)]_m = \sigma^2 \qquad (7.4.12)$$

where σ^2 is a conveniently chosen constant, there then follows

$$(a_r - \bar{a}_r)_m^2 = \sigma^2 \sum_{i=0}^{N} w(x_i)\Phi_r^2(x_i) \qquad (7.4.13)$$

Here the subscript m indicates the *mean* over the set of observations.

The quantity $(a_r - \bar{a}_r)_m^2$ may be referred to as the *estimated variance* of a_r or of its error $\delta a_r \equiv a_r - \bar{a}_r$, and the square root of that quantity as the corresponding *estimated standard deviation* (see Sec. 1.7).

This result can be put into a more convenient form. For this purpose, we notice first that if $\bar{f}(x_i)$ is identified with $\phi_s(x_i)$ in (7.4.1), where $0 \leqq s \leqq n$, there follows $\bar{a}_r = \delta_{rs}$, where $\delta_{rj} = 1$ when $r = j$ and 0 otherwise. Thus we deduce from (7.4.9) that

$$\sum_{i=0}^{N} w(x_i)\Phi_r(x_i)\phi_s(x_i) = \delta_{rs} \qquad (7.4.14)$$

so that the two sets of functions $\{\phi_r(x)\}$ and $\{\Phi_r(x)\}$ are *biorthogonal* (actually "biorthonormal") on S_N relative to $w(x)$. Reference to (7.4.8) then gives

$$\sum_{i=0}^{N} w(x_i)[\Phi_r(x_i)]^2 = \sum_{i=0}^{N} w(x_i)\Phi_r(x_i)\left[\sum_{k=0}^{n} \tilde{C}_{rk}\phi_k(x_i)\right]$$

$$= \sum_{k=0}^{n}\left[\sum_{i=0}^{N} w(x_i)\Phi_r(x_i)\phi_k(x_i)\right]\tilde{C}_{rk}$$

$$= \sum_{k=0}^{n} \delta_{rk}\tilde{C}_{rk} = \tilde{C}_{rr} \tag{7.4.15}$$

Thus (7.4.13) can be written in the form

$$(a_r - \bar{a}_r)_m^2 = \frac{C_{rr}}{D}\sigma^2 \tag{7.4.16}$$

where C_{rr} is the cofactor of c_{rr} in the matrix $\mathbf{C}$ and D is the determinant of that matrix.

It may happen that the individual values of $[E^2(x_i)]_m$ for $i = 0, 1, \ldots, N$ are not known but that their *ratios* can be estimated, that is, that the *relative* dependability of the measurements at $x_0, x_1, \ldots, x_N$ is known. If $w(x_0), \ldots, w(x_N)$ again are assumed to be so chosen that (7.4.12) holds, where σ^2 now is an *unknown* constant, and if it is now assumed (hypothetically) that the *true* function can be fitted exactly at the $N + 1$ points involved, it is possible to obtain an estimate of σ^2 in terms of the *calculable* residuals $\bar{R}(x_0), \ldots, \bar{R}(x_N)$, such that

$$\bar{R}(x_i) \equiv \bar{f}(x_i) - \bar{y}(x_i) = \bar{f}(x_i) - \sum_{k=0}^{n} \bar{a}_k\phi_k(x_i) \qquad (i = 0, \ldots, N) \tag{7.4.17}$$

For this purpose, we notice first that, since $E(x_i) = f(x_i) - \bar{f}(x_i)$, we have also

$$f(x_i) - \bar{y}(x_i) = E(x_i) + \bar{R}(x_i) \tag{7.4.18}$$

From (7.4.2) and (7.4.17), there follows

$$E(x_i) + \bar{R}(x_i) = \sum_{k=0}^{n} (a_k - \bar{a}_k)\phi_k(x_i) \tag{7.4.19}$$

or, after using (7.4.10),

$$E(x_i) + \bar{R}(x_i) = \sum_{k=0}^{n} \phi_k(x_i) \sum_{\nu=0}^{N} w(x_\nu)\Phi_k(x_\nu)E(x_\nu) \tag{7.4.20}$$

If we multiply both members of (7.4.19) by $w(x_i)\bar{R}(x_i)$ and sum over i, making use of the fact that $\Sigma w(x_i)\phi_k(x_i)\bar{R}(x_i) = 0$, in accordance with (7.2.8), there follows

$$\sum_{i=0}^{N} w(x_i)E(x_i)\bar{R}(x_i) = -\sum_{i=0}^{N} w(x_i)\bar{R}^2(x_i) \tag{7.4.21}$$

Also, by multiplying both members of (7.4.20) by $w(x_i)E(x_i)$, summing over i, and making use of (7.4.21), there follows

$$\sum_{i=0}^{N} w(x_i)E^2(x_i) - \sum_{i=0}^{N} w(x_i)\bar{R}^2(x_i) = \sum_{k=0}^{n} \sum_{i=0}^{N} \sum_{v=0}^{N} w(x_i)\phi_k(x_i)\Phi_k(x_v)w(x_v)E(x_i)E(x_v) \tag{7.4.22}$$

If we now average the equal members of (7.4.22) over a set of observations, and again assume that $[E(x_i)E(x_j)]_m = 0$ when $i \neq j$ and $[w(x_i)E^2(x_i)]_m = \sigma^2$ for $i = 0, \ldots, N$, only the terms for which $v = i$ will remain in the right-hand member, and there follows

$$(N + 1)\sigma^2 - \sum_{i=0}^{N} w(x_i)[\bar{R}^2(x_i)]_m = \sum_{k=0}^{n} \left[\sum_{i=0}^{N} w(x_i)\phi_k(x_i)\Phi_k(x_i)\right] \sigma^2 = (n + 1)\sigma^2 \tag{7.4.23}$$

since the sum in brackets in the first right-hand member is *unity* by (7.4.14). Thus we deduce that

$$\sigma^2 = \frac{1}{N - n} \sum_{i=0}^{N} w(x_i)[\bar{R}^2(x_i)]_m \tag{7.4.24}$$

We may notice also that, from (7.4.18) and (7.4.21), there follows

$$\sum_{i=0}^{N} w(x_i)[f(x_i) - \bar{y}(x_i)]^2 = \sum_{i=0}^{N} w(x_i)E^2(x_i) - \sum_{i=0}^{N} w(x_i)\bar{R}^2(x_i)$$

so that the mean of this weighted sum over the available set of observations is given by the right-hand member of (7.4.23). Thus (7.4.24) permits us to deduce also that

$$\sum_{i=0}^{N} w(x_i)[f(x_i) - \bar{y}(x_i)]_m^2 = \frac{n + 1}{N - n} \sum_{i=0}^{N} w(x_i)[\bar{R}^2(x_i)]_m \tag{7.4.25}$$

When only the residuals which correspond to a single set of given ordinates are available, the best estimate of the mean value of $\bar{R}^2(x_i)$ consists of the single calculated value, so that $[\bar{R}^2(x_i)]_m$ must be replaced by $\bar{R}^2(x_i)$ in (7.4.24) and (7.4.25).

It is convenient to summarize the preceding results in the case when $w(x) = 1$, which is of most common occurrence. Here we may write $\bar{\varepsilon}_{\text{RMS}}^2$ for the mean value of the squares of the $N + 1$ residuals $\bar{R}(x_i) = \bar{f}(x_i) - \bar{y}(x_i)$, so that

$$\bar{\varepsilon}_{\text{RMS}} = \sqrt{\frac{\Sigma[\bar{R}(x_i)]^2}{N + 1}} \tag{7.4.26}$$

measures the RMS departure between the observed function and its calculated approximation over the $N + 1$ points involved. Since here $\sigma^2 = [E^2(x_i)]_m$, the relation (7.4.24) then affords the *estimate*

$$E_{\text{RMS}} \approx \sqrt{\frac{\Sigma[\bar{R}(x_i)]^2}{N - n}} = \sqrt{\frac{N + 1}{N - n}}\, \bar{\varepsilon}_{\text{RMS}} \tag{7.4.27}$$

of the RMS departure between the true function and the observed function over those points and (7.4.25) yields the estimate

$$(f - \bar{y})_{\text{RMS}} \approx \sqrt{\frac{n + 1}{N - n}}\, (\bar{f} - \bar{y})_{\text{RMS}} = \sqrt{\frac{(n + 1)(N + 1)}{N - n}}\, \bar{\varepsilon}_{\text{RMS}} \tag{7.4.28}$$

for the RMS departure over that set of points between the true function and the least-squares approximation to the observed function.

Also, the combination of (7.4.16) and (7.4.25) gives

$$(\delta a_r)_{\text{RMS}} \equiv \sqrt{(a_r - \bar{a}_r)_m^2} \approx \sqrt{\frac{C_{rr}}{D}}\, E_{\text{RMS}} \approx \sqrt{\frac{C_{rr}}{D}} \sqrt{\frac{\Sigma[\bar{R}(x_i)]^2}{N - n}} \tag{7.4.29}$$

as an estimate of the RMS error in the rth calculated coefficient $\bar{a}_r$. In each case, $N + 1$ denotes the number of points employed and $n + 1$ the number of independent coordinate functions. The estimates (7.4.27) to (7.4.29) are essentially based on the assumption that sufficiently many ϕ's are used to ensure that the true function *can* be expressed in the form $\Sigma a_k \phi_k(x)$, apart from negligible deviations, for *some* choice of the a's, and are to be used in the more general case with a corresponding degree of caution. These estimates are properly meaningless when $N = n$ since then all the data are needed to determine the *approximation* and no data remain for the estimation of the error.

If the given data in the preceding example are empirical, the figure $\bar{\varepsilon}_{\text{RMS}} \doteq 0.042$ thus represents the RMS departure between the observed function and the "smoothed function" over the five relevant points. Since also $C_{00} = 30$, $C_{11} = 5$, and $D = 50$ in (7.3.7), (7.4.29) gives

$$(\delta a_0)_{\text{RMS}} \approx 0.8 E_{\text{RMS}} \qquad (\delta a_1)_{\text{RMS}} \approx 0.3 E_{\text{RMS}}$$

where E_{RMS} is the RMS value of the observational errors, under the assumption that the true function is linear. Also, if use is made of (7.4.27), with $n = 1$ and $N = 4$, we obtain the estimate

$$E_{\text{RMS}} \approx \sqrt{\tfrac{5}{3}}\,(0.042) \doteq 0.055$$

under the same assumption. Accordingly, the RMS errors in a_0 and a_1 then may be estimated as about 0.044 and 0.016, respectively, and also (7.4.25) then affords the estimate $\sqrt{\tfrac{10}{3}}\, \bar{\varepsilon}_{\text{RMS}} = \sqrt{2} E_{\text{RMS}} \doteq 0.076$ for the RMS value of the departure between the unknown *true* function $f(x)$ and the smoothed function $\bar{y}(x)$ over the five data points. On the other hand, if an *independent* estimate of the RMS value of the observational errors were available, a comparison of that estimate with the estimate of E_{RMS} obtained here would serve to indicate the validity of the assumption that the true function is indeed linear.

The solution of the normal equations and the evaluation of the relevant determinant and cofactors can be conveniently effected by the use of procedures described in Secs. 10.4 to 10.6. It should be noted, however, that unfortunately the set of normal equations often tends to be ill-conditioned in the sense that small errors in the coefficients or in the numerical solution process may lead to large errors in the solution of this set. This situation clearly can be avoided by the use of orthogonal coordinate functions (see Secs. 7.12 and 7.16), for which the normal equations are uncoupled, as was seen in Sec. 7.2.

The same methods are used more generally in dealing with sets of linear equations, in which there are more equations than unknowns, whether or not they arise from a problem (7.3.1) in "curve fitting." In general, the original set is inconsistent, and does not possess a solution. The normal equations then correspond to the result of minimizing the sum of the (weighted or unweighted) squared deviations between the two members of those equations. If the squared deviation associated with the kth equation is to be weighted by w_k, the same end result can be obtained alternatively by multiplying both sides of that equation by $\sqrt{w_k}$ and using a *unit* weight in forming the normal equations. In this connection, it should be noticed, for example, that the equations $x = 2.3$ and $5x = 11.5$ are *not* equivalent, if the right-hand members are known only to be correct to the places given, since the first assertion is equivalent to the condition $2.25 < x < 2.35$ and the second to the condition $2.29 < x < 2.31$.

In this more general case, the coefficients of the left-hand members of the original equations, as well as the right-hand members, may be subject to error. Here, if the normal equations are again represented by (7.4.1), and if ε_{RMS} represents the RMS error of each of the right-hand members of the original equations, whereas η_{RMS} denotes the RMS error of each *coefficient* in the original set, the estimate (7.4.29) is to be replaced by

$$(\delta a_r)_{\text{RMS}} \approx \sqrt{\frac{C_{rr}}{D}} \sqrt{\varepsilon_{\text{RMS}}^2 + (a_0^2 + a_1^2 + \cdots + a_n^2)\eta_{\text{RMS}}^2} \tag{7.4.30}$$

when $w(x) \equiv 1$, under the assumption that all errors are small, random, and independent, and that the RMS errors in the coefficients of the original equations are all equal.

7.5 Orthogonal Polynomials

We consider next the case when a least-squares approximation is to be effected over the *interval* $[a, b]$, and we attempt first to construct a set of polynomials $\phi_0(x), \phi_1(x), \ldots, \phi_r(x), \ldots$, such that each member is orthogonal to all others in the set over $[a, b]$ relative to a specified weighting function $w(x)$ which is *nonnegative* in that interval.† It is convenient to ask that $\phi_r(x)$ be a polynomial of degree r. The problem then will be solved, in particular, if we obtain a polynomial $\phi_r(x)$ which is orthogonal over $[a, b]$ to *all* polynomials of degree inferior to r.

Thus we require a polynomial $\phi_r(x)$, of degree r, such that

$$\int_a^b w(x)\phi_r(x)q_{r-1}(x)\,dx = 0 \qquad (7.5.1)$$

where w is specified and where q_{r-1} is an *arbitrary* polynomial of degree $r - 1$ or less. In order to express this requirement in a more useful form, we integrate by parts r times, making use of the fact that $q_{r-1}^{(r)} \equiv 0$. For this purpose, we first introduce the notation

$$w(x)\phi_r(x) \equiv \frac{d^r U_r(x)}{dx^r} \qquad (7.5.2)$$

so that (7.5.1) becomes

$$\int_a^b U_r^{(r)}(x)q_{r-1}(x)\,dx = 0$$

or, after r integrations by parts,

$$[U_r^{(r-1)}q_{r-1} - U_r^{(r-2)}q_{r-1}' + U_r^{(r-3)}q_{r-1}'' - \cdots + (-1)^{r-1}U_r q_{r-1}^{(r-1)}]_a^b = 0 \qquad (7.5.3)$$

The requirement that the function $\phi_r(x)$ defined by (7.5.2)

$$\phi_r(x) = \frac{1}{w(x)}\frac{d^r U_r(x)}{dx^r} \qquad (7.5.4)$$

be a polynomial of degree r implies that $U_r(x)$ must satisfy the differential equation

$$\frac{d^{r+1}}{dx^{r+1}}\left[\frac{1}{w(x)}\frac{d^r U_r(x)}{dx^r}\right] = 0 \qquad (7.5.5)$$

in $[a, b]$, whereas the requirement that (7.5.3) be satisfied for *any* values of

† For alternative derivations, see Sec. 7.10 and Prob. 40.

$q_{r-1}(a)$, $q_{r-1}(b)$, $q'_{r-1}(a)$, $q'_{r-1}(b)$, and so forth, is met by satisfaction of the $2r$ boundary conditions

$$U_r(a) = U'_r(a) = U''_r(a) = \cdots = U_r^{(r-1)}(a) = 0 \tag{7.5.6}$$

$$U_r(b) = U'_r(b) = U''_r(b) = \cdots = U_r^{(r-1)}(b) = 0 \tag{7.5.7}$$

Thus if, for each integer r, a solution of (7.5.5) which satisfies (7.5.6) and (7.5.7) can be obtained, the rth member of the required set of functions is given by (7.5.4). From the homogeneity of these conditions, it follows that each such solution will contain an arbitrary multiplicative constant. It is known (see Szego [1967]) that the problem thus formulated does indeed possess a (nontrivial) solution, even when a and/or b is infinite, under the assumptions that $w(x) \geqq 0$ in the interval and that $\int_a^b x^k w(x)\,dx$ exists for all nonnegative integral values of k.

In accordance with the results of Sec. 7.2, the coefficients in the expression

$$y(x) = \sum_{r=0}^{n} a_r \phi_r(x) \tag{7.5.8}$$

are then determined by the requirement

$$\int_a^b w(x)[f(x) - y(x)]^2\,dx = \min \tag{7.5.9}$$

in the form

$$a_r = \frac{\int_a^b wf\phi_r\,dx}{\int_a^b w\phi_r^2\,dx} \equiv \frac{\int_a^b wf\phi_r\,dx}{\gamma_r} \tag{7.5.10}$$

where

$$\gamma_r = \int_a^b w\phi_r^2\,dx \tag{7.5.11}$$

Although the numerator in (7.5.10) depends upon f, the denominator γ_r is independent of f and can be determined once and for all.

The calculation of γ_r is facilitated by the following considerations. If we write

$$\phi_r(x) = A_{r0} + A_{r1}x + \cdots + A_r x^r \tag{7.5.12}$$

so that A_{rk} is the coefficient of x^k in $\phi_r(x)$ and $A_r \equiv A_{rr}$ is its *leading* coefficient, there follows

$$\begin{aligned}\gamma_r &= \int_a^b w(x)\phi_r(x)\phi_r(x)\,dx \\ &\equiv \int_a^b w(x)\phi_r(x)[A_{r0} + A_{r1}x + \cdots + A_r x^r]\,dx\end{aligned}$$

and hence if we recall the relations

$$\int_a^b w(x)\phi_r(x)x^i\,dx = 0 \qquad (i = 0, 1, \ldots, r-1) \tag{7.5.13}$$

which are equivalent to (7.5.1), we may deduce that

$$\gamma_r = A_r \int_a^b x^r w(x)\phi_r(x)\,dx = A_r \int_a^b x^r U_r^{(r)}(x)\,dx$$

By integrating by parts r times and making use of (7.5.6) and (7.5.7), this relation takes the convenient form

$$\gamma_r \equiv \int_a^b w(x)\phi_r^2(x)\,dx = (-1)^r r!\, A_r \int_a^b U_r(x)\,dx \tag{7.5.14}$$

where A_r is the coefficient of x^r in $\phi_r(x)$.

It is useful to notice that the problem specified by (7.5.8) and (7.5.9) can be generalized in the following way. It may happen that $f(x)$ clearly *cannot* be satisfactorily approximated over $[a, b]$ by a polynomial of low degree, but that a certain function $v(x)$ is known such that the ratio $f(x)/v(x)$ *can* be so approximated. Thus, if we determine the coefficients of the polynomial

$$y(x) = \sum_{r=0}^{n} b_r\phi_r(x) \tag{7.5.15}$$

in such a way that

$$\int_a^b w(x)\left[\frac{f(x)}{v(x)} - \sum_{r=0}^{n} b_r\phi_r(x)\right]^2 dx = \min \tag{7.5.16}$$

the orthogonality of the ϕ's relative to w leads to the result

$$b_r = \frac{1}{\gamma_r}\int_a^b \frac{w}{v} f\phi_r\,dx \tag{7.5.17}$$

It is seen that (7.5.16) is equivalent to the result of minimizing the squared error $(f - vy)^2$ with the weighting function w/v^2. The choice $w(x) = v(x)$ is a frequently useful one. Several examples of such approximations are considered in the following sections.

7.6 Legendre Approximation

For least-squares approximation over an interval of finite length, it is convenient to suppose that a linear change in variables has transformed that interval into the interval $[-1, 1]$. We consider here the case when the weighting function is unity:

$$w(x) = 1 \tag{7.6.1}$$

The differential equation (7.5.5) then becomes

$$\frac{d^{2r+1}U_r}{dx^{2r+1}} = 0 \qquad (7.6.2)$$

and the boundary conditions (7.5.6) and (7.5.7) take the form

$$U_r(\pm 1) = U_r'(\pm 1) = \cdots = U_r^{(r-1)}(\pm 1) = 0 \qquad (7.6.3)$$

from which there follows (analytically or by inspection)

$$U_r = C_r(x^2 - 1)^r \qquad (7.6.4)$$

where C_r is an arbitrary constant. Hence, from (7.5.4), it follows that the rth relevant orthogonal polynomial is of the form

$$\phi_r(x) = C_r \frac{d^r}{dx^r}(x^2 - 1)^r \qquad (7.6.5)$$

With $C_r = 1/(2^r r!)$, the polynomial so obtained is known as the rth *Legendre polynomial* and is usually denoted by $P_r(x)$. The relation

$$P_r(x) = \frac{1}{2^r r!}\frac{d^r}{dx^r}(x^2 - 1)^r \qquad (7.6.6)$$

is often called the *Rodrigues formula* for $P_r(x)$. From the preceding derivation, it follows that

$$\int_{-1}^{1} P_r(x)P_s(x)\,dx = 0 \qquad (r \neq s) \qquad (7.6.7)$$

where r and s are nonnegative integers. The value assigned to C_r is such that $P_r(1) = 1$, and it is true also that $|P_r(x)| \leqq 1$ when $|x| \leqq 1$.

The first six of these polynomials may be obtained in the forms

$$\begin{aligned} P_0(x) &= 1 & P_1(x) &= x \\ P_2(x) &= \tfrac{1}{2}(3x^2 - 1) & P_3(x) &= \tfrac{1}{2}(5x^3 - 3x) \\ P_4(x) &= \tfrac{1}{8}(35x^4 - 30x^2 + 3) & P_5(x) &= \tfrac{1}{8}(63x^5 - 70x^3 + 15x) \end{aligned} \qquad (7.6.8)$$

and additional ones can be determined from the recurrence formula†

$$P_{r+1}(x) = \frac{2r + 1}{r + 1}xP_r(x) - \frac{r}{r + 1}P_{r-1}(x) \qquad (7.6.9)$$

It may be noted that the polynomials of even and odd degrees are even and odd functions of x, respectively.

† For the derivation of this formula and of other similar formulas to be listed in following sections, see Sec. 7.10 and Szego [1967].

In order to evaluate the factor (7.5.14), we notice first that (7.6.6) gives

$$P_r(x) = \frac{1}{2^r r!} \frac{d^r}{dx^r}(x^{2r} - rx^{2r-2} + \cdots) = \frac{(2r)!}{2^r(r!)^2} x^r - \cdots$$

so that

$$A_r = \frac{(2r)!}{2^r(r!)^2}$$

Hence (7.5.14) gives

$$\begin{aligned} \gamma_r \equiv \int_{-1}^{1} P_r^2(x)\,dx &= \frac{(2r)!}{2^r r!} \frac{1}{2^r r!} \int_{-1}^{1} (1 - x^2)^r\,dx \\ &= \frac{(2r)!}{2^{2r}(r!)^2} \frac{2^{2r+1}(r!)^2}{(2r+1)!} = \frac{2}{2r+1} \end{aligned} \tag{7.6.10}$$

Thus the nth-degree least-squares polynomial approximation to $f(x)$ over $[-1, 1]$ relevant to a constant weighting function is defined by

$$y(x) = \sum_{r=0}^{n} a_r P_r(x) \qquad (-1 \leqq x \leqq 1) \tag{7.6.11}$$

where

$$a_r = \frac{2r+1}{2} \int_{-1}^{1} f(x) P_r(x)\,dx \tag{7.6.12}$$

It has the property that, for all polynomials $y_n(x)$ of degree n or less, the integrated squared error

$$\int_{-1}^{1} [f(x) - y_n(x)]^2\,dx$$

is least when $y_n(x)$ is identified with the polynomial defined by (7.6.11). By virtue of (7.2.14), the minimum value is given by

$$\Delta_n = \int_{-1}^{1} f^2\,dx - \sum_{r=0}^{n} \frac{2a_r^2}{2r+1} \tag{7.6.13}$$

In accordance with (7.5.15) to (7.5.17), it follows also that the least-squares approximation to $f(x)$ of the form $v(x)y(x)$, with $y(x)$ a polynomial

$$y(x) = \sum_{r=0}^{n} b_r P_r(x) \qquad (-1 \leqq x \leqq 1) \tag{7.6.14}$$

and with the weighting function $1/[v(x)]^2$, where $v(x)$ is a specified function, is that for which

$$b_r = \frac{2r+1}{2} \int_{-1}^{1} \frac{f(x)}{v(x)} P_r(x)\,dx \tag{7.6.15}$$

7.7 Laguerre Approximation

For least-squares polynomial approximation over a semi-infinite interval, it is convenient to first transform that interval into the interval $[0, \infty)$ by a translation of the origin. A frequently used approximation makes use of a weighting function of the form

$$w(x) = e^{-\alpha x} \tag{7.7.1}$$

where α is a positive constant, taken to be sufficiently large to ensure the existence of the integral of the squared error over the semi-infinite interval (when this is possible).

From the results of Sec. 7.5, the relevant orthogonal polynomials are such that

$$\phi_r(x) = e^{\alpha x} \frac{d^r U_r}{dx^r} \tag{7.7.2}$$

where

$$\frac{d^{r+1}}{dx^{r+1}} \left[e^{\alpha x} \frac{d^r U_r}{dx^r} \right] = 0 \tag{7.7.3}$$

and

$$U_r(0) = U_r'(0) = \cdots = U_r^{(r-1)}(0) = 0 \tag{7.7.4}$$

$$U_r(\infty) = U_r'(\infty) = \cdots = U_r^{(r-1)}(\infty) = 0 \tag{7.7.5}$$

The general solution of (7.7.3) is readily found to be

$$U_r = e^{-\alpha x}(c_0 + c_1 x + \cdots + c_r x^r) + d_0 + d_1 x + \cdots + d_{r-1} x^{r-1}$$

where the c's and d's are arbitrary constants. The conditions (7.7.5) require that all d's vanish, and (7.7.4) gives $c_0 = c_1 = \cdots = c_{r-1} = 0$, so that there follows

$$U_r(x) = C_r x^r e^{-\alpha x} \tag{7.7.6}$$

and hence

$$\phi_r(x) = C_r e^{\alpha x} \frac{d^r}{dx^r} (x^r e^{-\alpha x}) \tag{7.7.7}$$

With $C_r = 1$ and $\alpha = 1$, this polynomial is called the rth *Laguerre polynomial* and is usually denoted by $L_r(x)$:

$$L_r(x) = e^x \frac{d^r}{dx^r} (x^r e^{-x}) \tag{7.7.8}$$

It follows that, again taking $C_r = 1$, the polynomial (7.7.7) can be expressed in the form

$$\phi_r(x) = L_r(\alpha x) \tag{7.7.9}$$

and that we have the orthogonality property

$$\int_0^\infty e^{-\alpha x} L_r(\alpha x) L_s(\alpha x)\, dx = 0 \qquad (r \neq s) \qquad (7.7.10)$$

when r and s are nonnegative integers.

The first six of the Laguerre polynomials can be obtained in the form

$$L_0(x) = 1 \qquad L_1(x) = 1 - x \qquad L_2(x) = 2 - 4x + x^2$$
$$L_3(x) = 6 - 18x + 9x^2 - x^3 \qquad L_4(x) = 24 - 96x + 72x^2 - 16x^3 + x^4$$
$$L_5(x) = 120 - 600x + 600x^2 - 200x^3 + 25x^4 - x^5 \qquad (7.7.11)$$

and additional ones can be determined from the recurrence formula

$$L_{r+1}(x) = (1 + 2r - x)L_r(x) - r^2 L_{r-1}(x) \qquad (7.7.12)$$

The value assigned to C_r is such that the coefficient of x^r in $L_r(\alpha x)$ is $(-\alpha)^r$. Hence, from (7.5.14), there follows

$$\gamma_r \equiv \int_0^\infty e^{-\alpha x} L_r^2(\alpha x)\, dx = \alpha^r r! \int_0^\infty x^r e^{-\alpha x}\, dx = \frac{1}{\alpha}(r!)^2 \qquad (7.7.13)$$

Thus the nth-degree least-squares polynomial approximation to $f(x)$ over $[0, \infty)$, relevant to the weighting function $w(x) = e^{-\alpha x}$, is defined by

$$y(x) = \sum_{r=0}^{n} a_r L_r(\alpha x) \qquad (0 \leqq x < \infty) \qquad (7.7.14)$$

where

$$a_r = \frac{\alpha}{(r!)^2} \int_0^\infty e^{-\alpha x} f(x) L_r(\alpha x)\, dx \qquad (7.7.15)$$

It has the property that, for all polynomials $y_n(x)$ of degree n or less, the integrated weighted squared error

$$\int_0^\infty e^{-\alpha x}[f(x) - y_n(x)]^2\, dx$$

is least when $y_n(x)$ is identified with the right-hand member of (7.7.14). In order for this integral to *exist*, it is generally necessary that $|f(x)|$ grow less rapidly than $e^{\alpha x/2}$ as $x \to \infty$.

Another type of approximation employing Laguerre polynomials is obtained if we require the coefficients in the relation

$$y(x) = \sum_{r=0}^{n} b_r L_r(\alpha x) \qquad (0 \leqq x < \infty) \qquad (7.7.16)$$

such that

$$\int_0^\infty e^{\alpha x}[f(x) - e^{-\alpha x}y(x)]^2\,dx$$

$$\equiv \int_0^\infty e^{\alpha x}\left[f(x) - e^{-\alpha x}\sum_{k=0}^{n} b_k L_k(\alpha x)\right]^2 dx = \text{min} \qquad (7.7.17)$$

This is a special case of the problem specified by (7.5.15) to (7.5.17) in which $v(x) = w(x)$, and the coefficients are thus obtained in the form

$$b_r = \frac{\alpha}{(r!)^2}\int_0^\infty f(x)L_r(\alpha x)\,dx \qquad (7.7.18)$$

In order that the integrals in (7.7.17) exist, it is generally necessary that $f(x)$ tend to zero more rapidly than $e^{-\alpha x/2}$ as $x \to \infty$.

7.8 Hermite Approximation

Over the doubly infinite interval $(-\infty < x < \infty)$, a frequently used weighting function is of the form

$$w(x) = e^{-\alpha^2 x^2} \qquad (7.8.1)$$

In this case the relevant orthogonal polynomials are defined by

$$\phi_r(x) = e^{\alpha^2 x^2}\frac{d^r U_r}{dx^r} \qquad (7.8.2)$$

where U_r satisfies the equation

$$\frac{d^{r+1}}{dx^{r+1}}\left[e^{\alpha^2 x^2}\frac{d^r U_r}{dx^r}\right] = 0 \qquad (7.8.3)$$

and where U_r and its first $r - 1$ derivatives are to tend to zero as $x \to \pm\infty$. Since the function

$$U_r(x) = C_r e^{-\alpha^2 x^2} \qquad (7.8.4)$$

has the property that its rth derivative is the product of itself and a polynomial of degree r, it satisfies these conditions and there follows

$$\phi_r(x) = C_r e^{\alpha^2 x^2}\frac{d^r}{dx^r}(e^{-\alpha^2 x^2}) \qquad (7.8.5)$$

The *Hermite polynomial* of degree r is usually defined by taking $C_r = (-1)^r$ and, in addition, either $\alpha^2 = 1$ or $\alpha^2 = \frac{1}{2}$ in (7.8.5). Both definitions are used in the literature. We adopt the former one and write

$$H_r(x) = (-1)^r e^{x^2}\frac{d^r}{dx^r}(e^{-x^2}) \qquad (7.8.6)$$

so that, with the choice $C_r = (-\alpha)^{-r}$, (7.8.5) becomes†

$$\phi_r(x) = H_r(\alpha x) = (-\alpha)^{-r} e^{\alpha^2 x^2} \frac{d^r}{dx^r}(e^{-\alpha^2 x^2}) \tag{7.8.7}$$

Thus Hermite polynomials possess the orthogonality property

$$\int_{-\infty}^{\infty} e^{-\alpha^2 x^2} H_r(\alpha x) H_s(\alpha x)\, dx = 0 \qquad (r \neq s) \tag{7.8.8}$$

when r and s are nonnegative integers. The first six of the polynomials defined by (7.8.6) are obtained in the form

$$\begin{aligned} &H_0(x) = 1 && H_1(x) = 2x \\ &H_2(x) = 4x^2 - 2 && H_3(x) = 8x^3 - 12x \\ &H_4(x) = 16x^4 - 48x^2 + 12 && H_5(x) = 32x^5 - 160x^3 + 120x \end{aligned} \tag{7.8.9}$$

and additional ones can be determined from the recurrence formula

$$H_{r+1}(x) = 2xH_r(x) - 2rH_{r-1}(x) \tag{7.8.10}$$

With $A_r = (2\alpha)^r$ and $U_r = (-1/\alpha)^r e^{-\alpha^2 x^2}$, Eq. (7.5.14) gives

$$\gamma_r \equiv \int_{-\infty}^{\infty} e^{-\alpha^2 x^2} H_r^2(\alpha x)\, dx = 2^r r! \int_{-\infty}^{\infty} e^{-\alpha^2 x^2}\, dx = \frac{2^r r!}{\alpha}\sqrt{\pi} \tag{7.8.11}$$

Thus the nth-degree least-squares polynomial approximation to $f(x)$ over $(-\infty, +\infty)$, relevant to the weighting function $w(x) = e^{-\alpha^2 x^2}$, is defined by

$$y(x) = \sum_{r=0}^{n} a_r H_r(\alpha x) \qquad (-\infty < x < \infty) \tag{7.8.12}$$

where

$$a_r = \frac{\alpha}{2^r r! \sqrt{\pi}} \int_{-\infty}^{\infty} e^{-\alpha^2 x^2} H_r(\alpha x) f(x)\, dx \tag{7.8.13}$$

It has the property that, for all polynomials $y_n(x)$ of degree n or less, the integrated squared error

$$\int_{-\infty}^{\infty} e^{-\alpha^2 x^2}[f(x) - y_n(x)]^2\, dx$$

is least when $y_n(x)$ is identified with the right-hand member of (7.8.12). It must be assumed that the behavior of $f(x)$ is such that this integral exists.

† With the definition $H_r(x) = (-1)^r e^{x^2/2}\, d^r(e^{-x^2/2})/dx^r$ and the corresponding choice $C_r = (-1)^r 2^{-r/2} \alpha^{-r}$, there would follow $\phi_r(x) = H_r(\sqrt{2}\, \alpha x)$.

It should be noticed that, since the weighting function $e^{-\alpha^2 x^2}$ becomes small very rapidly as x increases in magnitude, the least-squares criterion here requires that the magnitude of the deviation $f(x) - y(x)$ be small when x is small, but tolerates large values of that deviation when x is large in magnitude. A similar remark applies somewhat less strongly to the approximation of the preceding section. Thus, such approximations should not be used unless this situation is an acceptable one.

Another type of approximation, of particular importance in the theory of statistics, is obtained if we require the coefficients in the relation

$$y(x) = \sum_{r=0}^{n} b_r H_r(\alpha x) \qquad (-\infty < x < \infty) \tag{7.8.14}$$

such that

$$\int_{-\infty}^{\infty} e^{\alpha^2 x^2}[f(x) - e^{-\alpha^2 x^2} y(x)]^2 \, dx \equiv \int_{-\infty}^{\infty} e^{\alpha^2 x^2} \left[f(x) - e^{-\alpha^2 x^2} \sum_{k=0}^{n} b_k H_k(\alpha x) \right]^2 dx = \min \tag{7.8.15}$$

The conditions governing the b's are obtained directly, or by reference to (7.5.15) to (7.5.17), with $v(x) = w(x)$, in the form

$$b_r = \frac{\alpha}{2^r r! \sqrt{\pi}} \int_{-\infty}^{\infty} f(x) H_r(\alpha x) \, dx \tag{7.8.16}$$

assuming that the behavior of $f(x)$, for large values of $|x|$, is such that the integrals involved exist. In particular, the approximation (7.8.14) is often used in situations when $f(x)$ vanishes for all values of $|x|$ which exceed a certain value.

If the ith *moment* of $f(x)$ is defined as

$$m_i = \int_{-\infty}^{\infty} x^i f(x) \, dx \tag{7.8.17}$$

and use is made of the explicit forms of (7.8.9), we find that the leading coefficients in (7.8.14) are expressible in the forms

$$b_0 = \frac{\alpha}{\sqrt{\pi}} m_0 \qquad b_1 = \frac{\alpha^2}{\sqrt{\pi}} m_1 \qquad b_2 = \frac{\alpha}{4\sqrt{\pi}}(2\alpha^2 m_2 - m_0) \tag{7.8.18}$$

and that the remaining b's can be similarly expressed in terms of the moments.

7.9 Chebyshev Approximation†

In cases when errors near the ends of an interval $[a, b]$ are of particular importance, a weighting function which is of the form $1/\sqrt{(x - a)(b - x)}$ is

† The term "Chebyshev approximation" also is used frequently in the literature to refer to a "minimax approximation" (see Sec. 9.9).

often appropriate. It is supposed again that a linear change in variables has transformed the given interval into the interval $[-1, 1]$, so that the weighting function becomes

$$w(x) = \frac{1}{\sqrt{1 - x^2}} \tag{7.9.1}$$

In order to obtain the relevant orthogonal polynomials in this case, it is convenient to start with the basic condition (7.5.1) rather than with its consequences. Thus we require a polynomial $\phi_r(x)$, of degree r in x, such that

$$\int_{-1}^{1} \frac{\phi_r(x)q_{r-1}(x)}{\sqrt{1 - x^2}}\, dx = 0 \tag{7.9.2}$$

where $q_{r-1}(x)$ is an arbitrary polynomial of degree $r - 1$ or less in x. If we introduce the change in variables

$$x = \cos \theta \tag{7.9.3}$$

this requirement becomes

$$\int_0^{\pi} \phi_r(\cos \theta)q_{r-1}(\cos \theta)\, d\theta = 0 \tag{7.9.4}$$

Now, since $\cos k\theta$ is expressible as a polynomial of degree k in $\cos \theta$ and since, conversely, any polynomial of degree k in $\cos \theta$ can be expressed as a linear combination of $1, \cos \theta, \cos 2\theta, \ldots, \cos k\theta$, it follows that (7.9.4) will be satisfied if and only if

$$\int_0^{\pi} \phi_r(\cos \theta) \cos k\theta\, d\theta = 0 \qquad (k = 0, 1, \ldots, r - 1) \tag{7.9.5}$$

It is easily verified that the function

$$\phi_r(\cos \theta) = C_r \cos r\theta \tag{7.9.6}$$

has this property. Hence, returning to the variable x by using (7.9.3), we verify that the functions

$$\phi_r(x) = C_r \cos (r \cos^{-1} x) \tag{7.9.7}$$

are the required orthogonal polynomials. With $C_r = 1$, these polynomials are known as *Chebyshev polynomials*,† often denoted by $T_r(x)$, so that we may write

$$\phi_r(x) = T_r(x) = \cos (r \cos^{-1} x) \tag{7.9.8}$$

Thus, these polynomials possess the orthogonality property

$$\int_{-1}^{1} \frac{T_r(x)T_s(x)}{\sqrt{1 - x^2}}\, dx = 0 \qquad (r \neq s) \tag{7.9.9}$$

† The name of Chebyshev (or Tschebycheff) is associated with various sets of polynomials in the literature (see also Secs. 7.13 and 8.14).

when r and s are nonnegative integers. The first six are obtained in the form

$$
\begin{aligned}
T_0(x) &= 1 & T_1(x) &= x \\
T_2(x) &= 2x^2 - 1 & T_3(x) &= 4x^3 - 3x \\
T_4(x) &= 8x^4 - 8x^2 + 1 & T_5(x) &= 16x^5 - 20x^3 + 5x
\end{aligned}
\tag{7.9.10}
$$

and additional ones may be determined from the recurrence formula

$$
T_{r+1}(x) = 2xT_r(x) - T_{r-1}(x) \qquad (r \geqq 1) \tag{7.9.11}
$$

In order to evaluate the factor

$$
\gamma_r = \int_{-1}^{1} \frac{T_r^2(x)}{\sqrt{1 - x^2}}\, dx \tag{7.9.12}
$$

we again write $x = \cos\theta$ and $T_r(x) = \cos r\theta$, so that there follows directly

$$
\gamma_r = \int_0^{\pi} \cos^2 r\theta \, d\theta = \begin{cases} \pi & (r = 0) \\ \dfrac{\pi}{2} & (r \neq 0) \end{cases} \tag{7.9.13}
$$

Thus the nth-degree least-squares polynomial approximation to $f(x)$ over $[-1, 1]$, relevant to the weighting function $w(x) = 1/\sqrt{1 - x^2}$, is defined by

$$
y(x) = \sum_{r=0}^{n} a_r T_r(x) \qquad (-1 \leqq x \leqq 1) \tag{7.9.14}
$$

where

$$
a_0 = \frac{1}{\pi} \int_{-1}^{1} \frac{f(x)}{\sqrt{1 - x^2}}\, dx \qquad a_r = \frac{2}{\pi} \int_{-1}^{1} \frac{f(x)T_r(x)}{\sqrt{1 - x^2}}\, dx \quad (r \neq 0) \tag{7.9.15}
$$

It has the property that, for all polynomials of degree n or less, the integrated weighted squared error

$$
\int_{-1}^{1} \frac{1}{\sqrt{1 - x^2}} [f(x) - y_n(x)]^2 \, dx
$$

is least when $y_n(x)$ is identified with the right-hand member of (7.9.14).

On the other hand, if we wish to approximate $f(x)$ in the least-squares sense by the product of $1/\sqrt{1 - x^2}$ and a polynomial, over $(-1, 1)$, with the weighting function $\sqrt{1 - x^2}$, we are to determine the coefficients in the relation

$$
y(x) = \sum_{r=0}^{n} b_r T_r(x) \qquad (-1 < x < 1) \tag{7.9.16}
$$

such that

$$\int_{-1}^{1} \sqrt{1-x^2}\left[f(x) - \frac{y(x)}{\sqrt{1-x^2}}\right]^2 dx$$

$$\equiv \int_{-1}^{1} \sqrt{1-x^2}\left[f(x) - \frac{1}{\sqrt{1-x^2}} \sum_{k=0}^{n} b_k T_k(x)\right]^2 dx = \min$$

The conditions determining the b's are obtained in the form

$$\int_{-1}^{1} T_r(x)\left[f(x) - \frac{1}{\sqrt{1-x^2}} \sum_{k=0}^{n} b_k T_k(x)\right] dx = 0$$

and the use of (7.9.9) and (7.9.13), or of (7.5.15) to (7.5.17), yields the determination

$$b_0 = \frac{1}{\pi}\int_{-1}^{1} f(x)\,dx \qquad b_r = \frac{2}{\pi}\int_{-1}^{1} f(x)T_r(x)\,dx \quad (r \neq 0) \tag{7.9.17}$$

A great variety of other types of least-squares polynomial approximations can be formulated in terms of other weighting functions. In particular, for the weighting function

$$w(x) = (1-x)^{\alpha}(1+x)^{\beta} \qquad (\alpha > -1, \beta > -1) \tag{7.9.18}$$

over $[-1, 1]$, which reduces to the Legendre case when $\alpha = \beta = 0$ and to the Chebyshev case when $\alpha = \beta = -\frac{1}{2}$, the rth orthogonal polynomial is readily found to be of the form

$$\phi_r(x) = C_r(1-x)^{-\alpha}(1+x)^{-\beta}\frac{d^r}{dx^r}[(1-x)^{\alpha+r}(1+x)^{\beta+r}] \tag{7.9.19}$$

which may be identified with the rth *Jacobi polynomial* when C_r is suitably specified (see Sec. 8.9).

In particular, the factor C_r for $T_r(x)$ is given by $(-2)^r r!/(2r)!$, so that (7.9.8) can also be written in the form

$$T_r(x) = \frac{(-2)^r r!}{(2r)!}(1-x^2)^{1/2}\frac{d^r}{dx^r}(1-x^2)^{r-(1/2)} \tag{7.9.20}$$

Analogous polynomials $S_r(x)$, which are associated with the weighting function $w(x) = (1-x^2)^{1/2}$ and which can be expressed in the form

$$S_r(x) = \frac{\sin[(r+1)\cos^{-1}x]}{\sin(\cos^{-1}x)}$$

$$= \frac{(-2)^r(r+1)!}{(2r+1)!}(1-x^2)^{-1/2}\frac{d^r}{dx^r}(1-x^2)^{r+(1/2)} \tag{7.9.21}$$

are sometimes called *Chebyshev polynomials of the second kind* (considered in Prob. 33).†

For the weighting function

$$w(x) = x^{\beta} e^{-\alpha x} \qquad (\beta > -1, \alpha > 0) \qquad (7.9.22)$$

over $[0, \infty)$, there follows

$$\phi_r(x) = C_r x^{-\beta} e^{\alpha x} \frac{d^r}{dx^r} (x^{\beta + r} e^{-\alpha x}) \qquad (7.9.23)$$

and the resultant polynomials are frequently called *Sonine polynomials*, or *generalized Laguerre polynomials* (for additional information see Szego [1967]).

7.10 Properties of Orthogonal Polynomials. Recursive Computation

In order to provide a basis for computational techniques associated with the preceding approximations, as well as for later developments, we next exhibit some useful properties of orthogonal polynomials.

First, principally for later reference (Sec. 8.4), it is shown that if $\phi_r(x)$ is the rth member of the orthogonal set relative to the weighting function $w(x)$ over an interval $[a, b]$, then *if $w(x)$ does not change sign in $[a, b]$, the polynomial $\phi_r(x)$ possesses r distinct real zeros, all of which lie in the interval (a, b).* In order to establish this fact, we notice first that, since $\int_a^b w\phi_r\phi_0\,dx = A_0 \int_a^b w\phi_r\,dx = 0$ when $r \geqq 1$, and $w(x)$ is of constant sign, $\phi_r(x)$ must change sign at least *once* in (a, b) when $r \geqq 1$. Now let those real zeros of $\phi_r(x)$ which are of *odd* multiplicity, and which lie in (a, b), be denoted by $c_1, c_2, \ldots, c_m$ and *assume* that $m < r$. Then the product

$$(x - c_1)(x - c_2) \cdots (x - c_m)\phi_r(x)$$

does not change sign in $[a, b]$. But, since $m < r$, the coefficient of $\phi_r(x)$ is a polynomial of degree less than r, and hence, by (7.4.1), we must have

$$\int_a^b w(x)[(x - c_1)(x - c_2)\cdots(x - c_m)]\phi_r(x)\,dx = 0$$

However, since $w(x)$ does not change sign in $[a, b]$, the integrand therefore has the same property, and a contradiction follows. Hence there must follow $m = r$, and since the total multiplicity of *all* zeros is equal to r, all roots must be real and distinct and must lie in (a, b), as was to be shown.

Next it is shown that each set of orthogonal polynomials satisfies a simple

† The notation $U_r(x)$ is often used in place of $S_r(x)$.

three-term *recurrence formula*. In particular, a basis is provided for the derivation of the relations stated without proof in preceding sections (see Prob. 36). We notice first that

$$\phi_{k+1}(x) - \frac{A_{k+1}}{A_k} x\phi_k(x)$$

is a polynomial of maximum degree k. Hence, if we write

$$a_k \equiv \frac{A_{k+1}}{A_k} \tag{7.10.1}$$

it follows that $\phi_{k+1}(x) - a_k x\phi_k(x)$ can be expressed as a linear combination of $\phi_0(x), \phi_1(x), \ldots, \phi_k(x)$, in the form

$$\phi_{k+1}(x) - a_k x\phi_k(x) = b_k\phi_k(x) + c_k\phi_{k-1}(x) + \cdots \tag{7.10.2}$$

for some constant values of $b_k, c_k, \ldots$. But, since $x\phi_0(x), x\phi_1(x), \ldots$, and $x\phi_{k-2}(x)$ are polynomials of degree inferior to k, the two terms in the left-hand member of (7.10.2) are both orthogonal to $\phi_0, \phi_1, \ldots, \phi_{k-2}$ over $[a, b]$, relative to $w(x)$. Hence the same statement applies to the right-hand member, so that the omitted terms in (7.10.2) vanish, and we deduce that $\phi_k(x)$ satisfies a recurrence formula of the form

$$\phi_{k+1}(x) = (a_k x + b_k)\phi_k(x) + c_k\phi_{k-1}(x) \tag{7.10.3}$$

where a_k is defined by (7.10.1), and b_k and c_k are certain other constants. In order that (7.10.3) also hold when $k = 0$, the convention $\phi_{-1}(x) \equiv 0$ may be adopted.

If we multiply the equal members of (7.10.3) successively by $w\phi_{k+1}$, $w\phi_k$, and $w\phi_{k-1}$, and integrate each resultant equation over $[a, b]$, we obtain the additional relations

$$\gamma_{k+1} = a_k \int_a^b xw(x)\phi_k(x)\phi_{k+1}(x)\, dx \tag{7.10.4}$$

$$0 = a_k \int_a^b xw(x)[\phi_k(x)]^2\, dx + b_k\gamma_k \tag{7.10.5}$$

$$0 = a_k \int_a^b xw(x)\phi_{k-1}(x)\phi_k(x)\, dx + c_k\gamma_{k-1} \tag{7.10.6}$$

If k is replaced by $k - 1$ in the first equation, the result can be used to eliminate the unknown integral from the third equation and to establish the relation

$$c_k = -\frac{a_k\gamma_k}{a_{k-1}\gamma_{k-1}} \tag{7.10.7}$$

Hence (7.10.3) can be rewritten in the form

$$x\,\frac{\phi_k(x)}{\gamma_k} = \frac{\phi_{k+1}(x)}{a_k\gamma_k} + \frac{\phi_{k-1}(x)}{a_{k-1}\gamma_{k-1}} - \frac{b_k\phi_k(x)}{a_k\gamma_k} \tag{7.10.8}$$

For the purpose of deriving an important consequence of (7.10.8), it is noted next that if both members of (7.10.8) are multiplied by $\phi_k(y)$, where y is an arbitrary parameter, and the result of interchanging x and y in the result is subtracted from that result, the constant b_k is eliminated and the more symmetrical relation

$$(x-y)\,\frac{\phi_k(x)\phi_k(y)}{\gamma_k} = \frac{\phi_{k+1}(x)\phi_k(y) - \phi_k(x)\phi_{k+1}(y)}{a_k\gamma_k} - \frac{\phi_k(x)\phi_{k-1}(y) - \phi_{k-1}(x)\phi_k(y)}{a_{k-1}\gamma_{k-1}} \tag{7.10.9}$$

is obtained. The result of summing the equal members from $k = 0$ to $k = m$ and taking advantage of the "telescoping" of terms on the right is then the important relation

$$\sum_{k=0}^{m} \frac{\phi_k(x)\phi_k(y)}{\gamma_k} = \frac{\phi_{m+1}(x)\phi_m(y) - \phi_m(x)\phi_{m+1}(y)}{a_m\gamma_m(x-y)} \tag{7.10.10}$$

known as the *Christoffel-Darboux identity.* In addition, by considering the limiting form of that relation as $y \to x$, we obtain the equation

$$\sum_{k=0}^{m} \frac{[\phi_k(x)]^2}{\gamma_k} = \frac{A_m}{A_{m+1}\gamma_m}\,[\phi'_{m+1}(x)\phi_m(x) - \phi'_m(x)\phi_{m+1}(x)] \tag{7.10.11}$$

(The usefulness of these relations is illustrated in Probs. 37 and 38 as well as in Sec. 8.4.)

The fact that the set of the polynomials $\phi_r(x)$, which are orthogonal over $[a, b]$ relative to $w(x)$, satisfies a recurrence formula of form (7.10.3) is often useful for the purpose of generating members of the set recursively, without use of the generating function $U_r(x)$ and even without advance knowledge of the coefficients a_k, b_k, and c_k for a specific $w(x)$. For this purpose it is convenient first to determine the polynomials with *unit leading coefficients* (*monic* polynomials), so that

$$A_k = 1 \tag{7.10.12}$$

and hence also $a_k = 1$, and then to normalize the polynomials thus obtained in another way if this is desirable. The kth such polynomial is denoted here by $\tilde{\phi}_k(x)$.

Thus, with

$$\tilde{\phi}_0(x) = 1$$

(7.10.3) gives

$$\tilde{\phi}_1(x) = x + b_0$$

where b_0 is determined by (7.10.5) with $k = 0$:

$$b_0 = -\frac{\int_a^b xw\,dx}{\int_a^b w\,dx}$$

Next, (7.10.3) gives

$$\tilde{\phi}_2(x) = (x + b_1)\tilde{\phi}_1(x) + c_1$$

where, from (7.10.5) and (7.10.6),

$$b_1 = -\frac{\int_a^b xw\tilde{\phi}_1^2\,dx}{\int_a^b w\tilde{\phi}_1^2\,dx} \qquad c_1 = -\frac{\int_a^b xw\tilde{\phi}_1\,dx}{\int_a^b w\,dx}$$

In general, $\tilde{\phi}_{k+1}(x)$ is obtained from $\tilde{\phi}_k(x)$ and $\tilde{\phi}_{k-1}(x)$ by use of (7.10.3) with

$$a_k = 1 \qquad b_k = -\frac{\int_a^b xw\tilde{\phi}_k^2\,dx}{\int_a^b w\tilde{\phi}_k^2\,dx} \qquad c_k = -\frac{\int_a^b xw\tilde{\phi}_{k-1}\tilde{\phi}_k\,dx}{\int_a^b w\tilde{\phi}_{k-1}^2\,dx} \tag{7.10.13}$$

As was shown in Sec. 1.8, the fact that the polynomials $\phi_0, \ldots, \phi_n$ satisfy a three-term recurrence formula also can be used to simplify the numerical evaluation of a linear combination of them for a specific value of x. For this purpose, we identify (1.8.7) with (7.10.3) and obtain

$$\alpha_k = -(a_k x + b_k) \qquad \beta_k = -c_k$$

Hence, from the results of Sec. 1.8, we deduce that if the sequence $u_n(x)$, $u_{n-1}(x), \ldots, u_0(x)$ is generated by the recurrence formula

$$u_k = (a_k x + b_k)u_{k+1} + c_{k+1}u_{k+2} + C_k \tag{7.10.14}$$

for $k = n, n - 1, \ldots, 0$, with

$$u_{n+1} = u_{n+2} = 0 \tag{7.10.15}$$

then there follows

$$\sum_{k=0}^{n} C_k\phi_k(x) = \phi_0(x)u_0(x) + [\phi_1(x) - (a_0 x + b_0)\phi_0(x)]u_1(x) \tag{7.10.16}$$

In particular, if (7.10.3) holds when $k = 0$, with $\phi_{-1}(x) \equiv 0$, the right-hand side of (7.10.16) reduces to $\phi_0(x)u_0(x)$. (This situation could always be *made* to exist by suitably defining a_0 and/or b_0, but this is not always convenient.)

As an illustration, we consider the Chebyshev polynomials of Sec. 7.9, noticing that they are somewhat exceptional in that whereas they satisfy (7.9.11)

$$T_{k+1}(x) = 2xT_k(x) - T_{k-1}(x)$$

for $k = 1, 2, \ldots$, so that

$$a_k = 2 \qquad b_k = 0 \qquad c_k = -1 \qquad (k = 1, 2, \ldots)$$

the recurrence formula must take the form $T_1(x) = xT_0(x)$ when $k = 0$. Thus, if (for convenience) we also define

$$a_0 = 2 \qquad b_0 = 0 \qquad c_0 = -1$$

the recurrence formula is *not* satisfied when $k = 0$ with $T_{-1}(x) = 0$, in accordance with which the coefficient of $u_1(x)$ in (7.10.16) then does not vanish but becomes $-x$. Thus we deduce that

$$\sum_{k=0}^{n} C_k T_k(x) = u_0(x) - xu_1(x) \qquad (7.10.17)$$

where

$$u_k(x) = 2xu_{k+1}(x) - u_{k+2}(x) + C_k \qquad (7.10.18)$$

for $k = n, n - 1, \ldots, 0$, with

$$u_{n+1}(x) = u_{n+2}(x) = 0 \qquad (7.10.19)$$

This Chebyshev algorithm is associated with the name of Clenshaw [1955].

Before proceeding to a consideration of the use of orthogonal polynomials when *discrete* data are involved, it is desirable to establish certain analogies between integration and summation and to obtain certain special properties of the *binomial coefficient functions* and related functions, which play the same roles in summation and differencing as do the functions $1, x, \ldots, x^n$ in integration and differentiation.

7.11 Factorial Power Functions and Summation Formulas

The product $s(s - 1) \cdots (s - n + 1)$, where n is a positive integer, is often called the *factorial* nth *power* of s, and the notation

$$s^{(n)} = s(s - 1) \cdots (s - n + 1) \qquad (7.11.1)$$

is frequently used. It is related to the *binomial coefficient function* by the equation

$$\binom{s}{n} = \frac{s^{(n)}}{n!} \qquad (7.11.2)$$

In the more general case when n need not be a positive integer, (7.11.1) is generalized by the definition

$$s^{(n)} = \frac{\Gamma(s+1)}{\Gamma(s-n+1)} \qquad (7.11.3)$$

in accordance with which there follows, in particular,

$$s^{(0)} = 1 \qquad (7.11.4)$$

$$s^{(-1)} = \frac{1}{s+1} \qquad s^{(-2)} = \frac{1}{(s+1)(s+2)} \qquad (7.11.5)$$

and so forth.

In order to establish the usefulness of the notation (7.11.1) or (7.11.3), we notice that, from (7.11.3), there follows

$$\begin{aligned}\Delta_1 s^{(n)} &= \frac{\Gamma(s+2)}{\Gamma(s-n+2)} - \frac{\Gamma(s+1)}{\Gamma(s-n+1)} \\ &= \left(\frac{s+1}{s-n+1} - 1\right)\frac{\Gamma(s+1)}{\Gamma(s-n+1)} \\ &= n\frac{\Gamma(s+1)}{(s-n+1)\Gamma(s-n+1)} = n\frac{\Gamma(s+1)}{\Gamma(s-n+2)}\end{aligned}$$

or

$$\Delta_1 s^{(n)} = n s^{(n-1)} \qquad (7.11.6)$$

where Δ_1 denotes the forward-difference operator with unit spacing, acting on s, and where use is made of the fundamental property of the gamma function,†

$$\Gamma(u+1) = u\Gamma(u) \qquad (7.11.7)$$

Thus the factorial power $s^{(n)}$ is related to the operator Δ_1 just as the ordinary power x^n is related to the operator $\mathbf{D} \equiv d/dx$. In this connection, it is of interest to notice that Newton's forward-difference formula (4.3.5), with an error term, can be written in the form

$$f_s = f_0 + \frac{\Delta f_0}{1!}s^{(1)} + \frac{\Delta^2 f_0}{2!}s^{(2)} + \cdots + \frac{\Delta^n f_0}{n!}s^{(n)} + \frac{f^{(n+1)}(\xi)}{(n+1)!}s^{(n+1)} \qquad (7.11.8)$$

and is seen to be completely analogous to the Maclaurin expansion, with a remainder, expressed in the form

$$f(x) = f(0) + \frac{f'(0)}{1!}x + \frac{f''(0)}{2!}x^2 + \cdots + \frac{f^{(n)}(0)}{n!}x^n + \frac{f^{(n+1)}(\xi)}{(n+1)!}x^{n+1}$$

† It is often convenient to write $u!$ as an abbreviation for $\Gamma(u+1)$ even though u is not a positive integer.

We recall next that, from the telescoping of terms in the expansion

$$\sum_{s=M}^{N} \Delta_1 f_s = (f_{M+1} - f_M) + (f_{M+2} - f_{M+1}) + \cdots + (f_N - f_{N-1}) + (f_{N+1} - f_N)$$

it follows that

$$\sum_{s=M}^{N} \Delta_1 f_s = f_{N+1} - f_M \equiv f_s|_M^{N+1} \qquad (7.11.9)$$

This general relation is seen to be analogous to the relation

$$\int_a^b f'(x)\,dx = f(b) - f(a) = f(x)|_a^b$$

but careful notice should be taken of the fact that the limits on the right in (7.11.9) are not the same as those on the left. Thus, in particular, we may deduce from (7.11.9) and (7.11.6) the summation formula

$$\sum_{s=M}^{N} s^{(n)} = \left.\frac{s^{(n+1)}}{n+1}\right|_M^{N+1} \qquad (n \neq -1) \qquad (7.11.10)$$

which clearly corresponds to the integral formula

$$\int_a^b x^n\,dx = \left.\frac{x^{n+1}}{n+1}\right|_a^b \qquad (n \neq -1)$$

Since (7.11.8) permits any polynomial in s to be expressed as a linear combination of factorial powers of s, (7.11.10) then serves to effect the summation of that polynomial. In illustration, in order to express the sum

$$S_n \equiv 1 \cdot 3 + 2 \cdot 4 + \cdots + n(n+2)$$

in closed form, we could first obtain the relation

$$s^2 + 2s = 0 + 3s^{(1)} + s^{(2)}$$

by use of undetermined coefficients or by using (7.11.8), and then make the calculation

$$S_n = \sum_{s=1}^{n} [3s^{(1)} + s^{(2)}] = [\tfrac{3}{2}s^{(2)} + \tfrac{1}{3}s^{(3)}]_1^{n+1}$$
$$= \tfrac{3}{2}(n+1)n + \tfrac{1}{3}(n+1)n(n-1) = \tfrac{1}{6}n(n+1)(7+2n)$$

The summation could also be effected by making appropriate use of (5.8.4) or of the Euler-Maclaurin sum formula (5.8.12).

There exist a large number of useful identities involving either factorial power functions or, correspondingly, binomial coefficient functions. (See, for example, Probs. 46 and 47.) In particular, the relation

$$s^{(n)}(s-n)^{(k)} = s^{(n+k)} \qquad (7.11.11)$$

which follows immediately from the fact that the left-hand member is given by

$$[s(s-1)\cdots(s-n+1)][(s-n)(s-n-1)\cdots(s-n-k+1)]$$

is of frequent use.

In addition to the property

$$\binom{m}{n} = \binom{m}{m-n} \tag{7.11.12}$$

which follows immediately from the definition, many important relations are obtainable from the identity

$$\binom{m+p}{n} = \sum_{k=0}^{n} \binom{p}{k}\binom{m}{n-k} = \sum_{k=0}^{n} \binom{p}{n-k}\binom{m}{k} \tag{7.11.13}$$

when n is a nonnegative integer. In order to establish this relation, we multiply together the series expansions of $(1 + x)^p$ and $(1 + x)^m$, noticing that those series terminate when p and m are nonnegative integers and are absolutely convergent infinite series when $|x| < 1$ otherwise, to obtain the results

$$\begin{aligned}(1+x)^{m+p} &= \left[\sum_{j=0}^{\infty}\binom{m}{j}x^j\right]\left[\sum_{k=0}^{\infty}\binom{p}{k}x^k\right] = \sum_{j=0}^{\infty}\sum_{k=0}^{\infty}\binom{p}{k}\binom{m}{j}x^{k+j} \\ &= \sum_{n=0}^{\infty}\left[\sum_{k=0}^{n}\binom{p}{k}\binom{m}{n-k}\right]x^n \qquad (|x| < 1)\end{aligned}$$

But since also

$$(1+x)^{m+p} = \sum_{n=0}^{\infty}\binom{m+p}{n}x^n \qquad (|x| < 1)$$

and since the coefficient of x^n must be the same in these two forms, the first form of the desired result (7.11.13) follows. The second form results from interchanging p and m.

We may notice next that

$$\begin{aligned}\binom{-p}{q} &= \frac{(-p)(-p-1)\cdots(-p-q+1)}{q!} \\ &= (-1)^q\,\frac{(p+q-1)(p+q-2)\cdots p}{q!}\end{aligned}$$

and hence

$$\binom{-p}{q} = (-1)^q\binom{p+q-1}{q} \tag{7.11.14}$$

when q is a nonnegative integer. Hence we deduce from (7.11.13) and (7.11.14) the further relations

$$\binom{m-p}{n} = \sum_{k=0}^{n} (-1)^k \binom{p+k-1}{k}\binom{m}{n-k}$$
$$= \sum_{k=0}^{n} (-1)^{n+k} \binom{p+n-k-1}{n-k}\binom{m}{k} \qquad (7.11.15)$$

All these formulas can be expressed alternatively in terms of factorial power functions by making use of (7.11.2).

Finally, a general formula which will be needed in the sequel is the sum analogy to integration by parts. In order to derive the desired formula, we notice first that

$$\Delta_1 u_s v_s = u_{s+1}v_{s+1} - u_s v_s \equiv v_{s+1}(u_{s+1} - u_s) + u_s(v_{s+1} - v_s)$$
$$= u_s\,\Delta_1 v_s + v_{s+1}\,\Delta_1 u_s$$

Hence, by transposition and use of (7.11.9), we deduce a formula for *summation by parts* in the form

$$\sum_{s=M}^{N} u_s\,\Delta_1 v_s = u_s v_s|_M^{N+1} - \sum_{s=M}^{N} v_{s+1}\,\Delta_1 u_s \qquad (7.11.16)$$

Also, by replacing u_s by v_s and v_s by u_{s-1}, and transposing terms, we deduce an alternative form

$$\sum_{s=M}^{N} u_s\,\Delta_1 v_s = u_{s-1} v_s|_M^{N+1} - \sum_{s=M}^{N} v_s\,\Delta_1 u_{s-1} \qquad (7.11.17)$$

7.12 Polynomials Orthogonal over Discrete Sets of Points

For least-squares approximation to a function $F(x)$ over a discrete point set S_N, it is convenient to make use of a set of polynomials which are mutually orthogonal under *summation* over S_N with respect to a specified weighting function. We suppose that $N + 1$ points are to be employed in the approximation, with uniform separation h, and that the extreme points are at the ends of the interval $[a, b]$, where $b - a = Nh$. If we then write

$$x = a + sh \qquad (7.12.1)$$

the variable s takes on the values $s = 0, 1, 2, \ldots, N$ at those points, and we seek a set of polynomials $\phi_0(s, N), \phi_1(s, N), \ldots, \phi_N(s, N)$ such that ϕ_r is of degree r in s, and such that

$$\sum_{s=0}^{N} w(s)\phi_r(s, N)q_{r-1}(s) = 0 \qquad (7.12.2)$$

where $w(s)$ is a specified weighting function, assumed to be nonnegative on S_N, and where $q_{r-1}(s)$ is an *arbitrary* polynomial of degree $r - 1$ or less.

The procedure is analogous to that employed in Sec. 7.5. We first set

$$w(s)\phi_r(s, N) = \Delta_1^r U_r(s, N) \qquad (7.12.3)$$

so that (7.12.2) becomes

$$\sum_{s=0}^{N} [\Delta_1^r U_r(s, N)] q_{r-1}(s) = 0 \qquad (7.12.4)$$

and sum by parts r times, noticing that $\Delta_1^r q_{r-1}(s) \equiv 0$, to transform (7.12.4) to the relation

$$\begin{aligned}[\{\Delta_1^{r-1} U_r(s)\} q_{r-1}(s) - \{\Delta_1^{r-2} U_r(s+1)\}\, \Delta_1 q_{r-1}(s) + \cdots \\ + (-1)^{r-1}\{U_r(s+r-1)\}\, \Delta_1^{r-1} q_{r-1}(s)]_{s=0}^{s=N+1} = 0 \qquad (7.12.5)\end{aligned}$$

Since we require that ϕ_r be a polynomial of degree r, it follows from (7.12.3) that U_r must satisfy the *difference equation*

$$\Delta_1^{r+1}\left[\frac{1}{w(s)}\, \Delta_1^r U_r(s, N)\right] = 0 \qquad (7.12.6)$$

on S_N, and that (7.12.5) will be satisfied for arbitrary values of $q_{r-1}(s)$, $\Delta_1 q_{r-1}(s)$, and so forth, when $s = 0$ and when $s = N + 1$ if $U_r(s + r - 1, N)$, $\Delta_1 U_r(s + r - 2, N), \ldots,$ and $\Delta_1^{r-1} U_r(s, N)$ vanish when $s = 0$ and when $s = N + 1$. It is easily seen that these requirements are equivalent to the $2r$ conditions

$$U_r(0, N) = U_r(1, N) = U_r(2, N) = \cdots = U_r(r-1, N) = 0 \qquad (7.12.7)$$

and

$$U_r(N+1, N) = U_r(N+2, N) = U_r(N+3, N) = \cdots = U_r(N+r, N) = 0 \qquad (7.12.8)$$

Once U_r has been determined, necessarily with an arbitrary multiplicative constant, there follows, from (7.12.3),

$$\phi_r(s, N) = \frac{1}{w(s)}\, \Delta_1^r U_r(s, N) \qquad (7.12.9)$$

In consequence of the results of Sec. 7.2, the coefficients in the relation

$$y = \sum_{r=0}^{n} a_r \phi_r(s, N) \qquad (7.12.10)$$

are then determined by the requirement

$$\sum_{s=0}^{N} w(s)\left[F(a + sh) - \sum_{r=0}^{n} a_r \phi_r(s, N)\right]^2 = \min \qquad (7.12.11)$$

in the form

$$a_r = \frac{1}{\gamma_r(N)} \sum_{s=0}^{N} w(s)F(a + sh)\phi_r(s, N) \qquad (7.12.12)$$

where

$$\gamma_r(N) = \sum_{s=0}^{N} w(s)\phi_r^2(s, N) \qquad (7.12.13)$$

7.13 Gram Approximation

We restrict attention here to the case when $w(s) \equiv 1$. Equation (7.12.6) then requires merely that $U_r(s, N)$ be a polynomial in s of degree $2r$ and, since (7.12.7) and (7.12.8) determine its $2r$ zeros, there follows immediately

$$U_r(s, N) = C_{rN}[s(s-1)\cdots(s-r+1)] \\ \times [(s-N-1)(s-N-2)\cdots(s-N-r)]$$

or

$$U_r(s, N) = C_{rN}s^{(r)}(s-N-1)^{(r)} \qquad (7.13.1)$$

where C_{rN} is an arbitrary constant. Hence we have also

$$\phi_r(s, N) = C_{rN}\, \Delta_1^r[s^{(r)}(s-N-1)^{(r)}] \qquad (7.13.2)$$

In order to express this result in a more explicit form, we first expand $(s-N-1)^{(r)}$ in terms of factorial powers of $(s-r)$, then use (7.11.11) to express (7.13.1) in terms of factorial powers of s, and, finally, make use of (7.11.6). Thus, if we use the second form of (7.11.15), we obtain

$$(s-N-1)^{(r)} = r!\binom{(s-r)-(N+1-r)}{r} \\ = (-1)^r r! \sum_{k=0}^{r} (-1)^k \binom{N-k}{r-k}\binom{s-r}{k} \qquad (7.13.3)$$

and hence, by virtue of (7.13.1) and (7.11.11),

$$U_r(s, N) = (-1)^r r!\, C_{rN} \sum_{k=0}^{r} \frac{(-1)^k}{k!}\binom{N-k}{r-k} s^{(k+r)} \qquad (7.13.4)$$

Thus, by making use of (7.11.6), we obtain the result

$$\phi_r(s, N) = (-1)^r r!\, C_{rN} \sum_{k=0}^{r} (-1)^k \frac{(r+k)^{(r)}}{k!}\binom{N-k}{r-k} s^{(k)} \qquad (7.13.5)$$

which can be transformed to the more convenient form

$$\phi_r(s, N) = c_{rN} \sum_{k=0}^{r} (-1)^k \frac{(r+k)^{(2k)}}{(k!)^2} \frac{s^{(k)}}{N^{(k)}} \qquad (7.13.6)$$

where c_{rN} has been written for the arbitrary constant $(-1)^r r!\, N^{(r)} C_{rN}$. The expanded form appears as follows:

$$\phi_r(s, N) = c_{rN}\left[1 - \frac{r(r+1)}{(1!)^2}\frac{s}{N} + \frac{(r-1)r(r+1)(r+2)}{(2!)^2}\frac{s(s-1)}{N(N-1)} - \frac{(r-2)(r-1)r(r+1)(r+2)(r+3)}{(3!)^2}\frac{s(s-1)(s-2)}{N(N-1)(N-2)} + \cdots\right] \tag{7.13.7}$$

In most applications of least-squares methods, it is convenient to make use of an *odd* number of ordinates, so that N is *even*, and to write

$$N = 2M$$

In such cases, it is also convenient to make the change of variables

$$s = M + t \tag{7.13.8}$$

so that t represents distance from the *midpoint* of the set S_N in units of the spacing h (see Fig. 7.1) and takes on the values $0, \pm 1, \pm 2, \ldots, \pm M$ at the

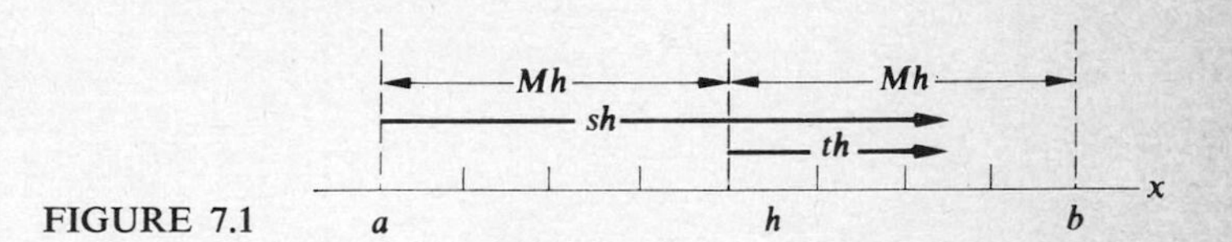

FIGURE 7.1

$2M + 1$ points of S_N. If also we choose

$$c_{rN} = (-1)^r \tag{7.13.9}$$

and write $p_r(t, 2M) \equiv \phi_r(s, 2M)$, the polynomials of degrees zero through five can be expressed explicitly as follows:

$$\begin{aligned}
p_0(t, 2M) &= 1 \\
p_1(t, 2M) &= \frac{t}{M} \\
p_2(t, 2M) &= \frac{3t^2 - M(M+1)}{M(2M-1)} \\
p_3(t, 2M) &= \frac{5t^3 - (3M^2 + 3M - 1)t}{M(M-1)(2M-1)} \\
p_4(t, 2M) &= \frac{35t^4 - 5(6M^2 + 6M - 5)t^2 + 3M(M^2-1)(M+2)}{2M(M-1)(2M-1)(2M-3)} \\
p_5(t, 2M) &= \\
&\frac{63t^5 - 35(2M^2 + 2M - 3)t^3 + (15M^4 + 30M^3 - 35M^2 - 50M + 12)t}{2M(M-1)(M-2)(2M-1)(2M-3)}
\end{aligned} \tag{7.13.10}$$

These polynomials thus possess the orthogonality property

$$\sum_{t=-M}^{M} p_i(t, 2M)p_j(t, 2M) = 0 \qquad (i \neq j)$$

and are usually known as *Gram polynomials* (or *Chebyshev polynomials,* although the latter name is usually reserved for either the polynomials considered in Sec. 7.9 or those to be considered in Sec. 8.14). It may be seen that the rth polynomial is an even function of t when r is even, and an odd function of t when r is odd. Also, each polynomial takes on the value *unity* when $t = M$. Further, if t is replaced by Mx and M is then increased without limit, it can be verified that $p_r(t, 2M)$ tends to the rth Legendre polynomial $P_r(x)$:

$$\lim_{M \to \infty} p_r(Mx, 2M) = P_r(x)$$

In accordance with the results of the preceding section, the nth-degree least-squares polynomial approximation to the function

$$f(t) \equiv F[\tfrac{1}{2}(a + b) + th] \qquad (7.13.11)$$

over the $(2M + 1)$-point set $t = -M, -M + 1, \ldots, -1, 0, 1, \ldots, M - 1, M$ is given by

$$y = \sum_{r=0}^{n} a_r p_r(t, 2M) \qquad (7.13.12)$$

where

$$a_r = \frac{1}{\gamma_r} \sum_{t=-M}^{M} f(t)p_r(t, 2M) \qquad (7.13.13)$$

and

$$\gamma_r = \sum_{t=-M}^{M} p_r^2(t, 2M) \qquad (7.13.14)$$

As in the earlier developments, the factors γ_r are independent of the function $f(t)$ which is to be approximated and can be calculated once and for all.

It should be noted that various conventions are adopted in the literature with regard to the value assigned to the arbitrary multiplicative constant c_{rN} in (7.13.7) in the general case. In particular, that constant sometimes is so defined that the coefficient of s^r in $\phi_r(s, N)$ is unity. Another choice is that for which the values taken on when $s = 0, 1, \ldots, N$ are *integers* without a common factor, so that tabulation is simplified. When $N + 1$ points are used, the sum of the squares of the $N + 1$ tabular values of the rth-degree polynomial corresponding to the normalization (7.13.9) used here is found to be

$$\gamma_r = \frac{(N + r + 1)!\,(N - r)!}{(2r + 1)(N!)^2} \qquad (7.13.15)$$

(see Prob. 53), whereas (7.13.7) shows that the leading coefficient in $\phi_r(s, N)$ is

$$A_r = \frac{(2r)!\,(N-r)!}{(r!)^2 N!} \tag{7.13.16}$$

These results permit tabulations relevant to other normalizations to be interpreted in terms of the one used here.

7.14 Example: Five-Point Least-Squares Approximation

In order to illustrate a method of using the preceding results, we consider here the case in which only five ordinates are used, so that $M = 2$. The relevant orthogonal polynomials of degrees zero through four are then obtained from (7.13.10) in the forms

$$\begin{aligned} p_0(t) &= 1 & p_1(t) &= \tfrac{1}{2}t \\ p_2(t) &= \tfrac{1}{2}(t^2 - 2) & p_3(t) &= \tfrac{1}{6}(5t^3 - 17t) \\ p_4(t) &= \tfrac{1}{12}(35t^4 - 155t^2 + 72) \end{aligned} \tag{7.14.1}$$

if we write $p_r(t)$ for $p_r(t, 4)$.

We may notice that $p_5(t)$, as defined by (7.13.10), is nonexistent when $M = 2$. This situation corresponds to the fact that the use of polynomials of degrees zero through five over five points would not lead to a determinate problem, since infinitely many fifth-degree polynomials would fit the data *exactly* at those points. Further, the use of polynomials of degrees zero through four over five points truly would not be a least-squares procedure since it necessarily would lead to the fourth-degree interpolation polynomial which fits the data exactly. Thus $p_4(t)$ is not needed when five points are used unless an exact fit at those points is desired, in which case the use of one of the methods given in earlier chapters usually is to be preferred.

Values of the polynomials at the five relevant points may be tabulated as in Table 7.1. According to (7.13.12) to (7.13.14), the coefficient a_r of each $p_r(t)$

Table 7.1

t	p_0	p_1	p_2	p_3	f
−2	1	−1	1	−1	f_{-2}
−1	1	−0.5	−0.5	2	f_{-1}
0	1	0	−1	0	f_0
1	1	0.5	−0.5	−2	f_1
2	1	1	1	1	f_2
$\gamma_r =$	5	2.5	3.5	10	

used in the approximation is obtained by multiplying each entry in its column by the corresponding entry in the column of values of $f(t)$, summing, and dividing the result by γ_r, which is listed at the foot of the p_r column. Once the a's are calculated, the least-squares polynomial $y(t)$ can be obtained *explicitly* by forming the corresponding combination of the polynomials listed in (7.14.1). If only the value of $y(t)$ at a tabular point is required, this explicit form of $y(t)$ is not needed, since the required value is obtained by merely multiplying the tabulated value of each p_r for that t by a_r, and summing the results.

In illustration, suppose that we are provided with the empirical data

x	0.0	0.2	0.4	0.6	0.8
$F(x)$	1.10	1.78	2.74	4.12	5.69

together with some assurance that the observed values are in error by no more than a few units in the last place given, and that the true function is "smooth." In order to obtain least-squares polynomial approximations by use of Table 7.1, we then set $x = 0.4 + 0.2t$, or $t = 5x - 2$, and write $F(0.4 + 0.2t) = f(t)$. Calculation then gives

$$a_0 \doteq 3.086 \qquad a_1 \doteq 2.304 \qquad a_2 \doteq 0.314 \qquad a_3 \doteq -0.009$$

so that least-squares approximations of degrees 1, 2, and 3 are obtained by retaining two, three, or four terms in the relation

$$f(t) \approx 3.086p_0(t) + 2.304p_1(t) + 0.314p_2(t) - 0.009p_3(t)$$

The corresponding smoothed values at the tabular points may be obtained, from Table 7.1, as follows:

t	-2	-1	0	1	2
f	1.10	1.78	2.74	4.12	5.69
y_3	1.105	1.759	2.772	4.099	5.695
y_2	1.096	1.777	2.772	4.081	5.704
y_1	0.782	1.934	3.086	4.238	5.390

The RMS value of the five deviations from the observed values is found to be 0.0198 for the third-degree approximation, 0.0235 for the second-degree approximation, and 0.264 for the linear approximation. The use of (7.3.34) then leads to corresponding estimates of 0.0443, 0.0372, and 0.341, respectively, for the RMS error in the observed values. Clearly, only the first two of these estimates are in accord with the given information.

If the smaller of these estimates is accepted as the more appropriate one, we may conclude that the additional smoothing afforded by the use of a parabolic approximation, in place of a cubic, probably represents a further removal of "noise" rather than a departure from the unknown true function. The use of a *linear* approximation could not be so justified.

If additional values of the least-squares polynomial are desired, they may be obtained conveniently by interpolation. However, if the *equation* of the parabolic approximation is required, it may be written down in the form

$$\begin{aligned} y &= 3.086p_0(t) + 2.304p_1(t) + 0.314p_2(t) \\ &= 2.772 + 1.152t + 0.157t^2 \end{aligned}$$

and reduced, if so desired, to the form

$$y = 1.096 + 2.620x + 3.925x^2$$

In particular, this result supplies the approximations 2.62, 5.76, and 8.90 to the *slope* of the unknown function at $x = 0.0, 0.4$, and 0.8, respectively, whereas the third-degree approximation would yield the values 2.30, 5.89, and 8.58. On the other hand, the result of differentiating the fourth-degree *interpolation* polynomial, which takes on the five observed values exactly, would give the respective values 3.40, 5.89, and 7.48.

By expressing the a's explicitly in terms of the observed values, we may obtain formulas which express the smoothed values directly in terms of the observed ones. Thus, corresponding to the *third*-degree least-squares approximation over five points, we obtain the formula

$$\begin{aligned} y_0 = a_0 - a_2 = \tfrac{1}{5}(f_{-2} + f_{-1} + f_0 + f_1 + f_2) \\ - \tfrac{2}{7}(f_{-2} - \tfrac{1}{2}f_{-1} - f_0 - \tfrac{1}{2}f_1 + f_2) \end{aligned}$$

or

$$y_0 = \tfrac{1}{35}(-3f_{-2} + 12f_{-1} + 17f_0 + 12f_1 - 3f_2) \qquad (7.14.2)$$

for the smoothed value at the midpoint $t = 0$, and the formulas

$$\begin{aligned} y_{-2} &= \tfrac{1}{70}(69f_{-2} + 4f_{-1} - 6f_0 + 4f_1 - f_2) \\ y_{-1} &= \tfrac{1}{35}(2f_{-2} + 27f_{-1} + 12f_0 - 8f_1 + 2f_2) \\ y_1 &= \tfrac{1}{35}(2f_{-2} - 8f_{-1} + 12f_0 + 27f_1 + 2f_2) \\ y_2 &= \tfrac{1}{70}(-f_{-2} + 4f_{-1} - 6f_0 + 4f_1 + 69f_2) \end{aligned} \qquad (7.14.3)$$

are obtained in a similar way. It is of interest to notice that these formulas can also be expressed in the compact forms

$$\begin{aligned} y_{-2} &= f_{-2} - \tfrac{1}{70}\delta^4 f_0 \qquad & y_{-1} &= f_{-1} + \tfrac{2}{35}\delta^4 f_0 \\ & & y_0 &= f_0 - \tfrac{3}{35}\delta^4 f_0 \\ y_1 &= f_1 + \tfrac{2}{35}\delta^4 f_0 \qquad & y_2 &= f_2 - \tfrac{1}{70}\delta^4 f_0 \end{aligned} \qquad (7.14.4)$$

The simplicity of these last forms is due to the fact that the degree of the least-squares polynomial is exactly *one* less than the degree of the polynomial which would be uniquely determined by the five data. In the cases when this difference exceeds unity, the formulas are less simply expressed in terms of differences, particularly for off-center points, as will be seen.

Explicit formulas which avoid the necessity of effecting the summations may also be obtained by first resolving the relations (7.14.1) in the forms

$$
\begin{aligned}
1 &= p_0(t) \qquad & t &= 2p_1(t) \\
t^2 &= 2p_0(t) + 2p_2(t) \qquad & t^3 &= \tfrac{1}{5}[34p_1(t) + 6p_3(t)] \\
t^4 &= \tfrac{1}{35}[238p_0(t) + 310p_2(t) + 12p_4(t)]
\end{aligned}
\tag{7.14.5}
$$

Now the interpolation polynomial (of degree 4) which agrees *exactly* with $f(t)$ when $t = 0, \pm 1$, and ± 2 can be expressed in the Stirling form (Sec. 4.5)

$$f_0 + (\mu\delta f_0)t + (\tfrac{1}{2}\delta^2 f_0)t^2 + (\tfrac{1}{6}\mu\delta^3 f_0)(t^3 - t) + (\tfrac{1}{24}\delta^4 f_0)(t^4 - t^2) \tag{7.14.6}$$

and hence, by introducing (7.14.5) into (7.14.6), we obtain the relation

$$
\begin{aligned}
f(t) = (f_0 + \delta^2 f_0 + \tfrac{1}{5}\delta^4 f_0)p_0(t) + (2\mu\delta f_0 + \tfrac{4}{5}\mu\delta^3 f_0)p_1(t) \\
+ (\delta^2 f_0 + \tfrac{2}{7}\delta^4 f_0)p_2(t) + (\tfrac{1}{5}\mu\delta^3 f_0)p_3(t) + (\tfrac{1}{70}\delta^4 f_0)p_4(t)
\end{aligned}
\tag{7.14.7}
$$

when t is restricted to the values 0, ± 1, and ± 2. For other values of t, the right-hand member represents the fourth-degree interpolation polynomial which coincides with $f(t)$ at those five points. The associated error $E(t)$ can be expressed in the familiar form

$$E(t) = \frac{1}{5!}\, t(t^2 - 1)(t^2 - 4)f^{\text{v}}(\tau) \qquad (|\tau| < 2)$$

when $f^{\text{v}}(t)$ exists and is continuous for $-2 \leqq t \leqq 2$.

Since the right-hand member of (7.14.7) accordingly is the polynomial which would be afforded by fourth-degree least squares, over the five points involved, and since the *coefficients* of the p's are independent of the number of p's retained, it follows that the *third*-degree least-squares polynomial relative to those points is then obtained by deleting the term involving $p_4(t)$. In particular, when attention is restricted to the five points themselves, the resultant formula can be expressed in the form

$$y(t) = f(t) - (\tfrac{1}{70}\delta^4 f_0)p_4(t) \qquad (t = 0, \pm 1, \pm 2) \tag{7.14.8}$$

in accordance with (7.14.4). Similarly, the first-degree least-squares polynomial relevant to five points may be obtained by retaining only $p_0(t)$ and $p_1(t)$ in the right-hand member of (7.14.7).

The methods of this section are readily generalized to cases in which more than five points are used in the least-squares calculation.

7.15 Smoothing Formulas

In place of approximating $f(t)$ by a single least-squares polynomial of degree n over the entire range of an extensive tabulation, it is frequently desirable to replace each tabulated value by the value taken on by a least-squares polynomial of degree n relevant to a subrange of $2M + 1$ points centered, if possible, at the point for which the entry is to be modified. Thus, except for points near the ends of the range of tabulation, each smoothed value is obtained from a distinct least-squares polynomial. In this section we list certain sets of smoothing formulas which are obtainable for this purpose by the methods of the preceding section.

For *first-degree* least-squares approximation relevant to *three points*, the formulas are of the form

$$\begin{aligned} y_{-1} &= \tfrac{1}{6}(5f_{-1} + 2f_0 - f_1) \equiv f_{-1} - \tfrac{1}{6}\delta^2 f_0 \\ y_0 &= \tfrac{1}{3}(f_{-1} + f_0 + f_1) \equiv f_0 + \tfrac{1}{3}\delta^2 f_0 \\ y_1 &= \tfrac{1}{6}(-f_{-1} + 2f_0 + 5f_1) \equiv f_1 - \tfrac{1}{6}\delta^2 f_0 \end{aligned} \qquad (7.15.1)$$

whereas the formulas relevant to *five points* are

$$\begin{aligned} y_{-2} &= \tfrac{1}{5}(3f_{-2} + 2f_{-1} + f_0 - f_2) \\ y_{-1} &= \tfrac{1}{10}(4f_{-2} + 3f_{-1} + 2f_0 + f_1) \\ y_0 &= \tfrac{1}{5}(f_{-2} + f_{-1} + f_0 + f_1 + f_2) \\ &\cdots\cdots\cdots\cdots\cdots\cdots \end{aligned} \qquad (7.15.2)$$

where the omitted formulas for y_1 and y_2 are obtained from the formulas for y_{-1} and y_{-2} by reversing the numbering of the ordinates. Thus, for example, if first-degree five-point least squares were to be used, the central formula would be used for all values except the first two and the last two, for which the off-center formulas would be used.

The formulas for *third-degree five-point* least squares were obtained in the preceding section and are listed again, for convenient reference, in the forms

$$\begin{aligned} y_{-2} &= \tfrac{1}{70}(69f_{-2} + 4f_{-1} - 6f_0 + 4f_1 - f_2) \equiv f_{-2} - \tfrac{1}{70}\delta^4 f_0 \\ y_{-1} &= \tfrac{1}{35}(2f_{-2} + 27f_{-1} + 12f_0 - 8f_1 + 2f_2) \equiv f_{-1} + \tfrac{2}{35}\delta^4 f_0 \\ y_0 &= \tfrac{1}{35}(-3f_{-2} + 12f_{-1} + 17f_0 + 12f_1 - 3f_2) \equiv f_0 - \tfrac{3}{35}\delta^4 f_0 \\ &\cdots\cdots\cdots\cdots\cdots\cdots \end{aligned} \qquad (7.15.3)$$

whereas the corresponding *seven-point* formulas are

$$\begin{aligned}
y_{-3} &= \tfrac{1}{42}(39f_{-3} + 8f_{-2} - 4f_{-1} - 4f_0 + f_1 + 4f_2 - 2f_3) \\
y_{-2} &= \tfrac{1}{42}(8f_{-3} + 19f_{-2} + 16f_{-1} + 6f_0 - 4f_1 - 7f_2 + 4f_3) \\
y_{-1} &= \tfrac{1}{42}(-4f_{-3} + 16f_{-2} + 19f_{-1} + 12f_0 + 2f_1 - 4f_2 + f_3) \\
y_0 &= \tfrac{1}{21}(-2f_{-3} + 3f_{-2} + 6f_{-1} + 7f_0 + 6f_1 + 3f_2 - 2f_3)
\end{aligned} \tag{7.15.4}$$

. . . .

Finally, the *fifth-degree seven-point* least-squares formulas may be listed as follows:

$$\begin{aligned}
y_{-3} &= \tfrac{1}{924}(923f_{-3} + 6f_{-2} - 15f_{-1} + 20f_0 - 15f_1 + 6f_2 - f_3) \\
&\equiv f_{-3} - \tfrac{1}{924}\delta^6 f_0 \\
y_{-2} &= \tfrac{1}{154}(f_{-3} + 148f_{-2} + 15f_{-1} - 20f_0 + 15f_1 - 6f_2 + f_3) \\
&\equiv f_{-2} + \tfrac{1}{154}\delta^6 f_0 \\
y_{-1} &= \tfrac{1}{308}(-5f_{-3} + 30f_{-2} + 233f_{-1} + 100f_0 \\
&\qquad - 75f_1 + 30f_2 - 5f_3) \\
&\equiv f_{-1} - \tfrac{5}{308}\delta^6 f_0 \\
y_0 &= \tfrac{1}{231}(5f_{-3} - 30f_{-2} + 75f_{-1} + 131f_0 + 75f_1 - 30f_2 + 5f_3) \\
&\equiv f_0 + \tfrac{5}{231}\delta^6 f_0
\end{aligned} \tag{7.15.5}$$

. . . .

The use of an nth-degree least-squares polynomial relevant to $2M + 1$ points essentially assumes that the true function can be approximated by some nth-degree polynomial over each subrange of $2M + 1$ points, but it admits the possibility that no single nth-degree polynomial may be satisfactory over the entire range. The amount of smoothing *increases* with the number of *points* used in the smoothing formula and *decreases* with increasing values of the *degree* n.

It is often desirable and convenient to employ a smoothing technique involving a relatively small number of points, so that the relevant formulas are of simple form, and to iterate the process as many times as appears to be desirable. The degree n is chosen to be as small as possible, in consistency with the assumption that differences of the *true* function, of order higher than n, are small. If such a process were iterated indefinitely, the sequence of smoothed functions would tend to the least-squares polynomial of degree n relevant to the *entire* range of tabulated values. The analyst can and generally must rely upon his judgment with regard to the stage at which the iteration is to be terminated, so that most of the noise is eliminated but essential characteristics of the function are not appreciably modified. The choice of n is often dictated by the fact that the first n differences of the observed function f are fairly regular, whereas the $(n + 1)$th differences fluctuate erratically and have a mean value near zero.

As an illustration, we consider the data listed in the second column of Table 7.2. A plot of the given data suggests that, whereas the true function is almost certainly not linear, it can be fairly approximated by a linear function

Table 7.2†

x	$f(x)$	5-point once	5-point twice	Spencer	W and R
0	431	402	405		419
1	409	423	422		422
2	429	444	439		435
3	422	459	456		454
4	530	469	472		473
5	505	483	485		487
6	459	504	499		496
7	499	510	516		508
8	526	527	536		526
9	563	554	557		550
10	587	584	585	582	578
11	595	612	616	614	610
12	647	649	650	648	646
13	669	683	684	682	685
14	746	720	720	716	724
15	760	756	752	749	758
16	778	792	784		787
17	828	810	815		812
18	846	841	847		837
19	836	876	880		868
20	916	914	922		910
21	956	960	966		961
22	1014	1019	1012		1016
23	1076	1061	1060		1069
24	1134	1106	1107		1112
25	1124	1152	1154		1141

† These data were taken from Spencer [1904] and have been analyzed in various ways by Spencer, by Whittaker and Robinson [1944], and others.

over any subrange of, say, three or five points. The smoothed data given by the first-degree five-point formulas of (7.15.2) are listed in the third column of the table. Each smoothed value except the two values at each end of the tabulation is obtained very simply as the average of the five values centered at the point considered. Off-center formulas are used for those points. A second application of this process leads to the values listed in the fourth column and is represented

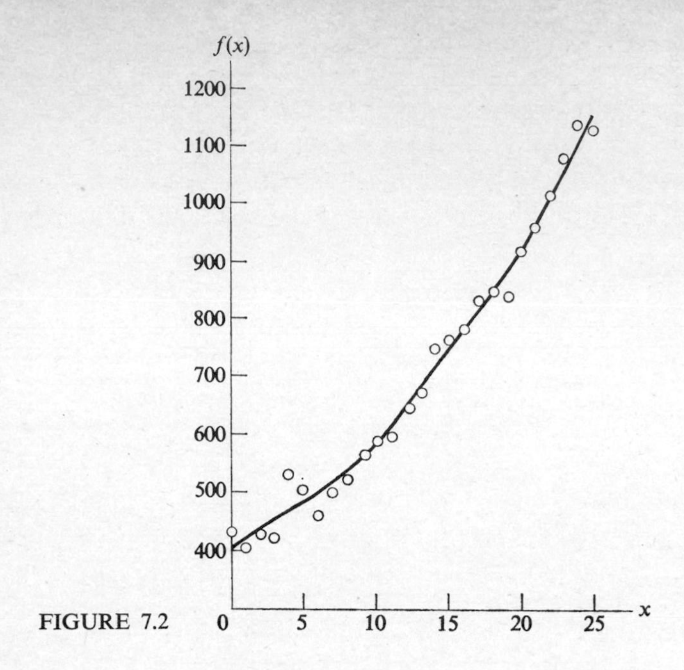

FIGURE 7.2

by a continuous curve in Fig. 7.2. A quantitative estimate of the degree of smoothing is afforded by the fact that the means of the absolute values of the second and third differences of the given data are 41 and 75, respectively, whereas the corresponding means for the results of the second smoothing are 2.1 and 2.6, respectively. At the same time, it appears that the characteristic trend of the data is preserved in the smoothing. The results of applying the first-degree three-point formulas of (7.15.1) *three* times are found to be quite similar to the results of using the five-point formulas twice in the present example.

In the fifth column of Table 7.2 are listed the results obtained by Spencer by use of an elaborate 21-point formula which yields smoothed values only at points which are more than 10 intervals away from the ends. The sixth column of the table lists results obtained by Whittaker and Robinson, by use of another 21-point formula combined with an appreciable amount of auxiliary calculation relevant to the smoothing of the first and last 10 entries. Whereas the smoothed values generally do not differ appreciably from those obtained (much more simply) in the fourth column, the advantage in smoothness actually belongs to the results of the simpler method, in the sense that the mean absolute second and third differences relevant to the data of the sixth column are found to be 5.2 and 3.4, respectively, as compared with 2.1 and 2.6 for the data of the fourth column.

It should be emphasized, however, that the smallness of certain mean absolute differences cannot *in itself* be taken as an indication of a satisfactory smoothing. By repeating the smoothing which led to column three indefinitely often, we eventually would be led to a "smoothed curve" which is represented by a *straight line* over the entire range, and hence for which *all* differences of order greater than one would *vanish.* This linear approximation would be obtained *directly* by use of first-degree 26-point formulas.

It is conceivable, of course, that the deviation from linearity of the smoothed curve is *still* predominantly noise and that a much more drastic smoothing is indeed called for. It is at this point that the judgment of the analyst (or the weight of additional evidence) must be brought into play.

As a further example, a plot of the data

x	0	1	2	3	4	5	6	7	8
$f(x)$	54	145	227	359	401	342	259	112	65

(see Fig. 7.3) suggests that the true function can be approximated by a third-

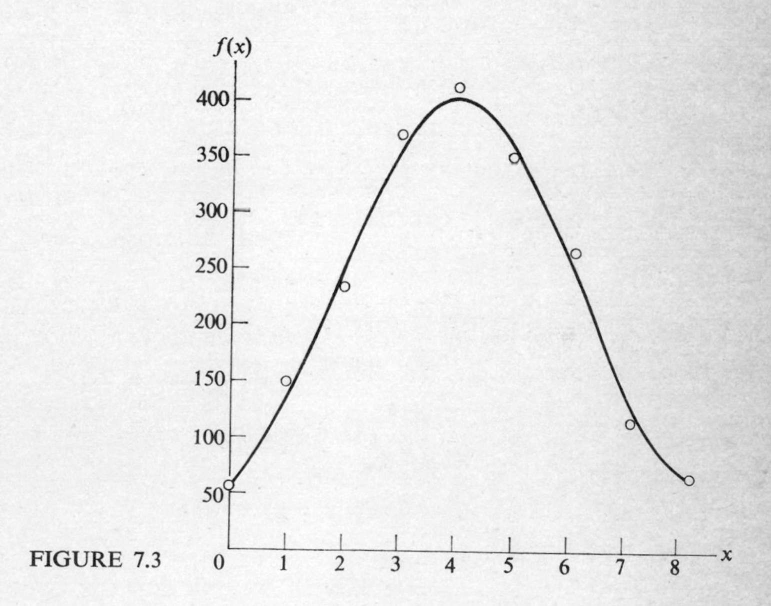

FIGURE 7.3

degree polynomial over each subrange of five points. The use of the formulas of (7.15.3) yields the smoothed values

x	0	1	2	3	4	5	6	7	8
$y(x)$	57	134	244	348	393	352	242	124	62

which are plotted and joined by a continuous curve in Fig. 7.3.

Whereas it is possible to determine a set of orthogonal polynomials over a discrete set of points relative to a specified weighting function w (see also Secs. 7.16 and 9.5) and derive corresponding smoothing formulas, a more convenient procedure which tends to accomplish about the same purpose, when w does not vary excessively, consists of applying the preceding smoothing formulas to the product wf and then dividing the result by w. More generally, the function f may be first transformed in an appropriate way to a new function g, and the new function g may be smoothed, after which the inverse transformation may be applied to the smoothed function. In particular, in the case of the last preceding example, the graph of the function $f(x)$ (Fig. 7.3) indicates a resemblance to a function of the form $\exp[-(Ax^2 + Bx + C)]$ and suggests that the smoothing be applied to $\log f(x)$, rather than to $f(x)$ itself.

Finally, it may be pointed out that the *central* smoothing formulas can be obtained rather simply without explicitly determining the least-squares polynomials involved. In this connection, we notice that the orthogonal polynomials $p_r(t, 2M)$ defined in Sec. 7.13 vanish at $t = 0$ when r is odd. Hence, if n is *even*, the central smoothing formula corresponding to a least-squares polynomial of degree n will be identical with the formula corresponding to that of degree $n + 1$.

For $n = 0$ or $n = 1$, there follows merely $y_0 = a_0$ and hence, since $p_0(t, 2M) = 1$,

$$y_0 = \frac{1}{2M+1} \sum_{r=-M}^{M} f_r \tag{7.15.6}$$

Thus, as in the special cases of (7.15.1) and (7.15.2), each smoothed value of f_0 is the average of the $2M + 1$ values centered about f_0. For $n = 2$ or $n = 3$, reference to (7.13.10) gives

$$y_0 = a_0 + a_2 p_2(0, 2M) = a_0 - \frac{M+1}{2M-1} a_2 \tag{7.15.7}$$

where

$$a_0 = \frac{1}{2M+1} \sum_{r=-M}^{M} f_r \tag{7.15.8}$$

and

$$a_2 = \frac{1}{\gamma_2} \frac{1}{M(2M-1)} \sum_{r=-M}^{M} [3r^2 - M(M+1)] f_r \tag{7.15.9}$$

with

$$\gamma_2 = \sum_{r=-M}^{M} p_2^2(r, 2M) \tag{7.15.10}$$

The value of γ_2 is given by (7.13.15) with $N = 2M$ and $r = 2$

$$\gamma_2 = \frac{(2M + 1)(2M + 2)(2M + 3)}{10M(2M - 1)} \tag{7.15.11}$$

and the insertion of (7.15.8), (7.15.9), and (7.15.11) into (7.15.7) leads immediately to the required formula

$$y_0 = \frac{3}{(4M^2 - 1)(2M + 3)} \sum_{r=-M}^{M} [(3M^2 + 3M - 1) - 5r^2]f_r \tag{7.15.12}$$

which specializes to the central formulas of (7.15.3) and (7.15.4) when $M = 2$ and $M = 3$, respectively. A similar analysis leads to the central smoothing formula

$$y_0 = \frac{15}{4(4M^2 - 1)(4M^2 - 9)(2M + 5)} \sum_{r=-M}^{M} [(15M^4 + 30M^3 - 35M^2 - 50M + 12) - 35(2M^2 + 2M - 3)r^2 + 63r^4]f_r \tag{7.15.13}$$

relevant to *fourth-* or *fifth-degree* least-square approximation using $2M + 1$ points, which specializes to the central formula of (7.15.5) when $M = 3$.†

As was pointed out earlier, the central smoothing formulas alone are generally useful only for smoothing values at points at least M intervals distant from the ends of the range of tabulation. However, they can be used throughout the entire range in the special cases when the true function is known to vanish outside the range of tabulation and to tend to zero smoothly as the ends of the range are approached from the interior, so that the zero values at exterior points can be used in smoothing values at interior points near the ends.

7.16 Recursive Computation of Orthogonal Polynomials on Discrete Sets of Points

Results analogous to those relevant to the Gram approximation can be obtained in other cases by similar methods. However, the determination of the associated orthogonal polynomials usually is more conveniently effected by recursive methods, particularly in those cases when the data points are not equally spaced. In this section, the necessary formulas are exhibited.

Here we denote by $\phi_r(s)$ the rth member of the set of polynomials which are orthogonal relative to a certain positive weighting function $w(s)$ over a discrete set S_N of $N + 1$ points, which are not necessarily uniformly spaced.

† The central formulas (7.15.12) and (7.15.13) are written out explicitly for $M \leqq 10$ in Whittaker and Robinson [1944].

Again we use the notation $\langle u(s) \rangle$ to represent the sum of $u(s)$ over S_N. Then, by proceeding just as in Sec. 7.10, again we find that the orthogonal polynomials satisfy a recurrence formula of the form

$$\phi_{k+1}(s) = (a_k s + b_k)\phi_k(s) + c_k \phi_{k-1}(s) \qquad (7.16.1)$$

where

$$a_k = \frac{A_{k+1}}{A_k} \qquad (7.16.2)$$

and where A_k is the coefficient of s^k in $\phi_k(s)$.

In complete analogy with the derivation of (7.10.5) and (7.10.6), it is found that b_k and c_k satisfy the relations

$$b_k = -a_k \frac{\langle sw\phi_k^2 \rangle}{\langle w\phi_k^2 \rangle} \qquad (7.16.3)$$

and

$$c_k = -a_k \frac{\langle sw\phi_{k-1}\phi_k \rangle}{\langle w\phi_{k-1}^2 \rangle} \qquad (7.16.4)$$

As in the continuous case, it is convenient to first generate the polynomials $\tilde{\phi}_0, \tilde{\phi}_1, \ldots, \tilde{\phi}_n$, for each of which the leading coefficient A_k is unity, so that also $a_k = 1$ in (7.16.1), (7.16.3), and (7.16.4). Hence there follows

$$\tilde{\phi}_0(s) = 1$$

$$\tilde{\phi}_1(s) = s + b_0 \qquad b_0 = -\frac{\langle sw \rangle}{\langle w \rangle}$$

and then

$$\tilde{\phi}_2(s) = (s + b_1)\tilde{\phi}_1(s) + c_1\tilde{\phi}_0(s)$$

with

$$b_1 = -\frac{\langle sw\tilde{\phi}_1^2 \rangle}{\langle w\tilde{\phi}_1^2 \rangle} \qquad c_1 = -\frac{\langle sw\tilde{\phi}_1 \rangle}{\langle w \rangle}$$

and so forth. Afterward, each $\tilde{\phi}_k(s)$ can be scaled in whatever way is convenient.

For the Gram polynomials defined in Sec. 7.13, with $c_{rN} = (-1)^r$, it can be shown that

$$\phi_{k+1}(s) = \frac{2(2k+1)}{(k+1)(N-k)}\left(s - \frac{N}{2}\right)\phi_k(s) - \frac{k(k+N+1)}{(k+1)(N-k)}\phi_{k-1}(s) \qquad (7.16.5)$$

and accordingly that

$$p_{k+1}(t) = \frac{2(2k+1)}{(k+1)(2M-k)}tp_k(t) - \frac{k(k+2M+1)}{(k+1)(2M-k)}p_{k-1}(t) \qquad (7.16.6)$$

7.17 Supplementary References

Szego [1967] contains the most comprehensive treatments of orthogonal polynomials. See also P. J. Davis [1963] and the bibliography of Shohat, Hille, and Walsh [1940].

Least-squares methods are presented in the classical manner in Whittaker and Robinson [1944] and in a more general and more modern setting in Guest [1961]. See also Aitken [1932*a*], Birge and Weinberg [1947], Lewis [1947], and Hayes and Vickers [1951].

The recursive generation of orthogonal polynomials, together with associated error analysis, is dealt with in Forsythe [1957]. The Gram polynomials, orthogonal relative to a unit weighting function on a discrete set of equally spaced points, are tabulated by Anderson and Houseman [1942] and by De Lury [1950]. Other such tabulations are listed in the index by Fletcher et al. [1962].

For additional smoothing techniques, see Rhodes [1921], Whittaker and Robinson [1944], Sard [1948*b*], Doodson [1950], Lanczos [1952], and Schoenberg [1952].

PROBLEMS

Section 7.2

1 Show that the functions $\phi_0(x) = 1$ and $\phi_1(x) = x$ are orthogonal under integration over $[-1, 1]$, and obtain the linear least-squares approximation $y_1(x)$ to a given function $f(x)$ over $[-1, 1]$

$$f(x) \approx y_1(x) \equiv a_0 + a_1 x \qquad (-1 \leqq x \leqq 1)$$

for which

$$\int_{-1}^{1} (f - y_1)^2 \, dx = \min$$

in the form

$$y_1(x) = \tfrac{1}{2} \int_{-1}^{1} (1 + 3xt) f(t) \, dt$$

Show also that the corresponding RMS error in $[-1, 1]$ is given by

$$\left[\tfrac{1}{2} \int_{-1}^{1} f^2 \, dx - a_0^2 - \tfrac{1}{3} a_1^2 \right]^{1/2}$$

2 Show that the functions $\phi_0(x) = 1$ and $\phi_1(x) = x$ are orthogonal under summation over the abscissas $x_0 = -1$, $x_1 = 0$, and $x_2 = 1$, and obtain the linear least-squares approximation $y_2(x)$ to a given function $f(x)$ over $[-1, 1]$

$$f(x) \approx y_2(x) \equiv A_0 + A_1 x \qquad (-1 \leqq x \leqq 1)$$

for which

$$\sum_{k=0}^{2} [f(x_k) - y_2(x_k)]^2 = \min$$

in the form

$$y_2(x) = \tfrac{1}{6}[(2 - 3x)f(-1) + 2f(0) + (2 + 3x)f(1)]$$

Show also that the corresponding RMS error over the three relevant points is given by

$$\left[\tfrac{1}{3}\sum_{k=0}^{2} [f(x_k)]^2 - A_0^2 - \tfrac{2}{3}A_1^2\right]^{1/2}$$

3 If $y_1(x)$ is the linear approximation to $f(x)$ obtained in Prob. 1, and if $f''(x)$ is continuous on $[-1, 1]$, show that

$$f(x) - y_1(x) = \int_{-1}^{1} g_1(x, s) f''(s)\, ds$$

where

$$g_1(x, s) = (x - s)_+ - \tfrac{1}{2}\int_{-1}^{1} (t - s)_+(1 + 3xt)\, dt$$

$$= \begin{cases} -\tfrac{1}{4}(1 + s)^2(1 - 2x + sx) & (s \leqq x) \\ -\tfrac{1}{4}(1 - s)^2(1 + 2x + sx) & (s \geqq x) \end{cases}$$

Show also that $g_1(x, s)$ is of constant sign for x and s in $[-1, 1]$ if and only if $|x| \leqq \frac{1}{3}$ or $|x| = 1$, and establish the relation

$$f(x) = y_1(x) - \tfrac{1}{6}(1 - 3x^2)f''(\xi) \qquad (|x| \leqq \tfrac{1}{3} \text{ or } |x| = 1)$$

where $-1 < \xi < 1$, showing, in particular, that

$$f(-1) - y_1(-1) = \tfrac{1}{3}f''(\xi_1) \qquad f(0) - y_1(0) = -\tfrac{1}{6}f''(\xi_2)$$
$$f(1) - y_1(1) = \tfrac{1}{3}f''(\xi_3)$$

4 If $y_2(x)$ is the linear approximation to $f(x)$ obtained in Prob. 2, and if $f''(x)$ is continuous on $[-1, 1]$, show that

$$f(x) - y_2(x) = \int_{-1}^{1} g_2(x, s) f''(s)\, ds$$

where

$$g_2(x, s) = (x - s)_+ - \tfrac{1}{3}(-s)_+ - \tfrac{1}{6}(1 - s)(2 + 3x)$$

Show also that $g_2(x, s)$ is of constant sign for x and s in $[-1, 1]$ if and only if $|x| \leqq \frac{2}{3}$ or $|x| = 1$, and establish the relation

$$f(x) - y_2(x) = -\tfrac{1}{6}(2 - 3x^2)f''(\xi) \qquad (|x| \leqq \tfrac{2}{3} \text{ or } |x| = 1)$$

where $-1 < \xi < 1$.

5 If $y_3(x)$ is the second-degree polynomial which agrees exactly with $f(x)$ when $x = -1$, 0, and 1, and if $y_2(x)$ is the linear approximation of Prob. 2, show that

$$y_2(x) = y_3(x) + \tfrac{1}{6}(2 - 3x^2)\,\delta^2 f(0)$$

where δ is the central-difference operator with unit spacing. In particular, show that

$$f(-1) - y_2(-1) = -\tfrac{1}{2}[f(0) - y_2(0)] = f(1) - y_2(1) = \tfrac{1}{6}\delta^2 f(0)$$

that the RMS error over the three points $x = -1$, 0, and 1 is

$$\frac{1}{3\sqrt{2}}\,|\delta^2 f(0)|$$

and that

$$y_2(x) = f(x) + \tfrac{1}{6}(2 - 3x^2)f''(x)$$

(for all values of x) if $f(x)$ is a polynomial of degree 2 or less.

Section 7.3

6 If the right-hand member of the rth normal equation associated with (7.3.4) is denoted by v_r ($r = 0, 1, \ldots, n$), show that the weighted sum of the squares of the $N + 1$ residuals is given by

$$\sum_{i=0}^{N} w(x_i)[f(x_i)]^2 - \sum_{r=0}^{n} a_r v_r$$

and use this relation to calculate that sum for the numerical example of Sec. 7.3.

7 Suppose that the following empirical data are available:

x	1.36	1.49	1.73	1.81	1.95	2.16	2.28	2.48
$\bar{f}(x)$	14.094	15.069	16.844	17.378	18.435	19.949	20.963	22.495

Determine least-squares polynomial approximations $y_1(x)$ and $y_2(x)$ of degrees 1 and 2, respectively, weighting all data equally, and calculate the RMS value of the eight residuals in both cases.

Section 7.4

8 Obtain estimated values of the RMS departure between the unknown true function $f(x)$ and the observed function $\bar{f}(x)$ in Prob. 7, based on the approximations $y_1(x)$ and $y_2(x)$; and also determine the approximate RMS errors in the calculated coefficients involved in those approximations. Would either (or both) of the approximations be acceptable if it were known, independently, that the RMS value of the observational errors is about 0.04?

9 The equations

$$\begin{aligned}
2.17x_1 + 0.86x_2 + 1.17x_3 &= 3.85 \\
1.06x_1 + 2.81x_2 - 1.21x_3 &= 3.03 \\
1.91x_1 - 1.02x_2 + 3.91x_3 &= 4.85 \\
1.07x_1 + 1.21x_2 + 1.06x_3 &= 3.27
\end{aligned}$$

are based on empirical data, and it is known that the RMS errors in the coefficients are each about 0.015 and the RMS errors in the right-hand members are each about 0.008. Use the method of least squares to obtain approximate values of x_1, x_2, and x_3, and give ranges within which each unknown lies with a probability of about 0.9. (See Table 1.4 in Sec. 1.7.)

Section 7.5

10 With the notation of Sec. 7.5, show that

$$a_r = \frac{1}{r!\,A_r} \frac{\int_a^b U_r(x) f^{(r)}(x)\,dx}{\int_a^b U_r(x)\,dx}$$

if $f^{(r)} \equiv d^r f/dx^r$ exists on $[a, b]$ and is continuous.

11 If $w(x) = (x - a)^\alpha (b - x)^\beta$, verify that

$$U_r(x) = C_r(x - a)^{r+\alpha}(b - x)^{r+\beta}$$

satisfies the conditions of Sec. 7.5 when C_r is a constant if $\alpha > -1$ and $\beta > -1$.

12 If $w(x) = x$ and $[a, b] = [0, 1]$, show that the rth orthogonal polynomial is given by

$$\phi_r(x) = C_r x^{-1} \frac{d^r}{dx^r} [x^{r+1}(1 - x)^r]$$

and that the arbitrary normalization $\phi_r(0) = 1$ requires that

$$C_r = \frac{1}{(r + 1)!}$$

Determine the polynomials of degrees 0 to 4, and prove that

$$A_r = (-1)^r \frac{(2r + 1)!}{[(r + 1)!]^2} \qquad \gamma_r = \frac{1}{2(r + 1)^3}$$

in consequence of the relation

$$\int_0^1 x^{p-1}(1 - x)^{q-1}\,dx = \frac{\Gamma(p)\Gamma(q)}{\Gamma(p + q)}$$

13 Use the results of Prob. 12 to show that the nth-degree least-squares polynomial approximation $f(x)$ over $[0, 1]$, relevant to the weighting function $w(x) = x$, is defined by

$$y(x) = \sum_{r=0}^{n} a_r \phi_r(x)$$

where

$$a_r = 2(r + 1)^3 \int_0^1 x f(x) \phi_r(x)\, dx$$

and where $\phi_r(x)$ is defined in Prob. 12. In particular, show that the linear approximation is of the form

$$y(x) = 6 \int_0^1 [(3t - 4t^2) - 2(2t - 3t^2)x] f(t)\, dt$$

Section 7.6

14 By expanding $(x^2 - 1)^r$ in descending powers of x^2, and appropriately differentiating term by term, show that (7.6.6) implies the relation

$$P_r(x) = \sum_{k=0}^{r} (-1)^k \frac{(2r - 2k)!}{2^r k!\,(r - k)!\,(r - 2k)!} x^{r-2k}$$

where the series terminates when $k = r/2$ if r is even and when $k = (r - 1)/2$ if r is odd.

15 Show that the coefficient of $P_r(x)$ in (7.6.11) can be expressed in the form

$$a_r = \frac{2r + 1}{2^{r+1} r!} \int_{-1}^{1} (1 - x^2)^r f^{(r)}(x)\, dx$$

if $f^{(r)}(x)$ exists on $[-1, 1]$ and is continuous.

16 Show that the leading terms in the Legendre expansion of $f(x) = \cos(\pi x/2)$ over $[-1, 1]$ are of the form

$$\cos\frac{\pi x}{2} = \frac{2}{\pi} P_0(x) - \frac{10}{\pi^3}(12 - \pi^2)P_2(x) + \frac{18}{\pi^5}(\pi^4 - 180\pi^2 + 1680)P_4(x) - \cdots$$

17 Compare numerically the approximations to $f(x) = \cos(\pi x/2)$ afforded by the least-squares polynomials of degrees 2 and 4, obtained in Prob. 16, with the approximating polynomials of corresponding degrees afforded by truncated power series and by fitting $f(x)$ exactly at three and at five equally spaced points in $[-1, 1]$.

18 If $f(x) = [(x+1)/2]^{1/2}$, show that the coefficient of $P_r(x)$ in the Legendre expansion of $f(x)$ over $[-1, 1]$ is given by

$$a_r = \frac{2r+1}{2^{r+1}r!}\int_{-1}^{1}(1-x^2)^r \frac{d^r}{dx^r}\left(\frac{x+1}{2}\right)^{1/2} dx$$

$$= \frac{2r+1}{r!}\int_0^1 s^r(1-s)^r \frac{d^r s^{1/2}}{ds^r}\, ds$$

$$= (-1)^{r+1}\frac{2}{(2r-1)(2r+3)}$$

so that

$$\left(\frac{x+1}{2}\right)^{1/2} = \frac{2}{3}P_0(x) + \frac{2}{1\cdot 5}P_1(x) - \frac{2}{3\cdot 7}P_2(x) + \frac{2}{5\cdot 9}P_3(x) - \cdots \qquad (|x| \leqq 1)$$

19 Assuming the results of Prob. 18, compare the least-squares polynomial approximations to $f(x) = [(x+1)/2]^{1/2}$ of degrees 2 and 4 over $[-1, 1]$ with the corresponding results of truncating power series and with the polynomials of degrees 2 and 4 which agree with $f(x)$ at three and at five equally spaced points in $[-1, 1]$.

20 Obtain the expansion

$$|x| = \tfrac{1}{2}P_0(x) + \tfrac{5}{8}P_2(x) - \tfrac{3}{16}P_4(x) + \tfrac{13}{128}P_6(x) - \cdots \qquad (|x| \leqq 1)$$

and compare the least-squares approximations of degrees 2 and 4 with the corresponding polynomial approximations which agree exactly with $f(x)$ at three and at five equally spaced points in $[-1, 1]$.

Section 7.7

21 By using the Leibnitz formula (3.3.11), show that (7.7.8) implies the relation

$$L_r(x) = \sum_{k=0}^{r} \frac{(-1)^k}{(r-k)!}\left(\frac{r!}{k!}\right)^2 x^k \equiv r!\sum_{k=0}^{r}(-1)^k\binom{r}{k}\frac{x^k}{k!}$$

22 Show that (7.7.15) can be expressed in the form

$$a_r = \frac{(-1)^r\alpha}{(r!)^2}\int_0^\infty x^r e^{-\alpha x} f^{(r)}(x)\, dx$$

if $f^{(r)}(x)$ exists for all $x \geqq 0$ and is continuous, and if $f(x)$ and its first r derivatives are dominated by $x^{-r-2}e^{\alpha x}$ as $x \to \infty$. Show also that

$$a_r = \frac{(-1)^r c^r \alpha}{r!\,(\alpha - c)^{r+1}} \qquad (\alpha > c)$$

when $f(x) = e^{cx}$, and that

$$a_r = \frac{(-1)^r[\Gamma(s+1)]^2}{\alpha^s(r!)^2\Gamma(s-r+1)} \qquad (s > -1)$$

when $f(x) = x^s$.

23 If $f(x) = [1 - (x/N)]^N$ when $0 \leqq x \leqq N$ and $f(x) = 0$ when $x \geqq N$, obtain the leading terms of the expansion (7.7.16), with $\alpha = 1$, in the form

$$f(x) = \frac{N}{N+1} e^{-x} \left[1 + \frac{2}{N+2} L_1(x) - \frac{N-6}{2(N+2)(N+3)} L_2(x) + \cdots \right] \qquad (x \geqq 0)$$

24 Show that the requirement that the best linear approximation to $e^{\alpha x} f(x)$ be a constant, in the sense of (7.7.17), determines α in the form

$$\alpha = \frac{\int_0^\infty f(x)\, dx}{\int_0^\infty x f(x)\, dx}$$

In particular, show that the most appropriate choice of α for Prob. 23 (in this sense) is $(N + 2)/N$.

Section 7.8

25 Obtain from (7.8.6) the relations

$$\frac{d}{dx} H_r(x) = 2rH_{r-1}(x)$$

$$\frac{d}{dx}\left[e^{-x^2} \frac{d}{dx} H_r(x)\right] = -2re^{-x^2} H_r(x)$$

and deduce also that $H_r(x)$ satisfies the differential equation

$$H_r'' - 2xH_r' + 2rH_r = 0$$

26 Use the first relation of Prob. 25, with the relation $H_0(x) = 1$, to show that the coefficient of x^r in $H_r(x)$ is 2^r. Also, by writing

$$H_r(x) = \sum_{k=0}^{\infty} a_k x^{r-2k}$$

in the differential equation of Prob. 25, show that

$$a_{k+1} = -\frac{(r-2k)(r-2k-1)}{4(k+1)} a_k$$

and deduce that

$$H_r(x) = (2x)^r - \frac{r(r-1)}{1!}(2x)^{r-2} + \frac{r(r-1)(r-2)(r-3)}{2!}(2x)^{r-4} - \cdots$$

where the series terminates with a multiple of x when r is odd and with a constant when r is even.

27 Show that (7.8.13) can be written in the form

$$a_r = \frac{\alpha^{1-r}}{2^r r!\sqrt{\pi}} \int_{-\infty}^{\infty} e^{-\alpha^2 x^2} f^{(r)}(x)\, dx$$

if $f^{(r)}(x)$ exists and is continuous for all x, and if $f(x)$ and its first r derivatives are dominated by $x^{-2}e^{\alpha^2 x^2}$ as $x \to \pm\infty$.

28 By taking $f(x) = e^{2cx}$ and $\alpha = 1$ in the result of Prob. 27, obtain the expansion

$$e^{2cx-c^2} = \sum_{k=0}^{\infty} \frac{c^k}{k!} H_k(x)$$

29 If $f(x) = 1 - |x|$ when $|x| \leqq 1$ and $f(x) = 0$ when $|x| \geqq 1$, obtain the expansion

$$f(x) = \frac{\alpha}{\sqrt{\pi}} e^{-\alpha^2 x^2} \left[H_0(\alpha x) - \frac{3 - \alpha^2}{12} H_2(\alpha x) \right.$$
$$\left. + \frac{45 - 30\alpha^2 + 4\alpha^4}{1440} H_4(\alpha x) + \cdots \right]$$

and show that, if α is chosen such that the coefficient of H_2 vanishes, there follows

$$f(x) = \sqrt{\frac{3}{\pi}}\, e^{-3x^2} \left[H_0(\sqrt{3}\, x) - \tfrac{1}{160} H_4(\sqrt{3}\, x) + \cdots\right]$$

30 If the origin is chosen such that $m_1 = 0$, and if α^2 is then taken to be $m_0/(2m_2)$, with the notation of (7.8.17), show that the expansion (7.8.14) becomes

$$f(x) = \frac{\alpha}{\sqrt{\pi}} e^{-\alpha^2 x^2} \left[m_0 + \frac{\alpha(2\alpha^2 m_3 - 3m_1)}{12} H_3(\alpha x) \right.$$
$$\left. + \frac{(4\alpha^4 m_4 - 12\alpha^2 m_2 + 3m_0)}{96} H_4(\alpha x) + \cdots \right]$$

Section 7.9

31 Verify that (7.9.8) satisfies the recurrence formula (7.9.11).

32 Obtain the expansion

$$|x| = \frac{4}{\pi} \left[\tfrac{1}{2} T_0(x) + \tfrac{1}{3} T_2(x) - \cdots + (-1)^{k+1} \frac{1}{4k^2 - 1} T_{2k}(x) + \cdots \right]$$

when $|x| \leqq 1$, and compare the least-squares approximations of degrees 2 and 4 with those obtained in Prob. 20.

33 Show that the function $S_r(x)$ defined by the relation

$$S_r(x) = \frac{1}{r+1} T'_{r+1}(x)$$

is a polynomial of degree r, expressible in the form

$$S_r(x) = \frac{1}{\sqrt{1 - x^2}} \sin [(r + 1) \cos^{-1} x] = \frac{\sin (r + 1)\theta}{\sin \theta} \qquad (\theta = \cos^{-1} x)$$

and that the polynomials $S_0(x), S_1(x), \ldots, S_r(x), \ldots$ are orthogonal over $[-1, 1]$ relative to the weighting function $w(x) = \sqrt{1 - x^2}$. Show also that

$$S_{r+1}(x) = 2rS_r(x) - S_{r-1}(x) \qquad (r \geqq 1)$$

that $A_r = 2^r$ and $\gamma_r = \pi/2$, and that the coefficients in the least-squares approximation

$$f(x) \approx y(x) = \sum_{r=0}^{n} a_r S_r(x) \qquad (|x| \leqq 1)$$

are given by

$$a_r = \frac{2}{\pi} \int_{-1}^{1} \sqrt{1 - x^2}\, f(x) S_r(x)\, dx$$

when the requirement

$$\int_{-1}^{1} \sqrt{1 - x^2}\, [f(x) - y(x)]^2\, dx = \min$$

is imposed.

34 Using the notation of Prob. 33, obtain the expansion

$$|x| = \frac{4}{\pi} \left[\tfrac{1}{3} S_0(x) + \tfrac{1}{5} S_2(x) - \cdots + (-1)^{k+1} \frac{1}{(2k - 1)(2k + 3)} S_{2k}(x) + \cdots \right]$$

and compare the least-squares approximations of degrees 2 and 4 with those considered in Prob. 32.

Section 7.10

35 If B_r is the coefficient of x^{r-1} in $\phi_r(x)$, show that the coefficient b_k in the recurrence formula (7.10.3) is given by

$$b_k = a_k \left(\frac{B_{k+1}}{A_{k+1}} - \frac{B_k}{A_k} \right)$$

so that (7.10.3) can be written in the form

$$\phi_{k+1}(x) = \frac{A_{k+1}}{A_k} \left[x + \left(\frac{B_{k+1}}{A_{k+1}} - \frac{B_k}{A_k} \right) \right] \phi_k(x) - \frac{A_{k+1} A_{k-1}}{A_k^2} \frac{\gamma_k}{\gamma_{k-1}} \phi_{k-1}(x)$$

36 Use the result of Prob. 35 to derive the relations (7.6.9), (7.7.12), and (7.8.10).

37 If $y_n(x)$ is the least-squares polynomial approximation of degree n to $f(x)$ over $[a, b]$, relative to the weighting function $w(x)$, and if $\phi_r(x)$ is the rth relevant orthogonal polynomial, use the Christoffel-Darboux identity to show that

$$y_n(x) = \frac{1}{a_n \gamma_n} \int_a^b \frac{w(t) f(t)}{x - t} [\phi_{n+1}(x) \phi_n(t) - \phi_n(x) \phi_{n+1}(t)] dt$$

38 Prove that *the zeros of* $\phi_m(x)$ *separate the zeros of* $\phi_{m+1}(x)$ by the following steps (or otherwise):

(*a*) Use (7.10.11) to deduce that if x_i and x_{i+1} are successive zeros of $\phi_{m+1}(x)$, then $\phi_m(x_i)\phi'_{m+1}(x_i)$ and $\phi_m(x_{i+1})\phi'_{m+1}(x_{i+1})$ have the same sign.

(*b*) From the established fact that the zeros of $\phi_{m+1}(x)$ are simple, deduce that $\phi'_{m+1}(x_i)$ and $\phi'_{m+1}(x_{i+1})$ are of opposite sign and hence complete the required proof.

39 Use the formulas (7.10.3) and (7.10.13), with $w(x) = 1$ and $[a, b] = [-1, 1]$, to generate the first five of the polynomials $\tilde{P}_k(x) \equiv P_k(x)/A_k$, where $P_k(x)$ is the kth Legendre polynomial and A_k its leading coefficient; compare the results with (7.6.8).

40 Show that the monic polynomials $\tilde{\phi}_k(x)$ which are orthogonal over $[a, b]$ relative to $w(x)$ can be generated recursively by use of the *Gram-Schmidt* formula

$$\tilde{\phi}_k(x) = x^k - \sum_{r=0}^{k-1} c_{kr}\tilde{\phi}_r(x)$$

where

$$c_{kr} = \frac{\int_a^b wx^k\tilde{\phi}_r\,dx}{\int_a^b w\tilde{\phi}_r^2\,dx}$$

starting with $\tilde{\phi}_0(x) = 1$. (First take $k = 1$; then use an inductive argument to derive the formula.) Also use this method in place of that used in Prob. 39 to determine $P_k(x)/A_k$ $(k = 0, 1, \ldots, 5)$ and compare the amounts of labor involved in the two processes.

41 Use Clenshaw's method to evaluate the sum

$$\sum_{k=0}^{10} \frac{(-1)^k}{2k+1} T_k(x)$$

to five places when $x = 0.10324$.

Section 7.11

42 Express the following sums in closed forms:

(*a*) $1 \cdot 2 + 2 \cdot 3 + \cdots + n(n+1) \equiv \sum_{s=2}^{n+1} s^{(2)}$

(*b*) $1 \cdot 2 \cdot 3 + 2 \cdot 3 \cdot 4 + \cdots + n(n+1)(n+2)$

(c) $1 \cdot 2 + 4 \cdot 5 + 7 \cdot 8 + \cdots + (3n-2)(3n-1)$

43 Express the following sums in closed forms, and determine the limit of each as $n \to \infty$:

(*a*) $\dfrac{1}{1 \cdot 2} + \dfrac{1}{2 \cdot 3} + \cdots + \dfrac{1}{n(n+1)} \equiv \sum_{s=0}^{n-1} s^{(-2)}$

(*b*) $\dfrac{1}{1 \cdot 2 \cdot 3} + \dfrac{1}{2 \cdot 3 \cdot 4} + \cdots + \dfrac{1}{n(n+1)(n+2)}$

(*c*) $\dfrac{1}{1 \cdot 2 \cdot 3} + \dfrac{3}{2 \cdot 3 \cdot 4} + \dfrac{5}{3 \cdot 4 \cdot 5} + \cdots + \dfrac{2n-1}{n(n+1)(n+2)}$

44 Show that

$$(-m)^{(n)} = (-1)^n(m + n - 1)^{(n)}$$

when n is a positive integer.

45 Show that

$$\binom{m}{n}\binom{m+p}{p} = \binom{n+p}{n}\binom{m+p}{n+p}$$

46 Use the binomial expansion of $(1 + x)^n$ to deduce the following relations:

(*a*) $\displaystyle\sum_{k=0}^{n} \binom{n}{k} = 2^n$

(*b*) $\displaystyle\sum_{k=0}^{n} (-1)^k \binom{n}{k} = 0$

47 Show that

$$\sum_{k=0}^{n} \binom{n}{k}^2 = \binom{2n}{n}$$

[Use (7.11.13) and (7.11.12).]

48 If m is a positive integer, show that

$$\binom{s}{m} = \sum_{k=0}^{m} \binom{n}{m-k}\binom{s-n}{k}$$

and hence that

$$s^{(m)} = \sum_{k=0}^{m} \frac{m!}{k!}\binom{n}{m-k}(s-n)^{(k)}$$

and deduce the relation

$$s^{(m)}s^{(n)} = \sum_{k=0}^{m} \frac{m!}{k!}\binom{n}{m-k} s^{(n+k)}$$

49 Show that

$$\sum_{s=M}^{N} s\, \Delta_1^2 u_s = [s\, \Delta_1 u_s - u_{s+1}]_M^{N+1}$$

and use this formula to obtain the result

$$\sum_{s=0}^{N} sa^s = \frac{1}{(a-1)^2}[Na^{N+2} - (N+1)a^{N+1} + a] \qquad (a \neq 1)$$

by taking $u_s = a^s/(a - 1)^2$ or otherwise.

50 Show that

$$\sum_{s=M}^{N} u_s\, \Delta_1^r v_s = [u_s\, \Delta_1^{r-1} v_s - (\Delta_1 u_s)(\Delta_1^{r-2} v_{s+1}) + (\Delta_1^2 u_s)(\Delta_1^{r-3} v_{s+2})$$

$$- \cdots + (-1)^{r-1}(\Delta_1^{r-1} u_s)v_{s+r-1}]^{N+1} + (-1)^r \sum_{s=M}^{N} (\Delta_1^r u_s)v_{s+r}$$

and also that

$$\sum_{s=M}^{N} u_s \, \Delta_1^r v_s = [u_{s-1} \, \Delta_1^{r-1} v_s - (\Delta_1 u_{s-2})(\Delta_1^{r-2} v_s) + (\Delta_1^2 u_{s-3})(\Delta_1^{r-3} v_s)$$

$$- \cdots + (-1)^{r-1}(\Delta_1^{r-1} u_{s-r}) v_s]_M^{N+1} + (-1)^r \sum_{s=M}^{N} (\Delta_1^r u_{s-r}) v_s$$

Section 7.12

51 Show that (7.12.13) can be written in the form

$$\gamma_r(N) = (-1)^r r! \, A_r(N) \sum_{s=0}^{N} U_r(s + r, N)$$

where $A_r(N)$ is the coefficient of s^r in $\phi_r(s, N)$.

52 Show that the minimum value of the left-hand member of (7.12.11) can be expressed in the form

$$\Delta_n(N) = \sum_{s=0}^{N} w(s)[F(a + sh)]^2 - \sum_{r=0}^{n} \gamma_r(N) a_r^2$$

Section 7.13

53 With the notation of Prob. 51, show that, when (7.13.9) is imposed on (7.13.6), there follows

$$\gamma_r(N) \equiv \sum_{s=0}^{N} [\phi_r(s, N)]^2 = \frac{(-1)^r (2r)!}{[r! \, N^{(r)}]^2} \sum_{s=0}^{N} (s + r)^{(r)} (s + r - N - 1)^{(r)}$$

By making use of appropriate summations by parts, show further that

$$\gamma_r(N) = \frac{1}{(N^{(r)})^2} \sum_{s=0}^{N} (s + r)^{(2r)}$$

and deduce the closed form

$$\gamma_r(N) = \frac{1}{2r + 1} \, \frac{(N + r + 1)^{(r+1)}}{N^{(r)}}$$

54 Use the results of Prob. 53 and of Eqs. (7.13.10) to express the leading terms of (7.13.12) in the form

$$y(t) = \frac{1}{2M + 1} \sum_{k=-M}^{M} f(k) + \frac{3t}{M(M + 1)(2M + 1)} \sum_{k=-M}^{M} k f(k)$$

$$+ \frac{15t^2 - 5M(M + 1)}{M(M + 1)(2M - 1)(2M + 1)(2M + 3)}$$

$$\times \sum_{k=-M}^{M} [3k^2 - M(M + 1)] f(k) + \cdots$$

Section 7.14

55 Prepare a table analogous to Table 7.1 in the case $M = 3$, when seven points are employed in the least-squares approximation, including only the orthogonal polynomials of degrees 5 and less.

56 By using the table prepared in Prob. 55, obtain least-squares polynomial approximations of degrees 1 to 5 to $f(t) \equiv F(\frac{3}{2} + \frac{1}{2}t)$ for $|t| \leqq 3$ from the following approximate data, calculate the respective smoothed values at the tabular points, and determine which approximation is probably most appropriate if the data are empirical, with errors having an estimated RMS value of about 0.07:

x	0.0	0.5	1.0	1.5	2.0	2.5	3.0
$F(x)$	15.564	18.059	20.548	23.554	26.348	29.498	32.830

57 Use the results of Prob. 56 to obtain approximate values of the following quantities from the smoothed data:

$$F(0.1) \qquad F(1.8) \qquad F'(1.0) \qquad F'(1.3) \qquad \int_0^3 F(x)\,dx \qquad \int_{1.2}^{2.3} F(x)\,dx$$

58 Obtain a formula analogous to (7.14.8) for fifth-degree seven-point least-squares approximation.

Section 7.15

59 Use (7.15.4) and (7.15.5) to obtain smoothed values of the data given in Prob. 56, corresponding to third-degree and fifth-degree least squares, and verify that the results agree with those obtained in Prob. 56.

60 The following data represent estimated world route mileages of scheduled air services in the years given in units of 1000 miles. Calculate smoothed values, using both first- and third-degree five-point formulas, and plot the two smoothed curves together with points representing the given data.

1919	3.2	1926	48.5	1933	200.3
1920	9.7	1927	54.7	1934	223.1
1921	12.4	1928	90.7	1935	278.2
1922	16.0	1929	125.8	1936	305.2
1923	16.1	1930	156.8	1937	333.5
1924	20.3	1931	185.1	1938	349.1
1925	34.0	1932	190.2		

Also use the two sets of smoothed data to obtain estimates of the annual rate of increase of mileage, at the end of the tabulation, to be used for long- and short-range predictions.

Section 7.16

61 Prove that if $w(s) \geqq 0$ for $s = 0, 1, \ldots, N$, then the polynomial $\phi_r(s)$ possesses r real zeros in the interval $0 < s < N$.

62 Prove that the zeros of $\phi_r(s)$ separate the zeros of $\phi_{r+1}(s)$. (See Prob. 38.)

63 Use the recursive method of Sec. 7.16 to generate the polynomials $\tilde{\phi}_r(s)$ $(r = 0, 1, 2, 3, 4)$ when $w(s) = 1$ and the set S_N comprises the five points $s = -2, -1, 0, 1, 2$, and check the results by reference to (7.14.1).

64 Modify the Gram-Schmidt formula of Prob. 40 to apply to the discrete case. Also use it recursively in Prob. 63 in place of the suggested method and compare the two methods with respect to amounts of labor involved.

8

GAUSSIAN QUADRATURE AND RELATED TOPICS

8.1 Introduction

The formulas given in Chaps. 3 and 5, for the purpose of numerical integration (with or without differences), each involve sets of ordinates which correspond to *equally spaced* abscissas. As might be expected, corresponding formulas which are generally capable of supplying comparable accuracy with fewer (about half as many) ordinates can be obtained by determining the *optimal* distribution of the abscissas, rather than prescribing them in an arbitrary way. It is found that the abscissas so determined are generally specified by irrational numbers and that the same is usually true of the weights by which the corresponding ordinates are to be multiplied.

As a specific, but typical, example, which may be helpful in motivating some remarks with regard to such formulas, the five-point Newton-Cotes formula (3.5.13), of closed type, is of the form

$$\int_{-1}^{1} f(x)\,dx = \tfrac{1}{45}[7f(-1) + 32f(-\tfrac{1}{2}) + 12f(0) + 32f(\tfrac{1}{2}) + 7f(1)] - \frac{f^{\mathrm{vi}}(\xi_1)}{15120} \qquad (8.1.1)$$

when related to the interval $[-1, 1]$, whereas the Legendre-Gauss three-point formula, to be derived in Sec. 8.5, is of the form

$$\int_{-1}^{1} f(x)\, dx = \frac{1}{9}\left[5f\left(-\frac{\sqrt{15}}{5}\right) + 8f(0) + 5f\left(\frac{\sqrt{15}}{5}\right)\right] + \frac{f^{\mathrm{vi}}(\xi_2)}{15750} \qquad (8.1.2)$$

where both ξ_1 and ξ_2 lie somewhere in $[-1, 1]$.

A comparison of the two error terms shows that the second formula, which requires the values of only three ordinates, generally may be expected to afford about the same accuracy as the first, which requires five ordinates, when the error terms are neglected. Also, since the weights are positive in both formulas, the error in the result, due to possible errors in the *ordinates*, cannot exceed (but can equal) twice the maximum of those errors in both cases. Moreover, if *random* errors in the ordinates are considered, the corresponding RMS errors in the approximations afforded by the first and second formulas are found to be given by about 0.48 and about 0.68 times the RMS ordinate error, respectively.

Thus the apparent advantage of the second formula consists of the fact that, aside from the central ordinate, which is needed in both, it involves only half as many ordinates as the first. However, unless $f(x)$ is a polynomial (in which case the formulas are not needed) the required ordinates are generally to be obtained by reference to a table of values of $f(x)$. It is then sometimes argued that, since two of the abscissas in (8.1.2) are irrational, interpolation involving *at least* two tabulated ordinates will be required for the determination of each of the two off-center ordinates, so that at least five ordinates truly will be involved in the use of (8.1.2). Thus, the apparent advantage is lost, and even reversed, since (8.1.1) involves the five ordinates needed in a simple and specific form.

For this reason, and also because of the fact that the *weights* in most gaussian formulas are also irrational (the present case is an exception), so that, in place of multiplying each ordinate by an integer, one must multiply it by a number with at least as many significant digits as are required in the final result, relatively little practical use was made of such formulas until recent years.

This situation was indeed unfortunate, since the second reason given, while an important one when calculations are necessarily effected by hand, slide rule, or use of tables of logarithms, is clearly of no significance when a computing device with even the relatively limited efficiency of a desk calculator is available, and the argument supplying the first reason is (rather obviously) generally fallacious. Specifically, it assumes that the ordinates denoted as $f(-1)$, $f(-\frac{1}{2})$, and so forth, are known or can be found directly in tables, without the need of interpolation, and is valid only then.

It is true that available tables of many functions, such as e^{-x^2} and $J_0(x)$, for example, include these arguments, and these are typical of the functions which most frequently appear in textbooks. But practical problems tend to deal instead with functions such as $e^{-\alpha x^2/L^2}$ and $J_0(\alpha x/L)$ over the interval $[-L, L]$ and, correspondingly, with $e^{-\alpha x^2}$ and $J_0(\alpha x)$ over the normalized interval $[-1, 1]$, where α is a function of certain physical quantities and is most unlikely to have an integral (or rational) value. Thus, in practical situations, it is probable that *each* of the ordinates appearing in either of the forms (8.1.1) and (8.1.2) will have to be determined by interpolation (or by direct calculation), and the interpolation for $J_0(\alpha\sqrt{15}/5)$ would be more difficult than that for $J_0(\alpha/2)$ *only* in that the determination of the numerical argument of the interpolate in the former case would involve a multiplication of *two* n-digit numbers. The necessary accuracy of the interpolation or calculation would be no higher in one case than in the other.

Further, it may be noted that, when use is made of a large-scale digital computer, and when $f(x)$ is defined analytically, approximate values of the integrand often are not obtained by interpolation in tables in any case, but are generated directly within the computer by subroutines incorporated in the program. Here, since the machine does not distinguish between rational and irrational arguments, the approximate evaluation of $f(\sqrt{15}/5)$ is in no way more complicated than that of $f(\frac{1}{2})$. Thus, formulas such as (8.1.2) are indeed advantageous when the determination of ordinates needed for the conventional formulas would involve either *direct calculation*, *physical measurement*, or *interpolation* and when the use of a *high-precision* formula is appropriate.†

The developments of this chapter relate these formulas to a method of *osculating interpolation*, associated with the name of Hermite, which is treated in Sec. 8.2, and to an associated quadrature formula (Sec. 8.3). Several of the classical quadrature formulas of the gaussian type, in which no abscissas are arbitrarily preassigned, are considered, together with their error terms, in the subsequent sections, which depend upon certain results from Chap. 7. Section 8.10 deals with the modifications necessary when certain of the abscissas are preassigned, and the results are illustrated in the next two sections. The convergence of sequences of approximations generated by the formulas considered, as the number of ordinates used is increased, is considered in Sec. 8.13. Section 8.14 deals with a special class of quadrature formulas in which the weights, rather than the abscissas, are preassigned; and Sec. 8.15 deals with algebraic

† Perhaps because of the fact that they are particularly useful when the integrand is defined analytically, they are usually called *quadrature formulas*, whereas formulas of the usual type are usually called *integration formulas*. There is no basic distinction between the terms.

methods for deriving quadrature formulas of the type considered in this chapter, without making use of the properties of orthogonal polynomials. The concluding section illustrates these procedures by obtaining a pair of simple formulas which approximate finite Fourier-transform integrals.

8.2 Hermite Interpolation

The interpolation formulas so far considered make use only of a certain number of values (approximate or exact) of the function to be approximated. Except in the case of the least-squares formulas of the preceding chapter, the approximating polynomial $y(x)$ has been defined as that polynomial of lowest degree which agrees with the approximated function $f(x)$ on a certain discrete set of points. In certain cases, values of both $f(x)$ and its derivative $f'(x)$ are available, say, at m points.† We next derive an interpolation formula which utilizes these $2m$ data and, in the remainder of this chapter, show that the result leads to useful formulas for numerical integration which do *not* depend upon knowledge of values of $f'(x)$.

Before proceeding to these matters, however, it is desirable to review the lagrangian interpolation formula treated in Chap. 3 and to write it in a slightly modified form. If the values of $f(x)$ are known at the m points $x = x_1, x_2, \ldots, x_m$, the auxiliary functions

$$\pi(x) = (x - x_1)(x - x_2)\cdots(x - x_m) \qquad (8.2.1)$$

and

$$l_i(x) = \frac{\pi(x)}{(x - x_i)\pi'(x_i)}$$
$$\equiv \frac{(x - x_1)\cdots(x - x_{i-1})(x - x_{i+1})\cdots(x - x_m)}{(x_i - x_1)\cdots(x_i - x_{i-1})(x_i - x_{i+1})\cdots(x_i - x_m)} \qquad (i = 1, 2, \ldots, m) \qquad (8.2.2)$$

are first defined, with the properties

$$\pi(x_j) = 0 \qquad (8.2.3)$$

† In the preceding chapters, integration formulas were obtained by integrating the interpolation polynomial of degree n which agrees with the integrand at $n + 1$ points, so that the principal emphasis was on the degree n of that interpolation polynomial, and it was convenient to number the relevant $n + 1$ abscissas from 0 to n. On the other hand, the derivations of the integration formulas which are to be treated in the present chapter are based on certain properties of the polynomial whose *zeros* are the abscissas of the points involved in the integration formula. It is thus more convenient to use a new symbol m to represent the degree of *that* polynomial, and hence also to represent the *number of ordinates* employed, and to number the ordinates from 1 to m.

and

$$l_i(x_j) = \delta_{ij} \tag{8.2.4}$$

where δ_{ij} is the Kronecker delta (zero when $i \neq j$ and unity when $i = j$). With these notations, the polynomial of degree $m - 1$ which takes on the values $f(x_1), f(x_2), \ldots,$ and $f(x_m)$ is expressible in the form

$$y(x) = \sum_{k=1}^{m} l_k(x) f(x_k) \tag{8.2.5}$$

Also, if $f^{(m)}(x)$ is continuous in the interval I limited by the largest and smallest of the $m + 1$ numbers $x_1, x_2, \ldots, x_m$, and x, the error

$$E(x) = f(x) - y(x)$$

is expressible in the forms

$$E(x) = \pi(x) f[x_1, \ldots, x_m, x] = \frac{f^{(m)}(\xi)}{m!} \pi(x) \tag{8.2.6}$$

where ξ is somewhere in I.

Now suppose that values of both $f(x)$ and $f'(x)$ are known for $x_1, \ldots, x_m$. Since a polynomial of degree $2m - 1$ is specified by $2m$ parameters, it is plausible that one such polynomial $y(x)$ can be determined in such a way that $y(x)$ and $f(x)$ possess the same value and the same derivative at each of these m points. We next attempt to determine such a polynomial by assuming that it is expressible in the form

$$y(x) = \sum_{k=1}^{m} h_k(x) f(x_k) + \sum_{k=1}^{m} \bar{h}_k(x) f'(x_k) \tag{8.2.7}$$

where $h_i(x)$ and $\bar{h}_i(x)$ $(i = 1, 2, \ldots, m)$ are polynomials of maximum degree $2m - 1$, to be determined.

The requirement that $y(x_j) = f(x_j)$ clearly will be satisfied if

$$h_i(x_j) = \delta_{ij} \qquad \bar{h}_i(x_j) = 0 \tag{8.2.8}$$

whereas the requirement $y'(x_j) = f'(x_j)$ will be satisfied if

$$h_i'(x_j) = 0 \qquad \bar{h}_i'(x_j) = \delta_{ij} \tag{8.2.9}$$

for $1 \leqq i \leqq m$ and $1 \leqq j \leqq m$. Now, since $l_i(x)$ is a polynomial of degree $m - 1$ which satisfies (8.2.4), the function $[l_i(x)]^2$ is a polynomial of degree $2m - 2$ which satisfies (8.2.4) and whose derivative vanishes at x_j when $i \neq j$. Hence, since $h_i(x)$ and $\bar{h}_i(x)$ are polynomials of degree $2m - 1$, there must follow

$$h_i(x) = r_i(x)[l_i(x)]^2 \qquad \bar{h}_i(x) = s_i(x)[l_i(x)]^2 \tag{8.2.10}$$

where $r_i(x)$ and $s_i(x)$ are *linear* functions of x, in order that the conditions of

(8.2.8) and (8.2.9) be satisfied when $i \neq j$. The other four conditions then give

$$r_i(x_i) = 1 \qquad r_i'(x_i) + 2l_i'(x_i) = 0 \tag{8.2.11}$$

and

$$s_i(x_i) = 0 \qquad s_i'(x_i) = 1 \tag{8.2.12}$$

from which there follows

$$r_i(x) = 1 - 2l_i'(x_i)(x - x_i) \qquad s_i(x) = x - x_i \tag{8.2.13}$$

Hence, by combining (8.2.7), (8.2.10), and (8.2.13), we obtain the desired polynomial in the form

$$y(x) = \sum_{k=1}^{m} h_k(x)f(x_k) + \sum_{k=1}^{m} \bar{h}_k(x)f'(x_k) \tag{8.2.14}$$

where

$$h_i(x) = [1 - 2l_i'(x_i)(x - x_i)][l_i(x)]^2 \tag{8.2.15}$$

and

$$\bar{h}_i(x) = (x - x_i)[l_i(x)]^2 \tag{8.2.16}$$

This result is known as *Hermite's interpolation formula* (Hermite [1878]) or, frequently, as the formula for *osculating interpolation.* (For a more general formula, which also uses values of higher derivatives of $f(x)$, see Fort [1948].)

An expression for the error $E(x) = f(x) - y(x)$ can be obtained by a method similar to that used in Sec. 2.6. Thus we notice that both $E(x)$ and $[\pi(x)]^2$ *vanish together with their first derivatives* at each of the m points $x = x_1, \ldots, x_m$. We then form a linear combination of these functions

$$F(x) = f(x) - y(x) - K[\pi(x)]^2 \tag{8.2.17}$$

which therefore has the same properties, and determine K in such a way that $F(x)$ also vanishes at an arbitrarily chosen additional point $x = \bar{x}$.

Now let $\bar{I}$ represent the closed interval limited by the smallest and largest of the numbers $x_1, x_2, \ldots, x_m$, and $\bar{x}$. Since $F(x)$ vanishes at these $m + 1$ distinct points, $F'(x)$ must vanish at at least m intermediate points inside $\bar{I}$. But since $F'(x)$ also vanishes at the m points $x_1, \ldots, x_m$, it vanishes at least $2m$ times in $\bar{I}$. Thus $F''(x)$ vanishes at least $2m - 1$ times *inside* $\bar{I}$, $F'''(x)$ at least $2m - 2$ times, ..., and hence, finally, $F^{(2m)}(x)$ vanishes at least *once* inside $\bar{I}$, assuming the existence of the derivatives considered. Let one such point be ξ. Then, recalling that $y(x)$ is a polynomial of degree $2m - 1$, and hence that $y^{(2m)}(x) \equiv 0$, we obtain from (8.2.17) the result

$$0 = F^{(2m)}(\xi) = f^{(2m)}(\xi) - K(2m)!$$

or

$$K = \frac{f^{(2m)}(\xi)}{(2m)!}$$

Thus, since $F(\bar{x}) = 0$, there follows

$$E(\bar{x}) \equiv f(\bar{x}) - y(\bar{x}) = \frac{f^{(2m)}(\xi)}{(2m)!} [\pi(\bar{x})]^2$$

where ξ is somewhere in $\bar{I}$. Since both sides of this relation vanish when $\bar{x}$ is identified with one of the points x_i, the relation is true also for such values of $\bar{x}$ and hence for *any* $\bar{x}$. Hence, suppressing the bars, we deduce that the error associated with approximating $f(x)$ by the right-hand member of (8.2.14) is of the form

$$E(x) = \frac{f^{(2m)}(\xi)}{(2m)!} [\pi(x)]^2 \tag{8.2.18}$$

where ξ is somewhere in the interval I.

It can be seen that the polynomial (8.2.14) could also be obtained as the limit of the polynomial of degree $2m - 1$ which agrees with $f(x)$ at $2m$ distinct points $x_1, x_1', x_2, x_2', \ldots, x_m, x_m'$, as each x_k' tends to x_k. From this fact we can conclude that the error $E(x)$ also can be expressed in the form

$$E(x) = [\pi(x)]^2 f[x_1, x_1, \ldots, x_m, x_m, x] \tag{8.2.19}$$

assuming only that $f'(x)$ exists in I.

From (8.2.18) it follows that the Hermite m-point formula yields exact results when $f(x)$ is identified with any polynomial of degree not exceeding $2m - 1$. In particular, we may deduce easily that the interpolation polynomial (8.2.14) is the *only* one having the specified properties.

8.3 Hermite Quadrature

From the Hermite interpolation formula we may deduce the formula

$$\int_a^b w(x)f(x)\,dx = \sum_{k=1}^{m} H_k f(x_k) + \sum_{k=1}^{m} \bar{H}_k f'(x_k) + E \tag{8.3.1}$$

with the weighting coefficients defined by the equations

$$H_i = \int_a^b w(x)h_i(x)\,dx = \int_a^b w(x)[1 - 2l_i'(x_i)(x - x_i)][l_i(x)]^2\,dx \tag{8.3.2}$$

and

$$\bar{H}_i = \int_a^b w(x)\bar{h}_i(x)\,dx = \int_a^b w(x)(x - x_i)[l_i(x)]^2\,dx \tag{8.3.3}$$

and with the error expressible in the form

$$E = \frac{1}{(2m)!} \int_a^b f^{(2m)}(\xi) w(x)[\pi(x)]^2\,dx \tag{8.3.4}$$

where $a < \xi < b$ if the points $x_1, x_2, \ldots, x_m$ lie in that interval. The result of neglecting the error term is called the *Hermite quadrature formula.*

If the *weighting function* $w(x)$ is nonnegative in $[a, b]$

$$w(x) \geqq 0 \qquad (8.3.5)$$

as will be assumed throughout this chapter, the coefficient of $f^{(2m)}(\xi)$ in the integrand of (8.3.4) is nonnegative. Hence the second law of the mean may be invoked to permit (8.3.4) to be written in the more convenient form

$$E = \frac{f^{(2m)}(\xi)}{(2m)!} \int_a^b w(x)[\pi(x)]^2 \, dx \qquad (8.3.6)$$

These results may be compared with the result which corresponds to lagrangian interpolation employing m points, which can be expressed in the form

$$\int_a^b w(x)f(x) \, dx = \sum_{k=1}^{m} W_k f(x_k) + E \qquad (8.3.7)$$

where

$$W_i = \int_a^b w(x)l_i(x) \, dx \qquad (8.3.8)$$

and

$$E = \frac{1}{m!} \int_a^b f^{(m)}(\xi) w(x)\pi(x) \, dx \qquad (8.3.9)$$

Since $\pi(x)$ changes sign at each of the points $x_1, \ldots, x_m$, the law of the mean cannot be applied directly to (8.3.9) to produce a form analogous to (8.3.6).

If a quadrature formula yields exact results when $f(x)$ is an arbitrary polynomial of degree r or less, but fails to give exact results for at least one polynomial of degree $r + 1$, it is said to possess a *degree of precision* equal to r (see Sec. 5.10). From the linearity of the process, it follows that this situation exists if and only if exact results are afforded for $1, x, x^2, \ldots, x^r$, but not for x^{r+1}.

From (8.3.6) we see that the degree of precision of the Hermite m-point formula is exactly $2m - 1$. It follows also from (8.3.9) that the degree of precision of the lagrangian quadrature formula, based on m points, is at least $m - 1$. Furthermore, if we take $f(x) = [\pi(x)]^2$, we see that all terms in the sum involved in (8.3.7) *vanish*, and hence, for this function, the lagrangian formula would give

$$\int_a^b w(x)[\pi(x)]^2 \, dx = 0$$

Under the assumption (8.3.5), this situation is impossible. Hence, since $[\pi(x)]^2$

is a polynomial of degree $2m$, it follows that the degree of precision of the lagrangian m-point formula cannot *exceed* $2m - 1$. Unless further information concerning the choice of the points $x_1, \ldots, x_m$ is available, no more specific statement can be made. However, it is shown in the following section that there exists a class of formulas of the simple lagrangian type (8.3.7) which actually have the maximum degree of precision $2m - 1$.

8.4 Gaussian Quadrature

An inspection of (8.3.1) shows that, if the points $x_1, \ldots, x_m$ can be chosen in such a way that the weighting coefficients $\bar{H}_k$ associated with the derivative terms vanish, then the Hermite formula will reduce to a formula of the simple type (8.3.7) while retaining the degree of precision $2m - 1$. With the notation of (8.2.1) and (8.2.2), the definition (8.3.3) can be expressed in the equivalent form

$$\bar{H}_i = \frac{1}{\pi'(x_i)} \int_a^b w(x)\pi(x)l_i(x)\,dx \tag{8.4.1}$$

where, as before,

$$\pi(x) = (x - x_1)(x - x_2)\cdots(x - x_m) \tag{8.4.2}$$

so that $x_1, \ldots, x_m$ are the m zeros of $\pi(x)$.

Thus $\bar{H}_i$ will vanish for $1 \leqq i \leqq m$, and the degree of precision $2m - 1$ will be preserved, if $\pi(x)$ is *orthogonal* to $l_1(x), \ldots, l_m(x)$ over $[a, b]$, relative to the weighting function $w(x)$. Since each $l_i(x)$ is a polynomial of degree $m - 1$, by virtue of (8.2.2), a *sufficient* condition is that $\pi(x)$ be orthogonal to *all* polynomials of degree inferior to m over $[a, b]$, relative to $w(x)$.

This condition is also *necessary*. To see this, assume that

$$\bar{H}_i = 0 \qquad (1 \leqq i \leqq m) \tag{8.4.3}$$

and that the formula has a degree of precision $2m - 1$. Let $f(x)$ be a polynomial, of degree $2m - 1$ or less, expressed in the special form

$$f(x) = \pi(x)u_{m-1}(x) \tag{8.4.4}$$

where $u_{m-1}(x)$ is an arbitrary polynomial of degree $m - 1$ or less. Then, since $\pi(x_i) = 0$ for $1 \leqq i \leqq m$, there follows $f(x_i) = 0$, and hence, for this polynomial, (8.3.1) becomes

$$\int_a^b w(x)f(x)\,dx \equiv \int_a^b w(x)\pi(x)u_{m-1}(x)\,dx = 0 \tag{8.4.5}$$

as was to be shown, since $\bar{H}_i = 0$ by assumption and $E = 0$ by virtue of the fact that here $f^{(2m)}(x) \equiv 0$.

Hence we deduce that *if and only if the polynomial* $\pi(x)$, *of degree* m, *is orthogonal to all polynomials of inferior degree over* $[a, b]$, *relative to* $w(x)$, *the Hermite quadrature formula reduces to the formula*

$$\int_a^b w(x)f(x)\,dx = \sum_{k=1}^{m} H_k f(x_k) + E \tag{8.4.6}$$

where

$$E = \frac{f^{(2m)}(\xi)}{(2m)!} \int_a^b w(x)[\pi(x)]^2\,dx \tag{8.4.7}$$

and where the m *abscissas* $x_1, \ldots, x_m$ *are the zeros of* $\pi(x)$.

A formula of this type is usually called a *gaussian quadrature* formula, although it appears that only the case in which $w(x) = 1$ was explicitly considered by Gauss, the generalization to other weighting functions being due to Christoffel [1858]. The weights H_i accordingly are sometimes called the *Christoffel numbers* of the formula.

Since (8.4.6) is a special case of both (8.3.1) and (8.3.7), the weighting coefficients H_i and W_i given by (8.3.2) and (8.3.8) must be equal in this case. (See also Prob. 7.) Thus we may write

$$H_i = \int_a^b w(x)[l_i(x)]^2\,dx = \int_a^b w(x)l_i(x)\,dx = W_i \tag{8.4.8}$$

the first form being obtained from (8.3.2) by writing that formula in the form

$$H_i = \int_a^b w(x)[l_i(x)]^2\,dx - 2l_i'(x_i)\overline{H}_i$$

and recalling that here $\overline{H}_i = 0$.

The polynomial $\pi(x)$ is precisely that numerical multiple of the polynomial $\phi_m(x)$, specified by Eqs. (7.5.4) to (7.5.7), for which the coefficient of the leading power of x is unity. Thus, as was shown in Sec. 7.10, its m zeros are indeed real and distinct and are all located in the interval (a, b). The interval need not be of finite length, as long as $w(x) \geqq 0$ and the integral $\int_a^b x^k w(x)\,dx$ exists for all nonnegative integral values of k. It is of particular importance to notice that, by virtue of the first form of (8.4.8), *the weighting coefficients in a gaussian quadrature formula are all positive.*

With the notation of Sec. 7.4, the error (8.4.7) can be expressed in the form

$$E = \frac{\gamma_m}{A_m^2} \frac{f^{(2m)}(\xi)}{(2m)!} \tag{8.4.9}$$

where γ_m is the normalizing factor corresponding to $\phi_m(x)$ and is defined by

(7.5.13), and where A_m is the coefficient of x^m in $\phi_m(x)$.† If use is made of (8.2.19), instead of (8.2.18), in (8.3.4), the error also can be expressed in the form

$$E = \frac{\gamma_m}{A_m^2} f[x_1, x_1, \ldots, x_m, x_m, \xi] \tag{8.4.10}$$

In order to determine explicitly the weights H_i defined by (8.4.8), we make use of the Christoffel-Darboux identity

$$\sum_{k=0}^{m} \frac{\phi_k(x)\phi_k(y)}{\gamma_k} = \frac{\phi_{m+1}(x)\phi_m(y) - \phi_m(x)\phi_{m+1}(y)}{a_m\gamma_m(x - y)} \tag{8.4.11}$$

with $a_m = A_{m+1}/A_m$, established in Sec. 7.10. If we notice that here

$$\phi_m(x) = A_m\pi(x) \tag{8.4.12}$$

and identify y with x_i, where x_i is a zero of $\pi(x)$, so that $\phi_m(x_i) = 0$, Eq. (8.4.11) specializes to the form

$$\frac{\phi_{m+1}(x_i)}{a_m\gamma_m} \frac{\phi_m(x)}{x - x_i} = -\sum_{k=0}^{m} \frac{\phi_k(x)\phi_k(x_i)}{\gamma_k} \tag{8.4.13}$$

The result of multiplying the equal members of (8.4.13) by $w(x)\phi_0(x)$, integrating the results over $[a, b]$, and making use of the orthogonality of the polynomials, relative to $w(x)$, is then

$$\frac{\phi_{m+1}(x_i)}{a_m\gamma_m} \int_a^b w(x) \frac{\phi_0(x)\phi_m(x)}{x - x_i}\, dx = -\phi_0(x_i)$$

or, since $\phi_0(x)$ is a constant,

$$\int_a^b w(x) \frac{\phi_m(x)}{x - x_i}\, dx = -\frac{a_m\gamma_m}{\phi_{m+1}(x_i)} \tag{8.4.14}$$

Finally, since

$$l_i(x) = \frac{\pi(x)}{\pi'(x_i)(x - x_i)} = \frac{\phi_m(x)}{\phi_m'(x_i)(x - x_i)} \tag{8.4.15}$$

reference to the second form of (8.4.8) leads to the desired result

$$H_i = \frac{1}{\phi_m'(x_i)} \int_a^b w(x) \frac{\phi_m(x)}{x - x_i}\, dx = -\frac{A_{m+1}\gamma_m}{A_m\phi_m'(x_i)\phi_{m+1}(x_i)} \tag{8.4.16}$$

† Whereas $\phi_m(x)$ could always be so defined that either γ_m or A_m is unity, this choice usually does not lead to a standard (tabulated) form. Hence, the formulas are given without such a restriction.

If use is made of the recurrence formula (7.10.3) in the form

$$\phi_{m+1}(x) = (a_m x + b_m)\phi_m(x) - \frac{a_m \gamma_m}{a_{m-1}\gamma_{m-1}} \phi_{m-1}(x)$$

there follows

$$\phi_{m+1}(x_i) = -\frac{A_{m+1}A_{m-1}}{A_m^2} \frac{\gamma_m}{\gamma_{m-1}} \phi_{m-1}(x_i)$$

so that H_i also can be expressed in the equivalent form

$$H_i = \frac{A_m \gamma_{m-1}}{A_{m-1}\phi_m'(x_i)\phi_{m-1}(x_i)} \tag{8.4.17}$$

In this connection, it may be noted that the result of setting $x = x_i$ in the confluent form (7.10.11) of the Christoffel-Darboux identity, where x_i is a zero of $\phi_m(x)$, is the curious relationship

$$\sum_{k=0}^{m} \frac{[\phi_k(x_i)]^2}{\gamma_k} = -\frac{A_m \phi_m'(x_i)\phi_{m+1}(x_i)}{A_{m+1}\gamma_m} = \frac{1}{H_i} \tag{8.4.18}$$

8.5 Legendre-Gauss Quadrature

In the case when a constant weighting function is to be used over a finite interval, it is convenient to suppose that a suitable change of variables has transformed that interval into the interval $[-1, 1]$. From the results of Sec. 7.6, we then have

$$\pi(x) = \frac{1}{A_m} P_m(x) \tag{8.5.1}$$

where $P_m(x)$ is the mth Legendre polynomial, and where

$$A_m = \frac{(2m)!}{2^m(m!)^2} \tag{8.5.2}$$

With the additional result

$$\gamma_m = \frac{2}{2m+1} \tag{8.5.3}$$

Equations (8.4.6), (8.4.16) or (8.4.17), and (8.4.9) reduce to the quadrature formula

$$\int_{-1}^{1} f(x)\,dx = \sum_{k=1}^{m} H_k f(x_k) + E \tag{8.5.4}$$

where x_i is the ith zero of $P_m(x)$, and where

$$H_i = -\frac{2}{(m+1)P_{m+1}(x_i)P'_m(x_i)} = \frac{2}{mP_{m-1}(x_i)P'_m(x_i)} \tag{8.5.5}$$

and

$$E = \frac{2^{2m+1}(m!)^4}{(2m+1)[(2m)!]^3} f^{(2m)}(\xi) \tag{8.5.6}$$

From the known relation†

$$\begin{aligned}(1-x^2)P'_m(x) &= -mxP_m(x) + mP_{m-1}(x) \\ &= (m+1)xP_m(x) - (m+1)P_{m+1}(x)\end{aligned} \tag{8.5.7}$$

there follows also

$$(1-x_i^2)P'_m(x_i) = mP_{m-1}(x_i) = -(m+1)P_{m+1}(x_i)$$

so that (8.5.5) can also be expressed in the forms

$$H_i = \frac{2}{(1-x_i^2)[P'_m(x_i)]^2} = \frac{2(1-x_i^2)}{(m+1)^2[P_{m+1}(x_i)]^2} \tag{8.5.8}$$

as well as in a variety of still other forms.

In illustration, when $m = 3$, there follows

$$\pi(x) = \tfrac{2}{5}P_3(x) = x(x^2 - \tfrac{3}{5})$$

and hence

$$x_1 = -\frac{\sqrt{15}}{5} \qquad x_2 = 0 \qquad x_3 = \frac{\sqrt{15}}{5}$$

and

$$H_1 = \tfrac{5}{9} \qquad H_2 = \tfrac{8}{9} \qquad H_3 = \tfrac{5}{9}$$

Thus

$$\int_{-1}^{1} f(x)\,dx = \frac{1}{9}\left[5f\left(-\frac{\sqrt{15}}{5}\right) + 8f(0) + 5f\left(\frac{\sqrt{15}}{5}\right)\right] + \frac{f^{\text{vi}}(\xi)}{15750} \tag{8.5.9}$$

where $|\xi| < 1$, as given in (8.1.2). When $m = 1$, a formula equivalent to the *midpoint rule* (3.5.18) results. The abscissas and weights corresponding to formulas for which $2 \leqq m \leqq 5$ are listed (to six digits) in Table 8.1. More elaborate tabulations are listed in the references (see Sec. 8.17).

† For derivations of (8.5.7), and of other similar differential recurrence formulas listed in this chapter, see Szego [1967].

Table 8.1

m	Abscissas	Weights
2	± 0.577350	1
3	0	$\frac{8}{9}$
	± 0.774597	$\frac{5}{9}$
4	± 0.339981	0.652145
	± 0.861136	0.347855
5	0	0.568889
	± 0.538469	0.478629
	± 0.906180	0.236927

8.6 Laguerre-Gauss Quadrature

In the case when the weighting function

$$w(x) = e^{-x} \tag{8.6.1}$$

is used over the semi-infinite interval $[0, \infty)$, the results of Sec. 7.7 give

$$\pi(x) = \frac{1}{A_m} L_m(x) \tag{8.6.2}$$

where $L_m(x)$ is the mth Laguerre polynomial and where

$$A_m = (-1)^m \tag{8.6.3}$$

In addition, there follows

$$\gamma_m = (m!)^2 \tag{8.6.4}$$

and hence the formulas of Sec. 8.4 become

$$\int_0^\infty e^{-x} f(x)\, dx = \sum_{k=1}^{m} H_k f(x_k) + E \tag{8.6.5}$$

where x_i is the ith zero of $L_m(x)$,† and where

$$H_i = \frac{(m!)^2}{L'_m(x_i) L_{m+1}(x_i)} = -\frac{[(m-1)!]^2}{L'_m(x_i) L_{m-1}(x_i)} \tag{8.6.6}$$

and

$$E = \frac{(m!)^2}{(2m)!} f^{(2m)}(\xi) \tag{8.6.7}$$

where $0 < \xi < \infty$. From the relation

$$xL'_m(x) = mL_m(x) - m^2 L_{m-1}(x) = (x - m - 1)L_m(x) + L_{m+1}(x) \tag{8.6.8}$$

there follows also

$$x_i L'_m(x_i) = -m^2 L_{m-1}(x_i) = L_{m+1}(x_i)$$

† The more general case in which e^{-x} is replaced by $e^{-\alpha x}$ in (8.6.5) is clearly reduced to the present case by a simple change of variables.

so that (8.6.6) can also be expressed in the forms

$$H_i = \frac{(m!)^2}{x_i[L'_m(x_i)]^2} = \frac{(m!)^2 x_i}{[L_{m+1}(x_i)]^2} \tag{8.6.9}$$

In illustration, when $m = 2$, there follows

$$\pi(x) = L_2(x) = x^2 - 4x + 2$$

and hence

$$x_1 = 2 - \sqrt{2} \qquad x_2 = 2 + \sqrt{2}$$

and

$$H_1 = \frac{2 + \sqrt{2}}{4} \qquad H_2 = \frac{2 - \sqrt{2}}{4}$$

Thus

$$\int_0^\infty e^{-x} f(x)\,dx = \tfrac{1}{4}[(2 + \sqrt{2})f(2 - \sqrt{2}) + (2 - \sqrt{2})f(2 + \sqrt{2})] + \frac{f^{\text{iv}}(\xi)}{6} \tag{8.6.10}$$

where $0 < \xi < \infty$ or, more generally,

$$\int_0^\infty e^{-\alpha x} f(x)\,dx = \frac{1}{4\alpha}\left[(2 + \sqrt{2})f\left(\frac{2 - \sqrt{2}}{\alpha}\right) + (2 - \sqrt{2})f\left(\frac{2 + \sqrt{2}}{\alpha}\right)\right] + \frac{f^{\text{iv}}(\xi)}{6\alpha^5} \tag{8.6.11}$$

The abscissas and weights corresponding to formulas for which $2 \leqq m \leqq 5$ are listed (to six digits) in Table 8.2. Other tabulations are listed in the references.

Table 8.2

m	Abscissas	Weights
2	0.585786	0.853553
	3.414214	0.146447
3	0.415775	0.711093
	2.294280	0.278518
	6.289945	0.0103893
4	0.322548	0.603154
	1.745761	0.357419
	4.536620	0.0388879
	9.395071	0.000539295
5	0.263560	0.521756
	1.413403	0.398667
	3.596426	0.0759424
	7.085810	0.00361176
	12.640801	0.0000233700

It may be noted that the use of Laguerre-Gauss quadrature is by no means restricted to cases in which the integrand is explicitly expressed as a product of the form $e^{-\alpha x}f(x)$. Its use may be appropriate in approximating an integral

$$\int_0^\infty F(x)\,dx$$

in which $F(x)$ can be approximated by a polynomial of moderate degree for small and moderately large positive x and is known to tend to zero as $x \to \infty$ like $e^{-\alpha x}$ times a polynomial, where α is a known positive constant. In this case, one would identify $f(x)$ with $e^{\alpha x}F(x)$ in using (8.6.5). As another example, we note that

$$\int_{-\infty}^{\infty} e^{-\sqrt{x^2+a^2}}F(x)\,dx = \int_0^\infty e^{-x}f(x)\,dx$$

with the definition

$$f(x) = \tfrac{1}{2}e^{-(\sqrt{x^2+a^2}-x)}[F(x) + F(-x)]$$

Similar comments apply to the other quadrature formulas to be considered.

When the weighting function (8.6.1) is generalized to the function

$$w(x) = x^\beta e^{-x} \qquad (\beta > -1) \tag{8.6.12}$$

it is easily found, by the methods of Sec. 7.7, that

$$\pi(x) = \frac{1}{A_m} L_m^\beta(x) \tag{8.6.13}$$

where $L_m^\beta(x)$ is the generalized Laguerre polynomial of degree m

$$L_m^\beta(x) = e^x x^{-\beta}\frac{d^m}{dx^m}(e^{-x}x^{\beta+m}) \tag{8.6.14}$$

and that

$$A_m = (-1)^m \qquad \gamma_m = m!\int_0^\infty x^{\beta+m}e^{-x}\,dx = m!\,\Gamma(m+\beta+1) \tag{8.6.15}$$

It can also be shown that the differential recurrence formula

$$\begin{aligned} xL_m^{\beta\prime}(x) &= mL_m^\beta(x) - m(m+\beta)L_{m-1}^\beta(x) \\ &= (x - m - \beta - 1)L_m^\beta + L_{m+1}^\beta \end{aligned} \tag{8.6.16}$$

is satisfied.

From these results, the corresponding quadrature formula is derived in the form

$$\int_0^\infty x^\beta e^{-x}f(x)\,dx = \sum_{k=1}^{m} H_k f(x_k) + E \tag{8.6.17}$$

where x_i is the ith zero of $L_m^\beta(x)$, and where

$$H_i = \frac{m!\,\Gamma(m+\beta+1)}{x_i[L_m^{\beta\prime}(x_i)]^2} = \frac{m!\,\Gamma(m+\beta+1)x_i}{[L_{m+1}^\beta(x_i)]^2} \tag{8.6.18}$$

and

$$E = \frac{m!\,\Gamma(m+\beta+1)}{(2m)!} f^{(2m)}(\xi) \tag{8.6.19}$$

8.7 Hermite-Gauss Quadrature

In the case when the weighting function

$$w(x) = e^{-x^2} \tag{8.7.1}$$

is used over the interval $(-\infty, \infty)$, the results of Sec. 7.8 give

$$\pi(x) = \frac{1}{A_m} H_m(x) \tag{8.7.2}$$

where $H_m(x)$ is the mth Hermite polynomial, and where

$$A_m = 2^m \tag{8.7.3}$$

In addition, there follows

$$\gamma_m = \sqrt{\pi}\, 2^m m! \tag{8.7.4}$$

so that the appropriate gaussian formula is of the form

$$\int_{-\infty}^{\infty} e^{-x^2} f(x)\, dx = \sum_{k=1}^{m} H_k f(x_k) + E \tag{8.7.5}$$

where x_i is the ith zero of $H_m(x)$, and where

$$H_i = -\frac{2^{m+1} m!\,\sqrt{\pi}}{H_m'(x_i)H_{m+1}(x_i)} = \frac{2^m(m-1)!\,\sqrt{\pi}}{H_m'(x_i)H_{m-1}(x_i)} \tag{8.7.6}$$

and

$$E = \frac{m!\,\sqrt{\pi}}{2^m(2m)!} f^{(2m)}(\xi) \tag{8.7.7}$$

for some ξ. From the relation

$$H_m'(x) = 2mH_{m-1}(x) = 2xH_m(x) - H_{m+1}(x) \tag{8.7.8}$$

there follows also

$$H_m'(x_i) = 2mH_{m-1}(x_i) = -H_{m+1}(x_i)$$

so that (8.7.6) can also be expressed in the forms

$$H_i = \frac{2^{m+1} m!\,\sqrt{\pi}}{[H_m'(x_i)]^2} = \frac{2^{m+1} m!\,\sqrt{\pi}}{[H_{m+1}(x_i)]^2} \tag{8.7.9}$$

In illustration, when $m = 3$, there follows

$$\pi(x) = \tfrac{1}{8}H_3(x) = x(x^2 - \tfrac{3}{2})$$

and hence

$$x_1 = -\frac{\sqrt{6}}{2} \qquad x_2 = 0 \qquad x_3 = \frac{\sqrt{6}}{2}$$

and

$$H_1 = \frac{\sqrt{\pi}}{6} \qquad H_2 = \frac{2\sqrt{\pi}}{3} \qquad H_3 = \frac{\sqrt{\pi}}{6}$$

Thus

$$\int_{-\infty}^{\infty} e^{-x^2} f(x)\,dx = \frac{\sqrt{\pi}}{6}\left[f\left(-\frac{\sqrt{6}}{2}\right) + 4f(0) + f\left(\frac{\sqrt{6}}{2}\right)\right] + \frac{\sqrt{\pi}\,f^{\text{vi}}(\xi)}{960} \tag{8.7.10}$$

or, more generally,

$$\int_{-\infty}^{\infty} e^{-\alpha^2 x^2} f(x)\,dx = \frac{\sqrt{\pi}}{6\alpha}\left[f\left(-\frac{\sqrt{6}}{2\alpha}\right) + 4f(0) + f\left(\frac{\sqrt{6}}{2\alpha}\right)\right] + \frac{\sqrt{\pi}\,f^{\text{vi}}(\xi)}{960\alpha^7} \tag{8.7.11}$$

The abscissas and weights corresponding to formulas for which $2 \leqq m \leqq 5$ are listed (to six digits) in Table 8.3. More extensive tabulations are listed in the references.

Table 8.3

m	Abscissas	Weights
2	± 0.707107	0.886227
3	0	1.181636
	± 1.224745	0.295409
4	± 0.524648	0.804914
	± 1.650680	0.0813128
5	0	0.945309
	± 0.958572	0.393619
	± 2.020183	0.0199532

It should be noticed that no restrictions are imposed on ξ in the error formula (8.7.7), other than that it be real. (Similarly, in the error formula of the preceding section, it is known only that ξ is real and positive.) Thus, in those cases when $f^{(2m)}(x)$ varies greatly in magnitude when m is large, as x takes on all real values, the imprecision associated with the use of the error formula (8.7.7) is correspondingly great.

In this connection, it may be recalled that the truncation error in the general gaussian quadrature formula (8.4.6) can be expressed in the form

$$E = \frac{\gamma_m}{A_m^2} f[x_1, x_1, x_2, x_2, \ldots, x_m, x_m, \xi_1] \qquad (8.7.12)$$

according to (8.4.10), where $a < \xi_1 < b$. This form reduces to the form of (8.4.9)

$$E = \frac{\gamma_m}{A_m^2} \frac{f^{(2m)}(\xi_2)}{(2m)!} \qquad (8.7.13)$$

when the divided difference is replaced by $f^{(2m)}(\xi_2)/(2m)!$, where $a < \xi_2 < b$. Since ξ_1 and ξ_2 generally cannot be estimated, one generally must replace either the divided difference or the derivative by its maximum absolute value for all ξ in $[a, b]$, to obtain an *upper bound* on $|E|$, and it may happen that the bound obtained from (8.7.12) is much less conservative than that obtained from (8.7.13).

In illustration, if $f(x) = 1/(x + a)$, there follows

$$f[x_1, x_1, \ldots, x_m, x_m, \xi_1] = \frac{1}{[(a + x_1)\cdots(a + x_m)]^2(a + \xi_1)}$$

and

$$\frac{f^{(2m)}(\xi_2)}{(2m)!} = \frac{1}{(a + \xi_2)^{2m+1}}$$

Thus, for example, if five-point Laguerre-Gauss quadrature were to be used to approximate the integral

$$\int_0^\infty \frac{e^{-x}}{x + 1}\, dx$$

the truncation-error terms corresponding to the use of (8.7.12) and (8.7.13) would be of the forms

$$\frac{(120)^2}{(2.39 \times 10^6)(1 + \xi_1)} \doteq \frac{0.0060}{1 + \xi_1} \qquad \text{and} \qquad \frac{1.44 \times 10^4}{(1 + \xi_2)^{11}}$$

respectively, where the abscissas are taken from Table 8.2 and where ξ_1 and ξ_2 are known only to be positive. Accordingly, the use of (8.7.12) here permits the determination of an error bound which is smaller than that obtainable from (8.7.13) in a ratio of about 2.4×10^6. The actual truncation error rounds to 0.0013.

Whereas this case is a rather extreme one, still, in those special cases when $f(x)$ is such that an upper bound on the magnitude of the relevant divided difference can be obtained or estimated practically, the use of (8.7.12) is usually preferable to that of (8.7.13).

8.8 Chebyshev-Gauss Quadrature

For the weighting function

$$w(x) = \frac{1}{\sqrt{1 - x^2}} \tag{8.8.1}$$

over the interval $[-1, 1]$, the results of Sec. 7.9 give

$$\pi(x) = \frac{1}{A_m} T_m(x) \tag{8.8.2}$$

where $T_m(x)$ is the mth Chebyshev polynomial. With the additional results

$$A_m = 2^{m-1} \qquad \gamma_m = \frac{\pi}{2} \tag{8.8.3}$$

the relevant gaussian formula is obtained in the form

$$\int_{-1}^{1} \frac{f(x)}{\sqrt{1 - x^2}}\, dx = \sum_{k=1}^{m} H_k f(x_k) + E \tag{8.8.4}$$

where x_i is the ith zero of $T_m(x)$ and where

$$H_i = -\frac{\pi}{T'_m(x_i) T_{m+1}(x_i)} \tag{8.8.5}$$

and

$$E = \frac{2\pi}{2^{2m}(2m)!} f^{(2m)}(\xi) \tag{8.8.6}$$

where $|\xi| < 1$.

Since

$$T_m(x) = \cos\,(m \cos^{-1} x) \tag{8.8.7}$$

the abscissas are obtainable in the explicit form

$$x_i = \cos\left[\frac{(2i - 1)\pi}{2m}\right] \qquad (i = 1, 2, \ldots, m) \tag{8.8.8}$$

Also, direct calculation shows that

$$T'_m(x_i) = \frac{(-1)^{i+1} m}{\sin \alpha_i} \qquad T_{m+1}(x_i) = (-1)^i \sin \alpha_i \tag{8.8.9}$$

where

$$\alpha_i = \frac{2i - 1}{2m}\,\pi \tag{8.8.10}$$

and hence (8.8.5) reduces to the remarkably simple form

$$H_i = \frac{\pi}{m} \tag{8.8.11}$$

Thus *the weights in* (8.8.4) *are all equal.*

The formula (8.8.4) hence can be written in the explicit form

$$\int_{-1}^{1} \frac{f(x)}{\sqrt{1-x^2}}\,dx = \frac{\pi}{m}\sum_{k=1}^{m} f\left(\cos\frac{2k-1}{2m}\pi\right) + \frac{2\pi}{2^{2m}(2m)!}f^{(2m)}(\xi) \tag{8.8.12}$$

where $|\xi| < 1$.

The differential recurrence formula

$$(1-x^2)T_m'(x) = -mxT_m(x) + mT_{m-1}(x) = mxT_m(x) - mT_{m+1}(x) \tag{8.8.13}$$

(which was not needed here) is included for reference purposes.

8.9 Jacobi-Gauss Quadrature

Many of the other gaussian quadrature formulas which have been investigated in the literature correspond to the use of a specialization of the weighting function

$$w(x) = (1-x)^\alpha(1+x)^\beta \qquad (\alpha > -1, \beta > -1) \tag{8.9.1}$$

over the interval $[-1, 1]$, or to the result of transforming this problem to the interval $[0, 1]$. The special cases $\alpha = \beta = 0$ and $\alpha = \beta = -\frac{1}{2}$ have been considered in Secs. 8.5 and 8.8.

In the general case, we may take $\pi(x)$ as the appropriate multiple of the polynomial

$$\begin{aligned}\phi_m(x) &= C_m(1-x)^{-\alpha}(1+x)^{-\beta}\frac{d^m}{dx^m}\left[(1-x)^{\alpha+m}(1+x)^{\beta+m}\right] \\ &= (-1)^m C_m 2^m m! \sum_{k=0}^{m}\binom{m+\alpha+\beta+k}{k}\binom{m+\alpha}{m-k}\left(\frac{x-1}{2}\right)^k \\ &\equiv C_m V_m(x)\end{aligned} \tag{8.9.2}$$

which, as was noted in Sec. 7.9, reduces with a certain (not universally agreed upon) choice of C_m to the mth Jacobi polynomial.†

The coefficient of x^m is found to be

$$A_m = (-1)^m \frac{\Gamma(2m+\alpha+\beta+1)}{\Gamma(m+\alpha+\beta+1)} C_m \tag{8.9.3}$$

† See Szego [1967]. The choice made in that reference is $C_m = (-1)^m/(2^m m!)$.

whereas the normalizing factor is obtained, from (7.5.14), in the form

$$\begin{aligned}\gamma_m &= C_m^2 m! \frac{\Gamma(2m+\alpha+\beta+1)}{\Gamma(m+\alpha+\beta+1)} \int_{-1}^{1} (1-x)^{\alpha+m}(1+x)^{\beta+m}\,dx \\ &= C_m^2 m! \frac{\Gamma(2m+\alpha+\beta+1)}{\Gamma(m+\alpha+\beta+1)} \\ &\qquad \times \frac{\Gamma(m+\alpha+1)\Gamma(m+\beta+1)}{\Gamma(2m+\alpha+\beta+2)} 2^{2m+\alpha+\beta+1} \\ &= C_m^2 \frac{2^{2m+\alpha+\beta+1} m!}{2m+\alpha+\beta+1} \frac{\Gamma(m+\alpha+1)\Gamma(m+\beta+1)}{\Gamma(m+\alpha+\beta+1)} \end{aligned} \qquad (8.9.4)$$

where use was made of the formula

$$\int_{-1}^{1} (1-x)^p(1+x)^q\,dx = 2^{p+q+1} \frac{\Gamma(p+1)\Gamma(q+1)}{\Gamma(p+q+2)} \qquad (p > -1,\ q > -1) \qquad (8.9.5)$$

The results (8.9.2) to (8.9.4) reduce to (8.5.1) to (8.5.3) when $\alpha = \beta = 0$ with the choice $C_m = (-1)^m/(2^m m!)$, corresponding to $\phi_m(x) = P_m(x)$, and to (8.8.2) and (8.8.3) when $\alpha = \beta = -\frac{1}{2}$ with the special choice $C_m = (-2)^m m!/(2m)!$, corresponding to $\phi_m(x) = T_m(x)$.

Thus we obtain the quadrature formula

$$\int_{-1}^{1} (1-x)^\alpha(1+x)^\beta f(x)\,dx = \sum_{k=1}^{m} H_k f(x_k) + E \qquad (8.9.6)$$

where x_i is the ith zero of $V_m(x)$, and where, from (8.4.16) or (8.4.17) and (8.4.10),

$$H_i = \frac{2m+\alpha+\beta+2}{m+\alpha+\beta+1} \frac{\Gamma(m+\alpha+1)\Gamma(m+\beta+1)}{\Gamma(m+\alpha+\beta+1)} \frac{2^{2m+\alpha+\beta+1} m!}{V_m'(x_i)V_{m+1}(x_i)} \qquad (8.9.7)$$

and

$$E = \frac{\Gamma(m+\alpha+1)\Gamma(m+\beta+1)\Gamma(m+\alpha+\beta+1)}{(2m+\alpha+\beta+1)[\Gamma(2m+\alpha+\beta+1)]^2} \frac{2^{2m+\alpha+\beta+1} m!}{(2m)!} f^{(2m)}(\xi) \qquad (8.9.8)$$

where $|\xi| < 1$. Integration based on (8.9.6) is sometimes known as *Mehler quadrature*.

It is possible to establish the relation (see Szego [1967])

$$\begin{aligned}(2m+\alpha+\beta+2)(1-x^2)V_m'(x) &= (m+\alpha+\beta+1)[(2m+\alpha+\beta+2)x + (\alpha-\beta)]V_m(x) \\ &\qquad + (m+\alpha+\beta+1)V_{m+1}(x)\end{aligned} \qquad (8.9.9)$$

from which there follows

$$V_{m+1}(x_i) = \frac{2m + \alpha + \beta + 2}{m + \alpha + \beta + 1}(1 - x_i^2)V_m'(x_i)$$

so that (8.9.7) can also be written in the somewhat simpler form

$$H_i = \frac{\Gamma(m + \alpha + 1)\Gamma(m + \beta + 1)}{\Gamma(m + \alpha + \beta + 1)} \frac{2^{2m+\alpha+\beta+1}m!}{(1 - x_i^2)[V_m'(x_i)]^2} \tag{8.9.10}$$

As an example, we consider the weighting function

$$w(x) = 1 - x^2 \tag{8.9.11}$$

in which case (8.9.2) gives

$$V_m(x) = (1 - x^2)^{-1} \frac{d^m}{dx^m}(1 - x^2)^{m+1}$$

By making use of the relationship

$$\frac{d^{r-k}(x^2 - 1)^r}{dx^{r-k}} = 2^r r! \frac{(r - k)!}{(r + k)!}(x^2 - 1)^k \frac{d^k P_r(x)}{dx^k} \tag{8.9.12}$$

this result can be written in the form

$$V_m(x) = \frac{(-1)^m 2^{m+1} m!}{m + 2} \frac{dP_{m+1}(x)}{dx}$$

Hence there follows

$$\int_{-1}^{1} (1 - x^2)f(x)\,dx = \sum_{k=1}^{m} H_k f(x_k) + E \tag{8.9.13}$$

where x_i is the ith zero of $P_{m+1}'(x)$, and where, from (8.9.10) and (8.9.8),

$$H_i = \frac{2(m + 1)(m + 2)}{(1 - x_i^2)[P_{m+1}''(x_i)]^2} \tag{8.9.14}$$

and

$$E = \frac{m!\,(m + 2)!}{(2m)!}\left[\frac{(m + 1)!}{(2m + 2)!}\right]^2 \frac{2^{2m+3}}{2m + 3} f^{(2m)}(\xi) \tag{8.9.15}$$

Since $P_{m+1}(x)$ satisfies the differential equation

$$(1 - x^2)P_{m+1}''(x) - 2xP_{m+1}'(x) + (m + 1)(m + 2)P_{m+1}(x) = 0$$

there follows

$$(1 - x_i^2)P_{m+1}''(x_i) = -(m + 1)(m + 2)P_{m+1}(x_i)$$

when $P_{m+1}'(x_i) = 0$, so that (8.9.14) can also be expressed in the form

$$H_i = \frac{2(1 - x_i^2)}{(m + 1)(m + 2)[P_{m+1}(x_i)]^2} \tag{8.9.16}$$

8.10 Formulas with Assigned Abscissas

In some applications it is desirable to *prescribe* one or more of the m abscissas to be involved in a quadrature formula. In particular, whereas none of the true gaussian formulas involves the values of $f(x)$ at the *ends* of the interval, it is sometimes important that one or both of these end values be used. It may be expected that, for each *arbitrarily* prescribed abscissa, the degree of precision generally will be reduced by unity below the maximum value of $2m - 1$. In particular, if *all* abscissas were prescribed, the maximum degree of precision would generally be reduced to $m - 1$. Naturally, exceptions occur when the abscissas are preassigned in favorable ways.

Whereas the gaussian formulas were derived in Sec. 8.4 from the hermitian formulas, by requiring that the m weights $\overline{H}_i$ vanish, a somewhat different approach (which also could have been used in the gaussian case) is desirable here. We recall first that the lagrangian quadrature formula

$$\int_a^b w(x)f(x)\,dx = \sum_{k=1}^{m} W_k f(x_k) + E \tag{8.10.1}$$

where

$$\pi(x) = (x - x_1)(x - x_2)\cdots(x - x_m) \tag{8.10.2}$$

and

$$W_k = \int_a^b w(x)l_i(x)\,dx \equiv \frac{1}{\pi'(x_i)}\int_a^b w(x)\frac{\pi(x)}{x - x_i}\,dx \tag{8.10.3}$$

always has a degree of precision of *at least* $m - 1$. Now any function $f(x)$ can be expressed as the sum

$$f(x) = p_{m-1}(x) + \pi(x)f[x_1, x_2, \ldots, x_m, x] \tag{8.10.4}$$

where $p_{m-1}(x)$ is the polynomial, of degree $m - 1$ or less, agreeing with $f(x)$ at the m points $x_1, \ldots, x_m$, and where $f[x_1, \ldots, x_m, x]$ is the mth divided difference of $f(x)$, relative to $x_1, \ldots, x_m$, defined in Chap. 2. Hence $f(x)$ can be replaced by that sum in (8.10.1). But since the two terms involving $p_{m-1}(x)$ in the result will cancel, and since $\pi(x_i) = 0$, we thus obtain the expression

$$E = \int_a^b w(x)\pi(x)f[x_1, \ldots, x_m, x]\,dx \tag{8.10.5}$$

for the error E in (8.10.1).†

If $f(x)$ is a polynomial of degree $m + r$, its divided difference of order m is a polynomial of degree r, and conversely. Hence we deduce from (8.10.5)

† This result is equivalent to (5.10.34). The derivation is repeated here, for completeness, in the modified notation of the present chapter.

that *the quadrature formula* (8.10.1) *has a degree of precision of at least* $m + r - 1$ *if and only if the polynomial* $\pi(x)$, *whose* m *zeros are the abscissas, is orthogonal, relative to* $w(x)$, *to all polynomials of degree less than* r. When $r = m$, this result reduces to the result of Sec. 8.4 and serves to specify the *gaussian* quadrature formulas, which were also derivable as special Hermite formulas for which $\overline{H}_i = 0$, and for which also $H_i = W_i$.

Now suppose that $m - r$ of the m abscissas are preassigned, leaving the r "free" abscissas $x_1, x_2, \ldots, x_r$ to be determined so that the degree of precision will be maximized. If we write

$$\begin{aligned}\pi(x) &= [(x - x_1) \cdots (x - x_r)][(x - x_{r+1}) \cdots (x - x_m)] \\ &\equiv \bar{\pi}(x)v(x)\end{aligned} \tag{8.10.6}$$

where

$$\bar{\pi}(x) = (x - x_1)(x - x_2) \cdots (x - x_r) \tag{8.10.7}$$

is a polynomial of degree r whose r zeros are the *free* abscissas, which are to be determined, and where

$$v(x) = (x - x_{r+1})(x - x_{r+2}) \cdots (x - x_m) \tag{8.10.8}$$

is a *known* polynomial, of degree $m - r$, whose zeros are the *preassigned* abscissas, the condition

$$\int_a^b w(x)\pi(x)u_{r-1}(x)\,dx = 0 \tag{8.10.9}$$

where $u_{r-1}(x)$ is an *arbitrary* polynomial of degree $r - 1$ or less, takes the form

$$\int_a^b [w(x)v(x)]\bar{\pi}(x)u_{r-1}(x)\,dx = 0 \tag{8.10.10}$$

Thus we may consider $\bar{\pi}(x)$ as the appropriate multiple of the rth member of a set of polynomials $\phi_0(x), \phi_1(x), \ldots, \phi_r(x), \ldots,$ of degrees $0, 1, \ldots, r, \ldots,$ respectively, which are orthogonal over $[a, b]$ relative to the *modified weighting function*

$$\bar{w}(x) = w(x)v(x) \tag{8.10.11}$$

and the methods of Sec. 7.5 are again available for its determination. However, if $v(x)$ changes sign in $[a, b]$, the modified weighting function $\bar{w}(x)$ will have the same property. Thus there is then no assurance that the zeros of $\phi_r(x)$ will be *real* or, if so, that they will lie inside $[a, b]$. In the important cases for which only one or both of the *end points* $x = a$ and $x = b$ are taken as preassigned abscissas, so that $v(x)$ is given by $x - a$, $x - b$, or $(x - a)(x - b)$, this difficulty does not arise since then $v(x)$ is of fixed sign in $[a, b]$. Attention will be restricted to these cases in what follows.

In order to evaluate the weights W_i, we write $\bar{\pi}(x) = \phi_r(x)/A_r$ to take into account the fact that the polynomial $\phi_r(x)$ which is most conveniently employed may not have unity as its leading coefficient, and notice that then

$$\pi(x) = \frac{1}{A_r} v(x)\phi_r(x)$$

where $v(x)$ is defined by (8.10.8). Equation (8.10.3) then becomes

$$W_i = \frac{1}{\{[v(x)\phi_r(x)]'\}_{x=x_i}} \int_a^b w(x) \frac{v(x)\phi_r(x)}{x - x_i} dx \quad (8.10.12)$$

and is, of course, independent of A_r. For $i = 1, 2, \ldots, r$ the abscissa x_i is a zero of $\phi_r(x)$. Hence there follows

$$W_i = \frac{1}{v(x_i)\phi_r'(x_i)} \int_a^b w(x)v(x) \frac{\phi_r(x)}{x - x_i} dx \quad (8.10.13)$$

for $i = 1, 2, \ldots, r$, and a comparison of this form with (8.4.16), with m replaced by r and $w(x)$ by $\bar{w}(x)$, leads to the desired result

$$W_i = -\frac{A_{r+1}\bar{\gamma}_r}{A_r v(x_i)\phi_r'(x_i)\phi_{r+1}(x_i)} \qquad (i = 1, 2, \ldots, r) \quad (8.10.14)$$

where A_r is the coefficient of x^r in $\phi_r(x)$, and where

$$\bar{\gamma}_r = \int_a^b \bar{w}(x)[\phi_r(x)]^2 \, dx \equiv \int_a^b w(x)v(x)[\phi_r(x)]^2 \, dx \quad (8.10.15)$$

Equation (8.10.14) determines all weights except those corresponding to the preassigned abscissas.

In the case when only the abscissa $x = a$ is preassigned, so that

$$v(x) = x - a$$

the corresponding weight is expressed by (8.10.12) in the form

$$W = \frac{1}{\phi_r(a)} \int_a^b w(x)\phi_r(x) \, dx \qquad (x_i = a) \quad (8.10.16)$$

whereas when only $x = b$ is fixed, so that

$$v(x) = x - b$$

there follows

$$W = \frac{1}{\phi_r(b)} \int_a^b w(x)\phi_r(x) \, dx \qquad (x_i = b) \quad (8.10.17)$$

In the case when both $x = a$ and $x = b$ are fixed, so that

$$v(x) = (x - a)(x - b)$$

there follows

$$W = \frac{1}{(b - a)\phi_r(a)} \int_a^b (b - x)w(x)\phi_r(x)\,dx \qquad (x_i = a) \qquad (8.10.18)$$

and

$$W = \frac{1}{(b - a)\phi_r(b)} \int_a^b (x - a)w(x)\phi_r(x)\,dx \qquad (x_i = b) \qquad (8.10.19)$$

Alternatively, the weights corresponding to the prescribed end ordinate or ordinates can be determined in terms of the remaining weights by use of one or both of the relations

$$\sum_{k=1}^{m} W_k = \int_a^b w(x)\,dx \qquad \sum_{k=1}^{m} x_k W_k = \int_a^b x w(x)\,dx \qquad (8.10.20)$$

which require that the error in (8.10.1) vanish when $f(x) = 1$ and when $f(x) = x$, respectively.

The special cases in which $w(x)$ is constant are treated in the two following sections.

In the general case, it is possible to show that the relevant quadrature formula can be obtained by replacing $f(x)$ in the integrand by the polynomial of degree $m + r - 1$ which agrees with $f(x)$ when $x = x_1, \ldots, x_m$ and whose derivative agrees with $f'(x)$ at the unassigned points $x = x_1, \ldots, x_r$. Thus the error can be expressed in the form

$$E = \int_a^b w(x)[(x - x_1)\cdots(x - x_r)]^2(x - x_{r+1})\cdots(x - x_m) \times f[x_1, x_1, \ldots, x_r, x_r, x_{r+1}, \ldots, x_m, x]\,dx \qquad (8.10.21)$$

In particular, if $w(x) \geqq 0$ in $[a, b]$, if no assigned abscissas lie *inside* $[a, b]$, and if $f^{(m+r)}(x)$ exists in (a, b), there follows also

$$\begin{aligned} E &= \frac{f^{(m+r)}(\xi)}{(m + r)!} \int_a^b w(x)[(x - x_1)\cdots(x - x_r)]^2\,(x - x_{r+1})\cdots(x - x_m)\,dx \\ &= \frac{f^{(m+r)}(\xi)}{(m + r)!} \int_a^b \bar{w}(x)[\bar{\pi}(x)]^2\,dx \\ &= \frac{\bar{\gamma}_r}{A_r^2}\,\frac{f^{(m+r)}(\xi)}{(m + r)!} \end{aligned} \qquad (8.10.22)$$

where ξ lies between the largest and smallest of $x_1, \ldots, x_m$, a, and b.

8.11 Radau Quadrature

In the case of a finite interval, with a unit weighting function, when one end of the interval is assigned as an abscissa, it is again convenient to suppose that the interval has been transformed to $[-1, 1]$, with $x = -1$ as the fixed abscissa, by an appropriate change in variables. We then have

$$v(x) = x + 1 \tag{8.11.1}$$

and

$$\pi(x) = (x + 1)\bar{\pi}(x) \tag{8.11.2}$$

where $\bar{\pi}(x)$ is a multiple of the rth member of a set of orthogonal polynomials $\phi_0(x), \phi_1(x), \ldots, \phi_r(x), \ldots$, which has the property

$$\int_{-1}^{1} (x + 1)\phi_r(x)u_{r-1}(x)\, dx = 0 \tag{8.11.3}$$

where $u_{r-1}(x)$ is an arbitrary polynomial of degree $r - 1$ or less.

If we follow the procedure of Sec. 7.5, by writing

$$(x + 1)\phi_r(x) = \frac{d^r}{dx^r} U_r(x)$$

and integrating the left-hand member of (8.11.3) by parts r times, we find that $U_r(x)$ must satisfy the equation

$$\frac{d^{r+1}}{dx^{r+1}}\left[\frac{1}{x + 1}\frac{d^r}{dx^r} U_r\right] = 0$$

and the requirements that $U_r, U_r', \ldots, U_r^{(r-1)}$ vanish when $x = \pm 1$, and hence that U_r must be of the form

$$U_r = C_r(x + 1)(x^2 - 1)^r$$

Thus it follows that

$$\phi_r(x) = \frac{C_r}{x + 1}\frac{d^r}{dx^r}[(x + 1)(x^2 - 1)^r] \tag{8.11.4}$$

which can be expressed in the form

$$\phi_r(x) = \frac{C_r}{x + 1}\left[(x + 1)\frac{d^r}{dx^r}(x^2 - 1)^r + r\frac{d^{r-1}}{dx^{r-1}}(x^2 - 1)^r\right]$$

or, by making use of the relationship (8.9.12), in the form

$$\phi_r(x) = 2^r r!\, C_r\left[P_r(x) + \frac{x - 1}{r + 1}P_r'(x)\right] \tag{8.11.5}$$

It is convenient to take

$$C_r = \frac{1}{2^r r!} \tag{8.11.6}$$

Then, noticing that here $r = m - 1$, since only one abscissa is preassigned, we conclude that the $m - 1$ free abscissas are the zeros of

$$\phi_{m-1}(x) = P_{m-1}(x) + \frac{x-1}{m} P'_{m-1}(x) = \frac{P_{m-1}(x) + P_m(x)}{1+x} \tag{8.11.7}$$

where the last form follows from the recurrence formula (8.5.7). The leading coefficient is found to be

$$A_{m-1} = \frac{(2m-1)!}{2^{m-1}m[(m-1)!]^2} \tag{8.11.8}$$

With this result, we notice next that

$$\begin{aligned}
\bar{\gamma}_{m-1} &= \int_{-1}^{1} (1+x)\phi_{m-1}(x)\phi_{m-1}(x)\,dx \\
&= A_{m-1} \int_{-1}^{1} x^{m-1}(1+x)\phi_{m-1}(x)\,dx \\
&= \frac{A_{m-1}}{2^{m-1}(m-1)!} \int_{-1}^{1} x^{m-1} \frac{d^{m-1}}{dx^{m-1}} [(1+x)(x^2-1)^{m-1}]\,dx
\end{aligned}$$

and an $(m - 1)$-fold integration by parts, followed by the use of (8.9.5), leads to the simple result

$$\bar{\gamma}_{m-1} = \frac{2}{m} \tag{8.11.9}$$

Thus, by introducing (8.11.8) and (8.11.9) into (8.10.14), we obtain the weights

$$W_i = -\frac{2(2m+1)}{m(m+1)} \frac{1}{(1+x_i)\phi'_{m-1}(x_i)\phi_m(x_i)} \qquad (x_i \neq -1) \tag{8.11.10}$$

corresponding to the $m - 1$ free abscissas. By making appropriate use of the formula (8.5.7), together with the fact that $\phi_{m-1}(x_i) = 0$ implies

$$P_{m-1}(x_i) = \frac{1-x_i}{m} P'_{m-1}(x_i) = -P_m(x_i)$$

we find, after some manipulation, that

$$\phi'_{m-1}(x_i) = \frac{2m}{1-x_i^2} P_{m-1}(x_i) \qquad \phi_m(x_i) = -\frac{2m+1}{m+1} P_{m-1}(x_i)$$

so that (8.11.10) reduces to

$$W_i = \frac{1}{m^2} \frac{1-x_i}{[P_{m-1}(x_i)]^2} = \frac{1}{1-x_i} \frac{1}{[P'_{m-1}(x_i)]^2} \qquad (x_i \neq -1) \tag{8.11.11}$$

The weight corresponding to the abscissa $x = -1$ follows from (8.10.16) in the form

$$W = \frac{1}{\phi_{m-1}(-1)} \int_{-1}^{1} \phi_{m-1}(x)\, dx \qquad (x_i = -1) \quad (8.11.12)$$

We obtain first

$$\int_{-1}^{1} \phi_{m-1}(x)\, dx = \int_{-1}^{1} \left[P_{m-1}(x) + \frac{x-1}{m} P'_{m-1}(x) \right] dx$$

or, after integrating the second member by parts and noticing that the first member integrates to zero (when $m > 1$),

$$\int_{-1}^{1} \phi_{m-1}(x)\, dx = \frac{2}{m} P_{m-1}(-1) = \frac{(-1)^{m-1}2}{m}$$

since $P_{m-1}(-1) = (-1)^{m-1}$. By making use of the additional fact that $P'_{m-1}(-1) = (-1)^m m(m-1)/2$, we obtain also

$$\phi_{m-1}(-1) = (-1)^{m-1} m \quad (8.11.13)$$

and hence (8.11.12) becomes

$$W = \frac{2}{m^2} \qquad (x_i = -1) \quad (8.11.14)$$

The error term is obtained, by use of (8.10.22), in the form

$$E = \frac{2^{2m-1} m[(m-1)!]^4}{[(2m-1)!]^3} f^{(2m-1)}(\xi) \qquad (|\xi| < 1) \quad (8.11.15)$$

Thus, in summary, we have obtained the quadrature formula

$$\int_{-1}^{1} f(x)\, dx = \frac{2}{m^2} f(-1) + \sum_{k=1}^{m-1} W_k f(x_k) + E \quad (8.11.16)$$

where x_i is the ith zero of the polynomial (8.11.7), and where the weights are defined by (8.11.11) and are *positive*. This formula is one of several attributed to *Radau*.

The first six of the polynomials are found to be of the form

$$\begin{aligned}
&\phi_0(x) = 1 \qquad\qquad \phi_1(x) = \tfrac{1}{2}(3x - 1) \\
&\phi_2(x) = \tfrac{1}{2}(5x^2 - 2x - 1) \qquad \phi_3(x) = \tfrac{1}{8}(35x^3 - 15x^2 - 15x + 3) \\
&\phi_4(x) = \tfrac{1}{40}(315x^4 - 140x^3 - 210x^2 + 60x + 15) \\
&\phi_5(x) = \tfrac{1}{16}(231x^5 - 105x^4 - 210x^3 + 70x^2 + 35x - 5)
\end{aligned} \quad (8.11.17)$$

and additional ones can be obtained from the recurrence formula

$$\phi_{r+1}(x) = \frac{1}{(r+2)(2r+1)}\{[(2r+1)(2r+3)x - 1]\phi_r(x) - r(2r+3)\phi_{r-1}(x)\} \qquad (8.11.18)$$

or by reference to (8.11.7).

In the simplest nontrivial case, $m = 2$, there follows $x_1 = \frac{1}{3}$. The weight W_1 is found to be $\frac{3}{2}$, and the weight relative to $x = -1$ to be $\frac{1}{2}$. Thus, the best two-point formula with $x = -1$ preassigned is of the form

$$\int_{-1}^{1} f(x)\,dx = \tfrac{1}{2}f(-1) + \tfrac{3}{2}f(\tfrac{1}{3}) + \tfrac{2}{27}f'''(\xi) \qquad (|\xi| < 1) \qquad (8.11.19)$$

By setting $x = 2t - 1$, and writing $f(2t - 1) = F(t)$, we may rewrite this formula in the form

$$\int_{0}^{1} F(t)\,dt = \tfrac{1}{4}F(0) + \tfrac{3}{4}F(\tfrac{2}{3}) + \tfrac{1}{216}F'''(\eta) \qquad (0 < \eta < 1) \qquad (8.11.20)$$

and similar forms can be obtained in the other cases. The abscissas and weights corresponding to formulas for which $2 \leqq m \leqq 5$ are listed (to six digits) in Table 8.4. More extensive tabulations are listed in the references.

Table 8.4

m	Abscissas	Weights
2	−1	$\frac{1}{2}$
	0.333333	$\frac{3}{2}$
3	−1	0.222222
	−0.289898	0.752806
	0.689898	1.024972
4	−1	0.125000
	−0.575319	0.657689
	0.181066	0.776387
	0.822824	0.440925
5	−1	0.080000
	−0.720480	0.446207
	−0.167181	0.623653
	0.446314	0.562712
	0.885792	0.287427

8.12 Lobatto Quadrature

In the case when both ends of the interval $[-1, 1]$ are preassigned as abscissas, the weighting function being unity, the derivation is quite similar to that of the preceding section. Thus, with

$$v(x) = x^2 - 1 \qquad (8.12.1)$$

it is found that

$$\phi_r(x) = \frac{C_r}{x^2 - 1}\frac{d^r}{dx^r}(x^2 - 1)^{r+1} \qquad (8.12.2)$$

and that, in accordance with (8.9.12), if we set

$$C_r = \frac{r + 2}{2^{r+1}r!} \qquad (8.12.3)$$

this result is of the form $\phi_r(x) = P'_{r+1}(x)$. Hence, since here $r = m - 2$, the free abscissas are the zeros of the polynomial

$$\phi_{m-2}(x) = P'_{m-1}(x) \qquad (8.12.4)$$

The additional results

$$A_{m-2} = \frac{(2m - 2)!}{2^{m-1}(m - 1)!\,(m - 2)!} \qquad (8.12.5)$$

and

$$\bar{\gamma}_{m-2} = -\frac{2m(m - 1)}{2m - 1} \qquad (8.12.6)$$

the negative sign in (8.12.6) being a consequence of the fact that here $v(x)$ is negative in $(-1, 1)$, are obtained by methods similar to those of the preceding section. Next the weights corresponding to the *free* abscissas are obtained in the form

$$W_i = -\frac{2m}{(1 - x_i^2)P''_{m-1}(x_i)P'_m(x_i)} \qquad (x_i \neq \pm 1)$$

which can be rewritten more conveniently as

$$W_i = \frac{2}{m(m - 1)[P_{m-1}(x_i)]^2} \qquad (x_i \neq \pm 1) \qquad (8.12.7)$$

The weights corresponding to the fixed abscissas $x = \pm 1$ are found to be equal and to have the value

$$W = \frac{P_{m-1}(1)}{P'_{m-1}(1)} = -\frac{P_{m-1}(-1)}{P'_{m-1}(-1)} = \frac{2}{m(m - 1)} \qquad (x_i = \pm 1) \qquad (8.12.8)$$

which is the same as that given by the right-hand member of (8.12.7) when $x_i = \pm 1$.

The error term is obtained from (8.10.22) in the form

$$E = -\frac{2^{2m-1}m(m - 1)^3[(m - 2)!]^4}{(2m - 1)[(2m - 2)!]^3} f^{(2m-2)}(\xi) \qquad (|\xi| < 1) \qquad (8.12.9)$$

and the corresponding quadrature formula

$$\int_{-1}^{1} f(x)\,dx = \frac{2}{m(m-1)}[f(1) + f(-1)] + \sum_{k=1}^{m-2} W_k f(x_k) + E \qquad (8.12.10)$$

where x_i is the ith zero of $P'_{m-1}(x)$, and W_i is given by (8.12.7) and is *positive*, is known as *Lobatto's quadrature formula.*

In the simplest nontrivial case, $m = 3$, the free abscissa is found from the equation $P'_2(x) \equiv 3x = 0$ to be $x = 0$, as would be expected from the symmetry. The corresponding weight is found to be $\frac{4}{3}$, whereas the weights corresponding to $x = \pm 1$ are each $\frac{1}{3}$. Hence, as also might have been anticipated, the Lobatto formula reduces in this simple case to *Simpson's rule.* The abscissas and weights corresponding to formulas for which $3 \leqq m \leqq 6$ are listed (to six digits) in Table 8.5. More elaborate tabulations are listed in the references.

Table 8.5

m	Abscissas	Weights
3	0	$\frac{4}{3}$
	± 1	$\frac{1}{3}$
4	± 0.447214	$\frac{5}{6}$
	± 1	$\frac{1}{6}$
5	0	$\frac{32}{45}$
	± 0.654654	$\frac{49}{90}$
	± 1	$\frac{1}{10}$
6	± 0.285232	0.554858
	± 0.765055	0.378475
	± 1	0.066667

When the Lobatto formula is applied to a function $f(x)$ which *vanishes* at both ends of the interval of integration, so that only $r \equiv m - 2$ ordinates are actually involved in the calculation, the degree of precision is $2m - 3 = 2r + 1$. Similarly, when the Radau formula is applied to an integrand which vanishes at the lower limit, so that $r \equiv m - 1$ ordinates are used, the degree of precision is $2m - 2 = 2r$. Thus, in such cases, a higher effective degree of precision is attained than that afforded by the formulas of gaussian type, in which the use of r ordinates leads to a degree of precision of $2r - 1$. However, in special cases where $f(x)$ has a singular derivative at an end point at which $f(x) = 0$, the associated advantage of an *open* formula may offset this consideration.

8.13 Convergence of Gaussian-Quadrature Sequences

For the purpose of considering the behavior of a *sequence* of approximations generated by a specific gaussian (or gaussian-type) formula as the number m of ordinates is increased, we express the m-point error in the form

$$E_m \equiv R_m[f(x)] = \int_a^b w(x)f(x)\,dx - \sum_{k=1}^{m} W_k^{(m)} f(x_k^{(m)}) \tag{8.13.1}$$

where the weights and abscissas now are supplied with the previously *implied* index m.

We first restrict attention to the cases when $[a, b]$ is finite. In order to include all such cases so far considered in this chapter as well as a class of other ones, we suppose here only that the weighting function $w(x)$ is nonnegative in (a, b) and all the weights $W_k^{(m)}$ are positive:

$$w(x) \geqq 0 \qquad (a \leqq x \leqq b) \qquad W_k^{(m)} > 0 \tag{8.13.2}$$

We also assume that the degree of precision of the formula in question is *positive* and *increasing with m.* (In the true gaussian cases that degree is $2m - 1$; for Radau or Lobatto quadrature it is $2m - 2$ or $2m - 3$, respectively.) We require of $f(x)$ only that it be *continuous* on $[a, b]$.

With these assumptions, given any positive number ε, no matter how small, the Weierstrass approximation theorem (Sec. 1.2) guarantees the existence of a polynomial $p(x)$ for which

$$|f(x) - p(x)| < \varepsilon \qquad (a \leqq x \leqq b) \tag{8.13.3}$$

If M is the degree of $p(x)$, we next take m sufficiently large that the degree of precision of the quadrature formula in question exceeds M, so that $R_m[p(x)] = 0$. Then, since we have

$$R_m[f] = R_m[f - p] + R_m[p]$$

from the linearity of the operator R_m, there follows

$$\begin{aligned} |R_m[f]| &= \left| \int_a^b w(x)[f(x) - p(x)]\,dx - \sum_{k=1}^{m} W_k^{(m)}[f(x_k^{(m)}) - p(x_k^{(m)})] \right| \\ &\leqq \int_a^b w(x)|f(x) - p(x)|\,dx + \sum_{k=1}^{m} W_k^{(m)}|f(x_k^{(m)}) - p(x_k^{(m)})| \\ &\leqq \varepsilon \left[\int_a^b w(x)\,dx + \sum_{k=1}^{m} W_k^{(m)} \right] = 2\varepsilon \int_a^b w(x)\,dx \end{aligned} \tag{8.13.4}$$

Here use was made of the assumed properties of $w(x)$ and $W_k^{(m)}$ and of the fact that $R_m[1] = 0$ implies the relation

$$\int_a^b w(x)\,dx = \sum_{k=1}^{m} W_k^{(m)} \tag{8.13.5}$$

Hence, since accordingly $|E_m|$ can be made smaller than any preassigned positive quantity by taking m to be sufficiently large, we may deduce that in fact $E_m \to 0$ as $m \to \infty$.

Although this conclusion thus follows for all the preceding formulas in this chapter for which $[a, b]$ is finite, some additional comments are in order. In the case of the Jacobi-Gauss quadrature of Sec. 8.9, over $(-1, 1)$, by making use of the Stirling approximation (3.9.6) to the factorial, and of the fact that $\Gamma(m + k + 1) \equiv (m + k)! \sim m^k m!$ as $m \to \infty$, when k is fixed, we find from (8.9.8) that

$$E_m \sim \frac{\pi}{2^{2m+\alpha+\beta}} \frac{f^{(2m)}(\xi_m)}{(2m)!} \qquad (8.13.6)$$

when m is large, in the general formula (8.9.6), where $-1 < \xi_m < 1$. This result holds, in particular, in the important special cases $\alpha = \beta = 0$ (Legendre-Gauss), $\alpha = \beta = -\frac{1}{2}$ (Chebyshev-Gauss), $\alpha = \beta = \frac{1}{2}$ (Prob. 24), and $\alpha = \beta = 1$ [Eq. (8.9.13)].

Thus, if $f(z)$ is an analytic function of the complex variable z in a region $\mathscr{R}$ of the complex z plane including the real interval $-1 \leqq x \leqq 1$, and if R is the shortest distance from a singular point of $f(z)$ to a point in that interval, then, as in Sec. 3.9, we can deduce from (8.13.6) only that there exists a constant C such that

$$|E_m| \leqq \frac{C}{(2R)^{2m}} \qquad (8.13.7)$$

when m is large. This fact would permit us to deduce convergence only when $f(z)$ is such that $R > \frac{1}{2}$, whereas actually convergence has just been established without any restriction on $f(x)$ except for continuity on $[-1, 1]$. In addition, however, it suggests that the *rate* of convergence when R is smaller than about $\frac{1}{2}$ will tend to be significantly less favorable than that described by (8.13.7) when $R > \frac{1}{2}$.

When the interval of integration is not finite, a less simple approach is needed since the Weierstrass theorem no longer is available. However, in the cases of Laguerre-Gauss and Hermite-Gauss quadrature, convergence has been established when $f(x)$ is continuous in every finite subinterval of $[0, \infty)$ [or $(-\infty, \infty)$] and also $f(x)$ is such that

$$w(x)|f(x)| < \frac{1}{|x|^{1+\rho}} \qquad (8.13.8)$$

for some $\rho > 0$ when x [or $|x|$] is sufficiently large.

It may be noted that for Hermite-Gauss quadrature the asymptotic bound corresponding to (8.13.7) would be proportional to $m!/(\sqrt{2}\, R)^{2m}$, whereas for

Laguerre-Gauss quadrature it would be proportional to $(m!)^2/R^{2m}$, from (8.6.7) and (8.7.7); both of these quantities increase unboundedly as $m \to \infty$ for *any* fixed R. Thus, in these cases an unfavorable *rate* of convergence may be feared unless $f(z)$ is an *entire function* of the complex variable z (that is, is analytic everywhere in the finite part of the complex z plane).

For example, if $f(x) = 1/(1 + x^2)$, it is seen from the Maclaurin expansion of $f(x)$ that $f^{(2m)}(x) = (-1)^m(2m)!$ when $x = 0$, so that in the evaluation of the integral

$$\int_{-\infty}^{\infty} e^{-x^2} \frac{dx}{1 + x^2} \doteq 1.34329$$

by m-point Hermite-Gauss quadrature, the formula (8.7.7) would admit the possibility of an error as large as $\sqrt{\pi}\,(m!/2^m)$ if the appropriate (but unknown) value of ξ_m were near zero. If this were indeed the case, the error would increase rapidly with m when $m > 2$. Thus the above-mentioned convergence implies that $|\xi_m|$ truly increases rapidly as m increases. The slowness of the convergence in this case was first noted by Rosser [1950], who reported that the errors corresponding to the use of 2, 10, and 16 points are about 0.16, 0.0016, and 0.00016, respectively. Similar slow convergence may be expected, more generally, whenever $f(z)$ is not an analytic function of z, or when $f(z)$ possesses singularities in the finite part of the complex z plane which are "fairly close" to the real axis.

8.14 Chebyshev Quadrature

By imposing various restrictions on the abscissas and/or weights in a formula of the type (8.10.1), various classes of quadrature formulas may be obtained in addition to those so far considered. In this connection, it may be noticed that, if the abscissas are required to be *equally spaced*, the Newton-Cotes formulas of Chap. 3 are obtained when $w(x) = 1$. In this case, m abscissas are fixed and the degree of precision may be expected to be reduced from $2m - 1$ to $m - 1$. However, as was seen in Chap. 3, when m is *odd*, so that the *midpoint* of the interval is one of the abscissas, the degree of precision is increased to m.†

Another interesting class of formulas, associated with the name of Chebyshev, is that in which all the *weights* are made equal. Whereas the significance of these formulas is perhaps more academic than practical, equality of the weighting coefficients is desirable, not only for convenience, but also in order

† It should be recalled that m here corresponds to $n + 1$ in Chap. 3.

that the effects of errors in the ordinates will be minimized. Here the common weight and the m abscissas are "free," and it may be expected that a formula with a degree of precision of at least m may be determinable. However, this expectation is not always to be realized.

We suppose again that the original interval has been transformed into $[-1, 1]$, so that the desired formula is of the form

$$\int_{-1}^{1} w(x)f(x)\,dx = W\sum_{k=1}^{m} f(x_k) + E[f(x)] \tag{8.14.1}$$

where W is the common weight. It may be noticed first that the weight W cannot be assigned if the degree of precision is to be positive, but is determined by the requirement that $E = 0$ when $f(x) = 1$ in the form

$$W = \frac{\lambda}{m} \qquad \text{where } \lambda = \int_{-1}^{1} w(x)\,dx \tag{8.14.2}$$

Now we *assume* that a set of m abscissas x_i exists in $[-1, 1]$ such that the degree of precision is indeed at least m and, as before, we write

$$\pi(x) = (x - x_1)(x - x_2)\cdots(x - x_m) \tag{8.14.3}$$

Then, following the derivation of Chebyshev, we identify $f(x)$ in particular with the special function

$$f(x) = \frac{1}{u - x} \qquad (u > 1) \tag{8.14.4}$$

in which case (8.14.1) becomes

$$\int_{-1}^{1} w(x)\frac{dx}{u - x} = \frac{\lambda}{m}\sum_{k=1}^{m}\frac{1}{u - x_k} + E\left[\frac{1}{u - x}\right] \tag{8.14.5}$$

The reason for choosing the special function $1/(u - x)$ is now seen if we notice that since

$$\log \pi(u) = \sum_{k=1}^{m} \log (u - x_k)$$

the finite sum in (8.14.5) can be expressed as

$$\frac{d}{du}\log \pi(u)$$

and hence that equation becomes

$$\int_{-1}^{1} w(x)\frac{dx}{u - x} = \frac{\lambda}{m}\frac{d}{du}[\log \pi(u)] + E\left[\frac{1}{u - x}\right] \tag{8.14.6}$$

or, after an integration with respect to u,

$$\int_{-1}^{1} w(x) \log (u - x)\, dx = \text{const} + \frac{\lambda}{m} \log \pi(u) - Q(u) \qquad (8.14.7)$$

where

$$Q(u) = \int_{u}^{\infty} E\left[\frac{1}{t - x}\right] dt \qquad (8.14.8)$$

Equation (8.14.7) can be resolved in the form

$$\pi(u) = C_m \exp\left[\frac{m}{\lambda} \int_{-1}^{1} w(x) \log (u - x)\, dx + \frac{m}{\lambda} Q(u)\right]$$

or, equivalently, in the form

$$\pi(u) = C_m u^m \exp\left[\frac{m}{\lambda} \int_{-1}^{1} w(x) \log \left(1 - \frac{x}{u}\right) dx\right] \exp\left[\frac{m}{\lambda} Q(u)\right] \qquad (8.14.9)$$

Now the error term in (8.14.6) is expressible in the form

$$E\left[\frac{1}{u - x}\right] = \int_{-1}^{1} G(s) \frac{d^{m+1}}{ds^{m+1}} \left(\frac{1}{u - s}\right) ds$$

$$= (m + 1)! \int_{-1}^{1} G(s) \frac{ds}{(u - s)^{m+2}} \qquad (8.14.10)$$

where $G(s)$ is the influence function defined by the relation

$$m!\, G(s) = \int_{s}^{1} w(x)(x - s)^m\, dx - \frac{\lambda}{m} \sum_{x_k > s} (x_k - s)^m \qquad (8.14.11)$$

in accordance with (5.10.15) and (5.10.16). Accordingly, there follows

$$Q(u) \equiv \int_{u}^{\infty} E\left[\frac{1}{t - x}\right] dt = m! \int_{-1}^{1} G(s) \frac{ds}{(u - s)^{m+1}} \qquad (8.14.12)$$

For present purposes, it is not necessary to evaluate this expression explicitly. However, it is important to notice that it can be expanded in the form

$$Q(u) = m! \frac{1}{u^{m+1}} \int_{-1}^{1} G(s) \left[1 + (m + 1) \frac{s}{u} + \frac{(m + 1)(m + 2)}{2!} \frac{s^2}{u^2} + \cdots\right] ds$$

$$= \frac{g_0}{u^{m+1}} + \frac{g_1}{u^{m+2}} + \cdots \qquad (8.14.13)$$

since $u > 1$, where the coefficient g_k is a certain multiple of $\int_{-1}^{1} s^k G(s)\, ds$.

Similarly, we see that

$$\int_{-1}^{1} w(x) \log\left(1 - \frac{x}{u}\right) dx = -\int_{-1}^{1} w(x)\left(\frac{x}{u} + \frac{x^2}{2u^2} + \cdots\right) dx$$

$$= -\frac{c_1}{u} - \frac{c_2}{2u^2} - \cdots \tag{8.14.14}$$

when $u > 1$, where

$$c_k = \int_{-1}^{1} x^k w(x)\, dx \tag{8.14.15}$$

Hence (8.14.9) can be expanded in the form

$$\pi(u) = C_m u^m \exp\left[-\frac{m}{\lambda}\left(\frac{c_1}{u} + \frac{c_2}{2u^2} + \cdots\right)\right]$$

$$\times \exp\left[\frac{m}{\lambda}\left(\frac{g_0}{u^{m+1}} + \frac{g_1}{u^{m+2}} + \cdots\right)\right]$$

$$= C_m u^m \left(1 - \frac{m}{\lambda}\frac{c_1}{u} + \cdots\right)\left(1 + \frac{m}{\lambda}\frac{g_0}{u^{m+1}} + \cdots\right) \tag{8.14.16}$$

where the two relevant power series in u^{-1} converge when $u > 1$. But, since $\pi(u)$ is a *polynomial* of degree m, the *product* of the two series will *terminate* before the term containing u^{-m-1}. Thus the terms in the second series, after the leading term, therefore do not enter into the determination of the terms which will remain in the product, but serve only to bring about the cancellation of all terms involving u^{-m-1}, u^{-m-2}, and so forth. Hence the second series can be disregarded, and the desired polynomial can be obtained by merely *terminating* the first series with the term involving u^{-m}. Also, since the coefficient of u^m in $\pi(u)$ is to be unity, we must take $C_m = 1$.

It thus follows that *if* $\pi(x)$ *exists such that* (8.14.1) *has a degree of precision of at least m, then* $\pi(x)$ *is defined by the expansion*

$$\exp\left[\frac{m}{\lambda}\int_{-1}^{1} w(t) \log(x - t)\, dt\right] \equiv x^m \exp\left[\frac{m}{\lambda}\int_{-1}^{1} w(t) \log\left(1 - \frac{t}{x}\right) dt\right]$$

$$\equiv x^m \exp\left[-\frac{m}{\lambda}\left(\frac{c_1}{x} + \frac{c_2}{2x^2} + \cdots\right)\right] \equiv x^m - \frac{mc_1}{\lambda} x^{m-1} - \cdots \tag{8.14.17}$$

where the last series is to be terminated with the last term having a nonnegative exponent.

In the special case when

$$w(x) = 1 \tag{8.14.18}$$

and hence also

$$\lambda = 2 \qquad W = \frac{2}{m} \tag{8.14.19}$$

the first four terms of the expansion of

$$x^m \exp\left[\frac{m}{2}\int_{-1}^{1} \log\left(1 - \frac{t}{x}\right) dt\right] = x^m \exp\left[-m\left(\frac{1}{6x^2} + \frac{1}{20x^4} + \frac{1}{42x^6} + \cdots\right)\right]$$

are found to be

$$\pi(x) = x^m - \frac{m}{6}x^{m-2} + \frac{m}{360}(5m - 18)x^{m-4} - \frac{m}{45360}(35m^2 - 378m + 1080)x^{m-6} + \cdots \tag{8.14.20}$$

where the series is to be terminated with the first term if $m = 0$ or 1, with the second if $m = 2$ or 3, and so forth. If the mth such polynomial is denoted here by $G_m(x)$, the first six such polynomials are thus obtained as follows:

$$\begin{aligned} G_0(x) &= 1 & G_1(x) &= x \\ G_2(x) &= \tfrac{1}{3}(3x^2 - 1) & G_3(x) &= \tfrac{1}{2}(2x^3 - x) \\ G_4(x) &= \tfrac{1}{45}(45x^4 - 30x^2 + 1) & G_5(x) &= \tfrac{1}{72}(72x^5 - 60x^3 + 7x) \end{aligned} \tag{8.14.21}$$

It is seen that the polynomials of even and odd degrees are even and odd functions of x, respectively, so that their zeros are symmetrically placed about $x = 0$.

It has been found that the zeros of the polynomials $G_1(x), G_2(x), \ldots, G_7(x)$ and $G_9(x)$ are all real, that they lie inside the interval $[-1, 1]$, and that the quadrature formula (8.14.1), with abscissas identified with a set of such zeros, accordingly does indeed have a degree of precision equal to or greater than the number of abscissas, when $w(x) = 1$. However, *six of the zeros of* $G_8(x)$ *are nonreal, and each* $G_m(x)$ *for* $m \geqq 10$ *possesses at least one pair of nonreal zeros* (see Bernstein [1937]). Thus, when $w(x) = 1$, the quadrature formula is useful only when $m \leqq 7$ and $m = 9$.

The abscissas corresponding to the formula

$$\int_{-1}^{1} f(x)\, dx = \frac{2}{m}\sum_{k=1}^{m} f(x_k) + E \tag{8.14.22}$$

for all relevant values of m, are listed to six digits in Table 8.6. Whereas the

Table 8.6

m	Abscissas	m	Abscissas
2	± 0.577350	7	0
3	0		± 0.323912
	± 0.707107		± 0.529657
			± 0.883862
4	± 0.187592		
	± 0.794654	9	0
			± 0.167906
5	0		± 0.528762
	± 0.374541		± 0.601019
	± 0.832497		± 0.911589
6	± 0.266635		
	± 0.422519		
	± 0.866247		

appropriate error term in each case can be expressed in the form

$$E = \int_{-1}^{1} G(s) f^{(m+1)}(s)\, ds \tag{8.14.23}$$

where $G(s)$ is defined by (8.14.11), with $w(x) = 1$, recourse to the third method of Sec. 5.10 leads more simply to the desired results. For this purpose, we may notice first that, since the coefficient of x^m in $G_m(x)$ is unity, there follows $w(x)\pi(x) = G_m(x)$. Further, by integrating the expressions given in (8.14.21), and determining the constant of integration in each case such that the integral vanishes at one (and hence both) of the limits ± 1, there follows

$$\begin{aligned}
G_1(x) &= [\tfrac{1}{2}(x^2 - 1)]' \\
G_2(x) &= [\tfrac{1}{3}(x^3 - x)]' = [\tfrac{1}{12}(x^2 - 1)^2]'' \\
G_3(x) &= [\tfrac{1}{4}x^2(x^2 - 1)]' \\
G_4(x) &= [\tfrac{1}{45}(9x^5 - 10x^3 + x)]' = [\tfrac{1}{90}(x^2 - 1)^2(1 + 3x^2)]'' \\
G_5(x) &= [\tfrac{1}{144}(x^2 - 1)(24x^4 - 6x^2 + 1)]'
\end{aligned} \tag{8.14.24}$$

and so forth. Thus, when m is *odd*, there follows $G_m(x) = V_m'(x)$, where V_m vanishes at the ends of the interval $[-1, 1]$ and is of constant sign inside that interval, whereas, when m is *even*, there follows $G_m(x) = V_m''(x)$, where V_m and V_m' vanish at the ends of the interval and V_m is of constant sign in the interior.

It follows from (5.10.38), with $n + 1 = m$, that the error E_m associated with an m-point formula is given by

$$E_m = \begin{cases} K_m \dfrac{f^{(m+1)}(\xi)}{(m+1)!} & (m \text{ odd}) \\[2ex] K_m \dfrac{f^{(m+2)}(\xi)}{(m+2)!} & (m \text{ even}) \end{cases} \tag{8.14.25}$$

where

$$K_m = \begin{cases} -\displaystyle\int_{-1}^{1} V_m(x)\,dx = \int_{-1}^{1} xG_m(x)\,dx & (m \text{ odd}) \\ 2\displaystyle\int_{-1}^{1} V_m(x)\,dx = \int_{-1}^{1} x^2G_m(x)\,dx & (m \text{ even}) \end{cases} \tag{8.14.26}$$

The first six of these values are found to be $K_1 = \frac{2}{3}$, $K_2 = \frac{8}{45}$, $K_3 = \frac{1}{15}$, $K_4 = \frac{32}{945}$, $K_5 = \frac{13}{756}$, $K_6 = \frac{16}{1575}$.

In the case $m = 2$, formula (8.14.22) reduces to the Legendre-Gauss two-point formula. It may be noticed that the degree of precision is m when m is odd, but is $m + 1$ when m is even. More generally, whenever $w(x)$ is an even function of x, it is apparent from the symmetry that both members of (8.14.1) will vanish when $f(x)$ is any polynomial of odd degree (or any odd function of x). Hence, in such cases, if m is even and if the degree of precision is at least m, then it is also at least $m + 1$.†

The Chebyshev-Gauss formula of Sec. 8.8, with $w(x) = (1 - x^2)^{-1/2}$, is a particularly notable member of the general class of formulas considered in this section, since in that case it was seen that *the degree of precision attains its maximum value* $2m - 1$, for all $m \geqq 1$, in spite of the fact that the weights are equal. It can be rederived here by noticing that (8.14.2) gives $\lambda = \pi$, and hence $W = \pi/m$, in accordance with (8.8.11). Equation (8.14.17) then gives

$$\begin{aligned} \exp\left[\frac{m}{\lambda}\int_{-1}^{1} w(t)\log(x-t)\,dt\right] &= \exp\left[\frac{m}{\pi}\int_{-1}^{1} \frac{\log(x-t)}{\sqrt{1-t^2}}\,dt\right] \\ &= \exp\left[m\log\left(\frac{x+\sqrt{x^2-1}}{2}\right)\right] = \left(\frac{x+\sqrt{x^2-1}}{2}\right)^m \\ &= \frac{x^m}{2^m}\left(1+\sqrt{1-\frac{1}{x^2}}\right)^m = x^m\left(1-\frac{1}{4x^2}+\cdots\right)^m \\ &= x^m - \frac{m}{4}x^{m-2} + \cdots \end{aligned}$$

when $x > 1$, and the *polynomial part* of the last indicated expansion can be shown to be identical with $T_0(x)$, when $m = 0$, and with the expanded form of

$$2^{1-m}T_m(x) = 2^{1-m}\cos(m\cos^{-1}x)$$

when $m \geqq 1$.

† The difference between this situation and that relevant to Newton-Cotes quadrature is a consequence of the fact that there the *minimum* degree of precision is $m - 1$, where m ordinates are used. Thus an increase of (at least) one degree occurs if $m - 1$ is even and hence m is *odd*.

8.15 Algebraic Derivations

Any specific one of the quadrature formulas considered in this chapter can be obtained directly by purely algebraic methods, without the use of properties of orthogonal functions. In cases in which the weighting function is given empirically, or in which only a single specific formula is desired, such methods are often to be preferred. For this reason, they are discussed briefly in this section.

We suppose here that the formula is to be of the form

$$\int_a^b w(x)f(x)\,dx = \sum_{k=1}^{m} W_k f(x_k) + E \tag{8.15.1}$$

where $w(x) \geqq 0$ in $[a, b]$, and that the abscissas and weights are to be chosen in such a way that the degree of precision is *at least* $m - 1$, so that $E = 0$ at least when $f(x) = x^r$ $(r = 0, 1, \ldots, m - 1)$. If we define the rth *moment* M_r, associated with $w(x)$ over $[a, b]$, by the equation

$$\int_a^b x^r w(x)\,dx = M_r \qquad (r = 0, 1, 2, \ldots) \tag{8.15.2}$$

the requirement that the degree of precision of (8.15.1) be at least N is represented by the $N + 1$ conditions

$$\sum_{k=1}^{m} W_k x_k^r = M_r \qquad (r = 0, 1, \ldots, N) \tag{8.15.3}$$

Whereas these equations are linear in the m weights W_i, they are nonlinear in the m abscissas x_i, and the purpose of this section is to indicate in what way the difficulties associated with this nonlinearity can be minimized.

The procedures to be used, in those situations in which no conditions are imposed on the *weights*, may be easily generalized from the simple case in which $m = 2$. Hence, in order to simplify the notation, we consider that case specifically, but describe the procedures in general terms. The *simplest case*, clearly, is that in which the m abscissas are preassigned. Then, unless they are chosen in a special way, we can require only that the degree of precision N be at least $m - 1$. When $m = 2$, the two conditions to be satisfied are then

$$\begin{aligned} W_1 + W_2 &= M_0 \\ W_1 x_1 + W_2 x_2 &= M_1 \end{aligned} \tag{8.15.4}$$

Since the abscissas are assigned, we have m simultaneous *linear* equations in the m unknown weights, and it can be shown that these equations always possess a unique solution.

On the other extreme, we have the *gaussian case*, in which no constraints are imposed and in which the degree of precision is to be $2m - 1$. In the case $m = 2$, the four conditions to be satisfied are then of the form

$$\begin{aligned} W_1 + W_2 &= M_0 \\ W_1 x_1 + W_2 x_2 &= M_1 \\ W_1 x_1^2 + W_2 x_2^2 &= M_2 \\ W_1 x_1^3 + W_2 x_2^3 &= M_3 \end{aligned} \tag{8.15.5}$$

representing four equations in the four unknown quantities x_1, x_2, W_1, and W_2. In order to solve these equations, we let x_1 and x_2 be the zeros of $\pi(x)$

$$\pi(x) = (x - x_1)(x - x_2) \equiv x^2 + \alpha_1 x + \alpha_2 \tag{8.15.6}$$

and attempt first to determine the *coefficients* α_1 and α_2. If we multiply the third equation of (8.15.5) by 1, the second by α_1, and the first by α_2, and add the results, making use of the fact that

$$x_1^2 + \alpha_1 x_1 + \alpha_2 = 0 \qquad x_2^2 + \alpha_1 x_2 + \alpha_2 = 0$$

we obtain the condition

$$M_2 + M_1 \alpha_1 + M_0 \alpha_2 = 0 \tag{8.15.7}$$

Similarly, from the fourth, third, and second equations we obtain the requirement

$$M_3 + M_2 \alpha_1 + M_1 \alpha_2 = 0 \tag{8.15.8}$$

The last two equations are *linear* in α_1 and α_2. If $M_1^2 \neq M_0 M_2$, they possess a unique solution. (This can be *guaranteed* when $w \geqq 0$ in $[a, b]$.) The abscissas x_1 and x_2 are then determined as the roots of the algebraic equation $\pi(x) = 0$, provided that $\pi(x)$ *has* real roots, and the weights W_1 and W_2 are finally determined from any two (say, the first two) equations of (8.15.5).

The generalization is obvious since, in the general case, $\pi(x)$ will be specified by m α's and the $2m$ equations replacing (8.15.5) will provide m sets of $m + 1$ successive equations, from each of which a *linear* equation in the α's may be obtained by the same general procedure as that which led to (8.15.7). These equations will (generally) determine the α's, after which the abscissas are obtained as the roots of $\pi(x) = 0$ and, finally, the first m of the basic equations determine the weights.

In the *intermediate cases*, in which, say, $m - r$ of the m abscissas are preassigned, we can hope only for a degree of precision $m + r - 1$ (unless those abscissas are assigned in a special way), and hence there will be $m + r$ basic equations replacing (8.15.5). If we again let $\pi(x)$ denote the product $(x - x_1)(x - x_2) \cdots (x - x_m)$, involving the fixed abscissas as well as the free ones, then $\pi(x)$ again will be specified by m α's. From the $m + r$ basic equations, we can proceed as in the derivation of (8.15.7) r times, and hence can obtain r linear equations in the α's. The $m - r$ additional linear equations needed for the determination of the m α's then follow from the requirements that the $m - r$ fixed abscissas satisfy the equation $\pi(x) = 0$.

Thus, in the case $m = 2$, $r = 1$, the three basic equations are

$$\begin{aligned} W_1 + W_2 &= M_0 \\ W_1 x_1 + W_2 x_2 &= M_1 \\ W_1 x_1^2 + W_2 x_2^2 &= M_2 \end{aligned} \tag{8.15.9}$$

Under the assumption that x_1 and x_2 satisfy the equation

$$x^2 + \alpha_1 x + \alpha_2 = 0 \tag{8.15.10}$$

we again obtain (8.15.7). By combining this condition with the requirement that the preassigned value x_1 satisfy (8.15.10), we deduce that α_1 and α_2 are determined uniquely by the two linear equations

$$\begin{aligned} M_2 + M_1\alpha_1 + M_0\alpha_2 &= 0 \\ x_1^2 + x_1\alpha_1 + \alpha_2 &= 0 \end{aligned} \tag{8.15.11}$$

under the assumption that $M_0 x_1 \neq M_1$.

There is no guarantee in this case, even though it be true that $w(x) \geqq 0$, that the zeros of $\pi(x)$ will be real and distinct or, if so, that they will lie in $[a, b]$. However, if a quadrature formula of the type sought exists, it can be obtained by the method outlined.

As a simple illustrative example, we suppose that a quadrature formula is required to be of the form

$$\int_0^1 x^{1/2} f(x)\,dx = W_1 f(x_1) + W_2 f(1) + E \tag{8.15.12}$$

where $x_2 = 1$ is preassigned. The expected degree of precision is then 2, corresponding to the fact that three free parameters x_1, W_1, and W_2 are available. We first calculate the relevant moments

$$M_r = \int_0^1 x^{(2r+1)/2}\,dx = \frac{2}{2r+3} \qquad (r = 0, 1, 2)$$

after which the three basic conditions (8.15.9) become

$$W_1 + W_2 = \tfrac{2}{3} \qquad W_1 x_1 + W_2 = \tfrac{2}{5} \qquad W_1 x_1^2 + W_2 = \tfrac{2}{7} \tag{8.15.13}$$

By writing $\pi(x) = (x - x_1)(x - 1) = x^2 + \alpha_1 x + \alpha_2$, we deduce from (8.15.11) that α_1 and α_2 must satisfy the equations

$$\tfrac{2}{7} + \tfrac{2}{5}\alpha_1 + \tfrac{2}{3}\alpha_2 = 0$$
$$1 + \alpha_1 + \alpha_2 = 0$$

and obtain $\alpha_1 = -\tfrac{10}{7}$, $\alpha_2 = \tfrac{3}{7}$, and hence $\pi(x) = (7x^2 - 10x + 3)/7$. Thus there follows

$$x_1 = \tfrac{3}{7} \qquad x_2 = 1 \tag{8.15.14}$$

With these results, the first two equations of (8.15.13) give

$$W_1 = \tfrac{7}{15} \qquad W_2 = \tfrac{1}{5} \tag{8.15.15}$$

Thus (8.15.12) becomes

$$\int_0^1 x^{1/2} f(x)\,dx = \tfrac{7}{15}f(\tfrac{3}{7}) + \tfrac{1}{5}f(1) + E \tag{8.15.16}$$

We verify that $E = 0$ for $f(x) = 1$, x, and x^2, and find that $E \neq 0$ when $f(x) = x^3$. Hence the degree of precision is indeed 2.

In order to obtain an expression for the error term E, we may make use of one of the methods of Sec. 5.10. In particular, the influence function (5.10.16) with $N = 2$ is readily determined in the form

$$G(s) = \begin{cases} -\tfrac{8}{105}s^{7/2} & (0 \leqq s \leqq \tfrac{3}{7}) \\ -\tfrac{1}{210}(16s^{7/2} - 49s^2 + 42s - 9) & (\tfrac{3}{7} \leqq s \leqq 1) \end{cases} \tag{8.15.17}$$

and is found to be negative throughout the interior of the interval [0, 1]. Thus the formula (5.10.31) can be used to give

$$E = \frac{f'''(\xi)}{3!}\left[\int_0^1 x^{7/2}\,dx - \tfrac{7}{15}(\tfrac{3}{7})^3 - \tfrac{1}{5}(1)^3\right] = -\tfrac{16}{6615}f'''(\xi) \tag{8.15.18}$$

where $0 < \xi < 1$.

The same result can be obtained somewhat more easily by use of the third method described in Sec. 5.10. For we find that

$$w(x)\pi(x) = \frac{d}{dx}\left[\tfrac{2}{7}x^{3/2}(x - 1)^2\right]$$

where the constant of integration is determined so that the content of the brackets vanishes when $x = 0$. Since it vanishes also when $x = 1$ and is positive for all intermediate values of x, we may make use of (5.10.38), with $n + 1 \equiv m = 2$, $r = 1$, and $V = \tfrac{2}{7}x^{3/2}(x - 1)^2$, to deduce that

$$E = -\frac{f'''(\xi)}{6}\int_0^1 \tfrac{2}{7}x^{3/2}(x - 1)^2\,dx = -\tfrac{16}{6615}f'''(\xi)$$

as before.

Finally, in the general case of *Chebyshev quadrature*, in which all the weights are to be equal, the formula is of the form

$$\int_a^b w(x)f(x)\,dx = W\sum_{k=1}^{m} f(x_k) + E \qquad (8.15.19)$$

and the $m + 1$ conditions requiring that the degree of precision be at least m are of the form

$$\begin{aligned} x_1^0 + x_2^0 + x_3^0 + \cdots + x_m^0 &= \overline{M}_0 \\ x_1 + x_2 + x_3 + \cdots + x_m &= \overline{M}_1 \\ x_1^2 + x_2^2 + x_3^2 + \cdots + x_m^2 &= \overline{M}_2 \qquad (8.15.20) \\ \cdots\cdots\cdots\cdots\cdots\cdots\cdots\cdots\cdots \\ x_1^m + x_2^m + x_3^m + \cdots + x_m^m &= \overline{M}_m \end{aligned}$$

where we have written

$$\overline{M}_r = \frac{1}{W}M_r = \frac{1}{W}\int_a^b x^r w(x)\,dx \qquad (8.15.21)$$

From the first equation, there follows immediately $\overline{M}_0 = m$, and hence

$$W = \frac{1}{m}\int_a^b w(x)\,dx = \frac{1}{m}M_0 \qquad (8.15.22)$$

Under the *assumption* that the problem possesses a (real) solution,† we again write

$$\begin{aligned} \pi(x) &= (x - x_1)(x - x_2)\cdots(x - x_m) \\ &\equiv x^m + \alpha_1 x^{m-1} + \alpha_2 x^{m-2} + \cdots + \alpha_{m-1}x + \alpha_m \qquad (8.15.23) \end{aligned}$$

and attempt to determine the m coefficients $\alpha_1, \ldots, \alpha_m$.

First, by multiplying the first equation in (8.15.20) by α_m, the second by $\alpha_{m-1}, \ldots$, the next-to-last by α_1, and the last by 1, adding the results, and using the fact that each x_i satisfies $\pi(x) = 0$, we obtain one linear equation relating the α's in the form

$$m\alpha_m + \overline{M}_1\alpha_{m-1} + \overline{M}_2\alpha_{m-2} + \cdots + \overline{M}_{m-1}\alpha_1 + \overline{M}_m = 0 \qquad (8.15.24)$$

In order to obtain $m - 1$ complementary relations, we make use of Newton's power-sum identities [see Theorem 13 of Sec. 1.9, with s_r, a_r, and n replaced by $\overline{M}_r$, α_r, and m] to deduce that

$$r\alpha_r + \overline{M}_1\alpha_{r-1} + \overline{M}_2\alpha_{r-2} + \cdots + \overline{M}_{r-1}\alpha_1 + \overline{M}_r = 0 \qquad (r = 1, 2, \ldots, m) \qquad (8.15.25)$$

† As was pointed out in Sec. 8.14, this assumption is not always valid.

This recurrence formula, which includes (8.15.24) when $r = m$, permits the expression of each of the α's in terms of the *reduced moments* $\overline{M}_1, \overline{M}_2, \ldots,$ and $\overline{M}_m$. The required abscissas then are the roots of the equation

$$x^m + \alpha_1 x^{m-1} + \alpha_2 x^{m-2} + \cdots + \alpha_{m-1} x + \alpha_m = 0 \qquad (8.15.26)$$

if those roots are real and distinct. Otherwise, the desired formula does not exist.

In illustration, in order to determine the Chebyshev abscissas for the quadrature formula

$$\int_{-1}^{1} (1 - x^2) f(x)\, dx = W[f(x_1) + f(x_2) + f(x_3)] + E \qquad (8.15.27)$$

we first calculate the common weight

$$W = \tfrac{1}{3} \int_{-1}^{1} (1 - x^2)\, dx = \tfrac{4}{9}$$

and then the relevant reduced moments

$$\overline{M}_1 = \tfrac{9}{4} \int_{-1}^{1} x(1 - x^2)\, dx = 0 \qquad \overline{M}_2 = \tfrac{9}{4} \int_{-1}^{1} x^2(1 - x^2)\, dx = \tfrac{3}{5}$$

$$\overline{M}_3 = \tfrac{9}{4} \int_{-1}^{1} x^3(1 - x^2)\, dx = 0$$

Next, from (8.15.25) with $r = 1$, 2, and 3, there follows

$$\alpha_1 = -\overline{M}_1 = 0 \qquad \alpha_2 = \tfrac{1}{2}(-\overline{M}_1 \alpha_1 - \overline{M}_2) = -\tfrac{3}{10}$$

$$\alpha_3 = \tfrac{1}{3}(-\overline{M}_1 \alpha_2 - \overline{M}_2 \alpha_1 - \overline{M}_3) = 0$$

Hence, the required abscissas then are obtained as the roots of the equation $x^3 - \frac{3}{10}x = 0$, in the form

$$x_1 = -\sqrt{\tfrac{3}{10}} \qquad x_2 = 0 \qquad x_3 = \sqrt{\tfrac{3}{10}} \qquad (8.15.28)$$

so that the desired formula is

$$\int_{-1}^{1} (1 - x^2) f(x)\, dx = \tfrac{4}{9}\left[f\left(-\sqrt{\tfrac{3}{10}}\right) + f(0) + f\left(\sqrt{\tfrac{3}{10}}\right)\right] + E \qquad (8.15.29)$$

It is easily verified that $E = 0$ when $f(x) = 1$, x, x^2, and x^3, but that $E \neq 0$ when $f(x) = x^4$, so that the degree of precision is 3.

An expression for the error term is obtained most readily by use of the third method of Sec. 5.10. Thus, we find that

$$\int_{-1}^{x} w(t)\pi(t)\, dt = \int_{-1}^{x} (t^3 - \tfrac{3}{10}t)(1 - t^2)\, dt$$

$$= -\tfrac{1}{120}(20x^6 - 39x^4 + 18x^2 + 1)$$

$$= -\tfrac{1}{120}(x^2 - 1)^2(20x^2 + 1)$$

so that $w(x)\pi(x) = V'(x)$, where $V(x) = -\frac{1}{120}(x^2 - 1)^2(20x^2 + 1)$. Since $V(x)$ vanishes at both ends of the interval of integration and is of constant sign inside that interval, use may be made of (5.10.38), with $n + 1 \equiv m = 3$ and $r = 1$, to give

$$E = -\frac{f^{\text{iv}}(\xi)}{4!}\int_{-1}^{1} V(x)\,dx$$

or

$$E = \tfrac{1}{700} f^{\text{iv}}(\xi) \tag{8.15.30}$$

where $|\xi| < 1$. The same result can be obtained, somewhat more laboriously, by use of the appropriate influence function (5.10.16) with $N = 3$. A check is afforded by an application of the formula to $f(x) = x^4$.

In applications of the methods of this section, when m is large and the relevant set of m linear equations determining the coefficients $\alpha_1, \ldots, \alpha_m$ which specify $\pi(x)$ is to be solved by *numerical* methods, proper account must be taken of the fact that small errors in the α's can propagate into large errors in the finally calculated zeros of $\pi(x)$ when $\pi(x)$ is an ill-conditioned polynomial (see Sec. 10.5).

8.16 Application to Trigonometric Integrals

To conclude this chapter, we illustrate the use of the methods of the preceding section in connection with the derivation of formulas for the approximate evaluation of the integrals

$$S_k \equiv \int_0^{\pi} F(\theta) \sin k\theta \, d\theta \qquad C_k \equiv \int_0^{\pi} F(\theta) \cos k\theta \, d\theta \tag{8.16.1}$$

when k is a positive integer.

In the case of S_k, if we consider $\sin k\theta$ as a weighting function w on the interval $[0, \pi]$, we notice that w changes sign inside that interval unless $k = 1$ and, in fact, that w oscillates rather violently if k is large. To deal with this oscillation, we may divide $[0, \pi]$ into the k subintervals $[0, \pi/k]$, $[\pi/k, 2\pi/k]$, $\ldots$, $[\pi - \pi/k, \pi]$, in each of which $\sin k\theta$ is of constant sign, and derive a gaussian formula for each subinterval. Thus we express S_k in the form

$$\begin{aligned} S_k &= \sum_{n=0}^{k-1} \int_{n\pi/k}^{(n+1)\pi/k} F(\theta) \sin k\theta \, d\theta \\ &= \frac{1}{k} \sum_{n=0}^{k-1} (-1)^n \int_0^{\pi} F\left(\frac{n\pi + x}{k}\right) \sin x \, dx \end{aligned} \tag{8.16.2}$$

with the substitution $x = k\theta - n\pi$, and we are led to consider the approximation of the integral

$$\int_0^{\pi} f(x) \sin x \, dx \qquad (8.16.3)$$

by gaussian quadrature, with the nonnegative weighting function $w = \sin x$.

For simplicity, we restrict attention here to the use of only two abscissas in this approximation. In this case (see Prob. 51) the methods of the preceding section give

$$W_1 = W_2 = 1 \qquad (8.16.4)$$

and require that x_1 and x_2 be the zeros of the polynomial

$$\pi(x) = x^2 - \pi x + 2 \qquad (8.16.5)$$

so that, with a convenient notation, there follows

$$x_1 = \frac{\pi}{4} + \alpha \qquad x_2 = \frac{3\pi}{4} - \alpha \qquad (8.16.6)$$

where

$$\alpha = \frac{\pi}{2}\left(\frac{1}{2} - \sqrt{1 - \frac{8}{\pi^2}}\right) \doteq 0.1017308 \qquad (8.16.7)$$

Also, according to (8.3.6), the relevant error term is

$$E = \frac{f^{\text{iv}}(\xi)}{24} \int_0^{\pi} (x^2 - \pi x + 2)^2 \sin x \, dx$$

or

$$E = K_1 f^{\text{iv}}(\xi) \qquad (8.16.8)$$

where

$$K_1 = \frac{10 - \pi^2}{6} \doteq 0.022 \qquad (8.16.9)$$

and where $0 < \xi < \pi$, so that the desired approximation to (8.16.3) takes the form

$$\int_0^{\pi} f(x) \sin x \, dx = f\left(\frac{\pi}{4} + \alpha\right) + f\left(\frac{3\pi}{4} - \alpha\right) + K_1 f^{\text{iv}}(\xi) \qquad (8.16.10)$$

Hence, if this result is used for each of the k integrals in (8.16.2), there follows

$$\int_0^{\pi} F(\theta) \sin k\theta \, d\theta = \frac{1}{k} \sum_{n=0}^{k-1} (-1)^n \left[F\left(\frac{n\pi + \pi/4 + \alpha}{k}\right) + F\left(\frac{n\pi + 3\pi/4 - \alpha}{k}\right)\right] + E \qquad (8.16.11)$$

where

$$E = K_1 k^{-5} \sum_{n=0}^{k-1} (-1)^n F^{\text{iv}}\left(\frac{\xi_n}{k}\right) \qquad (8.16.12)$$

and where $n\pi < \xi_n < (n + 1)\pi$.

The result of ignoring the error term in (8.16.11) accordingly yields an approximation to the integral S_k with a degree of precision equal to 3. It uses $2k$ values of F (two per half-period of $\sin k\theta$) and has an error term which is comparable (on the average, about twice as large) with the error term associated with a similar (but nongaussian) formula due to Price [1960], which employs $3k + 2$ values of F at *equally spaced* points.

In the case of C_k, a corresponding subdivision of $[0, \pi]$, defined now by the zeros of $\cos k\theta$, would yield the $k + 1$ subintervals $[0, \pi/2k]$, $[\pi/2k, 3\pi/2k]$, $\ldots$, $[\pi - \pi/2k, \pi]$, and the two end subintervals would require separate treatment. However, it happens that a piecewise-gaussian approximation to C_k can be obtained with the same uniform subdivision as that used for S_k in spite of the sign change of the weighting function $\cos k\theta$ at the midpoint of each subinterval. In fact, the approximation so obtained is found to possess a remarkable additional property.

Accordingly, we first write C_k in a form analogous to (8.16.2)

$$C_k = \frac{1}{k} \sum_{n=0}^{k-1} (-1)^n \int_0^{\pi} F\left(\frac{n\pi + x}{k}\right) \cos x \, dx \qquad (8.16.13)$$

and seek a two-point gaussian formula for the integral

$$\int_0^{\pi} f(x) \cos x \, dx \qquad (8.16.14)$$

with no assurance of success because of the sign change in $w = \cos x$ at $x = \pi/2$. The relevant equations (8.15.5) are found to yield the polynomial

$$\pi(x) = x^2 - \pi x + 6 - \tfrac{1}{2}\pi^2 \qquad (8.16.15)$$

whose zeros are to be the required abscissas, and there follows

$$x_1 = \frac{\pi}{8} - \beta \qquad x_2 = \frac{7\pi}{8} + \beta \qquad (8.16.16)$$

where

$$\beta = \frac{\sqrt{3\pi^2 - 24}}{2} - \frac{3\pi}{8} \doteq 0.0060494 \qquad (8.16.17)$$

Thus we are fortunate in that x_1 and x_2, not only are real, but also lie in $[0, \pi]$.

The associated weights are of equal magnitude and opposite sign (as might have been anticipated in view of the antisymmetry of the weighting function):

$$W_1 = -W_2 = \frac{1}{3\pi/8 + \beta} \doteq 0.8444900 \qquad (8.16.18)$$

Finally, the error term takes the form

$$E = \int_0^\pi [\pi(x)]^2 \cos x \, f[x_1, x_1, x_2, x_2, x] \, dx \qquad (8.16.19)$$

and, since $\cos x$ changes sign at $x = \pi/2$, the second law of the mean is not applicable. However, if we write

$$[\pi(x)]^2 \cos x = V'(x) \qquad V(x) = \int_0^x [\pi(s)]^2 \cos s \, ds \qquad (8.16.20)$$

in preparation for an integration by parts, we may notice that since $V'(x) \geqq 0$ when $0 \leqq x \leqq \pi/2$ and $V'(x) \leqq 0$ when $\pi/2 \leqq x \leqq \pi$, $V(x)$ increases steadily with increasing x from zero at $x = 0$ to a positive maximum at $x = \pi/2$ and then decreases steadily as x continues to increase from $\pi/2$ to π. But since clearly $V'(\pi - x) = -V'(x)$, the total decrease in $V(x)$ as x varies from $\pi/2$ to π is exactly equal to its increase as x varies from 0 to $\pi/2$. Hence we conclude that *the function* $V(x)$ *defined by* (8.16.20) *is positive for* $0 < x < \pi$ *and zero for* $x = 0$ *and* $x = \pi$.

Consequently, an integration by parts transforms (8.16.19) to the form

$$E = -\int_0^\pi V(x) f[x_1, x_1, x_2, x_2, x, x] \, dx$$

and hence, by use of the second law of the mean, to the form

$$E = -\frac{f^{\mathrm{v}}(\xi)}{120} \int_0^\pi V(x) \, dx \qquad (8.16.21)$$

where $0 < \xi < \pi$. Finally, the computation of the integral in (8.16.21) can be avoided, as in the transition from (5.10.30) to (5.10.31), by noticing that it must be equal to $-E$ when $f(x) = x^5$. In this way we find that

$$E = -K_2 f^{\mathrm{v}}(\xi) \qquad (8.16.22)$$

where

$$K_2 = \frac{336 - 24\pi^2 - \pi^4}{240} \doteq 0.0072 \qquad (8.16.23)$$

and where $0 < \xi < \pi$.

Thus the two-point formula

$$\int_0^\pi f(x)\cos x\,dx = \frac{1}{3\pi/8 - \beta}\left[f\left(\frac{\pi}{8} - \beta\right) - f\left(\frac{7\pi}{8} + \beta\right)\right] - K_2 f^{\text{v}}(\xi) \tag{8.16.24}$$

has in fact a degree of precision of 4, one greater than the *maximum* for a two-point gaussian formula with a *positive* weighting function; and the result of applying (8.16.24) to each integral in (8.16.13) is the formula

$$\int_0^\pi F(\theta)\cos k\theta\,d\theta = \frac{1}{k(3\pi/8 + \beta)}\sum_{n=0}^{k-1}(-1)^n\left[F\left(\frac{n\pi + \pi/8 - \beta}{k}\right) - F\left(\frac{n\pi + 7\pi/8 + \beta}{k}\right)\right] + E \tag{8.16.25}$$

where

$$E = -K_2 k^{-6}\sum_{n=0}^{k-1}(-1)^n F^{\text{v}}\left(\frac{\xi_n}{k}\right) \tag{8.16.26}$$

and where $n\pi < \xi_n < (n+1)\pi$.

In the special case when

$$F(\theta) = e^{a\theta} \tag{8.16.27}$$

the integrals S_k and C_k are elementary and also the sums in (8.16.11) and (8.16.25) can be expressed in closed form to give

$$S_k = \frac{k}{a^2 + k^2}(1 - e^{a\pi}\cos k\pi)$$
$$= \frac{\cosh[(\pi/4 - \alpha)a/k]}{k\cosh \pi a/2k}(1 - e^{a\pi}\cos k\pi) + E \tag{8.16.28}$$

where an expansion shows that the *relative* error is such that

$$\frac{E}{S_k} = \frac{K_1}{2}\left(\frac{a}{k}\right)^4 + O\left(\frac{a^6}{k^6}\right) \qquad \left(\frac{K_1}{2} \doteq 0.011\right) \tag{8.16.29}$$

and

$$C_k = -\frac{a}{a^2 + k^2}(1 - e^{a\pi}\cos k\pi)$$
$$= -\frac{\sinh[(3\pi/8 + \beta)a/k]}{k(3\pi/8 + \beta)\cosh \pi a/2k}(1 - e^{a\pi}\cos k\pi) + E \tag{8.16.30}$$

where

$$\frac{E}{C_k} = \frac{K_2}{2}\left(\frac{a}{k}\right)^4 + O\left(\frac{a^6}{k^6}\right) \qquad \left(\frac{K_2}{2} \doteq 0.0036\right) \tag{8.16.31}$$

Thus here the relative error becomes small of the order of $(a/k)^4$ in both cases when $|k/a|$ is large. For example, when $|k/a| = 10$, the approximation to S_k is in error by less than two units in the fifth significant figure and the approximation to C_k by less than four units in the sixth significant figure.

The preceding developments generalize without difficulty to the application of piecewise-gaussian quadrature employing more than two ordinates per half-period of the trigonometric weighting function. The use of a formula of Lobatto type over each half-period would afford obvious computational advantages. When the sum defined in (8.16.11) or (8.16.25), or in a generalization, involves a large number of terms with alternating signs, an accelerating process such as Euler's transformation (Sec. 5.9) may facilitate its evaluation.

8.17 Supplementary References

General references on numerical integration, dealing with methods of this chapter as well as with many others, include Krylov [1962], Davis and Rabinowitz [1967], and Ghizzetti and Ossicini [1970]. Davis and Rabinowitz include an extensive bibliography, together with a selection of Fortran programs, a list of Algol procedures, and a partial list of existent numerical tables, all relevant to numerical integration. Stroud [1961] presents an exhaustive bibliography of sources through 1960.

Classical references on gaussian (and gaussian-type) quadrature include Mehler [1864], Chebyshev [1874], and Radau [1880*a*, 1880b, 1883]. More recent contributions are by Shohat and Winston [1934], Winston [1934], and Bernstein [1937]. Stroud and Secrest [1966] provide a comprehensive collection of formulas of gaussian type, together with extensive tables of abscissas and weights for many of the formulas, including all the standard ones. Shorter tables are included in the NBS handbook [1964] and in Krylov [1962]. Other tables are listed in the index by Fletcher et al. [1962].

The relationship between gaussian quadrature and osculating (Hermite) quadrature is pointed out by Fort [1948]. Salzer [1954] presents a table of coefficients for osculating quadrature. Stieltjes [1895] related gaussian quadrature to continued fractions (see Cheney [1966], pp. 186–188), while Ghizzetti and Ossicini [1970] employ a quite different approach due to Radon [1935]. For methods of algebraic derivation, see Beard [1947] and Hamming [1962].

Convergence of gaussian-type quadrature sequences was established by Stieltjes [1884] on a finite interval, when $f(x)$ is continuous, and by Uspensky [1928] in special cases involving infinite intervals. For a stronger result when the interval is finite, see Davis and Rabinowitz [1967].

PROBLEMS

Section 8.2

1 Obtain the formula

$$f_s = (1 + 2s)(1 - s)^2 f_0 + (3 - 2s)s^2 f_1 + s(1 - s)^2 h f_0' - s^2(1 - s)hf_1' + \frac{h^4}{4!} f^{\text{iv}}(\xi)s^2(1 - s)^2$$

where $f_s \equiv f(x_0 + hs)$ and $x_0 < \xi < x_0 + h$, if $0 \leqq s \leqq 1$, and deduce the special formula

$$f_{1/2} = \tfrac{1}{2}(f_0 + f_1) + \frac{h}{8}(f_0' - f_1') + \frac{h^4}{384} f^{\text{iv}}(\xi)$$

2 Obtain the formula

$$f_s = \tfrac{1}{4}(4 + 3s)s^2(1 - s)^2 f_{-1} + (1 - s^2)^2 f_0 + \tfrac{1}{4}(4 - 3s)s^2(1 + s)^2 f_1 + \tfrac{1}{4}(1 + s)s^2(1 - s)^2 hf_{-1}' + s(1 - s^2)^2 hf_0' - \tfrac{1}{4}(1 - s)s^2(1 + s)^2 hf_1' + \frac{h^6}{6!} f^{\text{vi}}(\xi)s^2(1 - s^2)^2$$

where $f_s \equiv f(x_0 + hs)$ and $x_0 - h < \xi < x_0 + h$, if $|s| \leqq 1$, and deduce the special formula

$$f_{1/2} = \tfrac{1}{128}(11f_{-1} + 72f_0 + 45f_1) + \frac{3h}{128}(f_{-1}' + 12f_0' - 3f_1') + \frac{h^6}{5120} f^{\text{vi}}(\xi)$$

together with a corresponding formula for $f_{-1/2}$.

3 From the following tabular values of the function

$$\text{Si}(x) = \int_0^x \frac{\sin t}{t}\, dt$$

determine approximate values of Si(2.5) and Si(3.5) by use of the formulas of Probs. 1 and 2:

x	2.0	3.0	4.0
Si(x)	1.605	1.849	1.758

Section 8.3

4 From the results of Prob. 1, deduce the formulas

$$\int_{x_0}^{x_1} f(x)\, dx = \frac{h}{2}(f_0 + f_1) + \frac{h^2}{12}(f_0' - f_1') + \frac{h^5}{720} f^{\text{iv}}(\xi)$$

and

$$\int_{x_0}^{x_1} (x - x_0)f(x)\,dx = \frac{h^2}{20}(3f_0 + 7f_1) + \frac{h^3}{60}(2f'_0 - 3f'_1) + \frac{h^6}{1440}f^{\text{iv}}(\xi)$$

where $h = x_1 - x_0$ and $x_0 < \xi < x_1$ in each formula.

5 From the results of Prob. 2, deduce the formula

$$\int_{x_{-1}}^{x_1} f(x)\,dx = \frac{h}{15}(7f_{-1} + 16f_0 + 7f_1) + \frac{h^2}{15}(f'_{-1} - f'_1) + \frac{h^7}{4725}f^{\text{vi}}(\xi)$$

where $x_1 = x_0 + h = x_{-1} + 2h$ and $x_{-1} < \xi < x_1$.

6 Use the data given in Prob. 3 and the first formula of Prob. 4 to obtain approximate values of the integral $\int_a^b \text{Si}(x)\,dx$ for $[a, b] = [2, 3]$ and $[3, 4]$, and compare the sum with the result given by the formula of Prob. 5.

Section 8.4

7 By specializing $f(x)$ to $l_i(x)$ in (8.4.6), prove that $H_i = W_i$ for a gaussian quadrature formula.

8 With the notation of Sec. 8.4, suppose that $\phi(x)$ satisfies the recurrence formula

$$\phi_{k+1}(x) = (a_k x + b_k)\phi_k(x) + c_k\phi_{k-1}(x) \qquad (k \geqq 2)$$

with $\phi_0(x) = A_0$ and $\phi_1(x) = A_1 x + B_1$. If $\theta_k(x)$ is defined by the relation

$$\theta_k(x) = \int_a^b w(y)\frac{\phi_k(y) - \phi_k(x)}{y - x}\,dy$$

show that $\theta_k(x)$ satisfies the same recurrence formula

$$\theta_{k+1}(x) = (a_k x + b_k)\theta_k(x) + c_k\theta_{k-1}(x) \qquad (k \geqq 2)$$

with the modified starting values $\theta_0(x) = 0$ and $\theta_1(x) = A_1\gamma_0$. Show also that the gaussian weight H_i, defined by (8.4.16), can be expressed in the form

$$H_i = \frac{\theta_m(x_i)}{\phi'_m(x_i)}$$

[In the expression for $\theta_{k+1}(x)$, replace ϕ_{k+1} by the appropriate combination of ϕ_k and ϕ_{k-1}, write $a_k y = a_k(y - x) + a_k x$, and recall that $\int_a^b w\phi_k\,dx = 0$ when $k > 0$. Notice that if $\phi_k(x)$ is replaced by its monic multiple $\tilde{\phi}_k(x)$, as is convenient when the relevant orthogonal polynomials are *generated* recursively (Sec. 7.10), then a_k is to be replaced by 1 in *both* of the above recurrence formulas.]

9 If $[a, b] = [0, 1]$ and $w(x) = x$, show that the abscissas in (8.4.6) are the zeros of the polynomial

$$\phi_m(x) = \frac{1}{(m + 1)!}x^{-1}\frac{d^m}{dx^m}[x^{m+1}(1 - x)^m]$$

that the ith weight is given by

$$H_i = -\frac{2m+1}{m^2(m+1)^2\phi'_m(x_i)\phi_{m-1}(x_i)}$$

and that the error term is of the form

$$E = \frac{m+1}{2(2m+1)^2}\frac{(m!)^4}{[(2m)!]^3}f^{(2m)}(\xi)$$

In particular, obtain the formulas

$$\int_0^1 xf(x)\,dx = \tfrac{1}{2}f(\tfrac{2}{3}) + \tfrac{1}{72}f''(\xi)$$

and

$$\int_0^1 xf(x)\,dx = \frac{9-\sqrt{6}}{36}f\left(\frac{6-\sqrt{6}}{10}\right) + \frac{9+\sqrt{6}}{36}f\left(\frac{6+\sqrt{6}}{10}\right) + \frac{1}{14400}f^{\text{iv}}(\xi)$$

10 Rederive the abscissas and weights for the quadrature formulas obtained in Prob. 9 when $m = 1$ and when $m = 2$ by use of the methods of Sec. 7.10 and of Prob. 8.

Section 8.5

11 Rederive the Legendre-Gauss two- and three-point formulas by use of the methods of Sec. 7.10 and of Prob. 8.

12 Use a Legendre-Gauss three-point formula to show that

$$\int_{-1}^1 e^{-\alpha^2x^2}\,dx = \tfrac{2}{9}(4 + 5e^{-3\alpha^2/5}) + E$$

where

$$E = \alpha^6 e^{-\alpha^2\theta^2}H_6(\alpha\theta) \qquad (|\theta| < 1)$$

and where H_6 is a Hermite polynomial.

13 After making an appropriate linear change of variables, determine approximate values of the integral

$$\int_2^8 \frac{dx}{x}$$

by use of gaussian formulas involving two, three, four, and five ordinates, and compare the approximations with those afforded by corresponding Newton-Cotes formulas. In each case, obtain an upper bound on the error analytically and verify that it is conservative.

14 Proceed as in Prob. 13 with the integral

$$\int_0^{\pi/2} \sin x\,dx$$

15 Proceed as in Prob. 13 with the integral

$$\int_0^{\pi} \sin x \, dx$$

16 Obtain approximations to the integral

$$\int_{-4}^{4} \frac{dx}{1 + x^2}$$

using gaussian quadratures with two, three, four, and five points, and compare the results with the true value (see also Sec. 3.9).

Section 8.6

17 Use a Laguerre-Gauss two-point formula to show that

$$\int_0^{\infty} \frac{e^{-x}}{x + a} \, dx = \frac{a + 3}{a^2 + 4a + 2} + \frac{4\theta}{a^5} \qquad (0 < \theta < 1)$$

when $a > 0$.

18 Determine approximate values of the integral

$$\int_0^{\infty} e^{-x} \sin x \, dx$$

by use of Laguerre-Gauss quadratures employing two, three, four, and five ordinates. In each case, obtain an upper bound on the error and verify that it is conservative.

19 Proceed as in Prob. 18 with the integral

$$\int_0^{\infty} \frac{e^{-x}}{x + 4} \, dx \doteq 0.206346$$

20 Proceed as in Prob. 18 with respect to the integral

$$\int_0^{\infty} \frac{x^{1/2}}{x + 4} e^{-x} \, dx \doteq 0.16776$$

omitting the analytical determination of error bounds.

21 Derive the results of (8.6.17) to (8.6.19), and obtain the special formula

$$\int_0^{\infty} x^{\beta} e^{-x} f(x) \, dx = \frac{\Gamma(1 + \beta)}{2(2 + \beta)} [(2 + \beta - \sqrt{2 + \beta}) f(2 + \beta + \sqrt{2 + \beta})$$

$$+ (2 + \beta + \sqrt{2 + \beta}) f(2 + \beta - \sqrt{2 + \beta})] + \frac{\Gamma(3 + \beta)}{12} f^{\text{iv}}(\xi)$$

where $\beta > -1$ and $\xi > 0$, in the case when $m = 2$. Also use the two-point formula to approximate the integral in Prob. 20, and compare the result with that afforded by the two-point Laguerre-Gauss formula.

Section 8.7

22 Determine approximate values of the integral

$$\int_{-\infty}^{\infty} e^{-x^2} \cos x \, dx = \sqrt{\pi}\, e^{-1/4}$$

by use of Hermite-Gauss quadratures employing two, three, four, and five ordinates. In each case, obtain an upper bound on the error and verify that it is conservative.

23 Transform the integral of Prob. 20 to the form

$$\int_{-\infty}^{\infty} \frac{t^2}{t^2 + 4} e^{-t^2} \, dt$$

and determine approximate values by use of Hermite-Gauss quadratures employing two, three, four, and five ordinates. Also compare the results with the corresponding results in Prob. 20.

24 From the following tabulated rounded values of $J_0(x)$, together with the fact that $J_0(-x) = J_0(x)$, determine approximate values of the integral

$$\int_{-\infty}^{\infty} e^{-x^2} J_0(x) \, dx \doteq 1.570301$$

by use of Hermite-Gauss quadrature employing two, three, four, and five ordinates:

x	0.0	0.5	1.0	1.5	2.0	2.5
$J_0(x)$	1.000000	0.938470	0.765198	0.511828	0.223891	−0.048384

Section 8.8

25 Use a Chebyshev-Gauss three-point formula to show that

$$\int_{-1}^{1} \frac{e^{\alpha x}}{\sqrt{1 - x^2}} \, dx = \frac{\pi}{3}\left(1 + 2 \cosh \frac{\sqrt{3}\,\alpha}{2}\right) + \frac{\pi}{360}\left(\frac{\alpha}{2}\right)^6 e^{\alpha\theta} \qquad (|\theta| < 1)$$

26 Determine approximate values of the integral

$$\int_{-1}^{1} \frac{\cos x}{\sqrt{1 - x^2}} \, dx = \pi J_0(1) \doteq 2.40394$$

by use of Chebyshev-Gauss quadratures employing two, three, four, and five ordinates. In each case, obtain an upper bound on the error and verify that it is conservative.

27 Determine approximate values of the integral

$$\int_{-1}^{1} \frac{dx}{\sqrt{(1 - x^2)(16 - x^2)}} \doteq 0.794121$$

by use of Chebyshev-Gauss quadratures employing two, three, four, and five ordinates.

28 Use the results of Prob. 33 of Chap. 7 to deduce the quadrature formula

$$\int_{-1}^{1} \sqrt{1 - x^2}\, f(x)\, dx = \frac{\pi}{m+1} \sum_{k=1}^{m} \sin^2 \left(\frac{k\pi}{m+1}\right) f\left(\cos \frac{k\pi}{m+1}\right) + \frac{\pi}{2^{2m+1}} \frac{f^{(2m)}(\xi)}{(2m)!} \qquad (|\xi| < 1)$$

29 Proceed as in Prob. 26, using the formula of Prob. 28 to deal with the integral

$$\int_{-1}^{1} \sqrt{1 - x^2} \cos x\, dx \doteq 1.38246$$

Section 8.9

30 Determine, to six decimal places, the abscissas and weights in a formula

$$\int_{-1}^{1} (1 + x)^{1/2} f(x)\, dx = H_1 f(x_1) + H_2 f(x_2) + E$$

with degree of precision equal to 3, and obtain an expression for the error in terms of f^{iv}. Also transform the results into a formula of the form

$$\int_{0}^{1} x^{1/2} F(x)\, dx = H_1' F(x_1') + H_2' F(x_2') + E'$$

31 By making appropriate use of (8.9.12), obtain the quadrature formula

$$\int_{-1}^{1} (1 - x^2)^n f(x)\, dx = \sum_{k=1}^{m} H_k f(x_k) + E$$

when n is a nonnegative integer, with x_i the ith zero of the polynomial

$$\phi_m(x) = P_{m+n}^{(n)}(x) \equiv \frac{d^n}{dx^n} P_{m+n}(x)$$

and with

$$H_i = \frac{2(m + 2n)!}{m!\,(1 - x_i^2)[P_{m+n}^{(n+1)}(x_i)]^2}$$

and

$$E = \frac{m!\,(m + 2n)!}{(2m)!} \left[\frac{(m + n)!}{(2m + 2n)!}\right]^2 \frac{2^{2m+2n+1}}{2m + 2n + 1} f^{(2m)}(\xi) \qquad (|\xi| < 1)$$

Section 8.10

32 Suppose that a quadrature formula of the form

$$\int_{0}^{\infty} e^{-x} f(x)\, dx = W_1 f(0) + \sum_{k=2}^{m} W_k f(x_k) + E$$

is required, with the abscissa $x_1 = 0$ assigned. Show that the $m - 1$ free abscissas should then be zeros of the polynomial $\phi_{m-1}(x)$, where

$$\phi_r(x) = C_r x^{-1} e^x \frac{d^r}{dx^r} [x^{r+1} e^{-x}]$$

$$\equiv C_r x^{-1} e^x \left[x \frac{d^r}{dx^r} (x^r e^{-x}) + r \frac{d^{r-1}}{dx^{r-1}} (x^r e^{-x}) \right]$$

if the degree of precision is to be maximized. Verify the relation

$$\frac{d^{r-1}}{dx^{r-1}} (x^r e^{-x}) = -\frac{x}{r} e^{-x} \frac{d}{dx} L_r(x)$$

and hence, by taking $C_r = 1$, deduce that the free abscissas must be zeros of

$$\phi_{m-1}(x) = L_{m-1}(x) - L'_{m-1}(x)$$

where the prime denotes differentiation.

33 For the set of polynomials $\phi_r(x)$ obtained in Prob. 32, show that

$$A_r = (-1)^r \qquad \bar{\gamma}_r = r!\,(r+1)!$$

and hence deduce that the weights associated with the free abscissas are given by

$$W_i = \frac{(m-1)!\,m!}{x_i \phi'_{m-1}(x_i) \phi_m(x_i)} \qquad (i \neq 1)$$

Show also that

$$W_1 = \frac{-1}{\phi_{m-1}(0)} \int_0^\infty \frac{d}{dx} [e^{-x} L_{m-1}(x)]\, dx = \frac{1}{m}$$

34 Show that the error term in the formula of Prob. 32 can be obtained by the following steps:

$$E = (-1)^{m-1} \int_0^\infty e^{-x} x \phi_{m-1}(x) f[x_1, \ldots, x_m, x]\, dx$$

$$= (-1)^{m-1} \int_0^\infty f[x_1, \ldots, x_m, x] \frac{d^{m-1}}{dx^{m-1}} (x^m e^{-x})\, dx$$

$$= \int_0^\infty \left\{ \frac{d^{m-1}}{dx^{m-1}} f[x_1, \ldots, x_m, x] \right\} x^m e^{-x}\, dx$$

$$= \frac{(m-1)!}{(2m-1)!} f^{(2m-1)}(\xi) \int_0^\infty x^m e^{-x}\, dx \qquad \text{[see (5.10.36)]}$$

$$= \frac{(m-1)!\,m!}{(2m-1)!} f^{(2m-1)}(\xi) \qquad (\xi > 0)$$

Also verify that the same result follows directly from (8.10.22).

35 Show that the results of Probs. 32 to 34 reduce to the formulas

$$\int_0^\infty e^{-x} f(x)\,dx = \tfrac{1}{2}[f(0) + f(2)] + \tfrac{1}{3} f'''(\xi)$$

and

$$\int_0^\infty e^{-x} f(x)\,dx = \tfrac{1}{3} f(0) + \tfrac{1}{6}[(2 + \sqrt{3}) f(3 - \sqrt{3}) + (2 - \sqrt{3}) f(3 + \sqrt{3})] + \tfrac{1}{10} f^{\mathrm{v}}(\xi)$$

when $m = 2$ and 3, where $\xi > 0$ in both cases.

36 Use the second formula of Prob. 35 to approximate the integral

$$\int_0^\infty e^{-x} \sin x\,dx$$

and compare the result with that of using two (nonzero) ordinates in Prob. 18.

Section 8.11

37 Obtain approximate values of the integral

$$\int_0^{\pi/2} \sin x\,dx$$

by use of Radau quadratures employing two, three, four, and five ordinates, taking the vanishing ordinate as the assigned one. Also compare the results with corresponding ones (employing one, two, three, and four *nonvanishing* ordinates) in Prob. 14.

38 Proceed as in Prob. 37 with the integral

$$\int_0^1 x \cos x\,dx$$

and compare the results when $m = 2$ and 3 with those given by the two explicit formulas of Prob. 9 (in which x is to be considered as a weighting function).

Section 8.12

39 Obtain approximate values of the integral

$$\int_0^\pi \sin x\,dx$$

by use of Lobatto quadratures employing three, four, five, and six ordinates. Also compare the results with corresponding ones (employing one, two, three, and four *nonvanishing* ordinates) in Prob. 15.

40 Proceed as in Prob. 39 with the integral

$$\int_{-1}^{1} (1 - x^2) \cos x \, dx$$

and compare the results when $m = 3$ and 4 with corresponding ones obtained by using (8.9.13) and employing like numbers (one and two) of nonvanishing ordinates.

41 Derive the formula

$$\int_{-1}^{1} \frac{f(x)}{\sqrt{1 - x^2}} dx = \frac{\pi}{m}\left[\frac{1}{2}f(-1) + \frac{1}{2}f(1) + \sum_{k=1}^{m-1} f\left(\cos\frac{k\pi}{m}\right)\right] - \frac{2\pi}{2^{2m}(2m)!} f^{(2m)}(\xi)$$

making use of Prob. 33 of Chap. 7 and of Eq. (7.9.21), and noticing that $m + 1$ ordinates are employed.

Section 8.13

42 Determine Legendre-Gauss approximations to the integral

$$\int_{-1}^{1} \frac{dx}{x^2 + R^2}$$

for $R = 4, 2, 1$, and $\frac{1}{2}$. In each case obtain a sequence of four approximations, using successively $m = 2, 3, 4$, and 5 points, and calculate each error $E_m(R)$. Then, for each value of R, determine the successive ratios $|E_{m+1}(R)/E_m(R)|$ and compare the results with the asymptotic value $1/(4R^2)$ predicted by (8.13.7) as $m \to \infty$ (when $R > \frac{1}{2}$). (*Rounded true values:* 0.1224893, 0.4636476, 1.5707963, 4.4285949.)

43 Proceed as in Prob. 42, obtaining sequences of Hermite-Gauss approximations to the integral

$$\int_{-\infty}^{\infty} \frac{e^{-x^2}\, dx}{x^2 + R^2}$$

(*Rounded true values:* 0.1075993, 0.4011746, 1.3432934, 3.8684965.)

Section 8.14

44 Rework Prob. 13 by use of Chebyshev quadratures employing two, three, four, and five ordinates, and compare the results with the results of that problem.

45 Suppose that independent errors $\varepsilon_1, \varepsilon_2, \ldots, \varepsilon_m$ are associated with the m ordinates used in a Chebyshev quadrature over $[-1, 1]$, and that each of these errors is distributed about a zero mean with RMS deviation not exceeding ε_{RMS}. If R is the corresponding error in the approximate integral, show that

$$|R|_{\max} \leqq 2|\varepsilon|_{\max} \qquad R_{\text{RMS}} \leqq \frac{2}{\sqrt{m}}\varepsilon_{\text{RMS}}$$

Show also that the first relation holds for Legendre-Gauss quadrature while the factor $2/\sqrt{m}$ in the second relation is increased by only about 5 percent when $m = 3, 4$, and 5 (the same is true for $m = 6, 7$, and 9) but that somewhat larger increases in this factor occur when corresponding Newton-Cotes formulas are used over $[-1, 1]$. Also determine the values of this factor associated with the trapezoidal and parabolic rules when $m = 3, 5, 7$, and 9.

46 Verify that the interval $[-1, 1]$ can be replaced by $[a, b]$ in (8.14.1) and (8.14.17), when $[a, b] \neq (-\infty, \infty)$, and, in the *Laguerre-Chebyshev* case when $[a, b] = [0, \infty)$ and $w(x) = e^{-x}$, show that the m relevant abscissas are to be the zeros of the polynomial part of the formal expansion

$$x^m \exp\left[-m\left(\frac{1}{x} + \frac{1!}{x^2} + \frac{2!}{x^3} + \frac{3!}{x^4} + \cdots\right)\right]$$
$$= x^m - mx^{m-1} + \frac{m}{2}(m-2)x^{m-2} - \frac{m}{6}(m^2 - 6m + 12)x^{m-3}$$
$$+ \frac{m}{24}(m^3 - 12m^2 + 60m - 144)x^{m-4} + \cdots$$

if those zeros are real. Show further that, when $m = 2$, the quadrature formula is identical with the first formula of Prob. 35, and also that two of the abscissas are nonreal when $m = 3$, so that no three-point formula of the required type can exist. [The same has been shown to be true for all $m \geqq 3$ (see Krylov [1958]).]

47 Assuming the validity of (8.14.17) with $[-1, 1]$ replaced by $(-\infty, \infty)$ in the *Hermite-Chebyshev* case when $w(x) = e^{-x^2}$ (a modified derivation is necessary), show that the relevant abscissas are to be the zeros of the polynomial part of the formal expansion

$$x^m \exp\left[-m\left(\frac{1}{4x^2} + \frac{3}{16x^4} + \cdots\right)\right]$$
$$= x^m - \frac{m}{4}x^{m-2} + \frac{m}{32}(m-6)x^{m-4} - \cdots$$

that the zeros are real when $m = 2$ and $m = 3$, but that two of the zeros are nonreal when $m = 4$ and $m = 5$. [The presence of nonreal zeros has been established for all $m \geqq 4$ (see Krylov [1958]).]

48 Show that the two-point formula of Prob. 47 is identical with the Hermite-Gauss two-point formula, and that the three-point formula is of the form

$$\int_{-\infty}^{\infty} e^{-x^2} f(x)\,dx = \frac{\sqrt{\pi}}{3}\left[f\left(-\frac{\sqrt{3}}{2}\right) + f(0) + f\left(\frac{\sqrt{3}}{2}\right)\right] + E$$

Show also that the error term can be expressed in the form

$$E = \int_{-\infty}^{\infty} e^{-x^2}(x^3 - \tfrac{3}{4}x) f[x_1, x_2, x_3, x]\, dx$$

and transformed by integration by parts to give

$$E = \tfrac{1}{2}\int_{-\infty}^{\infty} e^{-x^2}(x^2 + \tfrac{1}{4}) f[x_1, x_2, x_3, x, x]\, dx = \frac{\sqrt{\pi}}{64} f^{\text{iv}}(\xi)$$

for some value of ξ.

Section 8.15

49 Determine algebraically the unknown abscissas and/or weights for the formula

$$\int_{-1}^{1} f(x)\, dx = \sum_{k=1}^{3} W_k f(x_k) + E$$

subject to the requirement that the degree of precision be as high as possible in consistency with each of the following sets of constraints, and determine the degree of precision in each case:

(*a*) $x_1 = -\tfrac{1}{2}, x_2 = 0, x_3 = \tfrac{1}{2}$
(*b*) No constraints
(*c*) $W_1 = W_2 = W_3$
(*d*) $x_1 = -1$

50 Suppose that the abscissas $x_1 = -1$ and $x_2 = \alpha$ are assigned and that the quadrature formula

$$\int_{-1}^{1} f(x)\, dx = W_1 f(-1) + W_2 f(\alpha) + W_3 f(x_3) + E \qquad (-1 < \alpha \leqq 1)$$

is to possess a degree of precision of at least 3. Determine x_3 and the three weights as functions of α, by algebraic methods, showing, in particular, that no such formula exists if $\alpha = \tfrac{1}{3}$, that x_3 is outside $[-1, 1]$ for all other α such that $0 < \alpha < \tfrac{1}{2}$, and that the ordinate at $x = -1$ is not involved if $\alpha = \pm 1/\sqrt{3}$.

51 Show that the two-point gaussian quadrature formula of the form

$$\int_{0}^{\pi} f(x) \sin nx\, dx = W_1 f(x_1) + W_2 f(x_2) + E$$

where n is a nonnegative integer, is such that

$$x_1 = \frac{\pi}{2} - \sqrt{\left(\frac{\pi}{2}\right)^2 - \frac{6}{n^2}} \qquad x_2 = \frac{\pi}{2} + \sqrt{\left(\frac{\pi}{2}\right)^2 - \frac{6}{n^2}}$$

$$W_1 = -W_2 = \frac{\pi}{n\sqrt{\pi^2 - 24/n^2}}$$

when n is *even*, and such that

$$x_1 = \frac{\pi}{2} - \sqrt{\left(\frac{\pi}{2}\right)^2 - \frac{2}{n^2}} \qquad x_2 = \frac{\pi}{2} + \sqrt{\left(\frac{\pi}{2}\right)^2 - \frac{2}{n^2}}$$

$$W_1 = W_2 = \frac{1}{n}$$

when n is *odd*. Show also that the degree of precision is 4 when n is even and 3 when n is odd.

52 Show that the error term in the quadrature formula of Prob. 51 can be expressed in the form

$$E = \int_0^\pi f[x_1, x_2, x]\pi(x) \sin nx \, dx = \tfrac{1}{2} \int_0^\pi f''(\eta)\pi(x) \sin nx \, dx \qquad (0 < \eta < \pi)$$

where $\pi(x) = x^2 - \pi x + 6n^{-2}$ when n is even and $\pi(x) = x^2 - \pi x + 2n^{-2}$ when n is odd, and that, when $n = 1$, this expression can be transformed to

$$E = \tfrac{1}{24} f^{\text{iv}}(\xi) \int_0^\pi x^2 \pi(x) \sin x \, dx = \frac{10 - \pi^2}{6} f^{\text{iv}}(\xi) \qquad (0 < \xi < \pi)$$

[Notice that here $w(x) \equiv \sin nx$ changes sign inside the range of integration when $n > 1$.]

53 Derive the gaussian integration formulas

$$\int_0^1 f(x) \log x \, dx = -f(\tfrac{1}{4}) + E$$

and

$$\int_0^1 f(x) \log x \, dx = -W_1 f(x_1) - W_2 f(x_2) + E$$

where

$$x_1 = \frac{15 - \sqrt{106}}{42} \doteq 0.112009 \qquad x_2 = \frac{15 + \sqrt{106}}{42} \doteq 0.602277$$

and

$$W_1 = \frac{212 + 9\sqrt{106}}{424} \doteq 0.718539 \qquad W_2 = \frac{212 - 9\sqrt{106}}{424} \doteq 0.281461$$

54 Show that the error terms associated with the two formulas of Prob. 53 are of the forms

$$E = \int_0^1 (x - \tfrac{1}{4})^2 f[\tfrac{1}{4}, \tfrac{1}{4}, x] \log x \, dx = -\tfrac{7}{288} f''(\xi)$$

and

$$E = \int_0^1 (x^2 - \tfrac{5}{7}x + \tfrac{17}{252})^2 f[x_1, x_1, x_2, x_2, x] \log x \, dx = -\tfrac{647}{5443200} f^{\text{iv}}(\xi)$$
$$\doteq -0.00012 f^{\text{iv}}(\xi)$$

respectively, where $0 < \xi < 1$ in each case.

Section 8.16

55 Derive the relations (8.16.15) to (8.16.18).

56 Derive the results of (8.16.28) and (8.16.30).

57 Use (8.16.25) to approximate the integral

$$\int_0^\pi \sin\theta \cos k\theta \, d\theta = \frac{2}{1 - k^2} \qquad (k = 0, 2, 4, \ldots)$$

for $k = 2$, 4, and 6, and verify that the relative error in each case is approximately $K_2/(2k^4) \doteq 0.0036/k^4$.

9

APPROXIMATIONS OF VARIOUS TYPES

9.1 Introduction

Whereas polynomials usually are convenient coordinate functions for the approximation of a continuous function (or for least-squares approximation of a function which is continuous except for finite "jumps") when the desired interval of approximation is finite, they are well adapted to the approximation of *periodic* functions only over relatively short ranges. When $f(x)$ is periodic and is to be approximated over one or more complete periods, it is desirable to make use of periodic coordinate functions, having the same period as $f(x)$, in constructing its approximation. The most convenient set of such functions (which, indeed, satisfies all the requirements of Sec. 1.2 when f is also continuous) is the composite set of all sines and cosines which possess that period. Although formulas analogous to Lagrange's formula exist for the determination of such an approximation (see Chap. 3, Prob. 7), they are seldom used, and resort is usually had to least-squares methods. The relevant analysis, due originally to Fourier and often known as *harmonic analysis*, is presented and illustrated for continuous domains in Sec. 9.2 and for discrete domains in Sec. 9.3.

When empirical data correspond to a simple decay or growth process, or to a combination of such processes, and an approximation is desired for a semi-infinite range of the independent variable (frequently representing time), real exponential functions are appropriate coordinate functions. On the other hand, when the superposition of two or more simple or damped harmonics, of unknown periods, is to be analyzed, complex exponential functions are appropriate. Prony's method of curve fitting, which includes both these cases when the data points are equally spaced, is presented in Sec. 9.4 and is specialized to the second case in Sec. 9.5.

Methods of *optimum* collocative polynomial interpolation are considered in Secs. 9.6 and 9.7, the Lanczos method of improving the efficiency of a given polynomial approximation is described in Sec. 9.8, and further approaches to uniform (minimax) polynomial approximation are considered in Sec. 9.9. Sections 9.10 to 9.13 are devoted to approximation by a class of coordinate functions, known as *cubic splines*, which may be described as sets of cubic polynomial segments with smooth joins.

A natural generalization of polynomial approximation consists of approximation by *ratios* of polynomials, that is, by rational functions. Such approximations can be expressed conveniently in terms of continued fractions and are treated in the concluding sections of this chapter (Secs. 9.14 to 9.18).

9.2 Fourier Approximation: Continuous Domain

We suppose here that a function $f(x)$ to be approximated is a *periodic* function, of known period, and that the scale of units has been so adjusted that the period is 2π, so that

$$f(x + 2\pi) = f(x) \qquad (9.2.1)$$

A particularly convenient class of coordinate functions is represented by the set

$$1, \cos x, \cos 2x, \ldots, \cos rx, \ldots; \sin x, \sin 2x, \ldots, \sin rx, \ldots$$

each member of which is of period 2π. This set has the useful property that the *product* of any two members is expressible as a linear combination of two members. Also, the *derivative* of each member is also a member, and the same is true of the *integral* of each member except the constant.

But the principal source of convenience is the verifiable fact that the set is *orthogonal* over any period interval, say, the interval $[-\pi, \pi]$, so that

$$\int_{-\pi}^{\pi} \sin jx \sin kx \, dx = 0 \qquad (j \neq k)$$

$$\int_{-\pi}^{\pi} \cos jx \cos kx \, dx = 0 \qquad (j \neq k) \qquad (9.2.2)$$

$$\int_{-\pi}^{\pi} \sin jx \cos kx \, dx = 0$$

when j and k are nonnegative integers. Clearly, negative integers need not be considered.

Suppose now that we require a so-called *Fourier approximation* to $f(x)$ of the form

$$f(x) \approx a_0 + \sum_{k=1}^{n} (a_k \cos kx + b_k \sin kx) \qquad (9.2.3)$$

involving harmonics through the nth, where the coefficients are to be determined in such a way that the integrated squared error over an interval of length 2π is least. From the periodicity of $f(x)$ and of the sine and cosine harmonics, it follows that attention may be restricted to *any* period interval, say, the interval $[-\pi, \pi]$. Then the requirement

$$\int_{-\pi}^{\pi} [f(x) - a_0 - \sum_{k=1}^{n} (a_k \cos kx + b_k \sin kx)]^2 \, dx = \min \qquad (9.2.4)$$

leads to the necessary conditions

$$\int_{-\pi}^{\pi} \left[f(x) - a_0 - \sum_{k=1}^{n} (a_k \cos kx + b_k \sin kx) \right] dx = 0$$

$$\int_{-\pi}^{\pi} \cos rx \left[f(x) - a_0 - \sum_{k=1}^{n} (a_k \cos kx + b_k \sin kx) \right] dx = 0$$

$$(r = 1, 2, \ldots, n) \qquad (9.2.5)$$

$$\int_{-\pi}^{\pi} \sin rx \left[f(x) - a_0 - \sum_{k=1}^{n} (a_k \cos kx + b_k \sin kx) \right] dx = 0$$

$$(r = 1, 2, \ldots, n)$$

when the partial derivatives of the left-hand member of (9.2.4) with respect to a_0, a_r, and b_r are equated to zero. Reference to the relations (9.2.2) and to the relations

$$\int_{-\pi}^{\pi} dx = 2\pi \qquad \int_{-\pi}^{\pi} \cos^2 kx\,dx = \int_{-\pi}^{\pi} \sin^2 kx\,dx = \pi \qquad (k \neq 0) \tag{9.2.6}$$

then leads to the determinations

$$\begin{gathered} a_0 = \frac{1}{2\pi}\int_{-\pi}^{\pi} f(x)\,dx \qquad a_k = \frac{1}{\pi}\int_{-\pi}^{\pi} f(x)\cos kx\,dx \qquad (k \neq 0) \\ b_k = \frac{1}{\pi}\int_{-\pi}^{\pi} f(x)\sin kx\,dx \end{gathered} \tag{9.2.7}$$

If $f(x)$ is an *even* function, so that $f(-x) = f(x)$, it is seen that $b_k = 0$, so that (9.2.3) then reduces to

$$f(x) \approx a_0 + \sum_{k=1}^{n} a_k \cos kx \tag{9.2.8}$$

where

$$\begin{aligned} a_0 &= \frac{1}{2\pi}\int_{-\pi}^{\pi} f(x)\,dx = \frac{1}{\pi}\int_{0}^{\pi} f(x)\,dx \\ a_k &= \frac{1}{\pi}\int_{-\pi}^{\pi} f(x)\cos kx\,dx = \frac{2}{\pi}\int_{0}^{\pi} f(x)\cos kx\,dx \qquad (k \neq 0) \end{aligned} \tag{9.2.9}$$

Similarly, if $f(x)$ is an *odd* function, so that $f(-x) = -f(x)$, there follows $a_0 = a_k = 0$ and (9.2.3) then becomes

$$f(x) \approx \sum_{k=1}^{n} b_k \sin kx \tag{9.2.10}$$

where

$$b_k = \frac{1}{\pi}\int_{-\pi}^{\pi} f(x)\sin kx\,dx = \frac{2}{\pi}\int_{0}^{\pi} f(x)\sin kx\,dx \tag{9.2.11}$$

If the periodic function $f(x)$ is fairly well behaved [in particular, if only $f(x)$ is bounded and piecewise differentiable], it is known that the approximation actually tends to $f(x)$ as $n \to \infty$ for all values of x at which $f(x)$ is continuous, and that it tends to the mean value $\frac{1}{2}[f(x+) + f(x-)]$ of the right- and left-hand limits at each point of discontinuity.

It is important to notice that, as is typical of least-squares approximations by orthogonal functions, each coefficient is determined independently of all others, and its value does not depend upon the number of harmonics to be retained in the approximation. As an example, suppose that $f(x)$ is defined in $[-\pi, \pi]$ in such a way that

$$f(x) = \begin{cases} 0 & (-\pi \leqq x \leqq 0) \\ x & \left(0 \leqq x \leqq \dfrac{\pi}{2}\right) \\ \dfrac{\pi}{2} & \left(\dfrac{\pi}{2} \leqq x < \pi\right) \end{cases} \tag{9.2.12}$$

and is defined elsewhere by the requirement that it be periodic, with period 2π (see Fig. 9.1). Since $f(x)$ is neither even nor odd, the presence of both sine and

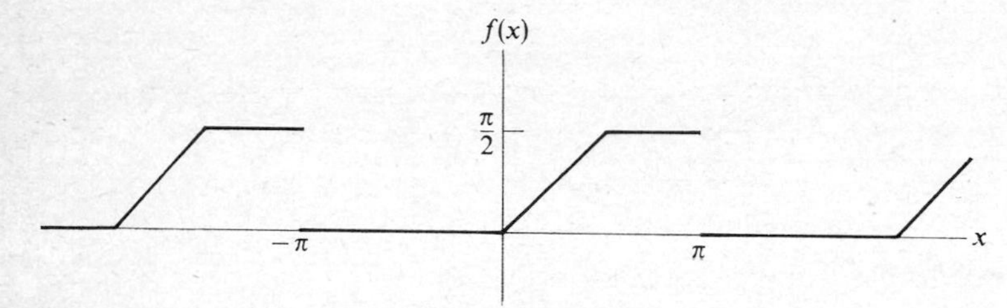

FIGURE 9.1

cosine harmonics may be anticipated. Equations (9.2.7) give

$$a_0 = \frac{1}{2\pi}\left[\int_{-\pi}^{0} 0\,dx + \int_{0}^{\pi/2} x\,dx + \int_{\pi/2}^{\pi} \frac{\pi}{2}\,dx\right] = \frac{3\pi}{16}$$

$$\begin{aligned} a_k &= \frac{1}{\pi}\left[\int_{-\pi}^{0} 0\cos kx\,dx + \int_{0}^{\pi/2} x\cos kx\,dx + \int_{\pi/2}^{\pi} \frac{\pi}{2}\cos kx\,dx\right] \\ &= -\frac{1}{\pi k^2}\left(1 - \cos\frac{k\pi}{2}\right) \qquad (k \neq 0) \end{aligned}$$

$$\begin{aligned} b_k &= \frac{1}{\pi}\left[\int_{-\pi}^{0} 0\sin kx\,dx + \int_{0}^{\pi/2} x\sin kx\,dx + \int_{\pi/2}^{\pi} \frac{\pi}{2}\sin kx\,dx\right] \\ &= \frac{1}{\pi k^2}\left(\sin\frac{k\pi}{2} - \frac{k\pi}{2}\cos k\pi\right) \end{aligned}$$

Thus there follows

$$\begin{aligned} f(x) &= \frac{3\pi}{16} - \frac{1}{\pi}\cos x - \frac{1}{2\pi}\cos 2x - \frac{1}{9\pi}\cos 3x - \cdots \\ &\qquad + \frac{2+\pi}{2\pi}\sin x - \frac{1}{4}\sin 2x + \frac{3\pi - 2}{9\pi}\sin 3x - \cdots \end{aligned} \tag{9.2.13}$$

when $x \neq \pm\pi, \pm 3\pi, \ldots$. If the best least-squares approximation to $f(x)$ over $[-\pi, \pi]$ involving only harmonics through the second were required, it would be obtained, by suppressing all higher harmonics in (9.2.13), in the form

$$f(x) \approx 0.589 - 0.318 \cos x - 0.159 \cos 2x + 0.818 \sin x - 0.250 \sin 2x$$

if (say) the coefficients were rounded to three places.

Because of the discontinuities in $f(x)$, a rather large number of terms would be needed, in this particular case, to afford a good approximation to $f(x)$, particularly near the discontinuities. However, there are in fact many practical situations in which only the *coefficients* of certain harmonics of low order are required and in which the degree of approximation afforded by a given number of harmonics is not of great interest.

It is clear that if $f(x)$ were *not* periodic, but were defined by (9.2.12) in the interval $[-\pi, \pi]$, the expansion (9.2.13) still would be valid *inside that interval* regardless of the behavior of $f(x)$ elsewhere. More generally, if the representation (9.2.3) were determined according to the formulas of (9.2.7) for *any* function $f(x)$ for which the integrals exist, the result would comprise the n-harmonic least-squares Fourier approximation to $f(x)$ *over the interval* $[-\pi, \pi]$. Outside that interval, the trigonometric expression would continue to define a periodic function regardless of the behavior of $f(x)$ itself outside that interval.

Further, if $f(x)$ is defined in $[0, \pi]$, and if the coefficients in the approximation

$$f(x) \approx a_0 + \sum_{k=1}^{n} a_k \cos kx \qquad (0 \leqq x \leqq \pi) \tag{9.2.14}$$

are determined by the equations

$$a_0 = \frac{1}{\pi}\int_0^{\pi} f(x)\,dx \qquad a_k = \frac{2}{\pi}\int_0^{\pi} f(x)\cos kx\,dx \qquad (k = 1, 2, \ldots, n) \tag{9.2.15}$$

the result will represent the n-harmonic least-squares Fourier cosine approximation to $f(x)$ over $[0, \pi]$. Similarly, the corresponding least-squares sine-harmonic approximation over that half-range is given by

$$f(x) \approx \sum_{k=1}^{n} b_k \sin kx \qquad (0 \leqq x \leqq \pi) \tag{9.2.16}$$

where

$$b_k = \frac{2}{\pi}\int_0^{\pi} f(x)\sin kx\,dx \tag{9.2.17}$$

These results are immediate consequences of the verifiable orthogonality relations

$$\int_0^\pi \sin jx \sin kx \, dx = \begin{cases} 0 & (j \neq k) \\ \dfrac{\pi}{2} & (j = k \neq 0) \end{cases} \tag{9.2.18}$$

$$\int_0^\pi \cos jx \cos kx \, dx = \begin{cases} 0 & (j \neq k) \\ \dfrac{\pi}{2} & (j = k \neq 0) \\ \pi & (j = k = 0) \end{cases} \tag{9.2.19}$$

where j and k are nonnegative integers. These relations can also be deduced directly from (9.2.2).

9.3 Fourier Approximation: Discrete Domain

We assume again that $f(x)$ is of period 2π, but now suppose that its values are known only on a discrete set of equally spaced points in a period interval, say at the $2N + 1$ points

$$-\pi, \; -\frac{(N-1)\pi}{N}, \ldots, -\frac{\pi}{N}, \; 0, \; \frac{\pi}{N}, \ldots, \frac{(N-1)\pi}{N}, \; \pi$$

of the interval $[-\pi, \pi]$. Since $f(-\pi) = f(\pi)$, from the assumed periodicity,† we then have $2N$ *independent* data, which may be expected to serve to determine the coefficients of $2N$ terms of an approximation of the form (9.2.3).

If we denote the rth abscissa as

$$x_r = r\frac{\pi}{N} \tag{9.3.1}$$

for $r = -N + 1, -N + 2, \ldots, -1, 0, 1, \ldots, N - 1, N,$ so that the $2N$ independent values $f_r \equiv f(x_r)$ are prescribed, we may verify that only the $2N$ functions

$$1, \cos x, \cos 2x, \ldots, \cos Nx; \sin x, \sin 2x, \ldots, \sin (N-1)x$$

of the set considered in the preceding section are independent over the domain comprising this set of abscissas; for the function $\sin Nx$ *vanishes* at each of these points, and each of the functions $\cos (N + 1)x, \ldots$ and $\sin (N + 1)x, \ldots$

† If $f(x)$ is *discontinuous* at the ends of the period interval $[-\pi, \pi]$ or, in other situations, is undefined outside that interval, the mean value

$$[f(\pi-) + f(\pi+)]/2 = [f(\pi-) + f(-\pi+)]/2$$

is to be assigned to $f(x)$ at both end points.

takes on the same values at points in the set as does one of the $2N$ functions listed above. For example, since $\sin Nx_r = 0$, we have

$$\cos (N + 1)x_r = \cos Nx_r \cos x_r = (-1)^r \cos x_r = \cos (N - 1)x_r$$

It is possible to show that this set of functions is orthogonal under *summation* over the set (9.3.1) (see Probs. 7 and 8), so that, with the notation of (9.3.1),

$$\begin{aligned} \sum_{r=-N+1}^{N} \sin jx_r \sin kx_r &= 0 \qquad (j \neq k) \\ \sum_{r=-N+1}^{N} \cos jx_r \cos kx_r &= 0 \qquad (j \neq k) \\ \sum_{r=-N+1}^{N} \sin jx_r \cos kx_r &= 0 \end{aligned} \tag{9.3.2}$$

when j and k are integers between 0 and N, inclusive, in analogy to (9.2.2). Furthermore, in the excluded cases for which $j = k$, the results

$$\begin{gathered} \sum_{r=-N+1}^{N} \sin^2 kx_r = \sum_{r=-N+1}^{N} \cos^2 kx_r = N \qquad (k \neq 0, N) \\ \sum_{r=-N+1}^{N} 1 = 2N \qquad \sum_{r=-N+1}^{N} \cos^2 Nx_r = 2N \end{gathered} \tag{9.3.3}$$

can be established in analogy to (9.2.6).

If now an approximation is assumed in the form

$$f(x) \approx A_0 + \sum_{k=1}^{n} (A_k \cos kx + B_k \sin kx) \tag{9.3.4}$$

where $n \leqq N$, and if the least-squares criterion

$$\sum_{r=-N+1}^{N} \left[f(x_r) - A_0 - \sum_{k=1}^{n} (A_k \cos kx_r + B_k \sin kx_r) \right]^2 = \min \tag{9.3.5}$$

is adopted, a derivation completely analogous to that leading from (9.2.4) to (9.2.7), making use of (9.3.2) and (9.3.3), yields the determinations

$$\begin{aligned} A_0 &= \frac{1}{2N} \sum_{r=-N+1}^{N} f(x_r) \qquad A_k = \frac{1}{N} \sum_{r=-N+1}^{N} f(x_r) \cos kx_r \qquad (k \neq 0, N) \\ A_N &= \frac{1}{2N} \sum_{r=-N+1}^{N} f(x_r) \cos Nx_r \qquad B_k = \frac{1}{N} \sum_{r=-N+1}^{N} f(x_r) \sin kx_r \end{aligned} \tag{9.3.6}$$

Thus the coefficients in (9.3.4) are easily obtained by summation, and the calculation of each coefficient is again independent of the calculation of the others, and is independent of n as long as $n \leqq N$. When $n = N$, the least-squares criterion becomes equivalent to the requirement that the two members of (9.3.4) agree exactly at the $2N$ points specified by (9.3.1).

The formulas (9.3.6) can be written in the more symmetrical forms

$$
\begin{aligned}
A_0 &= \frac{1}{2N}(\tfrac{1}{2}f_{-N} + f_{-N+1} + \cdots + f_{-1} + f_0 + f_1 + \cdots + f_{N-1} + \tfrac{1}{2}f_N) \\
A_k &= \frac{1}{N}(\tfrac{1}{2}f_{-N} \cos kx_{-N} + f_{-N+1} \cos kx_{-N+1} + \cdots \\
&\quad + f_{-1} \cos kx_{-1} + f_0 \cos kx_0 + f_1 \cos kx_1 + \cdots \\
&\quad\quad + f_{N-1} \cos kx_{N-1} + \tfrac{1}{2}f_N \cos kx_N) \qquad (k \neq 0, N) \\
A_N &= \frac{1}{2N}(\tfrac{1}{2}f_{-N} \cos Nx_{-N} + f_{-N+1} \cos Nx_{-N+1} + \cdots \\
&\quad + f_{-1} \cos Nx_{-1} + f_0 \cos Nx_0 + f_1 \cos Nx_1 + \cdots \\
&\quad\quad + f_{N-1} \cos Nx_{N-1} + \tfrac{1}{2}f_N \cos Nx_N) \\
B_k &= \frac{1}{N}(\tfrac{1}{2}f_{-N} \sin kx_{-N} + f_{-N+1} \sin kx_{-N+1} + \cdots \\
&\quad + f_{-1} \sin kx_{-1} + f_0 \sin kx_0 + f_1 \sin kx_1 + \cdots \\
&\quad\quad + f_{N-1} \sin kx_{N-1} + \tfrac{1}{2}f_N \sin kx_N)
\end{aligned}
\tag{9.3.7}
$$

in view of the relations $f_{-N} = f_N$.

If we notice that the spacing h is given by

$$h = \frac{\pi}{N} \tag{9.3.8}$$

we may observe the curious fact that Eqs. (9.3.7) are identical with the results of using the *trapezoidal rule* to approximate the right-hand members of (9.2.7), when $k < N$.†

For the purpose of numerical calculation, it is convenient to resolve $f(x)$ into its *even* and *odd* components by introducing the auxiliary functions

$$F(x) = \tfrac{1}{2}[f(x) + f(-x)] \qquad G(x) = \tfrac{1}{2}[f(x) - f(-x)] \tag{9.3.9}$$

so that

$$f(x) = F(x) + G(x) \tag{9.3.10}$$

† In this connection, it is interesting to recall that the Euler-Maclaurin sum formula, written in the form (5.8.17), reduces to the trapezoidal rule for *any* periodic function, with period equal to the length of the range of integration. That is, the "correction terms" in that formula all vanish in any such case. This fact obviously does not indicate that the trapezoidal rule is "exact" for periodic functions since the error term (5.8.14) remains, but may indeed serve to illustrate the dangers associated with lack of proper regard for the error term in such formulas.

If we recall that $x_{-r} = -x_r$, and that $x_0 = 0$, we find that Eqs. (9.3.6) or (9.3.7) may be reduced to the forms

$$A_0 = \frac{1}{N}(\tfrac{1}{2}F_0 + F_1 + F_2 + \cdots + F_{N-1} + \tfrac{1}{2}F_N)$$

$$A_k = \frac{2}{N}(\tfrac{1}{2}F_0 + F_1 \cos kx_1 + F_2 \cos kx_2 + \cdots$$
$$+ F_{N-1} \cos kx_{N-1} + \tfrac{1}{2}F_N \cos kx_N) \qquad (k \neq 0, N) \qquad (9.3.11)$$

$$A_N = \frac{1}{N}(\tfrac{1}{2}F_0 - F_1 + F_2 - \cdots + (-1)^{N-1}F_{N-1} + (-1)^N\tfrac{1}{2}F_N)$$

$$B_k = \frac{2}{N}(G_1 \sin kx_1 + G_2 \sin kx_2 + \cdots + G_{N-1} \sin kx_{N-1})$$

In order to illustrate the use of these formulas, we consider the case $N = 6$, corresponding to the use of 12 independent ordinates. The tabular forms given in Table 9.1*a* and *b* are then appropriate for desk calculation (although further systematization is possible). In Table 9.1*a*, the sum of the entries in the data column is $6A_0$, whereas the sum of products of corresponding entries in the data column and the column headed cos kx is $3A_k$ or $6A_6$. Similarly, the sum of products of corresponding entries in the data column of Table 9.1*b* and the column headed sin kx is $3B_k$.

Table 9.1*a*

x	Data	cos x	cos $2x$	cos $3x$	cos $4x$	cos $5x$	cos $6x$
0	$\tfrac{1}{2}f_0 = \tfrac{1}{2}F_0$	1	1	1	1	1	1
$\frac{\pi}{6}$	$\tfrac{1}{2}(f_1 + f_{-1}) = F_1$	$\tfrac{1}{2}\sqrt{3}$	$\tfrac{1}{2}$	0	$-\tfrac{1}{2}$	$-\tfrac{1}{2}\sqrt{3}$	-1
$\frac{\pi}{3}$	$\tfrac{1}{2}(f_2 + f_{-2}) = F_2$	$\tfrac{1}{2}$	$-\tfrac{1}{2}$	-1	$-\tfrac{1}{2}$	$\tfrac{1}{2}$	1
$\frac{\pi}{2}$	$\tfrac{1}{2}(f_3 + f_{-3}) = F_3$	0	-1	0	1	0	-1
$\frac{2\pi}{3}$	$\tfrac{1}{2}(f_4 + f_{-4}) = F_4$	$-\tfrac{1}{2}$	$-\tfrac{1}{2}$	1	$-\tfrac{1}{2}$	$-\tfrac{1}{2}$	1
$\frac{5\pi}{6}$	$\tfrac{1}{2}(f_5 + f_{-5}) = F_5$	$-\tfrac{1}{2}\sqrt{3}$	$\tfrac{1}{2}$	0	$-\tfrac{1}{2}$	$\tfrac{1}{2}\sqrt{3}$	-1
π	$\tfrac{1}{2}f_6 = \tfrac{1}{2}F_6$	-1	1	-1	1	-1	1
	$6A_0$	$3A_1$	$3A_2$	$3A_3$	$3A_4$	$3A_5$	$6A_6$

Table 9.1*b*

x	Data	$\sin x$	$\sin 2x$	$\sin 3x$	$\sin 4x$	$\sin 5x$
$\frac{\pi}{6}$	$\frac{1}{2}(f_1 - f_{-1}) = G_1$	$\frac{1}{2}$	$\frac{1}{2}\sqrt{3}$	1	$\frac{1}{2}\sqrt{3}$	$\frac{1}{2}$
$\frac{\pi}{3}$	$\frac{1}{2}(f_2 - f_{-2}) = G_2$	$\frac{1}{2}\sqrt{3}$	$\frac{1}{2}\sqrt{3}$	0	$-\frac{1}{2}\sqrt{3}$	$-\frac{1}{2}\sqrt{3}$
$\frac{\pi}{2}$	$\frac{1}{2}(f_3 - f_{-3}) = G_3$	1	0	-1	0	1
$\frac{2\pi}{3}$	$\frac{1}{2}(f_4 - f_{-4}) = G_4$	$\frac{1}{2}\sqrt{3}$	$-\frac{1}{2}\sqrt{3}$	0	$\frac{1}{2}\sqrt{3}$	$-\frac{1}{2}\sqrt{3}$
$\frac{5\pi}{6}$	$\frac{1}{2}(f_5 - f_{-5}) = G_5$	$\frac{1}{2}$	$-\frac{1}{2}\sqrt{3}$	1	$-\frac{1}{2}\sqrt{3}$	$\frac{1}{2}$
		$3B_1$	$3B_2$	$3B_3$	$3B_4$	$3B_5$

In illustration, for the empirical data

θ	0°	30°	60°	90°	120°	150°	180°	210°	240°	270°	300°	330°	360°
f	1.21	1.32	1.46	1.40	1.34	1.18	1.07	1.01	1.05	1.10	1.14	1.17	1.21
x	0	$\frac{\pi}{6}$	$\frac{\pi}{3}$	$\frac{\pi}{2}$	$\frac{2\pi}{3}$	$\frac{5\pi}{6}$	π	$-\frac{5\pi}{6}$	$-\frac{2\pi}{3}$	$-\frac{\pi}{2}$	$-\frac{\pi}{3}$	$-\frac{\pi}{6}$	0

the entries in the respective data columns of Tables 9.1*a* and *b* are

0.605	
1.245	0.075
1.300	0.160
1.250	0.150
1.195	0.145
1.095	0.085
0.535	

and calculation gives

$$A_0 = 1.204 \qquad A_1 = 0.084 \qquad A_2 = -0.062$$

$$A_3 = -0.012 \qquad A_4 = -0.009 \qquad B_1 = 0.165$$

$$B_2 = 0.001 \qquad B_3 = 0.003 \qquad B_4 = -0.007$$

for the coefficients of harmonics through the fourth.

In order to obtain a seven-point cosine approximation to a function $F(x)$ over the half-range $0 \leqq x \leqq \pi$, through harmonics of order not exceeding six, use would be made of Table 9.1*a* only, whereas for a five-point sine approximation to $G(x)$ over the same half-range, through harmonics of order not exceeding five, only Table 9.1*b* would be used. In any case, if all the *available* harmonics are retained, the resultant approximation takes on the prescribed value at each of the points employed in the calculation and accordingly represents a *collocative interpolation.* Retention of a smaller number of harmonics leads to the appropriate *least-squares approximation* relevant to that set of points.† Tables corresponding to those given here, but employing larger sets of data and further systematized in various ways, may be found in the literature (see Sec. 9.19). A related procedure is described in Sec. 9.7.

9.4 Exponential Approximation

In certain situations it is desired to determine an approximation of the form

$$f(x) \approx C_1 e^{a_1 x} + C_2 e^{a_2 x} + \cdots + C_n e^{a_n x} \tag{9.4.1}$$

or, equivalently, of the form

$$f(x) \approx C_1 \mu_1^x + C_2 \mu_2^x + \cdots + C_n \mu_n^x \tag{9.4.2}$$

where

$$\mu_k = e^{a_k} \tag{9.4.3}$$

It is somewhat more convenient here to work with the second form (9.4.2).

We suppose that values of $f(x)$ (exact or approximate) are specified on a set of N *equally spaced* points, and that a linear change of variables has been introduced in advance in such a way that the data points are $x = 0, 1, 2, \ldots, N - 1$. If (9.4.1) were to be an equality for these values of x, the equations

$$\begin{aligned} C_1 + C_2 + \cdots + C_n &= f_0 \\ C_1\mu_1 + C_2\mu_2 + \cdots + C_n\mu_n &= f_1 \\ C_1\mu_1^2 + C_2\mu_2^2 + \cdots + C_n\mu_n^2 &= f_2 \\ \cdots\cdots\cdots\cdots\cdots\cdots\cdots\cdots\cdots \\ C_1\mu_1^{N-1} + C_2\mu_2^{N-1} + \cdots + C_n\mu_n^{N-1} &= f_{N-1} \end{aligned} \tag{9.4.4}$$

necessarily would be satisfied, and the approximation (9.4.2) may be based on the result of satisfying these equations as nearly as possible. If the constants $\mu_1, \ldots, \mu_n$ were known (or preassigned), this set would comprise N linear equations in the n unknowns $C_1, \ldots, C_n$ and could be solved exactly if $N = n$ or approximately, by the least-squares method of Sec. 7.3, if $N > n$.

However, if the μ's are also to be determined, at least $2n$ equations are needed, and the difficulty consists of the fact that the equations are *nonlinear* in the μ's. This difficulty can be minimized by a method, similar to methods used in Sec. 8.15, to be described next.

† For the *cosine* approximation, the result corresponds to the use of one-half weights with respect to the errors at 0 and N (see Prob. 9).

Let $\mu_1, \ldots, \mu_n$ be the roots of the algebraic equation

$$\mu^n + \alpha_1\mu^{n-1} + \alpha_2\mu^{n-2} + \cdots + \alpha_{n-1}\mu + \alpha_n = 0 \qquad (9.4.5)$$

so that the left-hand member of (9.4.5) is identified with the product

$$(\mu - \mu_1)(\mu - \mu_2)\cdots(\mu - \mu_n)$$

In order to determine the coefficients $\alpha_1, \ldots, \alpha_n$, we multiply the first equation in (9.4.4) by α_n, the second equation by $\alpha_{n-1}, \ldots,$ the nth equation by α_1, and the $(n + 1)$th equation by 1, and add the results. If use is made of the fact that each μ satisfies (9.4.5), the result is seen to be of the form

$$f_n + \alpha_1 f_{n-1} + \cdots + \alpha_n f_0 = 0$$

A set of $N - n - 1$ additional equations of similar type is obtained in the same way by starting instead successively with the second, third, . . . , $(N - n)$th equations. In this way we find that (9.4.4) and (9.4.5) imply the $N - n$ *linear* equations

$$\begin{aligned} f_n + f_{n-1}\alpha_1 + f_{n-2}\alpha_2 + \cdots + f_0\alpha_n &= 0 \\ f_{n+1} + f_n\alpha_1 + f_{n-1}\alpha_2 + \cdots + f_1\alpha_n &= 0 \\ \cdots\cdots\cdots\cdots\cdots\cdots\cdots\cdots\cdots\cdots\cdots \\ f_{N-1} + f_{N-2}\alpha_1 + f_{N-3}\alpha_2 + \cdots + f_{N-n-1}\alpha_n &= 0 \end{aligned} \qquad (9.4.6)$$

Since the ordinates f_k are known, this set generally can be solved directly for the n α's if $N = 2n$, or solved approximately by the method of least squares if $N > 2n$.

After the α's are determined, the n μ's are found as the roots of (9.4.5). They may be real or imaginary. The equations (9.4.4) then become linear equations in the n C's, with known coefficients. The C's can be determined, finally, from the first n of these equations or, preferably, by applying the least-squares technique to the entire set. Thus the nonlinearity of the system is concentrated in the single algebraic equation (9.4.5). The technique described is known as *Prony's method.*

Obvious modifications are necessary when certain of the μ's (or a's) are prescribed and the remainder are to be determined. When such constraints are imposed, and are to be satisfied *exactly*, it is essential to satisfy them (by using them to eliminate unknowns from the set of equations to be solved) *before* applying the method of least squares.

The most common situation of this sort is that in which it is known that $f(x)$ tends to a finite limit (the value of which is generally unknown) as $x \to \infty$. The approximation

$$f(x) \approx C_0 + C_1 e^{a_1 x} + \cdots + C_n e^{a_n x} \qquad (9.4.7)$$

is then appropriate, where the a's are expected to have negative real parts. Since this approximation implies that

$$\Delta f(x) \approx C_1' e^{a_1 x} + \cdots + C_n' e^{a_n x}$$

where the coefficient C_k' is an unknown constant which is simply related to the unknown C_k, the equations (9.4.6) may be modified, in this case, by replacing each f_k by the difference $\Delta f_k \equiv f_{k+1} - f_k$, after which the α's and μ's are determined as before. The equations (9.4.4) are then modified by the insertion of the unknown C_0 in each left-hand member. At least $N = 2n + 1$ independent data are needed for the determination.

If one or more of the μ's satisfying (9.4.5) are not real and positive, the corresponding values of the a's in (9.4.1) will not be real. In particular, if μ_k is real and negative, say $\mu_k = -\rho_k$, where ρ_k is positive, the term $\mu_k^x = (-\rho_k)^x$ is real only when x takes on the (integral) values for which data are prescribed, or values which differ from those values by integral multiples of the (unit) spacing. However, we may notice that $(-1)^x = \cos \pi x$ *for any such value of* x. Hence, if we replace $(-\rho_k)^x$ by $\rho_k^x \cos \pi x$ or, equivalently, by $e^{x \log \rho_k} \cos \pi x$, we so obtain a suitable interpolating function which is real for all real values of x.

More generally, if one value of μ is *nonreal*, and hence expressible in the polar form $\rho e^{i\beta}$, where ρ and β are real and ρ is *positive*, then the conjugate $\rho e^{-i\beta}$ must also be involved, since the coefficients in (9.4.5) are necessarily real. The corresponding part of (9.4.2) can then be written as

$$\rho^x(A_1 e^{i\beta x} + A_2 e^{-i\beta x})$$

where A_1 and A_2 are constants which must be conjugate complex in order that the expression be real when x is real. Hence, by writing $A_1 = (C_1 - iC_2)/2$ and $A_2 = (C_1 + iC_2)/2$, this part of the approximation can be expressed in the more convenient form

$$\rho^x(C_1 \cos \beta x + C_2 \sin \beta x) \equiv e^{x \log \rho}(C_1 \cos \beta x + C_2 \sin \beta x) \qquad (9.4.8)$$

after the μ's are determined from (9.4.5) and (9.4.6) but before equations corresponding to (9.4.4) are formed and solved for the coefficients of the approximating functions.

In order to illustrate both the technique and the existence of unfavorable situations, we consider the attempt to recover the equation of the function

$$f(x) = 2.32 - 1.08e^{-x} + 1.20e^{-2x} \qquad (9.4.9)$$

from the values of that function for $x = 0, 1, 2, 3,$ and 4, under the hypothesis that the numerical coefficients in (9.4.9) are exact. These values are given, to four decimal places, in the following tabulation:

x	0	1	2	3	4
$f(x)$	2.4400	2.0851	2.1958	2.2692	2.3006

If the ordinates are arbitrarily rounded to *two* decimal places, the required differences of the rounded values are found to be -0.35, 0.11, 0.07, and 0.03, and Eqs. (9.4.6), with f_r replaced by Δf_r, become

$$\begin{aligned} 0.07 + 0.11\alpha_1 - 0.35\alpha_2 &= 0 \\ 0.03 + 0.07\alpha_1 + 0.11\alpha_2 &= 0 \end{aligned} \tag{9.4.10}$$

from which there follows $\alpha_1 = -\frac{91}{183} \doteq -0.497$ and $\alpha_2 = \frac{8}{183} \doteq 0.0437$. Equation (9.4.5) then becomes

$$183\mu^2 - 91\mu + 8 = 0$$

and yields $\mu_1 \doteq 0.383$ and $\mu_2 \doteq 0.114$ to three places. Thus the required approximation is to be of the form

$$\begin{aligned} f(x) &\approx C_0 + C_1(0.383)^x + C_2(0.114)^x \\ &\approx C_0 + C_1 e^{-0.96x} + C_2 e^{-2.18x} \end{aligned} \tag{9.4.11}$$

after which the C's may be determined by fitting the data at three points, or by use of a least-squares procedure over the five points for which data are provided. More nearly accurate determinations of the decay factors would have resulted from a reduction of inherent errors in the data employed, or from the result of using additional data to supply additional equations, and solving the resultant set approximately by least-squares methods.

Suppose, however, that the values $f(1) \doteq 2.0851$ and $f(2) \doteq 2.1958$ were incorrectly rounded to 2.08 and 2.19, respectively. We notice that the roundoff errors so introduced are only slightly greater than those effected by the correct rounding, and we may consider these additional errors as representative of observational errors which could result if the data were empirical. The four relevant differences are then -0.36, 0.11, 0.08, and 0.03, and the equations replacing (9.4.10) become

$$\begin{aligned} 0.08 + 0.11\alpha_1 - 0.36\alpha_2 &= 0 \\ 0.03 + 0.08\alpha_1 + 0.11\alpha_2 &= 0 \end{aligned} \tag{9.4.12}$$

from which there follows $\alpha_1 = -\frac{196}{409} \doteq -0.479$ and $\alpha_2 = \frac{31}{409} \doteq 0.0758$. The equation which determines approximations to μ_1 and μ_2 is then $409\mu^2 - 196\mu + 31 = 0$, which yields the *nonreal* roots $\mu_{1,2} \doteq 0.240 \pm 0.136i$. Since, accordingly $\mu_{1,2} \doteq e^{-1.29 \pm 0.515i}$, the form replacing (9.4.11) here becomes

$$f(x) \approx C_0 + e^{-1.29x}(C_1 \cos 0.515x + C_2 \sin 0.515x) \tag{9.4.13}$$

from which the C's may be determined by collocation or by least squares.

Whereas it is found that the coefficients in (9.4.11) and (9.4.13) can be determined in such a way that they *both* provide good approximations to the true function (9.4.9) for $0 \leqq x \leqq 4$ and, indeed, depart only slightly from $f(x)$ for all $x \geqq 0$, the latter approximation is oscillatory, while the true function and the former approximation are not. The slight additional errors introduced into the given data here lead to completely incorrect information concerning the *decay factors.*

While this example was selected deliberately to illustrate a particularly unfavorable situation, this type of "instability" is of common occurrence when it is necessary to determine the approximating *coordinate functions* themselves, in addition to the constants of combination to be associated with them. In such cases, it is particularly desirable that an error analysis be made.

Since here the true values of μ_1 and μ_2 are $e^{-1} \doteq 0.368$ and $e^{-2} \doteq 0.135$, the true values of α_1 and α_2 are $-(e^{-1} + e^{-2}) \doteq -0.503$ and $e^{-3} \doteq 0.0498$. Thus, in the second calculation, errors of magnitude smaller than 0.006 in the data employed lead to errors of about 0.024 and 0.026 in the calculation of α_1 and α_2, respectively, and these errors, in turn, lead to nonreal approximations of the real μ_1 and μ_2. The *possibility* of appreciably larger errors than those actually encountered in the calculation of α_1 and α_2 from *either* of the sets (9.4.10) and (9.4.12), assuming the coefficients to be correct to the places given, could have been predicted by an analysis of those sets.† Once such estimates are obtained, the maximum (or RMS) values of the errors $\delta\mu_1$ and $\delta\mu_2$ in the roots of $\mu^2 + \alpha_1\mu + \alpha_2 = 0$ may be estimated, by use of the differential relation

$$(2\mu + \alpha_1)\, d\mu + \mu\, d\alpha_1 + d\alpha_2 = 0$$

as

$$|\delta\mu_1|_{\max} \approx \frac{1 + |\mu_1|}{|\mu_2 - \mu_1|} |\delta\alpha|_{\max} \qquad |\delta\mu_2|_{\max} \approx \frac{1 + |\mu_2|}{|\mu_2 - \mu_1|} |\delta\alpha|_{\max} \tag{9.4.14}$$

or

$$(\delta\mu_1)_{\text{RMS}} \approx \frac{\sqrt{1 + \mu_1^2}}{|\mu_2 - \mu_1|} (\delta\alpha)_{\text{RMS}} \qquad (\delta\mu_2)_{\text{RMS}} \approx \frac{\sqrt{1 + \mu_2^2}}{|\mu_2 - \mu_1|} (\delta\alpha)_{\text{RMS}} \tag{9.4.15}$$

with μ_1 and μ_2 replaced by their calculated values, if those calculated values are real and if the errors are small. The reality of μ_1 and μ_2 depends upon the positivity of $\alpha_1^2 - 4\alpha_2$ and accordingly is in doubt if

$$|\alpha_1^2 - 4\alpha_2| < 2(2 + |\alpha_1|)|\delta\alpha|_{\max}$$

† The analysis of RMS errors relevant to the normal equations obtained in a least-squares procedure is described in Sec. 7.4 [see (7.4.30)]. For the corresponding analysis of maximum errors when least-squares methods are not used, as in the present case, see Sec. 10.7.

when α_1 and α_2 are estimated by their calculated values. Similar considerations apply to the more involved cases in which more coordinate functions are employed.

9.5 Determination of Constituent Periodicities

It frequently happens that an empirical function $f(x)$ is known to be expressible as a linear combination of two or more periodic terms whose periods are unknown and are not necessarily commensurable, and the approximate determination of these periods from empirical data is often of considerable importance.

If m distinct periods, denoted by $2\pi/\omega_1, \ldots, 2\pi/\omega_m$, are known (or assumed) to be present, then $f(x)$ correspondingly can be assumed to be approximated by an expression of the form

$$f(x) \approx A_1 \cos \omega_1 x + B_1 \sin \omega_1 x + \cdots + A_m \cos \omega_m x + B_m \sin \omega_m x \tag{9.5.1}$$

But such an approximation is a special case of (9.4.1), in which $n = 2m$ and in which the a's are identified, respectively, with $i\omega_1, -i\omega_1, \ldots, i\omega_m$, and $-i\omega_m$. Thus the desired values of ω may be obtained, by Prony's method, if we set $n = 2m$ and $\mu = e^{i\omega}$ in (9.4.5) and (9.4.6), again assuming $f(x)$ to be given for $x = 0, 1, 2, \ldots, N - 1$.†

Since, in this case, the roots of (9.4.5) are known (or required) to occur in reciprocal pairs $(e^{i\omega_k}, e^{-i\omega_k})$, it follows that (9.4.5) must be invariant under the substitution of $1/\mu$ for μ, so that we must have $\alpha_{2m} = 1$, $\alpha_{2m-1} = \alpha_1, \ldots,$ $\alpha_{m+1} = \alpha_{m-1}$. Thus, with $\mu = e^{i\omega}$, Eq. (9.4.5) becomes

$$e^{i2m\omega} + \alpha_1 e^{i(2m-1)\omega} + \cdots + \alpha_{m-1} e^{i(m+1)\omega} + \alpha_m e^{im\omega} + \alpha_{m-1} e^{i(m-1)\omega} + \cdots + \alpha_1 e^{i\omega} + 1 = 0$$

or

$$e^{im\omega}[(e^{im\omega} + e^{-im\omega}) + \alpha_1(e^{i(m-1)\omega} + e^{-i(m-1)\omega}) + \cdots + \alpha_{m-1}(e^{i\omega} + e^{-i\omega}) + \alpha_m] = 0$$

and hence, finally, since $e^{im\omega} \neq 0$, we find that the equation determining ω is of the form

$$2 \cos m\omega + 2\alpha_1 \cos (m - 1)\omega + \cdots + 2\alpha_{m-1} \cos \omega + \alpha_m = 0 \tag{9.5.2}$$

† In the more general case, the dimensionless variable x accordingly represents displacement from a reference point in units of the actual spacing h, as in the preceding section, and the calculated periods are also to be considered as expressed in units of h.

Since $\cos k\omega$ is expressible as a polynomial of degree k in $\cos \omega$, this equation can be expressed as an algebraic equation of degree m in $\cos \omega$. Indeed, it can be expressed in the form

$$T_m(\cos \omega) + \alpha_1 T_{m-1}(\cos \omega) + \cdots + \alpha_{m-1} T_1(\cos \omega) + \tfrac{1}{2}\alpha_m = 0$$

in terms of the Chebyshev polynomials of Sec. 7.9.

The N equations (9.4.6), which serve to determine the coefficients $\alpha_1, \ldots, \alpha_m$ in (9.5.2), reduce (again with $n = 2m$, $\alpha_{m+k} = \alpha_{m-k}$, and $\alpha_{2m} = 1$) to the forms

$$\begin{aligned}
&f_0 + f_{2m} + (f_1 + f_{2m-1})\alpha_1 + (f_2 + f_{2m-2})\alpha_2 + \cdots \\
&\qquad\qquad + (f_{m-1} + f_{m+1})\alpha_{m-1} + f_m\alpha_m = 0 \\
&f_1 + f_{2m+1} + (f_2 + f_{2m})\alpha_1 + (f_3 + f_{2m-1})\alpha_2 + \cdots \\
&\qquad\qquad + (f_m + f_{m+2})\alpha_{m-1} + f_{m+1}\alpha_m = 0 \qquad (9.5.3) \\
&\cdots\cdots\cdots\cdots\cdots\cdots\cdots\cdots\cdots\cdots \\
&f_{N-2m-1} + f_{N-1} + (f_{N-2m} + f_{N-2})\alpha_1 + (f_{N-2m+1} + f_{N-3})\alpha_2 + \cdots \\
&\qquad\qquad + (f_{N-m-2} + f_{N-m})\alpha_{m-1} + f_{N-m-1}\alpha_m = 0
\end{aligned}$$

In accordance with the fact that the approximation (9.5.1) involves $3m$ unknown constants, we must have $N \geqq 3m$. The set (9.5.3) then comprises $N - 2m \geqq m$ equations in the m unknowns α_k.

This set is to be solved (approximately, by least squares, if $N > 3m$) for the α's, and the ω's are then determined from (9.5.2), after which the coefficients in (9.5.1) are determined (if their values are desired) by writing down the conditions which would require (9.5.1) to be an equality for at least $2m$ of the N relevant values of x and solving that set approximately, by the method of least squares, if more than $2m$ conditions are used.

If, in addition, an unknown constant A_0 is present in the right-hand member of (9.5.1), Eqs. (9.5.3) are to be modified by replacing each f_k by Δf_k, and the constant A_0 will then appear only in the set of equations determining the *coefficients* in (9.5.1). Here we must have $N \geqq 3m + 1$ given data.

As a simple illustration, we attempt to determine the constituent periods of the function

$$f(x) = \cos \frac{\pi x}{3} + \sin \frac{\pi x}{4} \qquad (9.5.4)$$

assuming knowledge only of the following rounded values of that function:

x	0	1	2	3	4	5	6	7	8	9	10
$f(x)$	1.00	1.21	0.50	−0.29	−0.50	−0.21	0.00	−0.21	−0.50	−0.29	0.50

If we suppose that the vanishing of the overall mean value of $f(x)$ is not known in advance but that there is evidence (from physical considerations or otherwise) that the *deviation* from the mean is due to the superposition of two periodic processes, we first calculate the relevant *differences*:

x	0	1	2	3	4	5	6	7	8	9
$\Delta f(x)$	0.21	−0.71	−0.79	−0.21	0.29	0.21	−0.21	−0.29	0.21	0.79

Next, the six equations corresponding to (9.5.3) are written down

$$\begin{aligned} 0.50 - 0.92\alpha_1 - 0.79\alpha_2 &= 0 \\ -0.50 - 0.50\alpha_1 - 0.21\alpha_2 &= 0 \\ -1.00 + 0.29\alpha_2 &= 0 \\ -0.50 + 0.08\alpha_1 + 0.21\alpha_2 &= 0 \\ 0.50 - 0.08\alpha_1 - 0.21\alpha_2 &= 0 \\ 1.00 - 0.29\alpha_2 &= 0 \end{aligned} \tag{9.5.5}$$

and the relevant normal equations are obtained in the form

$$\begin{aligned} 1.1092\alpha_1 + 0.8654\alpha_2 &= 0.2900 \\ 0.8654\alpha_1 + 0.9246\alpha_2 &= 1.0800 \end{aligned} \tag{9.5.6}$$

The solution is found to be $\alpha_1 \doteq -2.4092$, $\alpha_2 \doteq 3.4230$, to four places, after which (9.5.2) becomes

$$2 \cos 2\omega - 4.8184 \cos \omega + 3.4230 = 0$$

or

$$4 \cos^2 \omega - 4.8184 \cos \omega + 1.4230 = 0 \tag{9.5.7}$$

and yields the values

$$\cos \omega_1 \doteq 0.5186 \qquad \cos \omega_2 \doteq 0.6860$$

from which the appropriate approximations to the true periods

$$P_1 = \frac{2\pi}{\omega_1} = 6 \qquad P_2 = \frac{2\pi}{\omega_2} = 8$$

are found to be

$$P_1 \approx \frac{2\pi}{1.0256} \doteq 6.12 \qquad P_2 \approx \frac{2\pi}{0.8147} \doteq 7.71 \tag{9.5.8}$$

Hence the roundoff errors introduced into the given data here lead to errors of about 2 and 4 percent in the calculations of P_1 and P_2, respectively. The corresponding approximation to the governing equation would then be obtained, if it were desired, by fitting the equation

$$f(x) \approx A_0 + A_1 \cos 1.026x + B_1 \sin 1.026x + A_2 \cos 0.815x + B_2 \sin 0.815x \tag{9.5.9}$$

to the data by use of the least-squares procedure.

Here, and in the general case, it may be noticed that only the value of $\cos \omega_k$ is determinate. Thus, if we denote by $\bar{\omega}_k$ the admissible value of ω_k which lies between 0 and π, we can conclude only that the proper approximate value of ω_k is one of the numbers $\pm\bar{\omega}_k + 2r\pi$ $(r = 0, 1, 2, \ldots)$, so that if the true physical spacing is h, the actual approximate period is known only to be one of the numbers

$$\frac{2\pi h}{\bar{\omega}_k + 2r\pi} \quad \text{or} \quad \frac{2\pi h}{2\pi - \bar{\omega}_k + 2r\pi} \qquad (r = 0, 1, \ldots)$$

Of these possibilities, only those corresponding to $r = 0$ can exceed the spacing h; the first $(2\pi h/\bar{\omega}_k)$ exceeds $2h$, whereas the second $[2\pi h/(2\pi - \bar{\omega}_k)]$ lies between h and $2h$. The data employed clearly cannot be expected to determine periods smaller than the spacing with any appreciable accuracy, in general. Whether or not either of the two remaining appropriate alternatives truly represents an approximate period could be determined mathematically by investigating whether a second calculation based on a set of additional data, with a spacing incommensurable with h, also yields that alternative. In practice, the decision frequently can be based more simply on an inspection of the graph of the data or on physical considerations.

Situations in which two or more of the constituent periods are nearly equal are of frequent practical occurrence and are the most troublesome ones. In such cases it is particularly important to retain sufficiently many terms in the approximation and to use a sufficiently large set of data when the data are inexact. An interesting example of this type is treated successfully in Whittaker and Robinson [1944] (Sec. 175) by a method which differs from the present one, and also in Willers [1950] (Sec. 30) by a method equivalent to that given here. In that case, 600 empirical data are available for the determination of two constituent periods, and all are employed in the former treatment. Whereas the latter treatment does not use all the 595 equations which could be formed, in analogy to (9.5.5), it first makes use of a selected set of 78 equations whose formation involves the use of most of the data, and then checks the results by a recalculation using a judiciously chosen similar set of 17 equations.

9.6 Optimum Polynomial Interpolation with Selected Abscissas

It has been shown in earlier chapters that, if a function $f(x)$ is approximated by the interpolation polynomial $y(x)$ of degree n which agrees with $f(x)$ at $n + 1$ points $x_0, x_1, \ldots, x_n$, we may write

$$f(x) = y(x) + \pi(x)\frac{f^{(n+1)}(\xi)}{(n+1)!} \tag{9.6.1}$$

where

$$\pi(x) = (x - x_0)(x - x_1)\cdots(x - x_n) \tag{9.6.2}$$

and where ξ lies in the interval I limited by the largest and smallest of x_0, $x_1, \ldots, x_n$, and x. We suppose here that an appropriate change of variables has reduced this interval to the interval $[-1, 1]$.

Furthermore, in the preceding chapter it was seen that appropriate choices of the $n + 1$ abscissas lead to *quadrature* formulas having certain desirable characteristics. In this section we investigate briefly a related class of interpolation formulas and single out a particular formula which is related to *trigonometric* approximation in the following section.

Whereas the parameter ξ in (9.6.1) depends upon the $n + 1$ abscissas and the variable x, the nature of that dependence will depend, in turn, upon the nature of the function $f(x)$. Thus, if we desire to choose the abscissas in such a way that the error $E(x) \equiv f(x) - y(x)$ will tend to be as small as possible over $[-1, 1]$, in some sense, for the set of *all* such functions having $n + 1$ continuous derivatives in $[-1, 1]$, we may attempt to make $|\pi(x)|$ as small as possible in the *same* sense in that interval, recalling that the coefficient of the highest power of x in $\pi(x)$ must be *unity* from (9.6.2).

In particular, we may simulate the condition

$$\int_{-1}^{1} w(x)[E(x)]^2\,dx = \min$$

by the requirement

$$\int_{-1}^{1} w(x)[\pi(x)]^2\,dx = \min \tag{9.6.3}$$

where $w(x)$ is a prescribed weighting function which is nonnegative in $[-1, 1]$. If we notice that $\pi(x)$ is expressible in the form

$$\pi(x) = x^{n+1} + c_n x^n + \cdots + c_2 x^2 + c_1 x + c_0 \tag{9.6.4}$$

and hence may be considered to be specified by the $n + 1$ coefficients c_0, $c_1, \ldots, c_n$, we deduce that (9.6.3) leads to the requirement that the partial derivative of the left-hand member with respect to each c_r must vanish. Since also

$$\frac{\partial \pi(x)}{\partial c_r} = x^r \qquad (r = 0, 1, \ldots, n) \tag{9.6.5}$$

this requirement becomes

$$2\int_{-1}^{1} w(x)\,\frac{\partial \pi(x)}{\partial c_r}\,\pi(x)\,dx \equiv 2\int_{-1}^{1} w(x)\pi(x)x^r\,dx = 0 \qquad (r = 0, 1, \ldots, n) \tag{9.6.6}$$

so that $\pi(x)$ *is to be that polynomial of degree* $n + 1$, *with leading coefficient unity, which is orthogonal to all polynomials of inferior degree over* $[-1, 1]$ *relative to* $w(x)$. The abscissas of the $n + 1$ points of collocation, at which the agreement between $f(x)$ and the polynomial approximation should be effected, are thus the zeros of that polynomial.

It is of interest to notice that, with the interpolation polynomial so determined, the integral approximation

$$\int_{-1}^{1} w(x)f(x)\,dx \approx \int_{-1}^{1} w(x)y(x)\,dx$$

is the corresponding *gaussian quadrature* formula of Chap. 8.

Thus, in particular, if we take $w(x) = 1$, and so tend "on the average" to minimize the integral of the square of the error $E(x)$ over the interval $[-1, 1]$, it follows from the results of Sec. 7.6 that the $n + 1$ abscissas should be the zeros of $P_{n-1}(x)$. Certain such sets of abscissas are listed in Table 8.1 (Sec. 8.5).

Further, if we take $w(x) = 1/\sqrt{1 - x^2}$, and so tend to minimize the integral of $[E(x)]^2/\sqrt{1 - x^2}$, the results of Sec. 7.9 show that the abscissas are to be the zeros of the $(n + 1)$th Chebyshev polynomial $T_{n+1}(x)$

$$T_{n+1}(x) = \cos\,[(n + 1)\cos^{-1} x] \tag{9.6.7}$$

and hence are given by

$$x_i = \cos\left(\frac{2i + 1}{n + 1}\,\frac{\pi}{2}\right) \qquad (i = 0, 1, \ldots, n) \tag{9.6.8}$$

Since the coefficient of x^r in $T_r(x)$ is 2^{r-1}, it follows also that then

$$\pi(x) = 2^{-n}T_{n+1}(x) \tag{9.6.9}$$

In addition, we may notice that the *extreme* values of $\pi(x)$ in $[-1, 1]$ are then $\pm 2^{-n}$ and are taken on (with successively alternating signs) at the end points $x = \pm 1$ and at n additional interior points, each of which separates a pair of adjacent abscissas. Thus, with this choice of the abscissas, the coefficient of

$f^{(n+1)}(\xi)/(n+1)!$ in the error term of (9.6.1) oscillates with *constant* amplitude 2^{-n} as x increases from -1 to 1.

On the other hand, since the coefficient of x^r in $P_r(x)$ is $2^{-r}(2r)!/(r!)^2$, the use of the zeros of $P_{n+1}(x)$ as the abscissas of collocation corresponds to the identification

$$\pi(x) = \frac{2^{n+1}[(n+1)!]^2}{(2n+2)!} P_{n+1}(x) \qquad (9.6.10)$$

Now the Legendre polynomial takes on the value $+1$ at $x = +1$ and the value $(-1)^{n+1}$ at $x = -1$, and $P_{n+1}(x)$ performs oscillations in $[-1, 1]$ in such a way that the n successive maxima and minima separating pairs of adjacent zeros inside the interval decrease in magnitude toward the center of that interval. Thus, in particular, the maximum absolute value of $\pi(x)$ in (9.6.10), over $[-1, 1]$, is given by the numerical factor in that equation, which is approximated by $2^{-n}\sqrt{\pi n/4}$ when n is large.

Hence it follows that, whereas use of the zeros of $P_{n+1}(x)$ minimizes the RMS value of $\pi(x)$ over $[-1, 1]$, the use of the zeros of $T_{n+1}(x)$ leads to a value of $|\pi(x)|_{\max}$ which is smaller than that corresponding to the former choice, by a factor which tends to increase in proportion to $n^{1/2}$ as n increases. Furthermore, the error will tend to oscillate uniformly over $[-1, 1]$ in the second case, whereas it will tend to oscillate with an amplitude increasing toward the ends of the interval in the first case, on the average. Thus, if it is desirable to control the *maximum* error, rather than the RMS error, the second choice generally will be preferable to the first.

Indeed, it was discovered by Chebyshev that this choice is the *best possible* one, when the maximum-error criterion is adopted. The proof follows most easily by assuming, on the contrary, that there exists a monic polynomial $\bar{\pi}(x)$ of degree $n + 1$ (with leading coefficient unity) whose maximum absolute value on $[-1, 1]$ is smaller than 2^{-n}. Then the difference $\bar{\pi}(x) - 2^{-n}T_{n+1}(x)$ is negative at the maxima of $T_{n+1}(x)$ and positive at its minima. Hence, since $2^{-n}T_{n+1}(x)$ takes on its extreme values ($\pm 2^{-n}$) at $n + 2$ points of $[-1, 1]$, including the ends, the difference $\bar{\pi}(x) - 2^{-n}T_{n+1}(x)$ must vanish at least $n + 1$ times. But, since this difference is a polynomial of degree n or less (the common leading term x^{n+1} being removed by the subtraction), this situation is impossible, and the desired contradiction is obtained.

It should not be forgotten that this result applies only to the minimization of the maximum value of $|\pi(x)|$ over $[-1, 1]$. For any *specific* function $f(x)$, the maximum absolute value of the *error* $\pi(x)f^{(n+1)}(\xi)/(n+1)!$ in (9.6.1) generally will not be minimized exactly since the maximum value of $|f^{(n+1)}(\xi)|$ generally will not be attained in correspondence with an abscissa for which $|\pi(x)|$ is greatest.

9.7 Chebyshev Interpolation

In this section, we consider in more detail the polynomial interpolation formula based on collocation at the zeros of $T_{n+1}(x)$. Since any polynomial of degree n can be expressed as a linear combination of Chebyshev polynomials of degrees zero through n, it is convenient to express the polynomial $y(x)$ which agrees with $f(x)$ when $x = x_0, x_1, \ldots, x_n$, where x_r is the rth zero of $T_{n+1}(x)$, in such a form, and so to write

$$f(x) = \sum_{k=0}^{n} C_k T_k(x) + \frac{1}{2^n(n+1)!} T_{n+1}(x) f^{(n+1)}(\xi) \tag{9.7.1}$$

in accordance with (9.6.1), where $|\xi| < 1$ under the assumption that x is in $[-1, 1]$. The C's are to be determined in such a way that the result of suppressing the error term is correct when $x = x_0, x_1, \ldots, x_n$, where

$$x_i = \cos\left(\frac{2i+1}{2n+2}\pi\right) \qquad (i = 0, 1, \ldots, n) \tag{9.7.2}$$

Whereas the desired interpolation polynomial could be expressed in the lagrangian form of Chap. 3, the following alternative procedure usually is more convenient for its determination. If we introduce the change of variables

$$x = \cos\theta \qquad (0 \leqq \theta \leqq \pi) \tag{9.7.3}$$

the requirement

$$f(x) \approx \sum_{k=0}^{n} C_k T_k(x) \qquad (-1 \leqq x \leqq 1) \tag{9.7.4}$$

becomes

$$F(\theta) \approx \sum_{k=0}^{n} C_k \cos k\theta \qquad (0 \leqq \theta \leqq \pi) \tag{9.7.5}$$

with the abbreviation

$$F(\theta) = f(\cos\theta) \tag{9.7.6}$$

The C's are now to be determined in such a way that (9.7.5) is an equality when $\theta = \theta_i$, where

$$\theta_i = \cos^{-1} x_i = \frac{2i+1}{2n+2}\pi \qquad (i = 0, 1, \ldots, n) \tag{9.7.7}$$

Thus the agreement is to occur at the *equally spaced* points $\pi/(2n+2)$, $3\pi/(2n+2), \ldots, (2n+1)\pi/(2n+2)$, which are seen to be midway between the successive points $0, \pi/(n+1), 2\pi/(n+1), \ldots, \pi$ which would have been employed in the procedure of Sec. 9.3, as the points of collocation for the determination of the C's in an approximation of the form (9.7.5).

In analogy to corresponding results of that section, it happens that $\cos j\theta$ and $\cos k\theta$ are orthogonal under summation over the $n+1$ points defined by (9.7.7) (see Prob. 23):

$$\sum_{r=0}^{n} \cos j\theta_r \cos k\theta_r = \begin{cases} 0 & (j \neq k) \\ \dfrac{n+1}{2} & (j = k \neq 0) \\ n+1 & (j = k = 0) \end{cases} \tag{9.7.8}$$

where j and k are nonnegative integers not exceeding n. Moreover, since the left-hand member of (9.7.8) is identical with

$$\sum_{r=0}^{n} T_j(x_r)T_k(x_r)$$

it follows that, whereas $T_0(x), T_1(x), \ldots$ are orthogonal under *integration* over $[-1, 1]$ relative to $w(x) = 1/\sqrt{1-x^2}$, the functions $T_0(x), T_1(x), \ldots, T_n(x)$ are orthogonal under *summation* over the zeros of $T_{n+1}(x)$, with a *unit* weighting function.

The truth of (9.7.8) permits us to deduce immediately that the required C's are expressible in the form

$$C_0 = \frac{1}{n+1}\sum_{r=0}^{n} F(\theta_r) \qquad C_k = \frac{2}{n+1}\sum_{r=0}^{n} F(\theta_r)\cos k\theta_r \qquad (k \neq 0) \tag{9.7.9}$$

where θ_i is defined by (9.7.7), or, alternatively, in the form

$$C_0 = \frac{1}{n+1}\sum_{r=0}^{n} f(x_r) \qquad C_k = \frac{2}{n+1}\sum_{r=0}^{n} f(x_r)T_k(x_r) \qquad (k \neq 0) \tag{9.7.10}$$

where x_i is defined by (9.7.2).

Thus, for example, we may construct Table 9.2 for desk calculation when $n = 5$. Here use has been made of the abbreviations

$$\begin{aligned} A &= \cos\frac{\pi}{12} = \tfrac{1}{2}\sqrt{2+\sqrt{3}} \doteq 0.96593 \\ B &= \cos\frac{5\pi}{12} = \tfrac{1}{2}\sqrt{2-\sqrt{3}} \doteq 0.25882 \end{aligned} \tag{9.7.11}$$

The dual headings permit the table to be used either with the function expressed as $f(x)$ over $-1 \leqq x \leqq 1$, with the unequally spaced abscissas listed in the third column, or with the function expressed as $F(\theta)$ over $0 \leqq \theta \leqq \pi$, with the equally spaced abscissas listed in the first column.

Table 9.2

θ	$f(x)$ / $F(\theta)$	$x = T_1(x)$ / $\cos\theta$	$T_2(x)$ / $\cos 2\theta$	$T_3(x)$ / $\cos 3\theta$	$T_4(x)$ / $\cos 4\theta$	$T_5(x)$ / $\cos 5\theta$
$\frac{\pi}{12}$	$F_1 = f_1$	A	$\frac{1}{2}\sqrt{3}$	$\frac{1}{2}\sqrt{2}$	$\frac{1}{2}$	B
$\frac{\pi}{4}$	$F_2 = f_2$	$\frac{1}{2}\sqrt{2}$	0	$-\frac{1}{2}\sqrt{2}$	-1	$-\frac{1}{2}\sqrt{2}$
$\frac{5\pi}{12}$	$F_3 = f_3$	B	$-\frac{1}{2}\sqrt{3}$	$-\frac{1}{2}\sqrt{2}$	$\frac{1}{2}$	A
$\frac{7\pi}{12}$	$F_4 = f_4$	$-B$	$-\frac{1}{2}\sqrt{3}$	$\frac{1}{2}\sqrt{2}$	$\frac{1}{2}$	$-A$
$\frac{3\pi}{4}$	$F_5 = f_5$	$-\frac{1}{2}\sqrt{2}$	0	$\frac{1}{2}\sqrt{2}$	-1	$\frac{1}{2}\sqrt{2}$
$\frac{11\pi}{12}$	$F_6 = f_6$	$-A$	$\frac{1}{2}\sqrt{3}$	$-\frac{1}{2}\sqrt{2}$	$\frac{1}{2}$	$-B$
	$6C_0$	$3C_1$	$3C_2$	$3C_3$	$3C_4$	$3C_5$

In illustration, the coefficient of $\cos 4\theta$ in (9.7.5) is given by

$$C_4 = \tfrac{1}{3}(\tfrac{1}{2}F_1 - F_2 + \tfrac{1}{2}F_3 + \tfrac{1}{2}F_4 - F_5 + \tfrac{1}{2}F_6)$$

whereas the coefficient of $T_4(x)$ in (9.7.4) is given by

$$C_4 = \tfrac{1}{3}(\tfrac{1}{2}f_1 - f_2 + \tfrac{1}{2}f_3 + \tfrac{1}{2}f_4 - f_5 + \tfrac{1}{2}f_6)$$

In order to obtain exact fit at the six relevant points, all the harmonics involved are to be used. The result of retaining a smaller number of harmonics would give the corresponding *least-squares* approximation over the six points. Once the C's are determined, the evaluation of the right-hand member of the approximation

$$f(x) \approx \sum_{k=0}^{n} C_k T_k(x) \qquad (9.7.12)$$

at intermediate points is conveniently effected by the method of Clenshaw (see Sec. 7.10).

9.8 Economization of Polynomial Approximations

It was seen in Sec. 7.9 that the nth-degree least-squares polynomial approximation to a function $f(x)$ over $[-1, 1]$, where the integral of the product of $1/\sqrt{1 - x^2}$ and the square of the error is to be minimized, is of the form

$$f(x) \approx y(x) = \sum_{k=0}^{n} a_k T_k(x) \qquad (|x| \leqq 1) \qquad (9.8.1)$$

where

$$a_0 = \frac{1}{\pi} \int_{-1}^{1} \frac{f(x)}{\sqrt{1 - x^2}} \, dx \qquad a_k = \frac{2}{\pi} \int_{-1}^{1} \frac{f(x)T_k(x)}{\sqrt{1 - x^2}} \, dx \qquad (k \geqq 1) \tag{9.8.2}$$

The approximation so determined generally will not be identified with that of (9.7.12) since the coefficients, determined in the one case by summation over a discrete set of points and in the other by integration over an interval, are generally unequal. However, the two approximations may be expected to be of similar nature, in the sense that the error associated with each will tend to oscillate with uniform amplitude over $[-1, 1]$, whereas that afforded by the finite Legendre series arising from least-squares approximation with uniform weighting (Sec. 7.6) will tend to oscillate with an amplitude which increases toward the ends of that interval, on the average. Accordingly, if the smallness of the *maximum* error is to be the governing criterion, it may be expected that a satisfactory approximation may be afforded by fewer terms of a Chebyshev series than would be required for a Legendre series.

Still another method of obtaining an approximation to a function $f(x)$ as a linear combination of Chebyshev polynomials is due to Lanczos. Its use presumes that a polynomial approximation to $f(x)$ is already available but that a more *efficient* one is desired. Specifically, it is supposed that one has the relation

$$f(x) = \sum_{k=0}^{n} A_k x^k + E_n(x) \tag{9.8.3}$$

where it is known that

$$|E_n(x)| < \varepsilon_1 \qquad (-1 \leqq x \leqq 1) \tag{9.8.4}$$

and that ε_1 is smaller than a prescribed error tolerance ε, whereas $|A_n| + \varepsilon_1$ is *not* a tolerable error, so that the last term in the approximation

$$f(x) \approx \sum_{k=0}^{n} A_k x^k \tag{9.8.5}$$

cannot be safely neglected.

Now let the right-hand member of (9.8.5) be expanded in a series of Chebyshev polynomials. Since that member is a polynomial of degree n, the resultant series will terminate with the term involving $T_n(x)$ and hence will be of the form

$$\sum_{k=0}^{n} A_k x^k \equiv \sum_{k=0}^{n} a_k T_k(x) \tag{9.8.6}$$

From the fact that the terms of highest degree in $T_r(x)$ are given by

$$T_r(x) = 2^{r-1}\left(x^r - \frac{r}{4} x^{r-2} + \cdots\right) \qquad (r \geqq 1) \tag{9.8.7}$$

it follows that the result of expressing the two members of (9.8.6) in terms of decreasing powers of x will be of the form

$$A_n x^n + A_{n-1}x^{n-1} + A_{n-2}x^{n-2} + \cdots = 2^{n-1}a_n\left(x^n - \frac{n}{4}x^{n-2} + \cdots\right)$$
$$+ 2^{n-2}a_{n-1}\left(x^{n-1} - \frac{n-1}{4}x^{n-3} + \cdots\right) + 2^{n-3}a_{n-2}(x^{n-2} - \cdots) + \cdots \tag{9.8.8}$$

so that there must follow

$$\begin{aligned} a_n &= 2^{-(n-1)} A_n \\ a_{n-1} &= 2^{-(n-2)} A_{n-1} \\ a_{n-2} &= 2^{-(n-3)}\left(A_{n-2} + \frac{n}{4}A_n\right) \end{aligned} \tag{9.8.9}$$

and so forth.

Thus, if n is sufficiently large, the coefficients of $T_n(x), T_{n-1}(x), \ldots, T_{n-m+1}(x)$ in (9.8.6) will be small relative to the respective coefficients of $x^n, x^{n-1}, \ldots, x^{n-m+1}$ in (9.8.3), for some m, and it may happen that

$$(|a_{n-m+1}| + |a_{n-m+2}| + \cdots + |a_n|) + \varepsilon_1$$

is smaller than ε and hence is a tolerable error in the desired approximation to $f(x)$. Since $|T_r(x)| \leqq 1$ on $[-1, 1]$, the last m terms in the right-hand member of (9.8.6) are then negligible, and the approximation (9.8.5) can then be replaced by

$$f(x) \approx \sum_{k=0}^{n-m} a_k T_k(x) \tag{9.8.10}$$

where $m > 0$, after which this approximation can be transformed back to an expression of the form

$$f(x) \approx \sum_{k=0}^{n-m} \bar{A}_k x^k \tag{9.8.11}$$

if this is desirable. In this way we obtain a polynomial approximation to $f(x)$ over $[-1, 1]$, involving fewer terms than would be required by the original approximation and tending to involve the smallest possible number of polynomial terms which will supply an accuracy within the prescribed tolerance limits.†

† It should be noticed that, in this procedure, the error $E_n(x)$ in (9.8.3) is accepted as a *fixed* error and an efficient approximation to $f(x) - E_n(x)$ is sought. Thus the approximation obtained is generally not the best possible one but would differ little from it if $|E_n(x)|$ were small relative to ε.

The transformations involved are facilitated by the use of the two following sets of relations, the second set being taken from the results of Sec. 7.9, and the first set being obtained by successively inverting the members of the second set:

$$
\begin{array}{ll}
1 = T_0 & T_0 = 1 \\
x = T_1 & T_1 = x \\
x^2 = \frac{1}{2}(T_0 + T_2) & T_2 = 2x^2 - 1 \\
x^3 = \frac{1}{4}(3T_1 + T_3) & T_3 = 4x^3 - 3x \\
x^4 = \frac{1}{8}(3T_0 + 4T_2 + T_4) & T_4 = 8x^4 - 8x^2 + 1 \\
x^5 = \frac{1}{16}(10T_1 + 5T_3 + T_5) & T_5 = 16x^5 - 20x^3 + 5x \\
x^6 = \frac{1}{32}(10T_0 + 15T_2 & T_6 = 32x^6 - 48x^4 \\
\qquad + 6T_4 + T_6) & \qquad + 18x^2 - 1 \\
x^7 = \frac{1}{64}(35T_1 + 21T_3 & T_7 = 64x^7 - 112x^5 \\
\qquad + 7T_5 + T_7) & \qquad + 56x^3 - 7x \\
x_8 = \frac{1}{128}(35T_0 + 56T_2 & T_8 = 128x^8 - 256x^6 + 160x^4 \\
\qquad + 28T_4 + 8T_6 + T_8) & \qquad - 32x^2 + 1 \\
x^9 = \frac{1}{256}(126T_1 + 84T_3 & T_9 = 256x^9 - 576x^7 + 432x^5 \\
\qquad + 36T_5 + 9T_7 + T_9) & \qquad -120x^3 + 9x
\end{array}
\tag{9.8.12}
$$

In illustration, suppose that a polynomial approximation to e^x is required over $[-1, 1]$, with a tolerance of 0.01. The truncation of a Maclaurin series gives a polynomial approximation of degree 5

$$e^x \approx 1 + x + \tfrac{1}{2}x^2 + \tfrac{1}{6}x^3 + \tfrac{1}{24}x^4 + \tfrac{1}{120}x^5 \equiv y(x) \tag{9.8.13}$$

with an error

$$|E(x)| = \left|\frac{e^\xi}{720}\,x^6\right| < \frac{e}{720} < 0.0038 \tag{9.8.14}$$

for which the neglect of the term $x^5/120$ would admit the possibility of an error exceeding the prescribed tolerance. The use of the first set of relations in (9.8.12) transforms (9.8.13) into the equivalent form

$$y(x) = \tfrac{81}{64}T_0 + \tfrac{217}{192}T_1 + \tfrac{13}{48}T_2 + \tfrac{17}{384}T_3 + \tfrac{1}{192}T_4 + \tfrac{1}{1920}T_5 \tag{9.8.15}$$

where $T_r \equiv T_r(x)$. Neglect of the last two terms will introduce an additional error not exceeding $\frac{11}{1920} < 0.0058$ for all x in $[-1, 1]$. Thus, with a total error smaller in magnitude than 0.0096, we have

$$e^x \approx \tfrac{81}{64}T_0 + \tfrac{217}{192}T_1 + \tfrac{13}{48}T_2 + \tfrac{17}{384}T_3 \tag{9.8.16}$$

or, after using the second set in (9.8.12),

$$e^x \approx \tfrac{1}{384}(382 + 383x + 208x^2 + 68x^3) \qquad (|x| \leqq 1) \tag{9.8.17}$$

For the purpose of comparison, it may be noted that a similar manipulation gives the form

$$y(x) = \tfrac{47}{40}P_0 + \tfrac{309}{280}P_1 + \tfrac{5}{14}P_2 + \tfrac{19}{270}P_3 + \tfrac{1}{105}P_4 + \tfrac{1}{945}P_5 \qquad (9.8.18)$$

in terms of the Legendre polynomials $P_r \equiv P_r(x)$. Here only the last term could be neglected, so that a polynomial approximation of *fourth* degree would be required.

The procedure described here, called the *economization of power series* by Lanczos, is useful in those situations when a minimization of the number of numerical operations is desirable. It clearly can be applied to *any* polynomial, whether that polynomial is obtained by truncating a power series or otherwise, once the interval of interest has been transformed to the interval $[-1, 1]$. An obvious alternative method consists of a stepwise elimination of powers, first eliminating x^n, then x^{n-1}, and so forth, until the total error is about to exceed the error tolerance. Other modifications and systematizations appear in the literature.

9.9 Uniform (Minimax) Polynomial Approximation

The problem of determining the approximation $y^*(x)$ of specified type to a given function $f(x)$ over a finite interval $[a, b]$, such that the maximum error over $[a, b]$ is minimized

$$\max_{a \leq x \leq b} |f(x) - y^*(x)| = \min \qquad (9.9.1)$$

is often referred to as a *minimax* problem.

We suppose here that $y^*(x)$ is to be a *polynomial* of degree n or less, where n is specified. In this case, we have seen that the procedures considered in Secs. 7.9, 9.7, and 9.8 provide approximations which *tend* to have this property for certain classes of functions but which generally are not optimum for a specific $f(x)$. The purpose of this section is to indicate an iterative process which can be used to determine a sequence of successive improvements to an initial approximation, permitting the determination of an approximation arbitrarily close to the optimum one.

The basis of this process is a theorem associated with Chebyshev (but apparently due to Borel), which states that when $f(x)$ is continuous on $[a, b]$, there is one and only one polynomial $y^*(x)$ of degree not exceeding n such that (9.9.1) is true, and that the optimum $y^*(x)$ is uniquely characterized by what is often called the *equal-ripple property: the deviation* $\delta^*(x) \equiv f(x) - y^*(x)$ *takes on its maximum absolute value at least* $n + 2$ *times on* $[a, b]$, *with alternating* *signs at successive extrema.*† Because of this property, a minimax approximation also often is referred to as a *uniform approximation* (or as a *Chebyshev approximation*, to further increase the ambiguity of that term).

The equal-ripple property of the minimax approximation suggests a crude iterative process in which a polynomial approximation $y_1(x)$ is first determined, perhaps by collocation with $f(x)$ at a certain appropriately chosen set of $n + 1$ points $x_0, x_1, \ldots, x_n$, and the corresponding deviation $\delta_1(x)$ is calculated and plotted over $[a, b]$. The points of collocation are then readjusted so that two points separated by a preponderant error ripple are brought a bit closer together while two points separated by an underdeveloped ripple are moved further apart, to yield a new approximation $y_2(x)$. A reasonably small number of successive adjustments frequently yields a satisfactory (nearly optimum) approximation in practice.

For a more systematic process of successive refinement, one can make use of the following algorithm, which is similar to one of several suggested by Rémès:

1 Choose a set of $n + 2$ successive points $c_0, c_1, \ldots, c_{n-1}$ in $[a, b]$ and determine $y(x)$ and a constant E such that

$$f(c_k) - y(c_k) = (-1)^k E \qquad (9.9.2)$$

for $k = 0, 1, \ldots, n + 1$.

2 Determine a set of $n + 2$ points $c_0', c_1', \ldots, c_{n+1}'$ in $[a, b]$ at which the deviation $\delta(x) = f(x) - y(x)$ has successive local extrema with alternating signs, including in that set the point at which the magnitude of $\delta(x)$ is *maximum* on $[a, b]$.

3 Repeat steps 1 and 2 with $c_0, \ldots, c_{n+1}$ replaced by $c_0', \ldots, c_{n+1}'$ and E replaced by a new unknown quantity E'.

The repetition of this process generates a sequence of approximations to $f(x)$ over $[a, b]$ which assuredly converges to the minimax approximation $y^*(x)$ when $f(x)$ is continuous on $[a, b]$. (See references listed in Sec. 9.19.)

It is seen that the set of conditions (9.9.2) requires that the values taken on by the deviation $f(x) - y(x)$ at the points $c_0, c_1, \ldots, c_{n+1}$ be of equal (but unknown) magnitude $\pm E$ and of alternating sign. The set comprises $n + 2$ linear algebraic equations in $n + 2$ unknown parameters, namely, the unknown

† Generally the optimum deviation has *exactly* $n + 2$ extrema on $[a, b]$. Exceptional situations usually are predictable because of the presence of symmetry. For example, the best *linear* approximation to $f(x) = x^2$ on $[-1, 1]$ in the minimax sense is the *constant* $y^*(x) = \frac{1}{2}$, which accordingly is also the best approximation of degree 0. Here there are *three* extreme deviations at -1, 0, and 1 of equal magnitude $\frac{1}{2}$.

deviation E and $n + 1$ parameters specifying $y(x)$, say, $a_0, a_1, \ldots, a_n$ in the relation

$$y(x) = a_0 + a_1 x + \cdots + a_n x^n \qquad (9.9.3)$$

The *initial* approximation need not be determined as in step 1 but may be more efficiently identified with an approximation obtained otherwise, say, as in Sec. 9.7 or 9.8.

Clearly the numerical determination of the local extrema of the deviation at each stage may present a substantial challenge. However, if the initial approximation to $y^*(x)$ is a good one, it may be expected that each of the new critical points $c'_0, \ldots, c'_{n+1}$ will be near to its predecessor. Also, except perhaps for the *final* approximation, the critical points need not be determined with high precision. Thus, if values of $\delta(x)$ are determined at a set of points near a preceding critical point c_k, a low-order process of inverse interpolation can be used to approximate c'_k. For this purpose, use may be made of the fact that if α_1, α_2, and α_3 are near a point α where $f'(\alpha) = 0$, an approximation to α is afforded by the formula

$$\alpha \approx \frac{\alpha_1 + 2\alpha_2 + \alpha_3}{4} - \frac{f[\alpha_1, \alpha_2] + f[\alpha_2, \alpha_3]}{4f[\alpha_1, \alpha_2, \alpha_3]} \qquad (9.9.4)$$

with the notation of divided differences (see Prob. 23 of Chap. 2). When $\alpha_3 - \alpha_2 = \alpha_2 - \alpha_1 = h$, this relation becomes

$$\alpha \approx \alpha_2 + \frac{h}{2}\left(\frac{f_1 - f_3}{f_1 - 2f_2 + f_3}\right) \qquad (9.9.5)$$

where $f_k = f(\alpha_k)$.

Care must be taken to introduce the point at which $|\delta(x)|$ is maximum as a *new* critical point in each cycle whether or not that point is a perturbation of a preceding one.

In illustration, we consider the example of Sec. 9.8 in which the third-degree approximation (9.8.17) was obtained for the function e^x over the interval $[-1, 1]$, by the economization process, in the form

$$y_1(x) = a_0 + a_1 x + a_2 x^2 + a_3 x^3 \qquad (9.9.6)$$

where

$$a_0 = 0.9948 \quad a_1 = 0.9974 \quad a_2 = 0.5417 \quad a_3 = 0.1771 \qquad (9.9.7)$$

when the coefficients are rounded to four places. When the deviation $\delta_1(x) = e^x - y_1(x)$ is tabulated by tenths over $[-1, 1]$, it is found that its five extrema are at the two end points and near $x = -0.7$, $x = 0$, and $x = 0.7$. The use of (9.9.5) then yields the following approximate data for the extrema:

c_k	$\delta_1(c_k)$
-1.000	0.0059
-0.676	-0.0047
0.030	0.0052
0.704	-0.0054
1.000	0.0073

Thus the ripples already are of roughly equal amplitude, the principal deviation being at $x = 1$.

The equations

$$e^{c_k} - y_2(c_k) = (-1)^k E \qquad (k = 0, 1, 2, 3, 4)$$

are then solved to determine the coefficients of $y_2(x)$ and the constant E; and the results

$$a_0' = 0.9946 \qquad a_1' = 0.9958 \qquad a_2' = 0.5430 \qquad a_3' = 0.1794 \tag{9.9.8}$$

and

$$E = 0.0055 \tag{9.9.9}$$

are obtained to four places. For the resultant modified approximation

$$y_2(x) = 0.9946 + 0.9958x + 0.5430x^2 + 0.1794x^3 \tag{9.9.10}$$

the approximate extrema of $\delta_2(x)$ then are determined as follows:

c_k'	$\delta_2(c_k')$
-1.000	0.0055
-0.677	-0.0055
0.048	0.0055
0.725	-0.0056
1.000	0.0055

Thus it appears that the first modification (9.9.10) is very nearly the desired third-degree minimax approximation to e^x over $[-1, 1]$.

9.10 Spline Approximation

Instead of approximating a given function $f(x)$ over an interval $[a, b]$ by a single polynomial, one may divide $[a, b]$ into n subintervals $[a, x_1]$, $[x_1, x_2]$, . . . , $[x_{n-1}, b]$ and approximate $f(x)$ by a different polynomial on each subinterval. For example, we may recall that the repeated midpoint, trapezoidal, and parabolic rules for approximate integration result from the process of replacing the integrand by *piecewise-polynomial* approximations of degree 0, 1, and 2 (or 3), respectively, with subintervals of uniform length. In the first case the approximation (a step function) generally is discontinuous at each division point x_k; in the other two cases this statement applies instead to the *derivative.*

For some purposes, particularly for numerical differentiation, it is highly desirable that the joins of the separate arcs be as "smooth" as possible. Specifically, if it is required that in each subinterval the approximation $s(x)$ be a polynomial of maximum degree 3, that $s(x)$ agree with $f(x)$ at each of the $n + 1$ points

$$x_0 = a, x_1, x_2, \ldots, x_{n-1}, x_n = b$$

and that the first and second derivatives $s'(x)$ and $s''(x)$ be continuous on $[a, b]$, then $s(x)$ is called a (cubic) *spline* (because of an analogy between its theory and the linearized theory of the so-called draftsman's spline).† The present section and Secs. 9.11 to 9.13 deal with some of its properties.

In the subinterval $[x_{k-1}, x_k]$ between two division points, or *nodes*, the equation of the spline is readily obtained by replacing f'_{k-1} and f'_k by s'_{k-1} and s'_k in the Hermite interpolation formula of Sec. 8.2 (or by direct derivation) in the form

$$s(x) = s'_{k-1}\frac{(x_k - x)^2(x - x_{k-1})}{h_k^2} - s'_k\frac{(x - x_{k-1})^2(x_k - x)}{h_k^2}$$
$$+ f_{k-1}\frac{(x_k - x)^2[2(x - x_{k-1}) + h_k]}{h_k^3} + f_k\frac{(x - x_{k-1})^2[2(x_k - x) + h_k]}{h_k^3} \tag{9.10.1}$$

with

$$h_k = x_k - x_{k-1} \tag{9.10.2}$$

The values $s'_{k-1} \equiv s'(x_{k-1})$ and $s'_k \equiv s'(x_k)$ are not known in advance but are to be determined, together with the unknown values of $s'(x)$ at the other nodes, so that $s''(x)$ is *continuous* at each of the interior nodes. For this purpose, we obtain from (9.10.1) the relations

$$s''(x_k-) = \frac{2}{h_k}(s'_{k-1} + 2s'_k) - 6\frac{f_k - f_{k-1}}{h_k^2} \qquad (k > 0) \tag{9.10.3}$$

and

$$s''(x_k+) = -\frac{2}{h_{k+1}}(2s'_k + s'_{k+1}) + 6\frac{f_{k+1} - f_k}{h_{k+1}^2} \qquad (k < n) \tag{9.10.4}$$

the last expression emerging as $s''(x_{k-1}+)$ with k replaced by $k + 1$.

† Splines of higher degree, as well as spline surfaces, have also been studied (see references listed in Sec. 9.19).

The requirement that s'' be continuous at each interior node x_k thus leads to the $n - 1$ linear equations

$$\frac{1}{h_k} s'_{k-1} + 2\left(\frac{1}{h_k} + \frac{1}{h_{k+1}}\right) s'_k + \frac{1}{h_{k+1}} s'_{k+1}$$
$$= 3\frac{f_k - f_{k-1}}{h_k^2} + 3\frac{f_{k+1} - f_k}{h_{k+1}^2} \qquad (k = 1, 2, \ldots, n-1) \qquad (9.10.5)$$

relating the $n + 1$ unknown spline slopes $s'_0, \ldots, s'_n$. Two additional conditions remain to be prescribed.

Alternatively, one can express $s(x)$ in $[x_{k-1}, x_k]$ in terms of y''_k, y''_{k-1}, f_k, and f_{k-1} (see Prob. 35) and then deduce the $n - 1$ conditions

$$\frac{h_k}{6} s''_{k-1} + \frac{h_k + h_{k+1}}{3} s''_k + \frac{h_{k+1}}{6} s''_{k+1}$$
$$= \frac{f_{k+1} - f_k}{h_{k+1}} - \frac{f_k - f_{k-1}}{h_k} \qquad (k = 1, 2, \ldots, n-1) \qquad (9.10.6)$$

relating the unknown quantities $s''_0, \ldots, s''_n$, by requiring continuity in $s'(x)$ at each interior node.

Once two appropriate auxiliary conditions (specifying, say, the values of s'_0 and s'_n or of s''_0 and s''_n) are prescribed, the solution of *either* (9.10.5) or (9.10.6) serves to determine $s(x)$ in each subinterval of $[a, b]$.

In order to exhibit a *minimal property* associated with spline approximations, we suppose that $s(x)$ is a spline approximation to $f(x)$ on $[a, b]$ satisfying a specified pair of auxiliary conditions. Furthermore, we consider any competing approximation $y(x)$ which also agrees with $f(x)$ at the $n + 1$ spline nodes x_k $(k = 0, 1, \ldots, n)$ and which also has two continuous derivatives on $[a, b]$. Starting with the identity

$$\int_a^b (y'')^2\, dx - \int_a^b (s'')^2\, dx = \int_a^b (y'' - s'')^2\, dx + 2\int_a^b s''(y'' - s'')\, dx \qquad (9.10.7)$$

we transform the last integral by use of integration by parts to obtain the relation

$$2\int_a^b s''(y'' - s'')\, dx = 2\sum_{k=0}^{n-1}\left\{\Big[s''(y' - s')\Big]_{x_k}^{x_{k+1}} - \int_{x_k}^{x_{k+1}} s'''(y' - s')\, dx\right\} \qquad (9.10.8)$$

A subdivision of the interval $[a, b]$ was necessary here since s''' generally is discontinuous at each interior node. Since s''' is *constant* over each subinterval and since $y = s$ at each node, each of the summed integrals in (9.10.8) vanishes;

and since s'', y', and s' are continuous, the other sum telescopes to yield the evaluation

$$2\int_a^b s''(y'' - s'')\,dx = 2[s''(y' - s')]_a^b \qquad (9.10.9)$$

Whenever $s(x)$ is such that (9.10.9) vanishes for every member $y(x)$ of a certain set of admissible competitors, the right-hand member of (9.10.7) is *positive* for any such $y(x)$ when $y(x) \not\equiv s(x)$, and hence the same statement applies to the left-hand member. It then follows that

$$\int_a^b (s'')^2\,dx \leqq \int_a^b (y'')^2\,dx \qquad (9.10.10)$$

with equality holding only when $y(x) \equiv s(x)$. Thus, in such cases, the spline approximation is "more smooth" than any member of the relevant set of competitors in the sense that the mean-square value of its second derivative over $[a, b]$ is minimal. Since the *curvature* $y''/(1 + y'^2)^{3/2}$ is approximated by y'' when the slope y' is small, this property of the spline approximation is sometimes termed a *minimum curvature* property. However, there is no implication that the slope is, in fact, to be small.

Among the situations in which (9.10.9) assuredly vanishes, so that (9.10.10) is valid, the following may be noted:

1 If the requirements

$$s''(a) = 0 \qquad s''(b) = 0 \qquad (9.10.11)$$

are selected as the auxiliary conditions completing the specification of $s(x)$, then (9.10.10) is true for *any* admissible $y(x)$. (A spline satisfying these conditions is sometimes called a "natural spline.")

2 If the auxiliary conditions

$$s'(a) = f'(a) \qquad s'(b) = f'(b) \qquad (9.10.12)$$

are selected, then (9.10.10) is true for any admissible $y(x)$ whose derivative also agrees with $f'(x)$ at the end points of $[a, b]$.

3 If the function $f(x)$ is such that

$$f(b-) = f(a+) \qquad f'(b-) = f'(a+) \qquad f''(b-) = f''(a+) \qquad (9.10.13)$$

and if $s(x)$ is required to have the same properties, so that the auxiliary conditions

$$s'(b-) = s'(a+) \qquad s''(b-) = s''(a+) \qquad (9.10.14)$$

are selected, then (9.10.10) holds for any admissible competitor $y(x)$ for which $y'(b-) = y'(a+)$. The conditions (9.10.13) will be satisfied, in particular, if $f(x)$ is a periodic function of period $b - a$ and are often referred to as *periodicity conditions* in the present context, whether or not $f(x)$ is in fact periodic. The conditions (9.10.14) can be conveniently and equivalently replaced by the requirement that the spline domain be extended over an additional subinterval $[x_n, x_{n+1}]$ of length $h_{n+1} = h_1$, and that (9.10.5) or (9.10.6) hold also for $k = n$, with

$$s'_n = s'_0 \qquad s'_{n+1} = s'_1 \qquad \text{or} \qquad s''_n = s''_0 \qquad s''_{n+1} = s''_1 \tag{9.10.14'}$$

The spline $s(x)$ so defined then is called a *periodic spline*.

Once a pair of auxiliary conditions has been selected, the spline approximation is to be determined by solving one of the two equation sets (9.10.5) and (9.10.6). When this solution is to be obtained by numerical methods, simplifications follow from the fact that the coefficient matrix in either case is of so-called *tridiagonal* form, with dominant diagonal elements, permitting the use of specially designed numerical procedures. (See, for example, Secs. 10.8 and 10.9.)

9.11 Splines with Uniform Spacing

When the nodes (division points) are equally spaced so that

$$h_k \equiv x_k - x_{k-1} = h \qquad (k = 1, 2, \ldots, n) \tag{9.11.1}$$

the preceding formulation simplifies. In particular, the equation set (9.10.5) takes the form

$$s'_{k+1} + 4s'_k + s'_{k-1} = \frac{6}{h}\,\mu\delta f_k \qquad (k = 1, 2, \ldots, n-1) \tag{9.11.2}$$

where $\mu\delta f_k$ is the mean central difference

$$\mu\delta f_k = \frac{f_{k+1} - f_{k-1}}{2h} \tag{9.11.3}$$

and the set (9.10.6) becomes

$$s''_{k+1} + 4s''_k + s''_{k-1} = \frac{6}{h^2}\,\delta^2 f_k \qquad (k = 1, 2, \ldots, n-1) \tag{9.11.4}$$

with the usual central-difference notation.

In order to obtain *analytical* solutions of the governing *difference equations*, we notice first that the associated homogeneous equation

$$u_{k+1} + 4u_k + u_{k-1} = 0 \tag{9.11.5}$$

is satisfied by $u_k = \beta^k$ if $\beta = -(2 \pm \sqrt{3})$. Hence, if account is taken of the fact that $2 - \sqrt{3} = 1/(2 + \sqrt{3})$, and if $2 + \sqrt{3}$ is denoted by e^α so that

$$e^\alpha = 2 + \sqrt{3} \qquad \sinh \alpha = \sqrt{3} \qquad \cosh \alpha = 2 \tag{9.11.6}$$

it can be deduced that the general solution of (9.11.5) is expressible in the form

$$u_k = (-1)^k(A \sinh k\alpha + B \sinh k\alpha) \tag{9.11.7}$$

when k takes on integral values.

Use then can be made of a method of variation of parameters (see, for example, Hildebrand [1968], Sec. 1.6), to obtain a particular solution of the modified equation in which a nonzero right-hand member is present, after which superposition yields the general solution of that equation. (See also Prob. 39.)

In this way, the solution of (9.11.2) can be obtained in the form

$$(-1)^k s_k' = s_0' \frac{\sinh (n - k)\alpha}{\sinh n\alpha} + (-1)^n s_n' \frac{\sinh k\alpha}{\sinh n\alpha} + \frac{6}{h} \sum_{r=1}^{n-1} (-1)^r G_{kr} \mu\delta f_r \tag{9.11.8}$$

where

$$G_{kr} = \begin{cases} \dfrac{\sinh r\alpha \sinh (n - k)\alpha}{\sinh \alpha \sinh n\alpha} & (r \leqq k) \\[2ex] \dfrac{\sinh k\alpha \sinh (n - r)\alpha}{\sinh \alpha \sinh n\alpha} & (r \geqq k) \end{cases} \tag{9.11.9}$$

It is seen that $G_{rk} = G_{kr}$, so that only those values of G_{kr} for which $r \leqq k$ need to be calculated, the other values then being obtained by symmetry. In fact, since also $G_{n-r,n-k} = G_{kr}$, the square matrix of values of G_{kr} is symmetric about *both* diagonals.

An additional simplification follows from the fact that $\sinh k\alpha$ satisfies the recurrence formula

$$\sinh (k + 1)\alpha = 4 \sinh k\alpha - \sinh (k - 1)\alpha$$

In particular, when k is an integer, it follows that $\sinh k\alpha$ is an integral multiple of $\sinh \alpha = \sqrt{3}$. Thus, if we write

$$g_k = \frac{\sinh k\alpha}{\sinh \alpha} = \frac{1}{\sqrt{3}} \sinh k\alpha \tag{9.11.10}$$

we can recast (9.11.8) and (9.11.9) in the more convenient form

$$(-1)^k s_k' = \frac{g_{n-k}}{g_n} s_0' + (-1)^n \frac{g_k}{g_n} s_n' + \frac{6}{h} \sum_{r=1}^{n-1} (-1)^r G_{kr} \mu\delta f_r \tag{9.11.11}$$

where

$$G_{kr} = \frac{g_r g_{n-k}}{g_n} \qquad (r \leqq k)$$
$$G_{rk} = G_{kr} \tag{9.11.12}$$

and where

$$g_0 = 0 \qquad g_1 = 1 \qquad g_2 = 4 \qquad g_3 = 15 \qquad \cdots$$
$$g_{k+1} = 4g_k - g_{k-1} \tag{9.11.13}$$

Analogously, from (9.11.4) there follows

$$(-1)^k s_k'' = \frac{g_{n-k}}{g_n} s_0'' + (-1)^n \frac{g_k}{g_n} s_n'' + \frac{6}{h^2} \sum_{r=1}^{n-1} (-1)^r G_{kr} \delta^2 f_r \tag{9.11.14}$$

with the same definitions of g_k and G_{kr}.

The constants s_0', s_n' or s_0'', s_n'' are to be determined by the chosen auxiliary conditions. It is useful to notice that if n is large, and if k is neither near zero nor near n, it follows that

$$\frac{g_{n-k}}{g_n} = \frac{\sinh (n-k)\alpha}{\sinh n\alpha} \sim e^{-k\alpha} \approx e^{-1.3k}$$
$$\frac{g_k}{g_n} = \frac{\sinh k\alpha}{\sinh n\alpha} \sim e^{-(n-k)\alpha} \approx e^{-1.3(n-k)} \tag{9.11.15}$$

Thus the end-effect terms in (9.11.11) and (9.11.14), which depend upon the selection of auxiliary conditions, damp out rather rapidly toward the interior of $[a, b]$ when n is large. Consequently, one might conclude that the selection of the auxiliary conditions generally is unimportant.

In accordance with this argument, the choice $s_0'' = 0$, $s_n'' = 0$ has been a popular one because of its simplicity and also because of the fact that then the mean-square value of s'' over $[a, b]$ will indeed be minimized under *all* admissible competition.

Additional insight is afforded by a consideration of this particular approximation in the special case of the function

$$f(x) = x^2 \tag{9.11.16}$$

Here it is found that†

$$s_k'' = 2\left\{1 - (-1)^k \left[\frac{(-1)^n g_k + g_{n-k}}{g_n}\right]\right\} \tag{9.11.17}$$

† In this simple case it is most convenient to attack the difference equation (9.11.4) directly rather than specialize the solution (9.11.14).

and also that

$$
\begin{aligned}
s_0' &= 2a + h - \frac{h}{3}\left[1 + \frac{g_{n-1} + (-1)^n}{g_n}\right] \\
s_k' &= 2x_k - (-1)^k \frac{h}{6}\left[\frac{(-1)^n(g_{k+1} - g_{k-1}) + (g_{n-k+1} - g_{n-k-1})}{g_n}\right] \\
&\qquad\qquad\qquad\qquad (0 < k < n) \\
s_n' &= 2b - h + \frac{h}{3}\left[1 + \frac{g_{n-1} + (-1)^n}{g_n}\right]
\end{aligned}
\tag{9.11.18}
$$

Thus, at the ends and at nearby nodes, the error in s_k' is small only of order h; that is, its magnitude decreases in proportion to $1/n$ as n is increased. The same is true of the error in $s'(x)$ at other points near the ends, while the error in $s(x)$ is found to be of order h^2 at such points in this case. This situation is particularly deplorable in the present case since here, in each subinterval, we are attempting to approximate a second-degree polynomial by one of third degree; but it is also typical of the end-effect contribution to the error distributions associated with the auxiliary conditions $s_0'' = s_n'' = 0$ in the general case, unless $f''(x)$ also vanishes at both end points.†

In the so-called *periodic* case described by (9.10.13), none of the quantities s_0', s_n', s_0'', and s_n'' are directly prescribed, but they are determined by the auxiliary conditions (9.10.14) which are consistent with the behavior of $f(x)$ at the end points, so that undesirable end effects are not introduced. (See Prob. 42.) In other cases, this introduction can be avoided without the sacrifice of a minimum curvature property if use is made of the auxiliary conditions (9.10.12), so that the value of $s'(x)$ at each end point is identified with the value of $f'(x)$ at that point, provided that $f'(a)$ and $f'(b)$ are known. Alternative procedures which do not require these data are considered in Sec. 9.13.

9.12 Spline Error Estimates

In order to estimate the errors which would be associated with a spline approximation at interior points if end effects were absent, we notice first that by averaging (9.10.3) and (9.10.4), with $h_k = h_{k+1} = h$, we obtain the relation

$$
s_k'' = \frac{3}{h^2}(f_{k+1} - 2f_k + f_{k-1}) - \frac{1}{h}(s_{k+1}' - s_{k-1}') \tag{9.12.1}
$$

when $1 \leqq k \leqq n - 1$. In a similar way (see Prob. 35), a formula relating

† Incorrect error bounds, which are inconsistent with this fact, can be found in the literature.

s'_k to s''_{k+1} and s''_{k-1} can be obtained in the form

$$s'_k = \frac{1}{2h}(f_{k+1} - f_{k-1}) - \frac{h}{12}(s''_{k+1} - s''_{k-1}) \qquad (9.12.2)$$

with the same restriction on k.

If it is assumed that (when f is sufficiently often differentiable and h sufficiently small) there exist representations

$$s'_k \sim f'_k + A_1 h f''_k + A_2 h^2 f'''_k + \cdots$$
$$s''_k \sim f''_k + B_1 h f'''_k + B_2 h^2 f^{\text{iv}}_k + \cdots$$

where omitted terms corresponding to a specific truncation are of higher order in h than those retained, we may deduce first from the symmetry of (9.12.1) and (9.12.2) that the terms involving odd powers of h must all vanish. For this purpose we recall, for example, that

$$\frac{f_{k+1} - f_{k-1}}{h} \equiv \frac{f(x_k + h) - f(x_k - h)}{h}$$

and hence conclude that this ratio (and similarly each other ratio present) is unchanged when h is replaced by $-h$.

If now the representations

$$\begin{aligned} s'_k &\sim f'_k + A_2 h^2 f'''_k + A_4 h^4 f^{\text{v}}_k + \cdots \\ s''_k &\sim f''_k + B_2 h^2 f^{\text{iv}}_k + B_4 h^4 f^{\text{vi}}_k + \cdots \end{aligned} \qquad (9.12.3)$$

are introduced into the equal members of (9.12.1) and (9.12.2), and if then each term of the form $f^{(r)}_{k\pm 1}$ is expanded as

$$f^{(r)}_{k\pm 1} \sim f^{(r)}_k \pm h f^{(r+1)}_k + \tfrac{1}{2} h^2 f^{(r+2)}_k \pm \cdots$$

the resultant relations become

$$f''_k + B_2 h^2 f^{\text{iv}}_k + B_4 h^4 f^{\text{vi}}_k + O(h^6)$$
$$\sim f''_k + h^2(-\tfrac{1}{12} - 2A_2) f^{\text{iv}}_k + h^4(-\tfrac{1}{120} - \tfrac{1}{3}A_2 - 2A_4) f^{\text{vi}}_k + O(h^6)$$

and

$$f'_k + A_2 h^2 f'''_k + A_4 h^4 f^{\text{v}} + O(h^6)$$
$$\sim f'_k + h^2(0) f'''_k + h^4(-\tfrac{7}{360} - \tfrac{1}{6}B_2) f^{\text{v}}_k + O(h^6)$$

By equating coefficients of corresponding powers of h, we then obtain a set of equations relating the A's and B's, which yields

$$A_2 = 0 \qquad B_2 = -\tfrac{1}{12} \qquad A_4 = -\tfrac{1}{180} \qquad B_4 = \tfrac{1}{360} \qquad \cdots$$

Accordingly there follows

$$f_k' - s_k' \sim \frac{h^4}{180} f_k^{\mathrm{v}} - \frac{h^6}{1512} f_k^{\mathrm{vii}} + O(h^8) \qquad (9.12.4)$$

and

$$f_k'' - s_k'' \sim \frac{h^2}{12} f_k^{\mathrm{iv}} - \frac{h^4}{360} f_k^{\mathrm{vi}} + O(h^6) \qquad (9.12.5)$$

where an additional term is included in (9.12.4) for reference purposes.

It is important to realize that these error relations merely *tend* to be valid at *interior* nodes as the number of subdivisions of $[a, b]$ increases. The portion of the total error which is due to end effects is not detected by the present methods since it dies out in proportion to $e^{-\alpha n} = e^{-\alpha(b-a)/h}$ when n is large, and hence it is not representable by a power series in h near $h = 0$.

From the relations (9.12.4) and (9.12.5) we can deduce other such relations at the interior nodes. In particular, we note that the right-hand limit of $s'''(x)$ at a node is given by

$$\begin{aligned} s'''(x_k+) &= \frac{1}{h}(s_{k+1}'' - s_k'') \\ &\sim f_k''' + \frac{h}{2} f_k^{\mathrm{iv}} + \frac{h^2}{12} f_k^{\mathrm{v}} - \frac{h^4}{360} f_k^{\mathrm{vii}} - \frac{67h^5}{60480} f_k^{\mathrm{viii}} + O(h^6) \end{aligned}$$

whereas the left-hand limit is given by

$$\begin{aligned} s'''(x_k-) &= \frac{1}{h}(s_k'' - s_{k-1}'') \\ &\sim f_k''' - \frac{h}{2} f_k^{\mathrm{iv}} + \frac{h^2}{12} f_k^{\mathrm{v}} - \frac{h^4}{310} f_k^{\mathrm{vii}} + \frac{67h^5}{60480} f_k^{\mathrm{viii}} + O(h^6) \end{aligned}$$

Hence there follows

$$f_k''' - \frac{1}{2}[s'''(x_k+) + s'''(x_k-)] \sim -\frac{h^2}{12} f_k^{\mathrm{v}} + O(h^4) \qquad (9.12.6)$$

and

$$f_k^{\mathrm{iv}} - \frac{1}{h}[s'''(x_k+) - s'''(x_k-)] \sim \frac{67h^4}{30240} f_k^{\mathrm{viii}} + O(h^6) \qquad (9.12.7)$$

Thus the jump in s''' at x_k would approximate $hf^{\mathrm{iv}}(x_k)$ and the mean of the constant values of s''' to the right and to the left of x_k would approximate $f'''(x_k)$ if n were sufficiently large and if end effects were not overpowering.

Finally, in any subinterval $[x_k, x_{k+1}]$ we may write

$$f(x) - s(x) = [f(x) - p(x)] + [p(x) - s(x)] \qquad (9.12.8)$$

where $p(x)$ is conveniently identified with the cubic Hermite polynomial approximation to $f(x)$ on that subinterval, having the property that agreement exists at x_k and at x_{k+1} between $f(x)$ and $p(x)$ and also between $f'(x)$ and $p'(x)$. If on that subinterval we write $x = x_k + \theta h$, where accordingly $0 \leqq \theta \leqq 1$, Eq. (8.2.18) takes the form

$$f(x_k + \theta h) - p(x_k + \theta h) = \frac{h^4}{24}\theta^2(1 - \theta)^2 f^{\text{iv}}(\xi_k) \qquad (9.12.9)$$

where $x_k < \xi_k < x_{k+1}$. Further, since $p(x)$ and $s(x)$ agree at x_k and at x_{k+1}, we must have

$$p(x) - s(x) = (p_k' - s_k')\frac{(x - x_k)(x_{k+1} - x)^2}{h^2} - (p_{k+1}' - s_{k+1}')\frac{(x - x_k)^2(x_{k+1} - x)}{h^2} \qquad (9.12.10)$$

And since also $p_k' = f_k'$ and $p_{k+1}' = f_{k+1}'$, there follows

$$p(x_k + \theta h) - s(x_k + \theta h) = h\theta(1 - \theta)^2(f_k' - s_k') - h\theta^2(1 - \theta)(f_{k+1}' - s_{k+1}') \qquad (9.2.11)$$

Finally, since (9.12.4) indicates that at the interior nodes the deviation $f' - s'$ is asymptotically small of order h^4, it follows that the spline approximation tends to differ from the cubic Hermite approximation only by terms which are small of order h^5 in interior subintervals. Consequently, we have also

$$f(x) - s(x) \sim \frac{h^4}{24}\theta^2(1 - \theta)^2 f^{\text{iv}}(x_k) + O(h^5) \qquad (9.12.12)$$

in any interior subinterval $[x_k, x_{k+1}]$, where $\theta = (x - x_k)/h$. It can be verified that the factor $\theta^2(1 - \theta)^2$ does not exceed $\frac{1}{16}$ when $0 \leqq \theta \leqq 1$.

We note again that the symbol $\sim$ is used in the present context to mean [in the case of (9.12.12)] that the error $f(x) - s(x)$ at a point in an interior subinterval differs from $(h^4/24)\theta^2(1 - \theta)^2 f^{\text{iv}}(x_k)$ by terms which depend upon the choice of the two auxiliary conditions but which decrease with $h = (b - a)/n$ in proportion to $e^{-c/h}$, where $c = \alpha(b - a) \approx 1.3(b - a)$, and by terms which decrease at least as rapidly as a multiple of h^5 as $h \to 0$.

9.13 A Special Class of Splines

Clearly, it is desirable to select auxiliary conditions which will ensure that the actual spline errors at the end points $x_0 = a$ and $x_n = b$ tend to zero as $h \to 0$ (and $n \to \infty$) at about the same rate as those predicted by the preceding asymptotic error formulas at interior points. In particular, in order to deal with situations in which the use of the ideal conditions $s_0' = f_0'$ and $s_n' = f_n'$ is not possible because $f'(a)$ and $f'(b)$ are not known, or is otherwise undesirable, we are led to the desirability of making $s_0' - f_0'$ and $s_n' - f_n'$ small of order h^4 for consistency with (9.12.4).

Reference to (3.11.5) and (3.11.7) shows that one method of accomplishing this end (with a minor invalidation of the minimum curvature property and under the assumption that $f^{\text{v}}(x)$ is continuous on $[a - h, b + h]$) consists of introducing the value of f at each of two additional nodes $x_{-1} \equiv a - h$ and $x_{n+1} \equiv b + h$ outside the original interval $[a, b]$ and imposing the auxiliary conditions

$$s_0' = \frac{1}{12h}(-3f_{-1} - 10f_0 + 18f_1 - 6f_2 + f_3) \tag{9.13.1}$$

and

$$s_n' = \frac{1}{12h}(-f_{n-3} + 6f_{n-2} - 18f_{n-1} + 10f_n + 3f_{n+1}) \tag{9.13.2}$$

There then follows

$$|f_0' - s_0'| \leqq \frac{h^4}{20} M_5 \qquad |f_n' - s_n'| \leqq \frac{h^4}{20} M_5 \tag{9.13.3}$$

where M_5 is a bound on $|f^{\text{v}}(x)|$ in $[a - h, b + h]$. Since also (9.10.4) then gives

$$\begin{aligned} s_0'' &= -\frac{2}{h}(2s_0' + s_1') + \frac{6}{h^2}(f_1 - f_0) \\ &= -\frac{2}{h}(2f_0' + f_1') + \frac{6}{h^2}(f_1 - f_0) \\ &\qquad + \frac{4}{h}(f_0' - s_0') + \frac{2}{h}(f_1' - s_1') \\ &= f_0'' - \frac{h^2}{12} f_0^{\text{iv}} + O(h^3) \end{aligned}$$

and a similar result is obtained for s_n'', the errors in s'' at the ends also will be small of the same order in h as the asymptotic interior errors predicted by (9.12.5).

Thus the end errors, the effects of which in any case damp out rather rapidly toward the interior of the interval $[a, b]$, are indeed of the same order of magnitude as the internal errors which would be present in their absence. This fact appears to justify the inconvenience associated with the introduction of the two supplementary ordinates when their values are available.† As the spacing is

† Another procedure, due to Curtis [1970], tends to accomplish about the same purpose by introducing additional *interior* nodes at the midpoints of the subintervals $[x_0, x_1]$, $[x_1, x_2]$, $[x_{n-2}, x_{n-1}]$, and $[x_{n-1}, x_n]$. Although a subsequent halving of h would require the introduction of these points as new basic nodes in any case, the fact that the $n + 4$ subintervals used at each stage are not of uniform length (when $n > 4$) complicates the formulation and adversely affects its accuracy. Such a modification would be particularly appropriate, however, if $f(x)$ were undefined outside $[a, b]$ or if nonuniform spacings were to be used in any case.

refined, perhaps by successive doubling of n, to yield a *sequence* of approximations, the two supplementary abscissas approach the end points of the interval $[a, b]$.

With these (or equally appropriate) auxiliary conditions satisfied, spline approximation is particularly useful for high-precision numerical differentiation when sufficiently accurate data are available, since the error $f'_k - s'_k$ decreases at least in proportion to h^4 as the spacing h is decreased (so that a halving of h tends to divide the error in s'_k by a factor of 16) when $f^{\text{v}}(x)$ is continuous on $[a - h, b + h]$. In fact, in this case the approximations to $f''(x_k)$ and $f'''(x_k)$ afforded by $s''(x_k)$ and by $\frac{1}{2}[s'''(x_k+) + s'''(x_k-)]$ are small of order h^2, and the approximation to $f^{\text{iv}}(x_k)$ afforded by $h^{-1}[s'''(x_k+) - s'''(x_k-)]$ is (surprisingly) small of order h^4 if $f^{\text{viii}}(x)$ is continuous.

These convergence properties are in sharp contrast with those associated with the derivatives of the polynomial $y(x)$ of degree n which agrees with $f(x)$ at all the $n + 1$ nodes $a, x_1, \ldots, x_{n-1}, b$, as n increases, since then even the *interpolation* error $f(x) - y(x)$ may not tend to zero on $[a, b]$ as $n \to \infty$ even though $f(x)$ has derivatives of *all* orders in that interval (Sec. 4.11).

However, the importance of suitably controlling the effects of roundoff errors (emphasized in Sec. 3.8) continues to be essential. Although the magnitudes of the coefficients of the individual ordinates in (9.11.11) are found to grow less rapidly as n increases than do the magnitudes of the corresponding coefficients in the sequences of formulas referred to above, still the necessity of strong sign variation in each formula remains and is inherently troublesome. In particular, if the data are empirical, smoothing of some sort is imperative before use is made of *any* formula for numerical differentiation and still a high degree of accuracy rarely can be anticipated.

The sum $K_k^{(n)}$ of the magnitudes of the coefficients of the ordinates in (9.11.11) would be of the form

$$K_k^{(n)} = \frac{3}{h}\left[\frac{g_{n-k}(g_k + g_{k-1}) + g_{|n-2k|} + g_k(g_{n-k} + g_{n-k-1})}{g_n}\right] \tag{9.13.4}$$

when $1 \leqq k \leqq n - 1$ if s'_0 and s'_n were taken to be zero. The imposition of the conditions (9.13.1) and (9.13.2) slightly *reduces* the coefficient of h^{-1} in $K_k^{(n)}$ when $1 \leqq k \leqq n - 1$ by an amount which rapidly tends to zero toward the interior of the interval $[a, b]$ and with increasing n. Thus, if all ordinates could possess errors of magnitude ε, the maximum possible associated error in s'_k would be slightly smaller than $K_k^{(n)}\varepsilon$. It is found further that the coefficient of h^{-1} in $K_k^{(n)}$ does not vary strongly with either k or n and, in addition, that

$$K_k^{(n)} < \frac{6}{h}(16 - 9\sqrt{3}) < \frac{2.5}{h} \tag{9.13.5}$$

for all $n \geqq 3$ and all k such that $1 \leqq k \leqq n - 1$.

When (9.13.1) and (9.13.2) are satisfied, the asymptotic relation (9.12.4) for the truncation error in $f'_k - s'_k$ can be replaced by the more useful estimate

$$f'_k - s'_k \approx \left[1 + (-1)^{k+1}\, 10\, \frac{g_{n-k} + (-1)^n g_k}{g_n}\right] \frac{h^4}{180} f^{\text{v}}_k \tag{9.13.6}$$

for $0 \leqq k \leqq n$, which incorporates the fact that end errors now known to be approximated by $-\frac{1}{20}h^4 f_0^{\text{v}}$ and $-\frac{1}{20}h^4 f_n^{\text{v}}$, together with their propagated effects, are superimposed on the errors considered in the derivation of (9.12.4).

When $n = 2$, the value of $s'(x)$ at the one interior node can be expressed in the form

$$s'_1 = \frac{1}{12h}(f_{-1} - 8f_0 + 8f_2 - f_3) \tag{9.13.7}$$

when (9.13.1) and (9.13.2) are to hold at the end points; and the corresponding formulas when $n = 4$ are found to be

$$\begin{aligned}
s'_1 &= \frac{1}{672h}(45f_{-1} - 390f_0 - 125f_1 + 588f_2 - 141f_3 + 26f_4 - 3f_5) \\
s'_2 &= \frac{1}{168h}(-3f_{-1} + 26f_0 - 127f_1 + 127f_3 - 26f_4 + 3f_5) \\
s'_3 &= \frac{1}{672h}(3f_{-1} - 26f_0 + 141f_1 - 558f_2 + 125f_3 + 390f_4 - 45f_5)
\end{aligned} \tag{9.13.8}$$

Finally, the formula for the midpoint derivative when $n = 8$ is listed for reference purposes:

$$\begin{aligned}
s'_4 = \frac{1}{2328h}(-3f_{-1} &+ 26f_0 - 126f_1 + 498f_2 - 1871f_3 \\
&+ 1871f_5 - 498f_6 + 126f_7 - 26f_8 + 3f_9)
\end{aligned} \tag{9.13.9}$$

Corresponding formulas relating instead to the imposition of the auxiliary conditions $s''_0 = s''_n = 0$ are considered in Prob. 47. In accordance with preceding analyses, the degree of precision for the formulas (9.13.7) to (9.13.9) is 4 (that is, these formulas would yield exact results if $f(x)$ were a polynomial of degree 4 or less). On the other hand, the degree of precision for the formulas corresponding to the auxiliary conditions $s''_0 = s''_n = 0$ is only 1 for off-center formulas and 2 for the midpoint formulas. (As has been pointed out before, however, the degree of precision of a formula is not *necessarily* a good measure of its usefulness.)

Some rounded numerical results obtained by use of only two- and four-division spline approximations from seven-digit values of the function $f(x) = x^{1/3}$ over the interval [1.0, 1.8] are listed in the following table:

	$s'(x)$			$-s''(x)$		
x	$n = 2$	$n = 4$	$f'(x)$	$n = 2$	$n = 4$	$-f''(x)$
1.0	0.33501	0.33349	0.33333	0.2267	0.2222	0.2222
1.2		0.29512	0.29518		0.1615	0.1640
1.4	0.26549	0.26636	0.26635	0.1209	0.1261	0.1268
1.6		0.24365	0.24367		0.1010	0.1015
1.8	0.22632	0.22530	0.22527	0.0750	0.0825	0.0834

	$\frac{1}{2}(s'''_{k+} + s'''_{k-})$			$-(1/h)(s'''_{k+} - s'''_{k-})$		
x	$n = 2$	$n = 4$	$f'''(x)$	$n = 2$	$n = 4$	$-f^{\text{iv}}(x)$
1.0			0.370			0.988
1.2		0.240	0.228		0.632	0.506
1.4	0.190	0.151	0.151	0.375	0.260	0.288
1.6		0.109	0.106		0.163	0.176
1.8			0.007			0.114

Whereas the asymptotic error estimates would suggest the propriety of iterated Richardson extrapolation (as in the Romberg integration process), this procedure usually is vitiated by the end effects still remaining, except near the midpoint of the interval when n is reasonably large.

When approximate values of $f''(x_k)$ are required, in place of using values of s''_k as above it is usually preferable to effect spline differentiation on the calculated values of $s'(x)$ at the nodes. (This process has been called *spline-on-spline* computation.) For this purpose, Eqs. (3.11.4) and (3.11.8) can be rewritten in the forms

$$\begin{aligned} s'_{-1} &= \frac{1}{12h}(-25f_{-1} + 48f_0 - 36f_1 + 16f_2 - 3f_3) \\ s'_{n+1} &= \frac{1}{12h}(3f_{n-3} - 16f_{n-2} + 36f_{n-1} - 48f_n + 25f_{n+1}) \end{aligned} \tag{9.13.10}$$

to supply the necessary additional data for the satisfaction of the auxiliary conditions. The results of this procedure in the present example are compared with the s'' values and with the rounded true values in the following table, and are seen to exhibit substantial improvement (except at $x = 1.0$):

x	$-s''(x)$	Spline on spline	$-f''(x)$
1.0	0.22220	0.22310	0.22222
1.2	0.16149	0.16431	0.16399
1.4	0.12607	0.12654	0.12684
1.6	0.10104	0.10152	0.10153
1.8	0.08252	0.08337	0.08343

Formulas for numerical integration also can be obtained by integrating spline approximations. In particular, from (9.10.1) there follows

$$\int_{x_k}^{x_{k+1}} s(x)\,dx = \frac{h_{k+1}}{2}(f_k + f_{k+1}) - \frac{h_{k+1}^2}{12}(s'_{k+1} - s'_k) \qquad (9.13.11)$$

Thus, when the spacing of the nodes is uniform we obtain, by summation, the simple result

$$\int_a^b s(x)\,dx = h(\tfrac{1}{2}f_0 + f_1 + f_2 + \cdots + f_{n-2} + f_{n-1} + \tfrac{1}{2}f_n) - \frac{h^2}{12}(s'_n - s'_0) \qquad (9.13.12)$$

In particular, for a periodic spline this formula reduces to the trapezoidal rule. When the spline satisfies the conditions $s'_0 = f'_0 \equiv f'(a)$, $s'_n = f'_n \equiv f'(b)$, Eq. (9.13.12) is the Euler-Maclaurin formula with one correction term.

Otherwise, if the end conditions (9.13.1) and (9.13.2) are used to define $s(x)$, the corresponding approximations to the integral in the cases $n = 2, 3$, and 4 are obtained as

$$\int_a^b f(x)\,dx \approx \frac{h}{72}(-f_{-1} + 28f_0 + 90f_1 + 28f_2 - f_3) \qquad (9.13.13)$$

$$\int_a^b f(x)\,dx \approx \frac{h}{48}(-f_{-1} + 21f_0 + 52f_1 + 52f_2 + 21f_3 - f_4) \qquad (9.13.14)$$

$$\int_a^b f(x)\,dx \approx \frac{h}{144}(-3f_{-1} + 62f_0 + 163f_1 + 132f_2 + 163f_3 + 62f_4 - 3f_5) \qquad (9.13.15)$$

where the error in each case differs from the Euler-Maclaurin error [see (5.8.14)] with $m = 1$ by h^6 terms if f^{vi} is continuous in $[a - h, b + h]$. In fact, it can be shown that the error is of the form

$$E = \frac{b-a}{720}h^4\left[f^{\text{iv}}(\xi_1) + \frac{h^2}{2}f^{\text{vi}}(\xi_2)\right] \qquad (9.13.16)$$

for any value of n, where $h = (b - a)/n$ as usual, and where ξ_1 and ξ_2 are in $(a - h, b + h)$.

Many of the corresponding approximations generated by integrating splines for which $s''_0 = s''_n = 0$ may be found in the literature. When $n = 2, 3$, and 4, the formulas are

$$\int_a^b f(x)\,dx \approx \frac{h}{8}(3f_0 + 10f_1 + 3f_2) \qquad (9.13.17)$$

$$\int_a^b f(x)\,dx \approx \frac{h}{10}(4f_0 + 11f_1 + 11f_2 + 4f_3) \qquad (9.13.18)$$

$$\int_a^b f(x)\,dx \approx \frac{h}{28}(11f_0 + 32f_1 + 26f_2 + 32f_3 + 11f_4) \qquad (9.13.19)$$

the simplicity of the forms being somewhat offset by the fact that the degree of precision of each such formula is only 1.

In the case of the preceding numerical example, the approximations to the integral

$$\int_{1.0}^{1.8} x^{1/3}\,dx \doteq 0.8921945$$

afforded by (9.13.13) and (9.13.15) round to

$$0.8922128 \qquad \text{and} \qquad 0.8921956$$

while those given by (9.13.17) and (9.13.19) round to

$$0.8918105 \qquad \text{and} \qquad 0.8921376$$

Here the Richardson extrapolation is indicated. The first pair yields the value 0.8921945, while the second pair yields 0.8921594.

9.14 Approximation by Continued Fractions

Newton's divided-difference polynomial interpolation formula (2.5.2), with an error term, can be considered as the *identity* which results from writing

$$f(x) = u_0(x) \tag{9.14.1}$$

and effecting the successive substitutions

$$u_k(x) = u_k(x_k) + (x - x_k)u_{k+1}(x) \qquad (k = 0, 1, \ldots, n-1) \tag{9.14.2}$$

with the abbreviation

$$u_k(x) = f[x_0, x_1, \ldots, x_{k-1}, x] \tag{9.14.3}$$

Thus, for example, when $n = 3$ there follows

$$\begin{aligned} f(x) &= u_0(x_0) + (x - x_0)\{u_1(x_1) + (x - x_1)[u_2(x_2) + (x - x_2)u_3(x)]\} \\ &= u_0(x_0) + (x - x_0)u_1(x_1) + (x - x_0)(x - x_1)u_2(x_2) + E(x) \\ &= f[x_0] + (x - x_0)f[x_0, x_1] \\ &\qquad\qquad + (x - x_0)(x - x_1)f[x_0, x_1, x_2] + E(x) \end{aligned} \tag{9.14.4}$$

where

$$\begin{aligned} E(x) &= (x - x_0)(x - x_1)(x - x_2)u_3(x) \\ &= (x - x_0)(x - x_1)(x - x_2)f[x_0, x_1, x_2, x] \end{aligned} \tag{9.14.5}$$

The algorithm for the calculation of the successive *divided differences* follows directly from (9.14.2) and (9.14.3), with $x = x_k$, in the form

$$f[x_0, \ldots, x_{k-2}, x_{k-1}, x_k] = \frac{f[x_0, \ldots, x_{k-2}, x_k] - f[x_0, \ldots, x_{k-2}, x_{k-1}]}{x_k - x_{k-1}} \tag{9.14.6}$$

The result of *assuming* that the $(n + 1)$st divided difference $u_{n+1}(x)$ is identically zero (or that the nth divided difference is constant) is the equation of the polynomial $y(x)$, of degree n or less, which agrees with $f(x)$ at the $n + 1$ points $x_0, x_1, \ldots, x_n$. If $u_{n+1}(x)$ *actually* vanishes identically, then $y(x) \equiv f(x)$.

A great variety of other identities can be obtained in a similar way and interpreted similarly as approximation formulas by making use of other sets of transformations in place of (9.14.2). In particular, the substitution sequence

$$\begin{aligned} f(x) &= v_0(x) \\ v_k(x) &= v_k(x_k) + \frac{x - x_k}{v_{k+1}(x)} \qquad (k = 0, 1, 2, \ldots) \end{aligned} \tag{9.14.7}$$

leads to an interesting and useful result. We see that the first three substitutions give

$$\begin{aligned} f(x) = v_0(x) &= v_0(x_0) + \frac{x - x_0}{v_1(x)} = v_0(x_0) + \cfrac{x - x_0}{v_1(x_1) + \cfrac{x - x_1}{v_2(x)}} \\ &= v_0(x_0) + \cfrac{x - x_0}{v_1(x_1) + \cfrac{x - x_1}{v_2(x_2) + \cfrac{x - x_2}{v_3(x)}}} \end{aligned} \tag{9.14.8}$$

More generally, we are thus led to the *continued-fraction* representation

$$f(x) = a_0 + \cfrac{x - x_0}{a_1 + \cfrac{x - x_1}{a_2 + \cfrac{x - x_2}{a_3 + \ddots}}} \tag{9.14.9}$$

where

$$a_k = v_k(x_k) \tag{9.14.10}$$

and where, when the fraction is terminated after n divisions, the constant a_n is to be replaced by $a_n + (x - x_n)/v_{n+1}(x)$ in the last denominator. If we then set $x = x_k$, where $0 \leqq k \leqq n$, the fraction terminates before the *residual* $(x - x_n)/v_{n+1}(x)$ is introduced. Thus, since (9.14.9) is an *identity*, the result of replacing $1/v_{n+1}(x)$ by zero (that is, terminating the fraction with a_n) will give a function $r_n(x)$ which agrees with $f(x)$ at the $n + 1$ points $x_0, \ldots, x_n$, under the assumptions that the constants $a_0, \ldots, a_n$ are actually existent and that the portion of the truncated fraction inferior to $x - x_k$ does not vanish when $x = x_k$, for $k = 0, \ldots, n - 1$. The result of this termination may be called the nth *convergent* (or *approximant*) of the representation, with the constant $a_0 = f(x_0)$ considered as the *zeroth convergent.*

If we introduce the notation

$$v_k(x) = \phi_k[x_0, x_1, \ldots, x_{k-1}, x] \tag{9.14.11}$$

so that (9.14.10) becomes

$$a_k = \phi_k[x_0, x_1, \ldots, x_{k-1}, x_k] \tag{9.14.12}$$

reference to (9.14.7) gives

$$\phi_0[x] = f(x)$$

$$\phi_1[x_0, x] = \frac{x - x_0}{\phi_0[x] - \phi_0[x_0]} = \frac{x - x_0}{f(x) - f(x_0)}$$

$$\phi_2[x_0, x_1, x] = \frac{x - x_1}{\phi_1[x_0, x] - \phi_1[x_0, x_1]}$$

and, in general,

$$\phi_k[x_0, x_1, \ldots, x_{k-1}, x] = \frac{x - x_{k-1}}{\phi_{k-1}[x_0, \ldots, x_{k-2}, x] - \phi_{k-1}[x_0, \ldots, x_{k-2}, x_{k-1}]} \tag{9.14.13}$$

Accordingly, we have also

$$\phi_k[x_0, \ldots, x_{k-1}, x_k] = \frac{x_k - x_{k-1}}{\phi_{k-1}[x_0, \ldots, x_{k-2}, x_k] - \phi_{k-1}[x_0, \ldots, x_{k-2}, x_{k-1}]} \tag{9.14.14}$$

Thus $\phi_1[x_0, x_1]$ is the *inverted* first divided difference of $f(x)$, relative to x_0 and x_1, $\phi_2[x_0, x_1, x_2]$ is the inverted divided difference of the inverted first divided difference $\phi_1[x_0, x]$, relative to x_1 and x_2, $\ldots$, and, in general, $\phi_k[x_0, \ldots, x_{k-2}, x_{k-1}, x_k]$ is the inverted divided difference of $\phi_{k-1}[x_0, \ldots, x_{k-2}, x]$, relative to x_{k-1} and x_k. For brevity, we will refer to the quantity defined by (9.14.13) as a kth *inverted difference* of $f(x)$.

Whereas the definition shows that always the inverted difference $\phi_k[x_0, \ldots, x_{k-2}, x_{k-1}, x_k]$ is symmetric in its *last two* arguments x_{k-1} and x_k, it is *not* generally symmetric in its other arguments.† Thus it *must* be formed

† A related quantity, which possesses complete symmetry, is considered in Sec. 9.17.

from the specific inverted differences $\phi_{k-1}[x_0, \ldots, x_{k-2}, x_{k-1}]$ and $\phi_{k-1}[x_0, \ldots, x_{k-2}, x_k]$, which possess its first $k - 1$ arguments in common. The following calculational arrangement is convenient for this purpose:

$$
\begin{array}{lllll}
x_0 & f(x_0) & & & \\
x_1 & f(x_1) & \phi_1[x_0, x_1] & & \\
x_2 & f(x_2) & \phi_1[x_0, x_2] & \phi_2[x_0, x_1, x_2] & \\
x_3 & f(x_3) & \phi_1[x_0, x_3] & \phi_2[x_0, x_1, x_3] & \phi_3[x_0, x_1, x_2, x_3] \\
\cdots & \cdots & \cdots & \cdots & \cdots
\end{array}
$$

Here, for example, we have

$$\phi_2[x_0, x_1, x_3] = \frac{x_3 - x_1}{\phi_1[x_0, x_3] - \phi_1[x_0, x_1]}$$

The *diagonal* elements thus are the desired constants $a_0, a_1, a_2, a_3, \ldots$ which appear in (9.14.9).

In illustration, for the given data

x	0	1	2	3	4	5	6
$f(x)$	2	$\frac{3}{2}$	$\frac{4}{5}$	$\frac{1}{2}$	$\frac{6}{17}$	$\frac{7}{26}$	$\frac{8}{37}$

we may number the abscissas in increasing algebraic order and, accordingly, form the array

x	f	ϕ_1	ϕ_2	ϕ_3	ϕ_4
0	2				
1	$\frac{3}{2}$	-2			
2	$\frac{4}{5}$	$-\frac{5}{3}$	3		
3	$\frac{1}{2}$	-2	∞	0	
4	$\frac{6}{17}$	$-\frac{17}{7}$	-7	$-\frac{1}{5}$	-5
5	$\frac{7}{26}$	$-\frac{26}{9}$	$-\frac{9}{2}$	$-\frac{2}{5}$	-5
6	$\frac{3}{37}$	$-\frac{37}{11}$	$-\frac{11}{3}$	$-\frac{3}{5}$	-5

for which the fourth inverted differences are equal. Thus, if we use only the first five points $x_0 = 0$, $x_1 = 1$, $x_2 = 2$, $x_3 = 3$, and $x_4 = 4$, we have $a_0 = 2$, $a_1 = -2$, $a_2 = 3$, $a_3 = 0$, and $a_4 = -5$, so that (9.14.9) becomes

$$f(x) \approx 2 + \cfrac{x}{-2 + \cfrac{x - 1}{3 + \cfrac{x - 2}{0 + \cfrac{x - 3}{-5}}}} \equiv r_4(x) \qquad (9.14.15)$$

where the approximation would become *exact* if the last denominator -5 were replaced by the (unknown) quantity $-5 + (x - 4)/\phi_5[0, 1, 2, 3, 4, x]$.

Thus $r_4(x)$ may be expected to agree with $f(x)$ at the five points employed in its determination. In addition, since the tabular array shows that the *same* approximation would be obtained if the abscissa $x = 4$ were replaced by either $x = 5$ or $x = 6$, it may be expected that $r_4(x)$ will agree with $f(x)$ at those two points as well.

Successive reductions will convert the right-hand member of (9.14.15) to the simpler form

$$r_4(x) = \frac{2 + x}{1 + x^2} \qquad (9.14.16)$$

if this reduction is desired, and the agreement can be verified directly. Furthermore, the respective approximating *convergents* corresponding to termination of the fraction with a_0, a_1, and a_2, and hence to collocation at one, two, and three successive points, are found to be

$$r_0(x) = 2 \qquad r_1(x) = \frac{4 - x}{2} \qquad r_2(x) = \frac{14 - 5x}{7 - x}$$

However, since the present example is exceptional to the extent that $a_3 = 0$, the third convergent does not agree with $f(x)$ at the four points $x = 0, 1, 2,$ and 3. Indeed, this convergent is identical with $r_1(x)$, which agrees with $f(x)$ at $x = 0, 1$, and also at $x = 3$, but does not do so at $x = 2$.

9.15 Rational Approximations and Continued Fractions

It is easily seen (inductively) that the nth convergent of the continued fraction (9.14.9) is expressible in the form

$$r_n(x) = \frac{\alpha'_0 + \alpha'_1 x + \cdots + \alpha'_p x^p}{\beta'_0 + \beta'_1 x + \cdots + \beta'_{p-1} x^{p-1}} \qquad (n = 2p - 1) \qquad (9.15.1a)$$

if n is *odd*, where $\alpha'_p \neq 0$, and in the form

$$r_n(x) = \frac{\alpha''_0 + \alpha''_1 x + \cdots + \alpha''_p x^p}{\beta''_0 + \beta''_1 x + \cdots + \beta''_p x^p} \qquad (n = 2p) \qquad (9.15.1b)$$

if n is *even*, where $\beta''_p \neq 0$. Thus the nth convergent affords an approximation to $f(x)$ by a *ratio of polynomials*, that is, by a *rational function* of x, which generally agrees with $f(x)$ at the $n + 1$ points $x_0, x_1, \ldots, x_n$ if $a_n \neq 0$ and if all preceding a's are finite.

This situation is in accordance with the fact that, since the numerator and denominator of either form of (9.15.1) can be divided through by any one of the nonzero constants, the first form involves $2p$ independent parameters and the second form $2p + 1$ such parameters, so that in either case $n + 1$ independent constants are available for the determination of the approximation.

Given a set of $n + 1$ distinct points, there cannot exist more than one *irreducible* rational function† of the form (9.15.1) which takes on prescribed values at those points. The proof follows simply by first writing (9.15.1) in the form $r_n(x) = M_n(x)/N_n(x)$, where M_n and N_n are polynomials, and supposing that *another* such ratio, $\overline{M}_n(x)/\overline{N}_n(x)$, takes on the same values as does $r_n(x)$ at $n + 1$ distinct points. In accordance with (9.15.1), the degrees of M_n and $\overline{M}_n$ cannot exceed $n/2$ when n is even or $(n + 1)/2$ when n is odd, whereas the degrees of N_n and $\overline{N}_n$ cannot exceed $n/2$ when n is even or $(n - 1)/2$ when n is odd. It then follows that the function $M_n(x)\overline{N}(x) - \overline{M}_n(x)N_n(x)$ also vanishes at those points. But, since this function is a polynomial of degree n or less, it must therefore vanish identically, so that $M_n(x)\overline{N}_n(x) \equiv \overline{M}_n(x)N_n(x)$. Under the assumption that $r_n(x)$ is irreducible, M_n and N_n possess no common linear factors. Thus all linear factors of M_n must also be factors of $\overline{M}_n$, and the converse is also true since $\overline{M}_n/\overline{N}_n$ is also assumed to be irreducible. The same argument applies to N_n and $\overline{N}_n$, so that the respective numerators and denominators can differ only to the extent of a common constant multiplicative factor, as was to be shown.

However, there may be *no* such function. For example, if we attempt to determine directly a function of the form

$$r_3(x) = \frac{\alpha_0 + \alpha_1 x + \alpha_2 x^2}{\beta_0 + \beta_1 x}$$

which takes on the values prescribed in the preceding example at the four points $x = 0$, 1, 2, and 3, we must solve the simultaneous equations

$$\begin{aligned} 2\beta_0 &= \alpha_0 \\ \tfrac{3}{2}(\beta_0 + \beta_1) &= \alpha_0 + \alpha_1 + \alpha_2 \\ \tfrac{4}{5}(\beta_0 + 2\beta_1) &= \alpha_0 + 2\alpha_1 + 4\alpha_2 \\ \tfrac{1}{2}(\beta_0 + 3\beta_1) &= \alpha_0 + 3\alpha_1 + 9\alpha_2 \end{aligned}$$

which result from clearing fractions and equating the resultant members at the four points, and we find that the general solution is given by the relations

$$\alpha_0 = 8\alpha_2 \qquad \alpha_1 = -6\alpha_2 \qquad \beta_0 = 4\alpha_2 \qquad \beta_1 = -2\alpha_2$$

where α_2 is arbitrary. Thus the assumed form becomes

$$r_3(x) = \frac{8 - 6x + x^2}{4 - 2x} = \frac{(4 - x)(2 - x)}{2(2 - x)}$$

† A rational function is said to be *reducible* if its numerator and denominator possess a common polynomial factor other than a constant.

and is *reducible* to $r_1(x) = (4 - x)/2$, in accordance with the result obtained from (9.14.15). The original form is indeterminate at $x = 2$, whereas the reduced form does not take on the prescribed value at that point. Thus the defect of the third convergent of (9.14.15) is due to the nonexistence of a form of the type required at that stage rather than to a failure of the determining process.

In the case of (9.14.15), a warning was served by the fact that $a_3 = 0$. It should be remarked, however, that the kth convergent may be defective, for the same reason as above, even though a_k does not vanish, although this situation is an unusual one. In illustration, the data

x	0	1	2	3
$f(x)$	2	1	1	0

lead to the inverted-difference array

x	f	ϕ_1	ϕ_2	ϕ_3
0	2			
1	1	-1		
2	1	-2	-1	
3	0	$-\frac{3}{2}$	-4	$-\frac{1}{3}$

in which no diagonal element vanishes. Whereas the corresponding approximation

$$f(x) \approx 2 + \cfrac{x}{-1 + \cfrac{x-1}{-1 + \cfrac{x-2}{-\frac{1}{3}}}} = \frac{12 - 13x + 3x^2}{6 - 4x}$$

is properly exact at the four tabular points, the second convergent of the fraction is seen to be

$$r_2(x) = 2 + \frac{x}{-1 - (x - 1)} = 2 - \frac{x}{x}$$

and is undefined at the tabular point $x = 0$. It is easily verified that there exists no irreducible fraction of the form $(\alpha_0 + \alpha_1 x)/(\beta_0 + \beta_1 x)$ which takes on the first three prescribed values.

On the other hand, even though the $n + 1$ given data serve to determine a rational approximation of the form (9.15.1), the continued-fraction expansion will fail to exist, in the form assumed, if $a_k = \infty$ for some $k \leq n$. Thus, whereas the data

x	0	1	2	3	4	5
$f(x)$	1	1	$\frac{3}{5}$	$\frac{2}{5}$	$\frac{5}{17}$	$\frac{3}{13}$

correspond to the function

$$f(x) = \frac{1 + x}{1 + x^2}$$

it is seen that $\phi_1[0, 1] = \infty$, so that there exists no expansion of the form

$$f(x) = a_0 + \cfrac{x}{a_1 + \cfrac{x - 1}{a_2 + \ddots}}$$

which takes on the prescribed values when $x_0 = 0$, $x_1 = 1$, and $x_2 = 2$. This difficulty can be averted here by reordering the abscissas in such a way that the equal ordinates are not consecutive. (See also Prob. 58.) Thus, if we take $x_0 = 0$, $x_1 = 2$, $x_2 = 1$, $x_3 = 3$, and $x_4 = 4$, we obtain the following array:

0	1				
2	$\frac{3}{5}$	-5			
1	1	∞	0		
3	$\frac{2}{5}$	-5	∞	0	
4	$\frac{5}{17}$	$-\frac{17}{13}$	-3	-1	-1
5	$\frac{3}{13}$	$-\frac{13}{2}$	-2	-2	-1

The additional line is included to illustrate the constancy of the fourth inverted difference in the present case.

From these results we deduce the approximation

$$f(x) \approx 1 + \cfrac{x}{-5 + \cfrac{x - 2}{0 + \cfrac{x - 1}{0 + \cfrac{x - 3}{-1}}}}$$

which properly reduces exactly to $(1 + x)/(1 + x^2)$. In this form, the successive convergents are 1, $(5 - x)/5$, 1, $(5 - x)/5$, and $(1 + x)/(1 + x^2)$. As was predicted by the presence of the zeros, the second and third convergents are both defective, in that the second takes on the prescribed values only at the first and third points while the third does so only at the first, second, and fourth points. If, for example, the abscissas are taken in the order 1, 2, 3, 4, 0, we obtain the form

$$f(x) \approx 1 + \cfrac{x - 1}{-\frac{5}{2} + \cfrac{x - 2}{-\frac{6}{5} + \cfrac{x - 3}{\frac{35}{2} + \cfrac{x - 4}{\frac{1}{5}}}}}$$

which naturally also reduces to $(1 + x)/(1 + x^2)$ but which possesses no defective convergents.

In the usual cases in practice, the ordinates can be introduced in any order. For calculation near the beginning of a tabulation, it is usually desirable to number the abscissas in increasing algebraic order, whereas near the end of a tabulation the reverse numbering is desirable, in analogy to the Newton forward- and backward-difference polynomial interpolation formulas. Inside the tabular range it often is desirable first to introduce the abscissa nearest the abscissa of the interpolant, and then successively to introduce abscissas at increasing distance from x_0, alternately forward and backward, in analogy to the central-difference interpolation formulas. These choices tend to maximize the effective initial rate of convergence of the sequence of successive convergents in practical situations for which the sequence generally does not terminate and for which the number of convergents *needed* to supply a specified accuracy usually cannot be predicted easily in advance.

9.16 Determination of Convergents of Continued Fractions

Since the direct evaluation of a truncated continued fraction necessarily "begins at the end," for the purpose of calculating values of a forward *sequence* of convergents or of expressing them as ratios of polynomials it is useful to proceed recursively by use of the formulas to be derived next.

From the definition (9.14.7), it can be seen that $f(x)$ is expressible as the ratio of two linear functions of any $v_r(x)$, say, in the form

$$f(x) = \frac{R_k(x) + v_{k+1}(x)M_k(x)}{S_k(x) + v_{k+1}(x)N_k(x)} \tag{9.16.1}$$

In order to determine M_k, N_k, R_k, and S_k, we may notice that $f(x)$ consequently also is given by the result of replacing k by $k - 1$ in (9.16.1) and then using (9.14.7) to express $v_k(x)$ in terms of $v_{k+1}(x)$, so that there must follow

$$\frac{R_k(x) + v_{k+1}(x)M_k(x)}{S_k(x) + v_{k+1}(x)N_k(x)} = \frac{(x - x_k)M_{k-1}(x) + v_{k+1}(x)[a_k M_{k-1}(x) + R_{k-1}(x)]}{(x - x_k)N_{k-1}(x) + v_{k+1}(x)[a_k N_{k-1}(x) + S_{k-1}(x)]} \tag{9.16.2}$$

This requirement is satisfied, for arbitrary v_{k+1}, if the desired functions satisfy the relations

$$\begin{aligned} M_{k+1}(x) &= a_{k+1}M_k(x) + (x - x_k)M_{k-1}(x) \\ N_{k+1}(x) &= a_{k+1}N_k(x) + (x - x_k)N_{k-1}(x) \\ R_k(x) &= (x - x_k)M_{k-1}(x) \\ S_k(x) &= (x - x_k)N_{k-1}(x) \end{aligned} \tag{9.16.3}$$

in accordance with which M_k, N_k, R_k, and S_k clearly will be *polynomials* in x if M_0, M_1, N_0, and N_1 are polynomials. Thus we may write (9.16.1) in the form

$$f(x) = \frac{M_k(x) + (x - x_k)M_{k-1}(x)/\phi_{k+1}[x_0, \ldots, x_k, x]}{N_k(x) + (x - x_k)N_{k-1}(x)/\phi_{k+1}[x_0, \ldots, x_k, x]} \tag{9.16.4}$$

Since, when $k = 0$, this form must reduce to the form given by (9.14.7)

$$f(x) = a_0 + \frac{x - x_0}{\phi_1[x_0, x]}$$

we must have $N_{-1}(x) = 0$, $M_0(x)/N_0(x) = a_0$, and $M_{-1}(x)/N_0(x) = 1$. It is convenient to take $N_0(x) = 1$.

Thus M_k and N_k can be determined by the recurrence formulas

$$\begin{gathered} M_{k+1}(x) = a_{k+1}M_k(x) + (x - x_k)M_{k-1}(x) \\ M_{-1}(x) = 1 \qquad M_0(x) = a_0 \end{gathered} \tag{9.16.5}$$

and

$$\begin{gathered} N_{k+1}(x) = a_{k+1}N_k(x) + (x - x_k)N_{k-1}(x) \\ N_{-1}(x) = 0 \qquad N_0(x) = 1 \end{gathered} \tag{9.16.6}$$

In particular, the kth convergent to $f(x)$ is given simply by

$$r_k(x) = \frac{M_k(x)}{N_k(x)} \tag{9.16.7}$$

The error associated with the approximation $f(x) \approx r_k(x)$ can be estimated by use of (9.16.4) if information with regard to $\phi_{k+1}[x_0, \ldots, x_k, x]$ is available, say, in the form of sample values of the $(k + 1)$th inverted differences formed with $x_0, \ldots, x_k$ as its first $k + 1$ arguments. For this purpose it is convenient to rewrite (9.16.4) in the equivalent form

$$f(x) - r_k(x) = -\frac{c_k(x)}{1 + c_k(x)}[r_k(x) - r_{k-1}(x)] \tag{9.16.8}$$

where

$$c_k(x) = \frac{(x - x_k)N_{k-1}(x)}{\phi_{k+1}[x_0, \ldots, x_k, x]N_k(x)} \tag{9.16.9}$$

(See also Prob. 63.)

When $f(x)$ is not a rational function, the sequence of convergents generally is infinite, and it may or may not tend to $f(x)$ as $n \to \infty$. However, it generally at least approaches $f(x)$ more and more closely, for any fixed value of x inside the range of the tabular values $x_0, \ldots, x_n$, as n increases up to a certain stage. The determination of successive convergents is desirable in order that the rate of *effective convergence* may be estimated.

It is useful to notice that the expansion (9.14.9) can be expressed in the alternative forms

$$f(x) = a_0 + \cfrac{x - x_0}{a_1 + \cfrac{x - x_1}{a_2 + \cfrac{x - x_2}{a_3 + \ddots}}} = a_0 + \cfrac{\cfrac{x - x_0}{a_1}}{1 + \cfrac{\cfrac{x - x_1}{a_1 a_2}}{1 + \cfrac{\cfrac{x - x_2}{a_2 a_3}}{1 + \ddots}}} \tag{9.16.10}$$

if none of the a_i vanish. The more compact symbolic arrangements

$$f(x) = a_0 + \frac{x - x_0}{a_1 +} \; \frac{x - x_1}{a_2 +} \; \frac{x - x_2}{a_3 +} \cdots$$

and

$$f(x) = a_0 + \frac{x - x_0|}{|a_1} + \frac{x - x_1|}{|a_2} + \frac{x - x_2|}{|a_3} + \cdots$$

of the first form are often used.

Approximation by rational functions is often useful in the neighborhood of a point α at which the function $f(x)$ becomes *infinite* in proportion to $1/(x - \alpha)^m$, where m is a positive integer whose value may or may not be known.† In illustration, the following calculation is for the purpose of determining an approximation to cot 0.15 ($\doteq$ 6.6166) from the given three-place values.

k	x_k	f_k	ϕ_1	ϕ_2	ϕ_3	$x - x_k$	M_k	N_k	r_k
0	0.1	9.967				0.05	9.967	1	9.967
1	0.2	4.933	−0.019865			−0.05	−0.14799	−0.019865	7.450
2	0.3	3.233	−0.029700	−10.168		−0.15	1.00641	0.15199	6.622
3	0.4	2.365	−0.039463	−10.205	−2.70	−0.25	−2.69511	−0.40739	6.6156

Here, for example, we have from (9.16.5) and (9.16.6) the computations

$$M_2 = (-10.168)(-0.14799) + (-0.05)(9.967) \doteq 1.00641$$

$$N_3 = (-2.70)(0.15199) + (-0.15)(-0.019865) \doteq -0.40739$$

In the calculation of the successive inverted differences, about one more digit was retained in each inverted difference than would be expected to be significant if all digits retained in the two preceding entries, from which it is

† The application of *polynomial* approximation to $1/f(x)$ or to $(x - \alpha)^m f(x)$ may be appropriate alternatives.

calculated, were correct, when account is taken of the loss of significant figures in the subtractions involved. The tabulated a's are then treated as though they were exact in the calculation of the M's and N's, so that, for example, (at least) five digits are retained in N_3, even though not more than three of its digits would be significant if the value -2.70 were correct to the places given, and only to those places.

Because of the fact that errors in the given ordinates and in the a's, M's, and N's enter into the determination of the required r's in a nonlinear way, it is difficult to estimate in advance the number of digits which should be retained in each intermediate calculation, but it is usually desirable to retain at least as many digits as are required by the preceding rule.† In the present case, the tabulated value of a_3 would be modified in its third digit if additional digits were retained in the calculation of preceding divided differences, and this modification would change M_3 and N_3 in the third digit. However, the calculated value of the *ratio* $r_3 = M_3/N_3$ would be modified by only two units in its *fifth* digit. The deviation of the calculated value r_3 from the true value, by one unit in its fourth digit, is due principally to the roundoff errors in the *given* data (see Prob. 65).

The fact that the value of the convergent r_3 itself is not sensitive to appreciable errors in a_3 can be seen more directly by inspection of the actual truncated continued fraction:

$$r_3(0.15) \approx 9.967 + \cfrac{0.05}{-0.019865 + \cfrac{-0.05}{-10.168 + \cfrac{-0.15}{-2.70}}}$$

In this particular example, the near linearity of the first inverted difference $\phi_1[0.1, x]$ suggests the use of *polynomial* extrapolation over the three available values for the determination of $\phi_1[0.1, 0.15]$. The use of Newton's forward-difference formula (retaining the second difference) gives $\phi_1[0.1, 0.15] \approx 0.014923$, so that there follows

$$\frac{0.15 - 0.10}{f(0.15) - 9.967} \approx 0.014923 \qquad f(0.15) \approx 6.6165$$

and the calculation of *inverted* differences of higher order is avoided. (See also Prob. 59.)

† In addition, it should be noted that possible rapid growth of M_k and N_k in special situations may occasion overflow in machine calculation, unless proper precautions are taken.

It can be shown (see Prob. 61) that the kth convergent of (9.16.10) can be expressed in the form

$$r_k = a_0 + \frac{x - x_0}{N_0 N_1} - \frac{(x - x_0)(x - x_1)}{N_1 N_2} + \frac{(x - x_0)(x - x_1)(x - x_2)}{N_2 N_3} - \cdots + (-1)^{k+1} \frac{(x - x_0)(x - x_1) \cdots (x - x_{k-1})}{N_{k-1} N_k} \tag{9.16.11}$$

In the case of the preceding example, the kth convergent r_k is thus obtained by retaining $k + 1$ terms in the sum

$$9.967 - \frac{0.05}{0.019865} - \frac{0.0025}{0.0030193} - \frac{0.000375}{0.06192}$$

and successive results agree, to three decimal places, with the values obtained previously.

If a specific rational approximation $r_n(x)$ is to be used repeatedly for numerical calculation, it is significant that usually neither the continued-fraction form (9.16.10) nor the form (9.16.7) [or (9.15.1)] provides the most efficient specification of the approximation in terms of the number of necessary "long operations" (multiplications and divisions). The evaluation of the nth convergent of (9.16.10) involves n divisions, whereas the most efficient evaluation of the ratio of polynomials in (9.15.1*a*) or (9.15.1*b*) generally requires $n - 1$ multiplications and one division [when one coefficient is reduced to unity in advance by preliminary division, and when the polynomials are evaluated recursively by Horner's method, as in (1.8.6)].

On the other hand, except in certain special cases which require individual treatment (see Probs. 69 and 70), it is possible to transform the expression for $r_n(x)$ to the form

$$r_n(x) = A_0 x + B_0 + \frac{A_1|}{|x + B_1} + \frac{A_2|}{|x + B_2} + \cdots + \frac{A_{p-1}|}{|x + B_{p-1}} \tag{9.16.12}$$

when $n = 2p - 1$, and

$$r_n(x) = B_0 + \frac{A_1|}{|x + B_1} + \frac{A_2|}{|x + B_2} + \cdots + \frac{A_p|}{|x + B_p} \tag{9.16.13}$$

when $n = 2p$. The second form involves only $n/2$ divisions, the first form $(n - 1)/2$ divisions and one multiplication.

9.17 Thiele's Continued-Fraction Approximations

Whereas the kth *inverted difference* $\phi_k[x_0, \ldots, x_{k-2}, x_{k-1}, x_k]$ of a function $f(x)$ is symmetric only in its last two arguments, it happens that the quantity

$$\rho_k[x_0, \ldots, x_k] = \phi_k[x_0, \ldots, x_k] + \phi_{k-2}[x_0, \ldots, x_{k-2}] + \phi_{k-4}[x_0, \ldots, x_{k-4}] + \cdots \tag{9.17.1}$$

is symmetric in all its $k + 1$ arguments. Here the last term on the right is $\phi_0[x_0]$ if k is even, and is $\phi_1[x_0, x_1]$ if k is odd. This quantity is often known as a kth *reciprocal difference* of $f(x)$.

In particular, we have

$$\rho_0[x_0] = \phi_0[x_0] = f(x_0)$$

$$\rho_1[x_0, x_1] = \phi_1[x_0, x_1] = \frac{x_1 - x_0}{f(x_1) - f(x_0)} \tag{9.17.2}$$

and calculation shows that

$$\begin{aligned} \rho_2[x_0, x_1, x_2] &= \phi_0[x_0] + \phi_2[x_0, x_1, x_2] \\ &= \frac{x_0 f_0(f_1 - f_2) + x_1 f_1(f_2 - f_0) + x_2 f_2(f_0 - f_1)}{x_0(f_1 - f_2) + x_1(f_2 - f_0) + x_2(f_0 - f_1)} \end{aligned} \tag{9.17.3}$$

in which cases the symmetry is apparent. Although an *inductive* generalization is possible, the following argument is considerably more simple.

It is easily verified that when use is made of (9.16.5) and (9.16.6), the nth convergent of (9.14.9) is given by (9.15.1a) when $n = 2p - 1$, with $\alpha_p' = 1$ and $\beta_{p-1}' = \rho_{2p-1}[x_0, \ldots, x_{2p-1}]$, and by (9.15.1$b$) when $n = 2p$, with $\alpha_p'' = \rho_{2p}[x_0, \ldots, x_{2p}]$ and $\beta_p'' = 1$. Thus it follows that ρ_{2p} is the ratio of the leading coefficients in the numerator and denominator of the rational function of form (9.15.1b) which agrees with $f(x)$ at the $2p + 1$ points $x_0, \ldots, x_{2p}$, whereas ρ_{2p-1} is the reciprocal of that ratio for the rational function of form (9.15.1a) which agrees with $f(x)$ at the $2p$ points $x_0, \ldots, x_{2p-1}$. Clearly these ratios are independent of any *ordering* of the points involved.

Since (9.17.1) implies that

$$\rho_k[x_0, \ldots, x_k] - \rho_{k-2}[x_0, \ldots, x_{k-2}] = \phi_k[x_0, \ldots, x_k] \tag{9.17.4}$$

reference to (9.14.13) shows that the successive reciprocal differences may be obtained by use of the recurrence formula

$$\begin{aligned} \rho_k&[x_0, \ldots, x_{k-2}, x_{k-1}, x_k] \\ &= \frac{x_k - x_{k-1}}{\rho_{k-1}[x_0, \ldots, x_{k-2}, x_k] - \rho_{k-1}[x_0, \ldots, x_{k-2}, x_{k-1}]} \\ &\qquad + \rho_{k-2}[x_0, \ldots, x_{k-2}] \end{aligned} \tag{9.17.5}$$

Although this formula is less simply applied than (9.14.13), the symmetry of the kth reciprocal difference permits its calculation from *any* two $(k - 1)$th reciprocal differences having $k - 1$ of its arguments in common, together with the $(k - 2)$th reciprocal difference formed with those arguments.

Thus, in particular, a *reciprocal-difference table* may be constructed in the convenient form

$$\begin{array}{llllll} x_0 & f(x_0) & & & & \\ & & \rho_1[x_0, x_1] & & & \\ x_1 & f(x_1) & & \rho_2[x_0, x_1, x_2] & & \\ & & \rho_1[x_1, x_2] & & \rho_3[x_0, x_1, x_2, x_3] & \\ x_2 & f(x_2) & & \rho_2[x_1, x_2, x_3] & & \\ & & \rho_1[x_2, x_3] & & & \\ x_3 & f(x_3) & & & & \end{array}$$

From this table we may determine the coefficients in (9.14.9) by combining (9.14.12) and (9.17.4), so that

$$a_0 = f(x_0) \qquad a_1 = \rho_1[x_0, x_1]$$

$$a_2 = \rho_2[x_0, x_1, x_2] - f(x_0) \qquad a_3 = \rho_3[x_0, x_1, x_2, x_3] - \rho_1[x_0, x_1]$$

and so forth. Thus the required coefficients are formed from (but are *not* identical with) reciprocal differences appearing in the forward diagonal beginning with $f(x_0)$. Furthermore, because of the symmetry, the data from the *same* table are available for the determination of formulas in which the ordinates are introduced in other orders, by choosing *difference paths* made up of suitable contiguous diagonal segments as was done in Sec. 2.5. Each such expansion is identical with the one which would be obtained *more simply* by the use of the *inverted*-difference array corresponding to an appropriate reordering of the abscissas, but only one array of *reciprocal* differences is needed for the formation of the entire set. Thus the use of reciprocal differences, rather than inverse differences, generally is advantageous only if several such formulas are required.

However, the definition of the reciprocal difference is particularly useful in the important limiting case when the abscissas $x_0, x_1, x_2, \ldots$ all become coincident, so that the requirement that the deviation between $f(x)$ and the kth convergent of the fraction vanish at $k + 1$ distinct points is replaced by the requirement that the deviation and its first k derivatives vanish at a single point x_0. The representation (9.14.9) then tends to the form

$$f(x) = a_0 + \cfrac{x - x_0}{a_1 + \cfrac{x - x_0}{a_2 + \cfrac{x - x_0}{a_3 + \ddots}}} \tag{9.17.6}$$

where

$$a_k = \phi_k(x_0) \tag{9.17.7}$$

with the function $\phi_k(x)$ defined by the limiting process

$$\phi_k(x) = \lim_{x_0, \ldots, x_k \to x} \phi_k[x_0, \ldots, x_k] \tag{9.17.8}$$

under the assumption that this limit exists for $k = 0, 1, \ldots$. Here, if the fraction is terminated after k divisions, it is necessary to replace $\phi_k(x_0)$ by $\phi_k(x_0) + (x - x_0)/\phi_{k+1}[x_0, \ldots, x_0, x]$ in order to restore true equality.

The consideration of this limit is complicated by the fact that the $k + 1$ arguments $x_0, \ldots, x_k$ are not symmetrically involved. Thus it is desirable to use (9.17.4) to express (9.17.8) in the form

$$\phi_k(x) = \lim_{x_0, \ldots, x_k \to x} \{\rho_k[x_0, \ldots, x_k] - \rho_{k-2}[x_0, \ldots, x_{k-2}]\} \tag{9.17.9}$$

so that both terms on the right are symmetric in their arguments. Accordingly, we have also

$$\phi_k(x) = \rho_k(x) - \rho_{k-2}(x) \tag{9.17.10}$$

with the additional abbreviation

$$\rho_k(x) \equiv \lim_{x_0, \ldots, x_k \to x} \rho_k[x_0, \ldots, x_k] \tag{9.17.11}$$

In addition, we have the relation

$$\phi_k(x) = \lim_{x_k \to x} \frac{x_k - x}{\rho_{k-1}[x, \ldots, x, x_k] - \rho_{k-1}[x, \ldots, x, x]} \tag{9.17.12}$$

from (9.17.4) and (9.17.5); and, if the limit on the right exists, it is given by

$$\left.\frac{1}{\dfrac{\partial \rho_{k-1}[x_0, \ldots, x_{k-1}]}{\partial x_{k-1}}}\right|_{x_0, \ldots, x_{k-1} = x} = \frac{k}{\dfrac{d\rho_{k-1}[x, \ldots, x]}{dx}} \tag{9.17.13}$$

in consequence of the symmetry in the arguments, so that (9.17.12) becomes

$$\phi_k(x) = \frac{k}{\rho'_{k-1}(x)} \tag{9.17.14}$$

Thus we may evaluate the coefficients $\phi_k(x_0)$ appearing in (9.17.6) successively, by using the formulas (9.17.10) and (9.17.14) in the form

$$\rho_k(x) = \rho_{k-2}(x) + \phi_k(x) \qquad \phi_{k+1}(x) = \frac{k+1}{\rho'_k(x)} \tag{9.17.15}$$

with the obvious starting values

$$\rho_{-2}(x) = \rho_{-1}(x) = 0 \qquad \phi_0(x) = f(x) \tag{9.17.16}$$

and evaluating the functions $\phi_k(x)$ at $x = x_0$. The function $\rho_k(x)$ is often called the kth *reciprocal derivative* of $f(x)$. In correspondence with the terminology of the preceding section, we may refer to $\phi_k(x)$ as the kth *inverse derivative* of $f(x)$. Naturally, Eqs. (9.16.5) and (9.16.6) and their implications [such as (9.16.11)] continue to hold here, with $x_1 = x_2 = \cdots = x_0$.

In order to illustrate the calculation in a simple case, we consider the function $f(x) = e^x$. By using successively the first and second relations in (9.17.15) with $k = 0, 1, 2, \ldots$, we obtain the functions

$$
\begin{aligned}
& & \phi_0 &= e^x \\
\rho_0 &= e^x & \phi_1 &= e^{-x} \\
\rho_1 &= e^{-x} & \phi_2 &= -2e^x \\
\rho_2 &= -e^x & \phi_3 &= -3e^{-x} \\
\rho_3 &= -2e^{-x} & \phi_4 &= 2e^x \\
\rho_4 &= e^x & \phi_5 &= 5e^{-x}
\end{aligned}
$$

and so forth. If we take $x_0 = 0$ in (9.17.6), we thus obtain the coefficients in the expansion

$$e^x = 1 + \frac{x|}{|1} + \frac{x|}{|-2} + \frac{x|}{|-3} + \frac{x|}{|2} + \cdots = 1 + \frac{x|}{|1} - \frac{x|}{|2} + \frac{x|}{|3} - \frac{x|}{|2} + \cdots$$

Inspection suggests that the inverse derivatives of e^x, of even and odd orders, are given by

$$
\begin{aligned}
\phi_{2n}(x) &= (-1)^n 2e^x \qquad (n \geqq 1) \\
\phi_{2n+1}(x) &= (-1)^n(2n+1)e^{-x}
\end{aligned}
$$

and the truth of this conjecture is readily established by induction.

The expansion (9.17.6) is attributed to Thiele. It is related to the more general expansions considered previously as the Taylor-series expansion is related to the divided-difference polynomial interpolation formulas. Whereas the nth convergent of the *confluent* expansion (9.17.6) generally affords a better approximation to $f(x)$ in the immediate neighborhood of x_0, the corresponding convergent of a development which yields exact results at $n + 1$ points of an interval including x_0 is usually to be preferred for approximation over that interval.

The nth convergent of the Thiele representation is expressible in the form $u_p(x)/v_{p-1}(x)$ when $n = 2p - 1$, where the numerator and denominator are polynomials of maximum degrees p and $p - 1$, respectively, whereas that convergent can be put in the form $u_p(x)/v_p(x)$ when $n = 2p$, in consequence of the results of Sec. 9.15. These ratios are sometimes denoted by $R_{p,p-1}(x)$ and $R_{p,p}(x)$, respectively. More generally, the notation $R_{m,k}(x)$ is sometimes used to denote a rational function, with numerator of degree m or less and denominator of degree k or less, which agrees with $f(x)$ at x_0 and whose first $m + k$ derivatives are respectively equal to the first $m + k$ derivatives of $f(x)$ at x_0, when such a function exists.

The approximation $R_{m,k}(x)$ is often called the (m, k) entry in the *Padé table* (Padé [1892]) of $f(x)$ relative to x_0. For a given value of the *sum* $n = m + k$ (which is variously called the *index*, *order*, or even *degree*, of the approximation), it is usually true that the entry for which $k = m - 1$ (n odd) or $k = m$ (n even) gives the most satisfactory approximation to $f(x)$ near x_0. This is the entry given by the nth convergent of the Thiele representation.

The entry $R_{n,0}$ is the sum of the first $n + 1$ terms of the *Taylor* expansion of $f(x)$ about $x = x_0$, while the entry $R_{0,n}$ is the reciprocal of the corresponding expansion for $1/f(x)$, assuming in each case the *existence* of the entry in question. Other entries, when they exist, could be determined as convergents of confluent forms of more general continued-fraction representations indicated in Prob. 57.

In addition, any Padé entry (if it exists) also can be obtained directly from the leading coefficients of the formal representation

$$f(x) = \sum_{r=0}^{\infty} c_r(x - x_0)^r \qquad (9.17.17)$$

if it is available, by determining $m + 1$ a's and $k + 1$ b's in such a way that

$$\frac{\sum_{r=0}^{m} a_r(x - x_0)^r}{\sum_{r=0}^{k} b_r(x - x_0)^r} - f(x) = O[(x - x_0)^{m+k+1}] \qquad (9.17.18)$$

so that the left-hand side and its first $m + k$ derivatives vanish at x_0. For this purpose, we may write $t = x - x_0$ and clear fractions to obtain the requirement

$$\sum_{r=0}^{m} a_r t^r - \left(\sum_{i=0}^{k} b_i t^i\right)\left(\sum_{j=0}^{\infty} c_j t^k\right) = O(t^{m+k+1})$$

With the conventions that $a_r = 0$ unless $0 \leqq r \leqq m$, $b_i = 0$ unless $0 \leqq i \leqq k$, and $c_j = 0$ unless $j \geqq 0$, this requirement can be put into the more convenient form

$$\sum_{r=0}^{\infty} \left(a_r - \sum_{i=0}^{k} b_i c_{r-i}\right) t^r = O(t^{m+k+1}) \qquad (9.17.19)$$

and hence imposes the conditions

$$\sum_{i=0}^{k} b_i c_{r-i} = a_r \qquad (r = 0, 1, \ldots, m + k)$$

with the same conventions, or, equivalently, the conditions

$$\sum_{i=0}^{k} b_i c_{r-i} = \begin{cases} a_r & (r = 0, 1, \ldots, m) \\ 0 & (r = m + 1, \ldots, m + k) \end{cases} \qquad (9.17.20)$$

with the single convention that

$$c_j = 0 \qquad \text{when } j < 0$$

The value of the constant b_0 can be arbitrarily assigned (say, taken to be 1), after which the conditions for which $r = m + 1, \ldots, m + k$ will determine the other k b's, and the remaining $m + 1$ conditions will determine the a's, if the assumed entry exists.

The expansion (9.17.6) can be generalized usefully as follows. If we replace the independent variable x by $G(x)$ and write

$$f(G(x)) = F(x) \qquad \rho_k(G(x)) = \mathrm{P}_k(x) \qquad \phi_k(G(x)) = \Phi_k(x)$$

the formulas of (9.17.15) become

$$\mathrm{P}_k(x) = \mathrm{P}_{k-2}(x) + \Phi_k(x) \qquad \Phi_{k+1}(x) = (k+1)\frac{G'(x)}{\mathrm{P}_k'(x)} \tag{9.17.21}$$

with the starting values

$$\mathrm{P}_{-2}(x) = \mathrm{P}_{-1}(x) = 0 \qquad \Phi_0(x) = F(x) \tag{9.17.22}$$

and (9.17.6) takes the form [compare the Bürmann-series expansion (1.9.10)]

$$F(x) = A_0 + \cfrac{G(x) - G(x_0)}{A_1 + \cfrac{G(x) - G(x_0)}{A_2 + \cfrac{G(x) - G(x_0)}{A_3 + \ddots}}} \tag{9.17.23}$$

where

$$A_k = \Phi_k(x_0) \tag{9.17.24}$$

Here, if the fraction is terminated with A_n, A_n is to be replaced by

$$A_n + \frac{G(x) - G(x_0)}{\phi_{n+1}[G(x_0), \ldots, G(x_0), G(x)]}$$

if strict equality is to be preserved.

The result of truncating (9.17.23), and neglecting the residual, is then an approximation to $F(x)$ in terms of a rational function of $G(x)$, which may be expected to be useful near x_0. It can be used, for example, to determine approximately the value of $F(x)$ when $G(x)$ takes on a prescribed value, if the corresponding value of x is unknown but is approximated by x_0. The first few Φ's are readily found to be governed by the equations

$$\Phi_0 = F \qquad \Phi_1 = \frac{G'}{F'} \qquad \Phi_2 = 2\frac{G'}{\Phi_1'} \qquad \Phi_3 = 3\frac{G'}{F' + \Phi_2'} \qquad \cdots \tag{9.17.25}$$

Thus, for example, if we take $F(x) = e^x$, $G(x) = \sin x$, $x_0 = 0$, we obtain the representation

$$e^x = 1 + \cfrac{\sin x}{1 + \cfrac{\sin x}{-2 + \ddots}}$$

near $x = 0$.

In particular, if we take $F(x) \equiv x$, we obtain a formula for *inverse interpolation* near $x = x_0$. For example, suppose that we require a zero $\bar{x}$ of $G(x)$ and that x_0 is a previously determined approximation to $\bar{x}$. Since then $A_0 \equiv F(x_0) = x_0$, (9.17.23) then reduces to the form

$$\bar{x} = x_0 + \cfrac{-G(x_0)}{A_1 + \cfrac{-G(x_0)}{A_2 + \cfrac{-G(x_0)}{A_3 + \ddots}}}$$

or to the more convenient equivalent form

$$\bar{x} = x_0 - \frac{\omega_1(x_0)|}{|1} - \frac{\omega_2(x_0)|}{|1} - \frac{\omega_3(x_0)|}{|1} - \cdots \tag{9.17.26}$$

where

$$\omega_1(x_0) = \frac{G(x_0)}{A_1} \qquad \omega_k(x_0) = \frac{G(x_0)}{A_k A_{k-1}} \qquad (k > 1) \tag{9.17.27}$$

and where (9.17.21), (9.17.22), and (9.17.24) apply with $F(x) = x$.

Here the relations of (9.17.25) reduce to

$$\Phi_0 = x \quad \Phi_1 = G' \quad \Phi_2 = 2\frac{G'}{G''} \quad \Phi_3 = 3\frac{G'}{1 + 2(G'/G'')'} \quad \cdots \tag{9.17.28}$$

and there follows

$$\omega_1 = \frac{G}{G'} \qquad \omega_2 = \frac{G}{G'}\frac{G''}{2G'} \qquad \omega_3 = \frac{G}{G'}\left(\frac{G''}{2G'} - \frac{G'''}{3G''}\right) \quad \cdots \tag{9.17.29}$$

The leading convergents of (9.17.26) are thus found to be

$$\bar{x}^{(0)} = x_0 \qquad \bar{x}^{(1)} = x_0 - \omega_1 \qquad \bar{x}^{(2)} = x_0 - \frac{\omega_1}{1 - \omega_2}$$

$$\bar{x}^{(3)} = x_0 - \frac{\omega_1(1 - \omega_3)}{1 - \omega_2 - \omega_3} \tag{9.17.30}$$

where the ω's are to be evaluated at x_0.†

† The approximation $\bar{x}^{(1)}$ is that given by the classical Newton-Raphson procedure (see Sec. 10.11), the approximation $\bar{x}^{(2)}$ is attributed to Halley, and the investigation of the entire sequence appears to have been initiated by Frame.

In illustration, the equation $x^3 - x - 1 = 0$ is easily seen to possess one real root $\bar{x}$, which lies between $x = 1$ and $x = 2$. If we choose the crude approximation $x_0 = 1$, and set $G(x) = x^3 - x - 1$, there follows $G_0 = -1$, $G_0' = 2$, $G_0'' = 6$, $G_0''' = 6$, and the successive convergents are found to be 1, $\frac{3}{2} = 1.5$, $\frac{9}{7} \doteq 1.29$, and $\frac{75}{56} \doteq 1.34$. If the process is iterated, starting now with $x_0 = 1.34$, the successive approximants round to 1.34, 1.3249, 1.324720, and 1.324718. The result yielded by the last approximation is in fact correct to more than the seven digits given.

Whereas expressions can be derived for the error of truncation (see Frame [1953]), they are too complicated to be generally useful, and one usually must attempt to estimate the error in a given approximant by inspecting the behavior of the *sequence* of preceding approximants.

9.18 Uniformization of Rational Approximations

Since the nth convergent $r_n(x)$ of the Thiele representation (9.17.6) of $f(x)$ has the property that $r_n(x)$ and $f(x)$, together with their first n derivatives, agree at x_0, it must be true that

$$f(x) - r_n(x) = O[(x - x_0)^{n+1}] \qquad (9.18.1)$$

as x tends to x_0. In fact, from Prob. 63 it follows that

$$f(x) - r_n(x) = (-1)^n \frac{(x - x_0)^{n+1}}{N_n(x_0)N_{n+1}(x_0)} + \cdots \qquad (9.18.2)$$

where omitted terms are of higher order in $x - x_0$.

Also, from (9.16.6), we may deduce that here

$$N_n(x_0) = p_n \qquad (9.18.3)$$

with the convenient abbreviation

$$p_k = \begin{cases} a_1 a_2 \cdots a_k & (k \geqq 1) \\ 1 & (k = 0) \end{cases} \qquad (9.18.4)$$

Thus, with this notation, it follows that

$$f(x) - r_n(x) = (-1)^n \frac{(x - x_0)^{n+1}}{p_n p_{n+1}} + \cdots \qquad (9.18.5)$$

near x_0.

Accordingly, as was to be expected, the magnitude of the error associated with $r_n(x)$ is small near x_0, and it tends to grow in proportion to $(x - x_0)^{n+1}$ with increasing distance from x_0. If we suppose that the use of the approximation is to be restricted to an interval $[x_0 - \varepsilon, x_0 + \varepsilon]$ centered at x_0, and if we write

$$x = x_0 + \varepsilon s \qquad s = \frac{x - x_0}{\varepsilon} \tag{9.18.6}$$

so that $-1 \leqq s \leqq 1$ in that interval, then we have

$$f(x) - r_n(x) = (-1)^n \frac{\varepsilon^{n+1}}{p_n p_{n+1}} s^{n+1} + \cdots \tag{9.18.7}$$

and hence also

$$\lim_{\varepsilon \to 0} \frac{f(x) - r_n(x)}{\varepsilon^{n+1}} = \frac{(-1)^n}{p_n p_{n+1}} s^{n+1} \tag{9.18.8}$$

with the notation of (9.18.4) and (9.18.6). This limiting relation essentially characterizes the *local behavior* of the approximation near x_0.

In analogy with preceding treatments of polynomial approximation, accordingly we may be led to seek a modified rational approximation $r_n^*(x)$ for which (9.18.8) is replaced by the relation

$$\lim_{\varepsilon \to 0} \frac{f(x) - r_n^*(x)}{\varepsilon^{n+1}} = \frac{(-1)^n}{p_n p_{n+1}} \frac{T_{n+1}(s)}{2^n} \tag{9.18.9}$$

where $2^{-n}T_{n+1}(s)$ is a monic Chebyshev polynomial, since then, at least over a sufficiently small interval centered at x_0, the error will tend to oscillate with uniform amplitude rather than to increase steadily in magnitude toward the ends of the interval.

One method of attempting this determination begins by writing $r_n^*(x)$ in the form

$$r_n^*(x) = \frac{M_n(x) + \sum_{k=0}^{n-1} \alpha_{k+1}\varepsilon^{n-k}M_k(x) + \alpha_0\varepsilon^{n+1}}{N_n(x) + \sum_{k=0}^{n-1} \alpha_{k+1}\varepsilon^{n-k}N_k(x)} \tag{9.18.10}$$

where the constants $\alpha_0, \alpha_1, \ldots, \alpha_n$ are to be determined, and where the appropriateness of the powers of ε is to be verified. There then follows

$$f(x) - r_n^*(x) = \frac{[N_n(x)f(x) - M_n(x)] + \sum_{k=0}^{n-1} \alpha_{k+1}\varepsilon^{n-k}[N_k(x)f(x) - M_k(x)] - \alpha_0\varepsilon^{n+1}}{N_n(x) + \sum_{k=0}^{n-1} \alpha_{k+1}\varepsilon^{n-k}N_k(x)} \tag{9.18.11}$$

Since (9.18.8) and (9.18.3) imply that

$$\lim_{\varepsilon \to 0} \frac{N_k(x)f(x) - M_k(x)}{\varepsilon^{k+1}} = \lim_{\varepsilon \to 0} \left\{ N_k(x_0 + \varepsilon s) \left[\frac{f(x) - r_k(x)}{\varepsilon^{k+1}} \right] \right\} = \frac{(-1)^k s^{k+1}}{p_{k+1}} \tag{9.18.12}$$

we then find that

$$\lim_{\varepsilon \to 0} \frac{f(x) - r_n^*(x)}{\varepsilon^{n+1}}$$

$$= \frac{(-1)^n (s^{n+1}/p_{n+1}) + \sum_{k=0}^{n-1} (-1)^k (\alpha_{k+1}/p_{k+1}) s^{k+1} - \alpha_0}{p_n}$$

$$= \frac{(-1)^n}{p_n p_{n+1}} \left\{ s^{n+1} + \sum_{k=0}^{n-1} \left[(-1)^{n+k} \frac{p_{n+1}}{p_{k+1}} \alpha_{k+1} \right] s^{k+1} + (-1)^{n+1} p_{n+1} \alpha_0 \right\} \tag{9.18.13}$$

Thus, if the coefficient of s^k in $2^{-n}T_{n+1}(s)$ is denoted by c_k (with $c_{n+1} = 1$),

$$2^{-n}T_{n+1}(s) = s^{n+1} + \sum_{k=0}^{n} c_k s^k \tag{9.18.14}$$

the relation (9.18.13) is identified with the desired relation (9.18.9) by taking

$$\alpha_k = (-1)^{n+k+1} \frac{p_k}{p_{n+1}} c_k$$

or, equivalently,

$$\alpha_k = \frac{(-1)^{n+k+1} c_k}{a_{k+1} a_{k+2} \cdots a_{n+1}} \tag{9.18.15}$$

Hence $r_n^*(x)$ is determined provided only that $a_1, a_2, \ldots, a_{n+1}$ all differ from zero.

In order to illustrate the process, we consider the function $f(x) = e^x$ with $x_0 = 0$, which was used as an example in the preceding section. Here we have

$$a_0 = 1 \qquad a_1 = 1 \qquad a_2 = -2 \qquad a_3 = -3 \qquad a_4 = 2 \qquad a_5 = 5 \qquad \cdots$$

and also, by use of (9.16.5) and (9.16.6), we find that

$$M_0(x) = 1 \qquad M_1(x) = 1 + x \qquad M_2(x) = -2 - x \qquad M_3(x) = 6 + 4x + x^2 \qquad M_4(x) = 12 + 6x + x^2 \qquad \cdots$$

and

$$N_0(x) = 1 \qquad N_1(x) = 1 \qquad N_2(x) = x - 2 \qquad N_3(x) = 6 - 2x \qquad N_4(x) = 12 - 6x + x^2 \qquad \cdots$$

The fourth Thiele convergent, in particular, is thus

$$r_4(x) = \frac{12 + 6x + x^2}{12 - 6x + x^2}$$

In order to determine $r_4^*(x)$, we note that since

$$T_5(x) = 16x^5 - 20x^3 + 5x$$

there follows

$$c_0 = 0 \qquad c_1 = \tfrac{5}{16} \qquad c_2 = 0 \qquad c_3 = -\tfrac{5}{4} \qquad c_4 = 0$$

Thus, with $n = 4$, (9.18.15) gives

$$\alpha_0 = 0 \qquad \alpha_1 = \tfrac{1}{192} \qquad \alpha_2 = 0 \qquad \alpha_3 = -\tfrac{1}{8} \qquad \alpha_4 = 0$$

and there follows

$$r_4^*(x) = \frac{12 + 6x + x^2 + (\varepsilon^2/8)(2 + x) + \varepsilon^4/192}{12 - 6x + x^2 + (\varepsilon^2/8)(2 - x) + \varepsilon^4/192}$$

Computation shows that the improvement in the approximation to e^x over the interval $[-\varepsilon, \varepsilon]$ afforded by $r_4^*(x)$ is by no means limited to "small" values of ε. For $\varepsilon = 1$, the maximum deviation on $[-1, 1]$ is in fact reduced from about 4×10^{-3} to about 2×10^{-4}, while for $\varepsilon = \frac{1}{2}$ the maximum deviation on $[-\frac{1}{2}, \frac{1}{2}]$ is reduced from about 7×10^{-5} to about 4×10^{-6}.

Once such an approximation has been obtained, iterative processes (again based on an equal ripple theorem) exist for the purpose of generating a sequence of additional improvements hopefully converging to an ideal *minimax* approximation. As might be expected from the fact that the data enter into the approximation nonlinearly, the calculation is involved and convergence can be guaranteed only when the initial approximation is sufficiently close to the ideal one. (See references listed in Sec. 9.19.)

A simple alternative "trial-and-error" process consists of first determining a rational approximation $r_n(x)$ of the desired type which agrees with $f(x)$ at $n + 1$ selected points inside the interval of interest, by the method of Sec. 9.14 or otherwise, and of calculating and plotting values of the deviation $f(x) - r_n(x)$. Then, as suggested in Sec. 9.9 for the polynomial case, the points of collocation may be successively adjusted for the purpose of more nearly attaining a deviation with at least $n + 2$ maximum extrema of alternating signs in the relevant interval.

Similar methods are applicable in the more general case when the rational function is to be such that the degree of the numerator does not exceed m and the degree of the denominator does not exceed k, where m and k are prescribed. This situation reduces to the preceding one, with $n = m + k$, when $k = m$ or when $k = m - 1$. In the general case, the presence of at least $m + k + 2$

maximum extrema with alternating signs in the relevant minimax approximation is guaranteed unless that approximation happens to be such that the degree of its numerator is smaller than m and that *also* the degree of its denominator is smaller than k. In this rather unusual event, the guaranteed number is $m + k + 2$ *reduced* by the *lesser* of the degree defects of the numerator and denominator.

9.19 Supplementary References

For more elaborate techniques of discrete harmonic analysis, see Whittaker and Robinson [1944], Danielson and Lanczos [1942], Willers [1950], Pollak [1947, 1949], and Krylov and Kruglikova [1969].

An efficient method of evaluating the coefficients in a complex discrete Fourier representation (or of evaluating a discrete Fourier transform) was devised by Cooley and Tukey [1965] and has been adapted to other related calculations.

The method of de Prony [1795] for determining exponential approximations is applied in Whittaker and Robinson [1944] to the numerical solution of certain integral equations. A related method is presented in Lanczos [1956]. Whittaker and Robinson [1944] also treat other methods for determining periodicities, giving collateral references, and Lanczos [1956] advocates a numerical "spectroscopic analysis" method for this purpose.

Optimum formulas for interpolation and approximate integration involving equally spaced abscissas are defined and studied by Sard [1949] and by Meyers and Sard [1950*a*, 1950*b*].

Approximation by Chebyshev polynomials was given its modern impetus by Lanczos [1938, 1952, 1956]. Tables of coefficients for approximating various functions are given by Clenshaw [1962], together with a treatment of associated theory and techniques, and by Clenshaw and Picken [1966] and Luke [1969]. See also Minnick [1957], Elliott [1964], and Fox and Parker [1968].

Texts dealing with the general theory of approximation and containing treatments of some or all of the topics considered in this chapter include Achieser [1956], P. J. Davis [1963], Sard [1963], Rice [1964, 1969], Cheney [1966], and Rivlin [1969]. Cheney, in particular, presents a 10-page section of bibliographic notes tracing the historical development of the principal theories and techniques, as well as an extensive bibliography. Classical treatments, available in reprint, include de la Vallée Poussin [1919] and Bernstein [1926].

Spline approximation in its present form apparently began with Schoenberg [1947]. The rapidly expanding literature in this area includes the works of Ahlberg, Nilson, and Walsh [1967], Greville [1969], Hayes [1970], Curtis

[1970], and Schoenberg [1971] among many others. Explicit forms of spline approximations, related to results of Sec. 9.11, are given by Schoenberg [1971]. The results of the error analysis in Sec. 9.12 are in accordance with corresponding results in Curtis [1970].

For general treatments of continued fractions, see Wall [1967] and Perron [1950]. Classic references include Padé [1892] and Thiele [1909]. See also Milne-Thomson [1951], Nörlund [1954], and Olds [1962].

The literature on minimax (and near minimax) approximation by polynomials or rational functions is (to quote someone) everywhere dense. The contributions of Maehly, Rémès, Loeb, and many others, as well as those of the several authors, may be found in Achiezer [1956], Cheney [1966], and most of the other general references cited above, together with useful bibliographies. See also Ralston [1965].

For associated compilations and techniques relative to computer approximations, see Hastings [1957], Kogbetliantz [1960], Hart et al. [1965], Handscomb [1966], and Fike [1968].

PROBLEMS

Section 9.2

1 If $f(x) = \sin x$ when $\sin x \geqq 0$ and $f(x) = 0$ when $\sin x \leqq 0$, obtain the expansion

$$f(x) = \frac{1}{\pi} + \tfrac{1}{2}\sin x - \frac{2}{\pi}\left(\frac{\cos 2x}{2^2 - 1} + \frac{\cos 4x}{4^2 - 1} + \frac{\cos 6x}{6^2 - 1} + \cdots\right)$$

Also compare graphically each of the three least-squares approximations corresponding to retention of harmonics through the second, fourth, and sixth with the true function over $[-\pi, \pi]$.

2 Obtain the expansion

$$x = \frac{\pi}{2} - \frac{4}{\pi}\left(\frac{\cos x}{1^2} + \frac{\cos 3x}{3^2} + \frac{\cos 5x}{5^2} + \cdots\right) \qquad (0 \leqq x \leqq \pi)$$

Assuming the validity of this expansion, show that the series represents a *triangular-wave* function of period 2π which coincides with $f(x) = |x|$ when $|x| \leqq \pi$, and sketch that function. Also compare graphically each of the three least-squares approximations corresponding to retention of harmonics through the first, third, and fifth, with the true function over $[0, \pi]$.

3 Obtain the expansion

$$x(\pi - x) = \frac{8}{\pi}\left(\frac{\sin x}{1^3} + \frac{\sin 3x}{3^3} + \frac{\sin 5x}{5^3} + \cdots\right) \qquad (0 \leqq x \leqq \pi)$$

and sketch the periodic function represented by the expansion. Also compare graphically each of the first three distinct least-squares approximations with the true function over $[0, \pi]$.

4 Show that the *squarewave* function $f(x)$, which is of period 2π and which is such that $f(x) = -1$ when $-\pi < x < 0$ and $f(x) = +1$ when $0 < x < \pi$, possesses the expansion

$$f(x) = \frac{4}{\pi}\left(\frac{\sin x}{1} + \frac{\sin 3x}{3} + \frac{\sin 5x}{5} + \cdots\right)$$

and verify that the expansion reduces to the average of the right- and left-hand limits of the function at its points of discontinuity. Also compare graphically the first three distinct least-squares approximations with the true function over $[-\pi, \pi]$. [Let $f(x) = 0$ when $x = 0, \pm\pi$.]

5 Show that the mean squared error associated with the approximation (9.2.3) over $[-\pi, \pi]$ is given by

$$\frac{1}{2\pi}\int_{-\pi}^{\pi} f^2\,dx - [a_0^2 + \tfrac{1}{2}\sum_{k=1}^{n}(a_k^2 + b_k^2)]$$

whereas the corresponding quantities associated with (9.2.8) and (9.2.10) over $[0, \pi]$ are given respectively by

$$\frac{1}{\pi}\int_0^{\pi} f^2\,dx - (a_0^2 + \tfrac{1}{2}\sum_{k=1}^{n} a_k^2)$$

and

$$\frac{1}{\pi}\int_0^{\pi} f^2\,dx - \tfrac{1}{2}\sum_{k=1}^{n} b_k^2$$

Also use these results to calculate the RMS errors in each of the least-squares approximations considered in Probs. 1 to 4.

6 Suppose that $y(x)$ is to be of period 2π and is to satisfy the differential equation

$$\alpha y^{\text{iv}}(x) + \beta y''(x) + \gamma y(x) = f(x)$$

where α, β, and γ are constants and $f(x)$ is a specified function of period 2π. Show that if an approximation to $y(x)$ is assumed in the form

$$y(x) \approx a_0 + \sum_{k=1}^{n}(a_k \cos kx + b_k \sin kx)$$

and is introduced into the differential equation, and if the coefficients are determined in such a way that the period integral of the square of the difference between the two sides of the resultant equation is as small as possible, then there follows

$$\gamma a_0 = \frac{1}{2\pi}\int_{-\pi}^{\pi} f(x)\,dx$$

$$(\alpha k^4 - \beta k^2 + \gamma)a_k = \frac{1}{\pi}\int_{-\pi}^{\pi} f(x)\cos kx\,dx \qquad (k \geqq 1)$$

$$(\alpha k^4 - \beta k^2 + \gamma)b_k = \frac{1}{\pi}\int_{-\pi}^{\pi} f(x)\sin kx\,dx$$

Also use this result to write down a series expansion of the solution of the equation $y''(x) + \lambda y(x) = f(x)$ which is of period 2π, when $f(x)$ is the squarewave function defined in Prob. 4 and λ is a constant such that $\lambda \neq 1^2, 3^2, \ldots, (2k+1)^2, \ldots$.

Section 9.3

7 By noticing that the relevant series is geometric, show that

$$\sum_{r=-N+1}^{N} e^{ir\alpha} = \begin{cases} e^{i\alpha/2} \dfrac{\sin N\alpha}{\sin \alpha/2} & (\alpha \neq 2\nu\pi) \\ 2N & (\alpha = 2\nu\pi) \end{cases}$$

where ν is any integer, and hence that

$$\sum_{r=-N+1}^{N} \cos r\alpha = \begin{cases} \cot \dfrac{\alpha}{2} \sin N\alpha & (\alpha \neq 2\nu\pi) \\ 2N & (\alpha = 2\nu\pi) \end{cases}$$

and

$$\sum_{r=-N+1}^{N} \sin r\alpha = \sin N\alpha$$

8 By taking $\alpha = m\pi/N$ in the results of Prob. 7, where m is an integer, and writing $x_r = r\pi/N$, show that

$$\sum_{r=-N+1}^{N} \cos mx_r = \begin{cases} 0 & (m \neq 2\nu N) \\ 2N & (m = 2\nu N) \end{cases}$$

and

$$\sum_{r=-N+1}^{N} \sin mx_r = 0$$

Then, by using the identity $2 \cos jx_r \cos kx_r = \cos (j-k)x_r + \cos (j+k)x_r$ and two similar identities, deduce the results of Eqs. (9.3.2) and (9.3.3).

9 Use the results of Probs. 7 and 8 to show that, if the range of summation is changed to $r = 0, 1, \ldots, N$ in (9.3.2) and (9.3.3), and if the weighting function w_r is inserted in each summand, where

$$w_r = \begin{cases} \frac{1}{2} & (r = 0) \\ 1 & (r = 1, 2, \ldots, N-1) \\ \frac{1}{2} & (r = N) \end{cases}$$

then the right-hand members of all formulas in those equations are to be divided by the factor 2.

10 Suppose that the ordinates $f_{-N+1}, \ldots, f_{N-1}, f_N$ are empirical, with $f_{-N} \equiv f_N$, and are subject to independent normal error distributions with zero means and a common RMS value σ. Show that the corresponding RMS errors associated with the coefficients calculated from (9.3.6) are given by

$$(\delta A_0)_{\text{RMS}} = (\delta A_N)_{\text{RMS}} = \frac{\sigma}{\sqrt{2N}}$$

$$(\delta A_k)_{\text{RMS}} = (\delta B_k)_{\text{RMS}} = \frac{\sigma}{\sqrt{N}} \qquad (k = 1, \ldots, N-1)$$

11 Suppose that the ordinates $F_0, F_1, \ldots, F_N$ and $G_1, G_2, \ldots, G_{N-1}$ are empirical and are subject to independent normal error distributions with zero means and a common RMS value σ. Use the result of Prob. 9 to show that the corresponding RMS errors associated with the coefficients of the cosine approximation to $F(x)$ over $[0, \pi]$, calculated from (9.3.11), are given by

$$(\delta A_0)_{\mathrm{RMS}} = (\delta A_N)_{\mathrm{RMS}} = \sqrt{\frac{2N-1}{2N^2}}\,\sigma$$

$$(\delta A_k)_{\mathrm{RMS}} = \sqrt{\frac{2(N-1)}{N^2}}\,\sigma \qquad (k = 1, \ldots, N-1)$$

whereas those associated with the coefficients of the sine approximation to $G(x)$ over $[0, \pi]$ are given by

$$(\delta B_k)_{\mathrm{RMS}} = \sqrt{\frac{2}{N}}\,\sigma$$

12 The following approximate values of a function $f(x)$, known to be of period 2π, are available:

x	$-\pi$	$-5\pi/6$	$-2\pi/3$	$-\pi/2$	$-\pi/3$	$-\pi/6$
$f(x)$	2.077	0.278	−1.014	−0.716	0.051	0.277

x	0	$\pi/6$	$\pi/3$	$\pi/2$	$2\pi/3$	$5\pi/6$	π
$f(x)$	1.015	3.031	4.759	4.680	3.689	3.032	2.077

Assuming first that the given values are correct to the number of places given, determine a trigonometric function of period 2π which agrees with $f(x)$, to those places, at all tabular points. If it is known that the magnitude of the errors in all given values cannot exceed 0.005, and that all higher harmonics are negligible, determine how many of the calculated harmonics can be neglected if the total error is nowhere to exceed about 0.01. If, instead, it is known only that the approximate ordinates are subject to error distributions with an RMS value of about 0.0025, and if all higher harmonics are again assumed to be negligible, use the result of Prob. 10 to estimate the RMS errors in the calculated coefficients.

13 Determine a seven-term cosine approximation of period 2π to the function $f(x)$ of Prob. 12 over $[0, \pi]$, and analyze the results as in Prob. 12.

14 Using the following data, determine a five-term sine approximation of period 2π to the function $f(x)$ over $[0, \pi]$, and analyze the results as in Prob. 12:

x	0	$\pi/6$	$\pi/3$	$\pi/2$	$2\pi/3$	$5\pi/6$	π
$f(x)$	0	1.136	0.864	4.002	6.059	2.868	0

Section 9.4

15 Repeat the calculations of the illustrative example using the given four-place values of $f(x)$. Also determine an approximation to the limiting value of $f(x)$ as $x \to \infty$ and compare it with the true value.

16 Suppose that approximate data are available for a function known to be of the form $F(t) = Ae^{bt}$, where A and b are unknown constants. Show that the change of notation

$$\log F(t) = f(t) \qquad \log A = c$$

leads to the linear relation $f(t) = c + bt$, after which the least-squares methods of Sec. 7.3, are available for the determination of c and b, and hence of A and b. Apply this method in the case when the following empirical values of $F(t)$ are given, and determine whether the result is consistent with the hypothetical fact that the errors in the given data do not exceed 0.0002 in magnitude:

t	9	12	15	18	21	24	27
$F(t)$	0.5820	0.4622	0.3672	0.2920	0.2320	0.1843	0.1463

17 Increase each of the given ordinates in Prob. 16 by unity, and suppose that the resultant ordinates correspond to a function $G(t)$. Show first that the assumption $G(t) \approx Ae^{bt}$ does not lead to an approximation consistent with the assumed error bounds. Then, assuming knowledge that the true function $G(t)$ is of the form $G(t) = A_0 + A_1e^{bt}$, but without making use of any other information, use Prony's method [with $x = (t - 9)/3$] to approximate A_0, A_1, and b.

18 Given the modified data of Prob. 17, assume an approximation of the form $G(t) \approx A_0 + A_1e^{b_1t} + A_2e^{b_2t}$, and use Prony's method to determine the approximation, showing that a negative value of e^{b_2} is obtained, so that the third term is of alternating sign at successive tabular points, and hence presumably is to be interpreted as "noise" in this case. (Take care to retain sufficiently many digits.)

Section 9.5

19 Repeat the calculations of the illustrative example, using given values of $f(x)$ rounded correctly to five decimal places.

20 The following data represent observed values of a certain physical quantity:

t	0	0.05	0.10	0.15	0.20
$F(t)$	0	0.954	1.527	1.502	0.913

t	0.25	0.30	0.35	0.40	0.45	0.50
$F(t)$	0.030	−0.752	−1.090	−0.833	−0.091	0.814

The errors in measurement are known not to exceed 0.001. Theory predicts that the true function $F(t)$ should satisfy the differential equation $MF''(t) + kF(t) = 0$, where $M \doteq 1.40$ and $k \doteq 248$, and hence should be of period $P \doteq 0.472$. There is reason to believe that the difference $G(t)$ between the true function and the function actually subject to measurement satisfies an equation of the form $G''(t) + c^2G(t) = 0$ for some constant c. Investigate the plausibility of this conjecture, and approximate the period of the perturbation $G(t)$.

Section 9.6

21 Plot the function $\pi(x) = x(x^2 - \frac{1}{4})(x^2 - 1)$ in $[-1, 1]$, relevant to a five-point interpolation employing data at equally spaced points, together with the corresponding functions associated with data prescribed at the zeros of $P_5(x)$ and $T_5(x)$, on a common graph. Also determine the maximum and RMS values of each of these functions over $[-1, 1]$.

22 Use the Lagrange interpolation formula to determine three parabolic approximations to $f(x) = e^x$ over $[-1, 1]$, such that $y_1(x)$ agrees with $f(x)$ at $x = -1$, 0, and 1, $y_2(x)$ agrees with $f(x)$ at the zeros of $P_3(x)$, and $y_3(x)$ agrees with $f(x)$ at the zeros of $T_3(x)$. Show also that the errors can be expressed in the forms $x(x^2 - 1)e^{\xi_1}/6$, $x(x^2 - \frac{3}{5})e^{\xi_2}/6$, and $x(x^2 - \frac{3}{4})e^{\xi_3}/6$, respectively, where each ξ is in $[-1, 1]$. Calculate the actual errors in the three approximations for $x = -1.0(0.2)1.0$, plot them on a common graph, and compare them with respect to approximate maximum and RMS values.

Section 9.7

23 Derive (9.7.8) by first obtaining the intermediate results

$$\sum_{r=0}^{n} e^{i(2r+1)\alpha} = \begin{cases} e^{i(n+1)\alpha}\dfrac{\sin(n+1)\alpha}{\sin\alpha} & (\alpha \neq \nu\pi) \\ (-1)^\nu(n+1) & (\alpha = \nu\pi) \end{cases}$$

and

$$\sum_{r=0}^{n} \cos m\theta_r = \begin{cases} 0 & [m \neq 2\nu(n+1)] \\ (-1)^\nu(n+1) & [m = 2\nu(n+1)] \end{cases}$$

where $\alpha = m\pi/[2(n+1)]$ and $\theta_r = (2r+1)\pi/(2n+2)$, and then using the identity $\cos j\theta_r \cos k\theta_r = \frac{1}{2}\cos(j-k)\theta_r + \frac{1}{2}\cos(j+k)\theta_r$.

24 Determine, to four decimal places, the coefficients in the approximation

$$e^x \approx \sum_{k=0}^{5} c_k T_k(x) \qquad (|x| \leqq 1)$$

if the approximation is to be exact at the zeros of $T_6(x)$, and show that the magnitude of the error is smaller than $e/23040 \doteq 0.00012$ everywhere in $[-1, 1]$. Also, recalling that $|T_k(x)| \leqq 1$ on $[-1, 1]$, obtain upper bounds on the errors relevant to the least-squares approximations of degrees 2, 3, and 4, obtained by truncation, and use Eqs. (7.8.10) to express these approximations in explicit polynomial form.

Section 9.8

25 Determine two third-degree approximations to e^x over $[-1, 1]$, in addition to (9.8.17), by truncating the Maclaurin expansion of e^x instead with the x^4 term and with the x^6 term, and proceeding by the Lanczos method. Also compare the error bounds associated with these approximations with each other and with the corresponding approximation obtained in Prob. 24.

26 Economize the fifth-degree Maclaurin approximation to e^x to a third-degree approximation over the interval $[-\varepsilon, \varepsilon]$. Also specialize to $\varepsilon = \frac{1}{2}$ and $\varepsilon = \frac{1}{4}$, in each case comparing the approximate maximum error with that corresponding to the third-degree Maclaurin approximation. (Set $x = \varepsilon s$, so that $-1 \leqq s \leqq 1$.)

27 Obtain an approximation of the form

$$\cos x \approx A_0 + A_2 x^2 + A_4 x^4$$

with an error smaller than 5×10^{-5} over $[-1, 1]$.

28 Show that the polynomials $T_k(2x - 1)$ play the same role over $[0, 1]$ as do the polynomials $T_k(x)$ over $[-1, 1]$ and, with the abbreviation

$$\bar{T}_k \equiv \bar{T}_k(x) \equiv T_k(2x - 1)$$

obtain the relations

$$\bar{T}_0 = 1 \qquad \bar{T}_1 = 2x - 1 \qquad \bar{T}_2 = 8x^2 - 8x + 1$$
$$\bar{T}_3 = 32x^3 - 48x^2 + 18x - 1$$
$$\bar{T}_4 = 128x^4 - 256x^3 + 160x^2 - 32x + 1$$
$$\bar{T}_5 = 512x^5 - 1280x^4 + 1120x^3 - 400x^2 + 50x - 1$$

and

$$1 = \bar{T}_0 \qquad 2x = \bar{T}_0 + \bar{T}_1 \qquad 8x^2 = 3\bar{T}_0 + 4\bar{T}_1 + \bar{T}_2$$
$$32x^3 = 10\bar{T}_0 + 15\bar{T}_1 + 6\bar{T}_2 + \bar{T}_3$$
$$128x^4 = 35\bar{T}_0 + 56\bar{T}_1 + 28\bar{T}_2 + 8\bar{T}_3 + \bar{T}_4$$
$$512x^5 = 126\bar{T}_0 + 210\bar{T}_1 + 120\bar{T}_2 + 45\bar{T}_3 + 10\bar{T}_4 + \bar{T}_5$$

29 Use the notation and results of Prob. 28 to obtain a third-degree polynomial approximation to e^{-x} with an error smaller than 0.001 in magnitude over $[0, 1]$.

30 After expressing a five-term (eighth-degree) truncation of the series representation

$$\frac{\sin x}{x} = 1 - \frac{x^2}{3!} + \frac{x^4}{5!} - \cdots$$

in terms of the polynomials $\bar{T}_k(x^2)$ $(0 \leqq k \leqq 4)$, by use of the results of Prob. 28, obtain a polynomial approximation to $\sin x$, involving as few terms as possible, with an error smaller than 10^{-5} over $[0, 1]$.

31 The modified Bessel function $K_0(x)$ possesses the asymptotic expansion

$$\sqrt{\frac{2x}{\pi}}\, e^{-x} K_0(x) \sim 1 - \frac{1^2}{1!\,8x} + \frac{(1 \cdot 3)^2}{2!\,(8x)^2} - \frac{(1 \cdot 3 \cdot 5)^2}{3!\,(8x)^3} + \cdots$$

in which the error of truncation in the right-hand member is smaller than the first neglected term. Show that truncation with the x^{-5} term corresponds to an error smaller than 0.0008 for all $x \geqq 3$. Then, after expressing the result of that truncation in terms of the functions $\overline{T}_k(3/x)$ $(0 \leqq k \leqq 5)$, obtain an approximation of the same type over $[3, \infty)$ involving as few terms as possible, with an error not exceeding 0.005 over that interval.

Section 9.9

32 Obtain a *linear* approximation to e^x on $[-1, 1]$ which is near to the minimax approximation, in the sense that the magnitudes of the three error extrema differ from each other by less than 0.1, by estimating suitable collocation abscissas x_0 and x_1, determining the resultant extrema, and modifying x_0 and x_1. (For the purpose of this problem, avoid shortcuts suggested by its simplicity.)

33 Proceed as in Prob. 32, but use a method which deals directly with the extrema.

34 Determine the minimax approximation involved in Prob. 32 analytically by use of calculus.

Section 9.10

35 Derive the equation of the spline approximation to $f(x)$ in the subinterval $[x_{k-1}, x_k]$, in the form

$$s(x) = \frac{(x_k - x)^3}{6h_k} s''_{k-1} + \frac{(x - x_{k-1})^3}{6h_k} s''_k + \frac{x_k - x}{h_k}\left(\frac{f_{k-1}}{h_k} - \frac{h_k}{6} s''_{k-1}\right) + \frac{x - x_{k-1}}{h_k}\left(\frac{f_k}{h_k} - \frac{h_k}{6} s''_k\right)$$

Also show that

$$s'(x_k-) = \frac{h_k}{6}(s''_{k-1} + 2s''_k) + \frac{f_k - f_{k-1}}{h_k}$$

$$s'(x_k+) = -\frac{h_{k+1}}{6}(2s''_k + s''_{k+1}) + \frac{f_{k+1} - f_k}{h_{k+1}}$$

and deduce (9.10.6).

36 Show that the end conditions (9.10.11)

$$s''(a) = s''(b) = 0$$

are equivalent to the conditions

$$s'_0 = -\frac{1}{2}s'_1 + \frac{3}{2}\frac{f_1 - f_0}{h_1}$$

$$s'_n = -\frac{1}{2}s'_{n-1} + \frac{3}{2}\frac{f_n - f_{n-1}}{h_n}$$

Also, in the case when $n = 3$ and the spacing is uniform, show that these end conditions lead to the spline derivative formulas

$$s'_0 = \frac{1}{15h}(-19f_0 + 24f_1 - 6f_2 + f_3)$$

$$s'_1 = \frac{1}{15h}(-7f_0 - 3f_1 + 12f_2 - 2f_3)$$

$$s'_2 = \frac{1}{15h}(2f_0 - 12f_1 + 3f_2 + 7f_3)$$

$$s'_3 = \frac{1}{15h}(-f_0 + 6f_1 - 24f_2 + 19f_3)$$

37 In the case when $n = 3$ and the spacing is uniform, show that the end conditions (9.10.12)

$$s'(a) = f'(a) \qquad s'(b) = f'(b)$$

lead to the spline derivative formulas

$$s'_0 = f'_0$$

$$s'_1 = \frac{1}{5h}(-4f_0 + f_1 + 4f_2 - f_3) + \tfrac{1}{15}(-4f'_0 + f'_3)$$

$$s'_2 = \frac{1}{5h}(f_0 - 4f_1 - f_2 + 4f_3) + \tfrac{1}{15}(f'_0 - 4f'_3)$$

$$s'_3 = f'_3$$

38 In the case when $n = 3$ and the spacing is uniform, show that the *periodicity* end conditions (9.10.13) lead to the spline derivative formulas

$$s'_0 = \frac{1}{h}(f_1 - f_2) \equiv \frac{1}{h}(f_1 - f_{-1})$$

$$s'_1 = \frac{1}{h}(f_2 - f_0)$$

$$s'_2 = \frac{1}{h}(f_0 - f_1) \equiv \frac{1}{h}(f_3 - f_1)$$

$$s'_3 = s'_0$$

Section 9.11

39 Deal with the difference equation

$$u_{k+1} + 4u_k + u_{k-1} = p_k$$

by the following steps.

(*a*) With the abbreviation

$$P_k = \sum_{r=1}^{k} (-1)^r p_r \frac{\sinh (k - r)\alpha}{\sinh \alpha}$$

verify that $u_k = (-1)^{k+1} P_k$ is a particular solution if $\sinh \alpha = \sqrt{3}$, so that, by superimposing the solution when $p_k \equiv 0$, the general solution is obtained in the form

$$u_k = (-1)^k (A \sinh k\alpha + B \cosh k\alpha - P_k)$$

(*b*) Show that when A and B are expressed in terms of u_0 and u_n, there follows

$$(-1)^k u_k = \frac{1}{\sinh n\alpha} [u_0 \sinh (k - n)\alpha + (-1)^n u_n \sinh k\alpha + P_n \sinh k\alpha - P_k \sinh n\alpha]$$

(*c*) Show that

$$(P_n \sinh k\alpha - P_k \sinh n\alpha) \sinh \alpha = \sum_{r=1}^{k} (-1)^r p_r \sinh r\alpha \sinh (n - k)\alpha + \sum_{r=k+1}^{n} (-1)^r p_r \sinh k\alpha \sinh (n - r)\alpha$$

(*d*) Deduce (9.11.8) and (9.11.9) from (9.11.4).

40 When $n = 4$, show that the matrix $[G_{kr}]$ is of the form

$$\frac{1}{56}\begin{bmatrix} 15 & 4 & 1 \\ 4 & 16 & 4 \\ 1 & 4 & 15 \end{bmatrix}$$

and deduce that when the end conditions

$$s'_0 = f'_0 \qquad s'_n = f'_n$$

are imposed, the five-point derivative formulas at the nodes are

$$-s'_1 = \frac{1}{56}\left[15 f'_0 + f'_4 - \frac{6}{h}(15\mu\delta f_1 - 4\mu\delta f_2 + \mu\delta f_3)\right]$$

$$s'_2 = \frac{1}{14}\left[f'_0 + f'_4 - \frac{6}{h}(\mu\delta f_1 - 4\mu\delta f_2 + \mu\delta f_3)\right]$$

$$-s'_3 = \frac{1}{56}\left[f'_0 + 15 f'_4 - \frac{6}{h}(\mu\delta f_1 - 4\mu\delta f_2 + 15\mu\delta f_3)\right]$$

41 When $f(x) = x^2$, show that (9.11.4) becomes

$$s''_{k+1} + 4s''_k + s''_{k-1} = 12$$

and derive the results of (9.11.17) and (9.11.18). (Use results of Prob. 35 to relate values of s' to values of s''.)

42 For a periodic spline, show that

$$-s_1' = \frac{g_{n-1} + (-1)^n g_1}{g_n} s_0' + \frac{6}{h} \sum_{r=1}^{n} (-1)^r G_{1r} \mu\delta f_r$$

and

$$-s_{n+1}' = \frac{g_{n+1} + (-1)^n g_{-1}}{g_n} s_0' + \frac{6}{h} \sum_{r=1}^{n} (-1)^{n+r} G_{n+1,r} \mu\delta f_r$$

Then use the relations

$$g_{-1} = -g_1 = -1 \qquad G_{1r} = \frac{g_{n-r}}{g_n} \qquad G_{n+1,r} = -\frac{g_r}{g_n}$$

to deduce that

$$s_0' = s_n' = \frac{6}{h} \sum_{r=1}^{n} (-1)^r \frac{g_{n-r} + (-1)^n g_r}{g_{n+1} - g_{n-1} - (-1)^n 2} \mu\delta f_r$$

43 Use the result of Prob. 42 to obtain the following periodic-spline derivative formulas:

$$n = 3 \qquad s_0' = \frac{1}{h}(f_1 - f_2) \equiv \frac{1}{h}(f_1 - f_{-1})$$

$$n = 4 \qquad s_0' = \frac{3}{4h}(f_1 - f_3) \equiv \frac{3}{4h}(f_1 - f_{-1})$$

$$n = 5 \qquad s_0' = \frac{3}{11h}(f_1 - 3f_2 + 3f_3 - f_4) \equiv \frac{3}{11h}(3f_{-2} - f_{-1} + f_1 - 3f_2)$$

$$n = 6 \qquad s_0' = \frac{1}{5h}(4f_1 - f_2 + f_4 - 4f_5) \equiv \frac{1}{5h}(f_{-2} - 4f_{-1} + 4f_1 - f_2)$$

[In each case the expression for the spline derivative at successive nodes is obtained by cyclic permutation of the subscripts and accordingly is of the same *form* at each node (see Prob. 38). Notice that in each case $h = (b - a)/n = P/n$, where P is the period of the spline.]

Section 9.12

44 Supply the omitted steps in the determination of A_2, A_4, B_2, and B_4 in (9.12.3).

45 Obtain the spline approximation to $f(x) = x^4$ over the interval $[a, b] = [0, 2h]$, so that $n = 2$, with the auxiliary conditions $s_0' = f'(0)$ and $s_2' = f'(2h)$, in the form

$$s(x) = \begin{cases} 2hx^3 - h^2x^2 & (0 \leqq x \leqq h) \\ 6hx^3 - 13h^2x^2 + 12h^3x - 4h^4 & (h \leqq x \leqq 2h) \end{cases}$$

Verify also that, in this case, $s(x)$ is identical with the third-degree Hermite interpolation polynomial in each subinterval.

46 Verify the validity of the error formulas (9.12.4) to (9.12.7) in the special case considered in Prob. 45. (Notice that here no end effects are present and the symbol $\sim$ can be replaced by $=$ in each formula.)

Section 9.13

47 Obtain the following spline derivative formulas corresponding to the auxiliary conditions $s_0'' = s_n'' = 0$:

$$n = 2 \qquad s_0' = \frac{1}{4h}(-5f_0 + 6f_1 - f_2)$$

$$s_1' = \frac{1}{2h}(-f_0 + f_2)$$

$$s_2' = \frac{1}{4h}(f_0 - 6f_1 + 5f_2)$$

$$n = 4 \qquad s_0' = \frac{1}{56h}(-15f_0 + 34f_1 - 24f_2 + 6f_3 - f_4)$$

$$s_1' = \frac{1}{28h}(-13f_0 - 6f_1 + 24f_2 - 6f_3 + f_4)$$

$$s_2' = \frac{1}{8h}(f_0 - 6f_1 + 6f_3 - f_4)$$

$$s_3' = \frac{1}{28h}(-f_0 + 6f_1 - 24f_2 + 6f_3 + 13f_4)$$

$$s_4' = \frac{1}{56h}(f_0 - 6f_1 + 24f_2 - 34f_3 + 15f_4)$$

$$n = 8 \qquad s_4' = \frac{1}{112h}(f_0 - 6f_1 + 24f_2 - 90f_3 + 90f_5 - 24f_6 + 6f_7 - f_8)$$

48 Apply the five-point ($n = 4$) derivative formulas of Prob. 47 to $f(x) = \sin x$ on $[0, \pi]$ and compare the results with those given by (9.13.8), together with (9.13.1) and (9.13.2). [Notice that here $f''(x) = 0$ at both ends of the interval.]

49 Determine the approximations to each of the integrals

$$\int_0^{\pi} \sin x\, dx = 2 \qquad \int_0^{\pi/2} \sin x\, dx = 1$$

afforded by (9.13.13) and (9.13.15) and also by (9.13.17) and (9.13.19).

Section 9.14

50 Show that an inverted difference of the sum $u(x) + v(x)$ generally is not equal to the sum of the individual inverted differences, that multiplication of $f(x)$ by a constant corresponds to multiplication of the nth inverted difference of $f(x)$ by c

if n is even and by $1/c$ if n is odd, and that the addition of a constant to $f(x)$ does not affect its inverted differences.

51 Calculate the successive inverted differences $\phi_1[1, x]$, $\phi_2[1, 2, x]$, $\phi_3[1, 2, 3, x], \ldots$ for the functions x^2, x^{-2}, and $x - x^{-1}$. Then deduce the identity

$$x^2 = 1 + \cfrac{x-1}{\cfrac{1}{3} + \cfrac{x-2}{-12 + \cfrac{x-3}{-\cfrac{1}{3}}}}$$

together with corresponding identities in the other two cases, and verify their correctness.

52 Form an inverted-difference array and use it to determine a function defined by a finite continued fraction which takes on the following values. Also express the result as a simple fraction.

x	0	1	2	3	4	5
$f(x)$	$\frac{3}{7}$	$\frac{7}{9}$	$\frac{13}{11}$	$\frac{21}{13}$	$\frac{31}{15}$	$\frac{43}{17}$

53 Replace the given ordinates in Prob. 52 by their three-place rounded values and repeat the determination, retaining an appropriate number of digits in the intermediate calculations and obtaining a function defined as a simple fraction whose values at the six given points round to the three-place values used. Compare the result with that of Prob. 52.

54 Proceed as in Prob. 52 with the following data:

x	0	1	2	3	4	5
$f(x)$	0	$-\frac{1}{3}$	∞	$\frac{3}{5}$	$\frac{1}{3}$	$\frac{5}{21}$

Section 9.15

55 If $f(x) = u_m(x)/v_k(x)$ is an irreducible rational function with $u_m(x)$ of exact degree m and $v_k(x)$ of exact degree k, show that the $(2m - 1)$th inverted difference of $f(x)$ is constant if $m > k$ and that the $(2k)$th inverted difference is constant if $k \geqq m$. [Use (9.15.1).]

56 Determine a rational function of the form (9.15.1), with $n \leqq 4$, which takes on the values

x	1	2	3	4	5
$f(x)$	1	2	3	3	2

or prove that no such function exists.

57 Show that the substitution sequence

$$f(x) = w_0(x) \qquad w_{2k}(x) = w_{2k}(x_{2k}) + (x - x_{2k})w_{2k+1}(x)$$

$$w_{2k+1}(x) = w_{2k+1}(x_{2k+1}) + \frac{x - x_{2k+1}}{w_{2k+2}(x)}$$

generates the representation

$$f(x) = f(x_0) + b_1(x - x_0) + \cfrac{(x - x_0)(x - x_1)}{a_2 + b_3(x - x_2) + \cfrac{(x - x_2)(x - x_3)}{a_4 + b_5(x - x_4) + \cdots}}$$

where $a_{2k} \equiv w_{2k}(x_{2k})$ and $b_{2k+1} \equiv w_{2k+1}(x_{2k+1})$, and that the a's and b's can be determined as leading elements of columns of first divided differences alternating with columns of first inverted differences, in each case relative to corresponding elements and leading elements of the preceding column. Investigate the form of the nth convergent and illustrate the procedure by use of ordinates of the function.

$$f(x) = \frac{6 - 6x + 3x^2}{6 - 5x + 2x^2}$$

at $x = 0, 1, 2, 3, 4$, and 5. Determine what representation would result if the definitions of $w_{2k}(x)$ and $w_{2k+1}(x)$ were interchanged. Determine what substitution sequence (and what mixed difference table) would generate the representation

$$f(x) = f(x_0) + b_1(x - x_0) +$$
$$\frac{(x - x_0)(x - x_1)}{a_2 + b_3(x - x_2) + b_4(x - x_2)(x - x_3) + b_5(x - x_2)(x - x_3)(x - x_4) + \cdots}$$

and also what sequence (and table) would generate the representation

$$f(x) = f(x_0) + b_1(x - x_0) + \cfrac{(x - x_0)(x - x_1)}{a_2 + \cfrac{x - x_2}{a_3 + \cfrac{x - x_3}{a_4 + \ddots}}}$$

58 For the data

x	0	1	2	3	4	5
$f(x)$	1	1	$\frac{3}{5}$	$\frac{2}{5}$	$\frac{5}{17}$	$\frac{3}{13}$

(see Sec. 9.15), form a column of first *divided* differences followed by columns of successive *inverted* differences (as in Prob. 57) and deduce that

$$f(x) \approx 1 + \cfrac{x(x - 1)}{-5 + \cfrac{x - 2}{-\frac{1}{5} + \cfrac{x - 3}{30 + \cfrac{x - 4}{\frac{1}{5}}}}}$$

Also verify that the right-hand member is expressible as $(1 + x)/(1 + x^2)$.

59 For the rounded data

x	0.1	0.2	0.3	0.4
$f(x)$	9.967	4.933	3.233	2.365

verify that the column of first *inverted* differences is nearly linear. Then determine following columns of successive *divided* differences, as in Prob. 57, and deduce that

$$f(x) \approx 9.967 + \frac{x - 0.1}{-0.019865 - 0.09835(x - 0.2) + 0.0036(x - 0.2)(x - 0.3)}$$

Also determine an approximation to $f(0.15)$. [Here $f(x) = \cot x$. See the example in Sec. 9.16.]

Section 9.16

60 Use the recurrence relations (9.16.5) and (9.16.6) to obtain the successive convergents $r_k \equiv M_k/N_k$ relevant to the illustrative example [$f(x) = \cot x$, $x_0 = 0.1$, $x_1 = 0.2$, $x_2 = 0.3$, $x_3 = 0.4$] as follows:

$$9.967 \qquad \frac{-0.29799 + x}{-0.019865} \qquad \frac{1.03656 - 0.20100x}{0.00199 + x} \qquad \frac{-2.70932 - 0.05529x + x^2}{0.00059 - 2.7199x}$$

Also verify that the kth convergent agrees appropriately with $\cot x$ at the $k + 1$ appropriate points, evaluate the successive convergents at $x = 0.05, 0.15, 0.25, 0.35$, and 0.45, and compare the predicted values with the rounded true values.

61 By eliminating a_{k+1} between (9.16.5) and (9.16.6), show that

$$\frac{M_{k+1}}{N_{k+1}} - \frac{M_k}{N_k} = -(x - x_k)\frac{N_{k-1}}{N_{k+1}}\left(\frac{M_k}{N_k} - \frac{M_{k-1}}{N_{k-1}}\right)$$

and deduce the relation

$$\frac{M_{k+1}}{N_{k+1}} - \frac{M_k}{N_k} = (-1)^k \frac{(x - x_0)(x - x_1)\cdots(x - x_k)}{N_k N_{k+1}}$$

Thus, with the notation $r_k \equiv M_k/N_k$, show that the kth convergent of the continued fraction (9.14.9) can be written in the form

$$r_k(x) = a_0 + \sum_{n=1}^{k} (-1)^{n+1} \frac{(x - x_0)(x - x_1)\cdots(x - x_{n-1})}{N_{n-1}(x)N_n(x)}$$

62 Use the result of Prob. 61 to show that the kth convergent obtained in Prob. 60 can be obtained also by terminating the expansion

$$\cot x \approx 9.967 + \frac{x - 0.1}{-0.019865} - \frac{(x - 0.1)(x - 0.2)}{(-0.019865)(x + 0.00199)} + \frac{(x - 0.1)(x - 0.2)(x - 0.3)}{(x + 0.00199)(0.00059 - 2.7199x)} - \cdots$$

with the $(k + 1)$th term.

63 Use results of Prob. 61 to show that the error expression (9.16.8) can be written in the form

$$f(x) - r_k(x) = (-1)^k \frac{\pi_k(x)/N_k(x)}{(x - x_k)N_{k-1}(x) + N_k(x)\phi_{k+1}[x_0, x_1, \ldots, x_k, x]}$$

where $\pi_k(x) = (x - x_0)(x - x_1)\cdots(x - x_k)$. Also show that this relation can be rewritten in the form

$$f(x) - r_k(x) = (-1)^k \frac{\pi_k(x)/N_k(x)}{N_{k+1}(x) + N_k(x)\{\phi_{k+1}[x_0, x_1, \ldots, x_k, x] - a_{k+1}\}}$$

64 Assuming knowledge of the fact that cot x becomes infinite at $x = 0$, verify that the introduction of $x = 0$ as a fifth abscissa in the text example leads to the approximate information

$$\phi_3[0.1, 0.2, 0.3, 0] \approx -3.00$$

so that $\phi_3[0.1, 0.2, 0.3, x]$ varies from about -3.00 when $x = 0$ to about -2.70 when $x = 0.4$. Under the assumption that that function increases steadily over that interval, use the result of Prob. 63 to show that the error in the calculation of cot 0.15 from the third convergent would be less than about 0.0006 if no roundoff errors were involved.

65 Repeat the calculations of the text example, using the following improved (four-place) approximate ordinates:

x	0.1	0.2	0.3	0.4
cot x	9.9666	4.9332	3.2327	2.3652

66 Deal as in Probs. 60, 62, and 64 with the results of Prob. 65.

67 Determine values of the approximate convergents to $K_1(0.3) \doteq 3.056$, where $K_1(x)$ is a modified Bessel function, which correspond to the successive introduction of the following rounded values at $x = 0.2$, 0.4, 0.6, and 0.8:

x	0	0.2	0.4	0.6	0.8
$K_1(x)$	∞	4.776	2.184	1.303	0.862

68 Deal as in Probs. 60, 62, and 64 with the results of Prob. 67, obtaining as much evidence as possible with regard to the accuracy afforded by the convergent $r_3(x)$ over the interval [0, 1]. Also verify the conclusions deduced by comparing calculated values with the following additional rounded true values:

x	0.1	0.3	0.5	0.7	0.9	1.0
$K_1(x)$	9.854	3.056	1.656	1.050	0.7165	0.5098

69 Suppose that a rational function $R(x)$ is expressible in the form

$$R(x) = \frac{A_1|}{|x + B_1} + \frac{A_2|}{|x + B_2} + \cdots + \frac{A_N|}{|x + B_N}$$

where none of the A's are zero.

(*a*) Show that A_1, B_1, and A_2 can be determined by the following sequence of operations:
Multiply $R(x)$ by x, take limit ($= A_1$) as $x \to \infty$, divide by A_1, take reciprocal, subtract 1, multiply by x, take limit ($= B_1$) as $x \to \infty$, subtract B_1, multiply by x, take limit ($= A_2$) as $x \to \infty$.

(*b*) Deduce that we can write

$$A_k = p_k(\infty) \qquad B_k = q_k(\infty)$$

where

$$q_k(x) = x\left[\frac{p_k(\infty)}{p_k(x)} - 1\right]$$

and

$$p_{k+1}(x) = x[q_k(x) - q_k(\infty)]$$

with

$$p_1(x) = xR(x)$$

provided that each $A_k \neq 0$. (Notice that x may be replaced by $1/t$, so that the limits are taken at $t \to 0$.)

70 Generalize the result of Prob. 69 as follows, supposing that $R(x) = u_m(x)/v_k(x)$, where u_m and v_k are polynomials of degree m and k, respectively:

(*a*) If $m \geqq k$, show that we can write

$$R(x) = C_0x^{m-k} + C_1x^{m-k-1} + \cdots + C_{m-k} + \frac{A_1|}{|x + B_1} + \cdots$$

if each $A_k \neq 0$. (Use long division.)

(*b*) If $k \geqq m + 2$, show that we can write

$$R(x) = \cfrac{D_0}{x^{k-m} + E_1x^{k-m-1} + \cdots + E_{k-m} + \cfrac{A_1|}{|x + B_1} + \cdots}$$

if each $A_k \neq 0$. [First take the reciprocal of $R(x)$.]

(*c*) If $A_j = 0$ in any one of the preceding representations, show that we can write the portion beginning with A_1 in the form

$$\frac{A_1|}{|x + B_1} + \cdots + \frac{A_{j-1}|}{|x + B_{j-1}} + \frac{F_0|}{|w(x)}$$

where

$$w(x) = x^r + G_1x^{r-1} + \cdots + G_r + \frac{K_1|}{|x + L_1} + \cdots$$

for some integer $r \geqq 2$. [Take the reciprocal of $p_j(x)$ in Prob. 69 and proceed as in part (*b*).]

71 Illustrate the methods of Probs. 69 and 70 by obtaining the following representations:

(*a*) $$\frac{6x^2 + 48x + 114}{x^3 + 9x^2 + 27x + 19} = \frac{6|}{|x + 1} + \frac{2|}{|x + 3} + \frac{4|}{|x + 5}$$

(*b*) $$\frac{x^4 + 8x^3 + 20x^2 + 10x + 29}{x^2 + 8x + 19} = x^2 + 1 + \frac{2|}{|x + 3} + \frac{4|}{|x + 5}$$

(*c*) $$\frac{2x + 12}{x^3 + 9x^2 + 22x + 29} = \frac{2|}{|x^2 + 3x + 4} + \frac{5|}{|x + 6}$$

(*d*) $$\frac{x^3 + 6x^2 + 4x + 29}{x^4 + 8x^3 + 16x^2 + 40x + 76} = \frac{1|}{|x + 2} + \frac{3|}{|x^2 + 4} + \frac{5|}{|x + 6}$$

Section 9.17

72 Construct the following reciprocal-difference table from the given ordinates:

x	f	ρ_1	ρ_2	ρ_3	ρ_4
0	2				
		-2			
1	$\frac{3}{2}$		5		
		$-\frac{10}{7}$		-2	
2	$\frac{4}{5}$		$-\frac{1}{4}$		0
		$-\frac{10}{3}$		14	
3	$\frac{1}{2}$		$-\frac{1}{13}$		0
		$-\frac{34}{5}$		66	
4	$\frac{6}{17}$		$-\frac{1}{28}$		
		$-\frac{442}{37}$			
5	$\frac{7}{26}$				

From this table, rederive (9.14.15) and also obtain the representation

$$r_4(x) = \frac{4}{5} + \cfrac{x-2}{-\cfrac{10}{3} + \cfrac{x-3}{-\cfrac{21}{20} + \cfrac{x-1}{\cfrac{52}{3} + \cfrac{x-4}{\cfrac{1}{4}}}}} = \frac{2+x}{1+x^2}$$

corresponding to a zigzag difference path launched from $x = 2$.

73 Obtain the formal expansion

$$\log(1+x) = \frac{x|}{|1} + \frac{x|}{|2} + \frac{x|}{|3} + \frac{x|}{|1} + \frac{x|}{|5} + \cdots$$

Also show that $\phi_{2n}(x) = 2/n$ $(n \geqq 1)$ and $\phi_{2n+1}(x) = (2n+1)(1+x)$, so that the $(2n)$th coefficient is $2/n$ and the $(2n+1)$th is $2n+1$.

74 Use (9.16.5) and (9.16.6) to show that the leading convergents of the expansion obtained in Prob. 73 are given by

$$0 \qquad x \qquad \frac{2x}{2+x} \qquad \frac{6x+x^2}{6+4x} \qquad \frac{6x+3x^2}{6+6x+x^2}$$

$$\frac{30x+21x^2+x^3}{30+36x+9x^2} \qquad \frac{60x+60x^2+11x^3}{60+90x+36x^2+3x^3}$$

and determine the successive approximations to $\log \frac{1}{2}$ and $\log 2$. Also use the result of Prob. 61 to show that these convergents are the partial sums of the formal expansion

$$\log(1+x) = 0 + x - \frac{x^2}{2+x} + \frac{x^3}{(2+x)(6+4x)}$$
$$- \frac{x^4}{(6+4x)(6+6x+x^2)} + \frac{x^5}{(6+6x+x^2)(30+36x+9x^2)}$$
$$- \frac{x^6}{(30+36x+9x^2)(20+30x+12x^2+x^3)} + \cdots$$

75 Obtain the formal representation

$$\tan^{-1} x = \frac{\pi}{4} + \frac{x-1|}{|2} + \frac{x-1|}{|1} + \frac{x-1|}{|-6} + \frac{x-1|}{|-\frac{1}{4}} + \cdots$$

and express the first five convergents as simple fractions. Also use the result of Prob. 61 to show that the representation can be expressed in the form

$$\tan^{-1} x = \frac{\pi}{4} + \frac{x-1}{2} - \frac{(x-1)^2}{2(1+x)} - \frac{(x-1)^3}{4(1+x)(2+x)} + \frac{(x-1)^4}{4(2+x)(1+x+x^2)} + \cdots$$

76 Obtain the formal representations

$$(x+c^2)^{1/2} = c + \frac{x|}{|2c} + \frac{x|}{|2c} + \frac{x|}{|2c} + \cdots$$

and

$$(x+c^2)^{-1/2} = \frac{1}{c} - \frac{x|}{|2c^3} + \frac{x|}{|2/3c} + \frac{x|}{|18c^3} + \frac{x|}{|2/15c} + \cdots$$

where c is a positive constant.

77 With the notation

$$A_k = \frac{1}{k!} f^{(k)}(x_0)$$

use (9.17.15) to show that the first five Thiele coefficients in (9.17.6) are given by

$$a_0 = A_0 \qquad a_1 = \frac{1}{A_1} \qquad a_2 = -\frac{A_1^2}{A_2}$$

$$a_3 = \frac{A_2^2}{A_1(A_1A_3 - A_2^2)} \qquad a_4 = -\frac{(A_1A_3 - A_2^2)^2}{A_2(A_2A_4 - A_3^2)}$$

where $a_k \equiv \phi_k(x_0)$. Deduce also that if a function possesses the formal Taylor expansion

$$A_0 + A_1u + A_2u^2 + \cdots$$

and the formal Thiele expansion

$$a_0 + \frac{u|}{|a_1} + \frac{u|}{|a_2} + \cdots$$

then the leading a's can be calculated from the leading A's by use of these relations.

78 Show that the expansion (9.17.6) is nonexistent when $f(x) = \cos x$ and $x_0 = 0$, but that, when $F(x) = \cos x$, $G(x) = x^2$, and $x_0 = 0$, the leading terms of the expansion (9.17.23) are of the form

$$\cos x = 1 + \frac{x^2|}{|-2} + \frac{x^2|}{|-6} + \frac{x^2|}{|\frac{10}{3}} + \cdots$$

Also obtain this form from the result of Prob. 77 with $u = x^2$.

79 Obtain the leading terms of an expansion of $\sin x$ analogous to that obtained in Prob. 78 by taking $F(x) = (\sin x)/x$, $G(x) = x^2$, and $x_0 = 0$, and also by using the result of Prob. 77.

80 Determine the Padé entries $R_{4,0}(x)$, $R_{3,1}(x)$, $R_{2,2}(x)$, $R_{1,3}(x)$, and $R_{0,4}(x)$ for $f(x) = \log(1 + x)$ at $x = 0$, by use of (9.17.17) and (9.17.20) or otherwise. Also verify that $R_{2,2}(x)$ agrees with the corresponding ratio in Prob. 74.

81 For each of the approximations obtained in Prob. 80, tabulate the error by tenths for $-0.5 \leqq x \leqq 0.5$. Then compare their (approximate) maximum errors as well as their (approximate) RMS values over that interval.

82 The equation $x - e^{-x} = 0$ possesses a real root between $x = 0.5$ and 0.6. Making use of the fact that $e^{-0.6} \doteq 0.548812$, determine that root to five decimal places.

83 Determine the root of the equation $x^4 - 3x + 1 = 0$ between $x = 1.3$ and 1.4 to five decimal places.

Section 9.18

84 Proceed as in the example of Sec. 9.18 with the approximation

$$\log(1 + x) \approx \frac{6x + 3x^2}{6 + 6x + x^2}$$

obtaining a more nearly uniform approximation over the interval $[-\varepsilon, \varepsilon]$ where $0 < \varepsilon < 1$. Also plot the error on that interval when $\varepsilon = 0.25$ and when $\varepsilon = 0.5$ and, in each case, compare the approximate maximum error with that associated with the original approximation on that interval.

85 Proceed as in Prob. 84 with the approximation

$$\cos x \approx 1 + \frac{x^2|}{|-2} + \frac{x^2|}{|-6} + \frac{3x^2|}{|10}$$

(see Prob. 78), taking x^2 as the basic variable and making the error approximately proportional to $\overline{T}_4(s)$ when $x^2 = \varepsilon s$ $(0 \leqq s \leqq 1)$. Specialize when $\varepsilon = 0.5$ and when $\varepsilon = 1$.

10

NUMERICAL SOLUTION OF EQUATIONS

10.1 Introduction

This chapter summarizes a number of methods which are available for the numerical solution of sets of linear algebraic equations (Secs. 10.2 to 10.9), nonlinear algebraic or transcendental equations in general (Secs. 10.10 to 10.13), and nonlinear algebraic equations in particular (Secs. 10.14 to 10.19). With only minor exceptions, the treatments are independent of the content of preceding chapters.

10.2 Sets of Linear Equations

A brief summary of terminologies and of some elementary results, relative to solutions of sets of linear algebraic equations, is presented in this section. We suppose first that we are concerned with a set of n equations relating n unknowns $x_1, x_2, \ldots, x_n$, and expressed in the form

$$\begin{aligned} a_{11}x_1 + a_{12}x_2 + \cdots + a_{1n}x_n &= c_1 \\ a_{21}x_1 + a_{22}x_2 + \cdots + a_{2n}x_n &= c_2 \\ \cdots\cdots\cdots\cdots\cdots\cdots\cdots\cdots\cdots \\ a_{n1}x_1 + a_{n2}x_2 + \cdots + a_{nn}x_n &= c_n \end{aligned} \tag{10.2.1}$$

where the n^2 coefficients a_{ij} and the n right-hand members c_i are prescribed. Here a_{ij} represents the coefficient of x_j in the ith equation of the set.

The left-hand members may be specified by the square array of the coefficients, known as the *coefficient matrix*,

$$\mathbf{A} = \begin{bmatrix} a_{11} & a_{12} & \cdots & a_{1n} \\ a_{21} & a_{22} & \cdots & a_{2n} \\ \cdots & \cdots & \cdots & \cdots \\ a_{n1} & a_{n2} & \cdots & a_{nn} \end{bmatrix} \tag{10.2.2}$$

whereas the complete set may be specified by the *rectangular* array

$$\mathbf{M} = \left[\begin{array}{cccc|c} a_{11} & a_{12} & \cdots & a_{1n} & c_1 \\ a_{21} & a_{22} & \cdots & a_{2n} & c_2 \\ \cdots & \cdots & \cdots & \cdots & \cdots \\ a_{n1} & a_{n2} & \cdots & a_{nn} & c_n \end{array}\right] \tag{10.2.3}$$

known as the *augmented matrix* and formed by adjoining the column $\mathbf{c}$ of right-hand members to the n columns in (10.2.2).

The *minor* of any element a_{ij} in the coefficient matrix $\mathbf{A}$ is defined as the value of the *determinant*† of the square array obtained by deleting the ith row and the jth column of the coefficient matrix. The *cofactor* of a_{ij}, to be denoted here by A_{ij}, is defined as the result of changing the sign of the minor $i + j$ times. It is then true that, *if each element of any column of the coefficient matrix is multiplied by its cofactor, then the sum of these n products is the value of the determinant of that matrix.* Furthermore, *if each element of any column is multiplied by the cofactor of the corresponding element of any other column, then the sum of these n products is zero.* Both statements also remain true if the word *column* is replaced by *row* throughout.

These facts permit the direct elimination of all unknowns except an arbitrarily chosen one, say, x_k, from (10.2.1). For if we multiply each equation by the cofactor in $\mathbf{A}$ of the coefficient of x_k in that equation and add the results, they lead immediately to the consequence that the result is of the form

$$Dx_k = c_1A_{1k} + c_2A_{2k} + \cdots + c_nA_{nk} \tag{10.2.4}$$

where D represents the value of the determinant of the coefficient matrix. Thus we may deduce that, if $D \neq 0$, and if (10.2.1) possesses a solution, then that solution is unique, and each x_k $(k = 1, 2, \ldots, n)$ is obtained from (10.2.4) by division by D.

† The definition and elementary properties of determinants are assumed.

It is then easily shown, by direct substitution, that the x's so obtained actually do satisfy (10.2.1). For since (10.2.1) is of the form

$$\sum_{k=1}^{n} a_{ik}x_k = c_i \qquad (i = 1, 2, \ldots, n)$$

and since (10.2.4) is expressible in the form

$$Dx_k = \sum_{j=1}^{n} A_{jk}c_j \qquad (k = 1, 2, \ldots, n)$$

the result of substituting (10.2.4) into the left-hand member of the ith equation of (10.2.1) is

$$\frac{1}{D}\sum_{k=1}^{n} a_{ik} \sum_{j=1}^{n} A_{jk}c_j = \frac{1}{D}\sum_{j=1}^{n}\left(\sum_{k=1}^{n} a_{ik}A_{jk}\right) c_j$$

Since the inner sum in the second form is the sum of the products of the elements in the ith row of the coefficient matrix and the cofactors of corresponding elements in the jth row, it is equal to D when $j = i$ and is zero when $j \neq i$, so that this quantity properly reduces to c_i.

Since the right-hand member of (10.2.4) would reduce to D if $c_1, c_2, \ldots, c_n$ were replaced by $a_{1k}, a_{2k}, \ldots, a_{nk}$, it follows also that *the right-hand member of* (10.2.4) *is the value of the determinant of the matrix obtained from the coefficient matrix by replacing the column of coefficients of* x_k *by the column of right-hand members of* (10.2.1). If we denote the value of this determinant by D_k, the solution of (10.2.1) can be written in the simple form

$$x_k = \frac{D_k}{D} \qquad (k = 1, 2, \ldots, n) \qquad (10.2.5)$$

if $D \neq 0$. This result is known as *Cramer's rule.*

It is convenient to write

$$\frac{A_{ij}}{D} = \tilde{A}_{ij} \qquad (10.2.6)$$

when $D \neq 0$, and to speak of this ratio as the *reduced cofactor* of a_{ij} in the matrix $\mathbf{A}$. With this notation, the result of writing out (10.2.4) for $k = 1, 2, \ldots, n$ is of the form

$$\begin{aligned} x_1 &= \tilde{A}_{11}c_1 + \tilde{A}_{21}c_2 + \cdots + \tilde{A}_{n1}c_n \\ x_2 &= \tilde{A}_{12}c_1 + \tilde{A}_{22}c_2 + \cdots + \tilde{A}_{n2}c_n \\ &\cdots\cdots\cdots\cdots\cdots\cdots\cdots\cdots \\ x_n &= \tilde{A}_{1n}c_1 + \tilde{A}_{2n}c_2 + \cdots + \tilde{A}_{nn}c_n \end{aligned} \qquad (10.2.7)$$

This set of relations is thus the result of "inverting" the relations (10.2.1), when $D \neq 0$.

The array of coefficients of the right-hand members in (10.2.7),

$$\mathbf{A}^{-1} = \begin{bmatrix} \tilde{A}_{11} & \tilde{A}_{21} & \cdots & \tilde{A}_{n1} \\ \tilde{A}_{12} & \tilde{A}_{22} & \cdots & \tilde{A}_{n2} \\ \cdots & \cdots & \cdots & \cdots \\ \tilde{A}_{1n} & \tilde{A}_{2n} & \cdots & \tilde{A}_{nn} \end{bmatrix} \tag{10.2.8}$$

thus may be called the *inverse* of the coefficient matrix of (10.2.1) in that case, in the sense that whereas the array (10.2.2) specifies a transformation of the x's into the c's, the array (10.2.8) specifies the *inverse transformation* of the c's back into the x's.

We notice that the inverse of (10.2.2) can be obtained by first replacing each element of (10.2.2) by its reduced cofactor (cofactor divided by D), and then interchanging rows and columns.

In order to describe the situation in the case $D = 0$, as well as the case in which there are m equations relating n unknowns, where $m \neq n$, it is desirable to define the *rank* of any rectangular matrix with, say, m rows and n columns. From any such array, we may form a number of *square* subarrays, by deleting certain rows and/or columns, and compressing the remaining elements into a compact arrangement. The largest such subarrays would be of an order equal to the smaller of the integers m and n; the smallest would be of order 1 and would consist of only a single element. The rank of the given matrix is defined as the order of the *largest* such subarray *whose determinant is not zero.*

The basic theorem relevant to the *existence* of a solution of a set of m equations in n unknowns states simply that *the system possesses a solution if and only if the rank of the coefficient matrix is equal to the rank of the augmented matrix.*

Suppose that the ranks *are* equal, and let the common rank be r. Then r is not greater than the smaller of m and n. Now there exists at least one square $r \times r$ subarray in the coefficient matrix whose determinant is not zero. If one such subarray is found, and if $r < m$, then it can be proved that the $m - r$ equations whose coefficients are not involved in that subarray are implied by the other m equations and can be suppressed. If $r = n$, then n equations in n unknowns remain and can be solved uniquely by Cramer's rule. However, if $r < n$, then the $n - r$ unknowns in the r remaining equations whose coefficients are not involved in the subarray can each be assigned completely arbitrary values, after which the remaining r unknowns can be determined *in terms of them* by Cramer's rule. Thus, the general solution of the system then involves $n - r$

arbitrary constants, and the system is said to be of *defect* $n - r$, since it fails to determine a unique solution by permitting $n - r$ degrees of freedom.

The cases of most frequent practical interest are those represented by (10.2.1), in which $m = n$. In particular, if all the right-hand members c_i are *zeros*, the set is said to be *homogeneous*. In this case, the coefficient matrix and augmented matrix are automatically of the same rank, so that the set has a solution. In fact, one solution is then always the *trivial* one, for which $x_1 = x_2 = \cdots = x_n = 0$. If $D \neq 0$, this is the only possible solution. Usually this solution is of no interest, and the important homogeneous sets are those for which $D = 0$, so that the system admits nontrivial solutions. On the other hand, if at least one of the right-hand members is *not* zero, so that the trivial solution is not admissible, the interest centers mainly on the cases when a nontrivial solution *exists and is unique*, so that $D \neq 0$. Attention will be restricted principally to this last case in the following sections.

Before proceeding, it may be pointed out that, whereas the preceding results are of basic theoretical importance and are essential to an understanding of the nature of sets of linear equations, the use of Cramer's rule in actual *numerical* cases is generally highly inefficient, because of the excessive labor involved in the evaluation of $n + 1$ determinants of order n, unless n is small. Many other methods have been devised, certain of which are described in the following sections.

10.3 The Gauss Reduction

In principle, the simplest practical method of solving the set (10.2.1) is one due to Gauss. In this reduction, the first equation is first divided by a_{11} (assuming that $a_{11} \neq 0$) and the result is used to eliminate x_1 from all succeeding equations. Next, the modified second equation is divided by the coefficient of x_2 in that equation, and the result is used to eliminate x_2 from the succeeding equations, and so forth. After this elimination has been effected n times, when $D \neq 0$, the resultant set, which is equivalent to the original one except for the effects of any roundoffs committed, is of the form

$$\begin{aligned} x_1 + a'_{12}x_2 + a'_{13}x_3 + \cdots + a'_{1n}x_n &= c'_1 \\ x_2 + a'_{23}x_3 + \cdots + a'_{2n}x_n &= c'_2 \\ \cdots\cdots\cdots\cdots\cdots\cdots\cdots\cdots\cdots & \\ x_{n-1} + a'_{n-1,n}x_n &= c'_{n-1} \\ x_n &= c'_n \end{aligned} \tag{10.3.1}$$

where a'_{ij} and c'_i designate specific numerical values, and the solution is completed by working backward from the last equation, to obtain successively

$x_n, x_{n-1}, \ldots, x_1$. It is convenient to work with the augmented *arrays* at each stage, rather than to write out each equation in full. A renumbering of equations and/or variables will be *necessary* if, at any stage, the coefficient of x_k in the kth equation is zero, and it is *desirable* if that coefficient is small relative to other coefficients in that equation, in order that the effects of roundoff errors may be minimized. (See Sec. 10.5.)

The exceptional cases in which $D = 0$ would evidence themselves through the fact that after r such eliminations, where r is the rank of the coefficient matrix, all coefficients in the $n - r$ succeeding equations would vanish (except for the errors due to roundoff). Unless all right-hand members of those equations *also* were reduced to zeros at that stage, the original set would be unsolvable. If all those members were zeros, the final $n - r$ equations would have been reduced to the form $0 = 0$, and hence would be ignorable. The rth equation would express x_r as the sum of a specified constant and a certain linear combination of $x_{r+1}, \ldots, x_n$, and the process of back substitution would finally express $x_1, x_2, \ldots, x_r$ in similar forms.

In illustration, we consider the three equations

$$\begin{aligned} 9.3746x_1 + 3.0416x_2 - 2.4371x_3 &= 9.2333 \\ 3.0416x_1 + 6.1832x_2 + 1.2163x_3 &= 8.2049 \\ -2.4371x_1 + 1.2163x_2 + 8.4429x_3 &= 3.9339 \end{aligned} \qquad (10.3.2)$$

The reduced equations, corresponding to (10.3.1), are obtained in the form

$$\begin{aligned} x_1 + 0.32445x_2 - 0.25997x_3 &= 0.98493 \\ x_2 + 0.38624x_3 &= 1.00246 \\ x_3 &= 0.61448 \end{aligned} \qquad (10.3.3)$$

if five decimal places are retained, and the "back solution" yields the values

$$x_1 = 0.89643 \qquad x_2 = 0.76512 \qquad x_3 = 0.61448 \qquad (10.3.4)$$

A discussion of the reliability of these results is deferred to later sections.

This method is known as the *Gauss reduction*. A modification, known as the *Gauss-Jordan reduction*, consists of using the kth equation, at the kth stage, to eliminate x_k from the preceding equations as well as the following ones, so that the solution is obtained after n (or less) eliminations and no back substitution is necessary. However, simple analysis shows that the Gauss-Jordan reduction involves about $\frac{1}{2}n^3$ multiplications and divisions whereas the Gauss reduction involves about $\frac{1}{3}n^3$. Hence the latter is to be preferred on this basis.

In practice, only the coefficients are recorded at the successive stages of the reduction, the array corresponding to the first stage intermediate between (10.3.2) and (10.3.3) thus being of the form

$$\begin{array}{rrrr} 1 & 0.32445 & -0.25997 & 0.98493 \\ & 5.19635 & 2.00702 & 5.20914 \\ & 2.00702 & 7.80933 & 6.33427 \end{array}$$

The necessity of introducing new arrays at each of the intermediate stages is time-consuming and conducive to gross errors, particularly when many equations are involved. In the following section, a more efficient technique is described.

10.4 The Crout Reduction

A modification of the Gauss reduction, which has the advantages that it is particularly well adapted to the use of desk calculators and of large-scale computers, and that the recording (or storage) of auxiliary data (such as the repeated rewriting of modified equations or arrays) is minimized, is due to Crout.†

Starting with the augmented matrix $\mathbf{M}$ of the original system

$$\mathbf{M} = \left[\begin{array}{cccc|c} \underline{a_{11}} & a_{12} & \cdots & a_{1n} & c_1 \\ a_{21} & \underline{a_{22}} & \cdots & a_{2n} & c_2 \\ \cdots & \cdots & \cdots & \cdots & \cdots \\ a_{n1} & a_{n2} & \cdots & \underline{a_{nn}} & c_n \end{array}\right] \equiv [\mathbf{A} \mid \mathbf{c}] \qquad (10.4.1)$$

which may be considered as being partitioned into the coefficient array $\mathbf{A}$ and the $\mathbf{c}$ column, one determines next the elements of an *auxiliary matrix* $\mathbf{M}'$ of the same dimensions

$$\mathbf{M}' = \left[\begin{array}{cccc|c} \underline{a'_{11}} & a'_{12} & \cdots & a'_{1n} & c'_1 \\ a'_{21} & \underline{a'_{22}} & \cdots & a'_{2n} & c'_2 \\ \cdots & \cdots & \cdots & \cdots & \cdots \\ a'_{n1} & a'_{n2} & \cdots & \underline{a'_{nn}} & c'_n \end{array}\right] \equiv [\mathbf{A}' \mid \mathbf{c}'] \qquad (10.4.2)$$

which may be considered as being partitioned, in the same way, into a square array $\mathbf{A}'$ and a $\mathbf{c}'$ column. From this matrix, one then obtains a solution column

† See Crout [1941], in which modifications which are convenient when the coefficients are *complex* are also given. Similar methods are attributed to Doolittle, Banachiewicz, Cholesky, and others.

$\mathbf{x}$ whose elements are the required values of $x_1, \ldots, x_n$,

$$\mathbf{x} = \begin{bmatrix} x_1 \\ x_2 \\ \vdots \\ x_n \end{bmatrix} \tag{10.4.3}$$

Each entry in (10.4.2) and (10.4.3) is obtained from previously calculated data by a continuous sequence of operations, which can be effected without the tabulation of intermediate data.

In order to describe the reduction in a simple way, it is convenient to introduce two definitions. First, the *diagonal elements* (or elements on the *principal diagonal*) of a matrix are those elements whose row and column indices are equal, and which are underlined in (10.4.1) and (10.4.2). Second, the *inner product* of a row into a column, each containing n elements, is defined as the sum of the n products of corresponding elements, the elements of a row being ordered from left to right, and the elements of a column from head to foot.

The n elements of the first column of the auxiliary matrix (10.4.2) are determined first, then the remaining n of the $n + 1$ elements of the first row. Next, the remaining $n - 1$ elements of the second column and of the second row are determined, then the remaining $n - 2$ elements of the third column and third row, and the process is continued until the array is filled.

The elements of the first column of $\mathbf{M}'$ are identical with the corresponding elements of $\mathbf{M}$; the remaining elements of the first row of $\mathbf{M}'$ (to the right of the diagonal element a'_{11}) are each obtained by dividing the corresponding element of $\mathbf{M}$ by the diagonal element a'_{11}. Thus, for example, $a'_{11} = a_{11}$, $a'_{21} = a_{21}$, and $a'_{12} = a_{12}/a'_{11}$. From this stage onward, the elements of $\mathbf{M}'$ are calculated, *in the order specified above*, according to two rules:

1 Each element on or below the principal diagonal in $\mathbf{M}'$ is obtained by subtracting from the corresponding element in $\mathbf{M}$ the inner product of its own column and its own row in the square subarray $\mathbf{A}'$, with all uncalculated elements imagined to be zeros.

2 Each element to the right of the principal diagonal in $\mathbf{M}'$ is calculated by the same procedure, followed by a division by the diagonal element in its row of $\mathbf{M}'$.

Finally, the elements of the solution column $\mathbf{x}$ are determined in the order $x_n, x_{n-1}, \ldots, x_2, x_1$, from foot to head. The element x_n is identical with c'_n. Each succeeding element above it is obtained as the result of subtracting from the corresponding element of the $\mathbf{c}'$ column the inner product of its row in $\mathbf{A}'$ and the $\mathbf{x}$ column, with all uncalculated elements of the $\mathbf{x}$ column imagined to be zeros.

The preceding instructions are summarized by the equations

$$a'_{ij} = a_{ij} - \sum_{k=1}^{j-1} a'_{ik} a'_{kj} \qquad (i \geqq j) \tag{10.4.4}$$

$$a'_{ij} = \frac{1}{a'_{ii}} \left[a_{ij} - \sum_{k=1}^{i-1} a'_{ik} a'_{kj} \right] \qquad (i < j) \tag{10.4.5}$$

$$c'_i = \frac{1}{a'_{ii}} \left[c_i - \sum_{k=1}^{i-1} a'_{ik} c'_k \right] \tag{10.4.6}$$

and

$$x_i = c'_i - \sum_{k=i+1}^{n} a'_{ik} x_k \tag{10.4.7}$$

where i and j range from 1 to n when not otherwise restricted. The derivation of these relations is included in Appendix A. It is seen that the process defined by (10.4.7) is identical with the back solution of the Gauss reduction, which determines $x_n, x_{n-1}, \ldots, x_1$ from (10.3.1).

In the important case when the coefficient array **A** is *symmetric*, so that each element a_{ij} in **A** above the principal diagonal is identical with the symmetrically placed element a_{ji} below the diagonal ($a_{ij} = a_{ji}$), as in the system (10.3.2), it can be shown that each element a'_{ij} in $\mathbf{A}'$ above the principal diagonal is given by the result of dividing the symmetrically placed element a'_{ji} below the diagonal by the diagonal element a'_{ii}. This fact leads to a considerable reduction in labor in such cases, particularly when n is large, since then each element below the diagonal thus can be recorded as the dividend involved in the calculation of the symmetrically placed element before the required division by the diagonal element is effected.

It can be shown that the elements to the right of the diagonal in $\mathbf{M}'$ are identical with the elements which appear in corresponding positions in the augmented matrix of (10.3.1), obtained by the Gauss reduction. The compactness of the tabulation is a consequence of the fact that all necessary intermediate data are tabulated in the remaining spaces, which would normally be occupied by ones and zeros.

Indeed, it is easily verified that the elements of the matrix $\mathbf{A}'$ can be associated with a left triangular matrix **L** and the right triangular matrix **R** just described

$$\mathbf{L} = \begin{bmatrix} a'_{11} & 0 & \cdots & 0 \\ a'_{21} & a'_{22} & \cdots & 0 \\ \cdots & \cdots & \cdots & \cdots \\ a'_{n1} & a'_{n2} & \cdots & a'_{nn} \end{bmatrix} \qquad \mathbf{R} = \begin{bmatrix} 1 & a'_{12} & \cdots & a'_{1n} \\ 0 & 1 & \cdots & a'_{2n} \\ \cdots & \cdots & \cdots & \cdots \\ 0 & 0 & \cdots & 1 \end{bmatrix} \tag{10.4.8}$$

such that

$$\mathbf{A} = \mathbf{LR} \tag{10.4.9}$$

with the conventional definition of matrix multiplication.†

The kth diagonal element a'_{kk} is the number by which the kth equation would be divided in the Gauss reduction before that equation is used to eliminate x_k from succeeding equations. In consequence of this fact, it is true that *the value of the determinant of the original coefficient matrix* **A** *is the product of the diagonal elements of* **A**′. In this connection, it is noted that the Crout reduction comprises an efficient method of specifically *evaluating* a determinant. When there is no associated set of equations to be solved, the **c**, **c**′, and **x** columns naturally are omitted.

A continuous check against gross errors is afforded by adjoining to the columns of **M** an additional column, each of whose elements is the sum of the elements in the corresponding row of **M**. If this column is treated in the same way as the **c** column, corresponding check columns are obtained and adjoined to **M**′ and **x**. The check consists of the fact that each element in the **M**′ check column should *exceed by unity* the sum of the elements in its row of **M**′ which lie *to the right* of the diagonal element, whereas each element in the **x** check column should exceed by unity the corresponding element in the **x** column itself. (See Prob. 4.) The sudden appearance of an appreciable discrepancy will generally indicate the commission of a gross calculational error. Small discrepancies generally correspond to the effects of intermediate roundoffs, effected in the steps of the reduction, which can be removed by retaining additional significant figures in the calculation or by an alternative procedure to be described in the following section.

For the set of Eqs. (10.3.2), the complete tabulation consists of the array of the elements of the given matrix (and its check column, if desired)

$$\left.\begin{array}{rrr|r|r|} 9.3746 & 3.0416 & -2.4371 & 9.2333 & 19.2124 \\ 3.0416 & 6.1832 & 1.2163 & 8.2049 & 18.6460 \\ -2.4371 & 1.2163 & 8.4429 & 3.9339 & 11.1560 \end{array}\right. \tag{10.4.10}$$

the array of the elements of the auxiliary matrix (and its check column)

$$\left.\begin{array}{rrr|r|r|} \underline{9.3746} & 0.32445 & -0.25997 & 0.98493 & 2.04941 \\ 3.0416 & \underline{5.19635} & 0.38624 & 1.00246 & 2.38870 \\ -2.4371 & 2.00702 & \underline{7.03414} & 0.61448 & 1.61448 \end{array}\right. \tag{10.4.11}$$

† The element in row i and column j of the product is the inner product of the ith row of the first factor into the jth column of the second.

and the solution column (and its check column)

$$\begin{array}{l|l|} 0.89643 & 1.89643 \\ 0.76512 & 1.76512 \\ 0.61448 & 1.61448 \end{array} \qquad (10.4.12)$$

if five decimal places are retained in the calculations.

10.5 Intermediate Roundoff Errors

In the preceding example the gross-error check columns display no discrepancies through the fifth place. This fact, however, does *not* guarantee that the results are correct to five decimal places. Indeed, the relationship between the sizes of such discrepancies and the effects of intermediate roundoffs or gross errors is not a simple one.

If the calculated solution values are substituted into the left-hand members of the original equations, then the presence of deviations between the resultant members and the original right-hand members serves to indicate the *presence* of errors due to intermediate roundoffs, or of gross errors; but again the magnitudes of the deviations are not dependable measures of the magnitudes of the solution errors.

The following oft-cited example, which illustrates this fact, was constructed by T. S. Wilson. It can be verified by direct calculation that the equations

$$\begin{aligned} 10x_1 + 7x_2 + 8x_3 + 7x_4 &= 32 \\ 7x_1 + 5x_2 + 6x_3 + 5x_4 &= 23 \\ 8x_1 + 6x_2 + 10x_3 + 9x_4 &= 33 \\ 7x_1 + 5x_2 + 9x_3 + 10x_4 &= 31 \end{aligned} \qquad (10.5.1)$$

are such that the approximate values

$$x_1 \approx -7.2 \qquad x_2 \approx 14.6 \qquad x_3 \approx -2.5 \qquad x_4 \approx 3.1 \qquad (10.5.2)$$

reduce the respective left-hand members to 31.9, 23.1, 32.9, and 31.1, whereas the approximate values

$$x_1 \approx 0.18 \qquad x_2 \approx 2.36 \qquad x_3 \approx 0.65 \qquad x_4 \approx 1.21 \qquad (10.5.3)$$

reduce these members to the more nearly correct values 31.99, 23.01, 32.99, and 31.01; but the true solution is

$$x_1 = x_2 = x_3 = x_4 = 1 \qquad (10.5.4)$$

Thus we see that a poor approximation to the true solution may "almost" satisfy the relevant equations.

The precise analytical treatment of roundoff effects is complicated, particularly when many equations are involved. Error bounds are difficult to determine and, when found, generally are extremely conservative. Statistical results are more realistic but are inherently incapable of affording error *bounds*. In practice, the following simple method of discovering and effectively removing roundoff-error effects usually suffices.

If substitution of the calculated values $\bar{x}_1, \bar{x}_2, \ldots, \bar{x}_n$ into the left-hand members of (10.2.1) yields $\bar{c}_1, \bar{c}_2, \ldots, \bar{c}_n$, so that

$$\begin{gathered} a_{11}\bar{x}_1 + a_{12}\bar{x}_2 + \cdots + a_{1n}\bar{x}_n = \bar{c}_1 \\ \cdots\cdots\cdots\cdots\cdots\cdots\cdots\cdots \\ a_{n1}\bar{x}_1 + a_{n2}\bar{x}_2 + \cdots + a_{nn}\bar{x}_n = \bar{c}_n \end{gathered}$$

whereas the true values are to satisfy the equations

$$\begin{gathered} a_{11}x_1 + a_{12}x_2 + \cdots + a_{1n}x_n = c_1 \\ \cdots\cdots\cdots\cdots\cdots\cdots\cdots\cdots \\ a_{n1}x_1 + a_{n2}x_2 + \cdots + a_{nn}x_n = c_n \end{gathered}$$

there follows, by subtraction,

$$\begin{gathered} a_{11}\,\delta x_1 + a_{12}\,\delta x_2 + \cdots + a_{1n}\,\delta x_n = \delta c_1 \\ \cdots\cdots\cdots\cdots\cdots\cdots\cdots\cdots \\ a_{n1}\,\delta x_1 + a_{n2}\,\delta x_2 + \cdots + a_{nn}\,\delta x_n = \delta c_n \end{gathered} \qquad (10.5.5)$$

where

$$\delta x_k \equiv x_k - \bar{x}_k \qquad \delta c_k \equiv c_k - \bar{c}_k \qquad (10.5.6)$$

Thus the necessary corrections $\delta x_1, \ldots, \delta x_n$ satisfy a set of equations which differs from the original set only in that each c_k is replaced by the *residual* $c_k - \bar{c}_k$.

If this set could be solved without roundoff, the corrections thus would be obtained exactly, provided that the δc_k were themselves computed without roundoff. But since this situation generally will not exist, the corrections obtained are themselves approximate. New residuals can then be calculated, and the process can be iterated, if so desired. Each such calculation is particularly simple in the Crout procedure, since only the **c** column of **M**, the **c**′ column of **M**′, and the solution column need be recalculated, all other data being unchanged.

Whereas there is no *certainty* that the iteration will converge rapidly (or at all), and whereas relevant criteria are difficult to apply in practical cases, both the presence and the rapidity of convergence generally are confirmed by actual calculation. Thus the method usually is an efficient one and, by successive iteration, without increasing the number of significant figures in the elements of $\mathbf{A}'$ but with the residuals δc_i and the corrections δx_i expressed in units of the leading decimal place in the δc's, it is usually possible to stabilize the desired number of significant figures in the approximate solution without an excessive amount of computation. Clearly, the same increase in accuracy could be obtained alternatively, but generally less conveniently, by repeating the original calculation with overall retention of additional significant figures.

In the case of the preceding example, the residuals corresponding to the approximate solution (10.4.12) are found to be

$$\delta c_1 = -1.2462 \times 10^{-5} \qquad \delta c_2 = 3.6504 \times 10^{-5}$$
$$\delta c_3 = -1.9095 \times 10^{-5}$$

and the approximate corrections are found to be

$$\delta x_1 = -0.59421 \times 10^{-5} \qquad \delta x_2 = 0.98893 \times 10^{-5}$$
$$\delta x_3 = -0.54016 \times 10^{-5}$$

if five significant figures are retained, yielding the improved values

$$x_1 = 0.8964240579 \qquad x_2 = 0.7651298893 \qquad x_3 = 0.6144745984$$

The new residuals are found (by substitution of the calculated values of δx_1, δx_2, and δx_3 into the equations which yielded them) to be of the order of 10^{-10}; and another iteration would supply 14-place accuracy, the rounded 10-place values agreeing with those given above except for a one-unit change in the tenth digit of x_1.

If the coefficients and right-hand members of the original set of equations are only four-decimal-place approximations to true values, the preceding retention of 10 or more decimal places may be expected to be foolish, since it is useless to strive for a higher degree of accuracy than that which is compatible with errors *inherent* in the given system. This problem is to be considered explicitly in Sec. 10.7.

Large relative errors and/or failure of this iterative correction process occur most often when a diagonal element a'_{kk} which is small relative to the elements to its right appears in $\mathbf{A}'$, in a position preceding the last one, since the necessary subsequent divisions by a'_{kk} then will propagate its error unfavorably into subsequent calculations. Such situations often can be avoided by use of the following procedures.

First, before the reduction is begun, the equal members of certain equations may be multiplied by constants as necessary in such a way that the *largest* coefficients in the several equations become of comparable magnitude. This process, as applied to the associated augmented matrix, is sometimes called *equilibration by rows*. Its principal purpose, in the present case, is to make the succeeding process effective.

Second, also in advance, the equations and/or unknowns may be reordered as necessary (with a relabeling of unknowns if they are, in fact, reordered) in such a way that each *diagonal* element in the coefficient matrix is dominant in its row, insofar as this is possible.

Finally, use may be made of a so-called *pivoting* (or "partial pivoting") process in which, at the kth stage of the *Gauss* reduction, instead of always using the kth transformed equation to eliminate x_k from all following equations, it is determined whether any of the $n - k$ following transformed equations have an x_k coefficient which is larger in magnitude than that in the kth equation. If so, then the equation with the *largest* x_k coefficient is interchanged with the current kth equation and is used instead for the elimination of x_k from the transformed equations which then follow it.

Correspondingly, in the *Crout* reduction, immediately after all elements in the kth column of $\mathbf{M}'$ have been determined, one may compare the diagonal element a'_{kk} in that column with all elements below it. If one or more of these elements exceeds a'_{kk} in magnitude, then one may interchange the row containing the largest such coefficient with the kth row of $\mathbf{M}'$, effect the same interchange on the two corresponding rows of $\mathbf{M}$, and then proceed as before. (For hand calculation on a desk calculator, the reordering of rows is somewhat time-consuming. Although it is not difficult to devise a procedure which accomplishes the same result without an actual physical interchange of rows, it is more easily discovered than described and hence is not detailed here.)

Clearly, generally there is little virtue in effecting a pivot interchange unless the magnitude of the incumbent is *substantially* smaller than that of a competitor. In addition, in those frequently occurring cases where all diagonal elements of $\mathbf{A}$ are rather strongly dominant in both their rows and their columns, as in (10.4.12), pivoting rarely is needed. On the other hand, there are unpleasant situations (considered in Sec. 10.7) which cannot be remedied by pivoting or any other available device.

As a simple illustration of the procedure, we deal with the system

$$\begin{aligned} 4.3x_1 + 1.6x_2 - 2.3x_3 &= 3.6 \\ 3.8x_1 + 1.4x_2 + 1.5x_3 &= 6.7 \\ 1.8x_1 + 1.2x_2 + 3.1x_3 &= 6.1 \end{aligned} \tag{10.5.7}$$

supposing for present purposes that the coefficients and right-hand members are exact, and retaining only three decimal places throughout the reduction process in order to emphasize the effects of intermediate roundoffs.

Here no significant preliminary modifications seem to be indicated. If the Crout reduction is used without pivoting, the auxiliary matrix is obtained in the form

$$\begin{array}{ccc:c} \underline{4.3} & 0.372 & -0.535 & 0.837 \\ 3.8 & \underline{-0.014} & -252.357 & -251.386 \\ 1.8 & 0.530 & \underline{137.812} & 1.000 \end{array} \tag{10.5.8}$$

and there results

$$x_1 \approx 1.011 \qquad x_2 \approx 0.971 \qquad x_3 \approx 1.000 \tag{10.5.9}$$

Substitution of these approximations into the left-hand members of (10.5.7) yields the three-place values 3.601, 6.701, and 6.085 and hence also the residuals -0.001, -0.001, and 0.015. Here the use of (10.5.5) produces the corrections $\delta x_1 \doteq -0.026$, $\delta x_2 \doteq 0.071$, and $\delta x_3 \doteq 0.000$; and, in correspondence with the "improved values" $x_1 \approx 0.985$, $x_2 \approx 1.041$, and $x_3 \approx 1.000$, the residuals are found to round to -0.001, 0.000, and 0.022 so that the iterative correction process appears to be ineffective.

Clearly, the small diagonal element $a'_{22} = -0.014$ is the probable source of the trouble. Indeed, the pivoting process would have dictated an interchange of rows 2 and 3 after the completion of the second column of $\mathbf{M}'$, in which case the auxiliary matrix would be obtained as

$$\begin{array}{ccc:c} \underline{4.3} & 0.372 & -0.535 & 0.837 \\ 1.8 & \underline{0.530} & 7.666 & 8.667 \\ 3.8 & -0.014 & \underline{3.640} & 1.000 \end{array} \tag{10.5.10}$$

and the resultant three-place approximate solution

$$x_1 \approx 1.000 \qquad x_2 \approx 1.001 \qquad x_3 \approx 1.000 \tag{10.5.11}$$

possesses only a small relative error which *can* be corrected by the iterative process.

10.6 Determination of the Inverse Matrix

From (10.2.7) and (10.2.8), it follows that the kth column of the matrix (10.2.8), which is the *inverse* of the coefficient matrix (10.2.2), is the solution column corresponding to the result of setting $c_k = 1$ and all other c's equal to zero in

(10.2.1). Thus if, in place of the single **c** column in (10.4.1), we insert the square array

$$\begin{matrix} 1 & 0 & \cdots & 0 \\ 0 & 1 & \cdots & 0 \\ \cdots & \cdots & \cdots & \cdots \\ 0 & 0 & \cdots & 1 \end{matrix} \qquad (10.6.1)$$

of n columns, and treat *each* column of this array as a **c** column, we will obtain finally the array (10.2.8) in place of the single **x** column. That is, the resultant solution array will supply the inverse of the coefficient matrix of the given set of equations. A check column can be included, if so desired, and the rules given for its use apply as stated.

Analysis shows that n^3 long operations (multiplications and divisions) are needed, in general, to determine the inverse of the $n \times n$ matrix **A**. The computation of a solution column **x** in correspondence to a **c** column by use of Eqs. (10.2.7) then requires n^2 additional long operations (assuming no zero elements). On the other hand, once the Crout auxiliary matrix **A**′ has been determined by $(n^3 - n)/3$ long operations, the derivation of the **x** column from a **c** column by its use also requires only n^2 additional long operations. Thus, from this point of view, the determination of $\mathbf{A}^{-1}$ for the purpose of computing the **x** column for a number of different **c** columns is not an efficient procedure.

However, the determination of the inverse matrix is necessary when $x_1, \ldots, x_n$ are to be expressed *explicitly* as combinations of $c_1, \ldots, c_n$ by means of (10.2.7), for the purpose of studying the analytical dependence of the x's on the c's. It is also desirable in certain statistical computations where the elements of the inverse have significance in the underlying theory (see Sec. 7.4).

In the case of the example (10.3.2), the auxiliary array corresponding to the given array

$$\begin{matrix} 1 & 0 & 0 \\ 0 & 1 & 0 \\ 0 & 0 & 1 \end{matrix}$$

is found to be

$$\begin{matrix} 0.106671 & 0 & 0 \\ -0.062438 & 0.192443 & 0 \\ 0.054773 & -0.054909 & 0.142164 \end{matrix}$$

and the solution array is obtained in the form

$$\begin{matrix} 0.148032 & -0.083594 & 0.054774 \\ -0.083594 & 0.213651 & -0.054909 \\ 0.054773 & -0.054909 & 0.142164 \end{matrix} \qquad (10.6.2)$$

Here six decimal places were retained, in order that five significant figures might be afforded after a rounding.† It may be noticed that the inverse matrix possesses the same symmetry as the given matrix. (The single discrepancy of one unit is due to roundoff.)

The result obtained is equivalent to the statement that, apart from the effects of roundoffs, the solution of the set (10.3.2) would be of the form

$$\begin{aligned} x_1 &= 0.148032c_1 - 0.083594c_2 + 0.054774c_3 \\ x_2 &= -0.083594c_1 + 0.213651c_1 - 0.054909c_3 \\ x_3 &= 0.054773c_1 - 0.054909c_2 + 0.142164c_3 \end{aligned} \qquad (10.6.3)$$

if the right-hand members of (10.3.2) were replaced by c_1, c_2, and c_3, respectively. In particular, the substitution of the actual right-hand members into (10.6.3) again leads to (10.4.12). The elements of (10.6.2) are the *reduced cofactors* defined in (10.2.6), in accordance with (10.2.7). Since, as stated in Sec. 10.4, the *determinant* D of the given matrix is the product of the diagonal elements of (10.4.11)

$$D = (9.3746)(5.19635)(7.03414) \doteq 342.66$$

the array of the cofactors themselves is obtained by multiplication by D (and interchange of rows and columns in the more general case).

10.7 Inherent Errors

In addition to the errors due to *intermediate* roundoffs, which usually can be detected and removed by the methods previously described, errors due to possible inaccuracies in the coefficients and right-hand members of the *given* equations themselves must be taken into account. For example, Wilson's equation set (10.5.1) shows that relatively small changes in the right-hand members of an apparently innocuous set of equations (with a symmetric and, in fact, positive

† For purposes of illustrating the technique, we again ignore the fact that *inherent* errors due to roundoff in the *given* data may adversely affect the significance of certain of the digits.

definite coefficient matrix) can bring about very severe changes in the components of the solution. (See also Prob. 23.)

In order to investigate these errors, we assume now that the errors due to *intermediate* roundoffs are negligible. We then suppose that the set actually solved is

$$\begin{aligned} a_{11}x_1 + \cdots + a_{1n}x_n &= c_1 \\ \cdots\cdots\cdots\cdots & \\ a_{n1}x_1 + \cdots + a_{nn}x_n &= c_n \end{aligned} \tag{10.7.1}$$

whereas the *true* values of the coefficients are $a_{ij} + \delta a_{ij}$ and the true values of the right-hand members are $c_i + \delta c_i$. If we denote the true value of the ith unknown by $x_i + \delta x_i$, there follows also

$$\begin{aligned} (a_{11} + \delta a_{11})(x_1 + \delta x_1) + \cdots + (a_{1n} + \delta a_{1n})(x_n + \delta x_n) &= c_1 + \delta c_1 \\ \cdots\cdots\cdots\cdots\cdots\cdots\cdots\cdots & \\ (a_{n1} + \delta a_{n1})(x_1 + \delta x_1) + \cdots + (a_{nn} + \delta a_{nn})(x_n + \delta x_n) &= c_n + \delta c_n \end{aligned} \tag{10.7.2}$$

and the result of subtracting Eqs. (10.7.1) from (10.7.2) is expressible in the form

$$\begin{aligned} a_{11}\,\delta x_1 + \cdots + a_{1n}\,\delta x_n &= \delta c_1 - (x_1\,\delta a_{11} + \cdots + x_n\,\delta a_{1n}) \\ \cdots\cdots\cdots\cdots\cdots\cdots\cdots\cdots & \\ a_{n1}\,\delta x_1 + \cdots + a_{nn}\,\delta x_n &= \delta c_n - (x_1\,\delta a_{n1} + \cdots + x_n\,\delta a_{nn}) \end{aligned} \tag{10.7.3}$$

if *products* of errors, of the form $\delta a_{ik}\delta x_k$, are *assumed* to be relatively negligible.

Thus, if the inherent errors δa_{ij} and δc_i were known, the corresponding solution errors δx_i would be obtained (to a degree of accuracy consistent with this assumption) by solving a set of equations which differs from the set actually solved only in that the right-hand member c_i is to be replaced by η_i, where

$$\eta_i = \delta c_i - (x_1\,\delta a_{i1} + \cdots + x_n\,\delta a_{in}) \tag{10.7.4}$$

However, in practice, it is usually known only that the errors δa_{ij} and δc_i do not exceed a certain positive number, say, ε, in magnitude, so that

$$-\varepsilon \leqq \delta a_{ij} \leqq \varepsilon \qquad -\varepsilon \leqq \delta c_i \leqq \varepsilon \tag{10.7.5}$$

Hence, in such cases, the value of η_i is not known and we are certain only that

$$|\eta_i| \leqq E \tag{10.7.6}$$

where

$$E \equiv (1 + |x_1| + |x_2| + \cdots + |x_n|)\varepsilon \tag{10.7.7}$$

The solution of the set (10.7.3) can be expressed in the form

$$\delta x_k = \tilde{A}_{1k}\eta_1 + \tilde{A}_{2k}\eta_2 + \cdots + \tilde{A}_{nk}\eta_n \qquad (k = 1, 2, \ldots, n) \tag{10.7.8}$$

with the notation of (10.2.6). Thus, if (10.7.6) is true, there follows

$$|\delta x_k| \leqq (|\tilde{A}_{1k}| + |\tilde{A}_{2k}| + \cdots + |\tilde{A}_{nk}|)E \qquad (10.7.9)$$

The reduced cofactors involved in this expression are the elements of the kth *row* of the inverse of the coefficient matrix. Thus, if the elements of the inverse matrix are calculated, approximate upper bounds on the effects of inherent errors are obtainable from (10.7.9). They are not *strictly* upper bounds, since they were derived under the assumption that terms of the form $\delta a_{ik}\delta x_k$ are small in magnitude relative to E, as defined in (10.7.7). However, unless the upper bounds predicted under this assumption are such that the truth of that assumption is contradicted, they may be accepted as close approximations to the true upper bounds (which could be attained, in any case, only when *all* the errors combined in the most unfavorable way).

In the case of (10.3.2), if it is supposed that the coefficients and right-hand members are merely rounded approximations to true values, there follows $E \approx 3.28\varepsilon = 1.64 \times 10^{-4}$, and reference to (10.6.2) yields the estimates

$$|\delta x_1|_{\max} \approx 0.29E = 0.48 \times 10^{-4} \qquad |\delta x_2|_{\max} \approx 0.35E = 0.57 \times 10^{-4}$$

$$|\delta x_3|_{\max} \approx 0.25E = 0.41 \times 10^{-4}$$

Thus, we could be confident only that the solution of the *true* equations is such that

$$0.89637 < x_1 < 0.89648 \qquad 0.76507 < x_2 < 0.76519$$
$$0.61443 < x_3 < 0.61452 \qquad (10.7.10)$$

so that we could write $x_1 = 0.8964$, $x_2 = 0.7651$, and $x_3 = 0.6145$, with the last digit in doubt by one unit in each case.

Unless the given system of equations is to be solved for *literal* values of the right-hand members, so that the determination of the inverse matrix is advisable in any case, an error estimate which does not involve the calculation of the elements of that matrix is desirable. It is clear that the upper bound (10.7.9) would correspond to the δx_k which is the kth element of the solution column satisfying a set of equations of the form

$$\begin{aligned} a_{11}\,\delta x_1 + \cdots + a_{1n}\,\delta x_n &= \pm E \\ \cdots\cdots\cdots\cdots&\cdots\cdots \\ a_{n1}\,\delta x_1 + \cdots + a_{nn}\,\delta x_n &= \pm E \end{aligned} \qquad (10.7.11)$$

for *some* choice of the ambiguous signs of each of the n right-hand members. However, the manner in which the signs are to be associated with successive equations cannot be determined unless the signs of the relevant cofactors of the coefficients of δx_k in these equations are known. Furthermore, a different combination of signs may be needed to maximize each of the δx's.

Reference to the Crout reduction shows that each of the elements in the solution column is obtained as a linear combination of the elements of the **c**′ column of the auxiliary matrix, each of which is, in turn, a linear combination of the elements of the **c** column (that is, of the original right-hand members). Thus, if each entry in the **c** column were $+E$, and if, in the calculation of each element of the **c**′ and **x** columns, we were to replace all subtractions by additions, it follows that no element of the resultant **x** column could be exceeded in magnitude by a corresponding element obtained by solving (10.7.11) with *any* prescribed combination of signs.

Hence, if an additional column with *unity* as *each* element is adjoined to the matrix **M** and is transformed just as the **c** column except for the fact that all subtractions are replaced by additions, the result of multiplying by E the elements of the final corresponding column, adjoined to the solution column, gives (approximate) upper bounds on the possible errors in the corresponding elements of the solution column, due to possible errors in the coefficients and right-hand members of the given equations. (This procedure appears to be due to Milne [1949].) The bounds obtained in this way usually exceed the more precise bounds afforded by (10.7.9), but are obtained much more simply.

In the case of the illustrative example, the inherent-error check columns adjoined to the given, auxiliary, and final arrays are found to be

1	0.10667	0.28640
1	0.25488	0.35215
1	0.25185	0.25185

Here it happens that the approximate upper bounds, obtained by multiplying the successive elements of the last column by E, are identical with those afforded by the preceding analysis.

The fact that this situation is not a general one may be illustrated, for example, by the case when only two equations are involved. Here, the estimates afforded by (10.7.9) are

$$|\delta x_1|_{\max} \approx \frac{|a_{12}| + |a_{22}|}{|D|} E \qquad |\delta x_2|_{\max} \approx \frac{|a_{11}| + |a_{21}|}{|D|} E$$

whereas it is readily verified that the estimates afforded by the simpler procedure are

$$|\delta x_1|_{\max} \approx \frac{|a_{11}a_{22} - a_{12}a_{21}| + |a_{12}a_{21}| + |a_{11}a_{12}|}{|a_{11}|\,|D|} E$$

$$|\delta x_2|_{\max} \approx \frac{|a_{11}| + |a_{21}|}{|D|} E$$

Although the estimates for $|\delta x_2|_{max}$ are identical, it is seen that the latter estimate for $|\delta x_1|_{max}$ exceeds the former estimate except in special cases, such as that in which $a_{11}a_{22} > a_{12}a_{21} > 0$.

Systems in which small relative errors in the coefficients or right-hand members, or in the process of solution, may correspond to large relative errors in the solution are often said to be *ill-conditioned* systems and are essentially characterized by the fact that the *determinant* of the coefficient matrix is small in magnitude relative to certain of the cofactors of elements of that matrix, when the matrix has been normalized such that its largest element is of the order of magnitude of 1.† When such a system is encountered, one must either make the inherent errors small by retaining a large number of significant figures in the given data (when this is possible) or control the effects of inherent errors as closely as possible and then accept the fact that the inaccuracies in the given data still may permit the solution to be determined only within relatively wide error limits. Wilson's example (10.5.1), with the coefficients and/or right-hand members modified "more realistically" by small perturbations, serves to illustrate this fact rather dramatically.

10.8 Tridiagonal Sets of Equations

Systems of equations of the special form

$$\begin{aligned}
&d_1x_1 + f_1x_2 &&= c_1 \\
&e_2x_1 + d_2x_2 + f_2x_3 &&= c_2 \\
&\qquad e_3x_2 + d_3x_3 + f_3x_4 &&= c_3 \\
&\cdots\cdots\cdots\cdots\cdots\cdots\cdots\cdots\cdots\cdots\cdots\cdots && \\
&\qquad\qquad e_{n-1}x_{n-2} + d_{n-1}x_{n-1} + f_{n-1}x_n &&= c_{n-1} \\
&\qquad\qquad\qquad e_nx_{n-1} + \quad d_nx_n &&= c_n
\end{aligned} \tag{10.8.1}$$

are of frequent occurrence in practice and are often said to be *tridiagonal*, since only the diagonal elements d_i and the adjacent elements e_i and f_i may differ from zero in the coefficient matrix **A** associated with any such system. It can be seen that the Crout reduction preserves this property, in the sense that the submatrix **A**′ in the auxiliary matrix **M**′ of (10.4.2) then also is of tridiagonal form. Furthermore, by virtue of the fact that the e's have no nonzero left neighbors and the f's no nonzero upper neighbors, it is easily verified that here the relation-

† More specific measures of the "condition" of a matrix have been proposed by von Neumann and Goldstine (see von Neumann and Goldstine [1947] and Goldstine and von Neumann [1951]), Turing [1948], and others. Whereas these measures are of marked theoretical importance, their usefulness in explicit numerical situations is limited by the amount of computation involved in their evaluation.

ships between the elements of $\mathbf{M}'$ and the corresponding elements of $\mathbf{M}$ take the simplified forms

$$e'_i = e_i \qquad (i = 2, 3, \ldots, n) \tag{10.8.2}$$

$$d'_1 = d_1 \qquad d'_i = d_i - e'_i f'_{i-1} \qquad (i = 2, 3, \ldots, n) \tag{10.8.3}$$

$$f'_i = \frac{f_i}{d'_i} \qquad (i = 1, 2, \ldots, n - 1) \tag{10.8.4}$$

and

$$c'_1 = \frac{c_1}{d'_1} \qquad c'_i = \frac{c_i - e'_i c'_{i-1}}{d'_i} \qquad (i = 2, 3, \ldots, n) \tag{10.8.5}$$

Finally, the relations of (10.4.7) reduce to the forms

$$x_n = c'_n \qquad x_i = c'_i - f'_i x_{i+1} \qquad (i = n - 1, n - 2, \ldots, 1) \tag{10.8.6}$$

For the purpose of compactness, it is convenient to record the coefficients and right-hand members of (10.8.1) in the array

$$\mathbf{P} \equiv \left[\begin{array}{ccc|c} & d_1 & f_1 & c_1 \\ e_2 & d_2 & f_2 & c_2 \\ \cdots & \cdots & \cdots & \cdots \\ e_{n-1} & d_{n-1} & f_{n-1} & c_{n-1} \\ e_n & d_n & & c_n \end{array}\right] \tag{10.8.7}$$

so that the diagonal elements of $\mathbf{A}$ are in the second column, the subdiagonal elements in the first column, and the superdiagonal elements in the third column. If a corresponding array $\mathbf{P}'$, with primed elements, is defined in correspondence with the Crout auxiliary matrix, its elements may be determined by use of (10.8.2) to (10.8.5), after which the elements of the solution $\mathbf{x}$ are obtained by use of (10.8.6).

For this purpose, we notice first that the *first column* of $\mathbf{P}'$ is identical with the first column of $\mathbf{P}$. To complete the *first row* of $\mathbf{P}'$, we next evaluate the quantities

$$d'_1 = d_1 \qquad f'_1 = \frac{f_1}{d'_1} \qquad c'_1 = \frac{c_1}{d'_1}$$

Then, to complete the *second row* of $\mathbf{P}'$, we determine successively

$$d'_2 = d_2 - e'_2 f'_1 \qquad f'_2 = \frac{f_2}{d'_2} \qquad c'_2 = \frac{c_2 - e'_2 c'_1}{d'_2}$$

The *third row* is completed by use of the same formulas with all subscripts advanced by unity and the remaining rows of $\mathbf{P}'$ are completed, in order, in the same way (except that f'_n is not needed in the last row). The elements of $\mathbf{x}$ then are determined successively, from the last two columns of $\mathbf{P}'$, in the reverse order

$$x_n = c'_n \qquad x_{n-1} = c'_{n-1} - f'_{n-1}x_n \qquad \cdots \qquad x_1 = c'_1 - f'_1 x_2$$

It may be noticed that the complete process requires a maximum of $3n - 3$ multiplications, $2n - 1$ divisions, and $3n - 3$ additions, when the system comprises n equations, and it compares favorably with other procedures in this respect.

As a simple illustration, the system

$$\begin{aligned} 2x_1 - x_2 &= 6 \\ -x_1 + 3x_2 - 2x_3 &= 1 \\ -2x_2 + 4x_3 - 3x_4 &= -2 \\ -3x_3 + 5x_4 &= 1 \end{aligned}$$

corresponds to the array

$$\mathbf{P} = \left[\begin{array}{rrr|r} & 2 & -1 & 6 \\ -1 & 3 & -2 & 1 \\ -2 & 4 & -3 & -2 \\ -3 & 5 & & 1 \end{array}\right]$$

and it may be verified that the auxiliary array $\mathbf{P}'$ and the solution column $\mathbf{x}$ are obtained in the forms

$$\mathbf{P}' = \left[\begin{array}{rrr|r} & 2 & -\frac{1}{2} & 3 \\ -1 & \frac{5}{2} & -\frac{4}{5} & \frac{8}{5} \\ -2 & \frac{12}{5} & -\frac{5}{4} & \frac{1}{2} \\ -3 & \frac{5}{4} & & 2 \end{array}\right] \qquad \mathbf{x} = \begin{bmatrix} 5 \\ 4 \\ 3 \\ 2 \end{bmatrix}$$

The check columns can be used exactly as before, if so desired, when account is taken of the fact that the *second column* of $\mathbf{P}$ (or of $\mathbf{P}'$) comprises the elements in the principal diagonal of $\mathbf{M}$ (or of $\mathbf{M}'$). It follows also that the determinant of the coefficient matrix associated with (10.8.1) is the product of the elements in the second column of $\mathbf{P}'$. In the preceding example, this determinant thus has the value 15.

10.9 Iterative Methods and Relaxation

In many sets of linear equations which arise in practice, the equations can be ordered in such a way that the coefficient of x_k in the kth equation is large in magnitude relative to all other coefficients in that equation. Such sets are often amenable to an iterative process in which the set is first rewritten in the form

$$
\begin{aligned}
x_1 &= \frac{1}{a_{11}}(c_1 - a_{12}x_2 - a_{13}x_3 - \cdots - a_{1n}x_n) \\
x_2 &= \frac{1}{a_{22}}(c_2 - a_{21}x_1 - a_{23}x_3 - \cdots - a_{2n}x_n) \\
&\cdots\cdots\cdots\cdots\cdots\cdots\cdots\cdots\cdots\cdots \\
x_n &= \frac{1}{a_{nn}}(c_n - a_{n1}x_1 - a_{n2}x_2 - \cdots - a_{n,n-1}x_{n-1})
\end{aligned}
\tag{10.9.1}
$$

The initial approximations may be taken to be

$$
x_1^{(0)} = \frac{c_1}{a_{11}} \qquad x_2^{(0)} = \frac{c_2}{a_{22}} \qquad \cdots \qquad x_n^{(0)} = \frac{c_n}{a_{nn}} \tag{10.9.2}
$$

The next approximations then are obtained by replacing the unknowns *in the right-hand members* of (10.9.1) by these initial approximations, and the feedback process is repeated in the hope that the input and output of a cycle eventually will agree within the specified error tolerance.

Thus, in this process, which is often called *Jacobi iteration*, in each step the current values $x_1, \ldots, x_n$ are all replaced by modified values $x_1^*, \ldots, x_n^*$ in accordance with the equations

$$
\begin{aligned}
x_1^* &= \frac{1}{a_{11}}(c_1 - a_{12}x_2 - \cdots - a_{1n}x_n) \\
x_2^* &= \frac{1}{a_{22}}(c_2 - a_{21}x_1 - \cdots - a_{2n}x_n) \\
&\cdots\cdots\cdots\cdots\cdots\cdots\cdots\cdots \\
x_n^* &= \frac{1}{a_{nn}}(c_n - a_{n1}x_1 - \cdots - a_{n,n-1}x_{n-1})
\end{aligned}
\tag{10.9.3}
$$

In the case of the system (10.3.2), about ten such iterations are required for three-place accuracy.

If the iteration is modified in such a way that each unknown in each right-hand member is replaced by its most recently calculated approximation rather than by the approximation afforded by the preceding cycle, the rate of convergence depends upon the *order* in which the x's are modified. In particular, if they are modified cyclically in their natural order, the modified values are related to the current values by the equations

$$
\begin{aligned}
x_1^* &= \frac{1}{a_{11}}(c_1 - a_{12}x_2 - \cdots - a_{1n}x_n) \\
x_2^* &= \frac{1}{a_{22}}(c_2 - a_{21}x_1^* - \cdots - a_{1n}x_n) \\
&\cdots\cdots\cdots\cdots\cdots\cdots\cdots\cdots\cdots \\
x_n^* &= \frac{1}{a_{nn}}(c_n - a_{n1}x_1^* - \cdots - a_{n,n-1}x_{n-1}^*)
\end{aligned}
\tag{10.9.4}
$$

In the case of (10.3.2), the number of cycles required to afford three-place accuracy is reduced to about six. The procedure described by (10.9.4) is often called *Gauss-Seidel* iteration, although its attribution to either Gauss or Seidel appears to be improper.

In the frequently occurring cases when the coefficient matrix is real and symmetric, so that $a_{ji} = a_{ij}$, and the diagonal elements are all positive, this iteration will converge if and only if all the n quantities

$$
a_{11} \qquad \begin{vmatrix} a_{11} & a_{12} \\ a_{12} & a_{22} \end{vmatrix} \qquad \begin{vmatrix} a_{11} & a_{12} & a_{13} \\ a_{12} & a_{22} & a_{23} \\ a_{13} & a_{23} & a_{33} \end{vmatrix} \qquad \cdots \qquad \begin{vmatrix} a_{11} & \cdots & a_{1n} \\ \cdots & \cdots & \cdots \\ a_{1n} & \cdots & a_{nn} \end{vmatrix}
$$

are positive (Reich [1949]).

Another useful theorem (Collatz [1950]) states that *both* the Jacobi and the Gauss-Seidel iterations will converge if the $n \times n$ matrix **A** has the two following properties:

1 The matrix **A** does not contain a $p \times q$ submatrix of *zeros*, with $p + q = n$.

2 The magnitude of each diagonal element of **A** is at least as large as the sum of the magnitudes of the other elements in its row and, in at least one case, is *larger* than that sum.

The first property of the matrix **A** is sometimes called *irreducibility*, and the second *diagonal row dominance*. It is important to notice that the two preceding conditions are *sufficient* but often are *not necessary*. In a wide class of situations (see Varga [1962]) the Gauss-Seidel iteration converges whenever the Jacobi iteration does and converges more rapidly. Although there are indeed cases in which the situation is reversed, the Gauss-Seidel iteration is almost always to be preferred.

There exist many other numerical techniques for solving sets of linear equations (see Sec. 10.20), some of which are direct methods (*reductions*) which would yield the exact solution of a set of n equations in n unknowns after a finite number of steps if no roundoffs were effected (as is true for the Gauss, Gauss-Jordan, and Crout reductions), and others which are basically iterative in the sense that generally an infinite sequence of approximations is generated, with convergence to the solution in a certain class of situations (as in the Jacobi and Gauss-Seidel processes).

Particular mention should be made of the *method of conjugate gradients* due to Hestenes and Stiefel [1952], which would terminate in the absence of roundoff, and of a rather extensive class of other gradient and related methods.

At the other extreme there exist the so-called *relaxation methods*, apparently invented by Gauss and revived and popularized by Southwell [1940, 1946, 1956], in which the rapidity (or existence) of convergence may depend upon the ingenuity of the user. In applying a relaxation process to the solution of a set of equations of the form

$$\begin{aligned} a_{11}x_1 + a_{12}x_2 + \cdots + a_{1n}x_n &= c_1 \\ \cdots\cdots\cdots\cdots\cdots\cdots &\cdots \\ a_{n1}x_1 + a_{n2}x_2 + \cdots + a_{nn}x_n &= c_n \end{aligned} \qquad (10.9.5)$$

we first define *residuals* $R_1, R_2, \ldots, R_n$ by the equations

$$\begin{aligned} c_1 - a_{11}x_1 - a_{12}x_2 - \cdots - a_{1n}x_n &= R_1 \\ \cdots\cdots\cdots\cdots\cdots\cdots &\cdots \\ c_n - a_{n1}x_1 - a_{n2}x_2 - \cdots - a_{nn}x_n &= R_n \end{aligned} \qquad (10.9.6)$$

The unknowns $x_1, x_2, \ldots, x_n$ are then *estimated*, and the corresponding residuals are calculated, after which the estimated values of the unknowns are to be successively modified (one or more at a time) in such a way that the magnitudes of all residuals are eventually reduced effectively to zero.

Reference to (10.9.6) shows that when x_i is *increased by unity*, and all other x's are held fixed, R_j *decreases by* a_{ji}. Thus the *transpose* of the coefficient matrix

$$\begin{matrix} a_{11} & a_{21} & a_{31} & \cdots & a_{n1} \\ a_{12} & a_{22} & a_{32} & \cdots & a_{n2} \\ \cdots & \cdots & \cdots & \cdots & \cdots \\ a_{1n} & a_{2n} & a_{3n} & \cdots & a_{nn} \end{matrix} \qquad (10.9.7)$$

in which the rows and columns of the original matrix are interchanged, serves as a *relaxation table*, in the sense that *the successive entries in the* kth *row of*

(10.9.7) *represent the decreases in the successive residuals which correspond to a unit increase in* x_k. If the original coefficient matrix is symmetric, the matrix (10.9.7) is then identical with it.

The situations which generally are most favorable to this process are those in which each of the diagonal elements $a_{11}, a_{22}, \ldots, a_{nn}$ is large in magnitude relative to the other elements in its row and column. For, in such cases, an increase in x_k of about R_k/a_{kk} will nearly reduce R_k to zero, but will affect the other residuals by relatively small amounts, and subsequent modifications of other x's generally will not seriously nullify the effect of this reduction.

It may be seen that the Gauss-Seidel iteration consists of determining successive corrections in exactly this way since, just before x_k is modified, the residual associated with the kth equation is

$$R_k = c_k - a_{k1}x_1^* - \cdots - a_{k,k-1}x_{k-1}^* - a_{kk}x_k - a_{k,k+1}x_{k+1} - \cdots - a_{kn}x_n$$

and the modified x_k is defined by the equation

$$x_k^* = \frac{1}{a_{kk}}(R_k + a_{kk}x_k) = x_k + \frac{R_k}{a_{kk}} \tag{10.9.8}$$

in accordance with (10.9.4). Also, the Jacobi iteration is one in which all entries are modified ("relaxed") simultaneously according to the formulas

$$x_1^* = x_1 + \frac{R_1}{a_{11}} \qquad \cdots \qquad x_n^* = x_n + \frac{R_n}{a_{nn}}$$

where the R's are given by (10.9.6). For these reasons, Gauss-Seidel iteration is sometimes referred to as *cyclic relaxation*, or *cyclic displacement*, and Jacobi iteration as *simultaneous relaxation*, or *simultaneous displacement*.

The advantages associated with the more general relaxation process follow from the fact that the values of the residuals are known at each stage. Thus it is possible to focus attention at each stage on the residual of *largest* magnitude and either to reduce its magnitude effectively to zero or to proceed otherwise if an alternative procedure appears to be desirable. At the same time, the fact that an efficient use of the process in its full generality requires a *decision* after each step, not only places a premium on the ingenuity of the user, but makes it useful for large-scale computers only when relatively limited latitude is provided for variation in the relaxation technique from step to step. In the two special cases just mentioned, the residuals are in fact usually not calculated at all since the decision as to what modification is to be effected at each step is made *in advance* independently of the values of the residuals.

In this connection, we note the existence of an important iteration closely related to the Gauss-Seidel process but in which the residuals *are* calculated and in which (10.9.8) is replaced by the formula

$$x_k^* = x_k + \omega \frac{R_k}{a_{kk}} \tag{10.9.9}$$

Here ω is a predetermined constant independent of k but dependent in a rather complicated way upon the coefficient matrix of (10.9.5), so chosen that the rate of convergence of the sequence of iterates is maximized relative to all other values of ω, including $\omega = 1$. Since it happens that always $\omega > 1$ (also $\omega < 2$), this process is known as *cyclic* (or *successive*) *overrelaxation.* (See Varga [1962] and Young [1971].)

A typical sequence of relaxations, as applied by hand to the result of first rounding all numerical coefficients and right-hand members in (10.3.2) to three digits, is included for the purpose of illustrating the more general procedure:

9.37	3.04	−2.44
3.04	6.18	1.22
−2.44	1.22	8.44
9	6	8

Δx_1	Δx_2	Δx_3	R_1	R_2	R_3	
0	0	0	9.23	8.20	3.93	
1			−0.14	5.16	6.37	
		1	2.30	3.94	−2.07	
	1		−0.74	−2.24	−3.29	
			−7.40	−22.40	−32.90	$\times 10^{-1}$
		−4	−17.16	−17.52	0.86	
	−3		−8.04	1.02	4.52	
−1			1.33	4.06	2.08	
	1		−1.71	−2.12	0.86	
			−17.10	−21.20	8.60	$\times 10^{-2}$
	−4		−4.94	3.52	13.48	
		2	−0.06	1.08	−3.40	
			−0.60	10.80	−34.00	$\times 10^{-3}$
		−4	−10.36	15.68	−0.24	
	3		−19.48	−2.86	−3.90	
−2			−0.74	3.22	−8.78	
		−1	−3.18	4.44	−0.34	
	1		−6.22	−1.74	−1.56	
−1			3.15	1.30	−4.00	

The relaxation table is written down immediately, and columns are provided for successive *changes* in the estimated unknowns and for the successive values of the three residuals. Values of the *diagonal* elements, rounded to the nearest integer, are encircled above the corresponding residuals, for convenience in estimating appropriate changes in the x's.

Starting arbitrarily with the crude approximation $x_1 = x_2 = x_3 = 0$, the *initial* residuals are then merely the right-hand members of the given equations and are listed in the first row of the calculation. Since the largest residual at this stage is $R_1 = 9.23$, we increase x_1 by the integer nearest $R_1/a_{11} \approx R_1/9$, and so enter a unit in the Δx_1 column and subtract unity times the first row of the relaxation table from the row of residuals. At this stage R_3 is largest in magnitude, and x_3 is increased by $1 \approx 6.37/8$, after which a unit increase in x_2 is called for. At this stage, each residual is less than one-half the corresponding rounded diagonal coefficient, and it is convenient to multiply the residuals by a factor of 10. The subsequent changes in the x's accordingly are then to be divided by 10 when all the changes are eventually accumulated.

The approximate solution at the last stage of the tabulation given is $x_1 = 0.897$, $x_2 = 0.764$, and $x_3 = 0.615$. It may be noticed that, with this arrangement of the calculations, the entries in the relaxation table need only be multiplied by *integers*. Also, it is possible to avoid all intermediate roundoff without carrying more decimal places than are involved in the given data. In particular, the residuals corresponding to the three-place approximations obtained at the last stage given would be *exactly* 0.00315, 0.00130, and -0.00400 if the *given* three-digit data were exact. However, it is desirable to accumulate the increments in the x's, from time to time, and then to calculate the corresponding residuals directly, in order to avoid the propagation of the effects of *gross* errors.

Iterative methods, in general, are appropriate for computer or desk use in solving sets of equations principally when the number of equations is rather large and when the coefficient matrix is "sparse," so that it contains relatively few nonzero elements. Even in such situations some users of computers tend to favor a direct method, particularly when that method takes advantage of the sparsity or can be made to do so.

10.10 Iterative Methods for Nonlinear Equations

Most of the useful methods for obtaining an approximate real solution of a real equation of the form

$$f(x) = 0 \qquad (10.10.1)$$

involve iterative processes in which an initial approximation z_0 to a desired real root $x = \alpha$ is obtained, by rough graphical methods or otherwise, and a certain recurrence relation is used to generate a sequence of successive approximations $z_1, z_2, \ldots, z_n, \ldots$ which converges (in a certain associated class of cases) to the limit α.

One such method is that of *successive substitutions*, in which (10.10.1) is first rewritten in an equivalent form

$$x = F(x) \qquad (10.10.2)$$

and use is then made of the simple recurrence relation

$$z_{k+1} = F(z_k) \qquad (10.10.3)$$

Generally there are many convenient ways of rewriting (10.10.1) in the form (10.10.2), and the convergence or divergence of the sequence of approximations to α may depend upon the particular form chosen.

In order to see why this is so, we first assume that $F(x)$ possesses a continuous derivative on the closed interval bounded by α and z_k and then notice that since

$$\alpha = F(\alpha)$$

Equation (10.10.3) implies the relation

$$\alpha - z_{k+1} = F(\alpha) - F(z_k) = (\alpha - z_k)F'(\xi_k) \qquad (10.10.4)$$

where ξ_k lies between z_k and α. If the iteration converges, so that $z_k \to \alpha$, then also $F'(\xi_k) \to F'(\alpha)$ as $k \to \infty$. Temporarily excluding the cases when $F'(\alpha) = 0$ and $F'(\alpha) = \pm 1$, we deduce that $\alpha - z_{k+1} \sim (\alpha - z_k)F'(\alpha)$ and hence also that

$$\alpha - z_k \sim A[F'(\alpha)]^k \qquad (k \to \infty) \qquad (10.10.5)$$

where A is a certain constant; and this deviation in fact would grow unboundedly in magnitude with increasing k if it were true that $|F'(\alpha)| > 1$. Thus it appears that in order that the iteration converge to $x = \alpha$ as an infinite sequence, it is *necessary* that $|F'(\alpha)| \leqq 1$.

If we define the *convergence factor* ρ_k as the ratio of the error in z_{k+1} to the error in z_k, it follows that if z_k is near α, then $\rho_k \approx F'(\alpha)$. The number $F'(\alpha)$ may be called the *asymptotic convergence factor*. Unless $|F'(\alpha)| \leqq 1$, a small error in z_k is *increased* in magnitude by the iteration, and we then say that the iteration is *asymptotically unstable* at α. When $|F'(\alpha)| > 1$, convergence to α could occur only in a finite number of steps, in consequence of an improbably fortunate choice of the initial approximation z_0 (such as $z_0 = \alpha$).

When $|F'(\alpha)| = 1$, the asymptotic behavior of the corresponding approximation sequence is unpredictable without further information. Finally, when $F'(\alpha) = 0$, a sufficiently small value of $|\alpha - z_k|$ certainly leads to a smaller value of $|\alpha - z_{k+1}|$, so that asymptotic stability is present, but (10.10.5) no longer describes the *nature* of the convergence when it exists, in this case.

If $|F'(\alpha)| < 1$, so that the iteration is asymptotically stable at α, *and if the initial approximation is sufficiently near to* α, the sequence of the iterates will indeed converge to α, in such a way that ultimately the successive approximations tend toward α from one direction if $0 < F'(\alpha) < 1$ and oscillate about α with decreasing amplitude if $-1 < F'(\alpha) < 0$. In the special cases when $F'(\alpha) = 0$, the nature of the convergence depends upon the behavior of the higher derivatives of $F(x)$ near $x = \alpha$.

In illustration, a simple analysis (or a rough plot) of the function $y = x^3 - x - 1$ shows that the real root of the equation

$$f(x) \equiv x^3 - x - 1 = 0 \qquad (10.10.6)$$

is between $x = 1$ and $x = 2$, and is near $x = 1.3$. This equation can be conveniently written in the form (10.10.2) in various ways, such as $x = x^3 - 1$, $x = 1/(x^2 - 1)$, and $x = (x + 1)^{1/3}$. However, only the third (and least convenient) of these particular forms is such that the derivative of the right-hand member is smaller than unity in absolute value near $x = 1.3$. Hence, we may use the recurrence formula

$$z_{k+1} = (z_k + 1)^{1/3}$$

and, with $z_0 = 1.3$, then obtain the sequence $z_1 \doteq 1.3200$, $z_2 \doteq 1.3238$, $z_3 \doteq 1.3245$, $z_4 \doteq z_5 \doteq 1.3247$, when four decimal places are retained. The true root is

$$\alpha \doteq 1.3247179573$$

to 10 places.

More generally, for any differentiable function $f(x)$, if an *interval* $[a, b]$ can be found such that $f(a)$ and $f(b)$ have opposite signs, and if $f'(x)$ is of constant sign in $[a, b]$, then certainly $f(x)$ has one and only one zero $x = \alpha$ inside $[a, b]$. If the equation $f(x) = 0$ is rewritten as $x = F(x)$ in such a way that

$$|F'(x)| \leqq M < 1 \qquad (10.10.7)$$

when $a \leqq x \leqq b$, then assuredly the iteration (10.10.3) is asymptotically stable at α. Furthermore, if z_0 is taken to be inside or at one end of $[a, b]$, it then follows from (10.10.4) that

$$|\alpha - z_1| \leqq M|\alpha - z_0| < |\alpha - z_0|$$

so that z_1 is closer to α than is z_0. Consequently, one may be led to conclude that also $|\alpha - z_2| \leqq M|\alpha - z_1| \leqq M^2|\alpha - z_0|$ and, by induction, that $|\alpha - z_k| \leqq M^k|\alpha - z_0|$, so that z_k will necessarily *converge* to α as $k \to \infty$.

The flaw in this argument is that there is no certainty that z_1 in fact will fall in $[a, b]$, so that (10.10.7) is true when $x = z_1$, but only that this will be the case *when z_0 is close enough to α.*

An additional condition which clearly is *sufficient* to ensure that z_1 and all subsequent z_k's remain in $[a, b]$, and hence that convergence to α will indeed follow in the preceding situation, is the requirement that $F(x)$ be such that $a \leqq F(x) \leqq b$ for *all* x such that $a \leqq x \leqq b$.

Another sufficient condition for convergence with any choice of z_0 in $[a, b]$, assuming again that α is known to lie in that interval, is that

$$0 < F'(x) < 1 \qquad \text{when } a < x < b \tag{10.10.8}$$

This follows easily from the fact that if z_k is in $[a, b]$, then

$$\begin{aligned} z_{k+1} - z_k &= F(z_k) - z_k \\ &= (\alpha - z_k) - [F(\alpha) - F(z_k)] \\ &= (\alpha - z_k)[1 - F'(\xi_k)] \end{aligned}$$

where $a < \xi_k < b$. Hence, since (10.10.8) guarantees that $0 < 1 - F'(\xi_k) < 1$, it follows that $z_{k+1} - z_k$ has the same sign as $\alpha - z_k$ and has a smaller magnitude. Thus z_{k+1} is between z_k and α and hence also in $[a, b]$.

In other cases, even though it is established that $|F'(\alpha)| < 1$, it may be difficult to determine in advance whether convergence is ensured for a particular z_0; and often one must determine by numerical calculation whether the initial approximation is in fact *sufficiently good.*

In view of (10.10.5), we may notice that if $0 < |F'(\alpha)| < 1$, and if the iteration (10.10.3) converges to α, then the relation

$$\alpha - z_k \sim A\beta^k \tag{10.10.9}$$

will be valid for *some* constants A and β, independent of k, when k is sufficiently large. If we rewrite this relation with k replaced by $k + 1$ and by $k + 2$, and eliminate the unknown A and β from the resultant three relations, we may deduce the approximation

$$\frac{\alpha - z_{k+2}}{\alpha - z_{k+1}} \approx \frac{\alpha - z_{k+1}}{\alpha - z_k}$$

which yields the estimate

$$\alpha \approx \frac{z_k z_{k+2} - z_{k+1}^2}{z_{k+2} - 2z_{k+1} + z_k}$$

or, equivalently,

$$\alpha \approx z_{k+2} - \frac{(z_{k+2} - z_{k+1})^2}{z_{k+2} - 2z_{k+1} + z_k} \equiv z_{k+2} - \frac{(\Delta z_{k+1})^2}{\Delta^2 z_k} \tag{10.10.10}$$

where

$$\Delta z_k \equiv z_{k+1} - z_k \qquad \Delta^2 z_k \equiv \Delta z_{k+1} - \Delta z_k = z_{k+2} - 2z_{k+1} + z_k$$

Thus, if three successive iterates z_k, z_{k+1}, and z_{k+2} are known, this relation affords an *extrapolation* which may be expected to provide an improved estimate of α, when the iteration converges. This procedure for accelerating convergence is often called *Aitken's* Δ^2 *process*. In the preceding example, with $z_3 = 1.3245$, $\Delta z_2 = 0.0007$, and $\Delta^2 z_1 = -0.0031$, to four places, (10.10.10) yields the extrapolation $\alpha \approx 1.3245 + 0.0002 = 1.3247$, which happens to agree with z_4 to four places and is correct to those four places. If additional digits had been retained in the calculation of the iterates z_1, z_2, and z_3, even though those digits were not of apparent importance to the iterates themselves, the approximate value of α obtained from them by an extrapolation based on (10.10.10) would have been found to be correct to additional places.

In a wide class of related methods for dealing with (10.10.1), a recurrence formula of the type

$$z_{k+1} = z_k - \frac{f(z_k)}{\gamma_k} \qquad (10.10.11)$$

is used, with a suitable definition of the auxiliary sequence $\gamma_0, \gamma_1, \ldots, \gamma_k, \ldots$. The relation (10.10.3) can be specialized to (10.10.11) by writing $F(x) = x - \phi(x)f(x)$, where $\phi(x)$ is a function such that $\phi(z_k) = 1/\gamma_k$. It should be noticed, however, that the function $F(x) - x$ relevant to the method of successive substitutions is not necessarily proportional to $f(x)$, but is required only to be a function which vanishes at the required point α for which f vanishes. Conversely, the explicit definition of a function $\phi(x)$ which takes on the chosen value $1/\gamma_k$ when $x = z_k$ obviously is not necessary in the present case.

Since $f(\alpha) = 0$, the recurrence relation (10.10.11) implies the relation

$$\begin{aligned} \alpha - z_{k+1} &= \alpha - z_k - \frac{f(\alpha) - f(z_k)}{\gamma_k} \\ &= (\alpha - z_k)\left[1 - \frac{1}{\gamma_k} f'(\xi_k)\right] \qquad (10.10.12) \end{aligned}$$

where ξ_k is between z_k and α. Thus the *convergence factor* ρ_k at the kth stage is given, to a first approximation, by $1 - [f'(\alpha)/\gamma_k]$ when z_k is near α, and, unless this factor is smaller than unity in magnitude, so that

$$0 < \frac{f'(\alpha)}{\gamma_k} < 2 \qquad (10.10.13)$$

when k is large, convergence of z_k to α generally cannot be obtained.

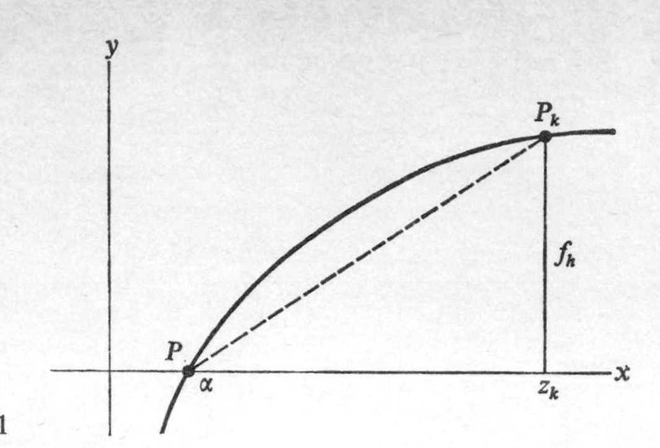

FIGURE 10.1

It is clear that if the z sequence converges, so that $z_{k+1} - z_k \to 0$, and if γ_k is bounded as $k \to \infty$, there then follows $f(z_k) \to 0$, so that z_k tends to a solution of (10.10.1). In particular, the requirement $z_{k+1} = \alpha$, where $f(\alpha) = 0$, would imply that

$$\gamma_k = \frac{0 - f(z_k)}{\alpha - z_k} \tag{10.10.14}$$

so that γ_k would then represent the slope of the secant line joining the points $P_k(z_k, f_k)$ and $P(\alpha, 0)$ in Fig. 10.1. Thus it is desirable to define the γ sequence in such a way that this situation is *approximated* at each stage of the calculation.

In the method of *false position* (*regula falsi*), the iteration is initiated by finding z_0 and z_1 such that f_0 and f_1 are of opposite signs, and by defining γ_1 as the slope of the secant P_0P_1 (Fig. 10.2), so that

$$z_2 = z_1 - \frac{z_1 - z_0}{f_1 - f_0} f_1 = \frac{f_1 z_0 - f_0 z_1}{f_1 - f_0} \tag{10.10.15}$$

In each following iteration, γ_k is taken as the slope of the line joining P_k and the most recently determined point at which the ordinate differs in sign from that at P_k. The procedure is seen to be merely iterated linear inverse interpolation and is clearly certain to converge, although the *rate* of convergence may be slow. In the case of the preceding example, in which $f(x) = x^3 - x - 1$, with the starting values $z_0 = 1.3$ and $z_1 = 1.4$, the next three iterates may be found as follows:

z_k	f_k	$1/\gamma_k$	$-f_k/\gamma_k$
1.3	−0.103	—	—
1.4	0.344	0.224	−0.077
1.323	−0.00731	0.219	0.0016
1.3246	−0.000503	0.219	0.000110
1.324710			

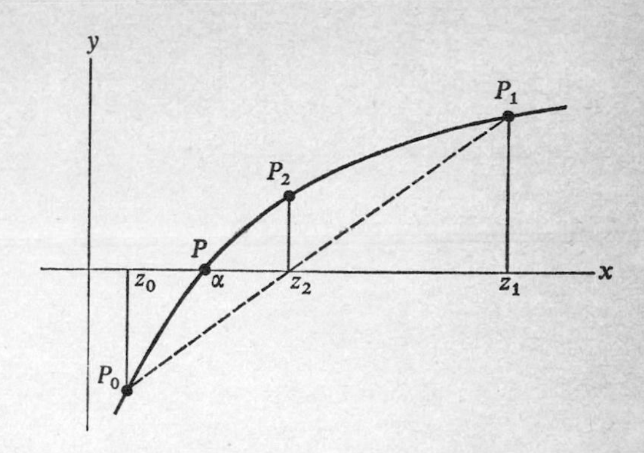

FIGURE 10.2

As this example illustrates, the factor γ_k often changes slowly after the first few steps, and the rate of convergence then is not significantly reduced if, from such a stage onward, γ_k is assigned a constant value.

In this illustration, the approximation z_4 was obtained by *interpolation* based on z_3 and z_1, in accordance with the preceding description of the procedure. If, instead, the *last two* abscissas available are used, so that here z_4 is obtained by *extrapolation* based on z_3 and z_2, with $1/\gamma_3 \doteq 0.235$, a better approximation (1.324718) is obtained. More generally, whereas the systematic use of the slope of the secant $P_{k-1}P_k$ cannot be guaranteed to yield a convergent sequence when it requires extrapolation, this modified procedure is usually advantageous when it does converge (see Sec. 10.12). It is often called the *secant method* and can be described by the iteration formula

$$z_{k+1} = z_k - \frac{z_k - z_{k-1}}{f_k - f_{k-1}} f_k \equiv z_k - \frac{f_k}{f[z_{k-1}, z_k]} \qquad (10.10.16)$$

with the divided-difference notation of Chap.2.

Another simple modification of the method of false position, with desirable properties, can be deduced by applying to it the Aitken acceleration process. It is noted first that, except in the special cases when $f''(\alpha) = 0$, the false-position iteration ultimately has one *stationary* end point (at which $ff'' > 0$), this situation occurring when f'' is of constant sign between $x = z_k$ and $x = z_{k+1}$. [See Prob. 42(*a*). In the preceding example, $z_1 = 1.4$ is stationary.] The sequence of the complementary end points (at which $ff'' < 0$) then tends to α monotonically, and the resultant one-sided approach may be slow.

To simplify the notation, we suppose here that z_0 is a stationary point and that z_1 is separated from z_0 by the desired root α, with $f''(x) \neq 0$ in $[z_0, z_1]$.

The false-position iteration then specializes to the form

$$z_{k+1} = z_k - \frac{z_k - z_0}{f_k - f_0} f_k \qquad (k \geqq 1) \quad (10.10.17)$$

Consequently, there follows also

$$z_{k+2} - z_{k+1} = -\frac{z_{k+1} - z_0}{f_{k+1} - f_0} f_{k+1} \qquad z_{k+1} - z_k = \frac{f_k}{f_0}(z_{k+1} - z_0)$$

and the introduction of these relations into the Aitken approximation (10.10.10) can be expressed in the convenient form

$$\alpha \approx z_{k+1} - \frac{z_{k+1} - z_0}{f_{k+1} - (1 - f_{k+1}/f_k)f_0} f_{k+1} \quad (10.10.18)$$

after a little manipulation. The right-hand member of this relation then can be used to provide a modified definition of z_{k+2} so that, when k then is replaced by $k - 1$, following approximations to α may be generated by the relation

$$z_{k+1} = z_k - \frac{z_k - z_0}{f_k - \mu_k f_0} f_k \qquad (k \geqq 2) \quad (10.10.19)$$

where

$$\mu_k = 1 - \frac{f_k}{f_{k-1}} \quad (10.10.20)$$

in place of (10.10.17).

The process so defined, which could be called an *accelerated false-position iteration*, differs from the classical process when z_0 is stationary only in that f_0 is replaced by $\mu_k f_0$ in the calculation of z_{k+1}.

Calculation shows that if $f'(\alpha)f''(\alpha) \neq 0$, there follows

$$|\alpha - z_{k+1}| \sim A|\alpha - z_k| \quad (10.10.21)$$

where the *asymptotic convergence factor* is

$$A = 1 - \sqrt{\frac{(z_0 - \alpha)f'(\alpha)}{f(z_0)}} \quad (10.10.22)$$

When $|z_0 - \alpha|$ is small, there follows also

$$A \approx \frac{(z_0 - \alpha)f''(\alpha)}{4f'(\alpha)} \quad (10.10.23)$$

so that A then also is small, as is desirable for rapid convergence. The corresponding factor in the classical case is given by $1 - [(z_0 - \alpha)f'(\alpha)/f(z_0)]$,

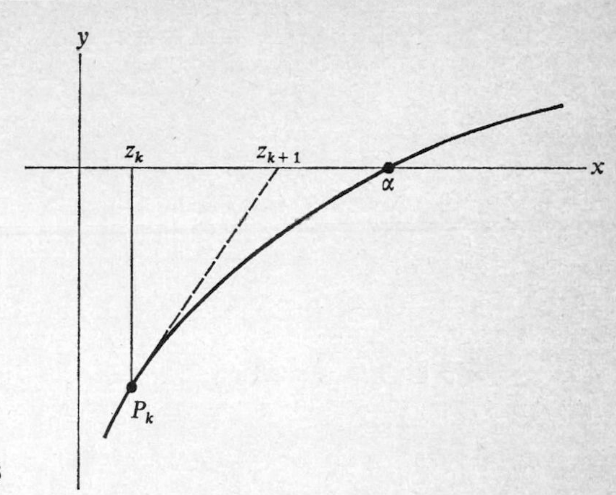

FIGURE 10.3

from which it follows that when $|z_0 - \alpha|$ is small, the acceleration approximately *halves* the asymptotic convergence factor.†

A particularly simple iterative process, often called the *bisection method*, consists of merely evaluating $f(x)$ at the *midpoint* x_3 of the interval $[x_1, x_2]$ at the ends of which $f(x)$ has opposite signs, discarding that one of x_1 and x_2 at which the ordinate has the same sign as $f(x_3)$ and repeating the process until half the length of the subinterval inside which α continues to be trapped is within the prescribed error tolerance. This process, in common with that of false position, is certain to converge; but here, in addition, the number of steps which will suffice is determinate in advance since after k steps the error will be smaller than $2^{-k-1}(x_2 - x_1)$ if the approximation to α at that stage is taken to be the abscissa of the midpoint of the remaining kth subinterval. Needless to say, the requisite number of steps may well turn out to be intolerably large.

10.11 The Newton-Raphson Method

An important method, associated with the names of Newton and Raphson, consists of taking γ_k in (10.10.11) as the slope of the curve $y = f(x)$ at the point z_k (Fig. 10.3), so that (10.10.11) becomes

$$z_{k+1} = z_k - \frac{f(z_k)}{f'(z_k)} \qquad (10.11.1)$$

† A similar process, in which μ_k is arbitrarily taken to be 2^{-k} (or to be c^{-k} for some c) is advocated by Hamming [1971]. Although successive approximations yielded by this process frequently approach α rather rapidly for a while, they ultimately will exhibit divergence except in very special cases.

This iteration is seen to be also the special case of (10.10.3) in which

$$F(x) = x - \frac{f(x)}{f'(x)}$$

and hence $F'(x) = f(x)f''(x)/[f'(x)]^2$. Thus, if $f'(\alpha) \neq 0$ and $f''(\alpha)$ is finite, there follows $F'(\alpha) = 0$, so that the convergence factor tends to zero when and if $z_k \to \alpha$.

In order to examine the behavior of the error $\alpha - z_k$, we rewrite (10.11.1) in the equivalent form

$$\alpha - z_{k+1} = \alpha - z_k - \frac{f(\alpha) - f(z_k)}{f'(z_k)} \tag{10.11.2}$$

and recall that

$$f(\alpha) - f(z_k) = (\alpha - z_k)f'(z_k) + \tfrac{1}{2}(\alpha - z_k)^2 f''(\xi_k)$$

where ξ_k lies between z_k and α, if $f''(x)$ is continuous in that interval, so that (10.11.2) becomes

$$\alpha - z_{k+1} = -\tfrac{1}{2}(\alpha - z_k)^2 \frac{f''(\xi_k)}{f'(z_k)} \tag{10.11.3}$$

Thus, if the iteration converges to α, there follows

$$\alpha - z_{k+1} \sim -\frac{f''(\alpha)}{2f'(\alpha)}(\alpha - z_k)^2 \qquad (k \to \infty) \tag{10.11.4}$$

provided that $f'(\alpha)$ and $f''(\alpha)$ are both finite and nonzero.

It is important to notice that here the error in z_{k+1} tends to be proportional to the *square* of the error in z_k, as $k \to \infty$, whereas in the other methods so far considered the two successive errors generally tend to be in a constant ratio, if the iteration converges. We say that such an iteration is a *second-order* process, whereas the preceding methods generally are *first-order* processes. If this method is applied to (10.10.6), the recurrence formula (10.11.1) becomes

$$z_{k+1} = z_k - \frac{z_k^3 - z_k - 1}{3z_k^2 - 1} = \frac{2z_k^3 + 1}{3z_k^2 - 1}$$

and, with $z_0 = 1.3$, the results of the first two iterations are $z_1 = 1.325$ and $z_2 = 1.324718$ when rounded to the places given.

Use can be made of (10.11.4) to predict in advance the probable number of correct digits in each iterate. For, since here f''/f' has a value of about 2 when $x = z_0 = 1.3$, it may be expected that the coefficient of $(\alpha - z_k)^2$ in (10.11.4) will have a value *approximating* -1, so that the error ε_k in the kth iterate will be of magnitude approximately the *square* of that of the preceding iterate and will be of negative sign, if the iteration converges to α. If convergence is assumed, and if initially it is known that the true value lies between 1.3 and

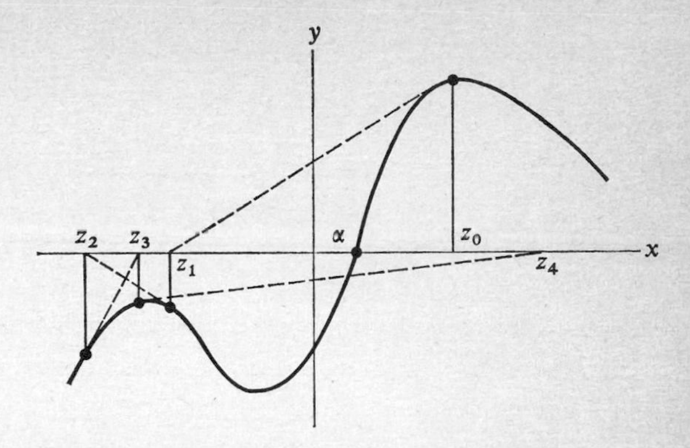

FIGURE 10.4

1.4 and hence that z_0 is in error by less than 0.1, it can be predicted that z_1 will be in error by less than about 0.01, so that three places would be retained. With ε_0 reestimated as $z_1 - z_0 \doteq 0.03$, there follows $|\varepsilon_1| \approx 10^{-3}$. Hence ε_2 may be expected to be less than about 10^{-6}, so that seven places might be retained at that stage. A comparison of z_1 and z_2 confirms the earlier prediction (although this method of error estimation may be undependable in early stages, in other cases) and suggests that the error in the next iterate z_3 will be in about the *twelfth* decimal place. Rigorous error *bounds* naturally would be obtained by use of (10.11.3) rather than the asymptotic formula (10.11.4).

If the curve representing $y = f(x)$ possesses turning points or inflections in the interval between the initial estimate $x = z_0$ and the true root $x = \alpha$, or between z_0 and z_1, the iteration may not converge to α, as is illustrated in Fig. 10.4, although it may well converge to some *other* root. However, if $f'(x)$ and $f''(x)$ do not change sign in the interval (z_0, α), and if $f(z_0)$ and $f''(z_0)$ have the same sign,† so that the iteration is initiated at a point at which the curve representing $y = f(x)$ is concave *away* from the x axis (as, for example, in Fig. 10.3), it is easily seen, by geometrical considerations or otherwise, that successive iterates must tend to $x = \alpha$ and that they all lie between z_0 and α. If $f(z_0)$ and $f''(z_0)$ have opposite signs, the first iterate z_1 is on the opposite side of α and convergence to α is uncertain unless $f'(x)$ and $f''(x)$ also do not change sign at $x = \alpha$ or in the interval (z_1, α), in which case convergence then follows as before.

If α is a zero of multiplicity m, where $m > 1$, so that

$$f(\alpha) = f'(\alpha) = \cdots = f^{(m-1)}(\alpha) = 0 \qquad f^{(m)}(\alpha) \neq 0 \qquad (10.11.5)$$

† These conditions are associated with Fourier. For less restrictive sets of conditions which also ensure convergence, see Henrici [1964], pp. 79–81, and Ostrowski [1966], p. 44.

the relation (10.11.4) no longer holds, but the Newton-Raphson method then can be shown to reduce to a *first*-order process. In such cases there are modifications which restore the order to two. (See Prob. 68.)

However, situations in which two (or more) zeros are coincident, or *nearly* so, generally are troublesome numerically since successive steps in the relevant iterative process then usually involve the evaluation of ratios of increasingly small quantities and high-precision computation is required (unless cancellation of factors or other simplification, by analytical methods, is possible). A method which often is useful in such cases is indicated in Prob. 55. When $f(x)$ is a *polynomial*, an alternative method consists of seeking a *quadratic* factor rather than a linear one (see Secs. 10.18 and 10.19).

The preceding methods can be combined and modified in various ways. In particular, if $f'(z_k)$ begins to change slowly with k after (say) r iterations, the Newton-Raphson procedure may be modified by taking $\gamma_k = f'(z_r)$ for all $k \geqq r$ or by recalculating the derivative (say) once out of every two or three steps. The method of false position may be modified, for example, by taking γ_k as the slope of the secant P_0P_1 for *all* k, where P_0 and P_1 are two fixed points on the curve $y = f(x)$ near to and separated by the point P at which $x = \alpha$, or by taking γ_k as the slope of the secant P_0P_k, where P_0 is an appropriately chosen fixed point on the curve. Whereas such modifications lead to reductions in labor, their use clearly may also adversely affect the nature and the rate of convergence.

10.12 Iterative Methods of Higher Order

Any iteration in which it is true that

$$|\alpha - z_{k+1}| \sim A|\alpha - z_k|^r \qquad (k \to \infty) \qquad (10.12.1)$$

with $A \neq 0$, when the iteration converges, is said to be a process of *order r*, which need not be an integer. The constant A is called the *asymptotic error constant*. It may be seen that any iterative process of order exceeding unity certainly will yield *convergence* to α if the iteration is initiated *sufficiently near to* α. On the other hand, when $r = 1$, this statement generally cannot be made unless the asymptotic convergence factor associated with α is smaller than 1 in absolute value. (The methods of false position and bisection are notable favorable exceptions.)

It is of some interest to determine the order of the *secant method* (that is, the variant of the method of false position in which linear extrapolation or interpolation is always based on the two most recently determined ordinates).

In this case it is true that

$$\alpha - z_{k+1} \sim C(\alpha - z_k)(\alpha - z_{k-1}) \qquad (10.12.2)$$

when the iteration converges, where $C = -\frac{1}{2}f''(\alpha)/f'(\alpha)$ (see Prob. 69). Assuming that $f'(\alpha)$ and $f''(\alpha)$ are finite and nonzero, and denoting the unknown order of the process by r, there must follow

$$A|\alpha - z_k|^r \sim |C||\alpha - z_k| \left(\frac{|\alpha - z_k|}{A}\right)^{1/r}$$

Thus r must be the positive root of the equation

$$r = 1 + \frac{1}{r}$$

and hence

$$r = \frac{1 + \sqrt{5}}{2} \doteq 1.62\dagger \qquad (10.12.3)$$

Since also

$$A = |C|^{1/r} = |C|^{r-1}$$

we conclude that here

$$|\varepsilon_{k+1}| \sim \left|\frac{f''(\alpha)}{2f'(\alpha)}\right|^{r-1} |\varepsilon_k|^r \qquad (10.12.4)$$

where r is given by (10.12.3).

In addition to the processes so far considered, a great variety of other iterative methods for the purpose of approximating the zeros of functions have been devised and studied. In particular, families of processes of higher order (integral or nonintegral) are available for computation. Although very often it is true that the theoretical advantages of such formulas are more than offset by the additional time and/or effort per iteration required by their use, this is not always the case. Accordingly, a few interrelated examples are considered briefly in this section; throughout the section it is assumed that the required zero is not repeated, so that $f'(\alpha) \neq 0$.

As a generalization of the secant method, which employs linear interpolation or extrapolation based on the values of $f(x)$ at two points z_k and z_{k-1} near the desired zero α, one may be led to investigate a *quadratic* process based on *three* such values. The second-degree polynomial $y(x)$ agreeing with $f(x)$ at the points z_k, z_{k-1}, and z_{k-2} could be expressed in the lagrangian form

† The reciprocal of this number is often called the *golden mean*.

(3.2.12) or in the newtonian divided-difference form (2.5.2). For present purposes we choose the latter form

$$y(x) = f_k + (x - z_k)f[z_k, z_{k-1}] + (x - z_k)(x - z_{k-1})f[z_k, z_{k-1}, z_{k-2}] \tag{10.12.5}$$

where

$$f[z_k, z_{k-1}] \equiv \frac{f_k - f_{k-1}}{z_k - z_{k-1}}$$

$$f[z_k, z_{k-1}, z_{k-2}] \equiv \frac{f[z_k, z_{k-1}] - f[z_{k-1}, z_{k-2}]}{z_k - z_{k-2}} \tag{10.12.6}$$

With the convenient abbreviations

$$\lambda_k = \frac{f_k}{\omega_k} \qquad \mu_k = \frac{f[z_k, z_{k-1}, z_{k-2}]}{\omega_k} \tag{10.12.7}$$

where

$$\omega_k = \begin{cases} f[z_k, z_{k-1}] + (z_k - z_{k-1})f[z_k, z_{k-1}, z_{k-2}] \\ f[z_k, z_{k-1}] + (f_k - f_{k-1})\dfrac{f[z_k, z_{k-1}, z_{k-2}]}{f[z_k, z_{k-1}]} \end{cases} \tag{10.12.8}$$

the equation $y(x) = 0$ takes the form

$$\mu_k(x - z_k)^2 + (x - z_k) + \lambda_k = 0 \tag{10.12.9}$$

and the identification of z_{k+1} with the proper root of this equation gives

$$z_{k+1} = z_k - \frac{2\lambda_k}{1 + \sqrt{1 - 4\mu_k\lambda_k}} \tag{10.12.10}$$

Here the ambiguous sign was chosen so that $z_{k+1} = z_k$ when $f(z_k) = 0$.

This formula is equivalent to a somewhat more complicated one obtained by Muller using the lagrangian form of $y(x)$, which has found some favor in practice. For this iteration it is found that

$$\varepsilon_{k+1} \sim -\frac{f'''(\alpha)}{6f'(\alpha)}\varepsilon_k\varepsilon_{k-1}\varepsilon_{k-2} \tag{10.12.11}$$

and, accordingly, that

$$|\varepsilon_{k+1}| \sim \left|\frac{f'''(\alpha)}{6f'(\alpha)}\right|^{(r-1)/2} |\varepsilon_k|^r \tag{10.12.12}$$

where r is the real root of the equation

$$r = 1 + \frac{1}{r} + \frac{1}{r^2}$$

and hence

$$r \doteq 1.84 \tag{10.12.13}$$

It may be noted that despite the relative complexity of the Muller iteration, its order is inferior to that of the Newton-Raphson iteration. In addition, its use requires a square-root extraction in each step and hence also may provide a nonreal approximation to a real zero. However, the fact that the value of the derivative $f'(z_k)$ is not needed in the kth step affords a computational advantage in those cases when the evaluation of $f'(z_k)$ is undesirable or impossible.

Now, just as the Newton-Raphson formula can be considered as the "tangent formula" obtained by confluence of the two data points for the secant formula, new formulas can be derived from (10.12.10) by letting either z_{k-1} or both z_{k-1} and z_{k-2} tend to z_k in (10.12.10). In either case there follows

$$\omega_k = f[z_k, z_k] = f'_k$$

Thus, in the former case, if we relabel the distinct abscissa as z_{k-1}, the iteration is again given by (10.12.10), where now

$$\lambda_k = \frac{f_k}{f'_k} \qquad \mu_k = \frac{f'_k - f[z_k, z_{k-1}]}{(z_k - z_{k-1})f'_k} \tag{10.12.14}$$

This process corresponds to interpolating or extrapolating from z_k along a *parabola* which is tangent to the curve at (z_k, f_k) and which also passes through the point (z_{k-1}, f_{k-1}). It is found that here

$$\varepsilon_{k+1} \sim -\frac{f'''(\alpha)}{6f'(\alpha)}\,\varepsilon_k^2\varepsilon_{k-1} \tag{10.12.15}$$

and accordingly that

$$|\varepsilon_{k+1}| \sim \left|\frac{f'''(\alpha)}{6f'(\alpha)}\right|^{(r-1)/2} |\varepsilon_k|^r \tag{10.12.16}$$

where

$$r = 1 + \sqrt{2} \doteq 2.41 \tag{10.12.17}$$

In the confluent case when z_{k-1} and z_{k-2} are *both* made to coincide with z_k, the iteration (10.12.10) becomes

$$z_{k+1} = z_k - \frac{2(f_k/f'_k)}{1 + \{1 - 2[f_k f''_k/(f'_k)^2]\}^{1/2}} \tag{10.12.18}$$

and it is found that

$$\varepsilon_{k+1} \sim -\frac{f'''(\alpha)}{6f'(\alpha)}\,\varepsilon_k^3 \tag{10.12.19}$$

so that the order is 3. Here the curve representing $y(x)$ and that representing $f(x)$ agree when $x = z_k$ and also have the same slope and the same curvature at the point (z_k, f_k).

From these formulas we may derive more simple ones by introducing approximations. For example, if (for small f_k) we introduce the approximation

$$\sqrt{1 - 2\frac{f_k f''_k}{(f'_k)^2}} \approx 1 - \frac{f_k f''_k}{(f'_k)^2}$$

in (10.12.18), we obtain the formula

$$z_{k+1} = z_k - \frac{f_k/f'_k}{1 - f_k f''_k/2(f'_k)^2} \tag{10.12.20}$$

for which

$$\varepsilon_{k+1} \sim \left\{\left[\frac{f''(\alpha)}{2f'(\alpha)}\right]^2 - \frac{f'''(\alpha)}{6f'(\alpha)}\right\}\varepsilon_k^3 \tag{10.12.21}$$

which is also of third order, but which does not require a square-root extraction. This is the frequently rediscovered formula of *Halley* (see footnote on page 513). Iterative approximation based on this formula is also sometimes called *Bailey's method*, or *Lambert's method.*

If we write

$$\left[1 - \frac{f_k f''_k}{2(f'_k)^2}\right]^{-1} \approx 1 + \frac{f_k f''_k}{2(f'_k)^2}$$

in Halley's formula, we obtain the iteration

$$z_{k+1} = z_k - \frac{f_k}{f'_k}\left[1 + \frac{f_k f''_k}{2(f'_k)^2}\right] \tag{10.12.22}$$

for which

$$\varepsilon_{k+1} \sim \left\{2\left[\frac{f''(\alpha)}{2f'(\alpha)}\right]^2 - \frac{f'''(\alpha)}{6f'(\alpha)}\right\}\varepsilon_k^3 \tag{10.12.23}$$

This is the third-order iteration supplied by the result of Prob. 60 and sometimes called *Chebyshev's formula.*

As a final illustration, which is also of practical significance, we start with (10.12.10), notice that λ_k will be small when f_k is small, and hence approximate $(1 - 4\mu_k\lambda_k)^{1/2}$ by $1 - 2\mu_k\lambda_k$, as was done in deriving (10.12.20). The resultant formula is

$$\begin{aligned} z_{k+1} &= z_k - \frac{\lambda_k}{1 - \mu_k\lambda_k} \\ &= z_k - \frac{f_k}{\omega_k - (f_k/\omega_k)f[z_k, z_{k-1}, z_{k-2}]} \end{aligned} \tag{10.12.24}$$

If we replace ω_k in its first appearance here by the second form of its definition (10.12.8) and replace it in its second (less significant) appearance by its approximation $f[z_k, z_{k-1}]$, the formula (10.12.24) takes the form

$$z_{k+1} = z_k - \frac{f_k}{f[z_k, z_{k-1}] - (f_{k-1}/f[z_k, z_{k-1}])f[z_k, z_{k-1}, z_{k-2}]} \tag{10.12.25}$$

for which

$$|\varepsilon_{k+1}| \sim \left|\frac{f'''(\alpha)}{6f'(\alpha)}\right|^{(r-1)/2} |\varepsilon_k|^r \tag{10.12.26}$$

where

$$r \doteq 1.84 \tag{10.12.27}$$

This formula, due to Traub [1964], can be considered, not only as a simplified approximation to the Muller formula, but also as a divided-difference simulation to the Halley formula. It has the virtue that its order and asymptotic error coefficient are the same as those of Muller's formula but that it does not require a root extraction. (The inviting replacement of f_{k-1} by f_k in Traub's formula, however, would in fact reduce the order.) In comparison with the Halley formula, we see that the replacement of derivatives by divided differences does cost an order reduction from 3 to about 1.84.

10.13 Sets of Nonlinear Equations

Some of the methods in preceding sections are readily generalized to the treatment of two or more simultaneous nonlinear equations (algebraic or transcendental). Thus, for example, the two simultaneous equations

$$f(x, y) = 0 \qquad g(x, y) = 0 \tag{10.13.1}$$

can be written (in various ways) in equivalent forms

$$x = F(x, y) \qquad y = G(x, y) \tag{10.13.2}$$

and the method of *successive substitutions* can be based on the recurrence formulas

$$x_{k+1} = F(x_k, y_k) \qquad y_{k+1} = G(x_k, y_k) \tag{10.13.3}$$

When the iteration converges to the true solution pair, say, $x = \alpha$ and $y = \beta$, it can be shown that the errors in the kth iterates tend to be described by the relations

$$\alpha - x_k \approx A_1\lambda_1^k + B_1\lambda_2^k \qquad \beta - y_k \approx A_2\lambda_1^k + B_2\lambda_2^k$$

where A_1, A_2, B_1, and B_2 are constants, independent of k, and where λ_1 and λ_2 are the roots of the equation

$$\begin{vmatrix} \lambda - F_x & -F_y \\ -G_x & \lambda - G_y \end{vmatrix} = 0$$

or

$$\lambda^2 - (F_x + G_y)\lambda + (F_xG_y - F_yG_x) = 0 \qquad (10.13.4)$$

with the partial derivatives evaluated at (α, β), if $F_xG_y \neq F_yG_x$ at that point. The constants A_1, B_1 and A_2, B_2 will be conjugate complex if the same is true of λ_1, λ_2. Thus the iteration will be asymptotically stable at (α, β) if and only if the roots λ_1 and λ_2 are smaller than unity in absolute value, the necessary and sufficient conditions for which are

$$|F_x + G_y| \leqq F_xG_y - F_yG_x + 1 < 2 \qquad (10.13.5)$$

A more stringent pair of conditions, which is *sufficient* (but generally not necessary) for asymptotic stability, is of the form

$$|F_x| + |F_y| < 1 \qquad |G_x| + |G_y| < 1 \qquad (10.13.6)$$

As before, these conditions are not sufficient for *convergence*, in that the iteration may fail to converge even though they are satisfied, unless the iteration is started with (x_0, y_0) sufficiently near (α, β).

The preceding discussion generalizes in the obvious way to the case when n simultaneous equations are involved, with $n \geqq 3$, except that no simple generalization of (10.13.5) to n dimensions is available.

The *Newton-Raphson iteration*, as applied to the solution of (10.13.1), is based on the result of replacing (α, β) by (x_{k+1}, y_{k+1}) in the right-hand members of the Taylor expansions

$$\begin{aligned} 0 = f(\alpha, \beta) &= f(x_k, y_k) + (\alpha - x_k)f_x(x_k, y_k) + (\beta - y_k)f_y(x_k, y_k) + \cdots \\ 0 = g(\alpha, \beta) &= g(x_k, y_k) + (\alpha - x_k)g_x(x_k, y_k) + (\beta - y_k)g_y(x_k, y_k) + \cdots \end{aligned} \qquad (10.13.7)$$

and neglecting nonlinear terms in $x_{k+1} - x_k$ and $y_{k+1} - y_k$, so that the recurrence formulas are of the form

$$\begin{aligned} (x_{k+1} - x_k)f_x(x_k, y_k) + (y_{k+1} - y_k)f_y(x_k, y_k) &= -f(x_k, y_k) \\ (x_{k+1} - x_k)g_x(x_k, y_k) + (y_{k+1} - y_k)g_y(x_k, y_k) &= -g(x_k, y_k) \end{aligned} \qquad (10.13.8)$$

Rather than resolve these equations for x_{k+1} and y_{k+1}, it is usually convenient to solve them, as written, for the *corrections* $\Delta x_k \equiv x_{k+1} - x_k$ and $\Delta y_k \equiv y_{k+1} - y_k$, which are to be added to x_k and y_k to yield the following iterates. When the iteration converges, the errors in the $(k + 1)$th iterates generally tend to become linear combinations of the *squares* and *products* of the errors in the kth iterates (that is, the iteration is a *second-order* process), whereas, in the method of successive substitutions, based on (10.13.3), the new errors generally tend to become linear combinations of the preceding errors themselves.

The generalization to n dimensions is obvious. In the two-dimensional case considered here, it can be seen that the Newton-Raphson process amounts to approximating the *surfaces* representing the relations

$$z = f(x, y) \qquad z = g(x, y) \tag{10.13.9}$$

in *three*-dimensional space by their *tangent planes* at (x_k, y_k, f_k) and (x_k, y_k, g_k), respectively, determining the common intersection (if it exists) of these planes with the plane $z = 0$, and finally identifying (x_{k+1}, y_{k+1}) with the x and y coordinates of that point.

When the so-called *Jacobian determinant* of f and g

$$J \equiv J(f, g) \equiv \begin{vmatrix} \dfrac{\partial f}{\partial x} & \dfrac{\partial f}{\partial y} \\ \\ \dfrac{\partial g}{\partial x} & \dfrac{\partial g}{\partial y} \end{vmatrix} \tag{10.13.10}$$

vanishes at the point (x_k, y_k), Eqs. (10.13.8) do not possess a unique solution. In this case, the lines in which the relevant tangent planes intersect the xy plane are either parallel or coincident. More generally, if J vanishes at or near a point (α, β) at which the curves $f = 0$ and $g = 0$ intersect, and if the higher derivatives of f and g exist in that neighborhood, then either the curves are *tangent* at their intersection or they possess, in fact, *two* (or more) nearly coincident intersections. In either case, a particularly unfavorable behavior of the Newton-Raphson sequence is to be anticipated.

One method of dealing with such a situation consists of first determining the common solution of the equations

$$f = 0 \qquad J(f, g) = 0 \tag{10.13.11}$$

(or of $g = 0$ and $J = 0$) near the desired point (α, β) by Newton-Raphson iteration or otherwise. The difficulty just discussed generally will not occur also in this iteration unless (α, β) is one of *three* or more coincident or nearly coincident zeros, in which case further modifications are needed. If the solution of (10.13.11) is denoted by $(\bar{x}, \bar{y})$, and if $f(x, y)$ and $g(x, y)$ are expanded in powers of $x - \bar{x}$ and $y - \bar{y}$, the conditions $f(\alpha, \beta) = 0$ and $g(\alpha, \beta) = 0$ then become

$$(\alpha - \bar{x})\bar{f}_x + (\beta - \bar{y})\bar{f}_y + \tfrac{1}{2}(\alpha - \bar{x})^2\bar{f}_{xx} + (\alpha - \bar{x})(\beta - \bar{y})\bar{f}_{xy} + \tfrac{1}{2}(\beta - \bar{y})^2\bar{f}_{yy} + \cdots = 0 \quad (10.13.12)$$

and

$$\bar{g} + (\alpha - \bar{x})\bar{g}_x + (\beta - \bar{y})\bar{g}_y + \tfrac{1}{2}(\alpha - \bar{x})^2\bar{g}_{xx} + (\alpha - \bar{x})(\beta - \bar{y})\bar{g}_{xy} + \tfrac{1}{2}(\beta - \bar{y})^2\bar{g}_{yy} + \cdots = 0 \quad (10.13.13)$$

where $\bar{f}, \bar{f}_x, \ldots$ abbreviate $f(\bar{x}, \bar{y}), f_x(\bar{x}, \bar{y}), \ldots$, and where account has been taken of the fact that $\bar{f} = 0$.

Equations (10.13.12) and (10.13.13) can be simplified by use of the fact that the condition $J(f, g) = 0$ at $(\bar{x}, \bar{y})$ implies the equality of the ratios $\bar{g}_x/\bar{f}_x$ and $\bar{g}_y/\bar{f}_y$, so that

$$\bar{g}_x = k\bar{f}_x \qquad \bar{g}_y = k\bar{f}_y \quad (10.13.14)$$

for some value of k. Thus the terms involving $\bar{g}_x$ and $\bar{g}_y$ in (10.13.13) can be eliminated by subtracting k times (10.13.12) from (10.13.13); and hence if only second-degree terms are retained in the results, we obtain the two approximate relations

$$(\alpha - \bar{x})\bar{f}_x + (\beta - \bar{y})\bar{f}_y + \tfrac{1}{2}(\alpha - \bar{x})^2\bar{f}_{xx} + (\alpha - \bar{x})(\beta - \bar{y})\bar{f}_{xy} + \tfrac{1}{2}(\beta - \bar{y})^2\bar{f}_{yy} \approx 0 \quad (10.13.15)$$

and

$$a(\alpha - \bar{x})^2 + 2b(\alpha - \bar{x})(\beta - \bar{y}) + c(\beta - \bar{y})^2 \approx -2\bar{g} \quad (10.13.16)$$

where

$$a = \bar{g}_{xx} - k\bar{f}_{xx} \qquad b = \bar{g}_{xy} - k\bar{f}_{xy} \qquad c = \bar{g}_{yy} - k\bar{f}_{yy} \quad (10.13.17)$$

Finally, if we write

$$\beta - \bar{y} = \lambda(\alpha - \bar{x}) \quad (10.13.18)$$

we can recast (10.13.15) and (10.13.16) in the forms

$$\lambda \approx -\frac{\bar{f}_x}{\bar{f}_y} - \frac{\alpha - \bar{x}}{2\bar{f}_y}(\bar{f}_{xx} + 2\lambda\bar{f}_{xy} + \lambda^2\bar{f}_{yy}) \quad (10.13.19)$$

$$\alpha - \bar{x} \approx \pm\left(\frac{-2\bar{g}}{a + 2\lambda b + \lambda^2 c}\right)^{1/2} \quad (10.13.20)$$

which generally are convenient for use of a process of successive substitutions for the determination of α and λ at each of two points, after which the corresponding values of β are given by (10.13.18). The approximations to (α_1, β_1) and (α_2, β_2) so obtained then generally will be sufficiently good to be suitable for further refinement by a conventional method based on the *original* pair of equations if this is desirable. This method appears to be due to Milne [1949].

In particularly difficult situations, when the two simultaneous equations of (10.13.1) are to be solved, it is possible to make use of either the false-position or the bisection method. For this purpose, suppose that the equation $f(x, y) = 0$ determines y as a continuous function $y_f(x)$ near (α, β) and the equation $g(x, y) = 0$ determines y as $y_g(x)$. Then, if x_1 and x_2 can be determined such that $\phi(x) \equiv y_f(x) - y_g(x)$ has opposite signs at these two points, the iterative procedures just mentioned are applicable to $\phi(x)$ with guaranteed convergence.

However, the probable slowness of the convergence here is coupled with the fact that generally the functions $y_f(x)$ and $y_g(x)$ cannot be obtained explicitly. Thus the equations $f(x_k, y) = 0$ and $g(x_k, y) = 0$ will have to be solved *numerically* for $y_f(x_k)$ and $y_g(x_k)$, respectively, at each stage of the iteration by use (say) of one of the preceding one-dimensional methods. Clearly, it may be desirable or necessary in a specific case to interchange the roles of x and y in the preceding description.

Although the other methods considered in this section generalize directly to higher dimensions, unfortunately it is true that no method completely analogous to the false-position or bisection methods exist in n-dimensional space when $n \geqq 3$. However, there does exist a usually convergent method known as the *method of steepest descent*. It is described here for the two-dimensional case, but the generalization to n dimensions is immediate. If a solution (α, β) of the simultaneous equations

$$f(x, y) = 0 \qquad g(x, y) = 0 \tag{10.13.21}$$

is desired, we first form the function $\phi(x, y)$ such that†

$$\phi = \tfrac{1}{2}(f^2 + g^2) \tag{10.13.22}$$

The desired solution then clearly is specified by a point at which ϕ is *minimized*.

We recall next that the gradient vector $\nabla\phi$, with components $\{\phi_x, \phi_y\}$, has the property that at a point in the xy plane it is normal to the curve $\phi =$ constant which passes through that point and, in addition, that its direction at that point is the direction along which ϕ changes most rapidly with distance from that point. Thus, if (x_k, y_k) is the current approximation to (α, β), we are led to define the next approximation in such a way that

$$x_{k+1} = x_k + tu_k \qquad y_{k+1} = y_k + tv_k \tag{10.13.23}$$

where

$$u_k = \phi_x(x_k, y_k) \qquad v_k = \phi_y(x_k, y_k) \tag{10.13.24}$$

so that the change $\{\Delta x_k, \Delta y_k\}$ is in the *direction* of most rapid variation of ϕ.

† Other definitions of ϕ can also be used.

The *amount* of change is then determined in such a way that

$$\phi(x_k + tu_k, y_k + tv_k) = \min \quad (10.13.25)$$

and hence t is to be determined by the equation

$$\frac{d}{dt}\phi(x_k + tu_k, y_k + tv_k) = 0 \quad (10.13.26)$$

after which (x_{k+1}, y_{k+1}) is known.

A somewhat simpler procedure consists of taking $u_0 = 1$, $v_0 = 0$, in the first step; then $u_1 = 0$, $v_1 = 1$, in the second; then $u_2 = 1$, $v_2 = 0$, in the third —and so forth—in (10.13.23) so that in each step only one of the components of the current approximation is modified. When n unknowns are involved, one may modify the components cyclically (in analogy to the Gauss-Seidel iteration) or at each stage one may modify the particular component of the current approximation which corresponds to the *largest* component of the current gradient vector (in partial analogy to the general relaxation process).

As might be anticipated, the convergence may be adversely affected by the presence of critical points where the Jacobian of f and g vanishes and by the fact that the direction of most rapid change of the function ϕ at a point unfortunately may differ significantly from the direction from that point to the desired *minimal* point. In addition, the computation per step may be excessive since the solution of (10.13.26) generally must be effected by an iterative method at each stage of the process.

10.14 Iterated Synthetic Division of Polynomials. Lin's Method

When $f(x)$ is a polynomial of degree n, so that the equation to be solved is an *algebraic* one, methods such as those of preceding sections can be systematized by the use of *synthetic division*. For this purpose, suppose that

$$f(x) = x^n + a_1x^{n-1} + \cdots + a_{n-1}x + a_n \quad (10.14.1)$$

and, first, let $f(x)$ be divided by the linear expression $x - z$, so that

$$\begin{aligned} f(x) &= x^n + a_1x^{n-1} + \cdots + a_{n-1}x + a_n \\ &= (x - z)(x^{n-1} + b_1x^{n-2} + \cdots + b_{n-2}x + b_{n-1}) + R \quad (10.14.2) \end{aligned}$$

where $x^{n-1} + \cdots + b_{n-1}$ represents the quotient, and R is the constant remainder. Here the coefficients $b_1, \ldots, b_{n-1}$ and the remainder R depend upon z. By setting $x = z$ in (10.14.2), it follows, in particular, that

$$R = f(z) \quad (10.14.3)$$

If now the quotient in (10.14.2) is again divided by $x - z$, so that

$$x^{n-1} + b_1 x^{n-2} + \cdots + b_{n-2}x + b_{n-1}$$
$$= (x - z)(x^{n-2} + c_1 x^{n-3} + \cdots + c_{n-3}x + c_{n-2}) + R' \qquad (10.14.4)$$

and hence

$$f(x) = (x - z)^2(x^{n-2} + c_1 x^{n-3} + \cdots + c_{n-2}) + (x - z)R' + R$$

there follows also

$$R' = f'(z) \qquad (10.14.5)$$

and, indeed, if the process is repeated k times, it is easily seen that the remainder $R^{(k)}$ is then $f^{(k)}(z)/k!$.

The method of synthetic division, often known as *Horner's method*, is based on the fact that, by equating coefficients of $x^{n-1}, x^{n-2}, \ldots, x$, and 1 in the two members of (10.14.2), we obtain the relations

$$a_1 = b_1 - z \qquad a_2 = b_2 - zb_1 \qquad \cdots$$
$$a_{n-1} = b_{n-1} - zb_{n-2} \qquad a_n = R - zb_{n-1}$$

Thus, if we introduce the recurrence formula

$$b_k = a_k + zb_{k-1} \qquad (k = 1, 2, \ldots, n) \qquad (10.14.6)$$

with

$$b_0 = 1 \qquad (10.14.7)$$

it follows that this formula will generate the coefficients of the quotient of (10.14.2) with $k = 1, 2, \ldots, n - 1$, and also that

$$R = f(z) = b_n = a_n + ab_{n-1} \qquad (10.14.8)$$

Further, the c's in (10.14.4) are related, for $k = 1, 2, \ldots, n - 2$, to the b's as the b's are related to the a's, and there follows also

$$R' = f'(z) = c_{n-1} = b_{n-1} + zc_{n-2} \qquad (10.14.9)$$

For desk calculation, it is convenient to arrange the entries in parallel columns (or rows), in the form

$$\begin{array}{l|ll|}
1 & 1 & 1 \\
a_1 & b_1 & c_1 \\
a_2 & b_2 & c_2 \\
\vdots & \vdots & \vdots \\
a_{n-2} & b_{n-2} & c_{n-2} \\
a_{n-1} & b_{n-1} & R' \\
a_n & R &
\end{array}$$

so that each element is obtained by adding to its left-hand neighbor z times its upward neighbor.

Thus, if the roots of the algebraic equation $f(x) = 0$ are $x = \alpha_1, \alpha_2, \ldots, \alpha_n$, and if the Newton-Raphson procedure is to be used to approximate one of the roots, starting with an initial approximation z, the next approximation z^* is given simply by†

$$z^* = z - \frac{R}{R'} \qquad (10.14.10)$$

and the process then can be repeated with z replaced by z^*. This method of computation tends to minimize the labor involved in evaluating the polynomials $f(z)$ and $f'(z)$ and is sometimes known as the *Birge-Vieta method.*

In the simple case of the cubic equation (10.10.6), for which

$$f(x) = x^3 - x - 1$$

the first two iterations (starting with $z = 1.3$) would be tabulated as follows:

$z =$	1.3		1.325	
1	1	1	1	1
0	1.3	2.6	1.325	2.65
−1	0.69	4.07	0.755625	4.267
−1	−0.103		0.001203	
$\Delta z =$	0.025		−0.000282	

The approximation obtained at this stage is thus 1.324718, in accordance with the results obtained in the preceding section.

Once the iteration is terminated, so that one zero of $f(x)$ is approximated and the last entry in the b column is effectively reduced to zero, the remaining entries in the b column are (approximately) the coefficients of the reduced polynomial, of degree $n - 1$, whose zeros are the remaining zeros of $f(x)$. Thus the remaining zeros can be obtained by solving equations of successively decreasing degree. Because of the errors propagated into the coefficients of those equations, however, it is desirable to use each zero so obtained as the *starting value* in a final correction run employing the coefficients of the *original* nth-degree polynomial.

† Here and henceforth, in dealing with polynomials, we minimize the number of indices used in a generic iteration formula by writing z for a typical member of an approximation sequence and z^* for the following member. Thus, if (10.14.10) were to be used to approximate the root α_r, it would abbreviate a more specific formula such as

$$z_{r,k+1} = z_{r,k} - \frac{R_{r,k}}{R'_{r,k}}$$

Other iterative methods involving the evaluation of derivatives, such as those based on (10.12.18), (10.12.20), and (10.12.22), can be systematized similarly when applied to algebraic equations.

A particularly simple procedure, due to S. N. Lin [1941, 1943], is based on the fact that, by virtue of (10.14.8), the condition $f(z) = 0$ is equivalent to the condition $a_n + zb_{n-1} = 0$. That is, if and only if the assumed value of z were a root of $f(x) = 0$, then the corresponding value of b_{n-1} (which depends upon z) would be such that $z = -a_n/b_{n-1}(z)$. *Lin's iteration* is the result of applying the method of successive substitutions to the equation written in this form, so that the revised estimate z^* is defined by the formula

$$z^* = -\frac{a_n}{b_{n-1}} \tag{10.14.11}$$

and hence

$$z^* - z = -\frac{a_n + zb_{n-1}}{b_{n-1}}$$

or, equivalently, by virtue of (10.14.8),

$$z^* = z - \frac{R}{b_{n-1}} \tag{10.14.12}$$

In this method, the formation of the c column is avoided, so that the labor per iteration is reduced by nearly one-half. However, if this method is applied to the example treated above, the first three iterations may be obtained as follows:

$z =$	1.3	1.45	0.91	−5.8 .
1	1	1	1	
0	1.3	1.45	0.91	
−1	0.69	1.102	−0.172	
−1	−0.103	0.598	−1.157	
$\Delta z =$	0.15	−0.54	−6.7	

Clearly, the iteration is not convergent in this case.

In order to investigate the Lin procedure more closely, we may notice that, since (10.14.8) gives

$$b_{n-1}(z) = \frac{f(z) - a_n}{z}$$

the recurrence relation written in the form (10.14.11) can also be put in the form

$$z^* = -\frac{a_n z}{f(z) - a_n}$$

Thus Lin's method is equivalent to applying the method of successive substitutions to the result of writing $f(x) = 0$ in the form

$$x = -\frac{a_n x}{f(x) - a_n} \equiv F(x) \quad (10.14.13)$$

In the example just considered, (10.14.13) becomes $x = 1/(x^2 - 1)$, which, as was seen in Sec. 10.10, is not suitable for successive substitutions since the convergence factor $F'(x)$ has a value of about -5 near the real root, whereas, for convergence, its absolute value should be smaller than unity. In confirmation, we may notice that the error in $z^* \doteq 1.45$ is indeed about five times the error in $z = 1.3$ and is of opposite sign.

More generally, we find from (10.14.13) that

$$F'(x) = a_n \frac{xf'(x) - f(x) + a_n}{[f(x) - a_n]^2}$$

and hence, at a zero α_r of $f(x)$, Lin's method possesses the asymptotic convergence factor

$$\rho_r = F'(\alpha_r) = 1 + \frac{\alpha_r}{a_n} f'(\alpha_r) \equiv 1 + \alpha_r \frac{f'(\alpha_r)}{f(0)} \quad (10.14.14)$$

Thus the result of applying Lin's iteration to a good approximation to α_r will lead to a *poorer* one unless $|\rho_r| \leqq 1$; that is, unless the condition

$$|\rho_r| = \left| 1 + \frac{\alpha_r}{a_n} f'(\alpha_r) \right| < 1 \quad (10.14.15)$$

is satisfied, the iteration generally will not converge to α_r.

This criterion is a useful one if a rough approximation to α_r is known initially, unless $f'(x)$ varies rapidly near $x = \alpha_r$. If we recall that $a_n = (-1)^n \alpha_1 \alpha_2 \cdots \alpha_n$ and that $f'(\alpha_r) = (\alpha_r - \alpha_1) \cdots (\alpha_r - \alpha_n)$, where the factor $(\alpha_r - \alpha_r)$ is to be omitted, we may deduce that (10.14.15) can also be expressed in the form

$$|\rho_r| = \left| 1 - \left[\left(1 - \frac{\alpha_r}{\alpha_1}\right)\left(1 - \frac{\alpha_r}{\alpha_2}\right) \cdots \left(1 - \frac{\alpha_r}{\alpha_n}\right) \right] \right| < 1 \quad (10.14.16)$$

in terms of the *remaining* roots of $f(x) = 0$.

In the case of the equation

$$x^4 - 8x^3 + 23x^2 + 16x - 50 = 0 \quad (10.14.17)$$

a real root is easily seen to lie between $x = 1$ and $x = 2$. If Lin's iteration is used, starting with $z = 1.5$, the results of the first three iterations are as follows:

$z =$	1.5	1.39	1.421	1.4125
1	1	1	1	
−8	−6.5	−6.61	−6.579	
23	13.25	13.8121	13.6512	
16	35.875	35.1988	35.3984	
−50	3.8125	−1.0737	0.3011	
$\Delta z =$	−0.11	0.031	−0.0085	

The true roots of (10.14.17) are $\pm\sqrt{2}$ and $4 \pm 3i$. The rate of convergence of the Lin iteration, in this case, might have been predicted in advance by approximating α_r by 1.5 in (10.14.14) to obtain $\rho \approx -\frac{1}{3}$. The *asymptotic* convergence factor is -0.25 to two places.†

It is of interest to notice that since

$$b_{n-1} = \frac{R - a_n}{z} = \frac{f(z) - f(0)}{z}$$

it follows that b_{n-1} is the slope of the secant joining the ordinate at $x = 0$ and the ordinate at $x = z$. Thus, (10.14.12) is equivalent to the result of taking γ as the slope of that secant in the more general recurrence relation (10.10.11), and the Lin iteration therefore amounts to determining z^* by linear interpolation (or extrapolation) based on the *fixed* ordinate $f(0)$ and the most recently calculated ordinate $f(z)$ (see Fig. 10.5). Also the requirement (10.14.15) is easily interpreted as demanding that the ratio of the slope of the curve at P to the slope of the secant P_0P be *positive and less than 2.*

From this fact it may be deduced that *when Lin's iteration is unstable for the determination of a zero α_r of a polynomial $f(x)$, stability can be attained by translating the origin to a new point $x = c$ if that point is sufficiently near to α_r.* Clearly, the process then would amount to using $f(c)$ as the fixed ordinate in place of $f(0)$ in Fig. 10.5. For example, if the origin is translated to $c = 1.3$ in the case of the previously considered equation $x^3 - x - 1 = 0$, the Lin iteration becomes convergent when initiated at that point (see Prob. 90).

A simple alternative method for introducing or improving the asymptotic stability of the Lin iteration at a required zero α_r, when a fair approximation to α_r is known in advance, is easily devised. For this purpose, we note that if the

† When the ratio of successive values of Δz_k becomes nearly constant, that ratio serves as an estimate of ρ, and a generally improved value of Δz_k then is given by $(\Delta z_k)(1 + \rho + \rho^2 + \cdots) = (\Delta z_k)/(1 - \rho)$.

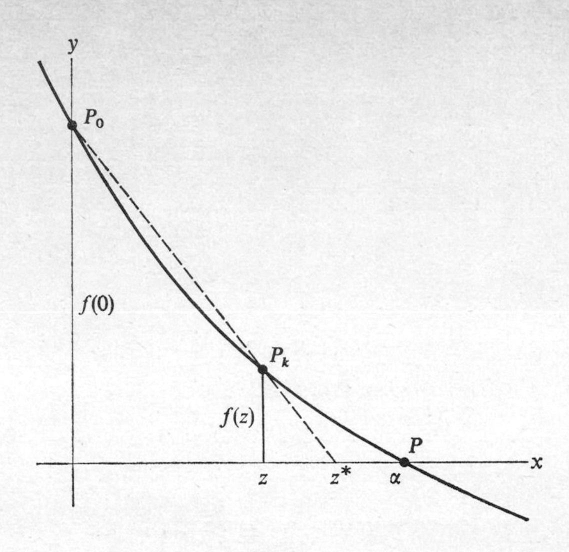

FIGURE 10.5

Lin formula (10.14.12) is modified by the insertion of a parameter λ into the form

$$z^* = z - \lambda \frac{R}{b_{n-1}} \qquad (10.14.18)$$

the asymptotic Lin convergence factor (10.14.14) at α_r is replaced by the factor

$$\rho_r = 1 + \lambda \alpha_r \frac{f'(\alpha_r)}{f(0)} \qquad (10.14.19)$$

If α_r were known exactly, λ generally could be determined so that $\rho_r = 0$, and the iteration process in fact then would be of higher order. Instead, we may replace α_r by a preliminary approximation $\bar{\alpha}_r$ and accordingly introduce the definition

$$\lambda = -\frac{f(0)}{\bar{\alpha}_r f'(\bar{\alpha}_r)} \qquad (10.14.20)$$

into (10.14.18).

In illustration, we again attempt the determination of the real root of Eq. (10.10.6):

$$x^3 - x - 1 = 0$$

With $\bar{\alpha} = 1.3$, (10.14.20) gives $\lambda \doteq 0.189$, so that the modified Lin formula is

$$z^* = z - 0.189 \frac{R}{b_{n-1}}$$

and the first three iterations yield the following results:

$z =$	1.3	1.328	1.3245	1.324733
1	1	1	1	
0	1.3	1.328	1.3245	
−1	0.69	0.7636	0.75430	
−1	−0.103	0.0141	−0.000930	
$\Delta z =$	0.028	−0.0035	0.000233	

An even simpler iterative procedure would replace (10.14.12) by the formula

$$z^* = z - \frac{R}{f'(\bar{\alpha}_r)} \qquad (10.14.21)$$

as a stationary simulation to the Newton-Raphson iteration, where the constant $f'(\bar{\alpha}_r)$ again is precalculated as an approximation to $f'(\alpha_r)$ and is not to be modified from step to step.

Both of these iterations generally are of first order, but they simulate second-order processes for a limited number of steps and require only about half as much computation per step as does the Newton-Raphson (Birge-Vieta) iteration. Ultimately, if the iterations were prolonged, both would become inferior to second-order processes.

The preceding methods are valid, in principle, for the determination of *complex* roots as well as real ones (see Prob. 86). However, since a real initial approximation leads necessarily to real iterates, when the coefficients are real, the process then must be initiated with a nonreal initial estimate, and operations with complex numbers are involved in each step of the process. When the coefficients are real, the complex roots occur in conjugate pairs, and it is generally preferable to exploit this fact by seeking *quadratic real factors* rather than linear complex ones. A generalized method of synthetic division for this purpose is considered in Secs. 10.18 and 10.19.

10.15 Determinacy of Zeros of Polynomials

In those cases when the coefficients $a_1, \ldots, a_n$ of $f(x)$ are inexact, it is desirable to have estimates of the corresponding inherent possible errors in the roots of $f(x) = 0$. If α_r is obtained as a root of the equation

$$f(\alpha_r) = \alpha_r^n + a_1\alpha_r^{n-1} + \cdots + a_{n-1}\alpha_r + a_n = 0 \qquad (10.15.1)$$

whereas the *true* coefficients are $a_1 + \delta a_1, \ldots, a_n + \delta a_n$, then the corresponding true root $\alpha_r + \delta\alpha_r$ must satisfy the equation

$$(\alpha_r + \delta\alpha_r)^n + (a_1 + \delta a_1)(\alpha_r + \delta\alpha_r)^{n-1} + \cdots + (a_{n-1} + \delta a_{n-1})(\alpha_r + \delta\alpha_r) + (a_n + \delta a_n) = 0 \quad (10.15.2)$$

If the first equation is subtracted from the second, and if it is assumed that the relative errors are sufficiently small to permit neglect of higher-order terms, it follows that, to a first approximation, $\delta\alpha_r$ must satisfy the equation

$$[n\alpha_r^{n-1} + (n-1)a_1\alpha_r^{n-2} + \cdots + a_{n-1}]\,\delta\alpha_r + \alpha_r^{n-1}\,\delta a_1 + \alpha_r^{n-2}\,\delta a_2 + \cdots + \delta a_n = 0$$

and hence

$$\delta\alpha_r \approx -\frac{\alpha_r^{n-1}\,\delta a_1 + \alpha_r^{n-2}\,\delta a_2 + \cdots + \delta a_n}{f'(\alpha_r)} \quad (10.15.3)$$

In particular, if each coefficient is known to be in error by no more than ε,

$$|\delta a_i| \leqq \varepsilon \qquad (i = 1, 2, \ldots, n) \quad (10.15.4)$$

there follows, within the same degree of approximation,

$$|\delta\alpha_r|_{\max} \approx \frac{1 + |\alpha_r| + |\alpha_r|^2 + \cdots + |\alpha_r|^{n-1}}{|f'(\alpha_r)|}\,\varepsilon$$

or

$$|\delta\alpha_r|_{\max} \approx \frac{|\alpha_r|^n - 1}{(|\alpha_r| - 1)|f'(\alpha_r)|}\,\varepsilon \quad (10.15.5)$$

when $|\alpha_r| \neq 1$. If $|\alpha_r| = 1$, this approximate bound becomes $n\varepsilon/|f'(\alpha_r)|$. Clearly, unless the right-hand member of (10.15.3) or (10.15.5) is sufficiently small to be consistent with the neglect of higher-order terms in its derivation, these results should be regarded with suspicion.

In the case of the real root of (10.10.6), it is found that errors of magnitude ε in the coefficients would correspond to a maximum error of very nearly the same magnitude in the root if ε is small. In the case of the root $\alpha = \sqrt{2}$ of (10.14.17), the maximum error in the approximation to the root is found to be about one-sixth of the maximum error in the coefficients.

In terms of *relative* errors, (10.15.3) yields the approximation

$$\frac{\delta\alpha_r}{\alpha_r} \approx -\sum_{k=1}^{n} c_{rk}\,\frac{\delta a_k}{a_k} \quad (10.15.6)$$

where

$$c_{rk} = \frac{\alpha_r^{n-k-1}a_k}{f'(\alpha_r)} \quad (10.15.7)$$

Hence, in particular, if the magnitude of the relative error in each coefficient does not exceed η,

$$\left|\frac{\delta a_i}{a_i}\right| \leqq \eta \qquad (10.15.8)$$

for $i = 0, 1, \ldots, n$, there follows

$$\left|\frac{\delta \alpha_r}{\alpha_r}\right|_{\max} \approx \eta \sum_{k=1}^{n} |c_{rk}| \qquad (10.15.9)$$

Unpleasant situations exist in which small changes in certain coefficients of a polynomial may lead to large changes in certain of its zeros. A well-known example of Wilkinson [1959] involves the equation

$$\begin{aligned} f(x) &= (x+1)(x+2)\cdots(x+20) \\ &= x^{20} + 210x^{19} + \cdots + 20! = 0 \quad (10.15.10) \end{aligned}$$

Here it happens that if the coefficient $a_1 = 210$ is changed by $2^{-23} \doteq 1.2 \times 10^{-7}$, the changes in the smaller roots are slight but, for example, the root α_{20} becomes -20.8 and the roots α_{16} and α_{17} become an imaginary root pair—approximately $-16.7 \pm 2.8i$. In fact, a total of five pairs of zeros become imaginary.

The relation (10.15.3) would in fact admit the possibility of considerably larger errors in this case but correspondingly would be useless for estimating them. In such a situation, the term *ill conditioned* is sometimes applied to the polynomial, just as it is applied to a set of linear equations with a similar sort of instability.

Even when the coefficients are exact, or nearly so, the process of *deflation*, in which a linear factor $x - \alpha$ is extracted from a polynomial $f(x)$ to yield a polynomial $g(x)$ of reduced degree, will introduce errors into the coefficients of $g(x)$ if α is inexactly known and a corresponding small residual remainder is ignored. As pointed out in Sec. 10.14, error propagation naturally also takes place in self-deflating iteration processes such as those of Birge-Vieta and Lin. In all cases, it is highly desirable that approximate roots obtained from $g(x)$ or from the results of subsequent deflations (by any method) be given final corrections by a method which uses the coefficients of the *original* polynomial.

Usually it is desirable to extract factors corresponding to zeros of successively increasing magnitude, in order to minimize the generated *relative* errors. However, this procedure may be inconsistent with the desirability of *beginning* with the calculation of any zero or zeros for which the polynomial is ill-conditioned (see Probs. 102 and 103).

10.16 Bernoulli's Iteration

A method, originally due to Daniel Bernoulli, for obtaining roots of the algebraic equation

$$x^n + a_1 x^{n-1} + \cdots + a_{n-1}x + a_n = 0 \qquad (10.16.1)$$

is based on the related recurrence formula

$$\mu_k + a_1\mu_{k-1} + \cdots + a_{n-1}\mu_{k-n+1} + a_n\mu_{k-n} = 0 \qquad (10.16.2)$$

having the *same coefficients* as (10.16.1).

If the roots of (10.16.1) are $\alpha_1, \alpha_2, \ldots, \alpha_n$, and if (10.16.2) is considered as a *difference equation*, its general solution is found to be†

$$\mu_k = C_1\alpha_1^k + C_2\alpha_2^k + C_3\alpha_3^k + \cdots + C_n\alpha_n^k \qquad (10.16.3)$$

where the n C's are constants, independent of k, which are determined by the values of $\mu_1, \mu_2, \ldots,$ and μ_n, if no roots are repeated. Under this assumption, let the roots be numbered in *decreasing* order of magnitude, so that α_1 here denotes the *largest* root in magnitude of (10.16.1). Then since (10.16.3) can be written in the form

$$\mu_k = C_1\alpha_1^k\left[1 + \frac{C_2}{C_1}\left(\frac{\alpha_2}{\alpha_1}\right)^k + \frac{C_3}{C_1}\left(\frac{\alpha_3}{\alpha_1}\right)^k + \cdots + \frac{C_n}{C_1}\left(\frac{\alpha_n}{\alpha_1}\right)^k\right] \qquad (10.16.4)$$

if $C_1 \neq 0$, it follows that, in any sequence generated by (10.16.2), the kth term is approximated by $C_1\alpha_1^k$ as $k \to \infty$ and, indeed, that the ratio

$$r_k = \frac{\mu_k}{\mu_{k-1}} \qquad (10.16.5)$$

tends to α_1 as $k \to \infty$ if the largest root α_1 is real and unrepeated and if no other root has equal magnitude, unless $\mu_1, \mu_2, \ldots, \mu_n$ are so chosen that the coefficient C_1 of α_1^k in (10.16.3) is zero.

If the largest root α_1 is *complex*, and the coefficients of (10.16.1) are real, then α_2 is the complex conjugate of α_1 and is of equal magnitude. If we write

$$\alpha_1 = \xi_1 + i\eta_1 = \beta_1 e^{i\phi_1} \qquad \alpha_2 \equiv \bar{\alpha}_1 = \xi_1 - i\eta_1 = \beta_1 e^{-i\phi_1} \qquad (10.16.6)$$

where $\beta_1 > 0$ and $\xi_1, \eta_1, \beta_1,$ and ϕ_1 are *real*, the terms corresponding to α_1 and α_2 in (10.16.3) can be expressed in the real form

$$\beta_1^k(C_1 \cos k\phi_1 + C_2 \sin k\phi_1)$$

† If a solution of (10.16.2) is assumed in the form $\mu_k = \alpha^k$, it is found that the characteristic equation determining admissible values of α is of the same form as (10.16.1). Thus $\alpha_1^k, \alpha_2^k, \ldots, \alpha_n^k$ are all solutions, and superposition leads to (10.16.3), which can be shown to represent the *most general* solution, if no roots are repeated, when only integral values of k are considered.

if C_1 and C_2 are replaced by $(C_1 - iC_2)/2$ and $(C_1 + iC_2)/2$, respectively, in (10.16.3).

Thus, if α_1 and $\bar{\alpha}_1$ are not repeated and if all other roots are smaller in magnitude than β_1, it follows that

$$\mu_k \approx \beta_1^k(C_1 \cos k\phi_1 + C_2 \sin k\phi_1) \qquad (k \to \infty) \qquad (10.16.7)$$

But, if μ_k were given exactly by the right-hand member of (10.16.7), it would satisfy the recurrence relation

$$\mu_{k+1} - 2\mu_k\beta_1 \cos \phi_1 + \beta_1^2\mu_{k-1} = 0 \qquad (10.16.8)$$

and conversely, as is easily verified. A second relation, involving the two real unknown quantities β_1 and ϕ_1, then would be obtained, by replacing k by $k - 1$, in the form

$$\mu_k - 2\mu_{k-1}\beta_1 \cos \phi_1 + \beta_1^2\mu_{k-2} = 0 \qquad (10.16.9)$$

The result of eliminating $\cos \phi_1$ from these two relations is

$$(\mu_{k-1}^2 - \mu_k\mu_{k-2})\beta_1^2 = \mu_k^2 - \mu_{k+1}\mu_{k-1} \qquad (10.16.10)$$

whereas the result of eliminating β_1^2 is

$$2(\mu_{k-1}^2 - \mu_k\mu_{k-2})\beta_1 \cos \phi_1 = \mu_k\mu_{k-1} - \mu_{k+1}\mu_{k-2} \qquad (10.16.11)$$

Thus, if we introduce the definitions

$$s_k = \mu_k^2 - \mu_{k+1}\mu_{k-1} \qquad t_k = \mu_k\mu_{k-1} - \mu_{k+1}\mu_{k-2} \qquad (10.16.12)$$

these approximate relations give

$$\beta_1^2 \equiv \xi_1^2 + \eta_1^2 \approx \frac{s_k}{s_{k-1}} \qquad 2\beta_1 \cos \phi_1 \equiv 2\xi_1 \approx \frac{t_k}{s_{k-1}} \qquad (10.16.13)$$

It follows that, unless it happens that $C_1 = C_2 = 0$ in (10.16.3), because of a very special choice of $\mu_1, \mu_2, \ldots, \mu_n$, the ratios s_k/s_{k-1} and t_k/s_{k-1} will *tend* to β_1^2 and $2\beta_1 \cos \phi_1$ as $k \to \infty$, from which limits the constants β_1 and ϕ_1, or ξ_1 and η_1, specifying the desired dominant complex root pair in (10.16.6), can be calculated.

If α_1 is a *repeated* real root, of multiplicity 2, so that $\alpha_2 = \alpha_1$, and all other roots are of smaller magnitude, then the combination of terms corresponding to α_1 and α_2 in (10.16.3) is of the form $\alpha_1^k(c_1 + c_2k)$. Since μ_k must then tend to such a form as $k \to \infty$, it follows that μ_k must tend to satisfy the relation

$$\mu_{k+1} - 2\mu_k\alpha_1 + \mu_{k-1}\alpha_1^2 = 0 \qquad (10.16.14)$$

as $k \to \infty$. Whereas an approximation to α_1, which tends to α_1 as $k \to \infty$, could be obtained as the appropriate one of the two roots of this equation, the solution of a quadratic equation can be avoided by rewriting (10.16.14) with k replaced by $k - 1$, and eliminating α_1^2 from the two relations, to give

$$2\alpha_1 \approx \frac{t_k}{s_{k-1}} \qquad (10.16.15)$$

with the notation of (10.16.12).

Other exceptional cases, in which several roots have the same maximum absolute value, can be treated in a similar way.

When the largest root α_1 is real and unrepeated and there are no other roots with the same absolute value, the ratio r_k tends to α_1, the rapidity of the convergence depending upon the magnitude of the ratio α_2/α_1 of the *two* largest roots. If α_1 and α_2 are conjugate complex, (10.16.7) shows that r_k will tend to oscillate about the value zero (although the period of the oscillation may comprise several iterations), whereas, if $\alpha_2 = \alpha_1$ [or $\alpha_2 \approx \alpha_1$], the convergence of the ratio r_k to α_1 will be slow; here the ratio t_k/s_{k-1} converges more rapidly to $2\alpha_1$ [or to $\alpha_1 + \alpha_2$]. Thus, after several iterations, the behavior of the sequence of r's generally will indicate the true situation, and recourse can be had to the appropriate choice between (10.16.13) and (10.16.15) when that sequence is not acceptable. The more complicated situations seldom occur in practice.

If α_1 is real and unrepeated, the *ideal* situation would be that in which $\mu_1, \mu_2, \ldots, \mu_n$ were so chosen that $C_2 = \cdots = C_n = 0$ in (10.16.3), so that $\mu_1, \mu_2, \ldots, \mu_n$ would be respectively proportional to $\alpha_1, \alpha_1^2, \ldots, \alpha_1^n$. The first calculated value of r, $r_{n+1} = \mu_{n+1}/\mu_n$, then clearly would be identical with α_1. In such cases, the starting values could be taken efficiently as successive powers of a previously determined *approximation* to α_1. If no information is easily available with regard to the nature of the largest root or roots, the starting values

$$\mu_1 = \mu_2 = \cdots = \mu_{n-1} = 0 \qquad \mu_n = 1$$

are often convenient. For this set of values it is easily seen that the undesirable case $C_1 = 0$ cannot occur.

A particularly notable set of n starting values having the same property is that determined by use of the formula

$$\mu_r = -(a_1\mu_{r-1} + a_2\mu_{r-2} + \cdots + a_{r-1}\mu_1 + ra_r) \qquad (r = 1, 2, \ldots, n) \qquad (10.16.16)$$

with $\mu_0 = \mu_{-1} = \cdots = 0$. For this set of starting values it can be shown†

† The proof follows directly from Newton's power-sum identities (see Theorem 13 of Sec. 1.9, with $s_r = \mu_r$).

that all the C's in (10.16.3) are *unity*, and hence that μ_k is then identified with the sum $\alpha_1^k + \alpha_2^k + \cdots + \alpha_n^k$ for all $k \geqq 1$. Thus, in particular, if $|\alpha_1| \gg |\alpha_2|, \ldots, |\alpha_n|$, there then follow both $\alpha_1 \approx \mu_k/\mu_{k-1}$ and $\alpha_1 \approx \mu_k^{1/k}$ when k is sufficiently large. With the convention that $a_r = 0$ when $r > n$, it is seen that the recurrence formula (10.16.16) is still applicable when $r > n$ since it then reduces to (10.16.2). This special procedure is closely related to the Graeffe procedure described in the following section.

In the case of the example (10.10.6), with $f(x) = x^3 - x - 1$, the Bernoulli recurrence relation is merely $\mu_k = \mu_{k-2} + \mu_{k-3}$. If the iteration is begun with the starting values 1.30, 1.69, and 2.20, about sixteen iterations are needed to establish the real root 1.3247 . . . to five significant figures, although here each iteration requires only a single addition. The remaining roots are complex, with an absolute value of about 0.9, so that the ratio of the magnitude of the dominant root to that of the subdominant root pair is about 1.5. The relative slowness of the convergence is due to the relative nearness of this ratio to unity. The fact that the subdominant roots are complex causes the sequence of iterates to tend to its limit in an oscillatory manner.

In the case of the example (10.14.17), the recurrence relation is

$$\mu_k = 8\mu_{k-1} - 23\mu_{k-2} - 16\mu_{k-3} + 50\mu_{k-4}$$

and, with the arbitrarily chosen starting values 0, 0, 0, 1, the ensuing calculation is as follows:

μ_k	r_k	s_k	t_k	s_k/s_{k-1}	t_k/s_{k-1}
8	8	23	8	—	—
41	5.12	657	200	28.565	8.696
128	3.12	16261	5224	24.750	7.951
3	0.02	406537	130600	25.001	8.031
−3176	−1059	10163401	3251272	25.000	7.997
−25475	8.02				

From the irregular behavior of the r sequence, it may be deduced that *either* the process has not yet begun to converge satisfactorily *or* there is a pair of dominant complex roots. To test the second hypothesis, the s and t sequences are constructed, and the convergence of the sequences of ratios in the last two columns is evident. The true dominant roots are $\xi_1 \pm i\eta_1 = 4 \pm 3i$, so that $\beta_1^2 \equiv \xi_1^2 + \eta_1^2 = 25$ and $2\xi_1 = 8$. The approximations afforded by the four successive pairs of ratios are $4.348 \pm 3.108i$, $3.976 \pm 2.990i$, $4.016 \pm 2.979i$, and $3.998 \pm 3.002i$.

The Bernoulli iteration has the useful property that it yields the dominant root (or roots) regardless of the starting values except in the unlikely (and avoidable) case when $C_1 = 0$ in (10.16.3), in which case another root or root pair will result; that is, it is not necessary to initiate the iteration with a *sufficiently accurate* approximation as is the case for many other iterative methods. This fact is of particular importance in those cases when *only* complex roots are present, since even rough approximations then are not readily obtained. The calculation is remarkably simple (and readily mechanized) when the dominant root is real and unequaled in absolute value, and is not unduly complicated otherwise.

Methods for improving the convergence of Bernoulli iteration are suggested in Probs. 106 and 109.

10.17 Graeffe's Root-squaring Technique

Graeffe's iterative method for determining roots of the algebraic equation

$$f(x) \equiv x^n + a_1 x^{n-1} + a_2 x^{n-2} + \cdots + a_{n-1} x + a_n = 0 \qquad (10.17.1)$$

consists of forming a sequence of *equations*, such that the roots of each equation are the *squares* of the roots of the preceding equation in the sequence, for the purpose of ultimately obtaining an equation whose roots are so widely separated in magnitude that they can be read approximately from the equation, by inspection.

The principle of the method can be illustrated by a consideration of the general equation of fourth degree, which can be written in the form

$$\begin{aligned} f(x) &= x^4 + a_1 x^3 + a_2 x^2 + a_3 x + a_4 \\ &= (x - \alpha_1)(x - \alpha_2)(x - \alpha_3)(x - \alpha_4) = 0 \qquad (10.17.2) \end{aligned}$$

or, equivalently,

$$\begin{aligned} f(x) = x^4 &- (\alpha_1 + \alpha_2 + \alpha_3 + \alpha_4)x^3 \\ &+ (\alpha_1\alpha_2 + \alpha_1\alpha_3 + \alpha_1\alpha_4 + \alpha_2\alpha_3 + \alpha_2\alpha_4 + \alpha_3\alpha_4)x^2 \\ &- (\alpha_1\alpha_2\alpha_3 + \alpha_1\alpha_2\alpha_4 + \alpha_1\alpha_3\alpha_4 + \alpha_2\alpha_3\alpha_4)x + \alpha_1\alpha_2\alpha_3\alpha_4 = 0 \qquad (10.17.3) \end{aligned}$$

where α_1, α_2, α_3, and α_4 are the roots.

If the roots are all real and are widely separated in magnitude, so that $|\alpha_1| \gg |\alpha_2| \gg |\alpha_3| \gg |\alpha_4|$, the result of retaining only the dominant part of each coefficient in (10.17.3) is

$$x^4 - \alpha_1 x^3 + \alpha_1\alpha_2 x^2 - \alpha_1\alpha_2\alpha_3 x + \alpha_1\alpha_2\alpha_3\alpha_4 \approx 0 \qquad (10.17.4)$$

Thus the four roots are given approximately, in this case, by equating to zero the four linear expressions $x + a_1$, $a_1 x + a_2$, $a_2 x + a_3$, and $a_3 x + a_4$.

If, say, α_1 and α_2 are conjugate complex, so that $\alpha_1 = \beta_1 e^{i\phi_1}$ and $\alpha_2 = \beta_1 e^{-i\phi_1}$, and if also $|\alpha_1| = |\alpha_2| \gg |\alpha_3| \gg |\alpha_4|$, the approximation replacing (10.17.4) is then

$$x^4 - 2\beta_1 x^3 \cos \phi_1 + \beta_1^2 x^2 - \beta_1^2 \alpha_3 x + \beta_1^2 \alpha_3 \alpha_4 \approx 0 \quad (10.17.5a)$$

The complex roots are then approximated by the zeros of the quadratic $x^2 + a_1 x + a_2$, and the remaining roots are found by equating $a_2 x + a_3$ and $a_3 x + a_4$ to zero. If, say, $\alpha_1 = \alpha_2$ and $|\alpha_1| = |\alpha_2| \gg |\alpha_3| \gg |\alpha_4|$, the approximate relation is

$$x^4 - 2\alpha_1 x^3 + \alpha_1^2 x^2 - \alpha_1^2 \alpha_3 x + \alpha_1^2 \alpha_3 \alpha_4 \approx 0 \quad (10.17.5b)$$

and the approximate roots are obtained in the same way. Other, more unusual situations can be analyzed similarly.

The root-squaring process itself can be based on the fact that the product

$$(-1)^n f(-x) f(x) = (x^2 - \alpha_1^2)(x^2 - \alpha_2^2) \cdots (x^2 - \alpha_n^2) \quad (10.17.6)$$

is a polynomial of degree n in x^2, whose zeros are the squares of the zeros of $f(x)$. Thus, if $f(x) = x^n + a_1 x^{n-1} + a_2 x^{n-2} + \cdots + a_{n-1} x + a_n$ is multiplied, term by term, by

$$(-1)^n f(-x) = x^n - a_1 x^{n-1} + a_2 x^{n-2} - \cdots + (-1)^{n-1} a_{n-1} x + (-1)^n a_n$$

and x^2 is then replaced by x, the result $f_2(x)$ is a polynomial of degree n with zeros $\alpha_1^2, \ldots, \alpha_n^2$. By repeating the process, a polynomial $f_4(x)$ with zeros $\alpha_1^4, \ldots, \alpha_n^4$ is obtained, then $f_8(x)$ with zeros α_k^8, and so forth.

If all roots are real, unrepeated, and of distinct magnitudes, the iteration may be concluded when the magnitude of each coefficient in an equation is the square of the magnitude of the corresponding coefficient in the preceding equation, within the tolerance adopted. Suppose that the *original* roots are $\alpha_1, \ldots, \alpha_n$, and that k root squarings are needed, so that the roots of the final equation are $\alpha_1^m, \ldots, \alpha_n^m$, where $m = 2^k$. If the final equation is of the form

$$f_m(x) = x^n - A_1 x^{n-1} + A_2 x^{n-2} - \cdots + (-1)^{n-1} A_{n-1} x + (-1)^n A_n = 0 \quad (10.17.7)$$

there then follows

$$\alpha_1^m \approx A_1 \qquad \alpha_2^m \approx \frac{A_2}{A_1} \qquad \alpha_3^m \approx \frac{A_3}{A_2} \qquad \cdots \qquad \alpha_n^m \approx \frac{A_n}{A_{n-1}} \quad (10.17.8)$$

Each of the right-hand members will be positive, and the proper sign must be chosen for the real mth root of each of these expressions by substitution of the two possibilities into the original equation or otherwise.

A dominant *double* original root α_1 would evidence itself by the fact that, after k root squarings, the equation would be approximately of the form

$$f_m(x) \approx x^n - 2\alpha_1^m x^{n-1} + \alpha_1^{2m} x^{n-2} - (\alpha_1^2\alpha_3)^m x^{n-3} + \cdots = 0 \quad (10.17.9)$$

where again $m = 2^k$, so that the magnitude of the coefficient of x^{n-1} would tend to be *half* the square of the magnitude of the corresponding coefficient in the preceding equation. Similarly, if α_r were a double real root, and if no other root were of equal magnitude, the coefficient of x^{n-r} would have this property. Thus α_r then would satisfy both of the relations

$$\alpha_r^{2m} \approx \frac{A_{r+1}}{A_{r-1}} \qquad \alpha_r^m \approx \frac{A_r}{2A_{r-1}} \quad (10.17.10)$$

and would be determined as the real root, with appropriate sign, of either equation.

A dominant conjugate complex root pair $\alpha_{1,2} = \beta_1 e^{\pm i\phi_1}$ would cause the kth equation to be approximately of the form

$$f_m(x) \approx x^n - 2\beta_1^m x^{n-1} \cos m\phi_1 + \beta_1^{2m} x^{n-2} - (\beta_1^2\alpha_3)^m x^{n-3} + \cdots = 0 \quad (10.17.11)$$

where $m = 2^k$, so that the coefficient of x^{n-1} in the kth equation would tend to fluctuate in magnitude and sign in the same way as $-2\beta_1^m \cos m\phi_1$, as k and $m = 2^k$ increased, and hence again would not tend to be the square of the corresponding coefficient in the $(k - 1)$th equation. The same sort of oscillation would occur in the coefficient of x^{n-r} if α_r and α_{r+1} were a complex root pair, and, for k sufficiently large, β_r and ϕ_r could be determined from the relations

$$\beta_r^{2m} \approx \frac{A_{r+1}}{A_{r-1}} \qquad 2\beta_r^m \cos m\phi_r \approx \frac{A_r}{A_{r-1}} \quad (10.17.12)$$

if no other root were also of magnitude β_r. The magnitude β_r thus would be the positive real $(2m)$th root of A_{r+1}/A_{r-1}, whereas the appropriate one of the values of ϕ_r obtained from the second relation would have to be selected by trial and error or otherwise.

When only one such pair of complex roots is present, say,

$$\beta_r e^{\pm i\phi_r} \equiv \xi_r \pm i\eta_r$$

the selection of the appropriate value of ϕ_r satisfying this relation can be avoided by noticing that, since the sum of all roots of (10.17.1) is given by $-a_1$, there follows

$$\alpha_1 + \alpha_2 + \cdots + \alpha_{r-1} + 2\xi_r + \alpha_{r+2} + \cdots + \alpha_n = -a_1 \quad (10.17.13)$$

Hence ξ_r is given immediately when the remaining $n - 2$ roots are known, after which η_r is given by $\sqrt{\beta_r^2 - \xi_r^2}$.

If *two* pairs of complex roots are present, say,

$$\beta_r e^{\pm i\phi_r} \equiv \xi_r \pm i\eta_r \qquad \text{and} \qquad \beta_s e^{\pm i\phi_s} \equiv \xi_s \pm i\eta_s$$

the corresponding relation is

$$2(\xi_r + \xi_s) = -(a_1 + \alpha_1 + \cdots + \alpha_{r-1} + \alpha_{r+2} + \cdots + \alpha_{s-1} + \alpha_{s+2} + \cdots + \alpha_n) \quad (10.17.14)$$

A second linear relation between ξ_1 and ξ_2 is then obtained by recalling that the sum of the *reciprocals* of the roots is $-a_{n-1}/a_n$, so that

$$\frac{1}{\alpha_1} + \cdots + \frac{1}{\xi_r + i\eta_r} + \frac{1}{\xi_r - i\eta_r} + \cdots + \frac{1}{\xi_s + i\eta_s} + \frac{1}{\xi_s - i\eta_s} + \cdots + \frac{1}{\alpha_n} = -\frac{a_{n-1}}{a_n}$$

or, after rationalizing the reciprocals of the complex numbers and transposing terms,

$$2\left(\frac{\xi_r}{\beta_r^2} + \frac{\xi_s}{\beta_s^2}\right) = -\left(\frac{a_{n-1}}{a_n} + \frac{1}{\alpha_1} + \cdots + \frac{1}{\alpha_n}\right) \quad (10.17.15)$$

where the reciprocals of the four complex roots are to be omitted in the right-hand member. Since the magnitudes β_r and β_s are known, the relations (10.17.14) and (10.17.15) comprise two linear equations for the determination of ξ_r and ξ_s if $\beta_r^2 \neq \beta_s^2$, after which $\eta_r = \sqrt{\beta_r^2 - \xi_r^2}$ and $\eta_s = \sqrt{\beta_s^2 - \xi_s^2}$.

In other situations where two or more of the original roots are of equal (or nearly equal) magnitude, it may be possible to exploit available information (for example, with regard to reality) for the purpose of determining which of the admissible ϕ's are appropriate. Otherwise, one may apply the Graeffe method both to $f(x) = 0$ and also to the equation $f(x + \varepsilon) = 0$, with a fixed value of ε, and select the appropriate candidates as points of intersection of circles of radii $|\alpha_r|$ and $|\alpha_r + \varepsilon|$ in the complex plane (see Prob. 116). A procedure using information corresponding to the limiting situation where $\varepsilon \to 0$ was proposed by Brodetsky and Smeal [1924] and systematized by Lehmer [1945, 1963].

In place of actually multiplying together the polynomials $f(x)$ and $(-1)^n f(-x)$ to obtain the function $f_2(x)$, it is desirable to work with detached coefficients and to obtain formulas recursively relating the new coefficients to the original ones. For this purpose, it is convenient to write

$$f(x) = A_0 x^n - A_1 x^{n-1} + A_2 x^{n-2} - \cdots = \sum_{i=0}^{\infty} (-1)^i A_i x^{n-i} \quad (10.17.16)$$

with the convention that $A_i = 0$ when $i > n$. If we use this convention, there follows

$$(-1)^n f(x) f(-x) = \sum_{i=0}^{\infty} (-1)^i A_i x^{n-i} \sum_{j=0}^{\infty} A_j x^{n-j}$$

$$= \sum_{i=0}^{\infty} \sum_{j=0}^{\infty} (-1)^i A_i A_j x^{2n-(i+j)}$$

Since clearly only the even powers of x will remain, we then may write $i + j = 2k$ and, after changing the limits appropriately, we have

$$(-1)^n f(x) f(-x) = f_2(x^2) = \sum_{k=0}^{\infty} (-1)^k A_k^*(x^2)^{n-k}$$

where

$$A_k^* = \sum_{i=0}^{2k} (-1)^{i+k} A_i A_{2k-i}$$

Thus $f_2(x)$ is given by

$$f_2(x) = \sum_{k=0}^{\infty} (-1)^k A_k^* x^{n-k} \tag{10.17.17}$$

where

$$A_k^* = A_k^2 - 2A_{k-1}A_{k+1} + 2A_{k-2}A_{k+2} - 2A_{k-3}A_{k+3} + \cdots \tag{10.17.18}$$

and where the series of products terminates when either the first subscript reduces to zero or the second increases to n. This formula is convenient because of the fact that the coefficients A_{k-r} and A_{k+r} involved in each product are symmetrically placed about A_k.

The procedure may be illustrated by the simple case of the cubic $f(x) = x^3 - x - 1$ considered in (10.10.6), for which $A_0 = 1$, $A_1 = 0$, $A_2 = -1$, and $A_3 = +1$, in accordance with (10.17.16). By making use of (10.17.18), the coefficients of the successive equations, again written in the form $x^3 - A_1x^2 + A_2x - A_3 = 0$, are obtained as follows in the first six iterations:

	A_0	A_1	A_2	A_3
f	1	0	−1	1
f_2	1	2	1	1
f_4	1	2	−3	1
f_8	1	10	5	1
f_{16}	1	90	5	1
f_{32}	1	8090	−155	1
f_{64}	1	65448410	7845	1

The coefficients A_0 and A_3 here remain fixed, whereas the coefficient A_1 in f_{64} is the square of that in f_{32} to five significant figures. The persistent fluctuation of A_2 indicates that the roots α_2 and α_3 are conjugate complex.

Thus the sequence of approximations to α_1 is 0, $\sqrt{2}$, $\sqrt[4]{2}$, $\sqrt[8]{10}$, $\sqrt[16]{90}$, $\sqrt[32]{8090}$, $\sqrt[64]{65448410}$, . . . , or 0, 1.4, 1.2, 1.33, 1.3248, 1.3247, ... The fact that the *positive* sign is correct would be determined most easily by noticing that $f(x)$ changes sign between $x = 1$ and $x = 2$. Reference to the first equation of (10.17.12), taking into account the fact that here $A_3 \equiv 1$, shows that the corresponding approximations to the magnitude β_2 of the complex root pair are the reciprocal square roots of the approximations to α_1, so that the best available approximation is $\beta_2 \approx 0.86884$ to five places. Rather than use the second relation of (10.17.12), which would involve choosing the appropriate value of $\cos \phi_2$ for which $\cos 64\phi_2 = 0.68263$ from among 64 possibilities, we use (10.17.13) to obtain $2\xi_2 = -\alpha_1$, and hence $\xi_2 \approx -0.6624$. Finally, there follows $\eta_2 = \sqrt{\beta_2^2 - \xi_2^2} \approx 0.5622$, so that the approximate roots are 1.3247 and $-0.6624 \pm 0.5622i$.

In order to illustrate the calculation involved in less simple cases, we display the results of five iterations as applied to the equation

$$x^4 - 10x^3 + 35x^2 - 50x + 24 = 0$$

when only three digits are retained:

	A_0	A_1	A_2	A_3	A_4
f	1.00	1.00(1)	3.50(1)	5.00(1)	2.40(1)
f_2	1.00	3.00(1)	2.73(2)	8.20(2)	5.76(2)
f_4	1.00	3.54(2)	2.65(4)	3.58(5)	3.32(5)
f_8	1.00	7.23(4)	4.49(8)	1.11(11)	1.10(11)
f_{16}	1.00	4.33(9)	1.86(17)	1.22(22)	1.21(22)
f_{32}	1.00	1.84(19)	3.45(34)	1.49(44)	1.46(44)

Here an integer in parentheses following a number represents the power of 10 by which that number is to be multiplied to give the relevant coefficient. The entries are obtained simply by use of (10.17.18). For example, the coefficients in f_{16} may be calculated as follows:

$$A_1 = 10^8[(7.23)^2 - 2(1.00)(4.49)]$$
$$A_2 = 10^{16}[(4.49)^2 - 2(7.23)(1.11)(10^{-1}) + 2(1.00)(1.10)(10^{-5})]$$
$$A_3 = 10^{22}[(1.11)^2 - 2(4.49)(1.10)(10^{-3})]$$
$$A_4 = 10^{22}(1.10)^2$$

In a sixth iteration, the squared term in (10.17.18) obviously would not be modified to three digits by the product terms in any case, so that the iteration is terminated. Here all roots are clearly real, and the application of (10.17.8) to f_{32} yields the approximations $\alpha_1 \approx 4.000$, $\alpha_2 \approx 3.001$, $\alpha_3 \approx 2.000$, and $\alpha_1 \approx 0.999$. The correctness of the positive signs is assured here by the fact that the expression for $f(-x)$ involves only positive coefficients, so that no negative real roots can be present.

It is of some interest to notice that the use of (10.17.8) at earlier stages of the iteration would yield the following sequences of approximate roots:

	1	2	3	4
f	10.000	3.500	1.429	0.480
f_2	5.477	3.017	1.733	0.838
f_4	4.338	2.941	1.917	0.981
f_8	4.049	2.979	1.991	0.999
f_{16}	4.002	3.000	2.000	1.000
f_{32}	4.000	3.001	2.000	0.999

The Graeffe method possesses the theoretical advantages that the iteration leads to *all* zeros of $f(x)$ at the same time, and that (as in the Bernoulli iteration) there is no question of the existence of ultimate convergence if appropriate attention is paid to the control of roundoff errors. As the preceding example illustrates, this control normally does not present difficulties in the Graeffe iteration. However, the process itself is often rather laborious for desk calculation, and the extraction of algebraic roots of high order, which is involved in the process, is conveniently effected in machine calculation only by an iterative process (see Prob. 47). The possibility of rapid growth of the relevant coefficients in a prolonged sequence of root squarings leads also to the danger of overflow when use is made of a computer, unless appropriate precautions are taken.

A serious disadvantage follows from the fact that a *gross* error committed at any stage of the calculation invalidates all subsequent calculations, whereas the other iterative methods considered here would suffer only a temporary reduction in the rate of convergence.

Rather than use this method for the complete determination of the roots, it is often convenient merely to iterate sufficiently to obtain crude approximations, when such approximations are not easily obtained by other methods, and then to improve these approximations by simpler or more rapidly convergent methods.

The root-squaring process is also useful in connection with the Bernoulli iteration, in cases when that iteration appears to converge slowly, since the rate of convergence increases with increasing values of the ratio of the magnitudes of the dominant and subdominant roots. Thus the convergence will be improved if the original equation is replaced by one whose roots are, say, the squares or fourth powers of the original roots.

10.18 Quadratic Factors. Lin's Quadratic Method

Among the most troublesome algebraic equations, in practice, are those which possess two or more pairs of nonreal roots. Whereas the methods of Secs. 10.16 and 10.17 can be used in such cases, and will always generate convergent sequences of approximations, the convergence is often slow and the time or labor involved may be excessive. We next treat two methods which are similar to those considered in Sec. 10.14, but in which successive approximations to a *quadratic* factor are generated. Both methods have the property that the iteration may not converge unless the initial approximation is sufficiently good and, in fact, one of them may require a modification to yield a convergent sequence even in that case. Thus, in troublesome cases, the use of the Bernoulli or Graeffe iteration may be desirable in order to afford a reasonably good initial estimate.

If the polynomial

$$f(x) = x^n + a_1 x^{n-1} + \cdots + a_{n-1}x + a_n \qquad (10.18.1)$$

is divided by the quadratic expression $x^2 + px + q$, so that

$$\begin{aligned} f(x) &= x^n + a_1 x^{n-1} + \cdots + a_{n-1}x + a_n \\ &= (x^2 + px + q)(x^{n-2} + b_1 x^{n-3} + \cdots + b_{n-3}x + b_{n-2}) \\ &\qquad + Rx + S \end{aligned} \qquad (10.18.2)$$

the requirement that this expression be a factor of $f(x)$ imposes the two conditions

$$R = 0 \qquad S = 0 \qquad (10.18.3)$$

where R and S are the coefficients of the linear remainder and are certain functions of the parameters p and q.

In order to obtain a recursive method for calculating R and S without actually effecting the long division, we equate coefficients of like powers of x in the two members of (10.18.2) and thus obtain the relations

$$\begin{gathered} a_1 = b_1 + p \qquad a_2 = b_2 + pb_1 + q \qquad a_3 = b_3 + pb_2 + qb_1 \\ \cdots \qquad a_k = b_k + pb_{k-1} + qb_{k-2} \qquad \cdots \\ a_{n-2} = b_{n-2} + pb_{n-3} + qb_{n-4} \qquad a_{n-1} = R + pb_{n-2} + qb_{n-3} \\ a_n = S + qb_{n-2} \end{gathered} \qquad (10.18.4)$$

Thus, if we introduce the recurrence formula

$$b_k = a_k - pb_{k-1} - qb_{k-2} \qquad (k = 1, 2, \ldots, n) \qquad (10.18.5)$$

with

$$b_{-1} = 0 \qquad b_0 = 1 \qquad (10.18.6)$$

it follows that this formula will generate the coefficients of the quotient in (10.18.2) with $k = 1, 2, \ldots, n - 2$, and also that

$$R = b_{n-1} = a_{n-1} - pb_{n-2} - qb_{n-3} \tag{10.18.7}$$

$$S = b_n + pb_{n-1} = a_n - qb_{n-2} \tag{10.18.8}$$

Hence the expression $x^2 + px + q$ will factor $f(x)$ if and only if the conditions

$$R \equiv a_{n-1} - pb_{n-2} - qb_{n-3} = 0 \qquad S \equiv a_n - qb_{n-2} = 0 \tag{10.18.9}$$

are satisfied.

Lin's quadratic iteration consists of applying the method of successive substitutions to the result of rewriting (10.18.9) in the form

$$p = \frac{a_{n-1} - qb_{n-3}}{b_{n-2}} \qquad q = \frac{a_n}{b_{n-2}}$$

so that "improved" values of p and q are defined by the formulas

$$p^* = \frac{a_{n-1} - qb_{n-3}}{b_{n-2}} \qquad q^* = \frac{a_n}{b_{n-2}} \tag{10.18.10}$$

and hence

$$p^* - p = \frac{a_{n-1} - pb_{n-2} - qb_{n-3}}{b_{n-2}} \qquad q^* - q = \frac{a_n - qb_{n-2}}{b_{n-2}}$$

or, equivalently, by virtue of (10.18.7) and (10.18.8),

$$p^* = p + \frac{R}{b_{n-2}} \qquad q^* = q + \frac{S}{b_{n-2}} \tag{10.18.11}$$

In analogy to (10.14.14), it is known that, if p and q are to be such that the zeros of $x^2 + px + q$ approximate the true zeros α_1 and α_2 of $f(x)$, then the two relevant asymptotic convergence factors are (see Prob. 120)

$$\rho_1 = 1 + \frac{\alpha_1\alpha_2}{\alpha_2 - \alpha_1}\frac{f'(\alpha_1)}{a_n} \qquad \rho_2 = 1 - \frac{\alpha_1\alpha_2}{\alpha_2 - \alpha_1}\frac{f'(\alpha_2)}{a_n} \tag{10.18.12}$$

That is, if either or both of these factors exceeds unity in absolute value, then one or both of the zeros of the modified expression $x^2 + p^*x + q^*$ generally will afford *poorer* approximations to α_1 and α_2 than the zeros of the expression $x^2 + px + q$. Thus, if $x^2 + px + q$ is to converge to $(x - \alpha_1)(x - \alpha_2)$, it is generally necessary that

$$|\rho_1| < 1 \qquad |\rho_2| < 1 \tag{10.18.13}$$

In addition, it is necessary that the *initial* estimates of p and q not differ excessively from $-(\alpha_1 + \alpha_2)$ and $\alpha_1\alpha_2$, respectively. The result (10.18.12) is useful only if fair approximations to a pair of roots can be obtained in advance. In analogy to (10.14.16), the conditions (10.18.13) can also be expressed in the form

$$\left|1 - \left[\left(1 - \frac{\alpha_k}{\alpha_3}\right)\left(1 - \frac{\alpha_k}{\alpha_4}\right)\cdots\left(1 - \frac{\alpha_k}{\alpha_n}\right)\right]\right| < 1 \qquad (k = 1, 2) \quad (10.18.14)$$

In the absence of preliminary information, the iteration may be started with arbitrarily chosen values of p and q, in the hope that convergence to *some* root pair (real or complex) will ensue. With the convenient initial choice $p = 0$, $q = 0$, the *first* iteration always yields the quadratic

$$x^2 + \frac{a_{n-1}}{a_{n-2}} x + \frac{a_n}{a_{n-2}}$$

whose zeros will approximate the two *smallest* roots of $f(x) = 0$ if those roots are sufficiently small relative to the others. It is seen that the initial choice $p = a_1$, $q = a_2$, corresponding to the quadratic $x^2 + a_1 x + a_2$, whose zeros would approximate the two *largest* roots of $f(x) = 0$ if those roots were sufficiently separated in magnitude from the others, leads always to $b_{n-2} = 0$ in the first iteration when $n = 3$ or $n = 4$, so that the following iteration then is undefined. This fact suggests that convergence of the Lin quadratic iteration to the *largest* root pair generally cannot be obtained *when that pair is widely separated in magnitude from the others*, as can be seen also by noticing that (10.18.14) then will tend to be violated.

In the more general case, however, (10.18.14) shows that the possibility of convergence to the largest root pair, or to any other chosen root pair, depends in a fairly complicated way upon the configuration of all the roots.

For desk calculation, the data can be arranged in parallel columns, as follows:

1	1
a_1	b_1
$\vdots$	$\vdots$
a_{n-2}	b_{n-2}
a_{n-1}	R
a_n	S

Here each entry in the b column *except the last* is obtained by subtracting from its left-hand neighbor p times its first upward neighbor and q times its second upward neighbor. (In calculating b_0 and b_1, the missing entries are taken to be zero.) The last element (S) is calculated in the same way except that its *first upward neighbor* is imagined to be replaced by zero. Finally, there follows

$$\Delta p \equiv p^* - p = \frac{R}{b_{n-2}} \qquad \text{and} \qquad \Delta q \equiv q^* - q = \frac{S}{b_{n-2}}$$

In illustration, the quartic equation

$$f(x) \equiv x^4 - 8x^3 + 39x^2 - 62x + 50 = 0 \quad (10.18.15)$$

possesses the complex roots $1 \pm i$ and $3 \pm 4i$, and $f(x)$ is factorable in the form $f(x) = (x^2 - 2x + 2)(x^2 - 6x + 25)$. The first steps in a Lin iteration, assuming ignorance of this information, and starting with $p = q = 0$, may be tabulated as follows:

$p =$	0	−1.6	−1.95	−2.009	−2.008	−2.003	−2.0007
$q =$	0	1.3	1.82	1.970	2.001	2.003	2.0012
1	1	1	1	1	1	1	
−8	−8	−6.4	−6.05	−5.991	−5.992	−5.997	
39	39	27.5	25.38	24.994	24.967	24.985	
−62	−62	−9.7	−1.50	0.015	0.124	0.057	
50	50	14.2	3.81	0.762	0.041	−0.045	
$\Delta p =$	−1.6	−0.35	−0.059	0.001	0.005	0.0023	
$\Delta q =$	1.3	0.52	0.150	0.031	0.002	−0.0018	

Thus, at this stage, the approximate factorization is

$$f(x) \approx (x^2 - 2.001x + 2.001)(x^2 - 5.997x + 24.985)$$

The Lin iteration technique is perhaps the simplest known method for the numerical solution of algebraic equations, when two or more pairs of complex roots are present. However, it possesses the disadvantage that convergence is not certain, even though the starting values are good approximations to true values, and that the rate of convergence, when present, is often rather slow. The following section describes a somewhat more elaborate method which usually has better convergence properties, as well as a less elaborate modification of the present method.

Use of (10.18.11) shows that the relation (10.18.2) is equivalent to the relation

$$f(x) = (x^2 + px + q)(x^{n-2} + b_1 x^{n-3} + \cdots + b_{n-3}x) + b_{n-2}(x^2 + p^*x + q^*) \quad (10.18.16)$$

Thus it follows that if $f(x)$ is divided by the trial factor $x^2 + px + q$, and if the steps in the division are terminated when the remainder is *quadratic* (rather than linear), the new Lin trial factor $x^2 + p^*x + q^*$ can be obtained by dividing that remainder by its leading coefficient. For this reason, Aitken [1952] refers to the new Lin trial factor as the *reduced penultimate remainder* and to Lin's quadratic method as the *RPR method.*

It may be noted that Lin also suggested an alternative technique in which the new value q^* is calculated first from the second relation of (10.18.10), after which q^* is used in place of q in the first relation for the calculation of an "improved" value of p^*. In some cases this alternative affords improved convergence; in others (including the preceding example) the reverse is true.

10.19 Bairstow Iteration

Another iterative method for solving algebraic equations, apparently first devised by Bairstow, but rediscovered by Hitchcock and others, differs from the Lin method in that the equations

$$R(p, q) = 0 \qquad S(p, q) = 0 \tag{10.19.1}$$

are solved by Newton-Raphson iteration, rather than by the method of successive substitutions used by Lin, so that it is a *second-order* process.

By virtue of the relations (10.18.7) and (10.18.8), we have

$$R = b_{n-1} \qquad S = b_n + pb_{n-1} \tag{10.19.2}$$

and hence the Newton-Raphson recurrence relations (10.13.8) become

$$\frac{\partial b_{n-1}}{\partial p}\,\Delta p + \frac{\partial b_{n-1}}{\partial q}\,\Delta q + b_{n-1} = 0$$

and

$$\left(\frac{\partial b_n}{\partial p} + p\,\frac{\partial b_{n-1}}{\partial p} + b_{n-1}\right)\Delta p + \left(\frac{\partial b_n}{\partial q} + p\,\frac{\partial b_{n-1}}{\partial q}\right)\Delta q + b_n + pb_{n-1} = 0$$

where $\Delta p \equiv p^* - p$ and $\Delta q \equiv q^* - q$. If the second relation is simplified, by subtracting from it p times the first equation, the two relations become

$$\begin{aligned} \frac{\partial b_{n-1}}{\partial p}\,\Delta p + \frac{\partial b_{n-1}}{\partial q}\,\Delta q + b_{n-1} &= 0 \\ \left(\frac{\partial b_n}{\partial p} + b_{n-1}\right)\Delta p + \frac{\partial b_n}{\partial q}\,\Delta q + b_n &= 0 \end{aligned} \tag{10.19.3}$$

If we recall that the b's are defined in terms of the coefficients of $f(x)$ by the recurrence formula (10.18.5),

$$\begin{aligned} b_k &= a_k - pb_{k-1} - qb_{k-2} \qquad (k = 1, 2, \ldots, n) \\ b_{-1} &= 0 \qquad b_0 = 1 \end{aligned} \tag{10.19.4}$$

it remains only to determine the partial derivatives involved in (10.19.3).

For this purpose, we obtain from the relation (10.19.4) the additional relations

$$-\frac{\partial b_k}{\partial p} = b_{k-1} + p\frac{\partial b_{k-1}}{\partial p} + q\frac{\partial b_{k-2}}{\partial p} \qquad (k = 1, 2, \ldots, n)$$
$$\frac{\partial b_{-1}}{\partial p} = 0 \qquad \frac{\partial b_0}{\partial p} = 0 \tag{10.19.5}$$

and

$$-\frac{\partial b_k}{\partial q} = b_{k-2} + p\frac{\partial b_{k-1}}{\partial q} + q\frac{\partial b_{k-2}}{\partial q} \qquad (k = 1, 2, \ldots, n)$$
$$\frac{\partial b_{-1}}{\partial q} = 0 \qquad \frac{\partial b_0}{\partial q} = 0 \tag{10.19.6}$$

Hence, if we introduce a new recurrence formula

$$c_k = b_k - pc_{k-1} - qc_{k-2} \qquad (k = 1, 2, \ldots, n-1)$$
$$c_{-1} = 0 \qquad c_0 = 1 \tag{10.19.7}$$

it follows, from (10.19.5), that

$$\frac{\partial b_k}{\partial p} = -c_{k-1} \qquad (k = 1, 2, \ldots, n) \tag{10.19.8}$$

and, from (10.19.6), that

$$\frac{\partial b_k}{\partial q} = -c_{k-2} \qquad (k = 1, 2, \ldots, n) \tag{10.19.9}$$

where the c's are obtained from the b's just as the b's are obtained from the a's.

Thus the first $n - 4$ of the c's are the coefficients in the relation

$$x^{n-2} + b_1x^{n-3} + \cdots + b_{n-3}x + b_{n-2}$$
$$= (x^2 + px + q)(x^{n-4} + c_1x^{n-5} + \cdots + c_{n-5}x + c_{n-4})$$
$$+ R'x + S' \tag{10.19.10}$$

and also

$$R' = c_{n-3} \qquad S' = c_{n-2} + pc_{n-3} \tag{10.19.11}$$

In particular, we have

$$\frac{\partial b_{n-1}}{\partial p} = -c_{n-2} \qquad \frac{\partial b_{n-1}}{\partial q} = -c_{n-3} \qquad \frac{\partial b_n}{\partial q} = -c_{n-2} \tag{10.19.12}$$

so that three of the four desired coefficients in (10.19.3) are now identified, and are calculable from (10.19.7). When $k = n$, Eq. (10.19.8) gives

$$\frac{\partial b_n}{\partial p} = -c_{n-1} \quad (10.19.13)$$

and hence the remaining coefficient in (10.19.3) is given by

$$\frac{\partial b_n}{\partial p} + b_{n-1} = -c'_{n-1} \quad (10.19.14)$$

where, in accordance with (10.19.7),

$$c'_{n-1} = c_{n-1} - b_{n-1} = -pc_{n-2} - qc_{n-3} \quad (10.19.15)$$

The basic equations of the Bairstow iteration then take the simple form

$$\begin{aligned} c_{n-2}\,\Delta p + c_{n-3}\,\Delta q &= b_{n-1} \\ c'_{n-1}\,\Delta p + c_{n-2}\,\Delta q &= b_n \end{aligned} \quad (10.19.16)$$

and the principal calculation involved in an iteration can be arranged as follows:

1	1	1
a_1	b_1	c_1
$\vdots$	$\vdots$	$\vdots$
a_{n-4}	b_{n-4}	c_{n-4}
a_{n-3}	b_{n-3}	c_{n-3}
a_{n-2}	b_{n-2}	c_{n-2}
a_{n-1}	b_{n-1}	c'_{n-1}
a_n	b_n	

Here each element in the b column (*including* b_n), and each element of the c column *except the last one* (c'_{n-1}), is calculated as in the Lin iteration, as the result of subtracting from the element to its left p times the last calculated element above it and q times the next-to-last element above it. The element c'_{n-1} is calculated in the same way except that the element to its *left* is imagined to be replaced by zero.

In addition, it is necessary to solve the simultaneous linear equations (10.19.16) for the corrections to be added to p and q to give p^* and q^*. For this purpose, the quantities

$$D = c_{n-2}^2 - c'_{n-1}c_{n-3} \quad (10.19.17)$$

and

$$\begin{aligned} D_p &= b_{n-1}c_{n-2} - b_n c_{n-3} \\ D_q &= -b_{n-1}c'_{n-1} + b_n c_{n-2} \end{aligned} \quad (10.19.18)$$

may be recorded, after which there follows

$$\Delta p = \frac{D_p}{D} \qquad \Delta q = \frac{D_q}{D} \quad (10.19.19)$$

The first three stages of the result of applying the Bairstow iteration to the equation (10.18.15), again starting with $p = q = 0$, appear as follows:

$p, q =$	0, 0		−1.3, 1.3		−1.9, 1.9		−1.998, 1.998
1	1	1	1	1	1	1	
−8	−8	−8	−6.7	−5.4	−6.10	−4.20	
39	39	39	29.0	20.7	25.51	15.63	
−62	−62	0	−15.6	33.9	−1.941	37.68	
50	50		−8.0		−2.157		
$D =$	1521		612		403		
$D_p, D_q =$	−2018, 1950		−366, 363		−39.4, 39.4		
$\Delta p, \Delta q =$	−1.3, 1.3		−0.6, 0.6		−0.098, 0.098		

The next (fourth) iteration gives $p \approx -1.9999992$ and $q \approx 1.9999992$ if sufficiently many digits are retained in the calculation. A comparison of these results with those obtained in the preceding section illustrates the fact that whereas the Bairstow iteration may converge more slowly than the Lin iteration in the *early* stages, when both iterations converge, its *ultimate* rate of convergence in such cases is far superior. This is due to the fact that it is a second-order process, whereas the Lin iteration is a first-order process.

Furthermore, the Bairstow iteration *will* converge if the starting values of p and q are sufficiently close to true values, whereas in the Lin iteration this is not always the case. On the other hand, the Bairstow iteration appears to be somewhat more sensitive to the choice of starting values than the Lin iteration, in the sense that, if the Lin iteration *is* asymptotically stable at (α_1, α_2), it may converge with starting values which correspond to cruder approximations to (α_1, α_2) than are required for convergence of the Bairstow iteration.

Various modifications of both the Lin and Bairstow procedures are possible (see Prob. 125).

In particular, a modification of the Lin quadratic method which is analogous to that defined by (10.14.18) and (10.14.20) can be obtained by first applying the Newton-Raphson formulas (10.13.8) to the functions

$$\phi(p, q) = \frac{R}{b_{n-2}} = \frac{qR}{a_n - S} \qquad \psi(p, q) = \frac{S}{b_{n-2}} = \frac{qS}{a_n - S} \tag{10.19.20}$$

with the partial derivatives of ϕ and ψ relative to p and q evaluated in the limiting (ideal) situation when $R = S = 0$. (Compare Prob. 91.) For example, there follows

$$\frac{\partial \phi}{\partial p} = \frac{(a_n - S)q(\partial R/\partial p) + qR(\partial S/\partial p)}{(a_n - S)^2} \sim \frac{q}{a_n}\frac{\partial R}{\partial p} = -\frac{qc_{n-2}}{a_n}$$

and, similarly,

$$\frac{\partial \phi}{\partial q} \sim -\frac{qc_{n-3}}{a_n}$$

$$\frac{\partial \psi}{\partial p} \sim -\frac{q}{a_n}(c_{n-1} - b_{n-1} + pc_{n-2})$$

$$\frac{\partial \psi}{\partial q} \sim -\frac{q}{a_n}(c_{n-2} + pc_{n-3})$$

Here p and q, as well as the c's which depend upon them, are imagined to have their ideal values, such that $x^2 + px + q$ is indeed a factor of $f(x)$. If, instead, approximate values $\bar{p}$ and $\bar{q}$ are used, and if the corresponding values of the c's are also indicated by bars, the associated simulated Newton-Raphson formulas become

$$\begin{aligned} \bar{c}_{n-2}\,\Delta p + \bar{c}_{n-3}\,\Delta q &= \frac{a_n}{\bar{q}}\frac{R}{b_{n-2}} \\ \bar{c}'_{n-1}\,\Delta p + \bar{c}_{n-2}\,\Delta q &= \frac{a_n}{\bar{q}}\frac{S - \bar{p}R}{b_{n-2}} \end{aligned} \tag{10.19.21}$$

after a slight rearrangement.

The solution of these equations can be written in the form

$$\Delta p = \frac{\lambda_{11}R + \lambda_{12}S}{b_{n-2}} \qquad \Delta q = \frac{\lambda_{21}R + \lambda_{22}S}{b_{n-2}} \tag{10.19.22}$$

where

$$\begin{aligned} \lambda_{11} &= \frac{a_n}{\bar{q}}\frac{\bar{c}_{n-2} + \bar{p}\bar{c}_{n-3}}{D} & \lambda_{12} &= -\frac{a_n}{\bar{q}}\frac{\bar{c}_{n-3}}{D} \\ \lambda_{21} &= -\frac{a_n}{\bar{q}}\frac{\bar{c}'_{n-1} + \bar{p}\bar{c}_{n-2}}{D} & \lambda_{22} &= \frac{a_n}{\bar{q}}\frac{\bar{c}_{n-2}}{D} \end{aligned} \tag{10.19.23}$$

and where

$$D = \bar{c}_{n-2}^2 - \bar{c}'_{n-1}\bar{c}_{n-3} \tag{10.19.24}$$

Here the $\bar{c}$'s and the λ's are evaluated once only, in correspondence with the approximations $\bar{p}$ and $\bar{q}$ to the desired parameters, after which only *one* quadratic synthetic division is required per iteration, the calculation differing from that in the Lin method only in that $\Delta p = p^* - p$ and $\Delta q = q^* - q$ are determined by (10.19.22) rather than by (10.18.11). In particular, here again *the last entry in the b column is S* (rather than b_n, as in the Bairstow process).

If $\bar{p}$ and $\bar{q}$ are fair approximations to true values, this modified Lin process simulates a second-order process for a limited number of steps. As in the case of the modified linear-factor method, ultimately the superiority of a truly second-order process (such as Bairstow's method) would be evident if a sufficiently large number of iterations were required.

A somewhat simpler process consists of using (10.19.16) with the c's replaced by their *initial* values throughout the iteration.

To illustrate the modified Lin quadratic process, we again consider (10.18.15) and suppose that the approximate values $\bar{p} = -1.9$ and $\bar{q} = 1.9$ have been obtained in advance. From the two-column array

$p =$	-1.9	
$q =$	1.9	
1	1	1
-8	-6.10	-4.20
39	25.51	15.63
-62	-1.941	37.68
50	1.531	

which differs from the corresponding array in the Bairstow tabulation only in that the entry $S = 1.531$ replaces the entry $b_4 = -2.157$, we obtain the rounded data

$$\bar{c}_1 \doteq -4.20 \qquad \bar{c}_2 \doteq 15.63 \qquad \bar{c}_3' \doteq 37.68 \qquad D \doteq 403$$

Hence, by use of (10.19.23), the modified iteration formulas follow in the form

$$\Delta p = \frac{1.542R + 0.274S}{b_{n-2}} \qquad \Delta q = \frac{-0.521R + 1.021S}{b_{n-2}}$$

and the tabulation of the first two iterations is obtained as follows:

$p =$	-1.9		-2.00088	-1.999994
$q =$	1.9		2.00092	1.999992
1	1	1	1	
-8	-6.10	-4.20	-5.99912	
39	25.51	15.63	24.99556	
-62	-1.941	37.68	0.016875	
50	1.531		-0.014116	
$\Delta p =$	-0.10088		0.000886	
$\Delta q =$	0.10092		-0.000928	

In this case the simulated (stationary) Bairstow process, using (10.19.16) with the same fixed values of the c's, is somewhat less effective.

10.20 Supplementary References

For treatments of the great variety of available methods for solving sets of linear equations, see Wilkinson [1960, 1961, 1964, 1965] on direct methods and associated error analysis, and Varga [1962], Householder [1964], and Young [1971] on iterative methods. Forsythe [1953] still provides a useful commentary.

The method of Crout [1941] was originally devised for use on desk calculators and sometimes is advocated only for that purpose in the current literature, in spite of the frequent use of this and related *compact elimination* methods in computer calculation when the processes of pivoting and equilibration are suitably incorporated. See, for example, the relevant Algol procedures of Forsythe [1960] and McKeeman [1962] and the treatments in Forsythe and Moler [1967] and in Wilkinson [1967]. The material in Sec. 10.8 was taken from Hildebrand [1965]. Doolittle's method, which is somewhat similar to that of Crout, is described in "Modern Computing Methods" [1961] and Ralston [1965].

Matrix eigenvalue (characteristic-value) problems, which are not treated here, are considered in detail in Householder [1964] and Wilkinson [1965], both of which include useful bibliographies.

The notion of the order of an iterative process was introduced by Schröder [1870]. Later contributions include those of Hamilton [1946], Bodewig [1949], Ehrmann [1959], and Traub [1964].

General treatments of numerical methods for solving nonlinear equations include Traub [1964], which presents an exhaustive study, classification, and compilation of iterative methods for determining a single zero, and Householder [1970], which deals with general methods intended principally for algebraic (polynomial) equations.

Traub [1964] also considers measures of relative efficiency of iterative processes, which attempt to predict the ratio of the labor, time, or "cost" totals associated with determinations effected by two different iterative processes, in terms of the orders and relative complexities of those processes. Other references are cited.

The accelerative Δ^2 process, which apparently is over a century old, was popularized by Aitken [1926] and has been generalized by Shanks [1955] and Wynn [1956] (see Sec. 5.12) and by others. See also Householder [1970].

Muller's method (Muller [1956]) is formulated in terms of divided differences by Traub [1964], who also points out that other iterative processes can be conveniently recast or modified by use of the notation of divided differences and of their recursive properties.

The iterative solution of sets of nonlinear equations is treated by Ortega and Rheinboldt [1970]. For specific numerical methods, see Booth [1949], Wolfe [1959], Kincaid [1961], Ostrowski [1966], and Rabinowitz [1970], the last of which also includes methods for single equations as well as a bibliography of methods for the solution of nonlinear sets.

Useful methods of obtaining suitable "starting values" for the iterative solution of nonlinear algebraic equations are given by Derwidué [1957] and Durand [1960]. Householder [1970] includes a chapter on the location of zeros of polynomials and lists many references, including Marden [1966].

Some interesting relationships between standard methods of solving algebraic equations and special properties of matrix eigenvalue problems are pointed out and exploited by Wilf [1960].

The method of false position is studied by Ostrowski [1966], and the secant method by Ostrowski [1966] and Barnes [1965]. Aitken [1926], in his analysis of Bernoulli's method, includes his first advocation of the Δ^2 process.

Graeffe's method is analyzed by Ostrowski [1940], Bodewig [1946], and Hoel and Wall [1947], and is modified and systematized for computer use by Bareiss [1960, 1967]. The modifications of Brodetsky and Smeal [1924] and of Lehmer [1945, 1963] are included in Householder [1970].

Lin's methods are studied by Aitken [1951, 1952] and are generalized to the extraction of higher-degree factors by Luke and Ufford [1951].

Among the many methods for solving algebraic equations which are not considered in this text are the following: Laguerre's method, treated in Derwidué [1957], Durand [1960, 1961], Ostrowski [1966], and Householder [1970]; the always-convergent method of Lehmer and Schur, presented in Lehmer [1961]; and the QD (quotient-difference) method of Rutishauser [1956] and Henrici [1958, 1963, 1964], which also has many other applications. In addition, Traub [1966] gives a procedure which constructs a sequence of iterative processes, in correspondence with any given algebraic equation and any given initial starting value, which terminates with a process yielding convergence to a root of the equation. This and related "global" procedures also are considered in Householder [1970].

The evaluation of zeros of ill-conditioned polynomials is studied in Wilkinson [1959] and Rice [1965], and the effects of roundoff errors in the more general case are dealt with in Wilkinson [1964].

Methods similar to those of Bairstow [1914] and Hitchcock [1944] for the iterative extraction of quadratic factors of polynomials are given by Friedman [1949] and McAuley [1962]. Generalizations to a second-order process for quartic factors and to a third-order process for quadratic factors were effected by Salzer [1961] and are being repeatedly rediscovered.

PROBLEMS

Section 10.2

1 Solve the following set of equations by use of determinants, without introducing roundoffs:

$$\begin{aligned} 1.4x_1 + 2.3x_2 + 3.7x_3 &= 6.5 \\ 3.3x_1 + 1.6x_2 + 4.3x_3 &= 10.3 \\ 2.5x_1 + 1.9x_2 + 4.1x_3 &= 8.8 \end{aligned}$$

2 Determine D times the inverse of the coefficient matrix in Prob. 1 without introducing roundoffs, where $D = -0.249$ is the determinant of that matrix. Then use this matrix to obtain explicit expressions for Dx_1, Dx_2, and Dx_3 when the respective right-hand members are replaced by c_1, c_2, and c_3, and check the results when the c's are assigned the values given. Also use this result to investigate the significance of the solution if it is supposed that the given coefficients are exact, but that the given right-hand members are only rounded numbers.

3 Show that the equations

$$\begin{aligned} \omega x_1 + 3x_2 + x_3 &= 5 \\ 2x_1 - x_2 + 2\omega x_3 &= 3 \\ x_1 + 4x_2 + \omega x_3 &= 6 \end{aligned}$$

possess a unique solution when $\omega \neq \pm 1$, that no solution exists when $\omega = -1$, and that infinitely many solutions exist when $\omega = 1$. Also, investigate the corresponding situation when the right-hand members are replaced by zeros.

Section 10.3

4 By considering the result of increasing each x by unity in each equation of (10.2.1), establish the validity of the following *gross-error check:*

If to each equation is adjoined an entry representing the sum of the coefficients and the right-hand member of that equation, and if the column of those entries is transformed under the Gauss (or Gauss-Jordan) reduction in the same way as the column of right-hand members, then, at each succeeding step, the transformed entry associated with any transformed equation will equal the sum of the coefficients and the right-hand member of that equation, except for the effects of intermediate roundoffs or gross errors.

5 Solve the set of equations in Prob. 1 by the Gauss reduction, retaining only five decimal places in the intermediate calculation and using the gross-error check of Prob. 4.

6 Proceed as in Prob. 5 with the following set of equations:

$$\begin{aligned}
8.467x_1 + 5.137x_2 + 3.141x_3 + 2.063x_4 &= 29.912\\
5.137x_1 + 6.421x_2 + 2.617x_3 + 2.003x_4 &= 25.058\\
3.141x_1 + 2.617x_2 + 4.128x_3 + 1.628x_4 &= 16.557\\
2.063x_1 + 2.003x_2 + 1.628x_3 + 3.446x_4 &= 12.690
\end{aligned}$$

7 Repeat the calculation of Prob. 6, using the Gauss-Jordan reduction.

Section 10.4

8 Verify (10.4.9) numerically in the case of the coefficient matrix in (10.4.10).

9, 10 Proceed as in Probs. 5 and 6 using the Crout reduction.

Section 10.5

11, 12 Assuming the given data to be exact, and starting with the approximate solutions of Probs. 9 and 10, obtain the solutions of those problems with 10-place accuracy by use of the Crout reduction.

13 Solve the equations [compare (10.5.1)]

$$\begin{aligned}
10.01x_1 + 6.99x_2 + 8.01x_3 + 6.99x_4 &= 32\\
6.99x_1 + 5.01x_2 + 5.99x_3 + 5.01x_4 &= 23\\
8.01x_1 + 5.99x_2 + 10.01x_3 + 8.99x_4 &= 33\\
6.99x_1 + 5.01x_2 + 8.99x_3 + 10.01x_4 &= 31
\end{aligned}$$

approximately, by the Crout method, rounding all entries in the auxiliary matrix to only *three* decimal places; compare the results with the true solution $x_1 = x_2 = x_3 = x_4 = 1$. Then calculate the residuals and verify the effectiveness of the iterative process of Sec. 10.5 in this case, obtaining the solution to three places.

14 Obtain a five-place solution of the set

$$\begin{aligned}
4.18x_1 + 2.87x_2 + 3.03x_3 + 2.11x_4 &= 27.45\\
6.81x_1 + 4.67x_2 + 4.09x_3 + 1.63x_4 &= 34.94\\
26.15x_1 + 17.96x_2 + 18.96x_3 + 19.94x_4 &= 198.71\\
1.23x_1 + 2.06x_2 + 1.19x_3 + 6.32x_4 &= 34.20
\end{aligned}$$

by the Crout method, assuming all coefficients and right-hand members to be exact.

Section 10.6

15, 16 Determine the inverse of the coefficient matrix in Probs. 9 and 10 by the Crout reduction, retaining five decimal places. Also evaluate the determinant of the coefficient matrix in each case.

17 Show that if **A** is the coefficient matrix of the equation set (10.5.1), then

$$\mathbf{A}^{-1} = \begin{bmatrix} 25 & -41 & 10 & -6 \\ -41 & 68 & -17 & 10 \\ 10 & -17 & 5 & -3 \\ -6 & 10 & -3 & 2 \end{bmatrix}$$

and also $D \equiv \det \mathbf{A} = 1$.

18 Determine the inverse of the coefficient matrix of the equation set in Prob. 13 approximately, by the Crout method, rounding all entries in the auxiliary matrix to only *three* decimal places. Then determine the matrix of the residuals and investigate the effectiveness of the iterative process of Sec. 10.5. (Assume that the coefficients of the equation set are exact.)

Section 10.7

19, 20 Use the results of Probs. 15 and 16 to obtain approximate upper bounds on the inherent errors relevant to the solutions of Probs. 9 and 10, assuming (*a*) that the coefficients are exact and the errors in the right-hand members cannot exceed ε in magnitude and (*b*) that the coefficients as well as the right-hand members may be in error by as much as $\pm\varepsilon$. In each case, determine what can be said about the solution if the errors in the given data are due to roundoff.

21, 22 Reestimate the error bounds considered in Probs. 19 and 20 by use of the inherent-error check column.

23 Use the result of Prob. 17 to obtain the information required in Probs. 19 and 20 for the equation set (10.5.1). Also verify that the results related to (10.5.2) and (10.5.3) are consistent with the bounds so obtained.

24 If $x_1, \ldots, x_n$ satisfy the equations

$$\sum_{k=1}^{n} a_{ik} x_k = c_i \qquad (i = 1, 2, \ldots, n)$$

show that there follows

$$\sum_{k=1}^{n} a_{ik} \frac{\partial x_k}{\partial c_r} = \begin{cases} 0 & (i \neq r) \\ 1 & (i = r) \end{cases}$$

and

$$\sum_{k=1}^{n} a_{ik} \frac{\partial x_k}{\partial a_{rs}} = \begin{cases} 0 & (i \neq r) \\ -x_s & (i = r) \end{cases}$$

and deduce the relations

$$\frac{\partial x_k}{\partial c_r} = \tilde{A}_{rk} \qquad \frac{\partial x_k}{\partial a_{rs}} = -\tilde{A}_{rk} x_s$$

25 Use the results of Prob. 24 and the data of Prob. 15 to obtain approximations to the changes in the values of x_1, x_2, and x_3 in Prob. 1 corresponding (*a*) to an increase of 0.05 in $c_3 \equiv 8.8$ and (*b*) to a decrease of 0.05 in $a_{23} \equiv 4.3$.

26 Use the results of Prob. 24 and the data of Prob. 16 to obtain approximations to the changes in the values of x_1, x_2, x_3, and x_4 in Prob. 6, corresponding (*a*) to an increase of 0.001 in $c_3 \equiv 16.557$ and (*b*) to a decrease of 0.001 in $a_{23} \equiv 2.617$.

Section 10.8

27 Determine a four-place solution of the following set of equations:

$$\begin{aligned} 3.955x_1 - 1.013x_2 &= 0.3068 \\ -1.007x_1 + 3.926x_2 - 1.023x_3 &= 0.8669 \\ -1.013x_2 + 3.887x_3 - 1.038x_4 &= 1.3168 \\ -1.021x_3 + 3.841x_4 &= 2.7997 \end{aligned}$$

28 If the interval $[0, 1]$ is subdivided by the points $x_0 = 0$, $x_1 = 0.1$, $x_2 = 0.2$, $x_3 = 0.3$, $x_4 = 0.5$, $x_5 = 0.7$, $x_6 = 0.8$, $x_7 = 0.9$, and $x_8 = 1.0$, and if the function $f(x) = e^x$ is approximated by a spline $s(x)$ with nodes at these points, the spline slopes at the interior nodes are related by the equation set

$$\frac{1}{h_k} s'_{k-1} + 2\left(\frac{1}{h_k} + \frac{1}{h_{k+1}}\right) s'_k + \frac{1}{h_{k+1}} s'_{k+1} = 3\frac{f_k - f_{k-1}}{h_k^2} + 3\frac{f_{k+1} - f_k}{h_{k+1}^2}$$

for $k = 1, 2, \ldots, 7$, where $h_k = x_k - x_{k-1}$ [see (9.10.5)]. Determine five-place values of the spline slopes at the seven interior nodes if the end conditions $s'_0 = f'(0)$ and $s'_8 = f'(1)$ are imposed.

Section 10.9

29 Determine a four-place solution of the equation set in Prob. 27 by use of Gauss-Seidel iteration.

30 Proceed as in Prob. 29 by Jacobi iteration.

31 Investigate (empirically) the efficiency of the Gauss-Seidel iteration in the case of the equations in Prob. 1.

32 Determine the solution of the equations in Prob. 27 to four places by use of a relaxation procedure.

33 Experiment with the application of relaxation methods to the equations in Prob. 1.

Section 10.10

34 Suppose that the equation $x^2 + a_1x + a_2 = 0$ possesses real roots α and β. Show that the iteration $z_{k+1} = -(a_1z_k + a_2)/z_k$ is stable at $x = \alpha$ if $|\alpha| > |\beta|$, the iteration $z_{k+1} = -a_2/(z_k + a_1)$ is stable at $x = \alpha$ if $|\alpha| < |\beta|$, and the iteration $z_{k+1} = -(z_k^2 + a_2)/a_1$ is stable at $x = \alpha$ if $2|\alpha| < |\alpha + \beta|$.

35 With the notation of Prob. 34, show that the iteration

$$z_{k+1} = z_k - (z_k^2 + a_1z_k + a_2)\phi(z_k)$$

is stable at $x = \alpha$ if $0 < (\alpha - \beta)\phi(\alpha) < 2$, that the asymptotic convergence factor is $\rho = 1 - (\alpha - \beta)\phi(\alpha)$, and that the three iterations of Prob. 34 are the special cases in which $\phi(x) = 1/x$, $1/(x + a_1)$, and $1/a_1$.

36 Show that if the asymptotic convergence factor ρ of an iteration can be estimated in any way, then the formula

$$\alpha \approx z_{k+1} + \frac{\rho}{1 - \rho}(z_{k+1} - z_k)$$

can be used to accelerate the convergence of the iteration in place of the Aitken Δ^2 process, and also that the latter process is equivalent to estimating ρ by the ratio $(z_{k+2} - z_{k+1})/(z_{k+1} - z_k)$.

37 The real root α of the equation $x + \log x = 0$ lies between 0.56 and 0.57. Show that the iteration $z_{k+1} = -\log z_k$ is unstable at $x = \alpha$, and verify this fact by calculation. Then show that the iteration $z_{k+1} = e^{-z_k}$ is stable at $x = \alpha$, and determine α to five places.

38 Suppose that the solution of Prob. 37 is required but that only values of $\log_{10} x$ are to be used. Determine a convenient value of the constant c for which the iteration

$$z_{k+1} = z_k - c(z_k + \log z_k) \equiv (1 - c)z_k - c(\log 10)\log_{10} z_k$$

is stable at $x = \alpha$, and use the result to determine α to five places.

39 Consider the application of the iteration

$$z_{k+1} = \frac{z_k^2 + 2}{3}$$

to the equation $x^2 - 3x + 2 = 0$.

(*a*) Show that this iteration is asymptotically stable at $\alpha = 1$ but unstable at $\alpha = 2$.

(*b*) Show that $z_k \to 1$ as $k \to \infty$ if $-2 < z_0 < 2$. (Prove that then z_{k+1} is between z_k and 1 when $k \geqq 1$.)

(*c*) Show that $z_{k+1} = 2$ if $z_0 = \pm 2$, but that convergence to $\alpha = 2$ for any other value of z_0 is impossible.

40 Consider the iterative solution of the equation $\tan x = x$.

(*a*) By superimposing the graphs of $y = x$ and $y = \tan x$, or otherwise, show that the rth positive root of this equation is in the interval $[r\pi, (r + \frac{1}{2})\pi]$.

(*b*) Show that the iteration

$$z_{k+1} = r\pi + \tan^{-1} z_k$$

is stable for the determination of the rth positive root α_r.

(*c*) With $[a, b] = [r\pi, (r + \frac{1}{2})\pi]$ and $F(x) = r\pi + \tan^{-1} x$, show that when $a \leqq x \leqq b$ it is true that both $a < F(x) < b$ and $0 < F'(x) < 1$. Hence deduce in two ways that convergence to α_r is assured if $a \leqq z_0 \leqq b$.

(*d*) Use the iteration of part (*b*) to determine both α_1 and α_2 to five decimal places.

(The *principal value* of $\tan^{-1} x$, for which $-\frac{1}{2}\pi < \tan^{-1} x < \frac{1}{2}\pi$, is to be presumed.)

41 Consider the polynomial $f(x) = x^5 + 5x - 1$.
(*a*) Prove that $f(x)$ has exactly one real zero α and that $0.1 < \alpha < 0.2$.
(*b*) Without more closely locating α, prove that the iteration

$$z_{k+1} = z_k - cf(z_k)$$

will converge to α if $0 < c < 1/5.008$ and if $0.1 \leqq z_0 \leqq 0.2$.
(*c*) With the choice $c = 1/5.01$, show that the asymptotic convergence factor is between 4×10^{-4} and 2×10^{-3}, so that ultimately each iteration will provide three or four additional correct decimal places.
(*d*) Verify that, with $c = 1/5.01$, two iterations provide 10-place accuracy when $z_0 = 0.2$, while three are needed when $z_0 = 0.1$.

42 Indicate by geometrical (or analytical) arguments why the following assertions are valid:
(*a*) If $f''(\alpha) \neq 0$, the false-position iteration ultimately has one stationary point. This situation occurs when $f''(x) \neq 0$ between z_k and z_{k+1}, and the stationary point is that one of z_k and z_{k+1} at which $ff'' \geqq 0$. (See Fig. 10.2.)
(*b*) If at some stage the secant method is based on abscissas z_k and z_{k+1} which are on the same side of a zero $x = \alpha$, if $ff'' \geqq 0$ at z_k and at z_{k+1}, and if f' and f'' are of constant sign in the interval spanned by α, z_k, and z_{k+1}, then convergence to α is certain.

43 Compare sequences generated by the false-position and secant methods when $f(x) = x^5 + 5x - 1$. Take $z_0 = 0$ and $z_1 = 0.5$ and obtain a five-place approximation to α by each method.

44 Deal with the "accelerated false-position" iteration process as follows:
(*a*) Complete the indicated derivation of (10.10.18).
(*b*) By writing $z_r = \alpha - \varepsilon_r$ and $f_r = f(\alpha - \varepsilon_r) = f(\alpha) - \varepsilon_r f'(\alpha) + \cdots = -\varepsilon_r f'(\alpha) + \cdots$ in (10.10.19), show that

$$\frac{\varepsilon_{k+1}}{\varepsilon_k} \sim 1 + \frac{\varepsilon_0 f'(\alpha)}{(1 - \varepsilon_{k+1}/\varepsilon_k)f_0}$$

when $f'(\alpha) \neq 0$, and deduce (10.10.21) and (10.10.22).
(*c*) Apply (10.10.19) to the equation of Prob. 43, obtaining a five-place approximation to the real root and comparing the process with the two processes used in that problem in terms of efficiency in this case.

45 Determine the number of steps which would be required if the bisection method were used for the calculation of Prob. 43, and obtain the results of the first five steps.

Section 10.11

46 Repeat the determination of Probs. 37 and 38 using the Newton-Raphson iteration both with $f(x) = x + \log x$ and with $f(x) = x - e^{-x}$.

47 Show that the Newton-Raphson iterations, as applied to $f(x) = x^n - a$ and to $f(x) = 1 - (a/x^n)$, for the determination of $\alpha \equiv a^{1/n}$, are of the respective forms

$$z_{k+1} = \frac{1}{n}\left[(n-1)z_k + \frac{a}{z_k^{n-1}}\right]$$

and

$$z_{k+1} = \frac{1}{n}\left[(n+1)z_k - \frac{z_k^{n+1}}{a}\right]$$

and that, if $\varepsilon_k \equiv \alpha - z_k$, there follows approximately

$$\varepsilon_{k+1} \approx -\frac{n-1}{2\alpha}\varepsilon_k^2$$

and

$$\varepsilon_{k+1} \approx \frac{n+1}{2\alpha}\varepsilon_k^2$$

respectively, when $z_k \approx \alpha$. (Notice that the second iteration formula possesses a constant denominator and that the two sequences approach α from above and from below, respectively, when $\alpha > 0$.) Also use both iterations to determine $(3.4765)^{1/5}$ and $(0.049672)^{1/2}$ to five places.

48 By applying the Newton-Raphson procedure to $f(x) = 1 - 1/(ax)$, obtain the recurrence formula

$$z_{k+1} = z_k(2 - az_k)$$

for the iterative determination of the reciprocal of a without effecting division, and show that, if ε_k denotes the error in z_k, there follows $\varepsilon_{k+1} \approx a\varepsilon_k^2$ when $z_k \approx 1/a$. Also show that the iteration will converge to $1/a$ if $0 < z_0 < 2/a$. Does it converge when $z_0 = 0$ or $z_0 = 2/a$?

49 Determine the smallest root of the equation $\tan x = cx$ to five places, with $c = 1.01$, $c = 2$, and $c = 30$.

Determine all real roots of the following equations to five places:

50 $x^3 - 2x - 5 = 0$†

51 $x^3 - 9x^2 + 18x - 6 = 0$

52 $x^4 - 16x^3 + 72x^2 - 96x + 24 = 0$

53 $x^4 - 3x + 1 = 0$

54 $x^2 - 3x - 4\sin^2 x = 0$

55 Suppose that $f(x)$ possesses two zeros α_1 and α_2 which are nearly coincident, so that $f'(x)$ vanishes at a point β between α_1 and α_2. By making use of the relation

$$f(\alpha) = f(\beta) + (\alpha - \beta)f'(\beta) + \frac{(\alpha - \beta)^2}{2}f''(\beta) + \cdots$$

†This equation was used by Wallis in 1685 to illustrate the Newton-Raphson method and has been included as an example in most subsequent works dealing with the numerical solution of equations.

show that if β is determined first, then initial approximations to the nearby zeros of $f(x)$ are given by

$$\alpha_{1,2} \approx \beta \pm \left[-\frac{2f(\beta)}{f''(\beta)}\right]^{1/2}$$

if $f''(\beta) \neq 0$, and are real if $f(\beta)$ and $f''(\beta)$ are of opposite sign, after which improved values may be obtained by an appropriate iterative method. [Note that the case of a *double* root α is also included since then $f(\beta) = 0$ and $\beta = \alpha$.] Also use this procedure to determine the two real roots of the equation

$$3x^4 + 8x^3 - 6x^2 - 25x + 19 = 0$$

(which are near $x = 1$) to five places.

56 The equation

$$2x^4 + 24x^3 + 61x^2 - 16x + 1 = 0$$

has two nearly coincident zeros near $x = 0.1$. Determine them to five decimal places by the method of Prob. 55.

57 Proceed as in Prob. 55 with the root pair of the equation

$$x^6 - 16x^3 + x^2 + 59 = 0$$

which is near $x = 2$.

58 The equation

$$x^4 - 8.2x^3 + 39.41x^2 - 62.26x + 30.25 = 0$$

has a double root near $x = 1$. Determine it to five places by the method of Prob. 55.

Section 10.12

59 If $x = \alpha$ is a root of $f(x) = 0$, if successive approximations to α are generated by the iteration $z_{k+1} = F(z_k)$, and if $F(x)$ possesses r continuous derivatives and is such that

$$F(\alpha) = \alpha \qquad F'(\alpha) = F''(\alpha) = \cdots = F^{(r-1)}(\alpha) = 0 \qquad F^{(r)}(\alpha) \neq 0$$

for some r, show that

$$\alpha - z_{k+1} = (-1)^{r-1} \frac{(\alpha - z_k)^r}{r!} F^{(r)}(\xi_k)$$

where ξ_k lies between z_k and α, and that the iteration is a process of order r. (This is the way in which *order* was first *defined* by Schröder [1870]. Here r must be an *integer*.)

60 With the notation of Prob. 59, show that the iteration corresponding to the definition

$$F(x) = x - \phi_1(x)f(x) - \phi_2(x)[f(x)]^2 - \phi_3(x)[f(x)]^3 - \cdots$$

is at least of second order if

$$1 - \phi_1 f' = 0$$

at least of third order if also

$$2\phi_1' f' + \phi_1 f'' + 2\phi_2 f'^2 = 0$$

and at least of fourth order if further

$$3\phi_1'' f' + 3\phi_1' f'' + \phi_1 f''' + 6\phi_2' f'^2 + 6\phi_2 f' f'' + 6\phi_3 f'^3 = 0$$

under the assumption that the ϕ's and an appropriate number of their derivatives are finite at $x = \alpha$ and that $f'(\alpha) \neq 0$. Thus deduce that the formula

$$z_{k+1} = z_k - \frac{f_k}{f_k'} - \frac{f_k''}{3f_k'}\left(\frac{f_k}{f_k'}\right)^2 - \left(\frac{f_k''^2}{2f_k'^2} - \frac{f_k'''}{6f_k'}\right)\left(\frac{f_k}{f_k'}\right)^3 - \cdots$$

with $f_k^{(n)} \equiv f^{(n)}(z_k)$, then yields a process of order equal to the number of terms retained in the right-hand member.

61 Rederive the formula of Prob. 60 by writing $z_{k+1} - z_k = h$ and

$$f(z_k + h) = f_k + hf_k' + \frac{h^2}{2} f_k'' + \frac{h^3}{6} f_k''' + \cdots$$

assuming an expansion of the form $h = -(\phi_1 f_k + \phi_2 f_k^2 + \phi_3 f_k^3 + \cdots)$, requiring that the coefficients of successive powers of f_k vanish in the result of substituting the second expansion into the first, and so obtaining the conditions $1 - \phi_1 f_k' = 0$, $2\phi_2 f_k' - \phi_1^2 f_k'' = 0$, $6\phi_3 f_k' - 6\phi_1\phi_2 f_k'' + \phi_1^3 f_k''' = 0$, ...

62 Consider the convergence of the sequence $\{z_k\}$ for which

$$z_{k+1} = \tfrac{1}{4}(10 - 19z_k + 14z_k^2 - 3z_k^3)$$

as follows:

(*a*) If the sequence converges to a limit α, determine all possible values of α.

(*b*) For each such possible value of α, find the order of the iteration process.

(*c*) Determine to which of the possible limits the convergence of an *infinite* sequence is in fact possible.

63 Use the formula of Prob. 60 to approximate the real root of $x^3 - x - 1 = 0$, taking $z_0 = 1.3$ and calculating separately the approximations to z_1 afforded by retention of one, two, and three correction terms. Also investigate the approximations corresponding to the choice $z_0 = 1$.

64 Use the results of Prob. 60, with $f(x) = x^2 - a$ and with $f(x) = 1 - (a/x^2)$, to obtain third-order iterations leading to $\alpha \equiv a^{1/2}$ in the forms

$$z_{k+1} = \frac{1}{2}\left(z_k + \frac{a}{z_k}\right) - \frac{1}{8z_k}\left(z_k - \frac{a}{z_k}\right)^2$$

and

$$z_{k+1} = \frac{1}{2} z_k \left(3 - \frac{z_k^2}{a}\right) + \frac{3}{8} z_k \left(1 - \frac{z_k^2}{a}\right)^2$$

and use them to determine $(16.324)^{1/2}$ and $(0.049672)^{1/2}$ to four places.

65 Use the results of Prob. 60 with $f(x) = 1 - 1/(ax)$ to obtain the third-order iteration

$$z_{k+1} = z_k(3 - 3az_k + a^2 z_k^2)$$

for the approximate calculation of $\alpha = 1/a$, and show that if $\varepsilon_k = \alpha - z_k$, then $\varepsilon_{k+1} = a^2\varepsilon_k^3$. Also account for the behavior of the iteration sequence when $z_0 = 2/a$.

66 By writing $z_k = \alpha - \varepsilon_k$ in the iteration formula and expanding the result in powers of ε_k, show that in the *Newton-Raphson* iteration there follows

$$\varepsilon_{k+1} = \varepsilon_k - \varepsilon_k \frac{f'(\alpha) - \frac{1}{2}f''(\alpha)\varepsilon_k + \frac{1}{6}f'''(\alpha)\varepsilon_k^2 - \cdots}{f'(\alpha) - f''(\alpha)\varepsilon_k + \frac{1}{2}f'''(\alpha)\varepsilon_k^2 - \cdots}$$

when ε_k is sufficiently small, assuming appropriate differentiability of $f(x)$. Thus deduce the following facts when convergence is present:

(*a*) If $f'(\alpha) \neq 0$ and $f''(\alpha) \neq 0$, then

$$\varepsilon_{k+1} \sim -\frac{f''(\alpha)}{2f'(\alpha)}\varepsilon_k^2$$

in accordance with (10.11.4).

(*b*) If $f'(\alpha) \neq 0, f''(\alpha) = 0$, and $f'''(\alpha) \neq 0$, then

$$\varepsilon_{k+1} \sim \frac{f'''(\alpha)}{3f'(\alpha)}\varepsilon_k^3$$

so that the process then is of order 3.

(*c*) If $f'(\alpha) = f''(\alpha) = \cdots = f^{(m-1)}(\alpha) = 0$ and $f^{(m)}(\alpha) \neq 0$, with $m > 1$, so that α is of multiplicity m, then

$$\varepsilon_{k+1} \sim \left(1 - \frac{1}{m}\right)\varepsilon_k$$

so that the process is of *first* order, with an asymptotic convergence factor $(m - 1)/m$.

67 Rederive the result of Prob. 66(*c*) by writing $f(x) = (x - \alpha)^m g(x)$, where $g(\alpha) \neq 0$, and $F(x) = x - f(x)/f'(x)$ in the result of Prob. 59. [Show that $r = 1$ and $F'(\alpha) = (m - 1)/m$.]

68 If α is a zero of $f(x)$ of multiplicity $m > 1$, show that the two *modified* Newton-Raphson iterations

$$z_{k+1} = z_k - \frac{mf(z_k)}{f'(z_k)}$$

and

$$z_{k+1} = z_k - \frac{h(z_k)}{h'(z_k)} \qquad \text{with } h(x) = \frac{f(x)}{f'(x)}$$

are both of order 2 or greater. (Use either the method of Prob. 66 or the method of Prob. 67. The first modification has the disadvantage that the value of m must be known; and both are somewhat objectionable for computation since the numerator and denominator of the correction term both tend to zero as $z_k \to \alpha$ unless an analytical simplification is possible.)

69 By use of the method of Prob. 66 (or otherwise), establish the following facts for the *secant* method:

(*a*) If $f'(\alpha) \neq 0$ and $f''(\alpha) \neq 0$, then

$$\varepsilon_{k+1} \sim -\frac{f''(\alpha)}{2f'(\alpha)}\varepsilon_k\varepsilon_{k-1}$$

in accordance with (10.12.2), and hence (10.12.4) follows.

(*b*) If α is a zero of $f(x)$ of multiplicity $m > 1$, then

$$\frac{\varepsilon_{k+1}}{\varepsilon_k} \sim \frac{(\varepsilon_k/\varepsilon_{k-1})^{m-1} - 1}{(\varepsilon_k/\varepsilon_{k-1})^m - 1}$$

so that the process is of order 1, with

$$\varepsilon_{k+1} \sim A\varepsilon_k$$

where the asymptotic convergence factor A is the real root of the equation

$$A^m + A^{m-1} - 1 = 0$$

In particular, $A \doteq 0.618$ when $m = 2$ and $A \doteq 0.755$ when $m = 3$.

70 Calculate two iterates approximating the zero $\alpha = 1$ of the equation $x^5 - 1 = 0$ using Muller's method, taking $z_0 = 1.05$, $z_1 = 0.95$, and $z_2 = 0.98$; compare them with the Newton-Raphson values 1.00083 and 1.0000014.

71 Proceed as in Prob. 70, using the semiconfluent process associated with (10.12.10) and (10.12.14), taking $z_0 = 0.95$ and $z_1 = 0.98$.

72 Use the confluent formula (10.12.18) in Prob. 70, taking $z_0 = 0.98$.

73 Use Halley's formula (10.12.20) in Prob. 70, taking $z_0 = 0.98$.

74 Use Chebyshev's formula (10.12.22) in Prob. 70, taking $z_0 = 0.98$.

75 Use Traub's formula (10.12.25) in Prob. 70, taking $z_0 = 1.05$, $z_1 = 0.95$, and $z_2 = 0.98$.

76 *Ostrowski's method.* Verify that the equation

$$\frac{y - f_{k-1}}{y - f_k} = \frac{1}{f_k'}\frac{f_k - f_{k-1}}{z_k - z_{k-1}}\frac{x - z_{k-1}}{x - z_k}$$

defines $y(x)$ as a *rational* function of x such that $y = f$ when $x = z_k$ and $x = z_{k-1}$ and also $y' = f'$ when $x = z_k$. By setting $y(x) = 0$ and identifying the resultant x with z_{k+1}, deduce the iteration formula

$$z_{k+1} = \frac{z_k - \mu z_{k-1}}{1 - \mu}$$

where

$$\mu = \frac{1}{f_k'}\frac{f_k - f_{k-1}}{z_k - z_{k-1}}\frac{f_k}{f_{k-1}}$$

[Here it is true that

$$\varepsilon_{k+1} \sim \left\{\frac{f'''(\alpha)}{6f'(\alpha)} - \left[\frac{f''(\alpha)}{2f'(\alpha)}\right]^2\right\} \varepsilon_k^2 \varepsilon_{k-1}$$

and the order of the process is $r = 1 + \sqrt{2} \doteq 2.414$. (See Ostrowski [1966], Sec. 11.2.) The confluent form of this formula is Halley's formula, as was seen in Sec. 9.17.]

77 Use Ostrowski's method (Prob. 76) in Prob. 70, taking $z_0 = 0.95$ and $z_1 = 0.98$.

Section 10.13

78 Determine $F(x, y)$ and $G(x, y)$ such that the Newton-Raphson iteration for a solution (α, β) of the equations $f(x, y) = 0$ and $g(x, y) = 0$ is expressed in the form

$$x_{k+1} = F(x_k, y_k) \qquad y_{k+1} = G(x_k, y_k)$$

and show that F_x, F_y, G_x, and G_y vanish when $(x, y) = (\alpha, \beta)$ in nonexceptional cases.

79 Determine to five places the real solution of the equations

$$x = \sin(x + y) \qquad y = \cos(x - y)$$

80 Determine to five places the real solution of the equations

$$4x^3 - 27xy^2 + 25 = 0 \qquad 4x^2y - 3y^3 - 1 = 0$$

in the first quadrant.

81 Determine to five places the real solution of the equations

$$\sin x \sinh y = 0.2 \qquad \cos x \cosh y = 1.2$$

nearest the origin.

82 Determine approximate coordinates of the intersections of the curves $y^2 - x^2 = 4$ and $(x - 1)^2 + (y - 1)^2 = 1.34$ as follows:

(*a*) Plot the curves, together with the jacobian locus $J(f, g) = 0$, where f and g are the left-hand members of the given equations; verify that the two curves have two intersections near the point (0.7, 2.1).

(*b*) Use the method leading to (10.13.19) and (10.13.20) to determine first the coordinates of the intersection of $f = 4$ and $J = 0$ to five decimal places and then corresponding approximations to the coordinates of the required intersections. Finally, use an iterative method to improve these approximations, as is necessary, so that five-place values are obtained.

(For purposes of this problem, avoid shortcuts or alternative procedures permitted by the fact that f and g are quadratic.)

83 Use the method of false position to determine one solution of the equation pair in Prob. 82 to four decimal places, starting with points at which $x = 0.63$ and 0.64.

84 Calculate two false-position iterates in the case of Prob. 80, starting with $x = 1.0$ and 1.1.

85 Determine the result of one step by the method of steepest descent for the solution of the simultaneous equations

$$x^2 + y^2 = 2 \qquad x - y = 0$$

taking $\phi = \frac{1}{2}[(x^2 + y^2 - 2)^2 + (x - y)^2]$ with $x_0 = 0.9$ and $y_0 = 1.1$. Also compare the initial and modified values of ϕ.

Section 10.14

86 Verify that if the coefficients of the polynomial $f(x)$ and/or the parameter z are complex, and if the notation

$$a_k = \tilde{a}_k + i\hat{a}_k \qquad b_k = \tilde{b}_k + i\hat{b}_k \qquad z = u + iv$$

is used, the recurrence formula (10.14.6) is replaced by the formulas

$$\tilde{b}_k = \tilde{a}_k + u\tilde{b}_{k-1} - v\hat{b}_{k-1} \qquad \hat{b}_k = \hat{a}_k + u\hat{b}_{k-1} + v\tilde{b}_{k-1}$$

with $\tilde{b}_0 = 1$ and $\hat{b}_0 = 0$. Also use this procedure to evaluate

$$f(x) = x^3 + (2 + 3i)x^2 + (1 - i)x + (4 - 3i)$$

when $x = 2.24 + 1.38i$.

87 Determine to five decimal places the minimum value of the function

$$f(x) = x^4 - 2.2x^3 + 2.24x^2 - 2.24x + 1.23$$

[*Suggestion:* Use the Birge-Vieta procedure to solve the equation $f'(x) = 0$. Also consider the sign of $f''(x)$.]

88 Suppose that a polynomial $f(x)$ is such that $f'(x) = 0$ at a point β near which $f(x)$ has (or may have) two nearly equal zeros.

(*a*) With the notation of Sec. 10.14, show that if β is evaluated first, the values of two nearby zeros are given approximately by

$$\alpha_{1,2} \approx \beta \pm \left(-\frac{R}{R''}\right)^{1/2}$$

if $R'' \neq 0$, and are real if $RR'' \leqq 0$. (Compare with Prob. 55.)

(*b*) Apply the result of part (*a*) to the determination of five-place values of the real zeros of the polynomial

$$f(x) = x^4 - 2.2x^3 + 2.24x^2 - 2.24x + 1.21$$

(Compare with Prob. 87.)

89 Show that the result of replacing x by $t + c$ in

$$f(x) \equiv x^n + a_1x^{n-1} + \cdots + a_{n-1}x + a_n$$

is of the form $\bar{f}(t) \equiv t^n + R^{(n-1)}t^{n-1} + R^{(n-2)}t^{n-2} + \cdots + R't + R$, where the coefficients can be determined by continuing the process leading to (10.14.5) until it terminates, with α replaced by c. Also illustrate this procedure in the case when

$$f(x) = x^3 - x - 1$$

and $c = 1.3$, showing that the calculations may be arranged as follows:

$$\begin{array}{r|llll} 1 & 1 & 1 & 1 & \underline{1} \\ 0 & 1.3 & 2.6 & \underline{3.9} & \\ -1 & 0.69 & \underline{4.07} & & \\ -1 & \underline{-0.103} & & & \end{array}$$

90 Apply the Lin iteration to the equation

$$t^3 + 3.9t^2 + 4.07t - 0.103$$

obtained in Prob. 89, starting with $t_0 = 0$. Thus obtain the real root of the equation $x^3 - x - 1 = 0$ to five places.

91 Verify that the modified Lin method specified by (10.14.18) and (10.14.20) can be derived by applying the Newton-Raphson formula to R/b_{n-1}

$$z^* = z - \frac{R/b_{n-1}}{(R/b_{n-1})'}$$

evaluating the derivative $(R/b_{n-1})'$ at $z = \alpha_r$ and then replacing α_r by $\bar{\alpha}_r$ in the result. [Note that $b_{n-1} = (R - a_n)/z$ and that $R = f(z)$ and $a_n = f(0)$.]

92 Show that the recurrence formula (10.14.21) takes the approximate form $z^* = z - 0.246R$ when applied to the iterative solution of (10.10.6) with $\bar{\alpha} = 1.3$, and verify its efficiency in this case by obtaining the real zero to seven places.

93 For the equation

$$(x - 1)(x - 2)(x - 3) \equiv x^3 - 6x^2 + 11x - 6 = 0$$

show that the Lin iteration is stable for the determination of $\alpha_1 = 1$ and $\alpha_3 = 3$, but is unstable for $\alpha_2 = 2$. Then calculate the results of two iterations using (*a*) Lin's method, (*b*) the modified Lin method of (10.14.18), and (*c*) the formula (10.14.21), starting with the initial approximations $\bar{\alpha}_1 = 0.9$, $\bar{\alpha}_2 = 1.9$, and $\bar{\alpha}_3 = 2.9$. Show that the modified Lin method is best for the determination of α_1 and α_2, but that the *original* Lin method is best for α_3; account for the last fact. [Determine the *order* of the Lin method in this case (when $\alpha = 3$).]

94 Use the result of Prob. 60 to devise a third-order iteration process extending (10.14.10), of the form

$$z^* = z - \frac{R}{R'} - \frac{R''}{R'}\left(\frac{R}{R'}\right)^2$$

where $R'' = d_{n-2}$ and $d_k = c_k + zd_{k-1}$. Also determine the real zero of $x^3 - x - 1$ to six decimal places by this method, starting with $z = 1.3$.

95 to *98* Determine the real roots of the equations in Probs. 50 to 53 to five places by use of any method exploiting synthetic division.

Section 10.15

99 Show that the formula (10.15.3) would provide the estimate $\delta\alpha \approx -5$ when $\alpha = -20$ and $\delta a_1 \approx 1.2 \times 10^{-7}$ in the Wilkinson example cited in Sec. 10.16. Also verify that with this estimate, the second-order term neglected in the approximation

$$(\alpha + \delta\alpha)^{20} = \alpha^{20} + 20\alpha^{19}\delta\alpha + 190\alpha^{18}(\delta\alpha)^2 + \cdots$$
$$\approx \alpha^{20} + 20\alpha^{19}\delta\alpha$$

[which is one of several approximations underlying (10.15.3) in this case] is about equal in magnitude to the first-order term retained in that case, so that the estimate (appropriately) should not be accepted *quantitatively* in this case, but only as a strong danger signal.

100 The equation

$$x^4 - 15.2x^3 + 59.7x^2 - 81.6x + 36.0 = 0$$

has roots approximated by 1.0, 1.2, 3, and 10. Determine approximately the maximum inherent error in each root, assuming (*a*) that each coefficient in the equation (except the leading one) may be in error by ± 0.1 and (*b*) that each such coefficient is correct within 1 percent.

101 If $\bar{\alpha}$ is an approximation to a root α of the equation $f(x) = 0$, and if $f(\bar{\alpha}) = \varepsilon$, show that $\alpha - \bar{\alpha} = -\varepsilon/f'(\xi)$, where ξ is between $\bar{\alpha}$ and α if $f'(x)$ is continuous.

102 Suppose that an approximation $\bar{\alpha}_1$ to a zero α_1 of a polynomial $f(x)$ has been found, that the polynomial $g_1(x)$ is the consequence of dividing $f(x)$ by $x - \bar{\alpha}_1$ and ignoring the constant remainder, and that an approximation $\bar{\alpha}_2$ to another zero α_2 of $f(x)$ is obtained as an *exact* zero of $g_1(x)$. By writing

$$f(x) = (x - \bar{\alpha}_1)g_1(x) + f(\bar{\alpha}_1)$$

and using the result of Prob. 101, show that

$$\alpha_2 - \bar{\alpha}_2 = \frac{f'(\xi_1)}{f'(\xi_2)}(\alpha_1 - \bar{\alpha}_1)$$

where ξ_k is between α_k and $\bar{\alpha}_k$. Thus deduce that *when approximations to* $\alpha_1, \alpha_2, \ldots, \alpha_k, \ldots$ *are obtained in that order by successive deflations, there follows*

$$\alpha_{k+1} - \bar{\alpha}_{k+1} \approx \left[\frac{f'(\alpha_k)}{f'(\alpha_{k+1})}\right](\alpha_k - \bar{\alpha}_k) + \delta_k$$

where δ_k is of the order of the adopted error tolerance. [Hence a significant error *amplification* generally occurs if $|f'(\alpha_{k+1})| \ll |f'(\alpha_k)|$.]

103 Illustrate the result of Prob. 102 in the case of the polynomial $f(x) = (x - 1)(x - 4)(x - 4.1)$, showing that deflation with the approximation 4.09 to the largest zero yields the nearly equal smaller zero with about the same accuracy and the smallest zero with an error of about 3×10^{-4}, whereas deflation with 1.01 as as approximation to the smallest zero yields the other zeros with errors of magnitude exceeding 0.12; relate these facts to Prob. 102. [In the second case, the quantitative deviation is due to the rapid variation of $f'(x)$ near $x = 4$.]

Section 10.16

104 Determine the largest root in Prob. 51 to four places by Bernoulli iteration.

105 Determine the largest root in Prob. 52 to four places by Bernoulli iteration. Also, after replacing x by $1/x$, determine the smallest root in a similar way. Then determine all roots to four places.

106 Show that the Bernoulli iteration converges very slowly when applied to Prob. 50, and account for this fact. Then translate the origin to a convenient point near the root (see Prob. 89), replace x by $1/x$, and apply the iteration to determine the reciprocal of the real root to six places.

107 Show that the μ, s, and t sequences all behave unsatisfactorily when the Bernoulli iteration is applied to Prob. 53. Then replace x by $1/x$, use the iteration to determine the smallest root to four places, and determine the other real root after translating the origin to a nearby point and replacing x by $1/x$. Finally, determine the remaining roots and account for the original difficulty.

108 Apply the Bernoulli iteration to the equation

$$x^4 - 8x^3 + 39x^2 - 62x + 50 = 0$$

determining the larger pair of complex roots to three significant figures.

109 Show that if $|\alpha_1| > |\alpha_2| > |\alpha_3|$ and if $C_1 \neq 0$ and $C_2 \neq 0$ in (10.16.4), there follows

$$\alpha_1 \sim r_k + A\beta^k$$

where $\beta = \alpha_2/\alpha_1$, with the notation of (10.16.5), and where $0 < \beta < 1$, so that Aitken's Δ^2 process then is applicable for accelerating the convergence of the Bernoulli iteration. [Compare (10.10.9).]

110 Apply the Δ^2 process to the determination of the largest root in Prob. 51 to five places by Bernoulli iteration.

Section 10.17

111 to *114* Determine all roots of the equations in Probs. 50 to 53 to three decimal places by the Graeffe procedure.

115 Use the Graeffe procedure to determine all roots of the equation in Prob. 108 to three significant figures.

116 Suppose that the Graeffe iteration is applied to the polynomial $f(x) = x^4 - 1$.
(*a*) Show that $f_4(x) = (x - 1)^4$, $f_8(x) = (x - 1)^8$, and so on, and hence that after k root squarings with $k \geqq 2$ one would know only that the four zeros of $f(x)$ are among the (2^k)th roots of unity and, in particular, that they lie on the unit circle $|z| = 1$ in the complex plane.
(*b*) Show that if the same iteration were also applied to the polynomial $f(x + \varepsilon) = (x + \varepsilon)^4 - 1$, where $0 < \varepsilon < 1$, it would be found that the zeros of that function lie on circles of radii $1 + \varepsilon$, $1 - \varepsilon$, and $\sqrt{1 + \varepsilon^2}$ and center at $z = 0$, and hence that the zeros of $f(x)$ lie on the circles $|z - \varepsilon| = 1 + \varepsilon$, $1 - \varepsilon$, and $\sqrt{1 + \varepsilon^2}$.
(*c*) Verify (a sketch is sufficient) that the intersections of $|z| = 1$ with the circles specified at the end of part (*b*) uniquely determine the four zeros of $f(x)$.
(The same method succeeds in less contrived situations, provided that ε is taken to be sufficiently small.)

Section 10.18

117 With the notation of (10.18.2), show that if $x - z$ is a factor of $x^2 + px + q$, then $f(z) = Rz + S$. Thus deduce that *if $f(x)$ is a polynomial, then quadratic synthetic division can be used to evaluate $f(u + iv)$ in the form*

$$f(u + iv) = (u + iv)R + S$$

where R and S correspond to

$$p = -2u \qquad q = u^2 + v^2$$

Also illustrate this procedure by evaluating $f(2 + 3i)$ when

$$f(x) = x^5 - 4.17x^4 + 2.81x^3 + 1.09x^2 + 3.21x + 1.42$$

(Compare with the procedure of Prob. 86.)

118 Show that the equation

$$x^4 - 9.00x^3 + 29.08x^2 - 39.52x + 18.82 = 0$$

has roots near $x = 1$ and $x = 2$ and two roots near $x = 3$, and determine the roots to four decimal places by extracting an approximate quadratic factor by Lin iteration, starting with $(x - 3)^2$.

119 Determine all roots of the equation

$$x^4 + 9x^3 + 36x^2 + 51x + 27 = 0$$

to four decimal places by iterative Lin extraction of a quadratic factor.

120 With the notation of Sec. 10.18, show that the Lin iteration (10.18.11) can be written in the form

$$p^* - p = \frac{qR}{a_n - S} \qquad q^* - q = \frac{qS}{a_n - S}$$

and that

$$Rx_1 + S = f(x_1) \qquad Rx_2 + S = f(x_2)$$

where x_1 and x_2 are the zeros of $x^2 + px + q$, and hence deduce the relations

$$x_1(p^* - p) + (q^* - q) = \frac{qf(x_1)}{a_n - S}$$

$$x_2(p^* - p) + (q^* - q) = \frac{qf(x_2)}{a_n - S}$$

Then show that these relations can be written in the forms

$$(x_2 - x_1)(x_1^* - x_1) = \frac{qf(x_1)}{a_n - S} - (x_1^* - x_1)(x_2^* - x_2)$$

$$(x_2 - x_1)(x_2^* - x_2) = -\frac{qf(x_2)}{a_n - S} + (x_1^* - x_1)(x_2^* - x_2)$$

where x_1^* and x_2^* are the zeros of $x^2 + p^*x + q^*$, and deduce that, when (x_1, x_2) and (x_1^*, x_2^*) are near (α_1, α_2), there follows

$$\alpha_1 - x_1^* \approx \left[1 + \frac{\alpha_1\alpha_2}{\alpha_2 - \alpha_1} \frac{f'(\alpha_1)}{a_n}\right] (\alpha_1 - x_1) \equiv \rho_1(\alpha_1 - x_1)$$

$$\alpha_2 - x_2^* \approx \left[1 - \frac{\alpha_1\alpha_2}{\alpha_2 - \alpha_1} \frac{f'(\alpha_2)}{a_n}\right] (\alpha_2 - x_2) \equiv \rho_2(\alpha_2 - x_2)$$

where ρ_1 and ρ_2 are the convergence factors listed in (10.18.12). Thus show that, if the zeros x_1 and x_2 of $x^2 + px + q$ approximate two zeros α_1 and α_2 of $f(x)$, and if x_1^* and x_2^* are the zeros of $x^2 + p^*x + q^*$, then x_1^* is generally a poorer approximation to α_1 than x_1 unless $|\rho_1| \leqq 1$ and x_2^* a poorer approximation to α_2 than x_2 unless $|\rho_2| \leqq 1$.

121 In the case of a quartic equation, show that the asymptotic convergence factors for the Lin method relevant to the root pair α_1, α_2 are

$$\rho_1 = \frac{\alpha_1}{\alpha_3\alpha_4}(\alpha_3 + \alpha_4 - \alpha_1) \qquad \rho_2 = \frac{\alpha_2}{\alpha_3\alpha_4}(\alpha_3 + \alpha_4 - \alpha_2)$$

where α_3 and α_4 are the remaining roots. Show, in particular, that the Lin iteration should converge rather rapidly to the root pair near $x = 3$ in Prob. 118, but that convergence to the pair near $x = 1$ and $x = 2$ should be very slow; verify the last fact by direct calculation.

122 Determine the first five quadratics yielded by the Lin iteration as applied to the equation $x^4 - 4x^3 + 7x^2 - 16x + 12 = 0$, starting with $p = q = 0$, and show that one of the zeros of the sequence of quadratics tends to approximate the smallest zero ($x = 1$) of the given equation. Also, use the fact that the four zeros are $x = 1, 3,$ and $\pm 2i$ to show that this situation is in accordance with the results of Prob. 120.

Section 10.19

123, 124 Repeat the determinations of Probs. 118 and 119, using the Bairstow iteration.

125 Show that, if the Newton-Raphson iteration is applied to the equations $b_{n-1} = 0$ and $b_n = 0$, rather than to the equivalent equations $b_{n-1} = 0$ and $b_n + pb_{n-1} = 0$, then the Bairstow procedure is modified only to the extent that c'_{n-1} is replaced by c_{n-1} in (10.19.16), so that *all* elements in the c column then are to be calculated from elements in the b column by the same rule. Also, apply this procedure to the example in the text, showing that the modification appears to lead to somewhat slower convergence in that case.

126 With the notation of (10.18.2) and (10.19.11), show that the Bairstow iteration can be described by the equations

$$(S' - pR')\,\Delta p + R'\,\Delta q = R$$

$$-qR'\,\Delta p + S'\,\Delta q = S$$

(These are the forms originally given by Bairstow.)

127 Show that if Bairstow's method is modified so that the values of the c's determined for $\bar{p}$ and $\bar{q}$ are used in all steps, then in the case of Eq. (10.18.15), with $\bar{p} = -1.9$ and $\bar{q} = 1.9$, the iteration formulas can be written in the form

$$\Delta p = 0.0586R + 0.0104S$$

$$\Delta q = -0.0198R + 0.0388S$$

and determine the results of two iterations. Also compare them with the results of the corresponding Lin method obtained in Sec. 10.19.

128, 129 Use the modified Lin iteration in Probs. 118 and 119, starting with $(x - 3)^2$ in the first case and with $x^2 - 2.0x + 1.3$ in the second.

130, 131 Proceed as in Probs. 128 and 129 with the stationary modification of Bairstow iteration.

132 The equation

$$2x^4 + 24x^3 + 61x^2 - 16x + 1 = 0$$

has two nearly equal small roots (see Prob. 56). Determine them to seven places by any method employing quadratic synthetic division.

APPENDIX A
JUSTIFICATION OF THE CROUT REDUCTION

The Gauss reduction of Sec. 10.3 reduces the set of equations

$$\begin{aligned} a_{11}x_1 + a_{12}x_2 + a_{13}x_3 + \cdots + a_{1n}x_n &= c_1 \\ a_{21}x_1 + a_{22}x_2 + a_{23}x_3 + \cdots + a_{2n}x_n &= c_2 \\ \cdots\cdots\cdots\cdots\cdots\cdots\cdots\cdots\cdots\cdots & \\ a_{n1}x_1 + a_{n2}x_2 + a_{n3}x_3 + \cdots + a_{nn}x_n &= c_n \end{aligned} \tag{A.1}$$

to an equivalent set of the form

$$\begin{aligned} x_1 + a'_{12}x_2 + a'_{13}x_3 + \cdots + a'_{1n}x_n &= c'_1 \\ x_2 + a'_{23}x_3 + \cdots + a'_{2n}x_n &= c'_2 \\ \cdots\cdots\cdots\cdots\cdots\cdots\cdots & \\ x_n &= c'_n \end{aligned} \tag{A.2}$$

after which the required values of $x_n, x_{n-1}, \ldots, x_1$ are obtained simply by solving the set (A.2) successively, in reverse order.

Since, in the Gauss reduction, the kth equation of (A.2) is obtained by a sequence of operations which involve the subtraction of multiples of the first $k - 1$ equations of (A.2) from the kth equation of (A.1) and the division of the result by a constant, it follows also that the kth equation of (A.1) can be expressed as a linear combination of the first k equations of (A.2), so that a set of constants a'_{ij} exists, with $i \geqq j$, such that

$$\begin{aligned} a'_{11}c'_1 &= c_1 \\ a'_{21}c'_1 + a'_{22}c'_2 &= c_2 \\ \cdots\cdots\cdots\cdots\cdots\cdots & \\ a'_{n1}c'_1 + a'_{n2}c'_2 + a'_{n3}c'_3 + \cdots + a'_{nn}c'_n &= c_n \end{aligned} \tag{A.3}$$

The Crout reduction amounts to first determining the coefficients a'_{ij} in such a way that the elimination of $c'_1, \ldots, c'_n$ between (A.2) and (A.3) lead to (A.1), then determining $c'_1, \ldots, c'_n$ from (A.3), and finally resolving (A.2) for $x_1, \ldots, x_n$ by the "back solution" of the Gauss procedure. In order to simplify the derivation of formulas for the determination of the coefficients a'_{ij} involved in both (A.2) and (A.3), it is convenient to introduce the temporary notations

$$\alpha_{ij} = \begin{cases} a'_{ij} & (i \geqq j) \\ 0 & (i < j) \end{cases} \qquad \beta_{ij} = \begin{cases} 0 & (i \geqq j) \\ a'_{ij} & (i < j) \end{cases} \tag{A.4}$$

The three sets (A.1) to (A.3) can then be specified by the equations

$$\sum_{j=1}^{n} a_{ij}x_j = c_i \tag{A.1$'$}$$

$$x_k + \sum_{j=1}^{n} \beta_{kj}x_j = c'_k \tag{A.2$'$}$$

and

$$\sum_{k=1}^{n} \alpha_{ik}c'_k = c_i \tag{A.3$'$}$$

where all indices range from 1 to n.

The introduction of (A.2′) into (A.3′) then gives

$$\sum_{k=1}^{n} \alpha_{ik}x_k + \sum_{j=1}^{n} \left(\sum_{k=1}^{n} \alpha_{ik}\beta_{kj} \right) x_j = c_i$$

and this relation is equivalent to (A.1′) if

$$\alpha_{ij} + \sum_{k=1}^{n} \alpha_{ik}\beta_{kj} = a_{ij} \tag{A.5}$$

By virtue of (A.4), the first term α_{ij} is zero unless $i \geqq j$ and the summand in the second term vanishes unless both $k \leqq i$ and $k < j$. Thus, when $i \geqq j$, (A.5) becomes

$$a'_{ij} + \sum_{k=1}^{j-1} a'_{ik}a'_{kj} = a_{ij} \qquad (i \geqq j) \tag{A.6}$$

whereas, when $i < j$, it can be written in the form

$$a'_{ii}a'_{ij} + \sum_{k=1}^{i-1} a'_{ik}a'_{kj} = a_{ij} \qquad (i < j) \tag{A.7}$$

These relations, together with the relations

$$a'_{ii}c'_i + \sum_{k=1}^{i-1} a'_{ik}c'_k = c_i \tag{A.8}$$

and

$$x_i + \sum_{k=i+1}^{n} a'_{ik}x_k + c'_i \tag{A.9}$$

which are equivalent to (A.3′) and (A.1′), respectively, are identical with the relations (10.4.4) to (10.4.7), establishing the validity of the Crout reduction as described in Sec. 10.4.

Clearly, the compactness of the relevant tabulation follows from the fact that, after suppressing the diagonal 1s in the coefficient matrix of (A.2) and the right-hand members of (A.3), which are also contained in the matrix of (A.1), the remaining elements of the *two* matrices associated with (A.2) and (A.3) can be recorded in a *single* auxiliary matrix. In order to establish the relation

$$a'_{ii}a'_{ij} = a'_{ji} \qquad (i < j) \tag{A.10}$$

in the special case when the given coefficient array is *symmetric*, so that

$$a_{ji} = a_{ij} \tag{A.11}$$

we may verify that (A.6) and (A.7) imply the relation

$$a'_{ii}a'_{ij} - a'_{ji} = a_{ij} - a_{ji} + \sum_{k=1}^{i-1} (a'_{ik}a'_{kj} - a'_{jk}a'_{ki}) \qquad (i < j)$$

which can be written in the form

$$a'_{ii}a'_{ij} - a'_{ji} = \sum_{k=1}^{i-1} [a'_{ki}(a'_{kk}a'_{kj} - a'_{jk}) - a'_{kj}(a'_{kk}a'_{ki} - a'_{ik})] \qquad (i < j) \tag{A.12}$$

if (A.11) is true. When $i = 1$, the sum on the right is absent, so that (A.10) is established in that case. When $i > 1$, (A.12) expresses $a'_{ii}a'_{ij} - a'_{ji}$ as a linear combination of terms of the form $a'_{rr}a'_{rs} - a'_{sr}$, where $r < i$ and $r < s$, so that (A.10) is established by induction on i.

The validity of the *gross-error check* described in Sec. 10.4, follows from the fact that an increase of each solution element x_i by *unity* would correspond to an increase of c_i by $\sum_{k=1}^{n} a_{ik}$, according to (A.1′), and likewise to an increase of c'_i by $1 + \sum_{k=i+1}^{n} a'_{ik}$, according to (A.2′).

Since the kth equation of (A.2) can be obtained by subtracting from the kth equation of (A.1) a certain linear combination of the first $k - 1$ equations of (A.1), and dividing the result by a'_{kk}, it follows, from the elementary properties of determinants, that the determinant of any square array of order k formed from elements in the first k rows of the augmented matrix of (A.1) is given by the result of multiplying the determinant of the corresponding array in (A.2) by $a'_{11}a'_{22} \cdots a'_{kk}$. In particular, we thus obtain the useful results

$$a_{11} = a'_{11} \qquad \begin{vmatrix} a_{11} & a_{12} \\ a_{21} & a_{22} \end{vmatrix} = a'_{11}a'_{22} \qquad \begin{vmatrix} a_{11} & a_{12} & a_{13} \\ a_{21} & a_{22} & a_{23} \\ a_{31} & a_{32} & a_{33} \end{vmatrix} = a'_{11}a'_{22}a'_{33}$$

$$\cdots \qquad \begin{vmatrix} a_{11} & \cdots & a_{1n} \\ \cdots & \cdots & \cdots \\ a_{n1} & \cdots & a_{nn} \end{vmatrix} = a'_{11}a'_{22} \cdots a'_{nn} \tag{A.13}$$

When the matrix composed of the coefficients in (A.1) is *symmetric*, it is said to be also *positive definite* if and only if each of the *n principal minors* indicated in (A.13) is positive. It follows that this matrix is positive definite if and only if all the diagonal elements of the associated Crout auxiliary matrix are positive.

APPENDIX B
BIBLIOGRAPHY

ACHIESER, N. I. [1956]: "Theory of Approximation," Frederick Ungar Publishing Co., New York; translated by C. J. Hyman.

AHLBERG, J. H., E. N. NILSON, and J. L. WALSH [1967]: "The Theory of Splines and Their Applications," Academic Press, Inc., New York.

AITKEN, A. C. [1926]: On Bernoulli's Numerical Solution of Algebraic Equations, *Proc. Roy. Soc. Edinburgh*, **46**:289–305.

——— [1932a]: On the Graduation of Data by the Orthogonal Polynomials of Least Squares, *Proc. Roy. Soc. Edinburgh*, **53**:54–78.

——— [1932b]: On Interpolation by Iteration of Proportional Parts, without the Use of Differences, *Proc. Edinburgh Math. Soc.*, **3**(2):56–76.

——— [1951]: Studies in Practical Mathematics, VI, On the Factorization of Polynomials by Iterative Methods, *Proc. Roy. Soc. Edinburgh*, **63**:174–191.

——— [1952]: Studies in Practical Mathematics, VII, On the Theory of Methods of Factoring Polynomials by Iterated Division, *Proc. Roy. Soc. Edinburgh*, **64**: 326–335.

ANDERSON, R. L., and E. E. HOUSEMAN [1942]: Tables of Orthogonal Polynomial Values Extended to N = 104, *Iowa State Coll. Agr. Exp. Sta. Res. Bull. No.* 297, pp. 595–672.

Arden, B. W., and K. N. Astill [1970]: "Numerical Algorithms: Origins and Applications," Addison-Wesley Publishing Company, Inc., Reading, Mass.

Babuska, I., M. Prager, and E. Vitasek [1966]: "Numerical Processes in Differential Equations," Interscience Publishers, a division of John Wiley & Sons, Inc., New York.

Bairstow, L. [1914]: Investigations Relating to the Stability of the Aeroplane, *Repts. & Memo. No.* 154, *Advis. Comm. Aeronaut.*

Bareiss, E. H. [1960]: Resultant Procedure and the Mechanization of the Graeffe Process, *J. Assoc. Comput. Mach.*, **7**:346–386.

——— [1967]: The Numerical Solution of Polynomial Equations and the Resultant Procedures, in Ralston and Wilf [1967].

Barnes, J. G. P. [1965]: An algorithm for Solving Non-Linear Equations Based on the Secant Method, *Comput. J.*, **8**:66–72.

Bashforth, F., and J. C. Adams [1883]: "An Attempt to Test the Theories of Capillary Action . . . with an Explanation of the Method of Integration Employed," Cambridge University Press, New York.

Bauer, F. L., H. Rutishauser, and E. Stiefel [1963]: New Aspects in Numerical Quadrature, in "Experimental Arithmetic, High Speed Computing, and Mathematics," American Mathematical Society, Providence, R.I.

Beard, R. E. [1947]: Some Notes on Approximate Product-Integration, *J. Inst. Actuar.*, **73**:356–416.

Bernstein, S. [1912a]: Démonstration du théorème de Weierstrasse fondée sur le calcul de probabilités, *Comm. Kharkov Math. Soc.*, **13**:1–2.

——— [1912b]: Sur l'ordre de la meilleure approximation des fonctions continues par des polynomes, *Comm. Kharkov Math. Soc.*, **13**:49–194.

——— [1926]: "Leçons sur les propriétés extremales et al meilleure approximation des fonctions analytiques d'une variable réele," Gauthier-Villars, Paris.

——— [1937]: Sur les formules de quadrature de Cotes et Tchebycheff, *C. R. Acad. Sci. URSS* (new ser.), **14**:323–326.

——— and C. J. de la Vallée Poussin [1969]: Reprinting of Bernstein [1926] and de la Vallée Poussin [1919], Chelsea Publishing Company, New York.

Bickley, W. G. [1948]: Difference and Associated Operators, with Some Applications, *J. Math. and Phys.*, **27**:183–192.

——— and J. C. P. Miller [1936]: The Numerical Summation of Slowly Convergent Series of Positive Terms, *Phil. Mag.*, **22**(7):754–767.

Birge, R. T., and J. W. Weinberg [1947]: Least Squares Fitting of Data by Means of Polynomials, *Rev. Modern Phys.*, **19**:298–360.

Birkhoff, G., and G-C. Rota [1969]: "Ordinary Differential Equations," 2d ed., Ginn and Company, Boston, Mass.

Birkhoff, G. D. [1906]: General Mean Value and Remainder Theorems with Applications to Mechanical Differentiation and Quadrature, *Trans. Amer. Math. Soc.*, **7**:107–136.

BODEWIG, E. [1946]: On Graeffe's Method of Solving Algebraic Equations, *Quart. Appl. Math.*, **4**:177–190.

——— [1949]: On Types of Convergence and on the Behavior of Approximations in the Neighborhood of a Multiple Root of an Equation, *Quart. Appl. Math.*, **7**:325–333.

BOOLE, G. [1970]: "Calculus of Finite Differences," 5th ed., Chelsea Publishing Company, New York. (1st ed., Macmillan and Co., London, 1860.)

BOOTH, A. D. [1949]: An Application of the Method of Steepest Descent to the Solution of Systems of Nonlinear Simultaneous Equations, *Quart. J. Mech. Appl. Math.*, **2**:460–468.

BRODETSKY, S., and G. SMEAL [1924]: On Graeffe's Method for Complex Roots of Algebraic Equations, *Proc. Cambridge Philos. Soc.*, **22**:83–87.

BURNSIDE, W. S., and A. W. PANTON [1935]: "Theory of Equations," Dublin University Press, Dublin.

CHEBYSHEV [TCHEBICHEFF], P. L. [1874]: Sur les Quadratures, *J. Math. Pures Appl.*, **19**(2):19–34.

CHENEY, E. W. [1966]: "Introduction to Approximation Theory," McGraw-Hill Book Company, New York.

CHERRY, P. M. [1950]: Summation of Slowly Convergent Series, *Proc. Cambridge Philos. Soc.*, **46**:436–449.

CHRISTOFFEL, E. B. [1858]: Über die Gaussische Quadratur und eine Verallgemeinerung derselben, *J. Reine Angew. Math.*, **55**:61–82.

CLENSHAW, C. W. [1955]: A Note on the Summation of Chebyshev Series, *MTAC*, **9**:118–120.

——— [1962]: "Chebyshev Series for Mathematical Functions," Her Majesty's Stationery Office, London.

——— and S. M. Picken [1966]: "Chebyshev Series for Bessel Functions of Fractional Order," Her Majesty's Stationery Office, London.

CODDINGTON, E., and N. LEVINSON [1955]: "Theory of Ordinary Differential Equations," McGraw-Hill Book Company, New York.

COLLATZ L. [1950]: Über die Konvergenzkriterion bei Iterationverfahren für lineare Gleichungssysteme, *Math. Z.*, **53**:149–161.

——— [1960]: "The Numerical Treatment of Differential Equations," 3d ed., Springer-Verlag OHG, Berlin.

COMRIE, L. J. (ed.) [1956]: "Interpolation and Allied Tables," Her Majesty's Stationery Office, London.

COOLEY, J. W., and J. W. TUKEY [1965]: An Algorithm for the Machine Calculation of Complex Fourier Series, *Math. Comp.*, **19**:297–301.

CRAMÉR, H. [1946]: "Mathematical Methods of Statistics," Princeton University Press, Princeton, N.J.

CROUT, P. D. [1941]: A Short Method for Evaluating Determinants and Solving Systems of Linear Equations with Real or Complex Coefficients, *Trans. AIEE*, **60**:1235–1240.

CURTIS, A. R. [1970]: The Approximation of a Function of One Variable by Cubic Splines, in Hayes [1970].

DAHLQUIST, G. [1956]: Convergence and Stability in the Numerical Solution of Ordinary Differential Equations, *Math. Scand.*, **4**:33–53.

DANIEL, J. W., and R. E. MOORE [1970]: "Computation and Theory in Ordinary Differential Equations," W. H. Freeman and Company, San Francisco.

DANIELL, P. J. [1940]: Remainders in Interpolation and Quadrature Formulae, *Math. Gaz.*, **24**:238–244.

DANIELSON, G. C., and C. LANCZOS [1942]: Some Improvements in Practical Fourier Analysis, *J. Franklin Inst.*, **233**:365–380, 435–452.

DAVIS, H. T. [1963]: "Tables of Mathematical Functions," 2 vols., Trinity University Press, San Antonio, Tex. (formerly "Tables of the Higher Mathematical Functions").

DAVIS, P. J. [1963]: "Interpolation and Approximation," Blaisdell Publishing Company, a division of Ginn and Company, Waltham, Mass.

——— and P. Rabinowitz [1967]: "Numerical Integration," Blaisdell Publishing Company, a division of Ginn and Company, Waltham, Mass.

DE LURY, D. B. [1950]: "Values and Integrals of the Orthogonal Polynomials up to $n = 26$," University of Toronto Press, Toronto.

DERWIDUÉ, L. [1957]: "Introduction a l'algèbre supérieure et au calcul numerique algébrique," Masson et Cie, Paris.

DOODSON, A. T. [1950]: A Method for the Smoothing of Numerical Tables, *Quart. J. Mech. Appl. Math.*, **3**:217–224.

DURAND, E. [1960, 1961]: "Solutions numériques des équations algébriques," 2 vols., Masson et Cie, Paris.

EHRMANN, H. [1959]: Konstruktion und Durchführung von Iterationsverfahren höher Ordnung, *Arch. Rational Mech. Anal.*, **4**:65–88.

ELLIOTT, D. [1964]: The Evaluation and Estimation of the Coefficients in the Chebyshev Series Expansion of a Function, *MTAC*, **18**:274–284.

FELLER, W. [1968]: "An Introduction to Probability Theory and Its Applications," 3d ed., 2 vols., John Wiley & Sons, Inc., New York.

FETTIS, H. E. [1955]: Numerical Calculation of Certain Definite Integrals by Poisson's Summation Formula, *MTAC*, **9**:85–92.

——— [1958]: Further Remarks Concerning the Relative Accuracy of Simpson's and the Trapezoidal Rule for a Certain Class of Functions, *Z. Angew. Math. Mech.*, **38**:159–160.

FIKE, C. T. [1968]: "Computer Evaluation of Mathematical Functions," Prentice-Hall, Inc., Englewood Cliffs, N.J.

FILON, L. N. G. [1928]: On a Quadrature Formula for Trigonometric Integrals, *Proc. Roy. Soc. Edinburgh*, **49**:38–47.

FLETCHER, A., J. C. P. MILLER, L. ROSENHEAD, and L. J. COMRIE [1962]: "An Index of Mathematical Tables 1," 2d ed., Addison-Wesley Publishing Company, Inc., Reading, Mass.

FLINN, E. A. [1960]: A Modification of Filon's Method of Numerical Integration, *J. Assoc. Comput. Mach.*, **7**:181–184.

FLORES, I. [1960]: "Computer Logic," Prentice-Hall, Inc., Englewood Cliffs, N.J.

——— [1963]: "The Logic of Computer Arithmetic," Prentice-Hall, Inc., Englewood Cliffs, N.J.

FORSYTHE, G. E. [1953]: Solving Linear Algebraic Equations Can Be Interesting, *Bull. Amer. Math. Soc.*, **59**:299–329.

——— [1957]: Generation and Use of Orthogonal Polynomials for Data-fitting with a Digital Computer, *SIAM J. Appl. Math.*, **5**:74–88.

——— [1958]: Singularity and Near Singularity in Numerical Analysis, *Amer. Math. Monthly*, **65**:229–240.

——— [1960]: Crout with Pivoting (Algorithm 16), *Comm. ACM*, **3**:507–508.

——— and C. B. MOLER [1967]: "Computer Solution of Linear Algebraic Systems, Prentice-Hall, Inc., Englewood Cliffs, N.J.

FORT, T. [1948]: "Finite Differences and Difference Equations in the Real Domain," Oxford University Press, Fair Lawn, N.J.

FOX, L. [1957]: "The Numerical Solution of Two-point Boundary Problems in Ordinary Differential Equations," Oxford University Press, Fair Lawn, N.J.

——— (ed.) [1962]: "Numerical Solution of Ordinary and Partial Differential Equations," Addison-Wesley Publishing Company, Inc., Reading, Mass.

——— and I. PARKER [1968]: "Chebyshev Polynomials in Numerical Analysis," Oxford University Press, Fair Lawn, N.J.

FRAME, J. S. [1953]: The Solution of Equations by Continued Fractions, *Amer. Math. Monthly*, **60**:293–305.

FRIEDMAN, B. [1949]: Note on Approximating Complex Zeros of a Polynomial, *Comm. Pure Appl. Math.*, **2**:195–208.

FRÖBERG, C-E. [1969]: "Introduction to Numerical Analysis," 2d ed., Addison-Wesley Publishing Company, Inc., Reading, Mass.

GHIZZETTI, M., and A. OSSICINI [1970]: "Quadrature Formulae," Academic Press, Inc., New York.

GILL, S. [1951]: A Process for the Step-by-Step Integration of Differential Equations in an Automatic Digital Computing Machine, *Proc. Cambridge Philos. Soc.*, **47**: 96–108.

GOLDSTINE, H. H., and J. VON NEUMANN [1951]: Numerical Inverting of Matrices of High Order II, *Proc. Amer. Math. Soc.*, **2**:188–202.

GOOD, I. J., and R. A. GASKINS [1971]: The Centroid Method of Numerical Integration, *Numer. Math.*, **16**:343–359.

GREVILLE, T. N. E. (ed.) [1969]: "Theory and Applications of Spline Functions," Academic Press, Inc., New York.

GUEST, P. G. [1961]: "Numerical Methods of Curve Fitting," Cambridge University Press, New York.

HABER, S. [1970]: Numerical Evaluation of Multiple Integrals, *SIAM Rev.*, **12**:481–526.

HAMILTON, H. J. [1946]: Roots of Equations by Functional Iteration, *Duke Math. J.*, **13**:113–121.

HAMMER, P. C. [1958]: The Midpoint Method of Numerical Integration, *Math. Mag.*, **31**:193–195.

——— [1959]: Numerical Evaluation of Multiple Integrals, in Langer [1959].

HAMMING, R. W. [1959]: Stable Predictor-Corrector Methods for Ordinary Differential Equations, *J. Assoc. Comput. Mach.*, **6**:37–47.

——— [1962]: "Numerical Methods for Scientists and Engineers," McGraw-Hill Book Company, New York.

——— [1971]: "Introduction to Applied Numerical Analysis," McGraw-Hill Book Company, New York.

HANDSCOMB, D. C. (ed.) [1966]: "Methods of Numerical Approximation," Pergamon Press, New York.

HARDY, G. H. [1949]: "Divergent Series," Oxford University Press, Fair Lawn, N.J.

HART, J. F., et al. [1965]: "Computer Approximations," John Wiley & Sons, Inc., New York.

HARTREE, D. R. [1958]: "Numerical Analysis," 2d ed., Oxford University Press, Fair Lawn, N.J.

HASTINGS, C. [1957]: "Approximations for Digital Computers," 2d ed., Princeton University Press, Princeton, N.J.

HAYES, J. G. (ed.) [1970]: "Numerical Approximations to Functions and Data," The Athlone Press, London.

——— and T. VICKERS [1951]: The Fitting of Polynomials to Unequally-spaced Data, *Phil. Mag.*, **42**(7):1387–1400.

HENRICI, P. [1958]: The Quotient-Difference Algorithm, in "Further Contributions to the Solution of Simultaneous Linear Equations and the Determination of Eigenvalues," National Bureau of Standards Applied Mathematics Series, vol. 49, U.S. Government Printing Office, Washington, D.C.

——— [1962]: "Discrete Variable Methods in Ordinary Differential Equations," John Wiley & Sons, Inc., New York.

——— [1963]: "Error Propagation for Difference Methods," John Wiley & Sons, Inc., New York.

——— [1964]: "Elements of Numerical Analysis," John Wiley & Sons, Inc., New York.

HERMITE, C. [1878]: Sur la formule d'interpolation de Lagrange, *J. Reine Angew. Math.*, **84**:70–79.

HESTENES, M. R., and E. STIEFEL [1952]: Method of Conjugate Gradients for Solving Linear Systems, *J. Res. Nat. Bur. Standards*, **49**:409–436.

HEUN, K. [1900]: Neue Methode zur approximativen Integration der Differentialgleichungen einer unabhängigen Variable, *Z. Angew. Math. Phys.*, **45**:23–38.

HILDEBRAND, F. B. [1965]: "Methods of Applied Mathematics," 2d ed., Prentice-Hall, Inc., Englewood Cliffs, N.J.

——— [1968]: "Finite-Difference Equations and Simulations," Prentice-Hall, Inc., Englewood Cliffs, N.J.

HITCHCOCK, F. L. [1944]: An Improvement on the G. C. D. Method for Complex Roots, *J. Math. and Phys.*, **23**:69–74.

HOEL, P. G., and D. D. WALL [1947]: The Accuracy of the Root-squaring Method for Solving Equations, *J. Math. and Phys.*, **26**:156–164.

HOUSEHOLDER, A. S. [1953]: "Principles of Numerical Analysis," McGraw-Hill Book Company, New York.

——— [1964]: "The Theory of Matrices in Numerical Analysis," Blaisdell Publishing Company, a division of Ginn and Company, Waltham, Mass.

——— [1970]: "Numerical Treatment of a Single Nonlinear Equation," McGraw-Hill Book Company, New York.

INCE, E. L. [1952]: "Integration of Ordinary Differential Equations," 6th ed., John Wiley & Sons, Inc., New York.

IRWIN, J. O. [1923]: "On Quadrature and Cubature, Tracts for Computers, X," Cambridge University Press, New York.

JOLLEY, L. B. W. [1961]: "Summation of Series," 2d ed., Dover Publications, Inc., New York.

JORDAN, C. [1965]: "Calculus of Finite Differences," 3d ed., Chelsea Publishing Company, New York.

KELLER, H. B. [1968]: "Numerical Methods for Two-Point Boundary-Value Problems," Blaisdell Publishing Company, a division of Ginn and Company, Waltham, Mass.

KINCAID, W. M. [1961]: A Two-Point Method for the Numerical Solution of Systems of Simultaneous Equations, *Quart. Appl. Math.*, **18**:305–324.

KNOPP, K. [1956]: "Infinite Sequences and Series," Dover Publications, Inc., New York.

KOGBETLIANTZ, E. G. [1960]: Generation of Elementary Functions, in Ralston and Wilf [1960].

KOPAL, Z. [1961]: "Numerical Analysis," 2d ed., John Wiley & Sons, Inc., New York.

KROGH, F. T. [1970]: Efficient Algorithms for Polynomial Interpolation and Numerical Differentiation, *Math. Comp.*, **24**:185–190.

KRYLOV, V. I. [1958]: Mechanical Quadratures with Equal Coefficients for the Integrals $\int_0^\infty e^{-x}f(x)\,dx$ and $\int_{-\infty}^{\infty} e^{-x^2}f(x)\,dx$, *Dokl. Akad. Nauk. SSSR*, **2**:187–192.

——— [1962]: "Approximate Calculation of Integrals," The Macmillan Company, New York; translated by A. H. Stroud.

——— and L. G. KRUGLIKOVA [1969]: "Handbook of Numerical Harmonic Analysis," Israel Program for Scientific Translation, Ltd., Jerusalem; translated by D. Louvish.

KUNTZMANN, J. [1959]: "Méthodes numériques," Dunod, Paris.

KUTTA, W. [1901]: Beitrag zur näherungsweisen Integration totaler Differentialgleichungen, *Z. Angew. Math. Phys.*, **46**:435–453.

LANCZOS, C. [1938]: Trigonometric Interpolation of Empirical and Analytic Functions, *J. Math. and Phys.*, **17**:123–199.

——— [1952]: Analytical and Practical Curve Fitting of Equidistant Data, *Nat. Bur. Standards Rept.* 1591.

——— [1956]: "Applied Analysis," Prentice-Hall, Inc., Englewood Cliffs, N.J.

LANGER, R. E. (ed.) [1959]: "On Numerical Approximation," University of Wisconsin Press, Madison.

LAPIDUS, L., and J. H. SEINFELD [1971]: "Numerical Solution of Ordinary Differential Equations," Academic Press, Inc., New York.

LEHMER, D. H. [1945]: The Graeffe Process as Applied to Power Series, *MTAC*, **1**:377–383.

——— [1961]: A Machine Method for Solving Polynomial Equations, *J. Assoc. Comput. Mach.*, **8**:151–162.

——— [1963]: The Complete Root-squaring Method, *SIAM J. Appl. Math.*, **11**:705–717.

LEWIS, D. C. [1947]: Polynomial Least Square Approximations, *Amer. J. Math.*, **69**:273–278.

LIDSTONE, G. J. [1943]: Notes on Interpolation, *J. Inst. Actuar.*, **71**(2):68–95.

LIN, S. N. [1941]: A Method of Successive Approximations of Evaluating the Real and Complex Roots of Cubic and Higher-order Equations, *J. Math. and Phys.*, **20**: 231–242.

——— [1943]: A Method for Finding Roots of Algebraic Equations, *J. Math. and Phys.*, **22**:60–77.

LUKE, Y. L. [1953]: Coefficients to Facilitate Interpolation and Integration of Linear Sums of Exponential Functions, *J. Math. and Phys.*, **31**:267–275.

——— [1956]: Simple Formulas for the Evaluation of Some Higher Transcendental Functions, *J. Math. and Phys.*, **34**:298–307.

——— [1969] "The Special Functions and Their Applications," vol. 2, Academic Press, Inc., New York.

——— and D. UFFORD [1951]: On the Roots of Algebraic Equations, *J. Math. and Phys.*, **30**:94–101.

MADELUNG, E. [1931]: Über eine Methode zur schnellen numerischen Lösung von Differentialgleichungen zweiter Ordnung, *Z. Physik*, **67**:516–518.

MANGULIS, V. [1965]: "Handbook of Series for Scientists and Engineers," Academic Press, Inc., New York.

MARDEN, M. [1966]: "Geometry of Polynomials," rev. ed., American Mathematical Society, Providence, R.I.

MCAULEY, V. A. [1962]: A Method for the Real and Complex Roots of a Polynomial, *SIAM J. Appl. Math.*, **10**:657–667.

MCKEEMAN, W. M. [1962]: Crout with Equilibration and Iteration (Algorithm 135), *Comm. ACM*, **5**:553–555.

MEHLER, F. G. [1864]: Bemerkungen zur Theorie der mechanischen Quadraturen, *J. Reine Angew. Math.*, **63**:152–157.

MEYERS, L. F., and A. SARD [1950a]: Best Approximate Integration Formulas, *J. Math. and Phys.*, **29**:118–123.

——— and ——— [1950b]: Best Interpolation Formulas, *J. Math. and Phys.*, **29**: 198–206.

MICHEL, J. G. L. [1946]: Central Difference Formulae Obtained by Means of Operator Expansions, *J. Inst. Actuar.*, **72**:470–480.

MILLER, J. C. P. [1950]: Checking by Differences, *MTAC*, **4**:3–11.

——— [1960a]: Quadrature in Terms of Equally-Spaced Function Values, *MRC Tech. Summ. Rept.* 167, Madison, Wis.

——— [1960b]: Numerical Quadrature over a Rectangular Domain in Two or More Dimensions, *MTAC*, **14**:13–20, 130–138, 240–248.

MILNE, W. E. [1949]: "Numerical Calculus," Princeton University Press, Princeton, N.J.

——— [1950]: Note on the Runge-Kutta Method, *J. Res. Nat. Bur. Standards*, **44**: 549–550.

——— [1970]: "Numerical Solution of Differential Equations," 2d ed., Peter Smith Publisher, Gloucester, Mass.

MILNE-THOMSON, L. M. [1951]: "The Calculus of Finite Differences," St. Martin's Press, Inc., New York.

MINNICK, R. C. [1957]: Tchebysheff Approximations for Power Series, *J. Assoc. Comput. Mach.*, **4**:487–504.

VON MISES, R. [1936]: Über allgemeine Quadraturformeln, *J. Reine Angew. Math.*, **174**:56–67.

"Modern Computing Methods" [1961], Her Majesty's Stationery Office, London.

MOULTON, F. R. [1926]: "New Methods in Exterior Ballistics," University of Chicago Press, Chicago; reprinted [1962] as "Methods in Exterior Ballistics," Dover Publications, Inc., New York.

MOURSUND, D. G., and C. S. DURIS [1967]: "Elementary Theory and Application of Numerical Analysis," McGraw-Hill Book Company, New York.

MULLER, D. E. [1956]: A Method for Solving Algebraic Equations Using an Automatic Computer, *MTAC*, **10**:208–215.

"NBS Tables of Lagrangian Interpolation Coefficients" [1944]: National Bureau of Standards Columbia Press Series, vol. 4, Columbia University Press, New York.

"NBS Handbook of Mathematical Functions" [1964]: National Bureau of Standards Applied Mathematics Series, vol. 55, U.S. Government Printing Office, Washington, D.C.

VON NEUMANN, J., and H. H. GOLDSTINE [1947]: Numerical Inverting of Matrices of High Order, *Bull. Amer. Math. Soc.*, **53**:1021–1099.

NEVILLE, E. H. [1934]: Iterative Interpolation, *J. Indian Math. Soc.*, **20**:87–120.

NÖRLUND, N. E. [1926]: "Leçons sur les séries d'interpolation," Gauthier-Villars, Paris.

——— [1954]: "Vorlesungen über Differenzenrechnung," Chelsea Publishing Company, New York.

OBRECHKOFF, N. [1942]: Sur les quadratures méchaniques (Bulgarian, with French summary), *Spisanie Bulgar. Akad. Nauk.*, **65**:191–289.

OLDS, C. D. [1962]: "Continued Fractions," Random House, New York.

ORTEGA, J. M., and W. C. RHEINBOLDT [1970]: "Iterative Solution of Nonlinear Equations in Several Variables," Academic Press, Inc., New York.

OSTROWSKI, A. M. [1940]: Recherches sur la méthode de Graeffe et les zéros des polynomes et des séries de Laurent, *Acta Math.*, **72**:99–155.

——— [1952]: On the Rounding Off of Difference Tables for Linear Interpolation, *MTAC*, **6**:212–214.

——— [1966]: "Solution of Equations and Systems of Equations," 2d ed., Academic Press, Inc., New York.

PADÉ, H. [1892]: Sur la représentation approchée d'une fonction par des fractions rationelles, *Ann. Sci. École Norm. Sup.*, **9**(suppl.):1–93.

PEANO, G. [1913]: Resto nelle formule di quadratura expresso con un integrale definito, *Atti Accad. Naz. Lincei Rend.*, **22**(5):562–569.

——— [1914]: Residuo in formulas de quadratura, *Mathesis*, **34**:5–10.

PEARSON, K. [1920a]: "On the Construction of Tables and on Interpolation, I," "Univariate Tables, Tracts for Computers, II," Cambridge University Press, New York.

——— [1920b]: "On the Construction of Tables and on Interpolation, II," "Bivariate Interpolation, Tracts for Computers, III," Cambridge University Press, New York.

PENNINGTON, R. H. [1970]: "Introductory Computer Methods and Numerical Analysis," 2d ed., The Macmillan Company, New York.

PERRON, O. [1950]: "Die Lehre von den Kettenbrüchen," 2d ed., Chelsea Publishing Company, New York.

POLLAK, L. W. [1947]: "Harmonic Analysis and Synthesis Schedules for 3 to 100 Equidistant Values of Empiric Functions," Geophysical Publications, Stationery Office, Dublin.

——— [1949]: "All Term Guide for Harmonic Analysis and Synthesis," Geophysical Publications, Stationery Office, Dublin.

PRICE, J. F. [1960]: Discussion of Quadrature Formulas for Use on Digital Computers, *Boeing Sci. Res. Labs. Rept.* D1-82-0052.

DE PRONY, R. [1795]: Essai expérimentale et analytique, *J. École. Polytech.* (*Paris*), **1**(2):24–76.

RABINOWITZ, P. (ed.) [1970]: "Numerical Methods for Nonlinear Algebraic Equations," Gordon and Breach, Science Publishers, Inc., New York.

RADAU, R. [1880a]: Sur les formules de quadrature à coefficients égaux, *C. R. Acad. Sci. Paris*, **90**:520–529.

——— [1880b]: Étude sur les formules d'approximation qui servent à calculer la valeur d'une intégrale définie, *J. Math. Pures Appl.*, **6**(3):283–336.

——— [1883]: Remarque sur le calcul d'une intégrale définie, *C. R. Acad. Sci. Paris*, **97**:157–158.

RADEMACHER, H. [1948]: On the Accumulation of Errors in Processes of Integration on High-Speed Calculating Machines. *Proceedings of a Symposium on Large-Scale Digital Calculating Machinery, Ann. Comput. Lab. Harvard Univ.*, **16:** 176–187.

RADON, J. [1935]: Restausdrücke bei Interpolations- und Quadraturformeln durch bestimmte Integrale, *Monatsh. Math.*, **42**:389–396.

——— [1948]: Zur mechanischen Kubatur, *Monatsh. Math.*, **52**:286–300.

RALL, L. B. (ed.) [1965]: "Error in Digital Computation," 2 vols., John Wiley & Sons, Inc., New York.

RALSTON, A. [1965]: "A First Course in Numerical Analysis," McGraw-Hill Book Company, New York.

——— and H. S. WILF (eds.) [1960, 1967]: "Mathematical Methods for Digital Computers," 2 vols., John Wiley & Sons, Inc., New York.

REICH, E. [1949]: On the Convergence of the Classical Iterative Method of Solving Linear Simultaneous Equations, *Ann. Math. Statist.*, **20**:448–451.

RÉMÈS, E. J. [1940]: Sur les termes complémentaires de certaines formules d'analyse approximative, *C. R. Acad. Sci. URSS*, **26**:129–133.

RHODES, E. C. [1921]: "Smoothing, Tracts for Computers, VI," Cambridge University Press, New York.

RICE, J. R. [1964, 1969]: "Approximation of Functions," 2 vols., Addison-Wesley Publishing Company, Inc., Reading, Mass.

——— [1965]: On the Conditioning of Polynomials and Rational Forms, *Numer. Math.*, **7**:426–435.

RICHARDSON, L. F., and J. A. GAUNT [1927]: The Deferred Approach to the Limit, *Trans. Roy. Soc. (London)*, **226A**:299–361.

RIVLIN, T. J. [1969]: "An Introduction to the Approximation of Functions," Blaisdell Publishing Company, a division of Ginn and Company, Waltham, Mass.

ROMBERG, W. [1955]: Vereinfachte numerische Integration, *Norske Vid. Selsk. Forh. (Trondheim)*, **28**:30–36.

ROSSER, J. B. [1950]: Note on Zeros of the Hermite Polynomials and Weights for Gauss' Mechanical Quadrature Formula, *Proc. Amer. Math. Soc.*, **1**:388–389.

——— [1951]: Transformations to Speed the Convergence of Series, *J. Res. Nat. Bur. Standards*, **46**:56–64.

RUNGE, C. [1895]: Über die numerische Auflösung von Differentialgleichungen, *Math. Ann.*, **46**:167–178.

——— [1901]: Über empirische Functionen und die Interpolation zwischen äquidistanten Ordinaten, *Z. Math. Phys.*, **46**:224–243.

RUTISHAUSER, H. [1952]: Über die Instabilität von Methoden zur Integration gewöhnlicher Differentialgleichungen, *Z. Angew. Math. Phys.*, **3**:65–74.

——— [1956]: "Der Quotienten-Differenzen-Algorithmus," Mitt. Inst. Angew. Math. Zürich, no. 7, Birkhäuser Verlag, Basel and Stuttgart.

SALZER, H. E. [1943]: Table of Coefficients for Inverse Interpolation with Central Differences, *J. Math. and Phys.*, **22**:210–224.

——— [1944a]: Table of Coefficients for Inverse Interpolation with Advancing Differences, *J. Math. and Phys.*, **23**:75–102.

——— [1944b]: A New Formula for Inverse Interpolation, *Bull. Amer. Math. Soc.*, **50**:513–516.

——— [1945]: Inverse Interpolation for Eight-, Nine-, Ten-, and Eleven-point Direct Interpolation, *J. Math. and Phys.*, **24**:106–108.

——— [1948]: "Table of Coefficients for Obtaining the First Derivative without Differences," National Bureau of Standards Applied Mathematics Series, vol. 2, U.S. Government Printing Office, Washington, D.C.

——— [1949]: Coefficients for Facilitating Trigonometric Interpolation, *J. Math. and Phys.*, **27**:274–278.

——— [1951]: Formulas for Finding the Argument for Which a Function Has a Given Derivative, *MTAC*, **5**:213–215.

——— [1954]: New Formulas for Facilitating Osculatory Interpolation, *J. Res. Nat. Bur. Standards*, **52**:211–216.

——— [1959]: Some New Divided Difference Algorithms for Two Variables, in Langer [1959].

——— [1961]: Some Extensions of Bairstow's Method, *Numer. Math.*, **3**:120–124.

SARD, A. [1948a]: Integral Representations of Remainders, *Duke Math. J.*, **15**:333–345.

——— [1948b]: The Remainder in Approximations by Moving Averages, *Bull. Amer. Math. Soc.*, **54**:788–792.

——— [1949]: Best Approximate Integration Formulas; Best Approximation Formulas, *Amer. J. Math.*, **71**:80–91.

——— [1963]: "Linear Approximation," American Mathematical Society, Providence, R.I.

SCHOENBERG, I. J. [1947]: Contributions to the Problem of Approximation of Equidistant Data by Analytic Functions, *Quart. Appl. Math.*, **4**:45–99.

——— [1952]: On Smoothing Operations and Their Generating Functions, *Nat. Bur. Standards Rept.* 1734.

——— [1971]: On Equidistant Cubic Spline Interpolation, *Bull. Amer. Math. Soc.*, **77**:1039–1044.

SCHRÖDER, E. [1870]: Über unendliche viele Algorithmen zur Auflösung der Gleichungen, *Math. Ann.*, **2**:317–365.

SHANKS, D. [1955]: Nonlinear Transformations of Divergent and Slowly Convergent Sequences, *J. Math. and Phys.*, **34**:1–42.

SHOHAT, J. A., and C. WINSTON [1934]: On Mechanical Quadratures, *Rend. Circ. Mat. Palermo*, **58**:153–165.

———, E. HILLE, and J. L. WALSH [1940]: "A Bibliography on Orthogonal Polynomials," Bull. Natl. Res. Counc. No. 103, National Academy of Sciences, Washington, D.C.

SINGER, J. [1964]: "Elements of Numerical Analysis," Academic Press, Inc., New York.

SOUTHARD, T. H. [1956]: Everett's Formula for Bivariate Interpolation and Throwback of Fourth Differences, *MTAC*, **10**:216–223.

SOUTHWELL, R. V. [1940]: "Relaxation Methods in Engineering Science," Oxford University Press, Fair Lawn, N.J.

——— [1946, 1956]: "Relaxation Methods in Theoretical Physics," 2 vols., Oxford University Press, Fair Lawn, N.J.

SPENCER, J. [1904]: On the Graduation of the Rate of Sickness and Mortality Presented by the Experience of the Manchester Unity of Oddfellows During the Period 1893–97, *J. Inst. Actuar.*, **38**:334–343.

STANCU, D. D. [1964]: The Remainder of Certain Linear Approximation Formulas in Two Variables, *SIAM J. Numer. Anal.*, **1**:137–163.

STEFFENSEN, J. F. [1950]: "Interpolation," 2d ed., Chelsea Publishing Company, New York.

STEGUN, I. A., and M. ABRAMOWITZ [1956]: Pitfalls in Computation, *SIAM J. Appl. Math.*, **4**:207–219.

STIELTJES, T. J. [1884]: Quelques recherches sur la théorie des quadratures dites mécaniques, *Ann. Sci. École Norm. Sup.*, **1**:409–426.

——— [1894, 1895]: Recherches sur les fractions continues, *Ann. Fac. Sci. Univ. Toulouse*, **8**(1894):1–122; **9**(1895):1–47.

STÖRMER, C. [1921]: Méthode d'integration numérique des équations différentielles ordinaires, *C. R. Congr. Intern. Math. Strasbourg* 1920, *Toulouse, Priv.*, 243–257.

STROUD, A. H. [1961]: A Bibliography on Approximate Integration, *Math. Comp.*, **15**:143–150.

——— [1967]: Approximate Multiple Integration, in Ralston and Wilf [1967].

——— [1971]: "Approximate Calculation of Multiple Integrals," Prentice-Hall, Inc., Englewood Cliffs, N.J.

——— and D. SECREST [1966]: "Gaussian Quadrature Formulas," Prentice-Hall, Inc., Englewood Cliffs, N.J.

SZASZ, O. [1950]: Summation of Slowly Convergent Series, *J. Math. and Phys.*, **28**:272–279.

SZEGO, O. [1967]: "Orthogonal Polynomials," 3d ed., American Mathematical Society, Providence, R.I.

THACHER, H. C. [1960]: Derivation of Interpolation Formulas in Several Independent Variables, *N. Y. Acad. Sci. Ann.*, **86**:758–775.

——— and W. E. MILNE [1960]: Interpolation in Several Variables, *SIAM J. Appl. Math.*, **8**:33–42.

THIELE, T. N. [1909]: "Interpolationsrechnung," B. G. Teubner, Leipzig.

TODD, J. (ed.) [1962]: "A Survey of Numerical Analysis," McGraw-Hill Book Company, New York.

TRAUB, J. F. [1964]: "Iterative Methods for the Solution of Equations," Prentice-Hall, Inc., Englewood Cliffs, N.J.

——— [1966]: A Class of Globally Convergent Iteration Functions for the Solution of Polynomial Equations, *Math. Comp.*, **93**:113–138.

TURING, A. M. [1948]: Rounding-off Errors in Matrix Processes, *Quart. J. Mech. Appl. Math.*, **1**:287–308.

USPENSKY, J. V. [1928]: On the Convergence of Quadrature Formulas Related to an Infinite Interval, *Trans. Amer. Math. Soc.*, **30**:542–559.

DE LA VALLÉE POUSSIN, C. [1919]: "Leçons sur l'approximation des fonctions d'une variable reélle," Gauthier-Villars, Paris; reprinted in Bernstein and de la Vallée Poussin [1969].

VARGA, R. [1962]: "Matrix Iterative Analysis," Prentice-Hall, Inc., Englewood Cliffs, N.J.

WALL, D. D. [1956]: Note on Predictor-Corrector Formulas, *MTAC*, **10**:167.

WALL, H. S. [1967]: "Analytic Theory of Continued Fractions," Chelsea Publishing Company, New York.

WEIERSTRASS, K. [1885]: Über die analytische Darstellbarkeit sogenannter willkürlicher Functionen einer reelen Veränderlichen, *Sitzungsber. Akad. Berlin*, 633–639, 789–805.

WHITTAKER, E. T., and G. ROBINSON [1944]: "The Calculus of Observations," 4th ed., D. Van Nostrand Company, New York.

——— and G. N. WATSON [1927]: "Modern Analysis," 4th ed., Cambridge University Press, New York.

WILF, H. S. [1960]: The Numerical Solution of Polynomial Equations, in Ralston and Wilf [1960].

WILKINSON, J. H. [1959]: The Evaluation of Zeros of Ill-conditioned Polynomials, *Numer. Math.*, **1**:150–166.

——— [1960]: Rounding Errors in Algebraic Processes, in "Information Processing," UNESCO, Paris.

——— [1961]: Error Analysis of Direct Methods of Matrix Inversion, *J. Assoc. Comput. Mach.*, **8**:281–330.

——— [1964]: "Rounding Errors in Algebraic Processes," Prentice-Hall, Inc., Englewood Cliffs, N.J.

——— [1965]: "The Algebraic Eigenvalue Problem," Oxford University Press, Fair Lawn, N.J.

——— [1967]: The Solution of Ill-Conditioned Equations, in Ralston and Wilf [1967].

WILLERS, F. A. [1950]: "Methoden der praktischen Analysis," 2d ed., Walter de Gruyter and Company, Berlin; translated by R. T. Beyer, "Practical Analysis," Dover Publications, Inc., New York, 1948.

WINSTON, C. [1934]: On Mechanical Quadratures Involving the Classical Orthogonal Polynomials, *Ann. of Math.*, **35**:658–677.

WOLFE, P. [1959]: The Secant Method for Simultaneous Non-linear Equations, *Comm. ACM*, **2**:12–13.

WYNN, P. [1956]: On a Procrustean Technique for The Numerical Transformation of Slowly Convergent Sequences and Sums, *Proc. Cambridge Philos. Soc.*, **52**: 663–671.

YOUNG, D. M. [1971]: "Iterative Solution of Large Linear Systems," Academic Press, Inc., New York.

APPENDIX C
DIRECTORY OF METHODS

A INTERPOLATION

1 Based on polynomials
- (*a*) Using an arbitrary set of ordinates without differences
  - (*1*) Noniterative: Sec. 3.2
  - (*2*) Iterative: Sec. 2.7
- (*b*) Using differences formed from ordinates at equally spaced points
  - (*1*) Near beginning or end of tabulation: Sec. 4.3
  - (*2*) Inside tabular range
    - (*a*) Using both odd and even differences: Secs. 4.5 and 4.6
    - (*b*) Using only even differences or only odd differences: Sec. 4.7
  - (*3*) With throwback: Sec. 4.10
- (*c*) Using divided differences formed from an arbitrary set of ordinates: Sec. 2.5
- (*d*) Using ordinates and slopes: Sec. 8.2
- (*e*) Using ordinates at appropriately selected points: Secs. 9.6 and 9.7
- (*f*) In two-way tables: Probs. 3 to 5 of Chap. 5
- (*g*) Inverse: Sec. 2.8

2 Based on ratios of polynomials: Secs. 9.14 to 9.17
3 Based on sines and/or cosines: Prob. 7 of Chap. 3; see also *B4*†
4 Based on exponential functions: see *B5*†
5 Based on cubic splines: Secs. 9.10 to 9.13

B APPROXIMATION

1 By polynomials
  (*a*) Determined as truncated Taylor expansions: Secs. 1.3 and 1.9
  (*b*) Determined by exact fit over a discrete set of points: see *A1*
  (*c*) Determined by least-squares methods
    (*1*) Using an arbitrary finite set of ordinates: Sec. 7.3
    (*2*) Using ordinates at equally spaced points: Sec. 7.13
    (*3*) Using a continuous set of ordinates
      (*a*) Over a finite interval: Secs. 7.6 and 7.9; Prob. 33 of Chap. 7
      (*b*) Over a semi-infinite interval: Sec. 7.7
      (*c*) Over an infinite interval: Sec. 7.8
  (*d*) Economized by use of Chebyshev polynomials: Sec. 9.8
  (*e*) Determined by a minimax condition: Sec. 9.9
2 By products of exponential functions and polynomials: Secs. 7.7 and 7.8
3 By ratios of polynomials: see *A2*
  (*a*) Uniformized: Sec. 9.18
4 By sines and/or cosines
  (*a*) With prescribed periods
    (*1*) Using a finite set of ordinates: Secs. 9.3 and 9.7; Prob. 33 of Chap. 7
    (*2*) Using a continuous set of ordinates: Sec. 9.2
  (*b*) With periods to be determined: Sec. 9.5
5 By exponential functions: Sec. 9.4
6 By cubic splines: Secs. 9.10 to 9.13

C NUMERICAL DIFFERENTIATION

1 Using ordinates without differences: Secs. 3.3, 3.11, and 9.13
2 Using differences formed from ordinates at equally spaced points
  (*a*) Near beginning or end of tabulation: Sec. 5.3; Prob. 5 of Chap 4
  (*b*) Inside tabular range
    (*1*) Near tabular point: Sec. 5.3; Prob. 13 of Chap. 4
    (*2*) Near midpoint between tabular entries: Sec. 5.3; Prob. 16 of Chap. 4

† The approximating functions obtained by least squares incorporating a number of ordinates equal to the number of independent coordinate functions fit the data exactly at those points and thus are strictly interpolative functions.

D NUMERICAL INTEGRATION

1 Using ordinates at equally spaced points without differences
  (*a*) Without end corrections: Secs. 3.5, 3.6, 3.8, and 9.13
    (*1*) Integrals of the form $\int_a^b f(x) \cos kx\, dx$ or $\int_a^b f(x) \sin kx\, dx$: Sec. 3.10
  (*b*) With end corrections: Sec. 5.8; Prob. 18 of Chap. 5
2 Using differences based on ordinates at equally spaced points
  (*a*) Over range near beginning or end of tabulation: Sec. 5.4; Prob. 5 of Chap. 4
  (*b*) Over range centered at interior tabular point: Sec. 5.6; Prob. 13 of Chap. 4
  (*c*) Over range centered midway between successive tabular points: Probs. 16 and 18 of Chap. 4
3 Using ordinates at appropriately selected points
  (*a*) Integrals of the form $\int_a^b f(x)\, dx$: Sec. 8.5
  (*b*) Integrals of the form $\int_a^\infty e^{-\alpha x} f(x)\, dx$: Sec. 8.6
  (*c*) Integrals of the form $\int_a^\infty (x - a)^\beta e^{-\alpha x} f(x)\, dx$: Sec. 8.6
  (*d*) Integrals of the form $\int_{-\infty}^\infty e^{-\alpha^2 x^2} f(x)\, dx$: Sec. 8.7
  (*e*) Integrals of the form $\int_{-a}^a f(x)\, dx/\sqrt{a^2 - x^2}$: Sec. 8.8; Prob. 41 of Chap. 8
  (*f*) Integrals of the form $\int_{-a}^a f(x)\sqrt{a^2 - x^2}\, dx$: Prob. 28 of Chap. 8
  (*g*) Integrals of the form $\int_a^b (x - a)^\alpha (b - x)^\beta f(x)\, dx$: Sec. 8.9
  (*h*) Integrals of the form $\int_0^\pi f(x) \sin kx\, dx$ or $\int_0^\pi f(x) \cos kx\, dx$: Sec. 8.16
  (*i*) Integration formulas involving ordinates at one or both of the integration limits: Secs. 8.11 and 8.12; Probs. 32 and 41 of Chap. 8
  (*j*) Integration formulas employing equal weights: Sec. 8.14
  (*k*) Algebraic derivation of miscellaneous formulas: Sec. 8.15
4 Using ordinates and slopes: Sec. 8.3 (see also Sec. 6.12)
5 Repeated: Secs. 5.5 and 5.6; Prob. 15 of Chap. 5
6 Two-way: Probs. 40 and 41 of Chap. 3

E SUMMATION OF SERIES

1 Finite sums of polynomials: Secs. 5.8 and 7.11
2 Approximate summation of series: Secs. 5.8 and 5.9; Prob. 8 of Chap. 1 and Prob. 37 of Chap. 5
  (*a*) Terms of constant sign: Sec. 5.9 [Eqs. (5.9.1) and (5.9.2)]
  (*b*) Terms of alternating signs: Sec. 5.9 [Eqs. (5.9.7) and (5.9.8)]

F SMOOTHING OF DATA

1 By determining a smooth approximating function: Sec. 7.13 (see also *B1c* and *B4*)
2 By point-by-point modification of data: Sec. 7.15

G NUMERICAL SOLUTION OF ORDINARY DIFFERENTIAL EQUATIONS

1 Initial-value problems: see Sec. 6.17
2 Boundary-value problems: Sec. 6.15
3 Characteristic-value problems: Sec. 6.16

H NUMERICAL SOLUTION OF EQUATIONS

1 Sets of linear algebraic equations
  (*a*) By use of determinants: Sec. 10.2
  (*b*) By a finite sequence of reductions: Secs. 10.3, 10.4, and 10.8
  (*c*) By iteration: Sec. 10.9
2 Nonlinear equations
  (*a*) General iterative methods: Secs. 10.10 to 10.12 (see also Secs. 2.8 and 9.17)
  (*b*) Special iterative methods for algebraic equations
    (*1*) Approximate determination of largest or smallest root: Sec. 10.16
    (*2*) Simultaneous approximate determination of all roots: Sec. 10.17
    (*3*) Approximate determination of one root: Sec. 10.14
    (*4*) Simultaneous approximate determination of two roots: Secs. 10.18 and 10.19

I MISCELLANEOUS PROCESSES

1 Inversion of power series: Probs. 16 and 46 of Chap. 1
2 Expansion of one function in powers of another: Sec. 1.9
3 Locating maxima or minima: Prob. 23 of Chap. 2
4 Checking tables by use of differences: Sec. 4.9
5 Expression of differences in terms of derivatives: Sec. 5.3
6 Subtabulation: Sec. 5.7; Probs. 21 and 22 of Chap. 5
7 Calculation of mean values over given intervals from known mean values over other intervals: Prob. 23 of Chap. 5
8 Determination of unknown periodicities from empirical data: Sec. 9.5
9 Continued-fraction representations: Secs. 9.14 to 9.17
10 Evaluation of determinants: Sec. 10.4
11 Inversion of matrices: Sec. 10.6

Index

Italic numbers in parentheses following page references indicate problem numbers.